Third Edition

Statistical Reasoning

GARY SMITH
Fletcher Jones Professor of Economics
Pomona College

ALLYN AND BACON   Boston   London   Toronto   Sydney   Tokyo   Singapore

For my wife, Betsy,
several standard deviations above the mean.

Series Editor: *Rich Wohl*
Series Editorial Assistant: *Cheryl Ten Eick*
Cover Administrator: *Linda Dickinson*
Composition and Manufacturing Buyer: *Louise Richardson*
Text Designer: *Sylvia Dovner*
Editorial-Production Service: *Technical Texts, Inc.*
Editorial-Production Administrator: *Mary Beth Finch*

Allyn and Bacon
A Division of Simon and Schuster, Inc.
160 Gould Street
Needham Heights, Massachusetts 02194

Library of Congress Cataloging-in-Publication Data

Smith, Gary, 1945–
    Statistical reasoning / Gary Smith. — 3rd ed.
      p. cm.
    Includes bibliographical references and index.
    ISBN 0-205-12939-0
    1. Statistics. 2. Probabilities.   I. Title.
    QA276.12.S58   1991
    519.2—dc20   90-22092
                            CIP

ISBN: Text only   0-205-12939-0
         Text with 5¼″ disk   0-205-12944-7
         Text with 3½″ disk   0-205-12943-9

Printed in the United States of America
10 9 8 7 6 5 4 3 2 1   96 95 94 93 92 91

Contents

Preface

This textbook is intended for an introductory course on probability and statistics. Its primary objective is to help students learn to think clearly about some important, yet unfamiliar, ideas. That is why this book is titled *Statistical Reasoning*. In some texts, students are given a formula and some data and are told to plug in the numbers and chug out an answer. While such an approach gives practice at addition, subtraction, division, and multiplication, it misleads students into thinking that probability and statistics are just tedious subcategories of arithmetic.

Because necessary calculations are now easily done with pocket calculators or computers, thinking is what students must learn. To help them, this text provides hundreds of interesting and relevant examples and discusses not only the uses but also the abuses of statistics. Students will see statistical reasoning used correctly and will see the errors, deliberate or accidental, that people often make. As indicated by the list inside the front cover, the examples are drawn from many areas to show the breadth and variety of statistical reasoning—to show students that probability and statistics are not an esoteric abstraction, but a real part of everyday life.

Almost every chapter has 100 exercises of varying type and difficulty. Some are simple computations to reinforce the chapter material. Many are real-world examples, which offer students opportunities to apply important statistical principles. Those that use the computerized data set available with the textbook are identified by a computer disk logo. Exercises that are relatively difficult are prefaced by a single bullet; those that are even more challenging have two bullets.

I am very grateful for the continuing stream of enthusiastic telephone calls and letters from professors and students who have used *Statistical Reasoning*. Many of the best new examples and exercises in this third edition were supplied by professors using *Statistical Reasoning.* Their suggestions and encouragement have made this enterprise a unique collaborative effort. I hope that professors and students will continue to find this text enjoyable and educational and continue to send me material that others will enjoy using.

CHANGES IN THE THIRD EDITION

Increased Number of Exercises: The number of exercises has been doubled, from 50 per chapter to 100 per chapter. With the help of users, these 1000 new exercises are as good or better than the first 1000. The new exercises are numbered 51 to 100 in each chapter, for the convenience of instructors who have their long-standing favorites among the first 50 exercises.

New and Updated Examples: Another 50 detailed examples have been added. Again with the help of users, these are quality examples, interesting real-world situations that nicely illustrate important statistical principles; for example, Junk Bond Default Rates, Quality Control in the Soviet Union, and Did the *Washington Post* Profit from the Watergate Scandal?

Addition of Data Disk: A data disk is now available so that students can do exercises using substantial amounts of data without the tedious chore of entering these data in a computer program. The exercises that use these data are identified by a computer-disk logo.

Addition of Computer Output: Sample output from MINITAB and MYSTAT have been added to illustrate the usefulness of computers for numerical calculations.

Addition of Bivariate Probability Distributions: An appendix to Chapter 4 now explains and illustrates bivariate probability distributions.

Increased Coverage of Quality Control: The coverage of quality control has been expanded. Chapter 5 uses the binomial distribution to explain acceptance sampling; Chapter 9 applies the normal distribution to control charts.

Expanded Coverage of Regression: An optional section on regression diagnostics has been added to Chapter 14. This section explains how the scrutiny of regression residuals can help identify regression problems, including heteroskedasticity and serial correlation. The use of the Durbin-Watson *d*-statistic is discussed in the appendix to Chapter 14.

Increased Flexibility: Dummy variables are now explained in both Chapter 15 (Multiple Regression) and Chapter 16 (Analysis of Variance) for instructors who assign only one of these chapters. This material is written so that it is self-contained but not redundant.

THREE POSSIBLE COURSE OUTLINES

Several sections are labeled as optional so that the book can be adapted to a variety of interests and time constraints. For example, in a general survey course, the following sequence can be used:

Chapter 1	Introduction
Chapter 2	The Probability Idea
Chapter 3	Probability Rules
Chapter 4	Probability Distributions (omit Sections 4.4–4.8 if desired)
Chapter 5	Three Useful Distributions
Chapter 6	The Normal Distribution
Chapter 7	Using the Normal Curve
Chapter 8	Sampling
Chapter 9	Estimation
Chapter 10	Hypothesis Tests
Chapter 11	The Use and Abuse of Statistical Tests
Chapter 12	A Few More Tests
Chapter 13	The Idea of Regression Analysis
Chapter 16	Analysis of Variance
Chapter 17	Some Nonparametric Tests

For courses emphasizing decision making, the following sequence is suitable:

Chapter 1	Introduction
Chapter 2	The Probability Idea (omit permutations and combinations, Section 2.3, if desired)
Chapter 3	Probability Rules
Chapter 4	Probability Distributions
Chapter 19	Decision Theory Analysis
Chapter 5	Three Useful Distributions
Chapter 6	The Normal Distribution
Chapter 7	Using the Normal Curve
Chapter 8	Sampling
Chapter 9	Estimation
Chapter 10	Hypothesis Tests
Chapter 11	The Use and Abuse of Statistical Tests
Chapter 12	A Few More Tests
Chapter 13	The Idea of Regression Analysis
Chapter 18	Time Series Analysis and Indexes

Where emphasis on regression is desired, the following sequence applies:

SUPPLEMENTS

Software: *Statistical Reasoning* can be ordered with or without software.

- The DOS version of Business MYSTAT is available as a 3 1/2" or 5 1/4" disk. Business MYSTAT is an educational version of the popular, comprehensive statistical program SYSTAT. A user's manual for Business MYSTAT is included as an appendix to *Statistical Reasoning*.
- Smith's Statistical Package, a menu-driven program, simpler to use but not as powerful or comprehensive, is available to users, free of charge, from the author in IBM, Macintosh, or Apple II format. The statistical programs on this disk are intended to avoid the drudgery of hand calculations and the complexity of large computer packages, so that students can concentrate on understanding and applying statistical reasoning. The primary limitation of Smith's Statistical Package is that it does not allow students to transform and manipulate data freely. The offsetting virtue is that there are no commands to be learned. Students can immediately use Smith's Statistical Package to do any of the text exercises.

Study Guide: A detailed study guide further explains and elaborates the text discussion. The guide gives detailed step-by-step calculations for the exercises that are answered in the back of *Statistical Reasoning* and provides additional exercises for homework assignment or student practice.

Instructor's Manual: The instructor's manual, which is now nearly as long as the text itself, contains lecture suggestions (including several examples not covered in the text), provides detailed solutions to all of the text exercises, and gives additional exercises.

Data Disk: The data disk provides substantial amounts of data that students can use with the exercises.

A NOTE FOR STUDENTS

H. G. Wells once wrote that "Statistical thinking will one day be as necessary for efficient citizenship as the ability to read and write." And today it is. We live in a complex society in which informed decisions require an intelligent interpretation of all sorts of statistical data and probability judgments.

Probability and statistics can help you decide which foods are unhealthful, which careers are lucrative, what car to drive, how much insurance to have, and which stocks to purchase. Businesses routinely use statistics to gauge markets, design products, schedule activities, monitor production, and even more. Government statisticians measure things ranging from unemployment to corn production to births to pollution. Each year, the tip of this statistical iceberg is published in the *Statistical Abstract,* a surprisingly interesting—some say fascinating—source of information about the United States. Some government data, such as weather patterns and crop production, help citizens and businesses make intelligent plans. Other data, such as unemployment, inflation, and product safety tests, are used by government policy makers.

Government data also provide ammunition for election campaigns, and voters must decide whether politicians are using or abusing statistics. In 1984, Ronald Reagan was so upset by contrary data that he used the nineteenth century British politician Benjamin Disraeli's immortal retort, "There are three kinds of lies—lies, damn lies, and statistics." And then President Reagan proceeded to cite some statistics of his own!

In this course, you will learn how to use statistical reasoning to recognize the abuses of others and to judge the competing claims of advertisers and politicians. The chapters rely heavily on examples to teach their lessons because there are so many real-world situations that require statistical analysis. A sampling is on the inside cover. There are hundreds more in the book itself. These will show you the breadth of probability and statistics, illustrate how the tools are used, and even teach you something about the examples themselves. The students who have used this book invariably say that they learned many interesting things from the examples—you will too.

The exercises are also an absolutely essential part of this book. You can't become a good racquetball player just by watching, and you can't learn probability and statistics just by reading. You have to go out and try to hit a few balls yourself. You may miss a few at first; but soon you will acquire the necessary skills and experience. Later, if you begin to feel rusty, maybe shortly before an exam, do some more exercises. These are the surest ways to strengthen your statistical reasoning and to prepare you for life after the exam, when you will encounter probability and statistics almost daily.

ACKNOWLEDGMENTS

Many have contributed to this book. I sincerely appreciate your advice and encouragement:

Neil Alper, Northeastern Univerity; Sergio Antiochia, Eastern Michigan University; Dan Brick, College of St. Thomas; George Briden, University of Rhode Island; Philip Carlson, Bethel College; Allin Cottrell, Elon College; Anirban DasGupta, Purdue University; Stephen DeCanio, University of California, Santa Barbara; Roy V. Erickson, Michigan State University; H. E. Frech III, University of California, Santa Barbara; Myron F. Goodman, University of Southern California; Marvin Jay Greenberg, University of California, Santa Cruz; David Harrington, Kenyon College; Iftekhar Hasan, University of Wisconsin; Jackie Hoell, Virginia Polytechnic Institute and State University; John Hoffman, Humboldt State University; Burt S. Holland, Temple University; Bernard Isselhardt, Rochester Institute of Technology; Raj Jagannathan, The University of Iowa; Mario L. Juncosa, University of California, Los Angeles; Hanhan Kim, South Dakota State University; Ronald S. Koot, The Pennsylvania State University; Michael Kuehlwein, Pomona College; Edward Leamer, University of California, Los Angeles; Jerry Lenz, St. John's University; Lonnie Magee, McMaster University; Stephen Marks, Pomona College; Fred S. McChesney, Emory University; Don Miller, Virginia Commonwealth University; Michael Murray, Bates College; Harold Petersen, Boston College; John R. Pickett, Georgia Southern College; Frank W. Puffer, Clark University; Ronald E. Shiffler, University of Louisville; Bob Smidt, Cal Poly, San Luis Obispo; James Stapleton, Michigan State University; A. H. Studenmund, Occidental College; Paul A. Thompson, Ohio State University; James Vanderhoff, Rutgers University; Mary Beth Walker, Emory University; Thomas O. Wisley, Western Kentucky University; Barbara Yanoska, Humboldt State University; and Mary Sue Younger, University of Tennessee at Knoxville.

I am grateful to the Literary Executor of the late Sir Ronald A. Fisher, F.R.S., to Dr. Frank Yates, F.R.S., and to Longman Group Ltd, London, for permission to reprint Table III from their book *Statistical Tables for Biological, Agricultural, and Medical Research* (6th Edition, 1974). I also thank the many people at Allyn and Bacon and Technical Texts who helped turn my drafts into a successful book.

In the year of our Lord 1432, there arose a grievous quarrel among the brethren of a monastery over the number of teeth in the mouth of a horse. For thirteen days the disputation raged without ceasing. All the ancient books and chronicles were fetched out, and wonderful and ponderous erudition, such as was never before heard of in this region, was made manifest. At the beginning of the fourteenth day, a youthful friar of goodly bearing asked his learned superiors for permission to add a word.

Straightaway, to the wonder of the disputants whose deep wisdom he sore vexed, he beseeched them to unbend in a manner coarse and unheard of, and to look in the mouth of a horse to find an answer to their questionings. At this, their dignity being greatly hurt, they waxed exceedinging wroth; and joining in a mighty uproar, they fell upon him, hip and thigh, and cast him out forthwith. For, said they, surely Satan hath tempted this bold neophyte to declare unholy and unheard of ways of finding truth contrary to all the teachings of their fathers!

Francis Bacon (1561–1626)

1

Introduction

Life is a school of probability.

Walter Bagehot

TOPICS

A great French mathematician, Pierre Laplace, observed that probabilities are "only common sense reduced to calculation." Probabilities help us analyze uncertain situations. If it is true that only death and taxes are certain, then that leaves a lot of targets for probability analysis! When you hear that the odds are 100 to 1 that the Chicago Cubs will be in the World Series, those odds reflect a probability assessment. The Educational Testing Service applies probabilities when it calculates standardized scores on its SAT tests. Colleges use probabilities when they decide how many students to admit, how many professors to employ, and how many classrooms to build. Probabilities are involved when an army decides to attack or retreat, when a business decides to expand or contract, and when you decide whether or not to take an umbrella to school. Uncertainties are all around us, and so are probabilities.

Historically, the first rigorous application of probability theory involved games of chance, and games are still a fertile field for probability analysis. Gambling casinos use probabilities when they set payoffs for roulette, craps, and slot machines. Governments use probabilities when they set payoffs for public lotteries. Good poker players have learned probabilities from long and sometimes expensive experience. Probability calculations have been used to devise systems for winning at blackjack.[1] Backgammon has been extensively studied by mathematicians and devoted players, again using probabilities. Probabilities also were used to analyze the game of Monopoly.[2]

Probabilities also are used in gambles that don't really involve standard games of chance. When a bookie offers odds on the winner of the next Super Bowl, the next president of the United States, or the sex of a celebrity's baby, that bookie is using probabilities. These probabilities cannot be as precisely reasoned as those in roulette or craps, but it is certain that the bookie has thought very carefully about how probable the wagered event is.

Another early use of probabilities was in setting insurance rates. Consider, for example, a life insurance policy that will pay $100,000 if Mr. Jones dies. In deciding whether to offer this policy and how large a premium to charge, the insurance company would like to know how probable it is that Mr. Jones will die soon. They will use data from the experiences of other men of similar age, health, and occupation to estimate the probabilities that Mr. Jones will live another year, another 10 years, and another 50 years. Their enthusiasm for issuing the policy and the premium they charge will depend on whether Mr. Jones is 18 or 98, whether he is in good health or the recent recipient of an artificial heart, and whether he is a college professor or a soldier of fortune. Insurance companies also estimate probabilities when they offer disability insurance, medical insurance, car insurance, boat insurance, home insurance, and business insurance. They have fewer data to go on, but they still estimate probabilities when they offer insurance against a baseball strike or offer to insure some part of a starlet's anatomy.

1.1 PROBABILITY VERSUS STATISTICS

If an unbent coin is fairly flipped, then simple logic tells us that the probability of heads is 1/2 (one chance in two) and the probability of tails is 1/2. Similarly, in roulette, craps, and other games of chance, we can use logic alone to deduce probabilities, and will begin doing so in the next chapter. With insurance rates, however, logic alone is not enough. Common sense tells us that a 98-year-old person is more likely to die within a year than an 18-year-old. But how much more likely? This is where statistics come in. By examining the observed mortality rates of 18-year-olds and 98-year-olds, we may be able to estimate the respective probabilities and life expectancies, and even to test hypotheses. The use of specific data to estimate probabilities and test theories is termed **statistical inference.**

There is a close relationship between probability and statistics that can be brought out by distinguishing between a **population** and a **sample.** Most statistical data are a sample from a much larger population. To predict the outcome of a presidential election, for instance, pollsters interview a few thousand voters. The people who are polled are a sample from a population consisting of the tens of millions of people who will vote in the actual election. Statistical inference involves using the poll data to predict the outcome of the election and assess how much confidence we have in our prediction.*

Probabilities enter because there is chance involved in the selection of the people to be interviewed. Even if 52 percent of all voters favor the Democratic candidate, there is a chance that, by the luck of the draw, 51 percent of the people picked for the poll will favor the Republican candidate. Probability calculations can tell us the likelihood of such an unfortunate sampling error. This possibility of sampling error explains why poll results are given as "51 percent $\pm$ 3 percent." In this course, you will learn where the $\pm$ 3 percent comes from and exactly what it means.

> In general, probabilities are used to describe anticipated outcomes. Statistical data are the observed outcomes, and statistical inference is the use of data to draw inferences about the underlying population.

When an unbent coin is fairly flipped, we can anticipate that heads and tails will come up about equally often. We can say even more about the anticipated outcomes with some easy probability rules that you will soon learn. If the coin is flipped 5 times, the probability of 5 straight heads works out to be .03 (3 chances in 100). With 20 flips, the probability of 15 or more heads turns out to be .02. With 100 flips, the probability of 60 or more heads is .03. These are all probability calculations about the likelihood of forthcoming data (the outcome of the coin flips), given certain assumptions (that the coin is unbent and fairly flipped).

We can think of the forthcoming coin tosses as a sample from a hypothetical infinite population consisting of all of the outcomes, were this coin to be flipped endlessly. Probability theory tells us how often various samples (such as 16 heads in 20 tosses) should occur. The actual results, perhaps 52 heads and 48 tails in 100

---

*A statistical population need not be large, but it must contain every person, place, or thing that we're interested in. If we're only interested in how the two adults in my household will vote, then my wife and I are the entire population. A sample could consist of either of us or both of us. In practice, partial samples are used when the population is too large to be analyzed completely. Thus, the distinction between a sample and the entire population usually arises in cases where the population is large.

flips, are our statistical data. An example of a statistical inference is that these particular data are consistent with our assumption that the coin is unbent and fairly flipped. If heads had come up 87 times in 100 flips, we might infer that this is too improbable a result to be consistent with our assumption that the coin is fair. The observation of 87 heads in 100 flips suggests that heads and tails are not equally likely with this particular coin. Either the coin or the coin tosser may be crooked. A more specific statistical inference is an estimate that the probability of throwing heads with this unusual coin is .87 $\pm$ .10. In this course you will learn how to calculate probabilities, scrutinize data, and draw such inferences.

Election Polls

The collection of all the people who will vote in an upcoming election is a finite population. A poll of 100 voters is a sample from this population of all voters. The most useful polls are what statisticians call **random samples,** in which each member of the population is equally likely to be selected for the sample. A random sample is analogous to a fair deal from a well-shuffled deck of cards. Just as probability theory tells us the chances of flipping 16 heads in 20 tosses and the chances of being dealt a pair of aces in poker, so it tells us the chances that a random sample will turn out in a specified way.

For instance, if half of all voters favor candidate H and half favor candidate T, then (just as with our coin flips) there is a .03 probability that at least 60 of the 100 voters we poll will favor H. It's not very likely, but there is this slim chance of a very misleading poll result. In this course, you will learn how the chances of an inaccurate poll can be reduced by taking a larger sample. If, for example, the population of all voters is evenly divided and 400 voters are surveyed, it turns out that there is then only a .00003 probability that 60 percent or more of the sample will favor H.

In this way, probability theory tells us how often we can anticipate observing certain data, given certain assumptions about the underlying population. These observed data may be a winning streak at roulette, high sales of a new product, or a favorable election poll. Statistical inference works in the other direction, using observed data to gauge the characteristics of the underlying population. Statistical inference is used to estimate one's chances of winning a game, to project future sales, and to predict the outcome of an upcoming election. In each case, we have to allow for a margin of error because luck alone can cause the observed data to be atypical. Statistical inference tells us how to make good estimates and how much confidence we can put in our estimates. Naturally, as with the election poll between H and T, the reliability of our estimates depends on the amount of data we have. A poll of 400 people is less likely to be misleading than a poll of 100 people or 5 people. Methods of statistical inference can tell us how big a poll we need to obtain reliable estimates.

Statistical inference can also be used to confirm or discredit theories about the underlying population. One theory might be that candidate T is going to win the upcoming election. If a poll is taken and 60 percent of those polled favor candidate

H, then these data cast some doubt on the theory that T is the voter's choice. The degree of doubt depends, of course, on the size of the sample. If 5 voters are surveyed and 3 favor H, this is very flimsy evidence. If 1,000 voters are polled and 600 favor H, this evidence strongly discredits the theory that T will win the election. In this course, you will learn to use probabilities to make this sort of statistical inference. You will be able to use data to test such theories as: smoking is dangerous to your health; vitamin C fights colds; Volvos last longer than Fords; and the unemployment rate influences presidential elections.

Statistical inference is very interesting and useful. It is used in election polls and unemployment surveys; in marketing tests and tests of product safety; in constructing macroeconometric forecasting models and interpreting psychological test scores; and in testing and refining scientific theories about the human body and the nature of the universe. But, as explained earlier, the foundation of statistical inference is probability—deducing the likelihood of observing various data from an assumed population. And so, starting in the next chapter, you will learn about probabilities before you learn about the details of statistical inference. But, before we begin, I want to show you a little bit of descriptive statistics and some of the tools that are repeatedly used in both probability analysis and statistical inference. **Descriptive statistics** are simple techniques for briefly describing a population or sample. For this purpose, statisticians use graphs and numerical measures of the center and spread of the data. We'll quickly run through some of these methods in the next section.

EXERCISES

1.1 A poll conducted shortly before the 1984 Democratic presidential primary in California found that 29 percent of those considered likely voters favored Gary Hart, 28 percent favored Walter Mondale, 11 percent favored Jesse Jackson, and 32 percent expressed no preference. From what population is the poll intended to be a sample? Why, by chance alone, might the poll results differ from the actual election results?

1.2 *(continuation)* Can you think of any other reasons why the election and poll results might differ?

1.3 *(continuation)* In the primary election itself, 38 percent voted for Hart, 35 percent for Mondale, 21 percent for Jackson, and 6 percent for others. Would the errors be more surprising if the poll had involved 50 voters or 5,000 voters? Explain your reasoning.

1.4 *(continuation)* In fact, this poll surveyed 445 likely voters. If you were a pollster, why might you survey more than 50 voters? Why might you decide to survey fewer than 5,000 voters?

1.2 DIAGRAMS OF FREQUENCY DISTRIBUTIONS

Statisticians use graphs to portray characteristics of a population or sample. Suppose we are on the eve of a gubernatorial election between two candidates, Mary Jones and John Martin. The population consists of all people who will vote in this election. Perhaps there are, conveniently enough, exactly 1,000,000 voters, of

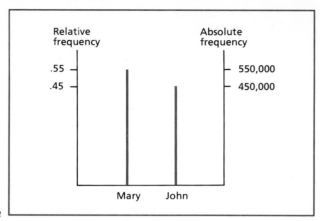

Figure 1.1 *A Voter Population*

whom 550,000 favor Mary and 450,000 favor John. Figure 1.1 illustrates this voter distribution. On the horizontal axis is a characteristic of interest, which here is just the names of the two candidates. On the vertical axis is a measure of the distribution of voter preferences. As shown in Figure 1.1, there are two equivalent alternatives for the vertical scale. One possibility is the number of voters who prefer each candidate, the **absolute frequency.** Another possibility is the **relative frequency,** which is the fraction of all voters who prefer each candidate. Either way, the graph looks exactly the same; it is just a matter of choosing to use either the absolute number or the relative frequency. In practice, the relative frequency is used more often because it is somewhat easier to digest and interpret, the fraction being easily converted into a percentage. When we say that a candidate receives 55 percent of the vote, we can easily compare that 55 percent figure to predictions and other elections:

> The incumbent, Mary Jones, scored a decisive victory today over her challenger, John Martin. In receiving 55 percent of the votes, she improved on the 52 percent of the vote received when elected four years ago. In the big cities, especially, she ran well ahead of her earlier pace. Jones also did better than her party yesterday, which received only 48 percent of the statewide presidential vote. With her stunning victory, Jones is now being mentioned as a future vice-presidential candidate.

The data in Figure 1.1 describe the entire population of 1,000,000 voters. Were we to ask 100 of these voters who they favored, this would be a sample. The results of this particular poll might be as shown in Figure 1.2; 54 favor Mary and 46 favor John. The vertical axis shows the relative frequencies: $54/100 = .54$ and $46/100 = .46$. Figures 1.1 and 1.2 have a similar appearance, but it is very important that you understand the conceptual distinction between a population and a sample. The population relative frequencies in Figure 1.1 give the fractions of all voters who favor each candidate. Before the actual election, these population frequencies are unknown and are an interesting subject of speculation. When we take a sample, we reach into this population and query only some of the voters. The sample relative

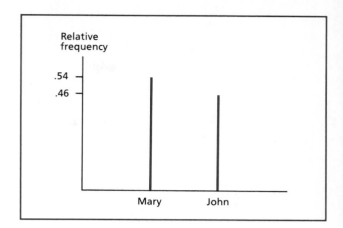

Figure 1.2 *A Voter Sample*

frequencies in Figure 1.2 are the actual result of one particular poll. Another poll will query a different sample of voters and, most likely, give somewhat different results. The sample frequencies will vary from poll to poll.

Probability theory can tell us, for an assumed population, the likely magnitudes of these sampling variations from poll to poll. Statistical inference can help us make informed judgments about the population from which the sample was drawn.

Figures 1.1 and 1.2 are examples of what statisticians call a frequency distribution or relative frequency distribution.

A **frequency distribution** shows the fraction of a population or sample having certain characteristics.

This first example involved political preferences. Often, there are numerical units on the horizontal axis, instead of qualitative distinctions like "John" and "Mary." For instance, here are some data on household incomes in the United States in 1985:

Income Category	Income in 1985	Number of Households	Relative Frequency
Low	$0– $9,999	17,780,000	.201
Middle	$10,000–$29,999	36,704,000	.415
Upper	$30,000–$74,999	30,046,000	.340
High	$75,000 and over	3,927,000	.044
Total		88,458,000	1.000

It is plainly impractical to list all 88,458,000 incomes. Instead, the incomes have been grouped into four categories: low ($0–$9,999), middle ($10,000–

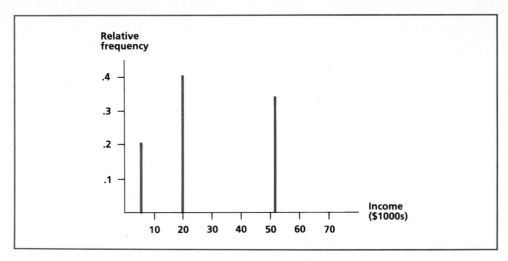

Figure 1.3 *A Relative Frequency Diagram of U.S. Income Distribution, 1985*

$29,999), upper ($30,000–$74,999), and high ($75,000 and up). By dividing the number of households in each category by the total number of households, we obtain the relative frequencies. For instance, the 17,780,000 households in the low-income group represent a fraction

$$\frac{17{,}780{,}000}{88{,}458{,}000} = .201$$

of the total population. If our arithmetic is correct, then the four relative frequencies should add up to 1, as they do here.

Figure 1.3 shows the relative frequency distribution of the income data. The heights of the three lines show the relative frequencies, with each relative frequency line placed in the middle of the appropriate interval. For instance, the .201 relative frequency of $0–$9,999 in household income is placed at the $5,000 point on the horizontal axis. The category "$75,000 and over" presents a special problem in that the lack of an upper limit precludes the specification of a midpoint. One possible approach, which I've taken in Figure 1.3, is just to ignore these people by omitting them from the graph. An alternative would be to use an estimate of the average income of this upper 4 percent of the households.

Bar Graphs and Histograms

Often, bars are used in place of relative frequency lines. Figure 1.4 is a simple *bar chart* representation of Figure 1.3. A bar chart has the advantage of showing the width of the interval represented by a frequency. In Figure 1.3, it is not at all easy

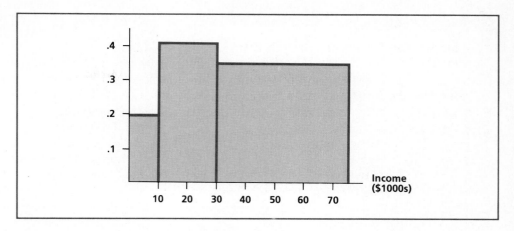

Figure 1.4   *A Bar Chart of U.S. Income Distribution, 1985*

to tell which incomes correspond to which relative frequencies. The bars in Figure 1.4, in contrast, show very clearly that the first frequency refers to $0 to $10,000 income; the second is for $10,000 to $30,000; and the third is for $30,000 to $75,000.

Statisticians use a superficially similar but actually quite different graphical device called a **histogram.** Figure 1.5 is a histogram representation of the household income data. A histogram has bars, like a bar chart, which show the width of each interval. The heights are determined differently, however, and so the bar chart in Figure 1.4 and the histogram in Figure 1.5 look quite different. As two statisticians, W. Allen Wallis and Harry Roberts, put it, "though bar charts and histograms are often confused, they have no relation except the fortuitous one that both involve bars—as do musical scores and saloons."[3]

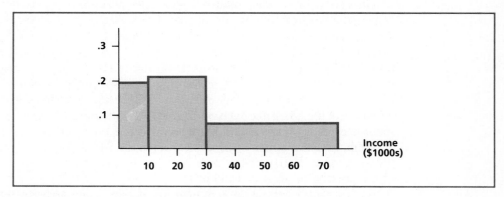

Figure 1.5   *A Histogram of U.S. Income Distribution, 1985*

Stretching Scales to Stretch the Truth

Sometimes the vertical axis of a bar graph is interrupted, in order to emphasize variations in the heights of the bars. For instance, the vertical axis in the advertisement in Figure 1(a) only runs from 89 to 100, rather than 0 to 100, to allow us to see clearly that this insurance company's 97.5 percent payoff rate is higher than that of nine competitors (and, not accidentally, to make 97.5 percent seem twice the size of 94 percent).[4] The alternative bar chart in Figure 1(b), with the vertical axis going from 0 to 100, seems to show that payoff rates are about the same for all ten companies, which is not the message this advertising agency had in mind.

In a simple *bar chart,* relative frequencies are shown by the *heights* of the bars. In a *histogram,* relative frequencies are shown by the *areas* inside the bars. Because the human eye gauges size by area, the bar chart in Figure 1.4 seems to say that the bulk of the population is in the upper-income group. Yet the data show that most households are actually in the low-income and middle-income categories. The simple bar chart in Figure 1.4 is misleading because it is the area inside a bar rather than its height that catches our attention. This discrepancy between height and area occurs because the three income intervals do not have the same width. The low-income category spans a $10,000 interval, while the middle-income interval spans $20,000, and the upper-income interval spans $45,000. The wider interval gives the upper-income bar a fatter base and a larger area, making it appear to hold more of the population than it really does.

A histogram corrects this optical illusion by using areas to represent relative frequencies. Let's measure income in $10,000 units. The width of the low-income interval is then 1 of these $10,000 units. A height of .201 for the low-income bar gives an area of (base)(height) = (1)(.201) = .201, as intended. What about the middle-income category? Its base is 2 units, because it spans the $20,000 interval from $10,000 to $30,000. To make the area equal to the relative frequency, we need to set the height equal to the relative frequency divided by 2:

$$\text{area} = (\text{base})(\text{height})$$

$$.415 = (2)(\text{height})$$

implies

$$\text{height} = \frac{.415}{2}$$

$$= .208$$

Similarly, because the upper-income interval spans 4.5 $10,000 units, the appropriate histogram height is

$$\text{height} = \text{relative frequency/width}$$

$$= \frac{.340}{4.5}$$

$$= .076$$

Figure 1.5 shows the resulting histogram, which does indeed give a very different visual image than the bar diagram in Figure 1.4

> If the intervals are of equal width, then a histogram is constructed in the same way as a bar chart. But, if the interval widths are unequal, then the height of a histogram block is calculated by dividing the relative frequency by the width of the interval.

Pluses and Minuses

One of the primary disadvantages of a bar chart is that its appearance is very sensitive to the largely arbitrary choice of intervals. How, exactly, did I decide where to draw the dividing lines between low, middle, and upper incomes? My particular choice was arbitrary in that there are other divisions that I could have used and there are no compelling criteria for my choice. Table 1.1 shows more detailed data on the income distribution. What if, for instance, I had let the middle-income category be from $15,000 to $40,000 instead of $10,000 to $30,000? Figure 1.6 shows the consequent bar chart and histogram, along with the appropriate arithmetic. First compare Figures 1.4 and 1.6(a). The appearance of the bar chart is drastically

Table 1.1   *Household Income, 1985*

Income in 1985	Number of Households	Relative Frequency
$0–$4,999	6,783,000	.0767
$5,000 –$9,999	10,997,000	.1243
$10,000–$14,999	10,149,000	.1147
$15,000–$19,999	9,674,000	.1094
$20,000–$24,999	8,839,000	.0999
$25,000–$29,999	8,042,000	.0911
$30,000–$34,999	6,965,000	.0787
$35,000–$39,999	5,739,000	.0649
$40,000–$49,999	8,208,000	.0928
$50,000–$74,999	9,134,000	.1033
$75,000 and over	3,927,000	.0444
Total	88,458,000	1.000

Note: If income groups are combined, the relative frequencies are added. For instance, the income range $0–$9,999 has a relative frequency .0767 + .1243 = .2010.

altered by this redefinition of the intervals. Now compare Figures 1.5 and 1.6(*b*). The histograms look pretty much the same.

Bar charts are also greatly affected by changes in the number of intervals. Why did I use three income categories? Why not four, five, or some other number? There is no compelling reason here, so I would hope that my graphs are not very sensitive to this arbitrary choice. Bar charts are, in fact, more sensitive than histograms, and this is another reason why statisticians prefer histograms. Figure 1.7 shows the bar chart and histogram when household incomes are divided into five categories. Again, the bar chart is strikingly transformed and seems to give a brand new picture of the income distribution. The appearance of the histogram, in contrast, is little changed and the changes that do occur just convey the additional information that is provided by five intervals instead of three.

Look at the incomes from $10,000 to $30,000. When this income range is divided into two intervals, as in Figure 1.7(*b*), the histogram shows that there are more people with incomes of $10,000 to $20,000 than with incomes of $20,000 to $30,000. When these two intervals are combined, as in Figure 1.5, the histogram gives us an average of the two separate intervals. The same thing occurs in the $30,000 to $75,000 range, but the results are more dramatic. The two intervals in Figure 1.7(*b*) show the preponderance of people in the $30,000 to $50,000 range, while the single interval in Figure 1.5 gives an average. If there are marked variations in the number of households in adjacent income levels, then it is usually a good idea to show that variation by using more intervals.

Bar charts can be useful, especially when the intervals are of equal width. But

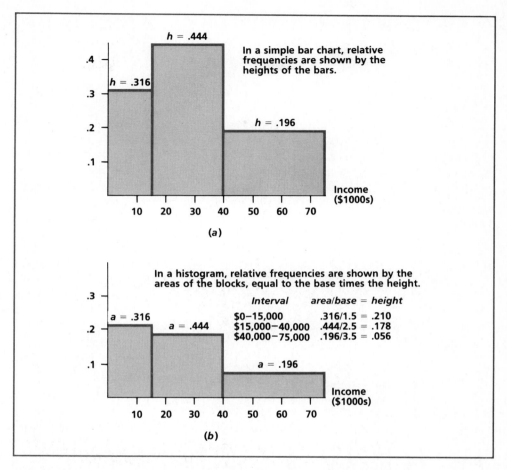

Figure 1.6 *(a) A Bar Chart and (b) a Histogram of U.S. Income Distribution, 1985, with Revised Intervals*

if the interval widths vary, bar charts can paint very misleading pictures, and which misleading picture is painted depends on the essentially arbitrary selection of the intervals. Histograms, however, give reasonably accurate pictures, which are not overly sensitive to the intervals picked. The histograms in Figures 1.5, 1.6(*b*), and 1.7(*b*) all show the distribution of households across income levels reasonably well. The only real difference is that five intervals provide a little more detail than three intervals.

Choosing Intervals

In general, histograms with only two or three intervals don't convey enough information to be worth the effort. If there are only two relative frequencies, these can

Disaggregating the Elderly

People over 65 years of age are often treated as a single group, perhaps because 65 is the traditional retirement age. However, the 6 million people in the United States who are 65 years of age and over are not a homogeneous group. For instance, life expectancies in the United States are now close to 80 years and people in their 80s may be very different from those in their 60s. Yet, with the exception of the decennial census, there are almost no data on the elderly that separate people in their 60s from those in their 80s. Even with the census data, analysts usually put everyone over 65 into a single category.

Figure 1 shows two overlapping graphs using 1980 census data for income by age.[5] In one graph all those over 65 are treated as a single group, with an average income of $7,505. In the second, the elderly are disaggregated into five-year intervals, showing that the annual income of those 65 to 70 is, on average, more than 50 percent higher than of those over 85. Medical expenses for people in their 80s are also very different from those for people in their 60s. In addition, the children of people in their 80s are close to 60 years of age themselves and may have difficulty providing financial and other resources to help care for their parents.

People in their 80s are, in many ways, very different from persons in their 60s. The U.S. Census Bureau has noticed this too. Barbara Boyle Torrey, the head of the Bureau's Center for International Research, explained: "When we disaggregated the data, breaking it down into five-year age groups, we found enormous differences among them. Right away we could see that one policy could never cover all of these groups. . . . Our views completely changed."[6]

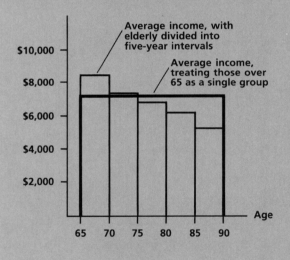

be told in words as easily as diagrams. A histogram is useful when there are several relative frequencies, because a figure is easier to absorb than a mass of numbers. At the other extreme, if the number of intervals gets too large, a histogram can turn into a jumble of jagged spikes, with each representing only a few observations. A rough rule of thumb is to divide the data into from five to fifteen intervals, using more intervals when you have more data. With 50 observations, five intervals is about as fine a gradation as you can have without spreading your data too thinly. With 50,000 observations, fifteen intervals could be supported. The exact number of intervals chosen is a matter of taste.

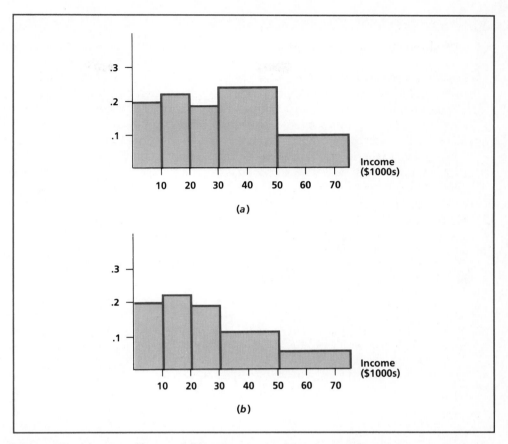

Figure 1.7   *(a) A Bar Chart and (b) a Histogram of U.S. Income Distribution, 1985, with Five Intervals*

The width and location of the intervals must also be chosen. The simplest decision is equally wide intervals covering all of the data. If the incomes range from $0 to $100,000 and five intervals are desired, then five $20,000 intervals are one possibility:*

$0–$20,000
$20,000–$40,000
$40,000–$60,000
$60,000–$80,000
$80,000–$100,000

*There is also a question of what to do with data right on an interval dividing line, such as a $20,000 income. Three alternatives are: (1) put them in the first interval (start the second interval at $20,000.01); (2) put them in the second interval (end the first interval at $19,999.99); and (3) alternate, putting half of the borderline data in the lower of the two intervals and half in the upper interval.

Misleading Picture Graphs

Graphic artists often use pictures instead of plain bars to enliven the bar graphs printed in newspapers and popular magazines. For example, Figure 1 shows a hypothetical comparison of the number of oranges that are produced, using two different fertilizers. Orange output is doubled using Harvey's Special Formula. This impressive accomplishment is made even more impressive by the use of oranges to construct the bar chart. Most peo-ple looking at such a chart will notice the areas occupied by the oranges, and not their heights. An orange with twice the diameter has four times the area. The visual impact of Figure 1 is that Harvey's Special Formula quadruples orange output. This dramatic impression may well have been the artist's intent, but it is still misleading.

These picturesque but exaggerated bar charts are quite common. I have seen charts using people, oil wells, dollar signs, and all sorts of things. Figure 2 shows an example

Figure 1   *A Misleading Comparison of Two Fertilizers*

Other choices are made for a variety of reasons. Often, the data are not evenly distributed over the entire range. There are many more people with incomes of $20,000 to $40,000 than with incomes of $60,000 to $80,000 or of $80,000 to $100,000. We may want to spend one of our five intervals distinguishing those with $20,000 and $30,000 incomes rather than separating people with $60,000 and $80,000:

$0–$20,000
$20,000–$30,000
$30,000–$40,000
$40,000–$60,000
$60,000–$100,000

from a newspaper advertisement boasting that "*The Times* has 2,244,500 readers every weekday—more than the next four area newspapers *combined*."[7] The height of *The Times* truck is about 3.6 times that of the nearest competitor, reflecting the fact that their circulation is 3.6 times that of *The Examiner*. But the width of *The Times* truck is also 3.6 times bigger, making the area of the truck $(3.6)^2 = 13$ times that of the nearest competitor. The circulation numbers say that *The Times* has somewhat more readers than the other four papers combined. The picture says that *The Times* dwarfs the competition. This visual image is very effective, but inaccurate.

One weekday issue	The Times	The Examiner	The Register	The Daily News	The Press-Telegram
Number of readers	2,244,500	624,000	485,500	350,500	310,500

Source: Harold R. Jacobs, *Mathematics: A Human Endeavor* (San Francisco: W.H. Freeman, 1982), p. 562.

Figure 2 *An Artist's Bar Chart*

Similarly, we may be more interested in seeing how many of those in the $0 to $20,000 range earn less than $10,000 than in seeing how many of those earning more than $40,000 are in the $40,000 to $60,000 range. If so, then we might choose these five intervals:

$0–$10,000
$10,000–$20,000
$20,000–$30,000
$30,000–$40,000
$40,000–$100,000

A rough rule of thumb is to choose the interval widths so that approximately equal amounts of data fall into each interval. The eleven income intervals in Table 1.1 have this characteristic.

Another consideration is that some intervals may have intrinsic interest. With income data, tax brackets or the official definition of the poverty level may dictate where an interesting interval begins or ends. The next example gives another illustration of how naturally interesting boundaries may influence the choice of intervals.

Years of Schooling

Table 1.2 shows the years of schooling received by 149,144,000 U.S. citizens who are age 25 and older. The Census Bureau has grouped these data into seven intervals to allow for natural breaks at the completion of elementary school (8 years), high school (12 years), and college (16 years).

Again, we have an endpoint problem with the unbounded category "16 years or more." We could omit these people, but it's difficult to ignore 20 percent of the population. So, I've just used 16–22 years as a "guesstimate."

Figure 1.8 shows the corresponding bar chart and histogram. The bar chart gives too much visual weight to college graduates and makes it appear that far more than a fifth of the population is in this category. The histogram puts things back into perspective. Figure 1.8(b) also shows the importance of the natural stopping points after 8 and 12 years of schooling. If we had more detailed dataon college graduates, there would undoubtedly be another large spike at 16 years.

Table 1.2   *Years of Schooling for Persons 25 Years Old and Over, 1987*

Years of Schooling	Number of Persons	Relative Frequency
0–4	3,579,000	.024
5–7	6,711,000	.045
8	8,650,000	.058
9–11	17,450,000	.117
12	57,570,000	.386
13–15	25,504,000	.171
16 or more	29,680,000	.199
Total	149,144,000	1.000

Source: U.S. Bureau of the Census, Current Population Survey, in *Statistical Abstract of the United States* (Washington, D.C.: U.S. Government Printing Office, 1989).

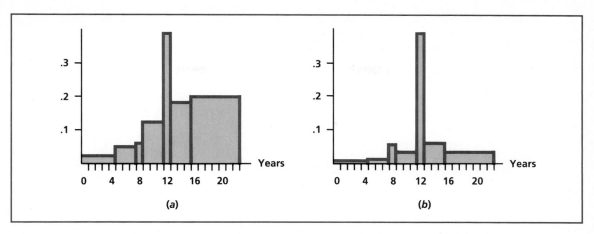

Figure 1.8   *(a) A Bar Chart and (b) a Histogram for Years of Schooling, 1987*

Identifying Authors

The authorship of important works is sometimes in doubt, because of the use of a pseudonym or inaccurate claims. Did Cervantes really write the novels about Don Quixote? Or, as many believe, did he take credit for the work of the servant, Sancho Panza? Did Conan Doyle write the Sherlock Holmes novels? Some suspect Dr. Watson. Were Shakespeare's plays written by Shakespeare or by Bacon, Marlowe, or someone else? Was Mark Twain the author of the letters signed "Quintus Curtius Snodgrass," chronicling a Southern soldier's adventures during the Civil War?

Two statisticians used *The Federalist* papers to show how such claims might be evaluated.[8] In 1787–1788, 77 essays urging New Yorkers to ratify the U.S. Constitution, signed with the pseudonym Publius, appeared in newspapers and then were published as a book, *The Federalist.* Five of these essays were written by John Jay, and the rest by Alexander Hamilton and James Madison, though in twelve cases, Hamilton and Madison (and later historians) disagreed as to which of the two was the author.

The statisticians looked at the frequency with which various words were used in papers known to be written by Hamilton and Madison and in the twelve disputed *Federalist* papers. For example, Hamilton almost always used "while" instead of "whilst" and Madison did the opposite; "whilst" appears in five of the disputed papers and "while" in none of them, suggesting that these five papers at least were authored by Madison. Similarly, "upon" was used frequently by Hamilton and seldom by Madison and appears in only one of the disputed papers. A less extreme case is the word "by," which was used by both, but more often by Madison, as shown in the frequency distributions in Figure 1.9. After applying similar data for 30 "marker" words to the twelve disputed papers, one by one, the statisticians concluded that the evidence strongly points to Madison as the author of each.

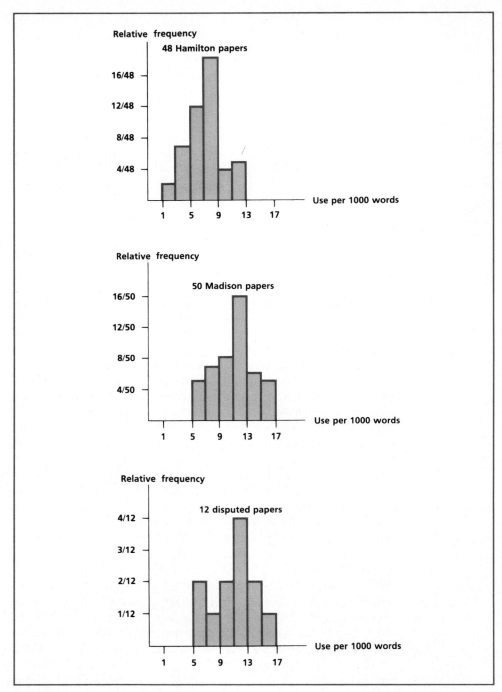

Figure 1.9   *Use of the Word "by"*

Rate per 1000 Words	Hamilton Papers	Madison Papers	Disputed Papers
1–3	2		
3–5	7		
5–7	12	5	2
7–9	18	7	1
9–11	4	8	2
11–13	5	16	4
13–15		6	2
15–17		5	1
17–19	—	3	—
Total	48	50	12

Stem-and-Leaf Diagrams (Optional)

In bar charts and histograms, the bars themselves do nothing more than show relative frequencies. Some well-meaning people add lines, blots, and splotches, apparently intending to enliven the graphs, but too often creating what Edward Tufte calls "Chartjunk," as in Figure 1.10—unattractive images that distract the reader and strain the eyes.[9]

John Tukey, a distinguished statistician, invented the stem-and-leaf diagram to allow the bars to convey more, not less, information about the underlying data: "If we are going to make a mark, it may as well be a meaningful one. The simplest—and most useful—meaningful mark is a digit."[10] In stem-and-leaf diagrams, as illustrated in Figure 1.11, the bar chart is turned sideways and each bar is made up of digits showing the details of the data in that interval. Each horizontal line is a stem and the digits on the stem are leaves. This particular figure is based on these data on the annual percentage rates of return for 24 mutual funds in 1986:

34.6 29.3 26.0 34.2 47.6   7.0 55.2 39.4 45.3 30.1 26.3 28.6
29.3 14.7 39.0 33.9 28.8 36.0 22.2 29.8 27.2 32.8 39.6   1.0

We're going to separate these data into the intervals 0–10 percent, 10–20 percent, and so on. So let's round off to the nearest percent:

35 29 26 34 48   7 55 39 45 30 26 29
29 15 39 34 29 36 22 30 27 33 40   1

and rank the data in order from smallest to largest, collecting the returns in the appropriate intervals:

0–10%:	1	7						
10–20%:	15							
20–30%:	22	26	26	27	29	29	29	29
30–40%:	30	30	33	34	34	35	36	39	39
40–50%:	40	45	48					
50–60%:	55							

Figure 1.10   *Chartjunk*

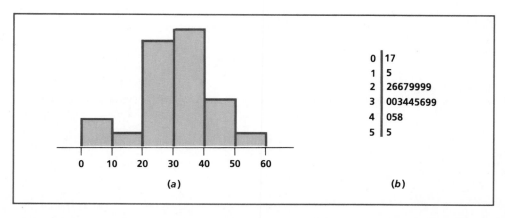

0	17
1	5
2	26679999
3	003445699
4	058
5	5

(a)   (b)

Figure 1.11   *Percentage Returns for 24 Mutual Funds, 1985*

The stem-and-leaf diagram in Figure 1.11 is a condensed, economical version of this display. The numbers to the right of the vertical line are the last digits of the numbers; the numbers on the left of the vertical line label the stem, showing all but the last digit. In comparison with the bar graph in Figure 1.11a, the stem-and-leaf diagram shows not only the relative frequencies in each interval, but also the last digit. Stem-and-leaf diagrams also can show the last two digits of each number and can be modified in a variety of clever ways to handle all sorts of data. Tukey's intention is to facilitate what he calls exploratory data analysis, "looking at data to see what it seems to say."[11]

EXERCISES

1.5 The U.S. Census Bureau reported that there were 116,649,000 males and 122,634,000 females living in the United States on July 1, 1985. Calculate the relative frequencies of male and female U.S. residents and draw a relative frequency diagram.

1.6 The 1980 census found that more than half of the people in Massachusetts identified with a single ancestry group. These single-ancestry people were divided as follows:

Irish	665,119
English	459,249
Italian	430,744
French	313,557
Portuguese	190,034
Polish	161,529
Other	947,003
Total	3,167,235

Calculate the relative frequencies and draw a relative frequency diagram.

1.7 Here is the actual age distribution of the U.S. civilian labor force in 1970 and the Bureau of Labor Statistics' projections for 1995. (All data are in millions.)

Age	1970	1995
16–20	7.2	7.8
20–25	10.6	13.3
25–35	17.0	34.4
35–45	16.5	36.8
45–55	16.9	24.9
55–65	11.3	11.0

Draw two histograms, one for 1970 and one for 1995, and describe any apparent differences.

1.8 The following data show the age distribution of deaths from scarlet fever in England and Wales in 1933.[12]

Age in Years	Number of Deaths
0–1	16
1–2	69
2–3	89
3–4	74
4–5	74
5–10	213
10–15	70
15–20	27
20–30	43
30–40	23
40–50	16
50–60	12
60–70	1
70–80	2

Construct a histogram for these data.

1.9 Regroup the household income data in Table 1.1 into the following three categories:

Low income:	$0–$24,999
Middle income:	$25,000–$34,999
Upper income:	$35,000–$74,999

Draw a bar chart and histogram using these three intervals and compare your graphs to Figures 1.4 and 1.5. Show any calculations that you find necessary in

Table 1.3   *Data for Exercise 1.11*

Fund	1984	1985	1986
State Bond Diversified	−2.2	27.3	34.6
Bull & Bear Capital Growth	−17.9	25.8	29.3
Composite Growth	−13.2	19.8	26.0
Keystone Custodian S-1	−14.8	17.7	34.2
AMEV Capital	−16.8	27.9	47.6
Valley Forge	7.7	10.7	7.0
Keystone International	−11.0	15.4	55.2
Bullock Dividend Shares	−3.1	31.0	39.4
Tri-Continental Corp.	−2.1	31.8	45.3
Hartwell Growth	−24.3	9.7	30.1
Fidelity Puritan	2.6	35.6	26.3
Columbia Growth	−18.0	34.7	28.6
Pennsylvania Mutual	−2.9	23.1	29.3
Vance, Sanders Special	−13.1	8.5	14.7
Value Line Leveraged Growth	−25.3	32.2	39.0
Exchange Fund of Boston	−12.8	30.2	33.9
IDS Progressive	−3.9	17.4	28.8
Smith, Barney Equity	−11.1	22.9	36.0
Mid-America Mutual	−9.0	32.4	22.2
Vanguard Windsor	2.8	42.6	29.8
Commonwealth Indenture A&B	−4.9	26.1	27.2
Acorn	−9.4	34.8	32.8
Keystone Custodian K-2	−16.5	21.8	39.6
ASA Limited	−9.5	−22.0	1.0

constructing these graphs. What specific differences do you notice in the appearance of this new bar chart? What specific differences do you notice in the new histogram?

1.10 Draw a bar chart and histogram for the income categories given in Table 1.1, omitting those earning $75,000 and over. Show any necessary cal-culations. Are these two graphs significantly different from Figures 1.4 and 1.5? What differences immediately catch your eye?

1.11 Table 1.3 gives the annual percentage returns for 24 randomly selected stock mutual funds. The 1986 data were used in Figure 1.11. Draw a stem-and-leaf plot using the 1985 data.[13]

1.3   THE CENTER OF A DISTRIBUTION

Statisticians often use one or two summary statistics other than graphs to charac-terize a frequency distribution. The most common of these summary statistics are the *mode, median,* and *mean* to identify the center of a distribution. The objective is usually to describe the "average" or "typical" value of the variable in question.

The **mode** identifies the most commonplace value of the variable being ana-lyzed. In Table 1.1, we can see that modal household income in 1985 is the $5,000–$9,999 interval. This is the "typical" household income in the sense that there are

Table 1.4 *Income Distribution in Smalland*

Income in 1985	Number of Households
$5,000	2
$10,000	1
$30,000	1
$40,000	1
$50,000	1
$60,000	1
$1,000,000	1

more people in this income range than in any other. However, the mode is often arbitrary in that it can be significantly altered by a simple reclassification of the intervals. If, for example, we put the $30,000–$34,999 and $35,000–$39,999 incomes into a single $30,000–$39,999 interval, then this would be the modal income interval.

In addition, the fact that some value is the mode is no guarantee that it is at all common. As an extreme example, consider the hypothetical country of Smalland. There are 8 people in Smalland, with the incomes shown in Table 1.4. The modal income is $5,000—yet most of the households earn considerably more than $5,000! The mode is seldom used in practice, except to answer the very special questions that it is designed to answer: What is the most watched television show? What is the best-selling automobile? What is the most common cause of death?

The **median** is the middle value of the data. Imagine that we arrange the data from lowest value to highest value and then, starting at each end of the data, count inward. When our two counts meet in the middle, we have located the median. (If there are an even number of observations, then we split the difference between the middle two.) Half of the data are less than or equal to the median and half are greater than or equal to the median. The median household income in Smalland is $35,000. This is the center of the income distribution, in that half of the population earned more than this and half earned less.

The arithmetic **mean** is the simple average value of the data.* If there are n values

$$X_1, X_2, \ldots, X_n$$

then the mean is the sum of these values, divided by n:

$$\text{mean} = \frac{X_1 + X_2 + \cdots + X_n}{n}$$

*There is also a *geometric mean,* which is found by multiplying rather than adding the data. The geometric mean is used, for example, in financial calculations of compounded rates of return.

The conventional symbol for the mean of a sample is $\overline{X}$ (pronounced "X bar"). Statisticians also use a shorthand notation when several numbers are to be added up:

$$\overline{X} = \frac{\sum_{i=1}^{n} X_i}{n}$$

The Greek letter Σ (capital "sigma") signifies the summation of a set of values X_1, $X_2, \ldots, X_n$ written as X_i for i ranging from 1 to n. If the range is obvious, then the limits can be omitted:

$$\overline{X} = \frac{\Sigma X_i}{n}$$

or even

$$\overline{X} = \frac{\Sigma X}{n}$$

There is an important conceptual distinction between a population and a sample—a sample is only part (and, often, a very small part) of a population and therefore subject to sampling error. To reinforce this distinction, different symbols are used for the sample and population means. A sample mean is labeled $\overline{X}$, while a population mean is labeled by the Greek symbol μ (pronounced "mu"):

$$\mu = \frac{\Sigma X}{N}$$

where the uppercase N is the number of values in the population, further reinforcing the sample-population distinction. Because the data in Table 1.3 are assumed to represent the entire population of Smalland, the symbol μ is appropriate here. The mean income in Smalland is

$$\mu = \frac{\$5{,}000 + \$5{,}000 + \$10{,}000 + \cdots + \$60{,}000 + \$1{,}000{,}000}{8}$$

$$= \frac{\$1{,}200{,}000}{8}$$

$$= \$150{,}000$$

The one millionaire pulls the arithmetic average far above the income levels of the other Smallanders. This is a good example of why you should not automatically interpret the statistical average as the typical value.

A national magazine once reported that a group of Colorado teachers had failed a history test, with an average score of 67. It turned out that four teachers had taken the test and one received a score of 20. The other three averaged 83.[14] Another example was provided in 1984 by Lawrence Simpson, director of the University of Virginia's office of career planning. Reporting on the jobs taken by the school's 1983 graduates, he noted that

> Our highest salaries were for graduates of the Department of Rhetoric and Communications Studies, where the beginning average pay was $55,000 a year. Of course, the average height was 6'5".
> Thanks, Ralph.[15]

Ralph Sampson, the first player picked in that year's National Basketball Association draft, pulled up the average salary and the average height.

The mean or average income tells us how much each person would earn if the total income were unchanged and everyone had the same income. If people do not, in fact, earn the same income, then there may be few people or even no one earning the "average" income. It need not be the case that half of the people earn more than the average income and half earn less. As in Smalland and at the University of Virginia, the mean can be affected greatly by a few extreme observations. It is for this reason that the U.S. Census Bureau reports both the mean and median household income. In 1985, the mean household income in the United States was $29,066 while the median was $23,618. Apparently, the mean was pulled up above the median by a few people with very high incomes.

Robust Estimation

The millionaire in Smalland is an "outlier," in that this income is very dissimilar from the other residents; Ralph Sampson's height was an outlier at the University of Virginia. In each case, the mean is distorted by such outliers in the sense that it gives a misleading description of the center of the distribution. Recognizing this potential distortion, some researchers discard data they consider to be outliers before calculating the mean.

Others routinely avoid the mean and use the median instead. The median is said to be "robust" or resistant to outliers, in that its value is little affected by a few extreme observations. Look again at the income distribution in Table 1.3. No matter whether the highest paid earns $100,000 or $100,000,000, the median is still $35,000. The mean, in contrast, is $37,500 if the highest income is $100,000, and $150,000 if it is $1,000,000.

An awkward illustration of this principle occurs whenever a large error is made in transcribing data. For example, in July 1986, the Joint Economic Committee of Congress released a report based on data compiled by researchers at the University of Michigan for the Federal Reserve Board. The report estimated that the share of the nation's wealth owned by the richest 0.5 percent of U.S. families had increased from 25 percent in 1963 to 35 percent in 1983. Politicians made speeches and newspapers nationwide reported the story with "Rich Get Richer" headlines. But some skeptics in Washington started poking through the numbers, rechecking the calculations, and a month later discovered that the reported increase was due almost entirely to the erroneous recording of one family's wealth as $200,000,000 rather than $2,000,000, an error that raised the mean wealth of the rich people surveyed by nearly 50 percent. Somewhere along the line, someone, most likely a keypuncher, typed two extra zeros and temporarily confused the nation.

Grouped Data

If all we have are grouped data, then the mean and the median can only be approximated. Let's use the eleven income intervals in Table 1.1 as an example. To calculate an average income, we need to estimate the incomes of the people within each interval. The simplest solution is to assume that everyone in an interval has an income equal to the midpoint of the interval.* Thus, the income for each group can be approximated as the product of the midpoint value and the number of households:

Income Interval	Approximate Midpoint	Number of Households	Total Income
$0 –$4,999	$2,500	6,783,000	$16,957,500,000
$5,000 –$9,999	$7,500	10,997,000	$82,477,500,000
$10,000–$14,999	$12,500	10,149,000	$126,862,500,000
$15,000–$19,999	$17,500	9,674,000	$169,295,000,000
$20,000–$24,999	$22,500	8,839,000	$198,877,500,000
$25,000–$29,999	$27,500	8,042,000	$221,155,000,000
$30,000–$34,999	$32,500	6,965,000	$226,362,500,000
$35,000–$39,999	$37,500	5,739,000	$215,212,500,000
$40,000–$49,999	$45,000	8,208,000	$369,360,000,000
$50,000–$74,999	$62,500	9,134,000	$570,875,000,000
$75,000 and over	$100,000(?)	3,927,000	$392,700,000,000
Total		88,458,000	$2,590,135,000,000

For the open-ended category "$75,000 and over," we can do little more than pluck a number (such as $100,000) out of the air, one that doesn't seem especially outrageous. The estimated mean is then given by dividing total income by the number of households,

$$\text{mean} = \frac{\$2,590,135,000,000}{88,458,000}$$

$$= \$29,281$$

which is gratifyingly close to the actual mean $29,066 that the Census Bureau calculated using data for all 88 million households.

---

*We would get the same answer, with a little more work, if we assumed incomes were evenly distributed throughout each interval.

The general formula is

$$\text{mean} = \frac{\Sigma(\text{midpoint})(\text{observations in interval})}{\text{total number of observations}}$$

There is an equivalent way of calculating a mean with grouped data, using relative frequencies. Suppose that we were trying to find the mean income of 10 people, 7 who earn \$20,000 and 3 who earn \$30,000. One way is the method used previously. That is, find the total income and then divide by the total number of people:

$$\text{mean} = \frac{7(\$20,000) + 3(\$30,000)}{10} = \frac{\$230,000}{10} = \$23,000$$

Alternatively, this computation can be rearranged as

$$\text{mean} = \left(\frac{7}{10}\right)\$20,000 + \left(\frac{3}{10}\right)\$30,000$$

$$= \$14,000 + \$9,000 = \$23,000$$

That is, the mean is a *weighted average* of the incomes, with the weights being equal to the relative frequencies. To turn this rule into a mathematical formula,

$$\text{mean} = \Sigma(\text{midpoint of interval}) \ (\text{relative frequency of observations in interval})$$

For the household income data, the first interval (\$0–\$4,999) has an approximate midpoint of \$2,500 and a relative frequency of 6,783,000/88,458,000 = .0767. The complete calculations give the same \$29,281 mean as before.

Income Interval	Approximate Midpoint	Relative Frequency	Product
\$0–\$4,999	\$2,500	.0767	\$191.70
\$5,000–\$9,999	\$7,500	.1243	\$932.39
\$10,000–\$14,999	\$12,500	.1147	\$1,434.16
\$15,000–\$19,999	\$17,500	.1094	\$1,913.85
\$20,000–\$24,999	\$22,500	.0999	\$2,248.27
\$25,000–\$29,999	\$27,500	.0911	\$2,500.11
\$30,000–\$34,999	\$32,500	.0787	\$2,558.98
\$35,000–\$39,999	\$37,500	.0649	\$2,432.93
\$40,000–\$49,999	\$45,000	.0928	\$4,175.54
\$50,000–\$74,999	\$62,500	.1033	\$6,453.63
\$75,000 and over	\$100,000(?)	.0444	\$4,439.39
Total		1.000	\$29,280.95

Now, let's estimate the median income using these grouped data. If we split 88,458,000 households down the middle, we have 44,229,000 in the lower half and 44,229,000 in the upper half. Counting in from either end of Table 1.1 puts us somewhere in the $20,000–$24,999 interval. For a more precise number, we can estimate where the median hides in this interval by assuming that incomes are evenly distributed throughout the interval. To start with, we know that the four lowest income intervals contain 37,603,000 households. To get the median 44,229,000th household, we need 6,626,000 more households out of the fifth interval, which contains 8,839,000 households. This represents about three-fourths of the people in that interval:

$$\frac{6,626,000}{8,839,000} = .750$$

If the incomes are evenly distributed, then the median income will be found about three-fourths of the way into the interval:

$$\text{median estimate} = \$20,000 + .750(\$25,000 - \$20,000)$$
$$= \$20,000 + \$3,750$$
$$= \$23,750$$

which is quite close to the actual median $23,618 that the Census Bureau found by having one of its giant computers count 44,229,000 households.

Midterm Scores

The nature of the mode, median, and mean, and a comparison among them can be brought out by an inspection of some relative frequency distributions. Here are some hypothetical data on midterm scores in a statistics class:

first test: 99   91   84   84   80   80   80   76   76   69   61

Figure 1.12 shows the relative frequency distribution. These scores were frankly rigged to be symmetrical with a single peak. Which score is the mode? The most common score is 80, which is the peak of the relative frequency distribution in Figure 1.12. What about the median? As there are 11 scores, we need to count in 5 from either end to locate the median. It, too, turns out to be 80. And the mean? If you add up these 11 scores and divide by 11, you will find that the data's symmetry implies that each score above 80 is balanced by a corresponding score below 80. Thus, the mean, too, is 80. These calculations are an illustration of this general principle.

If a relative frequency distribution is single peaked and perfectly symmetrical, then the mean, median, and mode coincide.

Conversely, if a relative frequency distribution has more than one peak or is asymmetrical, then the mean, median, and mode do not coincide.

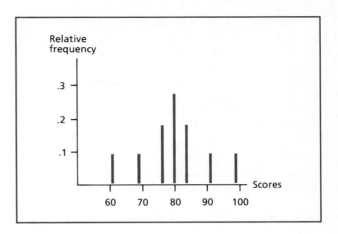

Figure 1.12 *Midterm Scores*

If the mean, median, and mode do differ, then we need to bear in mind what each measures. The mean tells us the arithmetic average value; the median tells us the middle value; and the mode tells us the most common value.

EXERCISES

1.12 Here are some data on United Airlines' annual profit, as a percentage of its net worth:

Year	Profit
1972	3.1
1973	7.2
1974	12.7
1975	−0.7
1976	2.4
1977	10.1
1978	25.6
1979	−6.3
1980	1.8
1981	−6.3
1982	1.0

Calculate the mean, median, and mode. Does each measure give a similar description of United Airlines' profits over this period? If they differ, which measures do you think are misleading?

1.13 Table 1.3 gives the annual returns for 24 randomly selected mutual funds. The average returns for the Standard & Poor's 500 index for these years were −4.8 percent in 1984, 30.9 percent in 1985, and 35.8 percent in 1986. Calculate the mean and median returns for these 24 mutual funds for each of these three years.

1.14 A Harvard Business School professor made a detailed study of 14 management successions, to see how these 14 new managers behaved during their first three years.[16] Here are his data on the timing of personnel changes made by these new managers.

Months After Taking Charge	Percentage of Personnel Changes
0–6	28%
6–12	9%
12–18	32%
18–24	15%
24–30	11%
30–36	4%

What is the modal interval? Estimate the median and mean number of months.

1.15 In estimating the average household income from Table 1.1, I assumed a $100,000 income for the households in the category "$75,000 and over." How much difference would it have made to my estimate of the mean if I had instead assumed an income of $90,000? Of $110,000?

1.16 A researcher is studying data on the financial wealth of 100 professors at a small liberal arts college. Their wealths range from $200 to $200,000, with a mean of $20,004 and a median of $12,500. When putting these data into a statistical computer program, the researcher mistakenly enters $2,000,000 for the person with the $200,000 wealth. How much does this error affect the mean and the median?

1.17 I used a computer program to simulate 600 rolls of a single six-sided die. The results were 96 ones, 94 twos, 110 threes, 91 fours, 103 fives, and 106 sixes. Calculate the relative frequencies for this sample and draw a relative frequency diagram. What are the mean, median, and modal die numbers? I used this computer program again and this time obtained 111 ones, 100 twos, 93 threes, 103 fours, 90 fives, and 103 sixes. What are the mean, median, and mode for this second sample? Now combine these two 600-roll samples into one 1200-roll sample. What are the mean, median, and mode for this combined sample? Which of these three measures of the center of a distribution is the least stable here? Is this a fluke, or do you think that this measure would probably be the least stable if we did experiment again?

1.4   THE SPREAD OF A DISTRIBUTION

The mean is, by far, the most commonly reported statistic. However, an average does not tell us the underlying variety that may, in fact, be of far greater interest. Sir Francis Galton once commented that

> It is difficult to understand why statisticians commonly limit their enquiries to Averages, and do not revel in more comprehensive views. Their souls seem as dull to the charm of variety as that of the native of one of our flat English counties, whose retrospect of Switzerland was that, if its mountains could be thrown into its lakes, two nuisances would be got rid of at once.[17]

For a more mundane example, imagine your state of mind a few weeks from now, when your statistics instructor hands back your midterms. You naturally will be interested in knowing how you did and, for comparison, how the class as a whole did. Perhaps the average score turns out to be 80. That's an interesting figure to know. But you may want to know a bit more about the scores. Did every student get 80? Or did half get 100 and half get 60? Or did most students get 100s and the rest get 0s?

The only way to know the complete distribution is to examine every single score. With a few tests, that's certainly feasible, but 300 tests would be trying. And no sane person wants to look at 88,458,000 household incomes. A starting place for organizing massive amounts of data is to sort the values into a small number of intervals. A relative frequency distribution or a histogram can then give a picture of the distribution of test scores, household incomes, or other data. However, graphs can also be difficult to digest and can, at times, be misleading. In addition, there are many occasions when graphs are just not sufficient or convenient, particularly when two sets of data, such as test scores for different classes or incomes in different years, are being compared. In these cases, statisticians use a few summary statistics to facilitate descriptions and comparisons.

While the mean, median, and mode describe the center of a distribution, other statistics gauge the spread of a distribution. One obvious choice is the **range,** which is simply the difference between the largest and the smallest values. In the 11 test scores discussed earlier,

first test: 99   91   84   84   80   80   80   76   76   69   61

the highest score is 99 and the lowest score is 61, giving a 38-point range. The professor might describe the test results by saying that, "The average score was 80, with a 38-point range from a high of 99 to a low of 61." The problem with the range is that it only looks at the extreme values, and the extremes are often atypical. In skating and diving contests, for example, the high and low scores of the judges are discarded to avoid their eccentricities.

Here is another set of 11 test scores that have a mean of 80 and a range of 38:

second test: 99   80   80   80   80   80   80   80   80   80   61

These scores are symmetrically centered at 80 with a 38-point range, from 99 to 61, the same as in the first set. But the range is now a very misleading indicator of the dispersion of the scores. Here, there are 9 students who did equally well and 2 outliers. Our common sense says that there is less dispersion in these scores than in the first test scores or in the following scores:

third test: 99   99   99   99   99   80   61   61   61   61   61

The problem with the range is that it only takes into account the spread between the two extreme values and ignores the dispersion among all the rest of the data. What we would really like is some measure of the overall or average spread of the data.

A natural attempt to measure the average spread in the data would be to compute the **deviation** of each data point from the mean and then calculate the average of these deviations. The deviation from the mean is simply each data point minus the mean. Those values greater than the mean give positive deviations; those values less than the mean give negative deviations. For the scores from the first test, as Table 1.5 shows, the average deviation about the mean turns out to be zero, which is hardly an informative statistic. In fact, the same result would have been obtained even if the data had not been symmetrical.

The average deviation from the mean is always zero.

The average deviation turns out to be zero because the positive and negative deviations from the mean offset each other. To eliminate this offsetting of positive and negative deviations, we could take the absolute value of each deviation and then calculate the *average absolute deviation,* which is the sum of the n absolute values of the deviations from the mean, divided by n. Table 1.5 shows that the average absolute deviation works out to be 6.9 for the first set of test scores. Do you think that the average absolute deviations for the second and third sets of test scores will be larger or smaller than 6.9? If you do the arithmetic, it will confirm your intuition that the average absolute deviation is less than 6.9 for the second test and

Table 1.5   *Three Measures of the Spread of a Distribution*

Test Score	Deviation from Mean	Absolute Deviation	Squared Deviation
99	19	19	361
91	11	11	121
84	4	4	16
84	4	4	16
80	0	0	0
80	0	0	0
80	0	0	0
76	−4	4	16
76	−4	4	16
69	−11	11	121
61	−19	19	361
Sum	0	76	1028
Average	0	6.9	93.5

greater than 6.9 for the third test. The average absolute deviation successfully gauges the dispersion in these data and it is easy to calculate and interpret. All in all, it is a very appealing measure of the spread of the data.

The only real problem with the average absolute deviation is that it presents formidable problems for mathematical analyses, and mathematicians love to analyze statistical measures—to see how they change as data are added, deleted, or combined. Mathematicians have developed differentiation, integration, and other powerful analytical tools, but these tools don't work very well with absolute values. It is largely for this reason that statisticians have developed another measure of the spread of the data.

Do you remember why we took absolute values in the first place? It was to keep the positive and negative deviations from offsetting each other. Another trick that accomplishes the same thing is to square each deviation from the mean. The squares of positive and negative deviations will both be positive. The average of these squared deviations is called the **variance:**

variance = average squared deviation from the mean

The population variance, using the Greek symbol σ (sigma), is

$$\sigma^2 = \frac{\Sigma(X - \mu)^2}{N}$$

The sample variance uses the sample mean $\overline{X}$ in place of the unknown population mean μ:

$$s^2 = \frac{\Sigma(X - \overline{X})^2}{n - 1}$$

Notice, too, that with a sample variance statisticians divide by $n - 1$, rather than n, because otherwise the sample variance tends to underestimate the population variance, as will be explained more fully in later chapters.

Table 1.5 shows the calculation of the variance for the first set of test scores, assumed to be the complete population. First, each deviation from the mean is calculated. Then these deviations are squared. Then the 11 squared deviations are added up and divided by 11. The result is a variance of 93.5.*

One problem with the variance is that the squaring of each deviation gives the variance a scale much larger than that of the original data. To offset this, statisticians also compute the **standard deviation:**

standard deviation = square root of the variance

With the exception of the scale difference, the variance and the standard deviation are interchangeable gauges of the spread of the data, in that a data set with a higher variance will also have a higher standard deviation. Although there are exceptions, it is also typically true that data with a large variance and standard deviation will also have a high average absolute deviation. The standard deviation of the first set of test scores, for instance, is $\sqrt{93.5} = 9.7$, while the average absolute deviation is 6.9. Computations of the average absolute deviation and the standard deviation both show that the second test scores are the least dispersed and that the third test scores are the most dispersed.

Another reason for the popularity of the standard deviation is that the histograms for many data, such as heights, IQ scores, and stock returns, follow a bell-shaped "normal" distribution, as illustrated in Figure 1.13 (and as you will see in later chapters). The standard deviation provides an informative gauge of the spread of a normal distribution: 68 percent of the data are within one standard deviation of the mean and 95 percent are within two standard deviations.† In Figure 1.13,

---

*When, as shown here, the calculations are made step by step, there are shortcut formulas:

$$\sigma^2 = \frac{\Sigma(X^2) - N\mu^2}{N} \qquad s^2 = \frac{\Sigma(X^2) - n\overline{X}^2}{n - 1}$$

In practice, variances are usually calculated by a computer program or a calculator that has these formulas built in.

†A famous theorem, Chebyshev's inequality, states that, no matter what the shape of the distribution, at least $1 - 1/k^2$ of the data are within k standard deviations of the mean; for example, at least three-fourths are within two standard deviations of the mean.

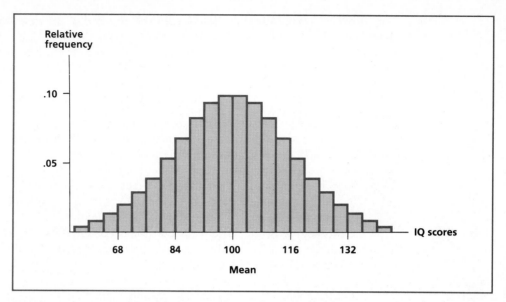

Figure 1.13   *A Bell-Shaped Histogram*

the data are IQ scores with a mean of 100 and a standard deviation of 16, so that 68 percent of the data lie between 84 and 116.

Often, data with a large mean also tend to have a large standard deviation. For instance, if there were 1,000 points possible on the test in Table 1.5 and the scores were 990, 910, and so on, the mean would be 800 instead of 80 and the standard deviation 97 instead of 9.7. Just looking at the standard deviation, it seems that there is more dispersion in the scores on a 1,000-point test than on a 100-point test. And there is, just looking at the absolute scores. But many find it more natural to scale the standard deviation by looking at its size relative to the mean:

$$\text{coefficient of variation} = \frac{\text{standard deviation}}{\text{mean}}$$

Here, the coefficient of variation is 9.7/80 = .12, whether the test is graded on a scale of 1 to 100 or 1 to 1,000.

Similarly, would you say that there is more dispersion in the incomes in Poorland, Midland, or Richland?

Poorland	Midland	Richland
$ 5,000	$20,000	$995,000
5,000	20,000	995,000
5,000	20,000	995,000
25,000	40,000	1,015,000

In absolute amounts, the dispersion is identical and the standard deviation of incomes in each country is the same ($8,660). But it is tempting to think in relative amounts, noting that one of the people in Poorland earns five times as much as everyone else, while the income of the highest-paid person in Midland is only twice that of the poorest. In Richland, everyone is a millionaire, give or take a millionaire's loose change. If a relative measure of dispersion makes the most sense to you, then so should the coefficient of variation.

The Profitability of Five Companies

Table 1.6 shows the annual profits (relative to net worth) for five very different companies over the ten-year period 1972–1981. These are a mass of data that is not easy to absorb. The mean and standard deviation can help summarize the data in a way that facilitates meaningful comparisons:

	Annual Profit, 1972–1981	
Company	*Mean*	*Standard Deviation*
American Water Works	7.61%	0.64%
Browne & Sharpe	7.62%	7.01%
Campbell Soup	13.56%	0.99%
McDonald's	20.04%	0.96%
Pan American	−0.98%	13.46%

American Water Works, which is the largest private water company in the United States, sells water in twenty states at prices set by regulatory commissions to give the company a "fair" rate of return. Over this ten-year period, the average annual profit rate was 7.61 percent. The small standard deviation accurately reflects the fact that the company's profit rate varied little from year to year.

Brown & Sharpe earned almost exactly the same average return as American

Table 1.6 *Five Companies' Profits, As a Percentage of Net Worth*

Year	American Water Works	Brown & Sharpe	Campbell Soup	McDonald's	Pan American
1972	7.2	2.1	11.0	17.9	−7.0
1973	6.6	5.8	13.6	19.9	−4.7
1974	6.8	6.5	13.7	20.4	−26.7
1975	7.3	−5.4	13.1	21.0	−18.0
1976	7.8	2.3	14.3	21.0	−2.3
1977	8.0	7.7	13.9	21.3	12.5
1978	7.7	15.2	14.2	20.4	18.3
1979	7.5	18.4	14.8	19.8	10.5
1980	8.7	16.8	14.0	19.4	10.0
1981	8.5	6.8	13.0	19.3	−2.4

The SAT Scores of College Applicants

Table 1 shows the distribution of verbal Scholastic Aptitude Test (SAT) scores for male and female applicants for admission to Pomona College in 1982. A natural question is whether the distribution of scores is similar for males and females, and an answer is provided by a computation of the mean and standard deviation for each sex. Because these are grouped data, we will assume that everyone in each interval received the same score, equal to the interval midpoint. The details of the female computation are shown in Table 2. The means and standard deviations turn out to be

Table 1   *Verbal SAT Scores of Female and Male Applicants for Admission to Pomona College in 1982*

Verbal SAT Score	Female Applicants	Male Applicants
750–800	9	10
700–750	48	48
650–700	114	98
600–650	157	152
550–600	186	159
500–550	187	143
450–500	105	98
400–450	87	40
350–400	20	23
300–350	17	15
250–300	5	9
200–250	5	5
Total	940	800

	Female	Male
Mean	558	564
Standard deviation	98	99

The average score of the male applicants is slightly higher than the average score of the female applicants. The standard deviations are virtually the same, indicating that there is the same spread in the female and male scores. Apparently, the distribution of verbal SAT scores is very similar for female and male applicants to this college.

Water Works, but with a much higher standard deviation. As a manufacturer of tools and equipment, their sales and profits are sensitive to the nation's economy. For instance, they lost money during the 1975 recession, but they made a lot of money in 1979, when the economy was relatively strong. Over this ten-year period, American Water Works and Brown & Sharpe were, on the average, equally profitable, but American Water Works's profits were much more stable.

Campbell Soup is an interesting contrast; their standard deviation is almost as low as that of American Water Works, but their average profit rate is nearly double that of the first two companies. Year in and year out, they made about the same 13 percent to 14 percent profit. McDonald's (the hamburger place) has done even better, earning an average profit of 20 percent, with the same low standard deviation as Campbell Soup. This is why so many investors like McDonald's stock, and

Table 2   *Computation of Mean and Standard Deviation for Female Applicants*

Interval Midpoint	Midpoint Squared	Number of Applicants	Midpoint Times Number	Midpoint Squared Times Number
775	600,625	9	6,975	5,405,625
725	525,625	48	34,800	25,230,000
675	455,625	114	76,950	51,941,250
625	390,625	157	98,125	61,328,125
575	330,625	186	106,950	61,496,250
525	275,625	187	98,175	51,541,875
475	225,625	105	49,875	23,690,625
425	180,625	87	36,975	15,714,375
375	140,625	20	7,500	2,812,500
325	105,625	17	5,525	1,795,625
275	75,625	5	1,375	378,125
225	50,625	5	1,125	253,125
Total		940	524,350	301,587,500

$$\text{mean} = \frac{\Sigma X}{N} = \frac{524,350}{940} = 557.82$$

$$\text{variance} = \frac{\Sigma(X^2) - N(\text{mean})^2}{N} = \frac{301,587,500 - 940(557.82)^2}{940} = 9,674.62$$

standard deviation = square root of variance = 98.35

why its price per share doubled over this ten-year period (from $24 a share at the beginning of 1972 to $48 a share at the end of 1981). At the other extreme is Pan American, the second largest U.S. airline, which lost money in six of these ten years. Their profits ranged from a high of 18.3 percent to a low of−26.7 percent. The average was −0.98 percent with a whopping standard deviation of 13.46 percent. No wonder the price of its stock fell from a high of $17.80 a share in 1972 to a low of $2.40 per share in 1981.

Box Plots (Optional)

The box plot is another creation of John Tukey, designed to emphasize some important features of the data in a clear, economical fashion. Table 1.7 shows the

Table 1.7   *Population of 15 Largest Metropolitan Areas, in 100,000*

	1960	1970	1980	1980 Rank	
New York	95	100	91	1	
Los Angeles	60	70	75	2	
Chicago	62	70	71	3	
Philadelphia	43	48	47	4	
Detroit	40	44	44	5	← fourth
San Francisco	26	31	33	6	
Washington, DC	21	29	31	7	
Dallas	17	24	30	8	← median
Houston	14	20	29	9	
Boston	27	29	28	10	
Nassau-Suffolk	20	26	26	11	
St. Louis	21	24	24	12	← fourth
Pittsburgh	24	24	23	13	
Baltimore	18	21	22	14	
Minneapolis	16	20	21	15	

fifteen largest metropolitan areas in the United States in 1980. Notice, first, that I have rounded off these data severely, to the nearest 100,000, leaving, with one exception, just two digits of data. The data in Tables 1.1 and 1.2, in contrast, are only rounded off to the nearest 1,000 and consequently show five nonzero digits. Tables 1.1 and 1.2 look cluttered, while Table 1.7 shows the essential features of the data that are of interest—the ranking of the cities and the level and changes in population to two digits. In addition, since the definition of a metropolitan area is subjective and the population counts are inexact, a reporting of the estimates to five digits gives an aura of precision that is not really warranted.

Let's look at the 1980 data, which are already ranked from the largest city to the smallest. The median city is Dallas, ranked 8 with a 1980 population of 30. We can also identify what Tukey calls the "fourths," which are halfway between the extremes and the median. Halfway between ranks 1 and 8, we find ranks 4 and 5. Splitting the difference between 44 and 47 and rounding off, we get a population of 45. The other fourth is between ranks 11 and 12 with a population of 25. The "fourth spread" is the difference between these fourth values: $45 - 25 = 20$:*

These values are shown graphically in Figure 1.14 in what is called a box plot or, even more descriptively, a box-and-whisker diagram. In Figure 1.14(*a*) the ends of the box are drawn at the fourths, 25 and 45, so that the box encompasses roughly

---

*There are slight technical differences between Tukey's fourths and the quartiles and interquartile ranges used by others.

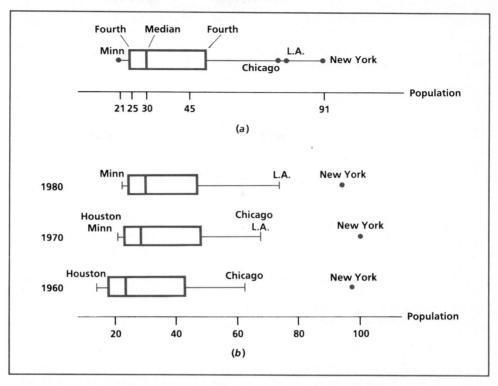

Figure 1.14 *Box-and-Whisker Diagrams of the 15 Largest Metropolitan Areas, in 100,000*

half the data. The median is denoted by the line inside the box. The single lines outside the box (the "whiskers") show the minimum and maximum values. Often, as here, it is helpful to label some of the extreme points. This figure then shows, in a relatively simple way, the median, the location of the center half of the data, and the extremes—information that is not conveyed very effectively just by the values of the mean (39.7), standard deviation (21.2), and skewness coefficient (1.3). A histogram would give additional detail, but is not as easily absorbed. A box plot is also relatively robust, in that its shape is resistant to the "wild" values of a few outliers. The same cannot be said for the mean, standard deviation, and skewness coefficient.

Figure 1.14(*b*) shows box plots for 1960, 1970, and 1980. Here I have followed Tukey's recommendation of stopping each whisker at the innermost identified value. One definition of an **outlier** is a value whose distance from the box is more than 1 1/2 times the length of the box. By this criterion, New York is the only outlier. Each whisker is drawn to the most extreme point that is not an outlier and ends in a "fence."

EXERCISES

1.18 Here are some annual data on profits (as a percentage of net worth) for Ford, General Motors, and Winnebago (a manufacturer of motor homes):

Year	Ford	General Motors	Winnebago
1972	14.7	18.7	25.0
1973	14.2	19.3	−8.9
1974	5.8	7.6	−10.3
1975	3.7	9.6	−3.7
1976	13.9	20.3	10.0
1977	19.8	21.4	4.1
1978	16.4	20.2	1.4
1979	11.2	15.2	−5.5
1980	−18.1	−4.3	−21.7
1981	−14.4	1.9	6.5

Calculate the mean and standard deviation of each of these companies' profits. Which of these three companies' stocks would you guess performed the best and the worst over this ten-year period?

1.19 Languages vary in the number of syllables in their words. Table 1.8 gives some relative frequencies for four different languages.[18]   **a.** Sketch a relative frequency diagram for each of these four languages.   **b.** Calculate the mean, median, and mode for each of these four languages.   **c.** Calculate the average absolute deviation, standard deviation, and coefficient of variation for each of these four languages.   **d.** Based on these descriptive statistics, describe (in words) the differences among these four languages.

1.20 Table 1.3 shows annual returns for 24 mutual funds. Draw three side-by-side box-and-whiskers diagrams, one for each year.

1.21 Here are some U.S. household income data, for whites and blacks:

Income in 1985	Relative Frequencies White	Black
$0–$4,999	.0641	.1762
$5,000–$9,999	.1166	.1879
$10,000–$14,999	.1115	.1405
$15,000–$19,999	.1080	.1242
$20,000–$24,999	.1014	.0852
$25,000–$29,999	.0933	.0738
$30,000–$34,999	.0819	.0553
$35,000–$39,999	.0673	.0452
$40,000–$49,999	.0975	.0539
$50,000–$74,999	.1101	.0467
$75,000 and over	.0482	.0110
Total	1.0000	1.0000

Assume that the average income for those in the category "$75,000 and over" is $100,000 for white households and $90,000 for black households. Calculate the mean, median, and modal income for white and black households. Calculate the average absolute deviation, the variance, the standard deviation, and the coefficient of variation of incomes for blacks and whites. What are the two pictures drawn by these various statistics?

Table 1.8   *Data for Exercise 1.19*

Number of Syllables	Relative Frequencies Arabic	English	German	Japanese
1	.23	.71	.56	.36
2	.50	.19	.31	.34
3	.22	.07	.09	.18
4	.05	.02	.03	.09
5	.00	.01	.01	.02
6	.00	.00	.00	.01
Total	1.00	1.00	1.00	1.00

1.22 Production costs usually decline over time, as a company or industry produces more of a product. Some of this decline is attributable to outside technological developments, some to improvements learned from production itself, and some are scale economies (e.g., if certain fixed expenses are required no matter how much is produced, then the cost per unit declines with the size of production). "Literally thousands of studies have shown that production costs usually decline by 10 percent to 30 percent with each doubling of cumulated output. For example, if the thousandth unit of a product costs $100, the two thousandth unit will normally cost between $70 and $90."[19] This decline often is labeled the *experience curve,* with the slope of the curve describing the relative cost as production doubles; in the quoted example, the slope would be between 70 percent and 90 percent. The quoted author summarized studies of 102 products:

Experience Curve Slope	Number of Cases
60–65%	3
65–70%	3
70–75%	10
75–80%	23
80–85%	30
85–90%	26
90–95%	6
95–100%	1

Draw a histogram and estimate the mean and standard deviation of the slope. Does this distribution seem to conform to the normal distribution rules that approximately 68 percent of the data are within one standard deviation of the mean and 95 percent are within two standard deviations?

1.5 THE USES AND MISUSES OF PROBABILITY AND STATISTICS

Over the years, the uses of probability and statistics have multiplied as people have recognized the prevalence of uncertainty and the power of this analysis. When Gregor Mendel first experimented with hybrid sweet peas, he used probabilities to explain his results. His experiments provided the basis for modern genetics, in which probabilities are still central. Scientists also use probability analysis to describe the movements of atoms, the decay of radioactive particles, and the life cycle of bacteria. Probabilities and statistics are used to evaluate medicines, judge the health effects of smoking, and explain longevity. One scientist has written that probability theory is now "a cornerstone of all of the sciences."[20]

Businesses use probability and statistics to gauge manufacturing precision, estimate demand, and design marketing strategies. Economists use probability and statistics to evaluate investments, describe financial markets, predict consumer spending, and formulate economic policies. Other social scientists use probability and statistics to conduct opinion polls, measure discrimination, and gauge the deterrence effects of capital punishment.

ASCAP (American Society of Composers, Authors, and Publishers) allocates hundreds of millions of dollars in performance-right royalties based on a statistical sampling of radio, television, jukebox, and even background music. Probability and statistics have long been used to decipher secret codes and are now being used in computer programs that translate spoken words into writing. These speech recognition programs use test periods to learn the user's pronunciation. Then, when the computer hears a sound, it selects a set of similar-sounding candidate words and, based on the context, computes probabilities and chooses the most likely word intended by this sound. For instance, if the user says, "Two computers are better than none," the machine will consider "to," "too," or "two" as the first word and "none" or "nun" as the last word. Using grammatical patterns and word frequencies, it will settle on "two" and "none" by deciding that, statistically, the sentence is most likely to make sense this way.

Probabilities and statistics abound in sports. Basketball players are told to take "high percentage shots," which are shots made from a distance and angle where a particular player has a high probability of making a basket. Football teams use detailed statistical models to evaluate prospective players. Football coaches ranging from Plano High to the Dallas Cowboys estimate their opponent's "tendencies," for instance, the plays a particular team is most likely to call when it has the ball inside its own 20-yard line on the third down with 3 yards to go for a first down. When a football coach must choose between punting, trying for a field goal, or trying for a first down, probabilities are racing through the coach's mind.

Baseball is probably the most statistical sport, starting with lifetime batting averages and running through a certain pitcher's record against Robin Yount during night games in August. Good baseball managers are continually "playing the percentages."[21] They bring in a left-handed pitcher to face a left-handed batter. They instruct a .245 hitter to make a sacrifice bunt to advance a runner to second base. They put their good hitters at the start of the lineup and their weak hitters at the end. In the late innings, they take out a good-hitting, weak-fielding player who has just batted and bring in a weak-hitting, good-fielding player. In all of these cases and many more, the good manager is trying to use statistics to make victory more probable.

Fallible humans also misuse probabilities and statistics. We see skill where there is luck. We see psychic phenomena where there is chance. We confirm flimsy theories with coincidences. We distort and misreport data to prove our prejudices. We take foolish chances, counting on an erroneous "law of averages" to bail us out. We waste money on worthless gambling systems. We play inferior backgammon, bridge, and poker. We buy too much insurance and too many different stocks. And we neglect many important risks.

I wrote this textbook to introduce you to probability and statistics. It is not encyclopedic and it is not just a collection of arcane formulas and crunched numbers. Instead, I've tried to emphasize some important principles that you can use all your life. You don't need to memorize 250 formulas or add up 2,500 six-digit numbers. You can always look up the formulas and have a computer do your arithmetic. What you really need is an understanding of what you are doing.

You will see that the world is filled with uncertainties. You will find out what probabilities are and what they can be used for. You will learn how to figure probabilities correctly and how to recognize the everyday errors in the calculations of others. You will see how data can be used and misused in estimating probabilities. You will learn why the unemployment rate is "± 0.2 percent" and why an election prediction is "± 3 percent." You will understand what it means to say that the evidence supports or rejects someone's theory. You will learn how to test theories yourself. You will find out what is meant by the commonplace report that researchers have found a *statistically significant relationship* between this and that. You will see why expert studies sometimes disagree and how statistical tests can be abused. By learning these things, you will make this course one of the most interesting and useful courses that you will ever take.

EXERCISES

1.23 Only 10 percent of all divorces involve a woman with a college degree and only 5 percent involve a woman with eight or fewer years of education. Do these data show that a man should marry a woman who either has a lot of education or very little? What other data would you like to see before jumping to this conclusion?

1.24 Professor Robinson is contemplating job offers in San Francisco and Albuquerque. Temperature is very important to Ms. Robinson because she cannot jog or write scholarly papers when the weather is either very hot or very cold. A check of the *Statistical Abstract* reveals that, over the thirty-year period of 1941–1970, the average temperature was 56.9°F in San Francisco and 56.8°F in Albuquerque. Should Professor Robinson therefore conclude that the temperatures are quite similar in these two cities?

1.25 An old joke is that a certain economics professor left Yale to go to Harvard and thereby improved the average quality of both departments. Is this possible?

1.26 Ann Landers once wrote, "Nothing shocks me anymore, especially when I know that 50 percent of the doctors who practice medicine graduated in the bottom half of their class."[22] Is Ann Landers using the mean, median, or mode to gauge doctor skills? Is there any relationship between the competency of the medical profession and the fact that 50 percent of all doctors are below average?

1.6 SUMMARY

Probabilities tell us how often we can anticipate observing certain outcomes, given certain assumptions about the underlying population. Statistical inference is the use of observed sample data to draw inferences about the characteristics of this population. For instance, probabilities tell us, if the voters are evenly divided, the chances that a poll of 400 voters will find 60 percent preferring one candidate. Statistical inference uses the observed 60 percent preference to estimate what fraction of the population prefer this candidate. Because a sample is only part of the population, this estimate must allow a margin for sampling error such as ± 5 percent.

Descriptive statistics are simple techniques for briefly describing a population or a sample. A frequency distribution shows the fraction of a population or sample that has certain characteristics. A histogram uses areas to show relative frequencies. The height of each block in a histogram is equal to the relative frequency divided by the width of the interval. In constructing histograms, the data are usually grouped so that there are between 5 and 15 intervals, with 15 intervals used only when there are thousands of observations and fewer intervals used when there are fewer data.

The mean, median, and mode are used to measure the center of a set of data.

The mean is the average value of the data; the median is the middle value; and the mode is the most common value. The formula for the mean is

$$\overline{X} = \frac{\Sigma X}{n} \quad \text{or} \quad \mu = \frac{\Sigma X}{N}$$

depending on whether the data represent a sample or complete population. With grouped data, the mean can be estimated by multiplying the midpoint of each interval by the number of observations in that interval:

$$\text{mean} = \frac{\Sigma (\text{midpoint}) \, (\text{observations in interval})}{\text{total number of observations}}$$

The variance and standard deviation gauge the spread of a distribution. The variance is calculated as the average squared deviation from the mean:

$$s^2 = \frac{\Sigma (X - \overline{X})^2}{n - 1} \quad \text{or} \quad \sigma^2 = \frac{\Sigma (X - \mu)^2}{N}$$

The standard deviation is the square root of the variance.

REVIEW EXERCISES

1.27 A recording of the degree of cloudiness at Greenwich each day during July in the years 1890–1904 obtained the following results:[23]

Degree of Cloudiness	Number of Days
0	320
1	129
2	74
3	68
4	45
5	45
6	55
7	65
8	90
9	148
10	676

Draw a relative frequency diagram and calculate the mean degree of cloudiness. Is the mean an accurate description of a typical Greenwich day?

1.28 Here is the age distribution of the cars in use in the United States in 1984:

Age (Years)	Number (Millions)
0–3	22.0
3–6	26.8
6–9	26.8
9–12	17.4
12 and up	19.0

The average age of the cars 12 years and older is about 16 years. Draw a histogram, letting the oldest interval run from 12 to 20 years. Estimate the mean age.

1.29 A study of the ages at which Australian men married during the years 1907–1914 found:[24]

Age	Number of Men
15–18	294
18–21	10,995
21–24	61,101

Age	Number of Men
24–27	73,054
27–30	56,501
30–33	33,478
33–36	20,569
36–39	14,281
39–42	9,320
42–48	11,006
48–54	5,810
54–60	2,755
60–66	1,459
66–78	1,143
78–90	119

Use a histogram to show the distribution of these marriage ages. What is the modal age interval? Estimate the mean and median age at marriage.

1.30 Here is the age distribution of adult U.S. stockholders in 1983:

Age	Number
21–35	10,552,000
35–45	8,346,000
45–55	6,586,000
55–65	6,850,000
65–75	7,277,000

Draw a bar chart and histogram, showing any necessary calculations. In what ways, if any, does the bar chart seem to give a misleading picture of the age distribution?

1.31 *(continuation)* Now regroup these data into three age intervals: 21–35, 35–55, and 55–75. Draw a bar chart and histogram for these three categories. Are there any striking differences between these graphs and the ones drawn in the preceding exercise?

1.32 Here are the number of U.S. residents, by age, below the poverty level in 1984:

Age	Number
0–16	12,132,000
16–22	3,954,000
22–45	9,886,000
45–65	4,398,000
65–80	3,330,000

Draw a histogram and estimate the mean and median age.

1.33 Regroup the schooling data in Table 1.2 into the three categories: 0–11 years, 12 years, and 13–22 years. Draw a bar chart and histogram using these intervals. Show any necessary calculations. Compare your two graphs with those in Figure 1.8. Did this interval regrouping cause any major changes in the appearance of the bar chart and the histogram?

1.34 The National Bureau of Economic Research (NBER) identifies the beginning of a recession (when economic activity peaks) and its end (when the economy hits its trough and begins to recover). The following table shows the beginning and end of each recession since 1920:

U.S. Business Cycles Peak to Trough	Recession (Months)	Preceding Expansion (Months)
Jan 1920–Jul 1921	18	10
May 1923–Jul 1924	14	22
Oct 1926–Nov 1927	13	27
Aug 1929–Mar 1933	43	21
May 1937–Jun 1938	13	50
Feb 1945–Oct 1945	8	80
Nov 1948–Oct 1949	11	37
Jul 1953–May 1954	10	45
Aug 1957–Apr 1958	8	39
Apr 1960–Feb 1961	10	24
Dec 1969–Nov 1970	11	106
Nov 1973–Mar 1975	16	36
Jan 1980–Jul 1980	6	58
Jul 1981–Dec 1982	18	12

Calculate the mean and median lengths of these 14 recessions and of these 14 expansions. Why does each of these means not equal the corresponding median?

1.35 Use the profits data in Table 1.6 to draw five box-and-whiskers diagrams, one for each company. For ease of comparison, draw all five diagrams on the same graph.

1.36 When, if ever, will a bar chart and histogram look the same?

• **1.37** The U.S. household income data in Table

1.1 includes the open-ended category "$75,000 and over." Estimate the average income of this upper 4 percent of the households using the Census Bureau's calculation that the average income for all 88,458,000 households was $29,066.

1.38 It has been said that, "The average stockholder is a 53-year-old white woman." What does "average" mean here?

• **1.39** What is your mean speed if (a) you drive 20 miles per hour for one hour and 40 miles per hour for one hour? (b) you drive 20 miles per hour for twenty miles and 40 miles per hour for twenty miles?

1.40 During the week ending February 8, 1987, the most popular television show in America was The Cosby Show. does this mean that most people in the United States watched the show that week?

1.41 Roll a pair of standard dice twenty times, and each time record the sum of the two numbers rolled. Calculate the mean, median, and mode. Repeat this experiment five times. Which of these three measures seems to be the least stable?

1.42 Most countries only have a handful of banks. The United States, in striking contrast, has nearly 15,000 banks, many of which are local "mom-and-pop" banks, as indicated by these 1984 data. (A bank with $50 million in assets need only have $2 million in net worth.)

Assets	Commercial Banks
Less than $250,000	2
$250,000–$1,000,000	9
$1,000,000–$5,000,000	320
$5,000,000–$10,000,000	1150
$10,000,000–$25,000,000	4064
$25,000,000–$50,000,000	3761
$50,000,000–$100,000,000	2741
$100,000,000–$500,000,000	1956
More than $500,000,000	478

(The average asset size of the 478 largest banks is $2.8 billion.) Estimate the overall mean and median asset size and explain why these two values are not equal.

1.43 The alumni of Clareville's Class of 1919 seem to be getting younger every year. In 1984, their average age was 87; in 1985 it was 86. How can this be?

• **1.44** The U.S. Census Bureau reported that the household income data shown in Table 1.1 have a mean of $29,066 and a median of $23,618. Why do you think that the mean and the median are not the same? What, specifically, pulls the mean up above the median?

1.45 Federal civilian white-collar employees are paid according to a General Schedule (GS) grade. Here are some 1984 data on the pay ranges and the number of employees within each of these five ranges:

	Annual Pay	Number
GS 1– 6	$8,676–$19,374	541,000
GS 7–10	$16,559–$29,003	360,000
GS 11–12	$24,508–$38,185	362,000
GS 13–15	$34,930–$63,115	220,000
GS 16–18	$59,945–$78,184	1,000

What is the modal pay interval? Estimate the mean and median pay.

1.46 You are going to roll a standard die four times and then calculate the mean and standard deviation for these rolls. What is **a.** the largest possible mean? **b.** the smallest possible mean? **c.** the largest possible standard deviation? **d.** the smallest possible standard deviation?

1.47 The following table compares five characteristics of privately owned one-family houses built in 1970 and in 1984: **a.** The average home built in 1984 had between 1,200 and 2,000 square feet, 3 bedrooms, 2 baths, electrical heat, and central air conditioning. Is this "average" a mean, median, or mode? **b.** Estimate the mean and median square feet in 1970 and 1984. The median price of a new one-family home was $23,400 in 1970 and $79,900 in 1984. Use these prices and the median square footage to compare the prices per square foot in 1970 and 1984. **c.** Does there appear to be any trend in the number of bedrooms or bathrooms between 1970 and 1984?

	1970	1984
Floor Area		
Under 1,200 sq. ft	36%	22%
1,200–2,000 sq. ft	43%	51%
More than 2,000 sq. ft.	21%	27%

	1970	1984
Bedrooms		
2 or less	13%	24%
3	63%	59%
4 or more	24%	18%
Bathrooms		
Fewer than 2	52%	24%
2	32%	48%
More than 2	16%	28%
Heating Fuel		
Electricity	28%	48%
Gas	62%	45%
Oil	8%	2%
Other	1%	5%
Central Air Conditioning		
With	34%	71%
Without	66%	29%

1.48 Construct a sample of dice rolls by rolling a standard six-sided die thirty times. Construct a table showing the absolute number and the relative frequency with which each of the six possible numbers appeared. Draw a relative frequency diagram for this sample.

• **1.49** At the end of each baseball season, newspapers publish the final batting averages for all players with at least 300 times at bat. Go to a library and find the 1989 National League batting averages. Group these averages into intervals .180–.199, .200–.219, and so on, and then draw a histogram.

•• **1.50** Table 1.5 shows that the average deviation from the mean of eleven symmetrical midterm scores works out to be zero. Show that the average deviation from a mean is always zero.

1.51 In January 1990, ten experts were asked to forecast the three-month Treasury bill rate on July 15, 1990. Calculate the mean, standard deviation, and coefficient of variation of their ten forecasts.

7.8%	8.1%	7.4%	7.6%	7.8%
8.0%	7.6%	7.5%	7.8%	8.0%

1.52 Are great discoveries generally made by people who are young and vigorous or by persons who are old and wise? The following is a tabulation of the ages at which 12 scientists made great discoveries.[25] Use these data to calculate the mean and median ages at which these great discoveries were made.

Scientist	Discovery	Age
Copernicus	Earth revolves around the sun	40
Galileo	Laws of astronomy	34
Newton	Motion, gravitation, calculus	23
Franklin	Nature of electricity	40
Lavoisier	Burning as oxidation	31
Lyell	Earth evolved gradually	33
Darwin	Natural selection in evolution	49
Maxwell	Equations for light	33
Curie	Radioactivity	34
Planck	Quantum theory	43
Einstein	Special relativity	26
Schroedinger	Equations for quantum theory	39

1.53 Entrepreneurs who start new companies generally work long hours trying to make their business succeed. Here are some 1989 data on the number of hours worked each week by new business owners:[26]

Hours Worked Each Week	Percentage of Entrepreneurs
Fewer than 50 hours	24%
50–60 hours	23%
60–70 hours	28%
70–80 hours	13%
More than 80 hours	12%

Estimate the mean and median number of hours. Assume that the average number of hours is 40 in the category "fewer than 50 hours" and 90 in the category "more than 80 hours."

1.54 Do entrepreneurs who start new companies generally do so when they are young or when they are middle-aged and have grown tired of working for someone else? Here are some data for answering that question:[27]

Age of Entrepreneur	Percentage of Entrepreneurs
20–25	9%
25–30	17%
30–35	21%
35–40	18%
40–45	15%
45–49	9%
50–60	11%

Use these data to draw a bar chart and a histogram. Is there any noticeable difference in their appearance?

1.55 *(continuation)* Estimate the mean and median age at which people start new businesses, using the data in the previous exercise. Explain why the mean is larger or smaller than the median.

1.56 The scores of ten students on a statistics midterm and final exam are listed next. Without doing any calculations, which test had the higher mean? Which had the higher standard deviation? Which had the higher coefficient of variation?

Student	Midterm	Final
Juan	98	92
Jessica	96	99
Laura	95	84
Bernard	92	94
Ed	85	92
Michael	85	86
Terri	80	75
Wei	76	88
Paul	72	85
Chris	65	74

1.57 *(continuation)* In Exercise 1.56, what would happen to the mean and standard deviation of final exam scores if Jessica's score on the final had been 90 instead of 99?

1.58 A stock's price-earnings (P/E) ratio is its price per share divided the company's annual profits. In January and February 1990, prices on the Tokyo Stock Exchange fell by 25 percent; yet Japanese price-earnings ratios were still several times higher than in other countries.[28]

Country	Price-Earnings Ratio
Japan	45.1
West Germany	17.9
United States	14.0
Australia	11.8
France	11.5
Britain	11.3
Hong Kong	10.4

Calculate the mean and median price-earnings ratio and explain why the mean is higher than the median. Is the mean P/E of the six countries other than Japan

higher or lower than the mean P/E of the seven countries including Japan? Why?

1.59 A pizza store delivers free anywhere within 10 miles of the store. One year's tabulation of the actual distances of each delivery found the following distribution:

Distance (Miles)	Number of Trips
0–1	2,504
1–3	6,130
3–5	6,517
5–7	6,266
7–10	7,085

Sketch a bar graph and a histogram. Are there significant differences in the appearances of these two graphs? Estimate the mean and median distance of a delivery.

1.60 John Maynard Keynes, the famous British economist, argued that consumption spending is less volatile than private investment spending. Use the data in Table 8 in the back of the book to calculate the annual percentage changes in nominal consumption and nominal private net investment for each of the 39 years, 1951–1989. Now calculate the standard deviation of the annual percentage change in consumption and the standard deviation of the annual percentage change in private net investment. Judging by the standard deviation, is consumption or investment more volatile?

1.61 Table 1.9 gives the annual percentage returns for four assets from 1950 through 1988.[29] Which asset looks as though it had the highest average return? Which looks as though it had the highest standard deviation? Calculate the mean and standard deviation for each of these four assets and see if you are right.

1.62 *(continuation)* Use the annual percentage returns in Table 1.9 to make a stem-and-leaf plot for the annual returns from high-grade corporate bonds. Do the same for New York Stock Exchange (NYSE) stock and describe the apparent differences in these two stem-and-leaf plots.

1.63 *(continuation)* Table 1.9 gives the annual percentage returns for four assets from 1950 through 1988. Draw four side-by-side box-and-whisker diagrams, one for each asset.

Table 1.9 *Data for Exercise 1.61*

Year	Short-Term Treasury Bills	Long-Term Treasury Bonds	High-Grade Corporate Bonds	NYSE Corporate Stocks
1950	1.2	0.1	2.1	31.7
1951	1.5	−3.9	−2.7	24.0
1952	1.7	1.2	3.5	18.3
1953	1.8	3.6	3.4	−1.0
1954	0.9	7.2	5.4	52.6
1955	1.6	−1.3	0.5	31.5
1956	2.5	−5.6	−6.8	6.6
1957	3.1	7.5	8.7	−10.8
1958	1.5	−6.1	−2.2	43.4
1959	3.0	−2.3	−1.0	12.0
1960	2.7	13.8	9.1	0.5
1961	2.1	1.0	4.8	26.9
1962	2.7	6.9	7.9	−8.7
1963	3.1	1.2	2.2	22.8
1964	3.5	3.5	4.8	16.5
1965	3.9	0.7	−0.5	12.4
1966	4.8	3.7	0.2	−10.0
1967	4.2	−9.2	−5.0	24.0
1968	5.2	−0.3	2.6	11.1
1969	6.6	−5.1	−8.1	−8.4
1970	6.5	12.1	18.4	3.9
1971	4.4	13.2	11.0	14.3
1972	3.8	5.7	7.3	19.0
1973	6.9	−1.1	1.1	−14.7
1974	8.0	4.4	−3.1	−26.4
1975	5.8	9.2	14.6	37.2
1976	5.1	16.8	18.6	23.8
1977	5.1	−0.7	1.7	−7.2
1978	7.2	−1.2	−0.1	6.6
1979	10.4	−1.2	−4.2	18.4
1980	11.2	−4.0	−2.6	32.4
1981	14.7	1.9	−1.0	−4.9
1982	10.5	40.4	43.8	21.4
1983	8.8	0.7	4.7	22.5
1984	9.9	15.4	16.4	6.3
1985	7.7	31.0	30.9	32.2
1986	6.2	24.4	19.9	18.5
1987	5.5	−2.7	−0.3	5.2
1988	6.4	9.7	10.7	16.8

r

1.64 The following are January–February 1990 revenue data for the five largest Atlantic City casinos:[30]

Casino	Jan.–Feb. 1990 (Millions)	Percentage Change from 12 Months Earlier
Harrah's Marina	$46.29	+ 3.3%
Caesar's	45.58	+ 2.3
Trump Castle	45.17	+15.0
Bally's Park Plaza	43.98	+ 6.8
Trump Plaza	43.47	− 10.0

Calculate the average revenue in January–February 1990 for these five casinos. Calculate the average percentage change in revenue (the average of the five percentage change figures shown). Now use the percentage change data to calculate each casino's revenue in January–February 1989 and the average revenue. Is the percentage change in the average revenue equal to the average percentage change in revenue?

1.65 The book value of a company is the value of its assets shown in its accounting books (generally historical cost less depreciation). The market value of a company, in contrast, is determined by the prices that investors are willing to pay for the company's bonds and stock, and these prices depend on the profitability of a company's assets, not their cost. A "white elephant" firm that loses money year after dreary year is worth little to its shareholders if its plant and equipment cost billions. Conversely, a computer company operating out of a $10,000 garage can be worth millions. Table 1.10 gives 1989 data on market value, book value, Tobin's q (the ratio of market value to book value), and return on equity (a measure of a firm's profitability) for the 20 largest firms in the computer and office equipment industry.[31] Calculate the mean and median market value and the mean and median book value for these 20 firms. Why, in each case, is the value of the mean substantially different from the median?

1.66 *(continuation)* Why might large firms be more profitable than small ones? Why might the

Table 1.10 *Data for Exercise 1.65*

Company	Market Value (Millions)	Book Value (Millions)	Tobin's q	Return on Equity
IBM	$94,644	$66,299	1.43	14.7%
Xerox	20,122	18,787	1.07	11.2
Hewlett-Packard	14,865	6,335	2.35	18.0
Digital Equipment	14,853	10,553	1.41	17.4
Unisys	11,394	12,203	0.93	9.0
Apple	6,692	1,764	3.79	39.9
Pitney-Bowes	6,016	3,925	1.53	18.7
Tandy	5,521	3,211	1.72	19.7
NCR	5,412	3,204	1.69	19.6
Automatic Data Processing	3,687	1,868	1.97	17.3
Compaq	3,453	1,248	2.77	30.5
Wang Laboratories	3,410	3,769	0.90	5.8
Control Data	2,633	2,872	0.93	3.5
Amdahl	2,397	1,388	1.73	21.2
Prime Computer	2,293	1,892	1.21	10.2
Tandem Computers	2,256	1,466	1.54	11.7
Sun	1,907	912	2.09	18.0
Cray Research	1,756	906	1.94	23.1
Intergraph	1,133	925	1.22	12.5
Seagate Technology	1,053	798	1.32	17.7

reverse be true? Divide the 20 firms in Table 1.10 into two groups of 10, based on their market value, and compare the average return on equity for these two groups. Repeat this calculation, dividing the firms into two groups based on their book value. Do these calculations tend to support the contention that large or small firms tend to be the more profitable?

1.67 *(continuation)* Tobin's *q* reflects an argument by Nobel Laureate James Tobin that the stock of very profitable firms tends to have a high market value relative to the firm's book value. Divide the 20 firms in Table 1.10 into two groups of 10 based on their return on equity. Calculate the mean and median value of *q* for the 10 firms with the highest return on equity and for the 10 firms with the lowest return on equity. Do these data support or contradict Tobin's theory?

1.68 *(continuation)* Draw two side-by-side box-and-whisker diagrams using the data in Table 1.10, one for the market values of these 20 firms and another for their book value.

1.69 *(continuation)* Draw a box-and-whisker diagram for the value of Tobin's *q* for the 20 firms in Table 1.10. Which, if any, of these firms seem to be outliers? Would the mean and median value of Tobin's *q* increase or decrease if these outliers were omitted from this data set?

1.70 Closed-end stock investment funds sell a fixed number of shares and use the proceeds to buy a portfolio of stocks. After issuance, the shares issued by the investment fund are traded among investors at a market price that need not equal the fund's net asset value (the value of its portfolio divided by the number of shares issued. Table 1.11 gives market price and net asset value data for 15 diversified stock funds on March 9, 1990. (Two dual-purpose funds, Gemini and Quest for Value, were excluded from these data.) Is the percent difference in this table calculated as a percentage of the fund's net asset value or as a percentage of its market price? Calculate the mean and median percent difference.

1.71 *(continuation)* Look in the "Money & Investing" section of the *Wall Street Journal* for the most recent Monday. Find the box titled "Publicly Traded Funds" and report the current percent differences for the 15 investment funds in Table 1.11. Calculate the current mean and median percent difference. Would you say that the percent difference has generally narrowed, widened, or is about the same as in March 1990?

1.72 *(continuation)* Use the data in Table 1.11 for March 9, 1990, and the recent data you collected for Exercise 1.71 to draw two stem-and-leaf diagrams. Do these distributions appear to be similar or dissim-

Table 1.11 *Data for Exercise 1.70*

Fund	Net Asset Value	Market Price	Percent Difference
Adams Express	17.72	16	− 9.71%
Baker Fentress	23.85	20	−16.14
Blue Chip Value	7.33	6⅝	− 9.62
Clemente Global	11.64	9½	−18.38
General American	21.29	17¾	−16.63
Growth Stock Outlook	10.05	9½	− 5.47
Lehman Corporation	15.01	12⅝	−15.89
Liberty All-Star Equity	9.30	8	−13.98
Niagara Share	16.80	14¾	−12.20
Nicholas-Applegate	11.04	9⅞	−10.55
Schafer Value Trust	10.67	9⅞	− 7.45
Source Capital	40.30	40⅜	+ 0.19
Tri-Continental	26.64	22⅞	−14.13
Worldwide Value	18.85	17¾	− 5.84
Zweig Fund	10.71	11⅜	+ 6.21

Table 1.12 *Data for Exercise 1.74*

Country	1960	1970	1980	1990
Canada	6.5%	5.7%	7.5%	7.5%
France	1.5	2.5	6.4	10.1
Italy	3.7	3.2	4.4	7.8
Japan	1.7	1.2	2.0	2.3
United Kingdom	2.2	3.1	7.0	5.9
United States	5.5	4.9	7.1	5.3
West Germany	1.1	.5	2.9	5.5

ilar? Which one of these 15 funds seems to be the farthest outlier in March 9, 1990? Is this same fund also the farthest outlier in the most recent data?

1.73 *(continuation)* Use the data for March 9, 1990, and the recent data to draw two side-by-side box-and-whisker diagrams.

1.74 Table 1.12 gives unemployment rates for seven countries in 1960, 1970, 1980, and 1990. Based on these data, which of these seven countries had the lowest average unemployment rate? Which had the highest?

1.75 *(continuation)* Use the data in Table 1.12 to calculate the average unemployment rate in 1960, 1970, 1980, and 1990. Did any of these seven nations consistently have a below-average unemployment rate? Did any have a consistently above-average unemployment rate?

1.76 During Ronald Reagan's 1980 campaign for the presidency of the United States, he repeatedly asked voters if they were better off in 1980 than they were four years ago, when the incumbent, Jimmy Carter, had been elected President. Here are some data on unemployment and inflation during Carter's four years in office (1977–1980) and Reagan's eight years (1981–1988):

Year	Unemployment Rate	Rate of Inflation
1977	7.1%	6.7%
1978	6.1	9.0
1979	5.8	13.3
1980	7.1	12.5
1981	7.6	8.9
1982	9.7	3.8

Year	Unemployment Rate	Rate of Inflation
1983	9.6	3.8
1984	7.5	3.9
1985	7.2	3.8
1986	7.0	1.1
1987	6.2	4.4
1988	5.5	4.4

Calculate the average unemployment rate and the average inflation rate during the Carter years and during the Reagan years. Is there any noticeable difference in unemployment or inflation between these two administrations?

1.77 A stock's price-earnings (P/E) ratio is its price per share divided by the company's annual profits. The P/E ratio for the stock market as a whole is used by some analysts as a measure of whether stocks are cheap or expensive in comparison with other historical periods. Here are some annual price-earnings ratios for New York Stock Exchange (NYSE) stocks:

Year	P/E
1970	15.5
1971	18.5
1972	18.2
1973	14.0
1974	8.6
1975	10.9
1976	11.2
1977	9.3
1978	8.3
1979	7.4

Year	P/E
1980	7.9
1981	8.4
1982	8.6
1983	12.5
1984	10.0
1985	12.3
1986	16.4

Calculate the mean and standard deviation. The stock market reached a peak in August 1987 when the Dow Jones Industrial Average topped 2700. The Dow slipped back to 2500 in October 1987 and then to 2250. Then on a single day, October 19, 1987, the Dow fell by 508 points. At its August 1987 peak, the market's price-earnings ratio was 23. Was this P/E value one standard deviation above the mean P/E since 1970? Was it two standard deviations above the mean?

1.78 The real interest rate is equal to the nominal interest rate minus the rate of inflation. For instance, in 1989 the average rate of return on 1-year Treasury bills was 9.3 percent and the average rate of inflation was 4.6 percent, giving a real rate of return of 9.3 percent − 4.6 percent = 4.7 percent. Use the data in Table 8 in the Appendix of this book to calculate the real rate of return on 1-year Treasury bills, in each year 1970–1989. Calculate the mean and median real rate of return for this 20-year period.

1.79 *(continuation)* Calculate the standard deviation of the annual real rates of return in Exercise 1.78. Are approximately 68 percent of the annual observations within one standard deviation of the mean, and 95 percent within two standard deviations?

1.80 The National Football League uses a weighted average of several performance data to rate quarterbacks:

$$\text{Rating} = \frac{12.5 + 5X_1 + 20X_2 - 25X_3 + 25X_4}{6}$$

where X_1 = percentage of passes completed ($30\% \leq X_1 \leq 77.5\%$)

X_2 = percentage of passes for touchdowns ($X_2 \leq 11.9\%$)

X_3 = percentage of passes intercepted ($X_3 \leq 9.5\%$)

X_4 = average yards gained per pass thrown ($X_4 \leq 12.5\%$)

(The limits indicate that, for example, if X_1 is less than 30, then $X_1 = 30$ is used in the equation.) In 1984 the highest-rated quarterback was Dan Marino of the Miami Dolphins, with $X_1 = 64.2$, $X_2 = 8.5$, $X_3 = 3$, and $X_4 = 9.01$. What was his rating? What is the highest possible rating? What is the lowest possible rating? Would you say that the NFL considers an interception more or less important than a touchdown pass?

1.81 Explain this argument, identifying the crucial assumption: "Here's how to interpret a standard deviation figure: Take the expected return and add one standard deviation to it. Then take the standard deviation and subtract one standard deviation. In two of three years, the actual result should fall within that range if all the assumptions were accurate."[32]

1.82 "At retirement, a person can choose to take a single lump sum payment or a fixed annual income until death. Those who choose the annual income live, on average, 2½ years longer than the general population. This shows that peace of mind is very important to one's health." Provide an alternative explanation.

1.83 A 1989 *Wall Street Journal* editorial criticized a ruling by New York State's highest court that called marriage "a fictitious legal distinction."[33] The *Wall Street Journal*'s promarriage editorial noted that "adult single men are five times more likely to commit violent crimes than married men," suggesting that marriage is a major factor in reducing violent crime. How else might these data be explained?

1.84 Observing that 40 percent of the 1989 freshman class at Pomona College is nonwhite,

> Dean of Admissions Bruce Poch noted in early October that the increase has not changed the school's high standards. "Combined median SAT (Scholastic Aptitude Test) scores," Mr. Poch stated of the 1989 crop of Pomona students, "increased slightly to 1320 at the same time that ethnic diversity increased substantially, clearly laying to rest the myth that academic excellence and ethnic diversity must be incompatible."[34]

Why, even if SAT scores are the appropriate measure of academic excellence, are the data cited of virtually no help in proving or disproving Poch's conclusion regarding Pomona College?

1.85 Between 1959 and 1983, the average number of hours that women worked outside the home increased by nearly 100 percent, while the average number of hours they worked inside the home fell by only 14 percent.[35] Do these data show that the sum of number of hours worked inside and outside the home definitely increased? Explain your answer, using an example.

1.86 A new golf ball design with an increased number of dimples was tested by having an amateur golfer hit 100 drives. The results are

Distance (Yards)	Number
100–150	10
150–200	30
200–250	40
250–275	20

Sketch a histogram and calculate the mean and standard deviation. If a professional golfer were to repeat this experiment, what do you predict will happen to the mean and standard deviation?

1.87 A new golf ball design with an increased number of dimples was tested by having an amateur golfer and a professional golfer each hit 100 shots on a 150-yard hole. Use the following data on their distances from the pin to calculate the mean and standard deviation for each:

Distance (Yards)	Amateur	Professional
0–10	50	80
10–20	30	20
20–30	20	0

1.88 Following are some 1985 data on the U.S. farm population (in 1,000s). (Assume that the oldest group runs from age 65 to 80.) Draw a bar graph and histogram for the total farm population by age. Is there any substantial difference in the appearance of these two graphs?

Age	Male	Female	Total
0–14	565	544	1,109
14–20	330	296	626
20–25	243	183	426
25–35	359	319	678
35–45	363	354	717
45–65	721	711	1,432
65+	393	372	765

1.89 *(continuation)* Draw histograms for the male and female farm populations by age, and estimate the mean and median age of each group. Do the ages of the male and female farm populations seem similar or dissimilar?

1.90 A 1989 radio commercial claimed, "If you have a LifeAlert security system, your chances of becoming a victim of a serious crime are 3,000 to 4,000 percent less than your neighbor's." Is this possible?

1.91 While driving 55 mph, you notice that the number of cars you are passing is equal to the number passing you. Is 55 mph the mean, median, or modal speed?

1.92 Sally's average score was 85 after four tests and 87 after five tests. What was her score on the fifth test?

1.93 "The average U.S. household has a clothes washer and clothes dryer, but no dishwasher or microwave oven." Is this "average" a mean, median, or mode?

1.94 The Public Health Service has tabulated data on the number of divorces according to the month in which the marriage ceremony was performed:

Month Married	Fraction of All Divorces
January	.064
February	.068
March	.067
April	.073
May	.080
June	.117
July	.087
August	.103
September	.090
October	.078
November	.079
December	.087
	1.0

Do these data show that it is dangerous to get married in June or August?

1.95 Here are data on the ages of U.S. mothers whose first child was born during the years 1975–1979:

Age of Mother	Relative Frequency
16–18	.117
18–20	.276
20–25	.369
25–30	.186
30–35	.052
	1.0

(The two end intervals are open-ended in the census data, but have been closed here for simplicity.) Draw a histogram for the age of these first-birth mothers and estimate the mean, median, and modal age.

1.96 A *Wall Street Journal* poll asked economic forecasters to predict the interest rate on 30-year Treasury bonds in June 1988. Here are ten of these predictions:[36]

$$9.25 \quad 7.50 \quad 9.25 \quad 9.40 \quad 8.60$$
$$10.30 \quad 8.25 \quad 9.00 \quad 9.15 \quad 9.25$$

Calculate the mean and standard deviation of these predictions. It is sometimes assumed that about two-thirds of the values will be within one standard deviation of the mean. What justifies this assumption? Is this rule of thumb approximately correct here?

1.97 The following are the age distributions of the Jewish population in Germany in 1928 and in 1959, and of a theoretical stable population.[37] Make two histograms, one comparing the 1928 population with the theoretical model, and one comparing the 1959 population with the model. (Assume that the interval 60+ is 60–75.)

Age	1928	1959	Stationary Model
0–20	26.9%	14.2%	33%
20–40	31.3	19.8	30
40–60	27.7	37.8	25
60+	14.3	28.2	12

1.98 The pavement condition of U.S. highways is rated on a scale of 0 to 5:

3.5 to 5.0	First class ride, few signs of wear?

2.0 to 3.5	Somewhat inferior ride, may show cracking
0.0 to 2.0	Extensive wear, may need resurfacing

Here is the 1983 distribution of ratings for rural interstate highways:

Rating	Percentage
0.0 to 2.0	9
2.0 to 3.5	33
3.5 to 5.0	58

Draw a histogram and estimate the mean and median road condition.

1.99 A cover story in the University of Texas at Austin publication *On Campus* boasted that the university was second in the nation in 1986 in the number of freshman National Merit Scholars with 270, behind Harvard's 297 but ahead of Yale's 183, Rice's 176, and Stanford's 172. If you were editor of the newspaper at one of these other colleges, how would you counter UT Austin's boast?

1.100 The Dow Jones Industrial Average is calculated by adding up the per-share prices of 30 stocks and dividing by the current value of the divisor (which is adjusted from time to time, for reasons that will be explained in a later chapter). Market value indexes, in contrast, multiply each stock's price by the number of shares outstanding. For example, on May 1, 1989, AT&T had 1,073.7 million shares outstanding and a per-share price of 34½, while IBM had 591.4 million shares outstanding and a per-share price of 113⅜. An unweighted index, such as the Dow, and a market value index would be calculated as

unweighted $\dfrac{34\frac{1}{2} + 113\frac{3}{8}}{k}$

market value $\dfrac{(1073.7)(34\frac{1}{2}) + (591.4)(113\frac{3}{8})}{h}$

where k and h are the respective divisors. If the divisors don't change, can this unweighted index ever increase while this market value index declined? Either show that this cannot happen or give an example or how it can happen.

PART

I

Probability

2

The Probability Idea

*It is remarkable that a science which began with the considera-
tion of games of chance should become the most important object
of human knowledge. . . . The most important questions of life
are, for the most part, really only problems of probability.*

Pierre Simon De Laplace

TOPICS

2.1   INTRODUCTION

There are very few things in our lives that are absolutely certain. This uncertainty makes life challenging and interesting. Think how boring it would be to know in advance everything that was going to happen to us. School, love, career, marriage, and family would all be so much less fun if the mystery were taken out. We wouldn't need elections because the winners and losers would be known beforehand. We wouldn't need much of a stock market because everyone would know what the stock prices would be tomorrow and every day after that. We wouldn't need sports events because we would already know the outcomes. How interesting do you find reruns of old football games? If the world were certain, life would be like a rerun of an old football game. We would be bored actors and actresses walking through a lifeless script.

Life is instead full of uncertainties, with all of the attendant thrills and frustrations. The idea behind probabilities is to try to quantify these uncertainties. Uncertainty means that a variety of outcomes are possible. We can better understand this uncertainty and be more prepared for the possibilities if we use probabilities to describe which outcomes are likely and which are unlikely. In this chapter, you will begin to see where these probabilities come from and what they are used for. We will begin with games of chance because, when viewed from a financially safe distance, these are an ideal vehicle for introducing probability logic.

2.2   THE GAMBLING ROOTS

The first systematic analysis of probabilities seems to have been conducted by Gerolamo Cardano "Cardan" (1501–1576). In his *Liber de Ludo Aleac (Book of Games of Chance),* Cardan explained the rules of various games, philosophized on the morality of gambling, and provided tips for catching cheats.* He also set forth some carefully thought-out probability principles.

The Flip of a Coin

When a fair coin is flipped, Cardan reasoned that there are two "equally likely" outcomes: heads or tails. As these outcomes are equally likely, each has a probability of 1/2 of occurring. If we roll a fair six-sided die, there are six equally likely outcomes. Therefore, each outcome has a probability of 1/6 of occurring. What about the probability of obtaining an even number when rolling a single die? Cardan reasoned that, as three of the six possible outcomes are even numbers, the probability must be 3/6, which reduces to 1/2. In general, if there are n equally

---

*Cardan suggested that gambling was "invented" during the Trojan wars to entertain the soldiers during a ten-year siege of Palamedes.

likely outcomes and m of these outcomes are "successes," then the probability of a successful outcome is m/n.

What is the probability of rolling a 5 or a 6? These are two of the six possibilities. So, the probability is 2/6, which reduces to 1/3. What is the probability of rolling a number greater than 1? Five of the six possible numbers are greater than 1, so the probability is 5/6. What is the probability of rolling a 1, 2, 3, 4, 5, or 6? These are six of the six possible outcomes, so the probability is 6/6, which reduces to 1.0. That answer, 1.0, is the largest a probability can ever be. If there are six possible outcomes, then the number of possible successes cannot be greater than six.

We can also use the idea of equally likely events to interpret probability statements. If I say that the probability of some event happening is 1/6, then we can interpret my assessment as meaning that it is as if this event were one of six equally likely possibilities. Or, even more concretely, we could say that this event is as likely to occur as is the number 3 when a fair die is rolled.

> The **classical interpretation of probabilities** is that the probability of any one of m events occurring, when there are n equally likely possible outcomes, is m/n.

The next milestone in probability theory was laid by the French mathematician Blaise Pascal in the 1600s. There is a (probably overdramatic) legend that the French nobleman Antoine Gombauld, the Chevalier de Mere, had won a considerable amount of money betting that he could roll at least one 6 in four throws of a single die. But he was losing money betting that he could roll at least one double-6 in twenty-four throws of a pair of dice. De Mere asked Pascal about this problem and, later, about other puzzling games of chance. De Mere's problem is tough and a little tricky, but you will soon be able to answer it with ease.

Pascal discussed these gambling puzzles with a number of European mathematicians, including Pierre de Fermet, and these discussions led to the first great books on the mathematical theory of probability: Christiaan Huygen's *De Ratiociniis in Ludo Aleae* (1657) and Pascal's *Treatise on Figurate Numbers* (1665). Huygens and Pascal, like Cardan before them, computed probabilities by counting the number of equally likely outcomes. However, Huygens and Pascal were much more systematic and rigorous than Cardan and were more successful in energizing other intellectuals, with their remarkable demonstration that chance was subject to mathematical laws. Pascal wrote that by "bringing together the rigor of scientific demonstration and the uncertainty of chance, and *reconciling those things which are in appearance contrary to each other,* this art can derive its name from both and justly assume the astounding title of the Mathematics of Chance."

The underpinning of their work is the identification of equally likely outcomes. But, how do we know if two events are equally likely? Another seventeenth-century French mathematician, Pierre Laplace, argued that we should presume that two events are equally likely unless we have good reason to believe otherwise.

Laplace's Principle of Insufficient Reason states that, in the absence of compelling evidence to the contrary, we should assume that events are equally likely.

If the coin is unbent and the die is symmetrical, then we should assume each possible outcome to be equally likely. We refrain from this assumption if the coin is two-headed, the die seems unbalanced, or the thrower is disreputable.

Several Coin Flips

So far, we have only looked at easy problems such as the flip of a fair coin and the roll of a fair die. More complicated problems may involve a more careful and lengthy enumeration of the possible outcomes. Consider a coin flipped twice. The possibilities are

> 2 heads
> 1 head and 1 tail
> 2 tails

It is tempting to apply Laplace's Principle of Insufficient Reason and assume that each of these three outcomes has a 1/3 probability of occurring. This was, in fact, the answer given by the prominent mathematician Jean d'Alembert. But that answer is wrong!

There is only one way to get two heads and only one way to obtain two tails, but there are two different ways of getting one head and one tail. We could get a head on the first flip and a tail on the second, or we could get a tail on the first flip and then a head on the second.

To enumerate the outcomes correctly, we can construct a **probability tree** to show the possibilities. On the first flip, there are two equally likely outcomes: a head or a tail. In either case, the second flip is also equally likely to be a head or a tail. If we use H for "heads" and T for "tails," then the possible outcomes are

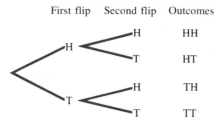

There are really four equally likely outcomes, with two involving one head and one tail. The probability of two heads is 1/4, the probability of two tails is 1/4, and the probability of one head and one tail is 2/4. The same logic would apply if we simultaneously flipped two coins, instead of flipping the same coin twice.

What about three flips? Now the probability tree looks like this:

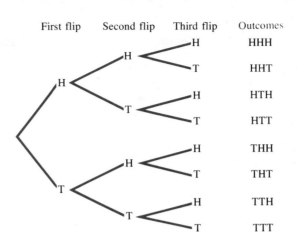

There are eight possible outcomes. By inspection, the probabilities are

Event	Number of Successes	Probability
Three heads	1	1/8
Two heads, one tail	3	3/8
One head, two tails	3	3/8
Three tails	1	1/8
Total	8	1

This logic can be extended to four flips, five flips, and more, ad nauseum.

Dice

Let's instead move on to some dice problems. If we roll two dice, what is the probability of getting double-6s? What is the probability of getting numbers that add up to 7? Which is more likely, numbers that add up to 7 or numbers that add up to 8? The classical, equally likely procedure is to enumerate the possible outcomes shown in Table 2.1. Thus, the probabilities are as shown in Table 2.2. The probability of double-6s is 1/36. The probability of numbers that add up to 7 is 1/6, which is slightly more likely than numbers that add up to 8.

In some professions or pastimes, it is helpful to remember the structure of dice probabilities shown in Table 2.2. You can see that 7 has the largest probability. The probabilities then decline symmetrically as we move away from 7. A sum of 7 is more likely than a sum of 6 or 8, which are more likely than a sum of 5 or 9, and so on.

Often, we are interested in just a single probability, rather than the entire Table

Why Are Credit Card Numbers So Long?

A single digit can have one of ten possible values: 0, 1, 2, . . . , 9. For a two-digit number, such as 47, there are ten possibilities for the first digit and, for each of these values, ten possibilities for the second digit, giving $10^2 = 100$ possible values in all. (Alternatively, we can reason directly that there are 100 possibilities, ranging from 00 to 99.) A three-digit number has $10^3 = 1,000$ possible values. A five-digit number has $10^5 = 100,000$ possible values.

The number of possible ten-digit numbers is $10^{10} = 10$ billion, enough to give a unique 10-digit identification number to every person who is now alive or will be born in the next 50 years. Yet, Visa credit cards have a minimum of 13 digits, American Express cards have 15 digits, and Master-Cards have 20 digits. Why do these credit cards have many, many more digits than they will ever need to keep track of their cardholders? Don't these extra, superfluous digits make it more likely that a clerk filling out a sales ticket, taking an order over the phone, or transferring numbers from sales receipts to

a computer will make a mistake, causing the charge to be billed to someone else's number?

In fact, the additional digits are a security device, intended to make erroneous billings *less likely,* by making it extremely improbable that an incorrectly entered number will belong to someone. MasterCard's 20-digit system creates 10^{20} possible numbers, of which approximately 75 million are randomly assigned to its active cardholders. If a clerk makes a mistake, the probability that this randomly mistaken number will belong to a cardholder is

$$\frac{75,000,000}{10^{20}} = \frac{7.5}{10^{13}}$$

which is less than 1 in a trillion. In virtually every case, a computer will reject an incorrectly entered number and ask the clerk to try again. Even with Visa's relatively modest 13-digit system, the probability that a randomly selected number will belong to a cardholder is less than 1 in 100,000 ($P < .00001$). Credit card companies use long numbers to put probabilities on their side, and on ours.

2.2, and there is a shortcut logic. Suppose we want to know the probability of rolling numbers that sum to 4. First, figure out the total number of possible outcomes. The first die has six possible outcomes and, for each of these outcomes, the second die has six possible outcomes. So, there are $6(6) = 36$ possibilities in all. Now, of these thirty-six possible outcomes, how many involve numbers that add up to 4? That is, how many different ways are there to roll numbers that sum to 4? Well, there's a 1 on the first die and a 3 on the second, 3 on the first and 1 on the second, or 2 on both dice. So, three possibilities out of thirty-six implies a probability 3/36, which reduces to 1/12. What about the probability of rolling numbers that add up to 5? There are four ways to roll a sum of 5: 1–4, 2–3, 3–2, and 4–1. So, the probability is 4/36, which reduces to 1/9.

Table 2.1 *Possible Outcomes When Two Dice Are Rolled*

First Die	Second Die	Outcome	Sum of Numbers
	1	1–1	2
	2	1–2	3
1	3	1–3	4
	4	1–4	5
	5	1–5	6
	6	1–6	7
	1	2–1	3
	2	2–2	4
2	3	2–3	5
	4	2–4	6
	5	2–5	7
	6	2–6	8
	1	3–1	4
	2	3–2	5
3	3	3–3	6
	4	3–4	7
	5	3–5	8
	6	3–6	9
	1	4–1	5
	2	4–2	6
4	3	4–3	7
	4	4–4	8
	5	4–5	9
	6	4–6	10
	1	5–1	6
	2	5–2	7
5	3	5–3	8
	4	5–4	9
	5	5–5	10
	6	5–6	11
	1	6–1	7
	2	6–2	8
6	3	6–3	9
	4	6–4	10
	5	6–5	11
	6	6–6	12

Backgammon

An understanding of probabilities is one of the keys to mastering the game of backgammon. Many of the world's best players have mathematical backgrounds. The others have learned the game's probabilities through long experience.

In backgammon, you roll two dice and use each number separately to move your pieces. If you roll a 5–3, for example, you move one piece 5 spaces and one piece (possibly the same piece) 3 spaces. An added wrinkle is that doubles are moved twice. A double-6, for example, lets you move four pieces 6 spaces.

Now, consider the following commonplace problem. You want to know whether it is safer to leave a piece 1 or 2 spaces from your opponent. Is your opponent more likely to be able to hit you by moving 1 space or by moving 2 spaces?

To move 1 space, your opponent must roll a 1 on one of the dice. The possible ways are 1–1, 1–2, 1–3, 1–4, 1–5, 1–6, 2–1, 3–1, 4–1, 5–1, or 6–1. (Notice that I did not count 1–1 twice.) Altogether, there are eleven ways to roll a 1, and so the probability of being hit from 1 space away is 11/36. What about from 2 spaces away? Your opponent must roll a 2 on one die (eleven possible ways) or double-1s (one way). Altogether, there are twelve ways, making the probability 12/36. Thus it is safer to be 1 space away than 2 spaces away.

If the piece is 3 spaces away, your opponent must roll a 3 on one die (eleven ways), a 1 on one die and a 2 on the other (two ways), or double-1s (one way). The total probability is 14/36. By similar calculations, the probabilities from 4, 5, and 6 spaces away are 15/36, 15/36, and 17/36.

When the piece is more than 6 spaces away, the probabilities drop off because it takes both dice to roll a 7 or higher. As shown in Table 2.2, the probability of rolling a sum of 7 is only 6/36, or 1/6. If the piece is 8 spaces away, your opponent must roll an 8 (five ways, according to Table 2.2) or use double-2s, giving a total probability of 6/36. To hit a piece 9 spaces away, there are four ways to roll a 9, plus double-3s, for a total probability of 5/36. There are three ways to roll a 10 and two ways to roll an 11. Twelve spaces away can be hit by double-6s, double-4s, or even double-3s. The probabilities are shown in the chart below.

We have derived one important backgammon rule of thumb. For 6 spaces or less, the closer you are to your opponent, the less likely you are to be hit. You are much safer if you are more than 6 spaces away and, after 6 spaces, the farther away you are, the safer you are. This easy-to-remember rule is a good approximation, but is not precisely correct. Twelve spaces away, for example, is more dangerous than 11 spaces, because doubles can be moved twice.

Number of spaces away	1	2	3	4	5	6	7	8	9	10	11	12
Probability of being hit	$\frac{11}{36}$	$\frac{12}{36}$	$\frac{14}{36}$	$\frac{15}{36}$	$\frac{15}{36}$	$\frac{17}{36}$	$\frac{6}{36}$	$\frac{6}{36}$	$\frac{5}{36}$	$\frac{3}{36}$	$\frac{2}{36}$	$\frac{3}{36}$

Table 2.2   *Probabilities When Two Dice Are Rolled*

Sum of Dice	Number of Successes	Probability
2	1	1/36
3	2	2/36
4	3	3/36
5	4	4/36
6	5	5/36
7	6	6/36
8	5	5/36
9	4	4/36
10	3	3/36
11	2	2/36
12	1	1/36
Total	36	1.0

Let's do one more. In many games, a special importance is attached to "doubles," with both dice showing the same number. Doubles can win a game, lose a game, give you extra moves, give you an extra roll, or get you out of jail. What is the probability of rolling doubles? By now, this should be easy for you. There are thirty-six possible outcomes and six ways to get doubles (1–1, 2–2, 3–3, 4–4, 5–5, and 6–6), so the probability of doubles is 6/36, which reduces to 1/6.

This framework is infinitely expandable. In any situation, just enumerate all the possible, equally likely outcomes and then count up the possible successes. This also grows infinitely complicated as the chain of outcomes grows longer. Remember de Mere's problem? He was interested in twenty-four throws of a pair of dice! Each pair has thirty-six possible outcomes, and each of these thirty-six possibilities has thirty-six different possibilities for the next pair. The total number of equally likely outcomes is $36^{24} = 2.25 \times 10^{37}$. It would be quite a chore to enumerate all of these possibilities. We might as well try counting the number of grains of sand on a beach. Fortunately, mathematicians have developed some probability rules that allow us to solve most problems, including de Mere's, very quickly. I'll show you a few counting tricks in the next section, but I want to save most of the rules for the next chapter.

EXERCISES

2.1 Your final exam consists of ten essay questions. Four of your answers are good enough to be graded "A," four are worth a "B," and two merit only a "C." There are over 100 students in this class, so your harried instructor decides to read just one essay per exam and use that grade as the final-exam score. If each of your essays is equally likely to be read, what is the probability that your instructor will choose one of your "A" answers? One of your "B" answers? One of your "C" answers?

2.2 A random-number generator is equally likely to produce any one of 100 numbers 0, 1, 2, . . . , 99. What is the probability that the number it generates will   **a.** be less than 10?   **b.** be odd?   **c.** be 99?   **d.** have 9 as the last digit?

2.3 In the card game Between the Sheets, each player is dealt two cards, face up. Ace is low and king is high. The player can then fold or bet that the value of a third dealt card will be between the values of the two original cards. The bet is lost if the third card is above, below, or matches the first two cards. You have been dealt two cards and there are fifty cards left in the deck. What is your probability of winning a bet if your cards are   **a.** a three and a nine?   **b.** a four and a ten?   **c.** a five and a queen?   **d.** an ace and a king?   **e.** a pair of jacks?

2.4 Mr. and Mrs. ZPG have decided to have two children. Let's assume that boy and girl babies are equally likely and that a baby's sex is unrelated to the sex of earlier babies. What is the probability that the ZPG's two children will be a boy and a girl?

2.5 A very successful football coach once explained why he preferred running the ball to passing it: "When you pass, three things can happen [completion, incompletion, or interception], and two of them are bad." Can we infer that there is a 2/3 probability that something bad will happen when a football team passes the ball?

2.6 You are playing draw poker and have been dealt four spades and a heart. You discard the heart and draw a new card. What is the probability that this new card will be a spade, giving you a flush?

2.7 Table 1.10 in Chapter 1 shows the 1989 market and book values of the 20 largest firms in the office equipment and computer industry. If these 20 names are written on slips of paper and dropped into a hat, with one slip picked at random, what is the probability that the market value of the company selected will be less than its book value? If this experiment is repeated using only those companies with a return on equity below 10 percent, what is the probability that the market value of the selected company will be less than its book value?

2.3   COUNTING PERMUTATIONS AND COMBINATIONS (OPTIONAL)

How many possible five-card hands are there in the game of poker? What is the probability of three of a kind? Is a flush less likely than a straight? These are probability questions that require a careful, complex counting of the possible number of outcomes. It turns out, though, that there is a standard counting procedure that we can use to handle these and a wide variety of other problems. The logic is clearest if we develop this counting procedure in three stages.

Permutations of *n* Distinct Items

Peter, Paul, and Mary are three distinct names. A rearrangement of the order in which these names appear is called a **permutation.** How many possible orderings are there? There are six permutations:

Peter, Paul, & Mary
Peter, Mary, & Paul
Paul, Peter, & Mary
Paul, Mary, & Peter
Mary, Peter, & Paul
Mary, Paul, & Peter

We can think of an ordering as three slots into which three names will be placed. Any one of the three names can be put into the first slot, and whichever name is put in the first slot leaves two possibilities for the second slot. Given the two names in the first and second slots, there is only one name left for the third slot. So, if there are three first-slot possibilities, each with two second-slot possibilities, and then one third-slot possibility, then there are $(3)(2)(1) = 6$ possible orderings. If there were four names (such as John, Paul, George, and Ringo), there would be $(4)(3)(2)(1) = 24$ possible orderings.

In general, the number of different permutations of n distinct items is

$$n! = n(n - 1)(n - 2) \ldots 1$$

The mathematical symbol $n!$ ("n factorial") signifies the product of the integers from n down to 1.

As another example, consider a horse race with 10 horses and an "exacta" prize for anyone who can pick the exact order of finish, first through tenth place. How many possible orderings are there? Any one of 10 horses could finish first. For each of these 10 possible first-place finishers, there are 9 possible second-place horses. We have 90 first-second orderings and, for each of these, there are 8 possible third-place finishers. And so, the possibilities multiply. Overall, the number of possible orderings of ten horses is

$$(10)(9)(8)(7)(6)(5)(4)(3)(2)(1) = 3,628,800$$

If the horses are evenly matched, it will not be easy to pick the exact order of finish. In fact, if each of these 3,628,800 possible orderings is equally likely, then your probability of picking the correct ordering is only $1/3,628,800$.

Permutations of r Out of n Distinct Items

Now consider a somewhat different problem. We again have n distinct items, but now we only want to use r of them. For example, in our race of 10 horses($n = 10$), the exacta prize may be awarded for picking the first 3 places in order ($r = 3$). If so, how many different orders of finish are there? Well, any of the 10 horses could finish first, and whichever finishes first, there are still 9 possible second-place finishers. For each of these 90 first-second pairs, there are 8 possible third-place finishers. So, there are $(10)(9)(8) = 720$ possible three-horse orderings in a ten-horse race.

In general, the number of different permutations of r out of n distinct items is

$$n(n - 1)(n - 2) \ldots (n - r + 1) = \frac{n!}{(n - r)!}$$

For another example, consider the Chief Executive Officer (CEO) of a company with offices in 20 different cities. This busy CEO decides to visit 5 of these 20 offices. How many different five-office sequences are possible? The first visit could be to any of 20 offices, followed by any of 19, and so on. The total number of possible sequences is

$$(20)(19)(18)(17)(16) = 1,860,480$$

Combinations of *r* Out of *n* Distinct Items

So far, we have looked at situations where the ordering makes a difference. Visiting Boston and then New York was assumed to be different from visiting New York and then Boston in the CEO's calculation. In many cases, though, order is unimportant. The CEO may not care which office is visited first and which second, but only which 5 offices are visited. If so, by how much did we overcount in our calculations?

Well, take any 5 selected offices, say Boston, New York, Philadelphia, New Haven, and Newark; this is one five-city combination. Our permutation count included all possible sequences of these 5 cities. There are $(5)(4)(3)(2)(1) = 120$ such ordered sequences. We counted 120 permutations, while our CEO considers it just one five-city combination; the same is true of every five-city combination. In arriving at 1,860,480 as the total number of possible ordered sequences, each five-city combination was counted 120 times. To find the number of different five-city combinations, we need to divide 1,860,480 by 120:

$$\frac{(20)(19)(18)(17)(16)}{(5)(4)(3)(2)(1)} = \frac{1,860,480}{120} = 15,504$$

Ignoring the order in which they are visited, there are 15,504 combinations of 5 cities to visit. For each of these 15,504 combinations, there are 120 different orders in which these 5 particular cities could be visited. If the sequence makes a difference, then there are 1,860,480 possible five-city itineraries.

In general, if there are *n* cities, of which *r* are to be selected, then there will be $n!/(n - r)!$ different itineraries. This calculation takes into account the fact that each combination of *r* cities could be visited in *r*! different sequences. If we do not care about the order in which the cities are visited, then there are $n!/(n - r)!r!$ different combinations of *r* cities.

When order is unimportant, we say that we are counting *r* **combinations,** rather than *r* permutations.

The number of different *r*-item combinations using *n* distinct items is

$$\frac{n!}{r!(n - r)!}$$

for which a shorthand notation is $\binom{n}{r}$.

A special word about $r = 0$ and $r = n$: We define $0! = 1$ so that

$$\binom{n}{0} = \binom{n}{n} = 1$$

This rule is in accordance with our common sense because, if we are going to visit no cities or every city, then there is only one possible choice.

Some Poker Hands

Card games are a fertile area for counting combinations. In five-card poker, for example, what is the probability of being dealt a royal flush: the ace, king, queen, jack, and ten of the same suit? First, let's count how many five-card hands are possible. We are taking $r = 5$ cards out of a deck of $n = 52$ distinct cards. As the order in which the cards are received is unimportant, the number of possible five-card hands is

$$\binom{52}{5} = \frac{(52)(51)(50)(49)(48)}{(5)(4)(3)(2)(1)} = 2,598,960$$

How many of these five-card hands are royal flushes? There are four, one in each suit. Therefore, if the cards are well shuffled and fairly dealt, so that every possible hand is equally likely to be dealt, then the probability of a royal flush is

$$P[\text{royal flush}] = \frac{4}{2,598,960} = .00000154$$

or a one-in-a-million shot.

What about the probability of being dealt a straight flush, other than a royal flush? You could have an ace through five, two through six, or all the way up to nine through king. There are 9 possible straight flushes in each suit, giving 36 altogether. Therefore,

$$P[\text{straight flush}] = \frac{36}{2,598,960} = .00001385$$

Let's do one more. What is the probability of being dealt four of a kind? There are 13 four-of-a-kinds and each could have any one of 48 other cards as the fifth card in the hand. Thus, $13(48) = 624$ hands contain four of a kind and

$$P[\text{four of a kind}] = \frac{624}{2,598,960} = .00024$$

Notice that these poker probabilities are inversely related to the hand's value. A royal flush is the most valuable hand and the least likely to occur. A straight flush

Table 2.3　*Five-Card Poker Hands*

Type of Hand	Number of Combinations	Probability
Royal flush	4	.00000154
Straight flush	36	.00001385
Four of a kind	624	.00024
Full house	3,744	.00144
Flush	5,108	.00197
Straight	10,200	.00392
Three of a kind	54,912	.02113
Two pair	123,552	.04754
One pair	1,098,240	.42257
Other hands	1,302,540	.50118
Total	2,598,640	1.0

beats four of a kind, and is less likely to be dealt. This is true through the complete ranking of poker hands: royal flush, straight flush, four of a kind, full house, flush, straight, three of a kind, two pairs, and a pair. In a comparison of any two of these types of poker hands, the more valuable hand is also the less probable.* I've shown you three of these probabilities and the remainder are given in Table 2.3. If your appetite is whetted, there are some good books that thoroughly analyze probabilities for poker, bridge, Monopoly, and other complex games.[1]

EXERCISES

2.8 In the United States, the telephone numbers within any area code contain seven digits. How many possible seven-digit numbers are there? (Use zeroes and allow numbers to repeat, such as 333-0202.)

2.9 As of 1986, there were 128 three-digit telephone area codes in use, all conforming to the restrictions that the first digit is not a 0 or 1, the middle digit is either a 0 or 1, the last digit is not 0, and the last two digits are not 11 (freeing 411 for directory assistance, 911 for emergencies, and so on). What is the maximum number of three-digit area codes meeting these restrictions?

2.10 The call letters of U.S. radio and television stations begin with W east of the Mississippi River and K west of the Mississippi. For example, the CBS radio station in New York is WCBS and the CBS station in Los Angeles is KCBS. How many different call letters are possible for the whole country if each station uses **a.** three letters? **b.** four letters? **c.** either three or four letters?

2.11 A television game show gives the contestant an opportunity to win a car by unscrambling the four digits 6789 to match the price of the car. How many possible arrangements of these four digits are there?

*This rule does not extend to specific hands. For example, three aces beats three kings, even though both are equally probable. Also, improbable hands are not necessarily valuable. The probability of drawing the seven of clubs, five of diamonds, four of hearts, three of spades, and two of clubs is 1/2,598,960, but this rare hand loses to any simple pair.

2.12 The *Daily Planet* has printed the pictures of 10 gurgling babies. The publisher has secretly judged their cuteness, ranking the babies first through tenth. A substantial prize will be given to any reader whose rankings match those of the publisher. What are one's chances of winning if the babies are all equally cute and the reader must correctly pick **a.** all 10 places in order? **b.** the first 5 places in order? **c.** the first 3 places in order? **d.** the 5 cutest babies, not necessarily in order? **e.** the 3 cutest babies, not necessarily in order?

2.13 Professor Smith likes to liven up his statistics course by telling 2 jokes during the semester. He repeats the jokes from year to year, but for the sake of variety, he has vowed never to tell the same pair of jokes twice. If Professor Smith knows 10 jokes, how many years can he last before breaking his vow?

• **2.14** A certain store sells 31 different flavors of ice cream. How many different 3-scoop cones are possible if **a.** each flavor must be different and the order of the flavors is unimportant? **b.** each flavor must be different and the order of the flavors is important? **c.** flavors need not be different and the order of the flavors is unimportant? **d.** flavors need not be different and the order of the flavors is important?

2.4 LONG-RUN FREQUENCIES

One of the difficulties with the classical approach to probabilities is that there may be compelling evidence that the possible events are not equally likely. A life insurance company is interested in the probability of a reasonably healthy 20-year-old woman dying within a year. The possible outcomes are "dead" or "alive" after one year, but we can hardly assume these to be equally likely. What can we do instead?

Another probability approach was developed to handle such situations. If a coin is equally likely to come up heads or tails, then we might suppose that in a large number of coin flips, heads and tails would each come up about half the time. Indeed, we might define *probability* in this way: If we say that some event has a probability of 1/2 of occurring, we mean that if the situation were repeated over and over, this event would occur about half of the time. Or, turning it around, if we observe an event consistently occurring half the time, we could say that its probability of occurring is 1/2.

> The **long-run frequency interpretation of probabilities** is that if, in *n* (a large number of) identical situations, a certain event occurs *m* times, then its probability is *m/n*.

The life insurance company might estimate the probability of a 20-year-old woman dying within a year by looking at the experiences of millions of other women when they were 20 years old. If they had data on 10 million such women and found that 12,000 died within a year, then they might estimate the probability as 12,000/10,000,000 = .0012, or slightly more than 1 in 1000. In fact, some thirty years after Pascal's *Treatise* appeared, the Royal Society of London published mortality tables that could be used to price life insurance.

The equally likely and long-run frequency approaches to probability are not inconsistent with each other. They are just two different ways of thinking about probabilities. If a coin is equally likely to land heads or tails, then we expect it to land heads about half the time. One statistician, Karl Pearson, actually whiled away the hours by flipping a coin 24,000 times and carefully recording the outcomes. There were 12,012 heads and 11,988 tails, which gives a probability of heads of 12,012/24,000 = .5005. Apparently, his coin was unbent.

Pearson's experiment is dwarfed by that of a Swiss astronomer named Wolf, who rolled dice over a forty-year period, from 1850 to 1893. In one set of experiments, Wolf tossed a pair of dice, one red and the other white, 20,000 times. An economist, John Maynard Keynes, commented on the results:

> . . . the records of the relative frequency of each face show that the dice must have been very irregular, the six face of the white die, for example, falling 38% more often than the four face of the same die. This, then, is the sole conclusion of these immensely laborious experiments—that Wolf's dice were very ill made. . . .
>
> Wolf recorded altogether . . . in the course of his life 280,000 results of tossing individual dice. It is not clear that Wolf had any well-defined object in view in making these records, which are published in curious conjunction with various astronomical results, and they afford a wonderful example of the pure love of experiment and observation.[2]

The equally likely and long-run frequency approaches are consistent with each other, but clearly they differ in how they go about obtaining probabilities. In the equally likely method, we sit down, enumerate the possible outcomes, and meditate on their equal likelihood. In the long-run frequency method, we get up, go out, and collect some real data. The equally likely method is most appropriate for well-structured gambling situations, where we are pretty sure that we have fair coins or fair dice and, thus, gathering data would just be a boring waste of time. The long-run frequency approach is best when we suspect that the possibilities are not really equally likely, and so we want to use some data to confirm or deny our suspicions. The long-run frequency approach is obviously needed for situations like insurance company mortality tables, where it is apparent that living and dying are not equally likely. It may also be useful in fine tuning the probabilities of events that are not quite equally likely.

Will It Be a Boy or a Girl?

Consider the probability that a newborn baby will be a boy. Laplace's Principle of Insufficient Reason suggests that we may as well assume that boy and girl babies are equally likely and, consequently, assign a probability of 1/2 to each. But, if we took the time to examine the data, we would find that boy babies slightly outnumber girl babies. During the years 1975–1984, there were 34,696,000 recorded births in the United States, of which 17,787,000 were gurgling boys and 16,909,000 were bouncing girls. With these data, the long-run frequency estimate of the probability of a boy is

$$P[\text{boy}] = \frac{17,787,000}{34,696,000} = .513$$

Mendel's Genetic Theory

Gregor Mendel (1822–1884), an Austrian monk, used probabilities to explain the inheritance of various traits. Mendel's theory laid the basis for modern genetics, including the development of many remarkable hybrid plants. One of his early experiments involved the cross-breeding of sweet peas, some with yellow seeds and some with green seeds. He noticed a number of statistical regularities that he explained by postulating that there are entities, now called genes, that determine seed color and are inherited from the parent plants.

Mendel found that a crossing of purebred yellow-seeded and green-seeded plants produced plants with yellow seeds. But when these hybrids were fertilized with their own pollen, a variety of peas emerged. One-fourth had green seeds and continued to produce green-seeded plants when they were self-fertilized. One-fourth had yellow seeds and continued to produce yellow-seeded plants when they were self-fertilized. The remaining half had yellow seeds and yielded the same offspring proportions as the original hybrids.

What theory could explain these data?

Mendel hypothesized that each plant has two genes, each of which could be either Y or G. Thus, a pea plant could be either YY, GG, or YG. A plant with YY genes produces yellow flowers, while a plant with GG genes produces green flowers. A plant with YG genes produces yellow seeds and, therefore, Y is said to be the dominant gene.

Probability comes in because one gene is inherited from each of the parent plants. When both parents are YY, the offspring will be YY and produce yellow flowers. When both parents are GG, the offspring will be GG, too. When a parent is YG, the offspring is equally likely to inherit the Y or the G gene. Thus, the offspring of two hybrid parents has probabilities 1/4 of being GG, 1/4 of being YY, and 1/2 of being YG—just as Mendel found in his experiments!

Matters are not always this simple. Some traits, such as hair color, depend on more than one pair of genes. Some genes, such as those for blood type, have more than two alternative kinds. And some traits, such as height, depend on environment as well as genes. However, Mendel's genetic theory is still an extraordinarily simple, yet powerful, application of probability theory.

A Scientific Study of Roulette

Outside of the United States, roulette wheels generally have 37 slots, numbered 0, 1, 2, . . . , 36. Those used in the United States have an additional 00 slot, giving 38 slots in all. If the wheel is balanced, clean, and fair, we might expect each slot to be equally likely to catch the spun ball; but in practice, a roulette wheel usually has a few imperfections that cause some numbers to win slightly more often than others.

In the late 1800s, an English engineer, William Jaggers, took dramatic advantage of these imperfections. He paid six assistants to spend every day for an entire month observing the roulette wheels at Monte Carlo and recording the numbers that won. Jaggers found that certain numbers did come up more often than others.

He then proceeded to bet heavily on these numbers and, in four days, won 1.5 million francs, nearly $200,000—a fortune in the late 1800s. In the 1960s, while their fellow students were studying or protesting, a group of Berkeley students were reported to have pulled off a similar feat in Las Vegas. Nowadays, unfortunately, casinos examine and rotate their roulette wheels frequently to frustrate the long-run frequency bettors.

EXERCISES

2.15 In 1977, 6101 oil wells were drilled in new fields and 1004 of these discovered significant oil or gas.[3] What is the probability that a randomly selected newly drilled well in 1977 was successful?

2.16 A study of monthly returns in the stock market over the 55-year period, 1926–1980, found that the return was positive in 403 months and negative in 257 months.[4] Based solely on these data, what is your estimate of the probability of a positive monthly return in the stock market?

2.17 The Swiss astronomer Wolf tossed two dice 100,000 times and found that pairs came up 16,466 times. Use these data to estimate the probability of rolling pairs. Is this empirical long-run frequency close to the theoretical probability implied by an equally likely assumption?

2.18 A 50-year-old has to pay more for life insurance than a 20-year-old. Is this because 50-year-olds have higher incomes and can therefore afford to pay more for insurance?

2.19 The U.S. Postal Service handled 56 billion pieces of mail in 1978, of which 4.5 billion pieces were initially misdelivered and 211,203 were reported lost completely. Based on these data, what is the probability that a letter will be misdelivered? Will be lost completely? Can you think of any reasons why these long-run frequency probabilities may understate or overstate the chances that the next letter you send will be misdelivered or lost?

2.20 The color of the flowers on Japanese four-o'clocks is determined by a pair of genes, each of which could be either R or W. The flower is red if the genes are RR, white if WW, and pink if RW. What are the respective probabilities of red, white, and pink flowers if **a.** the parent plants are RR and WW? **b.** the parent plants are RR and RW? **c.** the parent plants are RW and RW?

2.21 A book recently calculated the probabilities of "being injured by various items around your house."[5] The authors conclude that, "As the figures show, our homes are veritable booby traps. Even your cocktail table is out to get you. These accounted for almost 64,000 injuries [in 1977], more than all ladders [62,000 injuries]." Using data based on 74,050,000 households, they calculated that the probability of being injured by a cocktail table was 64,000/74,050,000 = .00086 and the probability of being injured by a ladder was 62,000/74,050,000 = .00084. Should we conclude that cocktail tables are indeed more dangerous than ladders? Do their probability calculations make sense?

2.5   SUBJECTIVE PROBABILITIES

The long-run frequency approach allows us to extend probabilities to events that are not equally likely. But its application is limited to situations where we have lots of repetitive data, while much of the uncertainty that surrounds us concerns virtually unique situations. There are no frequency data if the situation has never occurred before. For example, we might be keenly interested in the outcome of an upcoming presidential election. Perhaps the opposing candidates support policies that will have important effects on our lives. Decisions about buying or selling stocks, expanding or contracting a business, or even whether or not to enlist in the Army may well be affected by who you think is going to be elected president.

In 1946, the "Wizard of Odds" calculated presidential probabilities as follows:

> Miss Deanne Skinner of Monrovia, California, asks: Can the Wizard tell me what the odds are of the next President of the United States being a Democrat? . . . Without considering the candidates, the odds would be 2 to 1 in favor of a Republican because since 1861 when that party was founded, there have been 12 Republican Presidents and only 7 Democrats.[6]

But we can't very well predict an election winner simply by calculating the relative frequencies with which Democrats and Republicans have been elected president. In 1936, Franklin Roosevelt ran for reelection against Alf Landon. At that time, there had been 20 previous presidential contests between Republicans and Democrats, with the Republicans winning 14.* Should the 1936 forecasters have concluded that Landon had a 70 percent chance of winning? Or should they have somehow taken into account that Roosevelt was a popular president running for reelection against the unexciting governor of Kansas?

Similarly, by 1964, the tally was 16 election victories for the Republicans and 11 for the Democrats. Did Barry Goldwater then have a $16/27 = .59$ chance of thwarting Lyndon Johnson's reelection bid? Should we assume that the Republicans' odds were the same in 1964 and, eight years later, in 1972 when the Republican incumbent Richard Nixon ran against George McGovern? Clearly, the odds change from election to election, depending on the state of the nation and the personalities of the candidates.

The outcome of a presidential election is uncertain and it would be useful to quantify that uncertainty. We would like to have more than a shrug of the shoulders and sheepish "who knows?" Neither the equally likely nor the long-run frequency approach is satisfactory, however. The choice of president is not determined by the flip of a coin, bent or unbent, every four years.

In the same spirit, a good weather forecaster will use more than historical records and a lucky penny to predict tomorrow's weather. It would be hard to make rational plans if the forecaster simply said, "Maybe it will rain, and then again, maybe it won't." It's much more informative to hear, "There's a 90 percent chance of rain tomorrow." Similarly, a good doctor will base each prognosis on the individual patient, and not on some long-run average. To make rational decisions, the patient needs to know, "You have a 60 percent chance of surviving such an operation," and not "You will survive, unless you don't."

The same is true of many interesting situations. Who is going to win the next Super Bowl? Will this first-round draft choice be a successful pro quarterback? Will you be accepted by a certain graduate school? Will we find oil if we drill here? Is the unemployment rate going to increase or decrease? Did this person rob that

---

*These numbers differ from the Wizard's because he looked at presidents rather than elections. For example, Woodrow Wilson won in 1912 and 1916, and so I counted two Democratic Party victories, while the Wizard counted one Democratic president. Similarly, I ignored Chester Arthur, who became president in 1881, after James Garfield was assassinated; the Wizard counted Arthur as another Republican president.

bank? Will the Soviet Union attack China? Each example is uncertain and ripe for a probability assessment, but none fits into the equally likely or long-run frequency mold.

Bayes' Approach

In the eighteenth century, Reverend Thomas Bayes apparently wrestled with an even more challenging problem—the probability that God exists. The equally likely and long-run frequency approaches are useless, and yet, this is an uncertainty that is of great interest to many people, including Reverend Bayes. Such a probability is necessarily subjective. The best that any person can do is to weigh the available evidence and logical arguments and come up with a personal probability of God's existence. This idea of personal probabilities has been extended and refined by other *Bayesians,* who argue that many uncertain situations can only be analyzed by means of subjective probability assessments. Bayesians are ready, willing, and eager to assign probabilities to football games, oil drilling, war strategy, and God.

A Super Bowl Probability

These probabilities can be elicited by observing the choices people make. A few weeks before Denver played the New York Giants in Super Bowl XXI, Juan Guerra, a former student, called to ask if the stock market was headed up or down. In return, knowing that Juan was a football fan, I repeated a type of mental experiment that he had participated in during one of my statistics classes. Which of the following gambles would he choose, receiving

$10 if New York wins; or
$10 if a red card is drawn from a deck containing five red cards and five black cards?

He emphatically chose New York, indicating that he believes their probability of winning to be above .5. To learn more, I offered Juan more choices. Would he prefer

$10 if New York wins; or
$10 if a red card is drawn from a deck containing nine red cards and one black card?

Juan chose the card draw, indicating that he believed the probability of New York winning to be less than .9. I then offered him this choice:

$10 if New York wins; or
$10 if a red card is drawn from a deck containing eight red cards and two black cards

He chose New York, putting their probability of winning somewhere between .8 and .9. With further questioning, I pinned him down to a .85 probability by finding that he was indifferent between these choices:

$10 if New York wins; or
$10 if a red card is drawn from a deck containing 85 red cards and 15 black cards

Success Probabilities at Sandoz

Since 1970, Sandoz, a large Swiss pharmaceutical company, has been using subjective probabilities to gauge the potential success of its research and development (R&D) projects.[7] After a potentially useful chemical compound has been identified by exploratory research, it becomes a "project" and will be tested for a minimum of five years for efficacy and safety. If it passes these tests (on average, about one in ten do) then it is judged technically successful and will be registered and marketed.

The firm's decision to approve or reject a project for testing is of crucial importance, since the new products marketed five to ten years from now depend on which projects are approved for testing today. And even for those projects that are initially approved, as testing proceeds, the firm must decide whether to continue or abandon hope. A prime consideration for these important management decisions is each project's chance of eventually being judged technically successful. An equally likely assumption is of little use and not very plausible since all projects are not equally likely to be successful. And because each product is different, it is not reasonable to estimate its success probability from past relative frequencies.

Instead, Sandoz uses subjective probabilities, reflecting the consensus of a panel of its research and development experts. And once a project is launched, its success probability is updated semiannually. For some projects, the probability of success is revised upward and, as it approaches 1.0, the project is judged a technical success. Other projects—those for which the probability of success stays low or sinks as testing proceeds—are terminated by management.

Sandoz's subjective probabilities are no simple matter. We're talking about the chances, before testing, that a new compound will prove technically successful five to ten years hence. But as the elderly man says when asked about his health: "I've got problems; but life sure beats the heck out of the alternative." Sandoz reasons that subjective probabilities are better than none at all. And, even nicer, it has been pleased to report that its panel's subjective probability assessments have turned out to be a reliable guide to a project's potential success, in that of those projects given an X percent chance of succeeding, about X percent do succeed.

Juan knew a lot about these two teams. Denver had a good defense and an exciting quarterback, John Elway; but New York had an exceptional defense, with perhaps the best linebackers in NFL history, and had beaten its playoff opponents convincingly. Denver had played very well early in the season, including a close 16–19 loss to New York, but had seemed to tire as the season progressed. Denver's coach wears $700 suits to games, whereas the Giants players throw Gatorade on their coach. None of these varied tidbits of information can be directly translated into a probability, but Juan subjectively weighed what he knew and came up with a personal probability of .85 for New York winning.

The **Bayesian interpretation of probabilities** is that an event has a subjective probability of m/n of occurring if you are indifferent between a gamble hinging on this event and a gamble hinging on the draw of a red card from a deck in which a fraction m/n of the cards are red.

Interest Rate Uncertainty and Forecasting

The success of asset/liability decisions made by financial institutions depends critically on future interest rates. When interest rates rise unexpectedly, those who invested long-term at a fixed interest rate and those who borrowed short-term at a variable interest rate find that they made an expensive mistake. An unexpected drop in interest rates is costly for those who invested short-term or borrowed long-term. Yet, interest rates are notoriously difficult to forecast.

Up until 1971, the top managers at Morgan Guaranty Trust met regularly to determine the interest rate forecasts to use in all asset/liability decisions and, for each interest rate prediction, they would come up with a single number; for example, "our best estimate of the Treasury-bill (T-bill) rate three months from now is 6 percent." Yet, as the senior operations research officer explained,

> Considerable discussion normally preceded the managers' arrival at the single-valued expectation, but there was no formal procedure for quantifying the collective expectations of the participants, and the uncertainty surrounding these expectations.[8]

How confident were they in their 6 percent forecast? Are we talking 5 3/4 percent to 6 1/4 percent or 4 percent to 8 percent? If the T-bill rate isn't 6 percent, is it more likely to be somewhat higher or a bit lower? The answers to such questions require probabilities and, so, in 1971, for each forecast, they began reporting probabilities in place of a single interest rate number.

Twice a month, specialists write down interest rate probabilities individually and then meet as a group to compare notes. Instead of thinking of the most likely value for the interest rate, perhaps 6 percent, each member thinks of the possible values—5 percent, 5 1/2 percent, 6 percent, and so on—and assigns probabilities to each possibility. The resulting probabilities are graphed like a histogram and, indeed, called a histogram at Morgan Guaranty.

The first graph in Figure 2.1 reflects a committee member who is pretty confident that the T-bill rate will be 6 percent; the next graph shows considerable uncertainty. In the third and fourth graphs, 6 percent is the most likely value; but in the third case the person feels there is a good chance that the T-bill rate will be less than 6 percent and, in the fourth case, the opposite is true. None of these subtleties are revealed in a 6 percent forecast that is unaccompanied by probabilities. With probabilities, we can see not only the most likely value, but also the likely range, and we can see whether the individual believes that the rate is more likely to be above or below 6 percent.

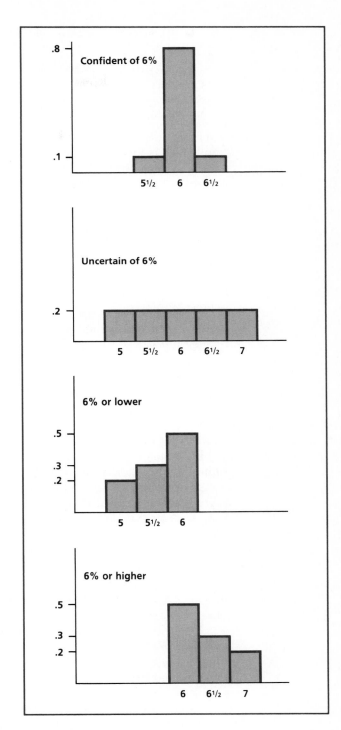

Figure 2.1   *Four Interest Rate Probability Histograms*

The senior research officer described the value of these probability graphs at a November 13, 1972 group meeting:

> . . . it was useful to know why both Griffin and Riefler felt the rate would stay essentially the same, whereas the others felt it would rise; and to find out why Engle felt that there was no chance the rate would drop below 5 1/8% and felt so strongly that it would rise over 5 5/8%, when nobody else in the group felt that way.

After such discussion, the committee constructs a consensus probability distribution, which is then presented to top management.

The probabilities used by Morgan Guaranty are necessarily subjective, but they accomplish their task of describing the beliefs of committee members:

> . . . the only "incorrect" histogram is one which does not correctly reflect its creators' expectations. While there is no guarantee that histograms will give an accurate picture of the future, they will give an accurate picture of an individual's view of the future.

Objectivist Objections

Many probability theorists are not comfortable with the Bayesian approach. They prefer objective probabilities derived from equally likely principles or long-run frequencies. It is not scientific to pull probabilities out of the air. It is especially disturbing that subjective probabilities vary from person to person. If a coin is fair, heads and tails are equally likely. If 17,787,000 out of 34,696,000 babies born were boys, then the probability of giving birth to a boy is .513. These are scientific "facts" on which we can all agree and draw common conclusions. How can we have a scientific discourse if one person says that the probability of something is .8 while another person says that it is .4?

Bayesians respond that these subjective disagreements are all about us and should not be ignored. As Mark Twain remarked, "It is a difference of opinion that makes horse races." And Super Bowls. After Juan settled on an 85 percent probability of New York winning, I told him that "On any given Sunday . . ." and expressed my own belief that New York's chances were closer to 60 percent. It is these very disagreements that make football games so interesting. If probability analysis ignores subjective probabilities, it will miss most of life's uncertainties.

Will we find oil here? Will interest rates go up or down? Will the Soviet Union attack China? These are all uncertainties about which informed and well-intentioned people disagree. Instead of ignoring these disagreements, Bayesians argue that it is better to quantify them by pinning down people's subjective probabilities.

You and I may be debating about whether or not we will find oil in a particular location. We can each present our arguments and call each other names. That might be fun, but it would be nice to have something more than me saying "maybe there is" and you cleverly retorting "maybe not." If we specify our subjective probabilities of finding oil, then we can see how much we really disagree. Perhaps we've been arguing viciously over a slight difference in probabilities, .48 versus .52. We can also use probabilities to calculate the expected costs and benefits of drilling, as we will see later in this book. Explicit probabilities allow us to see where we disagree about the value of drilling and by how much. Admittedly, it would be nicer to have an agreed-upon objective probability, but when there is none, Bayesians argue that subjective probabilities are better than nothing at all.

Can We Put Probabilities on Events That Have Already Occurred?

There is an even deeper dispute between the objectivists and the subjectivists. Some theoretical objectivists adhere very strictly to the coin-flip metaphor. We can imagine God flipping a coin (albeit a slightly bent coin) to determine whether a child will be a boy or a girl. Or, more exactly, we might imagine God reaching into a bag that contains 513 boys and 487 girls. These objectivists don't like to apply probabilities to situations that can't be fit into this metaphor. For example, they would not apply a .513 probability after a child's sex has been determined, even if the baby is not yet born and the sex is not yet known by the outside world. Once the child has been conceived, it is a boy either with probability 1.0 or with probability .0. Is my own oldest child a boy? A Bayesian who doesn't know the answer would freely assign a probability of about .5. An objectivist who doesn't know would refuse to answer, protesting that this is a misuse of probabilities.

By the same logic, these objectivists say that a coin has a .5 probability of landing heads before it is flipped, but not afterward. What if the coin has been flipped and then immediately covered before anyone could see whether it landed heads or tails? The strict objectivist argues that the coin is now either a head or a tail and, consequently, is no longer a fit subject for a .5 probability statement. The Bayesian, in contrast, argues that any uncertain situation is susceptible to a probability assessment. The Bayesian will hold to a .5 probability until there is some evidence of whether it is a head or a tail.

As another example, I was once asked to play the Bayesian devil's advocate in an academic objectivist-subjectivist debate. The objectivist asked me the probability that October 23, 1845 was a Tuesday. I replied, "Before or after I consult an almanac?" In the absence of any information, a Bayesian would undoubtedly say that the probability of this particular date being a Tuesday is 1/7. Once an almanac had been consulted, the probability would be revised to .0 because October 23, 1845 turns out to have been a Thursday. In this debate, the objectivist argued that 1/7 was not a proper use of probabilities. No matter how informed or ignorant I was, October 23, 1845 was a Thursday. Its probability of being a Tuesday always was and always will be zero.

Some theorists accept the peaceful coexistence of an objectively correct probability, which everyone would agree to if only they possessed all of the relevant information, and a subjective probability, which depends on the available information. The objective probability that October 23, 1845 was a Tuesday is zero; my subjective probability was 1/7 until I consulted an almanac. Thus a subjective probability is more a statement about a person's information than about the phenomenon under consideration. As in the Morgan Guaranty example, subjective probabilities are used to describe partially informed beliefs.

This debate may seem like a lot of academic hot air about nothing and, to a large extent, it is; but the objectivists and the subjectivists do look at many things a bit differently. Think of the question, "Is there oil where we are drilling?" The Bayesian would advise using the available information to come up with a subjective probability. The strict objectivist would object that there is either oil here or there isn't, so it is not appropriate to assign any probability other than zero or one. The same thing would happen with the question, "Did this person rob that bank?" The Bayesian is willing to assign probabilities; the objectivist isn't. Throughout this book, we will see that Bayesians assign probabilities to all sorts of uncertainties, whereas many objectivists keep a narrower perspective.

EXERCISES

2.22 A business executive believes that a new perfume for dogs is just as likely to fail as to succeed. What is this executive's implicit probability of the product's success?

2.23 A security analyst believes that it is twice as likely that the stock market will go up rather than down during the next twelve months. What is this analyst's implicit probability of the market going up during the next twelve months? (Assume there is no chance that the market will not change at all.)

2.24 Edward Yardeni, Director of Economics and Fixed Income Research at Prudential-Bache, is one of the best economic forecasters. In early 1987, it was very unclear whether the U.S. economy would boom, bust, or muddle in between. Yardeni explained his position as follows:

> We've been slumpers for the past few months. A slump is still possible, but now we feel more comfortable with a muddling scenario. Until recently, we assigned the slump scenario a probability of 50%, with muddling at 30% and with a boom given a 20% chance. Now, muddling gets 50%; bust gets 30%; and boom stays at 20%.[9]

Would you characterize his probabilities as being equally likely, long-run frequency, or subjective? Why are the other two kinds of probabilities inappropriate here?

2.25 About 20 percent of the students who take the GMAT test for admission to business school score above 600. Explain why you believe that your probability of scoring above 600 on the GMAT is either larger or smaller than .20.

2.26 A well-known automobile manufacturer claimed in a 1986 television commercial that 70 percent of all its cars registered in the U.S. since 1974 were still on the road. Does this imply that there is a 70 percent chance that a 1986 model will last at least twelve years?

● **2.27** A book recently answered the question, "What are the odds of being indicted or convicted for criminal tax activity?" by noting that, "Of the 136,718,000 tax returns of all kind filed in fiscal 1978, just 2,634 were referred for prosecution and 1,724 indictments issued."[10] Thus, they calculated the probability of being indicted as $1,724/136,718,000 = .000013$. Similarly, because 1,414 people were convicted, the book concludes that the probability of being convicted of criminal tax activity is $1,414/136,718,000 = .000010$.

Why are these probability calculations of dubious value? What is their correct interpretation?

2.6   SUMMARY

Probabilities are used to quantify life's uncertainties. There are three basic approaches: the enumeration of equally likely events, the computation of observed relative frequencies, and a subjective assessment based on a blending of facts and opinions.

The classical equally likely approach was devised to handle games of chance—the roll of the dice, spin of the roulette wheel, and deal of the cards. If there are m "successes" in n possible outcomes, with each being equally likely, then the probability of a success is m/n. Sometimes we have to use the permutation and combination formulas to enumerate the possible outcomes. Suppose that we are choosing r out of n distinct items.

When the order matters, the number of different permutations is $\dfrac{n!}{(n-r)!}$

When order is unimportant, the number of different combinations is $\dbinom{n}{r} = \dfrac{n!}{r!(n-r)!}$

The frequentist approach is designed for cases where there are lots of data and the possible outcomes are apparently not equally likely. If an event has occurred m times in n (a very large number) identical situations, then its probability is m/n. For instance, 17,787,000 boys in 34,696,000 births implies an $17,787,000/34,696,000 = .513$ probability of a boy.

The subjectivist approach is closely associated with Reverend Thomas Bayes. A subjective probability is based on an intuitive blending of a variety of information. Consider, for instance, the stock market. I may well resist making an equally likely assumption or using past relative frequencies. Instead, based on what I know about the stock market and the economy, I may believe that there is a .60 probability that the stock market will go up this year. Subjective probabilities such as this obviously vary from person to person.

REVIEW EXERCISES

2.28 The Braille writing system uses six dots, arranged in two columns of three dots. Each dot can be either raised or flat. How many different Braille characters are possible?

2.29 There are 40,000 students at Big State University (BSU). Of these 40,000 students, 5,000 love BSU, 10,000 hate BSU, and 25,000 never stop to think about it. *Timeweek* magazine decides to do a story on BSU and sends out a reporter to interview a student. If each student is equally likely to be interviewed, what is the probability that the reporter will talk to a student who hates BSU?

Now assume that the reporter is not interested in interviewing one of the 25,000 students who have no opinion of BSU. If one of these apathetic students is accidentally selected, the reporter will just move on to another student, until one is found who either loves or hates BSU. Now what is the probability that the person interviewed will be a student who hates BSU?

2.30 A prominent mathematician, Jean d'Alembert, argued that there is a 1/3 probability of two tails when a coin is flipped twice. He reasoned that we could obtain a head on the first flip, a head on the second flip, or no heads. Therefore, the probability of two tails (no heads) is 1/3. Similarly, he reasoned that a coin flipped three times could yield a head on the first flip, a head on the second flip, a head on the third flip, or no heads. Therefore, the probability of three tails (no heads) is 1/4. Explain the flaw in d'Alembert's reasoning.

2.31 Since World War II, the number of commercial banks in the United States has averaged about 14,000 and the number of bank failures each year has averaged about 65. On average, what fraction of U.S. banks fail during a year? Why might you be wary of interpreting this fraction as the probability that a specific bank will fail during the coming year?

2.32 Larry, Moe, and Joe are going to draw straws to decide who brings the beer. A neutral observer shows the tops of three straws and lets them pick, with the short straw losing. Joe insists on drawing last. That way, if Larry or Moe picks the short straw, Joe won't have any chance of drawing it. Is there an advantage to drawing last?

2.33 Do you think that a low-income or high-income tax return is more likely to be audited by the IRS? (Think about it carefully; there are arguments both ways.) Here are some data for 1983:

Total Income	Returns Filed	Audited by IRS
Less than $10,000	31,357,000	107,081
$10,000–$24,999	30,745,000	283,072
$25,000–$49,999	22,243,000	456,928
$50,000 and over	5,531,000	221,419

Calculate (and compare) the relative frequencies for each of these four income classes. Let's assume that in the year you graduate from college, your income falls into the second income category. Why should we be careful in interpreting the calculated relative frequency for this category as the probability that your return will be audited that year?

2.34 In 1984, 120,600 patent applications were filed with the U.S. Patent and Trademark Office, of which 72,700 patents were issued and 47,900 applications were denied. What percentage of the patent applications were granted? Does this mean that if you apply for a patent, you have this probability of being approved? For what question is your calculated relative frequency a valid probability?

2.35 You are playing draw poker and have been dealt a four, five, seven, eight, and king. You discard the king and draw a new card, hoping for a six to fill your inside straight. What is your probability of success? If you are instead dealt a four, five, six, seven, and king, what is your probability of drawing either a three or eight to complete your straight?

2.36 A Temple University mathematics professor used these figures to show that most Americans have an exaggerated fear of terrorists:

Without some feel for probability, car accidents appear to be a relatively minor problem of local travel while being killed by terrorists looms as a major risk of international travel. While 28 million Americans traveled abroad in 1985, 39 Americans were killed by terrorists that year, a bad year—1 chance in 700,000. Compare that with the annual rates for other modes of travel within the United States—1 chance in 96,000 of dying in a bicycle crash; 1 chance in 37,000 of drowning; and 1 chance in only 5,300 of dying in an automobile accident.[11]

How do you suppose the author calculated the probabilities of dying in a bicycle accident, of drowning, and of dying in a car accident? Do these calculations suggest that it is more dangerous to drive to school today than to fly to Paris for two weeks?

2.37 In craps, a bettor can make a "field" bet. The bet is lost if a 5, 6, 7, or 8 comes up on the next roll of a pair of dice. Any other number is a win. A $1 field bet gains $1 if a 3, 4, 9, 10, or 11 is rolled and gains $2 if a 2 or 12 is rolled: One author wrote that "field betting is a fascinating wager. . . . After all, how can they pass up a bargain that gives them seven numbers in their favor, compared to four in opposition?"[12]

What is the probability of winning a field bet? If you make 36 field bets and the numbers come up as frequently as they should (one 2, two 3s, and so on), how many dollars will you come out ahead?

• **2.38** *(continuation)* Explain why you agree or disagree with this advice on placing field bets:

One of the best ways to play the field is to go along with the dice. If a field number shows up, continue field betting. Thus, if a streak occurs, players will be in on it.

Or else you may attempt the opposite. When a field number comes up, stop betting and wait for a non-field number. Afterwards, bet the field again. In this way, players will be cutting down the percentage of numbers against them.[13]

• **2.39** There are three categories of professors at Metro U.:

	Male	Female	Total
Full professors	400	100	500
Associate professors	100	100	200
Assistant professors	150	150	300
Total	650	350	1,000

One of these three categories of professors is picked at random, and then one of the persons in that selected category is picked at random. What is the probability that the selected person is a male?

• **2.40** *(continuation)* If, using the data in the preceding exercise, a male is picked at random, what is the probability that the selected person is a full professor? If, in contrast, a female is picked at random, what is the probability that she is a full professor? Do these data suggest that gender and rank are related or unrelated?

• **2.41** Galileo wrote a short note on the probabilities of obtaining a sum of 9, 10, 11 or 12 when three dice are rolled. Someone else had concluded that these numbers are equally likely, because there are six ways to obtain a 9 (1–4–4, 1–3–5, 1–2–6, 2–3–4, 2–2–5, or 3–3–3), six ways to obtain a 10 (1–4–5, 1–3–6, 2–4–4, 2–3–5, 2–2–6, or 3–3–4), six ways to obtain an 11 (1–5–5, 1–4–6, 2–4–5, 2–3–6, 3–4–4, or 3–3–5), and six ways to obtain a 12 (1–5–6, 2–4–6, 2–5–5, 3–4–5, 3–3–6, or 4–4–4). Yet Galileo observed "from long observation, gamblers consider 10 and 11 to be more likely than 9 and 12." How do you think Galileo resolved this conflict between theory and observation?

•• **2.42** Here's a probability variant on three-card Monte. You are shown three cards: one black on both sides, one white on both sides, and one white on one side and black on the other. The three cards are dropped into an empty bag and you slide one out; it happens to be black on the side that is showing. The operator of the game says, "We know that this is not the double-white card. We also know that it could be either black or white on the other side. I will bet $5 against your $4 that it is, in fact, black on the other side." Can the operator make money from such bets without cheating somehow?

2.43 Explain why you think that the following strategy for winning coin tosses is sound or unsound:

Let the other man make the call, for if you are the one who calls, the chances are 3 to 2 against you. Explanation: 7 out of 10 people will cry heads, but heads will turn up only 5 times out of 10, so if you let your opponent call, you have the greater probability of winning.[14]

2.44 Raymond Queneau wrote an unusual book of poetry, *Cent mille milliards de poemes.* This book contains only 10 pages, but each page is sliced into 14 strips, with each strip of paper containing one line of poetry. The reader can then construct a 14-line poem by taking each of the 14 lines from any one of the 10 pages. For example, the first line of the poem can be any of the first lines on these 10 pages. How many lines of poetry did Queneau write? How many different 14-line poems can be constructed?

2.45 California license plates once used a number followed by a letter, followed by four more numbers. How many different license plates were possible? In 1956, California changed to a system used by several other states: three letters, followed by three numbers. If there are no forbidden combinations, how many different license plates of this type are possible? Then California also began issuing plates with three numbers followed by three letters. How many additional plates did this reversal allow? In 1978, California introduced license plates with a number followed by three letters and then three more numbers. How many different plates of this type are possible? Why do you think California keeps changing its license plate system?

2.46 In 1983, there were 83.9 million housing units in the United States, classified in Table 2.4 (in millions) by age and size. **a.** What fraction of all

Table 2.4   *Data for Exercise 2.46*

| When Built | *Heated Square Footage of Residence* | | | | |
	−1,000	*1,000–1,599*	*1,600–2,399*	*2,400 or More*	*Total*
Before 1940	8.9	7.1	4.7	3.0	23.7
1940–1949	2.8	2.1	1.4	0.7	7.0
1950–1959	4.0	4.8	3.2	1.4	13.4
1960–1969	6.0	4.9	3.6	2.2	16.7
1970–1979	7.5	5.4	4.4	3.0	20.3
1980–1982	1.2	0.8	0.5	0.3	2.8
Total	30.4	25.1	17.8	10.6	83.9

units were built before 1960? **b.** What fraction of all units were built before 1940? **c.** What fraction of all units with 2,400 or more square feet were built before 1940? **d.** What fraction of all units built before 1950 have less than 1,000 square feet? **e.** What fraction of all units built since 1950 have less than 1,000 square feet?

2.47 There are 11 positions on a soccer team. A team has just been formed for children who are five to six years old, with one player's parent as coach. This inexperienced coach decides to try every child in every position to find the best possible starting lineup. How many possible lineups are there **a.** if there are 11 children on the team? **b.** if there are 13 children on the team? **c.** if there are 13 children on the team and it is predetermined that the coach's child will start in the center-forward position?

2.48 The Morse code consists of a sequence of telegraphic "dot" or "dash" signals, with each sequence representing a letter of the English alphabet. For example, dot = E; dot-dash = A; and dot-dash-dot = R. Thus, dot dot-dash dot-dash-dot gives the message "EAR." The distress signal is dot-dot-dot dash-dash-dash dot-dot-dot = SOS. How many English letters can be represented by sequences of from one to three dot or dash signals? How long a sequence did Morse have to allow in order to include all 26 letters of the English alphabet?

2.49 The Center for UFO Studies recorded 3,126 UFO sightings by Californians in 1978 and only 79 sightings by Delaware residents.[15] John lives in California and Mary lives in Delaware. Can we conclude from these data that John is more likely to report a UFO sighting than Mary?

2.50 A book of probabilities advises

... next time you pat your little nephew on the head and ask him what he wants to be when he grows up don't expect a quick and intelligent answer. The poor kid has the odds stacked 17,452 against him. That's how many specified occupations there are listed in the *Dictionary of Occupational Trades.*[16]

Explain why you would or would not use 1/17,452 as the probability of your nephew correctly selecting his future occupation.

2.51 To decide who pays for lunch, Jack and Jill each flip a coin. If one coin is a head and the other a tail, then Jack pays. If the coins match, then Jill pays.

What is the probability that Jack pays? In the long run, about how frequently will Jack pay?

2.52 To decide who pays for lunch, three people simultaneously flip coins, with the odd person paying. What is the probability that one round will be enough to determine the odd person?

2.53 If you roll two standard dice, what is the probability that the two numbers rolled will not be the same?

2.54 You are taking a multiple-choice test and have run out of time, with one question left to answer. You arbitrarily mark C. If the five possible answers (A, B, C, D, and E) are equally likely, what is the probability that you have answered the last question correctly?

2.55 *(continuation)* Assume now that you are forced to guess the answers to two multiple-choice questions. You choose C for the first and D for the second. How many possible pairs of answers are there? If all pairs are equally likely to be correct, what is the probability that you answered both questions correctly?

• 2.56 *(continuation)* What is the probability that you answered both questions incorrectly? What is the probability that you got one right and one wrong?

2.57 You are one of ten people at a Christmas party. Each has brought a gift, which is placed in a large bag. After all ten gifts are in the bag, each person reaches into the bag and randomly withdraws a gift. You are the last person to pull out a gift. What is the probability that it is the gift that you brought?

2.58 The U.S. Public Health Service estimates that the chances of being murdered sometime during one's lifetime are 1 in 369 for a white female, 1 in 131 for a white male, 1 in 104 for a black female, and 1 in 21 for a black male. Should we conclude that if you are black or white, the appropriate one of these four figures is the probability that you will be murdered?

2.59 In 1983 there were 2,966,000 business firms in the United States, of which 31,334 failed that year. Calculate the percentage of firms that failed. Explain (with examples) why we should be cautious in using this relative frequency to estimate the probability of failure next year (a) as an individual starting a new business and (b) as a bank evaluating a loan application.

2.60 Give an equally likely, frequentist, and subjectivist interpretation of a weatherman's statement that "there is a 50 percent chance of rain tomorrow."

2.61 The Postal Service uses five-digit ZIP codes to sort mail. How many possible five-digit ZIP codes are there? In 1983 nine-digit ZIP codes were introduced. How many possible nine-digit codes are there? Why do you think that the Postal Service wanted to increase the number of digits in our ZIP codes? There are about 225 million people in the United States. How many digits would be needed to give every person a separate ZIP code?

2.62 In the game odds-or-evens, two players simultaneously take a hand from behind their backs, showing either one or two fingers. If each player is equally likely to show one or two fingers and their choices are made independently, what is the probability that the sum will be even?

2.63 *(continuation)* The traditional Indian game Tong is played similarly, except that players can show one, two, or three fingers. If each possibility is equally likely, what is the probability that the sum will be an even number?

2.64 The coach of a young girl's softball team has learned that there are three key positions: pitcher, shortstop, and first base. If this coach has five skilled players, how many different ways can he fill these three key positions with skilled players, assuming that it does not matter which person plays which position?

• 2.65 Greasy Pizza Parlor offers ten different extras on their pizzas: sausage, salami, pepperoni, mushrooms, peppers, anchovies, olives, Canadian bacon, onions, and bean sprouts. You can order your pizza plain or with one to ten extras. How many different pizzas are possible?

2.66 A book reported that 1,356 people were arrested in Newark, New Jersey, in 1973 for gambling.[17] Since there were 367,476 people in Newark, the authors reported that the probability "of being arrested in a gambling raid" in Newark is 1,356/367,476 = 0.0037. They advise: "If you have a hankering for the ponies and want to avoid the paddy wagon, head for Tucson. They didn't have a single [gambling] arrest in 1973." What's wrong with their reasoning?

2.67 In one carnival game, three balls are tossed into a tic-tac-toe arrangement of nine slots:

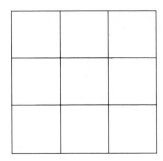

(The slots are surrounded with a fence, so that each ball has to land in one of the slots; only one ball fits in each slot.) If the balls land randomly, what are the chances of a tic-tac-toe (three balls in a row horizontally, vertically, or diagonally)?

2.68 Johnny has forgotten the three-digit combination to his school padlock. He decides to try all possible combinations. How many are there? What is the probability that his first guess will be correct? (Assume that there are 39 numbers on the padlock and that numbers can be repeated.)

• 2.69 You are playing draw poker and have been dealt three spades and two hearts. You discard the two hearts. What is the probability of drawing two spades?

2.70 A popular magazine prints the pictures of ten celebrities, as they look now and as they looked as babies. Prizes are given to readers who can correctly match all ten celebrities with their baby pictures. If the match is completely random, what is the probability of winning?

2.71 In the carnival game Under and Over Seven, the player can predict that the sum of a pair of dice will be either less than 7 or larger than 7. What is the probability of a correct prediction?

2.72 A television show gives the contestant a chance to win a trip to Buffalo by guessing the price of mystery package. The contestant is given a penny, nickel, dime, and quarter and told that a combination of some or all of these coins gives the exact price of the mystery package. How many possibilities are there?

2.73 A *Wall Street Journal* article asked:

Is Rhode Island a tax haven? Last year, the Internal Revenue Service audited only 0.47% of the individual income-tax returns filed in the Ocean State—the lowest audit rate in the country. By contrast, the IRS examined 2.61% of returns in Nevada.[18]

Does this disparity prove that moving to Rhode Island will reduce your chances of a tax audit?

• **2.74** Galileo was asked whether the numbers on three rolled dice will sum to 9 or 10 more often. He counted the possible outcomes and provided the answer. What is the probability of rolling a 9? A 10? Which is the more probable?

2.75 The four aces are removed from a standard deck of cards, shuffled, and placed face down on a table. If you select two cards, what is the probability that they are the same color?

2.76 An American Express card number has 15 digits. If each digit can be any number from 0 to 9, how many American Express cards can be issued without repeating a sequence of 15-digit numbers? There are, in fact, 11 million American Express cards outstanding. If a clerk makes a mistake entering an American Express number, what is the probability that the 15-digit number typed will be one of these 11 million active card numbers?

2.77 In 1978 there were 73,767 deaths in New York City and 217 deaths in Dover, Delaware. Do these data prove that New York City was a far more dangerous place than Dover?

2.78 In the board game Careers, there are two corner squares, labeled Hospital and Park Bench. To leave the hospital square, you must either pay half of your salary or roll a sum of 5 or less with two dice. What is the probability of doing so? To leave the Park Bench, you must pay half of your cash on hand or else roll a 7, 11, or doubles. What is this probability?

2.79 A market research firm asks a focus group to rank four different versions of a commercial; for instance, version 1 first, version 3 second, version 4 third, and version 2 fourth. How many possible rankings are there?

2.80 A beverage company is introducing a new cola, which it hopes will compete successfully against the three leading brands. One hundred cola drinkers are asked to rate each of these four colas on a scale of 1 to 10, and the average rating for each cola is then calculated. Based on this average rating, the four colas are ranked first, second, third, and fourth. If we label the four colas A, B, C, and D, how many possible ways can they be ranked. In how many of these possible rankings is A first? First or second? If the ranking is purely random, what is the probability that the new cola will be ranked first or second?

2.81 A computer store has ordered 20 hard drives from a new supplier. Two of these drives are defective. If the store tests two drives, what is the probability that the two drives it tests will be the defective ones?

2.82 The economics department at a small college is housed in a house built in 1929. There are five economics professors and five offices—all different, with one office clearly better than the other four. To reduce jealousy, the professors agree to rotate offices every year. How often will each professor have the best office? How many years will it take for these five professors to exhaust every possible arrangement of these five offices?

2.83 A survey of college teachers of English as a Second Language (ESL) obtained the following data:

	Males	Females
Part time	2	24
Full time	3	8

a. If a part-timer is picked at random, what is the probability that the selected person is female?   **b.** If a full-timer is picked at random, what is the probability that the selected person is female?   **c.** If a female is picked at random, what is the probability that she teaches part time?   **d.** If a male is picked at random, what is the probability that he teaches part time?   **e.** Do these data indicate to you that gender and part-time/full-time status are related or unrelated?

2.84 A medical researcher wants to investigate whether the order in which four prescribed medications are taken matters. How many volunteers does she need if she wants each possible sequence to be tried by 30 different patients? (Each patient will try only one sequence.)

2.85 The President of the United States, the president of the USSR, and the leaders of eight other countries stand in a straight line for a photograph commemorating the signing of an important peace agreement. If the placement of the leaders is random, what is the probability that the presidents of the United States and the USSR will stand next to each other?

2.86 When three people play the card game hearts, they are each dealt 17 cards from a standard deck of 52 cards; the last card is set aside. If you play this game, what is the probability that you will be dealt the queen of spades? That the queen of spades will be set aside?

• **2.87** *(continuation)* What is the probability that your hand will contain all 13 hearts? That it will contain 13 hearts and the queen of spades?

2.88 An anagram is a rearrangement of the letters of a word to spell a new word. How many different arrangements of a four-letter word are possible if each letter is different? If two of the letters are the same?

2.89 In the game Scrabble, a group of letters are arranged to spell words. How many arrangements are possible for constructing a seven-letter word using seven different letters? A four-letter word using any four of seven different letters?

• **2.90** *(continuation)* How many arrangements are possible for constructing a seven-letter word using seven letters, three of which are the same letter?

•• **2.91** *(continuation)* How many arrangements are possible for constructing a four-letter word using any four of seven letters, if three of the seven letters are the same letter?

2.92 A bank has targeted 20 cities in which it wants to open branches during the next five years. If it opens five of these branches this year, how many possible choices are there? If 2 cities are definite and the other 3 are still to be selected?

2.93 Characterize each of the following probabilities as equally likely, long-run frequency, or subjective. Explain your reasoning. **a.** The probability that the next president of the United States will be a Republican. **b.** The probability of winning a state lottery. **c.** The probability that a baby will be male. **d.** The probability that it will rain tonight.

2.94 A college has 30 trustees, of whom 5 will be assigned to a special committee to study the school's fraternity system. The committee members will be selected by writing the 30 names on separate pieces of paper, shuffling these papers, and selecting 5. How many possible outcomes are there? What is the probability that the head of the board of trustees will be on this committee?

• **2.95** *(continuation)* If four 5-person committees are chosen randomly, with no one allowed to serve on more than one committee, how many different outcomes are possible?

2.96 Value Line, a large investment advisory service, recommends that investors diversify their portfolios by buying at least 15 stocks. How many 15-stock portfolios are possible if you restrict yourself to the 30 stocks included in the Dow Jones Industrial Average? The 500 stocks in the S&P 500 average? The 5,000 stocks in the Wilshire 5,000?

2.97 Experiments with a chimpanzee (named Sarah) investigated whether nonhuman primates could be taught language skills.[19] In one experiment Sarah was shown a series of symbols that formed a question. To answer this question correctly, Sarah had to choose four out of eight symbols and arrange these four symbols in correct order. What is the probability that four symbols chosen and arranged randomly will be the correct four symbols arranged in proper order?

• **2.98** On the first day of class, Professor Smith conducts a brief experiment in which he flips a coin ten times and records the results, while each student writes down what he or she feels the outcomes to be. Often, a student will write down ten heads before the experiment even begins. Is this student more or less likely to get all ten flips correct than a student who writes down more varied guesses, such as H, T, H, H, T, H, T, T, H, T?

2.99 In a carnival game, three balls are rolled down a ramp and stop in three of six numbered slots, as shown below. Railing along the side ensures that each ball lands in a slot, while the small size of the slot

ensures that each slot can hold only one ball. The nar-
rowness of the slots and the condition of the balls and
the track ensure that this is a game of luck, not skill.
A stuffed animal is won if the sum of the three balls is
more than 14 or less than 7. What is the probability
of winning?

2.100 *(continuation)* What is the probability of
winning if each slot is large enough to hold three balls
and the sum has to be larger than 16 or smaller than
5?

3

Probability Rules

This branch of mathematics [probability] is the only one, I believe, in which good writers frequently get results entirely erroneous.

Charles Sanders Peirce

TOPICS

You have now seen what probabilities are and why there is a need for them. You have also seen that there is some disagreement among theoreticians about the interpretation of probabilities. Probabilities can also be treated as just another interesting mathematical problem. Many clever mathematicians have proven some useful probability theorems, without ever specifying what probabilities are and where they come from.[1] Both objectivists and subjectivists alike can consequently use these theorems to simplify their probability calculations.

3.1   VENN DIAGRAMS

We won't go through any elaborate mathematical proofs here, but we can use a simple picture to justify the theorems intuitively. Consider an uncertain situation, an "experiment," for which there are, say, ten possible outcomes. An outcome is sometimes called a "simple event" or an "elementary event," to emphasize the requirement that it cannot be decomposed into two or more possible events; for example, when a six-sided die is rolled, the possible outcomes are the numbers 1, 2, 3, 4, 5, and 6, and "an even number is rolled" is a compound event rather than a simple event.

In any experiment, one and only one outcome (i.e., simple event) must occur. The set of all possible outcomes is called the "sample space" and can be represented here by placing ten numbered points inside the box shown in Figure 3.1, an aid originated by the mathematician John Venn. The circle A in this figure represents the compound event "Outcomes 4, 5, or 6 will occur."

> A **Venn diagram** is a box representing the possible outcomes of an uncertain situation. It can be used to justify probability rules intuitively.

The probability that is assigned to an event is a measure of the likelihood that the event will occur. For the objectivist, probabilities describe the relative frequency with which events happen when the experiment is repeated a very large number of

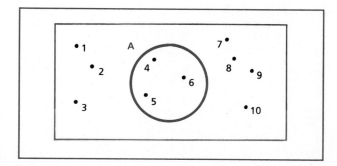

Figure 3.1   *A Venn Diagram*

times. For the subjectivist, probabilities reflect degrees of belief, a personal assessment of how likely each event is relative to others. Either way, probabilities must conform to these rules:

> The probability of a compound event A is the sum of the probabilities of the simple events in A; in Figure 3.1, $P[A] = P[4] + P[5] + P[6]$.
>
> The sum of the probabilities of all outcomes (or simple events) is 1.
>
> The event that A does not occur is called the "complement" of event A, what we will label "not A." Obviously, $P[\text{not } A] = 1 - P[A]$.
>
> The probability of an impossible event is 0.
>
> The probability of an event that is certain to occur is 1.
>
> The probability of an event cannot be negative or larger than 1.

$$0 \le P[A] \le 1 \qquad\qquad (3.1)$$

Venn diagrams can be used to show three useful probability theorems: The addition rule, the multiplication rule, and the subtraction rule.

EXERCISES

3.1 Consider the roll of a four-sided die that is numbered 1, 2, 3, and 4. Draw a Venn diagram showing these four possible outcomes. Now draw three circles showing the following events: **a.** 1 **b.** an even number **c.** 1 or an even number

Now draw a new Venn diagram with the numbers 1, 2, 3, and 4 and then draw three circles showing the following events: **d.** 4 **e.** 4 or an even number **f.** 4 and an even number

What are the probabilities of each of these six events?

3.2 In 1984, 83 percent of all new cars sold in the United States had air conditioning and 39 percent had power windows. Use Venn diagrams to determine **a.** the maximum possible percentage of these cars that had both air conditioning and power windows. **b.** the minimum possible percentage of these cars that had both air conditioning and power windows. **c.** the maximum possible percentage of these cars that had neither air conditioning nor power windows.

3.2   THE ADDITION RULE

You are going to flip two coins. What is the probability of getting at least one head, either on the first flip or on the second flip? It is tempting to simply add up the probability of a head on the first flip plus the probability of a head on the second flip, but that would give .5 + .5 = 1.0, which can't be right. It is not at all certain that you will get at least one head; you may well get tails on both flips.

The error is a double counting of the case of the two heads. Our .5 probability for heads on the first toss includes the case where we also get heads on the second toss, and our .5 probability for heads on the second toss also includes the case of two heads. If we add together these two .5 probabilities, we will count the two-heads case twice.

Let's list the four possible outcomes, as we have done before:

HT
HH
TH
TT

The first and second outcomes involve a head on the first flip. The second and third outcomes involve a head on the second flip. If we simply add two successful outcomes plus two successful outcomes, the second outcome, HH, is counted twice. The correct probability, counting HH only once, is 3/4.

Figure 3.2 shows a Venn diagram representation of this double-counting potential. Set A contains the outcomes in which the first flip is a head. Set B contains the outcomes in which the second flip is a head. The outcome TT is in neither set. The outcome HH is in both sets. We are interested in the probability of "A or B": the probability of a head on the first or second flip. If we simply count the two outcomes in A and the two outcomes in B and add them together to get four, we will double count the outcome HH that is in both A and B.

One way to correct this double counting is to enumerate all the possible outcomes and then count the successful ones carefully. Another way, which is often simpler, is to knowingly double count and then subtract the outcomes that are counted twice. This is the **addition rule:**

The probability that either A or B or possibly both will occur is

$$P[A \text{ or } B] = P[A] + P[B] - P[A \text{ and } B] \qquad (3.2)$$

In this two-coin case, the probability of a head on the first flip is 1/2, the probability of a head on the second flip is 1/2, and the probability of heads on both flips is 1/4. Therefore, the probability of a head on either the first or second flip is 1/2 + 1/2 − 1/4 = 3/4.

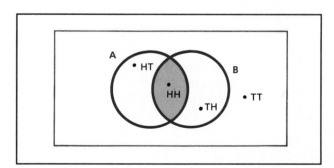

Figure 3.2 *Double Counting Double Heads*

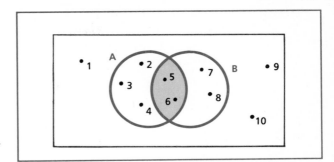

Figure 3.3 *Two Overlapping Sets*

Figure 3.3 shows another example. There are ten possible outcomes, of which five are in *A* and four are in *B*. Perhaps one of ten people is to be randomly selected to chair a committee. Group *A* might be people under 30 years of age, while group *B* is females. What is the probability that the selected person will be in either *A* or *B*? It depends on how much overlap there is—how many outcomes are included in both *A* and *B* (that is, how many of those under 30 are also female). In Figure 3.3, there are two outcomes that are in both *A* and *B*. Therefore,

$$P[A \text{ or } B] = P[A] + P[B] - P[A \text{ and } B]$$

$$= \frac{5}{10} + \frac{4}{10} - \frac{2}{10}$$

$$= \frac{7}{10}$$

If there are no outcomes that are in both *A* and *B*, and hence $P[A \text{ and } B] = 0$, then there is no double-counting problem. If this is the case, then *A* and *B* are said to be mutually exclusive.

A and *B* are **mutually exclusive** if $P[A \text{ and } B] = 0$. If *A* and *B* are mutually exclusive, then the addition rule (Equation 3.2) simplifies to

$$P[A \text{ or } B] = P[A] + P[B]$$

If, in the committee example, all five of the people under 30 were male, then *A* and *B* would be mutually exclusive. The probability of selecting someone who is female or under 30 would be 5/10 + 4/10 = 9/10.

Figure 3.4 shows two mutually exclusive sets. These six possible outcomes might be the number that comes up when a die is rolled. Set *A* contains the even numbers and set *B* contains the number 5. Sets *A* and *B* are mutually exclusive because the outcome cannot be both even and the number 5. The probability that

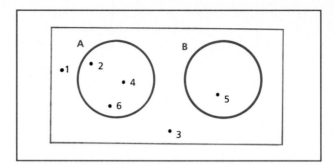

Figure 3.4 *Two Mutually Exclusive Sets*

the number that is rolled is either even or 5 is

$$P[\text{even number or 5}] = P[\text{even number}] + P[5]$$

$$= \frac{3}{6} + \frac{1}{6}$$

$$= \frac{4}{6}$$

What about the probability of either an even number or a 6? Here the sets overlap because 6 is an even number.

$$P[\text{even number or 6}] = P[\text{even number}] + P[6] - P[\text{even number and 6}]$$

$$= \frac{3}{6} + \frac{1}{6} - \frac{1}{6}$$

$$= \frac{3}{6}$$

Let's do one more that is a little more complicated. You are playing a card game in which spades and honors (ace, king, queen, or jack) are valuable. If you draw a card from a full 52-card deck, what is the probability that it is one of these valuable cards? Think about it for a minute and see if you can figure out the answer before reading on.

Let's define A and B as follows:

> A: a spade
> B: an honor card

There are 13 spades and 16 honor cards (four in each of the four suits). Therefore, $P[A] = 13/52$ and $P[B] = 16/52$. Now, what is $P[A \text{ and } B]$? There are four cards (the ace, king, queen, and jack of spades) that are both spades and honors. So,

$P[A \text{ and } B] = 4/52$. Our answer is

$$P[\text{spade or honor}] = P[\text{spade}] + P[\text{honor}] - P[\text{spade and honor}]$$

$$= \frac{13}{52} + \frac{16}{52} - \frac{4}{52}$$

$$= \frac{25}{52}$$

There is nearly a 50 percent chance of drawing a card that is either a spade or an honor.

As one more example, imagine that you own a small business and have contracted to supply your product at a fixed price—$20 apiece. You like the security provided by a guaranteed customer, but you are worried about two possible catastrophes: (*a*) the Federal Reserve (Fed) will engineer a tight-money "credit crunch" that dries up the financing you need to buy raw materials and manufacture your product; and (*b*) the Fed will cause an easy-money inflation that drives up your material and labor costs and prevents you from making a profit at a price of $20 apiece. Perhaps you believe that the probability of tight money is .20 and the probability of easy money is .10. As tight money and easy money are mutually exclusive, the probability of one or the other is .20 + .10 = .30.

EXERCISES

3.3 You draw two cards out of a standard 52-card deck. What is the probability that one or the other of these cards is the queen of spades?

3.4 You are playing backgammon and need to roll a 1 on either of your two dice to hit your opponent. What are your chances?

3.5 A security analyst tells you that there is a .3 chance that a certain company will be bought up by Exxon and a .3 chance that it will be purchased by Mobil. What is the probability that it will be acquired by one or the other of these two companies?

3.6 You run a small business and have two big fears: (*a*) you will be squeezed by a credit crunch; and (*b*) the Environmental Protection Agency (EPA) will decide that your product is a threat to the environment. You believe that there is a .20 probability of a credit crunch, a .05 probability of a damaging EPA decision, and a .01 chance that both of these disasters will occur. What is the probability that at least one of your fears will come true?

3.3 CONDITIONAL PROBABILITIES AND INDEPENDENT EVENTS

We are sometimes interested in the probability that one of the events in A will occur, given that one of the events in B occurs. This is written as $P[A \mid B]$ and is called a conditional probability.

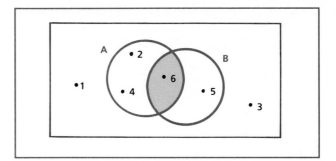

Figure 3.5   *A Conditional Probability*

The **conditional probability** $P[A \mid B]$ is the probability that A occurs, given that B has occurred.

Be careful that you don't interpret $P[A \mid B]$ as the probability of A divided by the probability of B:

$$P[A \mid B] \quad is \; not \quad \frac{P[A]}{P[B]}$$

The notation $P[A \mid B]$ is a special shorthand for "the probability of A, given that we know B has occurred."

Consider, for example, the roll of a standard six-sided die. The six possible outcomes are shown in the Venn diagram in Figure 3.5. The set A contains the even numbers 2, 4, and 6. Set B contains the numbers larger than 4. The conditional probability

$$P[A \mid B] \; = \; P[\text{even number} \mid \text{larger than } 4]$$

asks the question, "If we roll a number larger than 4, what is the probability that it is even?" To answer this question, we can think of B as a restricted-outcome space. If the number is larger than 4, then it must be either a 5 or a 6. Now, what fraction of these two outcomes are even? One-half is the answer. To obtain this answer, we compared the number of elements in both A and B (one)* with the number of elements in B (two). Given in terms of probability this translates into a ratio of the chances of both A and B happening to the chances of B happening. The conditional probability is given by this ratio of probabilities:

$$P[A \mid B] \; = \; \frac{P[A \text{ and } B]}{P[B]} \tag{3.3}$$

---

*The expression "both A and B" is used here (and elsewhere) to signify the area of intersection of the two sets, not their combined areas.

Here,

$$P[\text{even} \mid \text{larger than 4}] = P[\text{even and larger than 4}]/P[\text{larger than 4}]$$

$$= \frac{(1/6)}{(2/6)}$$

$$= \frac{1}{2}$$

What about the conditional probability

$$P[B \mid A] = P[\text{larger than 4} \mid \text{even number}]$$

This probability asks the question, "If the number is even, what is the probability that it is larger than 4?" Now, we want to know the number of elements that are in both A and B relative to the number that are in A. That is,

$$P[B \mid A] = \frac{P[A \text{ and } B]}{P[A]} \qquad (3.4)$$

Here

$$P[\text{larger than 4} \mid \text{even}] = P[\text{larger than 4 and even}]/P[\text{even}]$$

$$= \frac{(1/6)}{(3/6)}$$

$$= \frac{1}{3}$$

Notice the differences between Equations 3.3 and 3.4. Conditional probabilities are important, but they can be confusing. Go through this example again if you are not sure that you understand what $P[A \mid B]$ and $P[B \mid A]$ mean.

It is also important to recognize the difference between $P[A]$ and $P[A \mid B]$. The expression $P[A]$ asks, "Considering all possible outcomes, what is the probability that A occurs?" The expression $P[A \mid B]$, in contrast, asks, "Assuming that one of the outcomes in B occurs, what is the probability that A also occurs?" These two probabilities are not equal unless A and B are independent.

> A and B are **independent** if $P[A \mid B] = P[A]$. If $P[A \mid B]$ does equal $P[A]$, then $P[B \mid A] = P[B]$, too.

The conditional clause "given that B has occurred" means that you have this additional information that may persuade you to revise your probability of A occurring.

Two events are independent if the knowledge that one has occurred does not alter the probability that the other also occurs.

In the die example shown in Figure 3.5, A and B are, in fact, independent. The probability of an even number is 1/2. If the number is larger than 4, then the probability that it is even is still 1/2, so both $P[A]$ and $P[A\,|\,B]$ are 1/2. Similarly, both $P[B]$ and $P[B\,|\,A]$ are 1/3. The probability of a number larger than 4 is 1/3, and if the number is even (2, 4, or 6), then the probability that it is larger than 4 is still 1/3.

Age and Sex

To reinforce the ideas of conditional probabilities and independence, let's do a somewhat different example. Table 3.1 gives some 1985 data on age and sex in the United States. One possible Venn diagram that results from these data is shown in Figure 3.6.

Now imagine that we are going to reach into the United States and pick someone at random. What is the probability that this person will be a male? What is the probability that this person will be at least 25 years old? Are age and sex independent? Let's use M and F to denote male and female and use "25" and "not 25" to label age. Table 3.1 yields the following probabilities:

$$P[M \text{ and not } 25] = \frac{46,783,000}{239,283,000} = .196$$

$$P[M \text{ and } 25] = \frac{69,866,000}{239,283,000} = .292$$

$$P[F \text{ and not } 25] = \frac{44,973,000}{239,283,000} = .188$$

$$P[F \text{ and } 25] = \frac{77,661,000}{239,283,000} = .325$$

We can also see from the totals in Table 3.1 that

$$P[M] = \frac{116,649,000}{239,283,000} = .487$$

$$P[F] = \frac{122,634,000}{239,283,000} = .513$$

$$P[\text{not } 25] = \frac{91,756,000}{239,283,000} = .383$$

$$P[25] = \frac{147,527,000}{239,283,000} = .617$$

Table 3.1 *Age and Sex in the United States*

Age	Males	Females	Total
Less than 25 years old	46,783,000	44,973,000	91,756,000
25 years and older	69,866,000	77,661,000	147,527,000
Total	116,649,000	122,634,000	239,283,000

The population of the United States is 51 percent female and 62 percent at least 25 years old.

Now let's ask a trickier question. If the person picked is 25 or older, what is the probability that this person is male? Using our conditional probability rule,

$$P[M \,|\, 25] = \frac{P[M \text{ and } 25]}{P[25]}$$

$$= \frac{.292}{.617}$$

$$= .473$$

Similarly, the probability that someone under 25 is a male works out to be

$$P[M \,|\, \text{not } 25] = \frac{P[M \text{ and not } 25]}{P[\text{not } 25]}$$

$$= \frac{.196}{.383}$$

$$= .512$$

These conditional probabilities compare to an overall .487 probability of being male. Therefore, age and sex are not independent. The young are predominantly

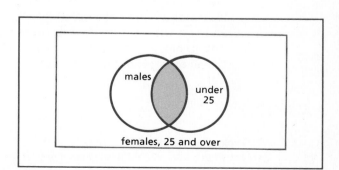

Figure 3.6 *One Representation of Data for Age and Sex in the U.S.*

male because babies are slightly more likely to be male. The old are predominantly female because auto accidents, murders, wars, and health problems give males a shorter life expectancy.

Winning Streaks

In any fair game of chance, a player will win some and lose some (or, it often seems, win some and lose many). The wins will at times be scattered and, at other times, be bunched together. Many gamblers attach a great deal of significance to these runs of luck. They apparently believe that luck is some sort of infectious disease that a player catches and then takes awhile to get over. For example, Clement McQuaid, "author, vintner, home gardener, and keen student of gambling games, most of which he has played profitably," offers this advice:

> There is only one way to show a profit. Bet light on your losses and heavy on your wins. Many good gamblers follow a specific procedure. a. *Bet minimums when you're losing.* . . . b. *Bet heavy when you're winning.* . . . c. *Quit on a losing streak, not a winning streak.* While the law of mathematic probability averages out, it doesn't operate on a set pattern. Wins and losses go in streaks more often than they alternate. If you've had a good winning streak and a loss follows it, bet minimums long enough to see whether or not another winning streak is coming up. If it isn't, quit while you're still ahead.[2]

You will indeed show a profit if you win your big bets and lose only your small ones. But how are you to know in advance whether you are going to win or lose your next bet? Suppose you are playing a dice game and have won three times in a row. You know that you have been winning and you are probably excited about it, but dice have no memories or emotions. Games were invented by people. Dice don't know the difference between a winning number and a losing number. Dice do not know what happened on the last roll and do not care whether you are betting heavily or lightly on the next roll. A die is just an inanimate celluloid cube and each of the six sides is still equally likely to come up. The outcomes are independent in that the probabilities are the same, roll after roll.

Sometimes you will win four times in a row, but at least as often, you will win three in a row and then lose the fourth. Sometimes you will win five times in a row, but at least as often, you will win four in a row and then lose the fifth. You may fondly recall those rare occasions when you won four, five, or more times in a row, and you will wish that you had bet more heavily. Unfortunately, there is just no way to predict when a winning streak will begin or end. Games of chance are the classic example of independent events.

EXERCISES

3.7 Use Table 3.1 to calculate $P[F|25]$ and $P[25|F]$. Explain what these probabilities mean. Use these probabilities to see whether sex and age are independent.

3.8 In 1983, there were 80,390,000 occupied housing units in the United States, classified below by when the structure was built and whether the person living there owned or rented the unit:

Year Built	Owner	Renter
1939 or earlier	12,949,000	10,580,000
1940–1983	38,846,000	18,015,000

If a unit is picked at random what is the probability that **a.** the occupant is a renter? **b.** the unit was built before 1939? **c.** the occupant is a renter if the structure was built before 1939? **d.** the structure was built before 1939 if the occupant is a renter?

3.9 Each season, Calvin Clever's new fashion designs are shown to a panel of experts—ten Los Angeles teenagers—and each secretly votes thumbs up or down. Here are historical data on the panel's rankings of 1500 designs and the subsequent market success, measured by sales:

	Number of Positive Panel Votes		
Sales	*7–10*	*4–6*	*0–3*
Very successful	380	140	80
Modest success	180	120	100
Disappointment	40	40	420

If we interpret these relative frequencies as probabilities, what is the probability that **a.** a design will be given 7–10 thumbs up? **b.** a design will be given 0–3 thumbs up? **c.** a design given 7–10 thumbs up will be very successful? **d.** a design given 0–3 thumbs up will be very successful? **e.** a very successful design had been given 7–10 thumbs up? **f.** a very successful design had been given 0–3 thumbs up? Are success and ranking independent?

3.10 Evaluate this advice for winning at craps at Las Vegas:

First of all, try to find a hot table. Never remain at a cold one. Always make it a policy to keep looking— move around! A tip-off might be the yelling crowd where a hot roll may be taking place. Another indication usually is a lot of chips spread every which way around the table by numerous players. If the action is continuous, you'll know you've caught a good table. Keep in mind that it is far more lucrative to tag along on the tail end of a hot roll than it is to go in fresh on a cold one. And no one moving from table to table will actually catch a hot roll from the beginning. If 65 percent of a streak is caught, it's enough![3]

• **3.11** Let's pick words at random from a typical book and see if there is any pattern in the arrangements of the vowels and consonants. In particular, we'll use the words on page 100 of this chapter. Let p_1 be the probability that the first letter of a word chosen at random from that page is a vowel, and let p_2 be the probability that the second letter of a word chosen at random is a vowel. Do you think that these two events are independent? Use the words on page 100 of this chapter to calculate p_1 and p_2 and to check your theory about independence.

3.4 THE MULTIPLICATION RULE

The equations for conditional probabilities show us how to calculate $P[A \mid B]$ and $P[B \mid A]$ if we know $P[A$ and $B]$, $P[A]$, and $P[B]$. But often, we are interested in the other way around; we know $P[A \mid B]$ or $P[B \mid A]$ and want to calculate $P[A$ and $B]$. We can make this calculation by rearranging Equations 3.4 and 3.3 to obtain the **multiplication rule:**

The probability that both A and B will occur is

$$P[A \text{ and } B] = P[A]P[B \mid A] \tag{3.5}$$

$$P[A \text{ and } B] = P[B]P[A \mid B] \tag{3.6}$$

For both A and B to occur, first one must occur and then, given that it has occurred, the other must occur, too. Notice that $P[A$ and $B]$ can be calculated in either of two ways, depending on the information at our fingertips.

Let's do an example. What is the probability of drawing two aces in a row out of a standard 52-card deck? Let A_1 stand for an ace on the first pick and A_2 for an ace on the second pick. The probability that we're after is

$$P[A_1 \text{ and } A_2] = P[A_1]P[A_2|A_1]$$

$P[A_1]$ is 4/52 because there are initially 4 aces scattered among the 52 cards. If we are lucky enough to pick an ace, then the odds of drawing another ace are diminished because there will be only 3 aces left among the 51 remaining cards, $P[A_2|A_1]$ = 3/51. The probability of drawing 2 aces in a row is

$$P[A_1 \text{ and } A_2] = \left(\frac{4}{52}\right)\left(\frac{3}{51}\right) = .0045$$

If we attempted this feat over and over again, we would succeed less than 5 times in every 1000 tries.

A $P[A \text{ and } B]$ question usually can be given a sequential structure and then solved by the multiplication rule. Even if we were picking two cards simultaneously, we could imagine that one was drawn slightly before the other and then use the formula given previously. The logic of an "A and B" problem is that two conditions, A and B, have to be met. The first condition will be met a certain fraction of the time. Then a fraction of these times, the second condition will be met. Both conditions will be met only a fraction of a fraction of the time.

We can visualize this logic by thinking of a probability tree in which the branches are not equally likely. Figure 3.7 shows such a tree. Here, we need to pick aces. We will pick the first ace 4/52 of the time and, then, 3/51 of these times we

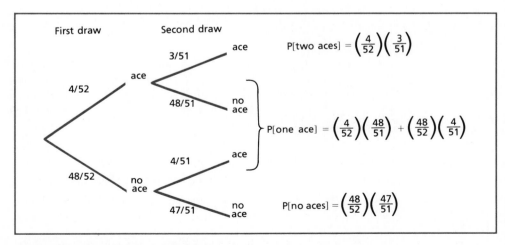

Figure 3.7   *A Probability Tree for Two Aces in a Row*

will successfully draw the second ace. Our probability of drawing 2 aces is a fraction 3/51 of the fraction 4/52, or (4/52) (3/51).

This logic can be extended indefinitely. Consider the probability of drawing 4 straight aces:

$P[A_1$ and A_2 and A_3 and $A_4]$

$$= P[A_1]P[A_2|A_1]P[A_3|A_1 \text{ and } A_2]P[A_4|A_1 \text{ and } A_2 \text{ and } A_3]$$

$$= \left(\frac{4}{52}\right)\left(\frac{3}{51}\right)\left(\frac{2}{50}\right)\left(\frac{1}{49}\right)$$

$$= .0000037$$

If you were to try again and again, you could anticipate drawing 4 straight aces about 4 times in every 1 million tries.

These draws are not independent of each other, because the odds change as cards are taken out of the deck. If, instead, the selected card is put back into the deck and the deck is shuffled thoroughly, then the probability of picking an ace remains 4/52 on each draw. These draws are independent of each other because the probability is not affected by the outcomes of other draws. This is an example of the distinction between sampling *without* replacement and sampling *with* replacement.

Sampling without replacement means that the selected item is removed and cannot be selected again.

Sampling with replacement means that the selected item is replaced and can be selected again.

With replacement, the probability of picking 4 straight aces is

$$P[A_1 \text{ and } A_2 \text{ and } A_3 \text{ and } A_4] = P[A_1]P[A_2]P[A_3]P[A_4]$$

$$= \left(\frac{4}{52}\right)\left(\frac{4}{52}\right)\left(\frac{4}{52}\right)\left(\frac{4}{52}\right)$$

$$= .0000350$$

The odds are somewhat improved because you now may be able to draw the same ace again, but it is still a very unlikely outcome.

Is the Multiplication Rule Legal?

Perhaps the earliest legal use of probability theory and statistical evidence in a U.S. court was in 1867 when Benjamin Peirce, a Harvard mathematics professor, testified that the many similarities between the signature on a will and the signature on a contested insert to the will strongly suggested that the second signature had been traced from the first.[4] Based on a detailed comparison of the downstrokes in 42 other uncontested signatures by the deceased, Peirce estimated the probability of a matched downstroke to be .2. Yet, in this will, all 30 downstrokes on the insert signature matched those on the valid signature. Using the multiplication rule, he calculated the probability of 30 such matches to be $.2^{30}$ and concluded that, "So vast an improbability is practically an impossibility."

Notice that his use of the multiplication rule implicitly assumes independence in all 30 downstrokes. There is no mention in his testimony of the reasonableness of this crucial assumption. Nor did he take into account the fact that the .2 estimate was based on 42 signatures made at very different times, while the original and contested signatures were reportedly written on the same day. But with little understanding of probabilities and great respect for Peirce's academic credentials, the opposing attorney did not challenge his calculations.

A more recent and notorious case is *People* v. *Collins* (1968),[5] in which a white woman with blond hair tied in a ponytail was seen fleeing a Los Angeles robbery in a yellow car driven by a black man with a beard and moustache. Four days later, the police found and subsequently arrested Malcolm Collins, a black man with a beard, moustache, and yellow Lincoln, and his common-law wife, a white woman with a blond ponytail. A mathematics professor calculated the probability that two people picked at random would have this combination of characteristics by multiplying estimates of the probability of each characteristic:

$$P[\text{black man with a beard}] = \frac{1}{10}$$

$$P[\text{man with a moustache}] = \frac{1}{4}$$

$$P[\text{owning a yellow car}] = \frac{1}{10}$$

$$P[\text{interracial couple}] = \frac{1}{1,000}$$

$$P[\text{blonde woman}] = \frac{1}{3}$$

$$P[\text{wearing a ponytail}] = \frac{1}{10}$$

so that, by the multiplication rule,

$$P[\text{all 6 characteristics}] = \left(\frac{1}{10}\right)\left(\frac{1}{4}\right)\left(\frac{1}{10}\right)\left(\frac{1}{1,000}\right)\left(\frac{1}{3}\right)\left(\frac{1}{10}\right)$$

$$= \frac{1}{12,000,000}$$

The Incentive to Know Slow Horses

Some racetracks offer "superfectas" in which a great deal of money can be won by correctly picking the first four finishers in a horse race. From time to time, there are allegations that jockeys are persuaded to slow their horses to keep them out of the top four places. Superfectas provide a considerable incentive for this chicanery because the chances of picking the first four finishers correctly are much higher if some horses can be excluded.

To illustrate the point simply, let's assume that the order of finish is completely random. If there are eight horses in the race, what is the probability of picking the first four finishers in correct order? This is a multiplication problem. You must pick first place, second, third, and fourth correctly. If eight horses are equally likely to win, then the probability that your selected horse will finish first is 1/8. If you get first place right, then there are seven possible second-place finishers, and the probability that your second-place pick will also be correct is 1/7. So far, so good. If you pick the first two right, there are six horses left that could finish third and the probability of a correct pick is 1/6. Similarly, if your first three selections are right, you have a 1/5 chance of picking fourth place correctly, too.

To put these numbers together, let "1st" represent the correct selection of the first-place horse, "2nd" the correct pick of the second-place horse, and so on:

$P[$1st and 2nd and 3rd and 4th$]$

$\quad = P[$1st$]P[$2nd$\,|\,$1st$]$

$\quad\quad \times P[$3rd$\,|\,$1st and 2nd$]$

$\quad\quad \times P[$4th$\,|\,$1st and 2nd and 3rd$]$

$\quad = \left(\dfrac{1}{8}\right)\left(\dfrac{1}{7}\right)\left(\dfrac{1}{6}\right)\left(\dfrac{1}{5}\right)$

$\quad = .0006$

What if you were certain that a particular horse will not be one of the first four finishers? Then, there would be only seven horses to worry about, and your probability of winning the superfecta is doubled:

$P[$1st and 2nd and 3rd and 4th$]$

$$= \left(\frac{1}{7}\right)\left(\frac{1}{6}\right)\left(\frac{1}{5}\right)\left(\frac{1}{4}\right)$$

$$= .0012$$

What if you could rule out three horses? There would be only five horses to consider, and

$P[$1st and 2nd and 3rd and 4th$]$

$$= \left(\frac{1}{5}\right)\left(\frac{1}{4}\right)\left(\frac{1}{3}\right)\left(\frac{1}{2}\right)$$

$$= .0083$$

If you know that three horses will run slowly enough to lose, then your probability of winning the superfecta is fourteen times greater than when the race is fair.

The remoteness of this probability helped convict Collins and his wife, the jurors apparently believing, in the words of the Supreme Court, that "there could be but one chance in 12 million that the defendants were innocent and that another equally distinctive couple actually committed the robbery."

As with the disputed signature 100 years earlier, the professor's calculation implicitly assumes independence. But, this time, the California Supreme Court was well enough informed about probabilities to question the appropriateness of this assumption. The probability of having a moustache is clearly affected by being a black man with a beard; while only 25 percent of the overall population has moustaches, perhaps 75 percent of the black men with beards do. Similarly, being a black man, a blond woman, and an interracial couple are interrelated, as perhaps are ponytails and blonde hair. And yellow cars? The Court also questioned the absence of evidence supporting the assumed probabilities and raised the possibility that the assumed distinctive characteristics of the guilty couple might be incorrect:

> Conceivably, for example, the guilty couple might have included a light-skinned Negress with bleached hair rather than a caucasian blonde; or the driver of the car might have been wearing a false beard as a disguise; or the prosecution's witnesses might simply have been unreliable.

Further, taking into account the population of Los Angeles and using the binomial probabilities discussed in Chapter 5, the court calculated that even if there really were only a 1/12,000,000 chance that a single, randomly picked couple would match these six characteristics, there was roughly a 40 percent chance of finding another such couple somewhere in Los Angeles. The court decided that this was not proof beyond a reasonable doubt of the Collins' guilt and reversed their conviction.

EXERCISES

3.12 On a Monte Carlo roulette wheel, there are 37 slots: 18 red, 18 black, and 1 green. If you were to sit down and spin such a wheel 26 times, what would be the probability of duplicating the 26 straight blacks that occurred on August 18, 1913 at Monte Carlo?

3.13 At the beginning of 1985, an investment advisor picked five stocks that he thought "should do especially well" in 1985. As it turned out, 75 percent of the stocks on the New York Stock Exchange increased in price that year, while all five of this advisor's recommended stocks went down. If a monkey had thrown five darts at the financial pages at the beginning of 1985, what is the probability that all five picks would have gone down in price?

3.14 Players of the California Lotto game select six different numbers, from 1 to 49, and win the grand prize if these match (not necessarily in order) the six numbered balls mechanically selected by a machine that Saturday. What is the probability of winning the grand prize?

3.15 Before the fatal 1986 explosion of the space shuttle Challenger, many government officials believed that the space shuttle program would *never* have a fatal failure.[6] Within NASA, estimates of the probability of a mission failure ranged from 1 in 100,000 (by management) to 1 in 100 (by engineers). On other U.S. rockets, failure rates have ranged from 1 percent (Thor) to 10 percent (Atlas). The tragic Challenger explosion came on its twenty-sixth mission. What is the probability of 25 successes in 25 missions if the probability of failure on each mission is 1 percent? 4 percent? 10 percent?

3.16 Many corporate takeovers in 1984 and 1985 were financed by "junk bonds," bonds backed by little more than the assets of the takeover target and considered too speculative to be rated by Standard & Poor's or Moody's. Suppose that 5 percent of these junk bonds default within three years of their issuance. If you buy two newly issued junk bonds, what is the probability that one or both will default within three years? Assume that the outcomes are independent.

3.17 You want to pick the first four finishers in a six-horse race. If the order of finish is random, what is your probability of success? If you can rule out one horse, what is your probability of success? Now, compare your answers with the probabilities given in the text for an eight-horse superfecta. Is there greater incentive to guarantee a slow horse in a six-horse or an eight-horse superfecta? If you were a statistical consultant to your state gambling commission, what recommendation could you make on the basis of these calculations?

We've seen how the multiplication rule can be extended easily to a long sequence of events. The addition rule is not so easily extended. For two sets of outcomes, the addition rule is

$$P[A \text{ or } B] = P[A] + P[B] - P[A \text{ and } B]$$

With three sets, the correction for double counting becomes more complicated. Figure 3.8 shows the relevant Venn diagram. Points 1 through 7 are representative outcomes that should each be counted once in calculating $P[A \text{ or } B \text{ or } C]$. If we simply add the three probabilities $P[A]$, $P[B]$, and $P[C]$, we will include the following:

Probability	Outcomes
$P[A]$	1, 4, 6, 7
$P[B]$	2, 4, 5, 7
$P[C]$	3, 5, 6, 7

The summation $P[A] + P[B] + P[C]$ counts outcomes 1, 2, and 3 once; it counts outcomes 4, 5, and 6 twice; and it counts outcome 7 three times. What if we now subtract $P[A \text{ and } B]$, $P[A \text{ and } C]$, and $P[B \text{ and } C]$? That will undo the double counting of outcomes 4, 5, and 6, but it will also leave outcome 7

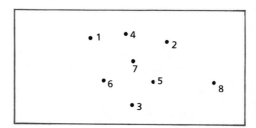

Figure 3.8 *The Addition Rule for Three Groups*

uncounted. So, the correct rule must be

$$P[A \text{ or } B \text{ or } C] = P[A] + P[B] + P[C]$$
$$- P[A \text{ and } B] - P[A \text{ and } C] - P[B \text{ and } C] \quad (3.7)$$
$$+ P[A \text{ and } B \text{ and } C]$$

This expression is complex but not unusable. However, as we go to more than three sets, the addition rule quickly becomes unwieldy. Fortunately, we have a subtraction rule to spare us such misery.

> The **subtraction rule** is a powerful ploy based on the simple observation that the probability of something occurring is equal to one minus the probability that it does not occur.
>
> $$P[A] = 1 - P[\text{not } A]$$
>
> This is very useful because it often is easier to figure out the probability of something not occurring than it is to figure out the probability that it will occur.

Consider, for example, the probability of getting at least one head in three flips of a coin. This is a three-set addition problem. We want to know the probability of a head on the first flip or the second flip or the third flip. We could coerce an answer out of Equation 3.7:

$$P[H_1 \text{ or } H_2 \text{ or } H_3] = P[H_1] + P[H_2] + P[H_3]$$
$$- P[H_1 \text{ and } H_2] - P[H_1 \text{ and } H_3] - P[H_2 \text{ and } H_3]$$
$$+ P[H_1 \text{ and } H_2 \text{ and } H_3]$$
$$= \frac{1}{2} + \frac{1}{2} + \frac{1}{2} - \frac{1}{4} - \frac{1}{4} - \frac{1}{4} + \frac{1}{8}$$
$$= \frac{7}{8}$$

The subtraction rule provides an easy alternative. The probability of at least one head is equal to one minus the probability of no heads, or one minus the probability of three straight tails. The probability of three straight tails is an easy multiplication rule: $(1/2)(1/2)(1/2) = 1/8$. Thus, the probability of at least one head in three flips is $1 - 1/8 = 7/8$, as simple as that.

De Mere's Problem

We're now ready to tackle de Mere's problem. De Mere was puzzled by his experience with two wagers:

1. Roll at least one 6 in four throws of a single die.

2. Roll at least one double-6 in twenty-four throws of a pair of dice.

The probability of a 6 on a single throw of a die is 1/6. A naive calculation of the probability of a 6 with four throws would be 4/6. This calculation is wrong because of the double-counting problem. Apparently de Mere was using it as a rough approximation to judge that the odds were in his favor, and he was winning more than half the time.

The probability of a double-6 on a single throw of a pair of dice is 1/36. A naive calculation of the probability of a double-6 with twenty-four throws is 24/36, which reduces to 4/6 again. There is still a double-counting problem, but de Mere apparently believed that, as a rough approximation, the odds should be in his favor; yet, he lost his second bet more often than he won. He turned to Pascal for an explanation.

We could try to construct a probability tree and enumerate the possible outcomes, but the second bet has 36^{24} branches, which is many more than I want to draw. We could try the addition rule, but the addition rule for twenty-four sets is more complicated than I want to think about. The easy path is, of course, the subtraction rule. The desired probabilities are, with remarkable simplicity,

1. $P[$ at least one 6$] = 1 - P[$ no 6s in 4 throws$]$

$$= 1 - \left(\frac{5}{6}\right)^4$$

$$= .518$$

2. $P[$at least one double-6$] = 1 - P[$no double-6s in 24 throws$]$

$$= 1 - \left(\frac{35}{36}\right)^{24}$$

$$= .491$$

De Mere's experience was consistent with the correct probabilities. The odds were with him on the first bet, but they were against him on the second bet.

The more general moral is, if you've got a problem that seems hopelessly complicated, try the subtraction rule. It just may be the trick you need.

The Probabilistic Case for Redundancy

Word processing programs must allow users to clear the text that is currently in the computer's memory so that the user can begin working on something new. This necessary memory-clearing command is potentially dangerous, however, in that if users invoke it accidentally, they will lose whatever they have been working on. Realizing this, computer programmers design memory-clearing commands with some redundancy, requiring users to take more than one action before the memory is cleared. These additional actions make failure less likely.

Matched Birthdays

In a class with 23 students, what is the probability that at least 2 students were born on the same day of the year? Surprisingly, the probability is greater than 1/2!

How would we go about figuring such a probability? This is a tough addition problem: the probability that one student has the same birthday as any of the other 22 students, plus the probability that a second student has the same birthday as any of the other students, and so on, in addition to a mind-boggling adjustment for all of the double counting. The easy alternative is, you guessed it, the subtraction rule. The probability of at least one matched birthday is one minus the probability of no matches.

It is traditional in this classic problem to ignore February 29 and to assume that all of the remaining 365 days of the year are equally likely to be a birthday. Now, pick any two students. The first has some birthday, say November 11. The probability that the second person has a different birthday is 364/365. Now add a third student. The probability of another different birthday, given that the first two didn't match, is 363/365. The probability that the first two don't match and that the third also doesn't match is the product of these probabilities:

$P[1 \ \& \ 2 \ \& \ 3 \text{ don't match}]$

$\qquad = P[1 \ \& \ 2 \text{ don't match}]P[3 \text{ doesn't}$
$\qquad \text{match 1 or 2} \mid 1 \ \& \ 2 \text{ don't match}]$

$\qquad = \left(\dfrac{364}{365}\right)\left(\dfrac{363}{365}\right)$

Extending this logic, the probability that none of the 23 students' birthdays match is

$$\left(\frac{364}{365}\right)\left(\frac{363}{365}\right)\left(\frac{362}{365}\right)\cdots\left(\frac{343}{365}\right)$$

and the probability of at least one match is

$$1 - \left(\frac{364}{365}\right)\left(\frac{363}{365}\right)\left(\frac{362}{365}\right)\cdots\left(\frac{343}{365}\right) = .51$$

Analogous computations show that the probability of at least one matched birthday is .71 with 30 students, .97 with 50 students, and .9999997 with 100 students. Unless you have an unusually small class, there are probably at least two students in your statistics class who share a birthday.

For instance, in one word processing program, the computer's memory is not cleared unless the user presses the CONTROL and N keys simultaneously, and then presses the Y key. Three keys are not really necessary; any single key could have been reserved for memory clearing. But then there would be a substantial chance that a user presses this single key accidentally and loses a good deal of work. The extra two keystrokes are redundancies that make disaster far less likely, in that the chances of accidentally hitting these three keys in this specific order are remote.

The probability principle is simple and very general: To reduce the chances of a system's failure, increase the number of things that must go wrong before the system fails. Another word processing rule is the wise advice to make backups of everything that you cannot afford to lose.

Similarly, any system can be designed with the component parts in a series so that the whole system fails if any part fails, or in parallel, with some parts redundant in that the system fails only if all parallel parts fail. Figure 3.9 shows one example, with the failure probabilities on each part. In the series design, though each part has a relatively small chance of failure, the probability is substantial that some part will fail, causing the entire system to fail:

$$P[\text{system failure}] = 1 - P[\text{no part fails}]$$
$$= 1 - (.90)(.95)(.95)$$
$$= .1878$$

In the parallel design, Part 1 has been replaced by three, perhaps smaller, parts; two are redundant since these are not necessary for the system to work. But this very redundancy means that there is a backup, in that even if two parts fail, the third can carry on. The probability that Parts 1*a*, 1*b*, and 1*c* will all fail is .10(.10)(.10)

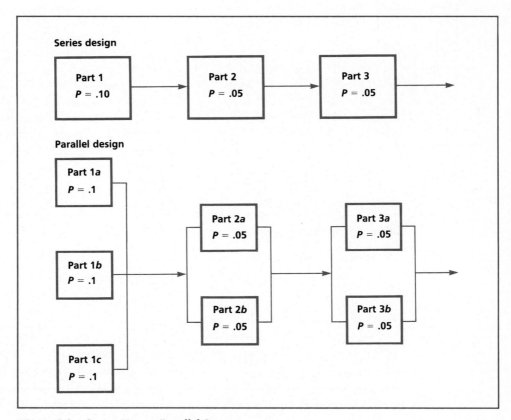

Figure 3.9 *Series Versus Parallel Systems*

Junk Bond Default Rates

Junk is a generic label for low-quality bonds with an unusually large probability of default. Some of these bonds are new junk sold to finance doubtful ventures; some are fallen angels, bonds issued by companies that once were financially strong. Default rates on junk bonds are usually computed by comparing the number (or value) of junk bonds that default each year with the total junk bonds outstanding. Calculated in this way, the annual default rates on junk bonds in the 1980s were around 2 percent. However, very few junk bonds default in the first few years after issuance, and the rapid growth of the junk bond market in the 1980s meant that, in any given year, most of the outstanding bonds were only a few years old.

An extreme case would be that every junk bond defaults in the third year but that 98 percent of all junk bonds were issued during the previous two years, giving a misleading annual default rate of only 2 percent. For a more realistic example, suppose that every junk bond is a ten-year bond that has no chance of default in the first two years after issuance and a 5 percent annual probability of default during each of the next eight years. The probability of default by the end of the nth year is given by the subtraction rule:

probability of default

$$= 1 - \text{probability of survival} = 1 - .95^{n-2}$$

The exponent is $n - 2$ since the first two years are assumed to be default-free. The calculated default probabilities are shown in the accompanying table. Two default-free years and eight years of a 5 percent default probability implies a 33.66 percent probability of default by the end of ten years (and a $1 - .3366 = .6634$ probability of survival), which is equivalent to a constant annual 4 percent default rate during all ten years ($.3366 \approx 1 - .96^{10}$)

Now suppose that 60 percent of the outstanding junk bonds have been issued within the last two years (and consequently have no chance of default). The probability that a ran-

$= .001$. At stages 2 and 3, there are two parallel parts, reducing the chances of failure at each stage to $.05(.05) = .0025$. The overall probability of a system failure is now

$$P[\text{system failure}] = 1 - P[\text{no stage fails}]$$
$$= 1 - (.999)(.9975)(.9975)$$
$$= .0060$$

about a thirtieth of the failure probability with a series design.

Examples of parallel designs abound: in parallel electronic circuitry, in the use of several smaller engines on airplanes in place of one large engine, in the launching of nuclear missiles, and even in the appearance of two kidneys in the human body.

domly selected bond defaults is given by the multiplication rule:

P[default]

$= P$[bond is > 2 years old]P[default by bond that is > 2 years old]

$= (.40)(.05) = .02$

Each bond has the equivalent of a 4 percent annual probability of default, but only 2 percent of the outstanding bonds, on average, default each year. This stylized example is roughly consistent with a study by Paul Asquith, David Mullins, and Eric Wolff, who found that even though, on average, only 2 percent of outstanding junk bonds default each year, 34 percent of these bonds issued in 1977 and 1978 had defaulted ten years later and 23 percent of the bonds issued between 1979 and 1983 had defaulted five to nine years later.[7]

Number of Years After Issuance, n	Probability of Default	
	During Year n	By End of Year n
1	.00	.00
2	.00	.00
3	.05	.05
4	.05	.0975
5	.05	.1426
6	.05	.1855
7	.05	.2262
8	.05	.2649
9	.05	.3017
10	.05	.3366

EXERCISES

3.18 Farmer Fran sends out 10,000 cartons of eggs every day. Two inspectors independently glance at each carton to see if there are any broken eggs, and each has a .8 probability of noticing when a carton has cracked eggs. What is the probability that a carton with broken eggs will be noticed by at least one inspector? What is the probability if there are three inspectors?

3.19 In parallel research and development efforts, instead of concentrating resources on a single project, several small teams work independently on different projects.[8] If each of fifteen projects has, say, a 20 percent chance of success (and assuming independence), what is the probability that at least one project will succeed?

3.20 In the carnival game Queens, there are six cards—two kings, two queens, and two jacks. The cards are turned face down and shuffled. The player picks two cards and wins if neither card is a queen. What is the probability of winning?

3.21 A dance studio once offered $25 worth of dancing lessons to anyone with a "lucky" dollar bill containing a 2, 5, or 7 in its eight-digit serial number.

What is the probability that a dollar bill will win this prize? If you have three dollar bills, what is the probability that at least one will win?

3.22 It has been estimated that the chances of a fatality in a parachute jump are 1/71,000 if the jump is from an airplane and 1/600 if the jump is from a tall building, bridge, or other stationary object.[9] What

is the probability of making 100 nonfatal jumps from an airplane? From a stationary object?

• **3.23** If you are in a class with 23 students, what is the probability that at least one other person has the same birthday as you? What if there are 100 students in the class? 250 students? (Be careful!)

3.6 BAYES' THEOREM

We've discussed the conditional probabilities $P[A|B]$ and $P[B|A]$, and we've seen how they can be used via the multiplication rule to find $P[A$ and $B]$. Sometimes we are interested in the special problem of converting knowlege of $P[A|B]$ into a calculation of $P[B|A]$. That is, we know the probability that A will occur, given that B has occurred, but what we want to know is the probability that B will occur, given that A has occurred. For instance, we may know the probability that a test will be "positive" if we have a certain disease, but what we are interested in is the probability of disease when the test turns out "positive." We can use the multiplication rule and a bit of logic to make such conversions successfully.

Reverend Bayes, as we saw in Chapter 2, was interested in a very special problem, using known facts to confirm the existence of God. He seemingly wanted to calculate the probability that God exists, given what we observe about the world: $P[\text{God}|\text{observations}]$. Our logic is concerned with the reverse question—the probability that we would observe what we do, if there were a god: $P[\text{observations}|\text{God}]$. Bayes' theorem tells us how to go from one conditional probability to the other.* Bayes did not prove the existence of God, but his conversion theorem has turned out to be extremely useful, nonetheless. It is also the foundation for the modern Bayesian approach to probability and statistics.

The multiplication rule, Equations 3.5 and 3.6, can be written in two ways,

$$P[A \text{ and } B] = P[A]P[B|A]$$

$$P[A \text{ and } B] = P[B]P[A|B]$$

---

*We must speculate about Bayes' motives because little is known about him. He left few published writings and these do not even contain what is now called "Bayes' theorem." It was Laplace[10] who actually wrote down the general rule and then labeled it "Bayes' theorem." Bayes did work out some specific problems of this sort, such as using the frequency with which some event has occurred to calculate the probability that the probability of this event occurring lies in a certain range. The friend who found Bayes' major paper and had an annotated version published posthumously argued that Bayes' approach could be used "to confirm the argument taken from final causes for the existence of the Diety."[11] It is not certain, though, that this is the usage that Bayes himself had in mind.

If we substitute one equation into the other to eliminate $P[A \text{ and } B]$, we obtain **Bayes' theorem:**

$$P[B \mid A] = \frac{P[B]P[A \mid B]}{P[A]} \qquad (3.8)$$

This equation is the vehicle for converting $P[A \mid B]$ in $P[B \mid A]$.

In practice, as we will soon see, $P[A]$ often is not immediately apparent in the kinds of questions that Bayes' theorem is designed to answer. But, with a subtle trick, $P[A]$ can be derived from information that is available. Figure 3.10 uses a Venn diagram to reveal the trick. The set A consists of two parts that are shaded light and dark. The dark portion is "A and B," which we have analyzed frequently. The remaining portion of A is "A and not B," which comprises those outcomes that are in A and are not in B. The multiplication rule tells us that the probabilities of an outcome being in these two separate parts of A are given by

$$P[A \text{ and } B] = P[B]P[A \mid B]$$

$$P[A \text{ and not } B] = P[\text{not } B]P[A \mid \text{not } B]$$

Logic tells us that an outcome in A must be in one part or the other, either in "A and B" or in "A and not B," and the addition rule tells us that, because the two parts are mutually exclusive, the probability of being in one part or the other is the simple sum of the separate probabilities:

$$P[A] = P[A \text{ and } B] + P[A \text{ and not } B] \qquad (3.9)$$
$$= P[B]P[A \mid B] + P[\text{not } B]P[A \mid \text{not } B]$$

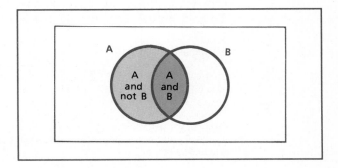

Figure 3.10 *A Tricky Way to Calculate P[A]*

Now, substituting Equation 3.9 into Equation 3.8, we obtain Bayes' theorem in its most frequently used form,*

$$P[B|A] = \frac{P[B]P[A|B]}{P[B]P[A|B] + P[\text{not } B]P[A|\text{not } B]} \qquad (3.10)$$

A Rare Disease

Enough theory! On to an example. Consider a rare disease that can be accurately detected by a medical test. Let "disease" and "no disease" signify the presence or absence of the disease, and let "yes" and "no" represent the test results. We'll assume that only one person in a hundred has this disease:

$$P[\text{disease}] = .01$$
$$P[\text{no disease}] = .99$$

In either case, whether the person does or does not have the disease, the test gives a correct diagnosis 95 percent of the time:

$$P[\text{"yes"}|\text{disease}] = .95$$
$$P[\text{"no"}|\text{no disease}] = .95$$

It is important to recognize precisely what these conditional probabilities mean. If you have the disease, there is a 95 percent chance that the test will register "yes" and a 5 percent chance that the test will register "no." If you do not have the disease, there is a .95 probability that the test will register "no" and a .05 probability that the test will register "yes."

Most of us are interested in the reverse question. If the test does register "yes," what is the probability that we have the disease? If the test registers "no," what is the probability that we do not have the disease? These probabilities are not .95! The probability that the test will say "yes" if you have the disease is not the same concept as the probability that you have the disease if the test does, in fact, say "yes." To go from the former to the latter, we need Bayes' theorem.

$$P[\text{disease}|\text{"yes"}]$$
$$= \frac{P[\text{disease and "yes"}]}{P[\text{"yes"}]}$$

*Sometimes, the outcomes are divided into several sets: $B_1, B_2, \ldots, B_n$ instead of just B and "not B." If, like B and "not B," these sets include all possible outcomes, then Equations 3.9 and 3.10 can be generalized by using

$$P[A] = P[B_1]P[A|B_1] + P[B_2]P[A|B_2] + \cdots + P[B_n]P[A|B_n]$$

$$= \frac{P[\text{disease}]P[\text{"yes"}\,|\,\text{disease}]}{P[\text{disease}]P[\text{"yes"}\,|\,\text{disease}] + P[\text{no disease}]P[\text{"yes"}\,|\,\text{no disease}]}$$

$$= \frac{(.01)(.95)}{(.01)(.95) + (.99)(.05)}$$

$$= .161$$

One percent of the entire population has the disease. Sixteen percent of those diagnosed "yes" have the disease. The "yes" reading makes it more likely that you have the disease, but it is still far from certain.

We can also calculate the probability of disease when the test result is negative:

$P[\text{disease}\,|\,\text{"no"}]$

$$= \frac{P[\text{disease}]P[\text{"no"}\,|\,\text{disease}]}{P[\text{disease}]P[\text{"no"}\,|\,\text{disease}] + P[\text{no disease}]P[\text{"no"}\,|\,\text{no disease}]}$$

$$= \frac{(.01)(.05)}{(.01)(.05) + (.99)(.95)}$$

$$= .0005$$

If you have not been tested, then there is a .01 probability that you have the disease. If you are tested and the results are negative, then the probability that you have the disease falls to .0005.

These results are typical of the effects of applying Bayes' theorem. We can often interpret the situation as one in which initial probabilities are revised in the light of empirical evidence. Here, the initial probability of disease is .01. After the test, this probability is either revised up to .16 or down to .0005, depending on whether the test reading is positive or negative. Academics call these initial and revised probabilities the *prior* and *posterior* probabilities.

> The **prior probability** $P[A]$ is the probability assigned to A before certain data are gathered. The **posterior probability** $P[A\,|\,\text{data}]$ is the revised probability, given these observed data.

As more data are collected, the probabilities can be revised again and again. For example, in the rare-disease problem, if the first diagnosis is positive, the doctor may recommend a second test. The .16 probability of disease will then be revised further, up or down, depending on the outcome of the second test.

Some Probabilistic Evidence Against Cigarette Smoking

If you put a hand on a hot stove, you will be burned. The effects of smoking cigarettes are neither as certain nor as immediate. The case against cigarette smoking is instead that smokers are more likely to suffer eventually from a variety of diseases. For instance, 6 percent of all deaths in the United States are due to lung cancer (about 100,000 deaths each year) and 85 percent of those who die of lung cancer are cigarette smokers. Since about 1/3 of the adult population smokes and 2/3 doesn't, we can use Bayes' theorem to figure the respective probabilities that smokers and nonsmokers die of lung cancer:[12]

$$P[\text{cancer}|\text{smoke}] = \frac{P[\text{cancer and smoke}]}{P[\text{smoke}]}$$

$$= \frac{P[\text{cancer}]P[\text{smoke}|\text{cancer}]}{P[\text{smoke}]}$$

$$= \frac{.06(.85)}{(1/3)}$$

$$= .153$$

while

$$P[\text{cancer}|\text{don't smoke}] = \frac{P[\text{cancer and don't smoke}]}{P[\text{don't smoke}]}$$

$$= \frac{P[\text{cancer}]P[\text{don't smoke}|\text{cancer}]}{P[\text{don't smoke}]}$$

$$= \frac{.06(.15)}{(2/3)}$$

$$= .014$$

Not all smokers die of lung cancer; but they are ten times more likely to die of lung cancer than nonsmokers.

Adjusting the Weather Forecast

Your favorite television weather forecaster may pretend to be a standup comic, an underweight starlet, or an overweight turkey. Some even pretend to be meteorologists. If you pay attention to their forecasts, you will find that they make mistakes, and you may find that their mistakes are disproportionately of certain kinds. If so, you can make your own forecasts by adjusting the television forecasts.

Perhaps you live in a city where there is .01 inch of precipitation (either rain or snow) on about one out of every three days. Yet your friendly television forecaster predicts "wet" or "dry" by flipping a coin. As a result, there are a disproportionate number of "wet" forecasts. You could do better with homemade forecasts that ignored the television forecaster and just used the historical data, giving a 1/3 probability of precipitation.

A more subtle bias occurs when the forecaster is informed, but tilts the forecasts one way or the other. There are some days when the forecast is clearcut; the forecaster sticks his or her head out the window and is soaked. There are other days when the weather is very uncertain. An optimistic forecaster might invariably forecast "dry" on these iffy days, to keep the show happy and upbeat. In contrast, a pessimistic forecaster may always call the close ones "wet," to avoid the wrath of angry viewers caught in the rain without an umbrella. In either case, you can adjust these slanted forecasts to get a better estimate of the chances of precipitation.

To illustrate this adjustment, we'll consider the case of the forecaster who, "when in doubt, calls it 'wet.'" The forecaster who, "when in doubt, calls it 'dry,'" is left as an exercise at the end of the chapter. Let's say that it is dry two out of three days and that one-fourth of these dry days are so obviously going to be dry that the forecaster predicts "dry." But on three-fourths of the dry days, the weather appears sufficiently uncertain that the forecaster predicts "wet" just to be safe. Thus, $P[\text{"dry"}|\text{dry}] = .25$ and $P[\text{"wet"}|\text{dry}] = .75$, where the quotation marks signify the forecast. When it is actually going to be wet, there is only a .05 chance that this cautious forecaster will be misled into making a "dry" forecast: $P[\text{"wet"}|\text{wet}] = .95$ and $P[\text{"dry"}|\text{wet}] = .05$.

If we are aware of this forecaster's conservative bias, how can we adjust the forecasts to obtain a more accurate weather gauge? Bayes' formula tells us how to combine the historical data with the forecaster's record to obtain revised weather probabilities:

$$P[\text{dry}|\text{"dry"}] = \frac{P[\text{dry}]P[\text{"dry"}|\text{dry}]}{P[\text{dry}]P[\text{"dry"}|\text{dry}] + P[\text{wet}]P[\text{"dry"}|\text{wet}]}$$

$$= \frac{(2/3)(.25)}{(2/3)(.25) + (1/3)(.05)}$$

$$= .91$$

$$P[\text{wet}|\text{"wet"}] = \frac{P[\text{wet}]P[\text{"wet"}|\text{wet}]}{P[\text{wet}]P[\text{"wet"}|\text{wet}] + P[\text{dry}]P[\text{"wet"}|\text{dry}]}$$

$$= \frac{(1/3)(.95)}{(1/3)(.95) + (2/3)(.75)}$$

$$= .39$$

You can be pretty confident that the weather will be dry when the forecast is "dry," because this cautious forecaster almost never says "dry" unless dry weather is almost certain. On the other hand, the "wet" forecasts are of little value because this forecaster makes them so readily. The revised .39 probability of precipitation is not much higher than the historical 1/3 probability. The "dry" forecast is informative, but the "wet" forecast is almost worthless.

These principles can be applied to the consequences of a federal judge's 1985 precedent-setting decision that the National Weather Service was liable for $1.2 million in damages because it did not forecast a freak winter hurricane that killed three lobstermen 100 miles off the Massachusetts coast. As might be expected, this court decision persuaded the Weather Service to change to very conservative forecasts, issuing storm warnings when storms were, in fact, several days away and their arrival strength uncertain. The fishermen soon noticed this bias, after they repeatedly cut short fishing trips only to find that the storm was mild and/or days away. Lamented one,

> This ruling puts fishermen and the weather service in a terrible position. How many times is a skipper going to take evasive action for nothing before he ignores the forecast, or just assumes he has a certain amount of time before he really needs to get into the lee? I think this ruling is really stupid.[13]

A Miracle or a Trick?

Let's do a second example. This one is a little more off the wall, but it's closer to Bayes' initial inquiry into the existence of God. We saw earlier that the probability of drawing 4 straight aces from a standard 52-card deck is

$$\left(\frac{4}{52}\right)\left(\frac{3}{51}\right)\left(\frac{2}{50}\right)\left(\frac{1}{49}\right) = .0000037$$

This is a very unlikely event. If some character were to walk up to you, announce that she is going to pick 4 aces in a row, and then do just that, you are likely to suspect some chicanery. Perhaps she is really an amateur magician performing a prepared trick. The 4 aces may be prearranged on top of the deck, the deck may contain 52 aces, or she may be using other gimmicks that I am pledged not to reveal. If you had any suspicions that the draw was going to be rigged, these suspicions would certainly be bolstered by the woman's success. Four aces in a row is just too improbable to attribute to luck alone.

Bayes' theorem can handle this kind of situation, too. Let's say that, beforehand, as a trusting person, you are 99 percent sure that this is a random draw from a standard deck. We can write this prior probability as

$$P[\text{standard deck}] = .99$$

$$P[\text{rigged deck}] = .01$$

What are your revised probabilities, after seeing the feat performed? The probabilities of picking 4 aces from a standard deck and from a perfectly rigged deck are

$$P[4 \text{ aces} \,|\, \text{standard deck}] = .0000037$$

$$P[4 \text{ aces} \,|\, \text{rigged deck}] = 1.0$$

Therefore, Bayes' theorem yields the posterior probability

$$P[\text{standard deck}|4\text{ aces}] = \frac{.99(.0000037)}{.99(.0000037) + .01(1.0)}$$

$$= .000366$$

You went in trusting, 99 percent sure that the deck was fair, but after witnessing a feat that would be miraculously improbable if the deck were fair, you are 99.96 percent certain that the deck is rigged somehow.

This example is very much in the spirit of Bayes' intended application of his theorem. You have some theory about the world, such as this being a fair deck or there being a god. You then gather evidence. If the evidence is consistent with your theory, then you are more confident that your theory is correct. If the evidence is not consistent with your theory, then you are more doubtful of the theory. This general methodological approach can be applied to virtually all of the physical and social sciences. Indeed, this is the very way science proceeds, although by more informal means. A theory is accepted as long as it is successful in explaining what we observe. When new evidence accumulates that cannot be explained by the accepted theory, then the conventional wisdom is discredited and replaced with a new theory that can explain the data. Later in this book we will look at some formal procedures for confirming or rejecting hypotheses.

Objectivist Objections

There can be no dispute about the mechanics of Bayes' theorem. Its mathematics are impeccable. Objectivists, however, are not comfortable with the way it is applied. They don't like the whole idea of applying subjective probabilities to theories. Such probabilities vary substantially from person to person and from time to time, and even more fundamentally, theories are not a proper subject for probabilities. A theory is either correct or it is incorrect; it is not correct half of the time, with its correctness being determined by the flip of a coin.

Let's go back to the doctor's office for the disease diagnosis. This is an example where the probabilities did not seem subjective or whimsical, yet strict objectivists would balk at assigning a probability of disease to a particular person. If we were about to pick one person at random from the population, we could legitimately say that there is a .01 probability that we will pick someone who has the disease. Once the person has been selected and is in the doctor's office being diagnosed, objectivists argue that we can no longer talk about a .01 or .16 probability of this person having the disease. The person either does or does not have the disease. We can say things like, "If a person does not have the disease, there is only a .05 probability of obtaining a positive diagnosis." But, objectivists argue that we should not turn this conditional probability around. A strict objectivist will not say, "Because the diagnosis is 'yes' for Mary Morgan, there is a .16 probability that Mary has the disease."

Some Guilty People Go Free

In the United States, defendants are presumed innocent until they are proven guilty "beyond a reasonable doubt." The intent is to protect the innocent fully, because it is "better to let nine guilty people go free, than to send one innocent person to prison."

For illustrative purposes, let's assume that 75 percent of all defendants brought to trial are guilty, 70 percent of all guilty defendants are convicted, and 90 percent of all innocent defendants are set free. These hypothetical numbers imply that, on the average, out of every 1000 defendants, 750 are guilty and 250 are innocent. Of the 750 who are guilty, .7(750) = 525 are convicted, while the remaining 225 go free. Of the 250 who are innocent, .9(250) = 225 go free while 25 are convicted. With these figures, it is indeed true that for every innocent person who goes to prison, nine guilty people go free (225/25 = 9).

Now let's ask two somewhat different questions. Of those who do go free, what fraction are actually guilty? Of those who are convicted, what fraction are innocent? We can find the answers in the previous data. If 225 guilty people go free and 225 innocent people go free, half of those who go free are, in fact, guilty. If 525 guilty people and 25 innocent people are convicted, 25/550 = .045 (4.5 percent) of the people convicted are, in fact, innocent. See Figure 1.

These calculations are actually an application of Bayes' rule. Here are the formulas, using "G" for guilty, "I" for innocent, "GF" for go free, and "C" for convicted:

Figure 1   *Of Those Who Go Free, Half Are Guilty!*

$$P[G|GF] = \frac{P[G]P[GF|G]}{P[G]P[GF|G] + P[I]P[GF|I]}$$
$$= \frac{.75(.3)}{.75(.3) + .25(.9)}$$
$$= \frac{.225}{.225 + .225}$$
$$= .5$$

and

$$P[I|C] = \frac{P[I]P[C|I]}{P[I]P[C|I] + P[G]P[C|G]}$$
$$= \frac{.25(.1)}{.25(.1) + .75(.7)}$$
$$= \frac{.25}{.25 + .525}$$
$$= .045$$

Bayesians argue that it is precisely this latter probability that is of interest. The .16 probability is used to convey our estimate of how likely it is that Mary has the disease. The interpretation is simply, and admittedly rather cruelly put in this instance, that we are indifferent between a wager in which we win $100 if Mary turns out to have the disease and a wager in which we win $100 if a red ball is drawn from a bag containing 16 red balls and 84 black balls.

The rigged-deck example is even more discomforting to objectivists because it requires a totally subjective assessment of whether a particular unexamined deck is rigged or fair. Where did that .99 probability of a fair deck come from? It was apparently plucked out of the air and has nothing to support it. It is not reliable and cannot be scientific. Bayesians respond that, as fragile as this probability is, it is necessary to answer the relevant question: Do we believe the four-ace demonstration to be a miracle or a trick? Indeed, people make such subjective assessments all the time. Bayes' theorem just shows us how to do the arithmetic correctly.

This objectivist-subjectivist debate may still seem like a lot of academic quarreling about nothing. You will soon see that it matters very much to the way one thinks about scientific theories and the use of data to test and revise these hypotheses.

EXERCISES

3.24 According to the U.S. Federal Highway Administration, 7.7 percent of all drivers are under the age of 20, and each year a randomly selected driver under the age of 20 has a .00082 probability of being involved in a fatal automobile accident, while the similar probability for a driver 20 or older is .00039. What is the probability that a randomly selected driver involved in a fatal accident is under the age of 20?

3.25 It has been suggested that Bayes' theorem can be used to help jurors assess the significance of statistical evidence.[14] Suppose that a palm print on a murder weapon is similar to that of the defendant and, in the rest of the population is similar to that of one person in a thousand. Let

E = the print has these distinctive features

D = the print was left by the defendant

Use Bayes' theorem to find $P[D|E]$ for each of two alternative prior probabilities: $P[D] = .25$ and $P[D] = .75$.

3.26 In three careful studies, polygraph (lie detector) experts examined several persons, some known to be innocent and the rest known to be guilty, to see if the experts could tell which were which.[15] Overall, 83 percent of the guilty people were pronounced "deceptive" and 57 percent of the innocent were judged "truthful." Based on these data, let us assume that $P[\text{"lies"}|\text{liar}] = .83$ and $P[\text{"truth"}|\text{truthful}] = .57$, where the quotation marks indicate the polygraph expert's assessment. And let's assume that, in practice, 80 percent of all people tested are truthful and 20 percent are liars. If the expert judges "truth," what is the probability that the person is a liar? If the expert judges "liar," what is the probability that the person is truthful? How would these probabilities be altered if half the people tested are truthful and half are liars? How would an objectivist interpret these probability calculations?

3.27 Econ Oil is interested in the possibility of an oil deposit under the Astrodome in Houston, Texas. The local expert believes that there is a 30 percent chance that a significant oil field will be found there. An exploratory drilling (on the fifty-yard line) is positive. Historically, such tests are 70 percent accurate: $P[\text{positive}|\text{oil}] = .7$ and $P[\text{negative}|\text{no oil}] = .7$.

What should Econ's revised probability of oil be? (Is this why the football team is called the Houston Oilers?)

- **3.28** Critically evaluate the following:

  A large metropolitan police department made a check of the clothing worn by pedestrians killed in traffic at night. About four-fifths of the victims were wearing dark clothes and one-fifth light-colored garments. This study points up the rule that pedestrians are less likely to encounter traffic mishaps at night if they wear or carry something white after dark so that drivers can see them more easily.[16]

- **3.29** Twenty-five percent of the designer jeans manufactured for Claudia La Claudia have imperfections. Each pair of jeans is independently examined by two inspectors before it leaves the plant. Each inspector has a .9 probability of correctly spotting an imperfect pair. Perfect pairs are never classified "imperfect." What is the probability that a pair of jeans that passes inspection is actually imperfect, if **a.** jeans are rejected if either or both inspectors spot a flaw? **b.** jeans are rejected only if both inspectors spot a flaw?

3.7 SUMMARY

A number of rules simplify the computation of probabilities. The addition rule gives us the probability that either A or B will occur, or both:

addition rule: $P[A \text{ or } B] = P[A] + P[B] - P[A \text{ and } B]$

The multiplication rule gives the probability that both A and B will occur:

multiplication rule: $P[A \text{ and } B] = P[A]P[B|A]$
$$= P[B]P[A|B]$$

where $P[B|A]$ is the conditional probability that B will occur, given that A has occurred.

The subtraction rule is handy because sometimes the easiest way to find the probability that A will occur is to reason out the probability that A will not occur:

subtraction rule: $P[A] = 1 - P[\text{not } A]$

Bayes' theorem tells us how to get from $P[A|B]$ to $P[B|A]$:

Bayes' theorem: $P[B|A] = \dfrac{P[B]P[A|B]}{P[A]}$

$$= \dfrac{P[B]P[A|B]}{P[B]P[A|B] + P[\text{not } B]P[A|\text{not } B]}$$

Two events are said to be mutually exclusive if $P[A \text{ and } B] = 0$. Two events are independent if $P[A \text{ and } B] = P[A]P[B]$. When sampling with replacement from a deck of cards, for example, the outcomes are independent. When sampling without replacement, the outcomes are not independent because the probabilities change as cards are removed from the deck.

REVIEW EXERCISES

3.30 "When the cards are dealt in bridge, there is a 1/4 probability that I will get the ace of spades and a 1/3 probability that my partner will get it if I don't. Therefore, we have a 7/12 probability of getting the ace of spades." Explain why you agree or disagree with this reasoning.

3.31 Here is a division of Nobel prize winners in chemistry, physics, and medicine according to when the prize was won and whether the winner was a U.S. citizen:

Winner	1901–1945	1946–1984
United States	20	115
Other Countries	122	106

If one of these 363 Nobel laureates is picked at random, what is the probability that the person selected is **a.** a U.S. citizen? **b.** not a U.S. citizen? **c.** a U.S citizen, given that the person won the prize before 1946? **d.** a U.S. citizen, given that the person won the prize after 1945?

3.32 Businesses commonly project revenues under alternative economic scenarios. For a stylized example, inflation could be high or low and unemployment could be high or low. There are four possible scenarios, with the assumed probabilities shown in Table 3.2. **a.** What is the probability of high inflation? **b.** What is the probability of high unemployment? **c.** What is the probability of high unemployment if inflation is high? **d.** Are inflation and unemployment independent?

3.33 Table 3.3 presents 1985 data on the age and sex distribution in the United States. Calculate the following probabilities: $P[M]$, $P[M|\text{under }18]$, $P[F|\text{under }18]$, $P[M|18\text{–}30]$, $P[M|30\text{–}55]$, and $P[M|\text{over }55]$. Are age and sex independent? If not, what dependencies do you detect in these data?

3.34 In Exercise 2.39, we looked at the distribution of males and females among three categories of professors at Metro U. Are rank and sex independent in these data? If not, what dependencies do you detect?

3.35 What are the sucker's chances of winning this game:

Take a small opaque bottle and seven olives, two of which are green, five black. The green ones are considered the "unlucky" ones. Place all seven olives in the bottle, the neck of which should be such a size that it will allow only one olive to pass through at a time. Ask the sucker to shake them and then wager that he will not be able to roll out three olives without getting an unlucky green one amongst them.[17]

• **3.36** Explain why you think that the following system does or does not have merit:

There are many systems of all kinds but in talking to gamblers you will find the one they believe in most (and it can be used for most gambling games) is the "watch" or "patience" system. It's simple. If you are playing dice and certain bets pay even money, before you put your first stake on any of the numbers watch until one chance shows up at least three or four consecutive times and then put your first stake on the opposite chance.[18]

Table 3.2 *Scenarios for Exercise 3.32*

Scenario	Inflation	Unemployment	Probability
1	high	high	.16
2	high	low	.24
3	low	high	.36
4	low	low	.24

Table 3.3 *Data for Exercise 3.33*

Age	Males	Females	Total
Under 18	32,256,000	30,757,000	63,013,000
18–30	25,530,000	25,097,000	50,627,000
30–55	36,835,000	37,943,000	74,778,000
Over 55	22,028,000	28,837,000	50,865,000
Total	116,649,000	122,634,000	239,283,000

• **3.37** Pepys asked Newton which of the following three events is most likely: **a.** at least one 6 when six dice are rolled, **b.** at least two 6s when twelve dice are rolled, or **c.** at least three 6s when eighteen dice are rolled? Answer Pepys's question.

• **3.38** In the game craps, a pair of dice are rolled. The roller immediately wins if a 7 or 11 comes up and loses with a 2, 3, or 12 ("craps"). If any other number comes up, then this becomes the "point" number and the dice are rolled again and again until either the point number or a 7 comes up. The roller wins if the point number comes up first and loses if a 7 is rolled first. What is the roller's probability of winning a game of craps?

3.39 A car was ticketed in Sweden for parking too long in a limited time zone when a policeman recorded the position of the two tire air valves on one side of the car (in the 1:00 and 8:00 positions) and returned hours later to find the car in the same space with the air valves in the same position.[19] The driver claimed that he had driven away and returned later to park in the same spot, and that the air valves happened to stop in the same positions. The court accepted the driver's argument, calculating the probability that both valves would stop at their earlier positions as (1/12) (1/12) = 1/144 and feeling that this was not a small enough probability to preclude reasonable doubt. The court advised, though, that had the policeman noted the position of all four tire valves and found these to be unchanged, the very slight (1/12)^4 = .00005 probability of such a coincidence would be accepted as proof that the car had not moved. As defense attorney for a four-valve client, how might you challenge this calculation?

3.40 At a certain selective college, 80 percent of the freshman class were ranked in the top quarter of their high school class. It turns out that 25 percent of the freshman class got good enough grades to make the dean's list their first year, and that 90 percent of the students on the dean's list were in the top quarter of their high school class. Of those freshmen in the top quarter of their high school class, what fraction made the dean's list their first year? What fraction of those not in the top quarter of their high school class made the dean's list their first year?

3.41 In 1974, Cutler-Hammer was offered a six-month option on the patent rights for a flight-safety system that they might be able to manufacture for the military.[20] They estimated the probability of deciding, after six months, to exercise the option as .71 and the probability, if they did exercise the option, of obtaining a defense contract as .15. What was their estimate of the probability that they would both exercise the option and obtain a defense contract?

• **3.42** There are four secretaries in my department: Bob, Carol, Ted, and Alice. Their typing speed and accuracy is quite dissimilar. Bob types 40 percent of the departmental correspondence, and 25 percent of his work contains serious mistakes (transposed words, omitted lines, and so on). Carol types 30 percent of the correspondence and 20 percent of her work has serious mistakes. Ted types 20 percent of the correspondence and 20 percent of his work has serious mistakes. Alice types only 10 percent of the correspondence, and she never makes a serious mistake.

I wrote a letter to the student newspaper, vigorously protesting the demise of the dress code. Fortunately, I decided to proofread the letter before signing it, because several of my choicest words were omitted or misplaced by the typist. What is the probability that this letter was typed by Bob? By Carol? By Ted? By Alice?

• **3.43** Half of the cars that come off of the assembly line at U.S. Motors have flaws in their appearance and half have flaws in their performance. These flaws are independent. There are two independent inspectors at U.S. Motors, one who checks each car's appearance and one who checks its performance. When a car is flawed, each inspector will spot the flaw 80 percent of the time and overlook the flaw 20 percent of the time. Neither inspector ever sees a flaw when there is none.

As a car comes off the assembly line, what is the probability **a.** that it has flaws in its appearance and performance? **b.** that it only has flaws in its appearance? **c.** that it only has flaws in its performance? **d.** that it is flawless?

Each car is then checked by the two inspectors. What are the probabilities **e.** that a car that has flaws in both its appearance and performance will pass both inspectors? **f.** that a car with flaws in both its appearance and performance will pass neither inspector? **g.** that a car with an acceptable appearance but flawed performance will pass both inspectors? **h.** that a car will pass both inspectors?

If you buy a car that has been passed by both of these inspectors, what is the probability that **i.** it has a flawed appearance and performance? **j.** only its appearance is flawed? **k.** only its performance is flawed? **l.** it has no flaws?

• **3.44** Jack Cooper lives in New York City where throughout the year there is at least .01 inch of precipitation on about one out of every three days. But Dr. Chuckles, Mr. Cooper's favorite weather forecaster, uses a Susan B. Anthony dollar to forecast "wet" or "dry" each day. Thus, $P[\text{"wet"}|\text{wet}] = P[\text{"wet"}|\text{dry}] = .5$. If Mr. Cooper knows Bayes' theorem and wants better forecasts, what should he use for $P[\text{wet}|\text{"wet"}]$?

How might he actually use this revised probability to make homemade forecasts that will be accurate more often than those of Dr. Chuckles? If he follows your advice, what is Jack's overall probability of making a correct forecast? What is Dr. Chuckles's probability of making a correct forecast?

3.45 A certain optimistic weather forecaster predicts "dry and sunny" unless there is clear evidence of imminent precipitation. When there is going to be precipitation, this is apparent one-fourth of the time and the forecaster predicts "wet." The other three-quarters of the wet days, the precipitation is not certain and this optimistic forecaster predicts "dry." Thus $P[\text{"wet"}|\text{wet}] = .25$ and $P[\text{"dry"}|\text{wet}] = .75$. When it is going to be dry, there is only a .01 chance that the weather will look threatening enough to provoke a "wet" forecast: $P[\text{"wet"}|\text{dry}] = .01$ and $P[\text{"dry"}|\text{dry}] = .99$. If, in fact, it is wet one-third of the days ($P[\text{wet}] = 1/3$), then what will be your adjusted weather probabilities $P[\text{wet}|\text{"wet"}]$ and $P[\text{dry}|\text{"dry"}]$?

•• **3.46** Consider a deadly disease that is transmitted genetically. Let's say that the gene H is healthy, while the gene L is lethal. The genetic pair LL causes death during birth or early childhood. The genetic pairs HH and HL are both healthy, in that H is the dominant gene. But individuals with HL genes are carriers of the lethal gene who, if they marry other carriers, may have children with the lethal genetic combination LL. Unfortunately, the only way to be certain that someone is a carrier is to discover this deadly disease in their children.

If 1 percent of the adult population have HL genes and 99 percent have HH genes, what fraction of their children will have LL genes? Of those children who do not have LL genes, what fraction will be HH and what fraction will be HL?

•• **3.47** *(continuation)* If a child has the lethal disease considered in the preceding exercise, then the parents must both be carriers. If these parents have another child, what is the probability that this next child, too, will have the lethal disease? Many societies prohibit brother-sister marriages. One explanation is that such marriages are bad genetic risks. If two people are picked at random, what is the probability that both will be carriers of this lethal gene? If a brother and sister marry, what is the probability that they will both be carriers? How much more likely is it that a child will be LL if the parents are brother and sister?

• **3.48** Bertram's matchbox contains four matches, two of which are burnt and two of which are fresh. If we draw two matches randomly, what is the probability that we will get one burnt match and one fresh match?

• **3.49** We have three boxes, each with two drawers and a coin in each drawer. The coins are both gold in one box, both silver in another box, and one gold and one silver in the third box. A drawer is chosen at random from a randomly selected box. If the coin in this drawer is gold, what is the probability that this box's other drawer also contains a gold coin?

•• **3.50** Larry, Moe, and Curly are three prisoners, kept in separate cells. Two of them will be hanged in the morning and the third will be set free. Part of their punishment is that the prisoners do not know which two will be hanged. As each is equally likely to be set free, Curly calculates his chances to be only 1/3. Knowing some probability analysis, he says to a guard, "Look, I know that Larry or Moe is going to hang in the morning, so it can't hurt to tell me the name of one man, either Larry or Moe, who is going to be hanged." Accepting this argument, the jailer reveals that Larry will be hanged. Curly sleeps easier that night because, now that he knows that only he or Moe will live, his odds of being set free have risen from 1/3 to 1/2. Yet Curly would have reasoned in exactly the same way if the jailer had revealed that Moe would be hanged, implying that this probability of being set free goes up to 1/2 no matter who the jailer names! Whose probability logic is flawed?

3.51 In 1944 an aircraft executive accused a movie actor of winning $18,500 from him in a crooked craps game. The executive claimed that the movie actor had made 13 consecutive passes, where the probability of a pass with fair dice is approximately .5. If the dice had been fair, what is the probability of 13 consecutive passes?

3.52 The four Simpsons are posing for a family portrait. Each Simpson has a .6 probability of looking acceptable at the moment the picture is taken. Assuming independence, what is the probability that all four Simpsons will look acceptable at the same time? How many pictures must the photographer take to be 80 percent certain that all four Simpsons will look acceptable simultaneously in at least one picture?

3.53 A television weather forecaster once said that there was a 50 percent chance of rain on Saturday and a 50 percent chance of rain on Sunday, and therefore a 100 percent chance of rain that weekend. Explain the error in this reasoning. If there is a 40 percent chance that it will rain on both Saturday and Sunday, what is the probability that it will rain on at least one of these days?

3.54 A research group estimated that college football players have a 23 percent chance of a knee injury each season and a 64 percent chance of a knee injury during a four-year career.[21] Does this 64 percent probability assume independence from one year to the next?

3.55 The *Wall Street Journal* runs a monthly stock picking contest involving four investment professionals and four stocks selected by tossing darts at the financial pages of the *Wall Street Journal*. During a five-month period between December and April 1989, one expert beat the other three experts four times and tied for first once.[22] If four outright wins and one tie occur in a five-month contest, what is the probability that these victories will be by a single person?

3.56 A card player claims that a thoroughly shuffled deck usually has an ace within nine cards of the top. To test this claim, a deck is shuffled and the cards turned over one at a time. What is the probability that an ace will be among the first nine cards turned up?

3.57 In one version of Russian roulette, the player chooses one of three guns, aims the gun at himself, and pulls the trigger. Each gun has five chambers, one gun with 1 bullet and 4 empty chambers, one gun with 2 bullets and 3 empty, and one gun with 3 bullets and 2 empty. What is the probability that the player will be unlucky enough to fire a bullet?

3.58 Explain why you agree or disagree with this statement: "The probability that one of these missiles will hit the target is .1; therefore, if we fire ten missiles we're certain to hit the target."

3.59 You are playing draw poker and have been dealt a pair of 4s, a jack, a 7, and a 3. You keep the pair and discard the other three cards. What is the probability of drawing at least one more 4?

• **3.60** Use a Venn diagram to confirm this addition rule:

$$P[A \text{ or } B \text{ or } C] = P[A] + P[B] + P[C]$$
$$- P[A \text{ and } B] - P[B \text{ and } C]$$
$$- P[A \text{ and } C] + P[A \text{ and } B \text{ and } C]$$

• **3.61** You are playing poker and have been dealt five cards. What is the probability of a hand with four aces? What is the probability of being dealt a hand with four of a kind, not necessarily aces?

• **3.62** Events A and B are mutually exclusive and neither $P[A]$ nor $P[B]$ is equal to zero. Show that A and B are not independent.

3.63 A single six-sided die is rolled. Let A contain the odd numbers and B the numbers larger than 3. What is $P[A]$? $P[B]$? $P[A \mid B]$? $P[B \mid A]$? $P[A$ and $B]$? Are A and B independent?

• **3.64** Six dice are rolled from a cup. What is the probability that the six numbers 1, 2, 3, 4, 5, and 6 will be obtained?

3.65 You flip a coin seven times. What is the probability of heads on the seventh flip, given that you've gotten heads on five of the first six flips?

3.66 One fifth of the people in this town leave their keys in their cars. Two percent of the cars in this town will be stolen sometime this year. If three fourths of these stolen cards have keys left in them, what is the probability that a car with keys left in it will be stolen?

3.67 Carefully explain the error in this logic:

Most automobile accidents involve people driving at moderate speed, and almost no accidents involve people driving 100 mph. Therefore, it is safest to drive 100 mph.

• **3.68** A certain textbook has 500 pages, with approximately 1,000 printed characters per page. There are typically 500 typographical errors (mistakenly printed characters) randomly scattered through a book of this size published by this company. If there are 1,000 characters on the first page, what is the probability of at least one typographical error on the first page?

3.69 You are playing draw poker and have been dealt three clubs and two spades. If you discard the two spades, what are the chances of drawing two clubs to give you a flush?

3.70 A bag contains 4 red balls, 5 white balls, and 6 blue balls. Determine the following probabilities: **a.** $P[R]$ if 1 ball is drawn.   **b.** $P[R$ or $W]$ if 1 ball is drawn.   **c.** $P[R, W,$ and B in that order] if 3 balls are drawn without replacement.   **d.** $P[R, W,$ and B in that order] if 3 balls are drawn with replacement.

3.71 In a family of four, what is the probability that at least two persons have the same astrological sign? Assume that each of the twelve astrological signs is equally likely to contain a person's birthday.

3.72 Each box of Cracker Jack candy contains a small prize. Each of the 10 children at Joanna's birthday party are given a box of Cracker Jack. If there are 50 equally likely prizes, what is the probability that every child will get a different prize?

3.73 Explain the difference between two events being independent and being mutually exclusive. Give an example to illustrate this distinction.

3.74 During World War II, 40,000 German soldiers served on U-boats, and 30,000 never returned. What is the probability that one of these soldiers, picked at random, didn't return? Now consider a father and two sons who served on German U-boats. Assuming independence, what is the probability that all three returned? That none returned?

3.75 Cheaters sometimes sneak gimmicked dice into friendly craps games. One is "tops and bottoms," dice that contain only three different numbers, with the matching numbers on opposite ends to avoid detection. If one die has a 3, 4, and 5 and the other die has a 1, 5, and 6, what are the probabilities of the different possible sums?

3.76 A card player claims that a thoroughly shuffled deck of cards, containing 4 aces and 48 other cards, usually has an ace within 9 cards off the top. To test this claim, a deck is shuffled and the cards turned over one at a time. What is the probability that an ace will be among the first 9 cards?

3.77 Many have said that the average person cannot distinguish among different brands of beers. Two self-proclaimed experts, Bob and Doug, claim that they can and that they both like brand A better than brand B and brand B better than brand C. In separate blindfold taste tests, they both rate brand A best, brand B next, and brand C lowest. If, in fact, they cannot tell one brand from another, what is the probability that, by luck alone, both would give these particular rankings?

3.78 A car rental company has 120 Japanese cars and 80 American cars. It has found that, on any given day, each Japanese car has about a .5 percent chance of being disabled while each American car has about a 4 percent chance. If a car is disabled, what is the probability it is American?

3.79 Here are some 1984 data on the ages of licensed drivers and of drinking drivers involved in fatal accidents:

Age	Licensed Drivers	Drinking Drivers in Fatal Accidents
16–18	4,372,000	850
18–25	27,331,000	8,000
25–35	39,316,000	7,140
35–55	49,182,000	3,790
55+	37,075,000	1,300
Total	157,305,000	21,080

Assume that all drivers are licensed. What is the probability that a randomly selected driver will be 18–25 years of age? Over 55? What is the probability that a randomly selected drinking driver involved in a fatal accident will be 18–25 years of age? Over 55? Is involvement as a drinking driver in a fatal accident independent of age?

3.80 When the available panel of jurors has been almost depleted, both sides can agree to a trial with 6 rather than 12 jurors. Consider a case in which each juror has a .05 probability of voting not guilty. What is the probability that a jury will not unanimously vote guilty if there are 12 people on the jury? If there are 6 people?

3.81 Some cars turn out to be "lemons" but, unfortunately, the quality of a new car is something that cannot be ascertained until after the car has been driven several hundred miles. The unlucky purchaser of a lemon may then try to unload it by resale to someone else. Assume that 10 percent of all new cars are lemons and that 90 percent of all lemons and 5 percent of all nonlemons are offered for sale within the first year of ownership. If so, what fraction of all cars offered for sale within the first year are lemons?

3.82 On long automobile trips, Mrs. Smith drives and Mr. Smith gives directions. When there is a fork in the road, his directions are correct 40 percent of the time and wrong 60 percent of the time. Having been misled many times. Mrs. Smith follows Mr. Smith's direction 40 percent of the time and does the opposite 60 percent of the time. Assuming independence, how often do they drive down the correct fork in the road? If Mrs. Smith wants to maximize the probability of choosing the correct road, how often should she follow Mr. Smith's directions?

• **3.83** The weatherwoman forecasts "fair," "cloudy," or "wet" each day. The weather is, in fact, equally likely to be fair, cloudy, or wet. Here is the weatherwoman's forecasting record:

$P[\text{"fair"}|\text{fair}] = 0.7$ $P[\text{"cloudy"}|\text{fair}] = 0.2$
$P[\text{"wet"}|\text{fair}] = 0.1$ $P[\text{"wet"}|\text{cloudy}] = 0.1$
$P[\text{"cloudy"}|\text{cloudy}] = 0.8$ $P[\text{"fair"}|\text{wet}] = 0.1$
$P[\text{"fair"}|\text{cloudy}] = 0.1$ $P[\text{"cloudy"}|\text{wet}] = 0.3$
$P[\text{"wet"}|\text{wet}] = 0.6$

What are the Bayesian adjusted weather probabilities $P[\text{fair}|\text{"fair"}]$, $P[\text{cloudy}|\text{"cloudy"}]$, and $P[\text{wet}|\text{"wet"}]$? Overall, what fraction of her forecasts are correct?

3.84 A vacation resort has a lip-synch contest in which staff members perform seven songs and a panel of judges and audience applause determine the rankings of these seven performances. Before the show, each member of the audience predicts the songs that will finish first, second, and third. Those whose predictions turn out to be correct win a bottle of champagne. If you pick three of the seven songs randomly, what is the probability that you will win a bottle of champagne? What is the probability that you will pick the top three songs, not necessarily in the right order?

3.85 It has been estimated that 5 percent of all adult males have coronary heart disease and that 47 percent of those with coronary heart disease have diagonal earlobe creases whereas 30 percent of those with no heart disease have earlobe creases. If so, what is the probability that an adult male with an earlobe crease has heart disease?

3.86 Scholars are uncertain whether a certain letter was written by A or by B, but they believe that the probability that it was written by A is .6. A historian now identifies a key phrase in this disputed letter that was used in 40 percent of the letters written by A and in only 20 percent of the letters written by B. Use Bayes' rule to compute the revised probability that this letter was written by A.

3.87 Explain the error in this argument: Almost no automobile accidents involve people driving 100 mph; therefore it is safest to drive 100 mph.

3.88 Strings of Christmas tree lights used to be designed so that if any bulb failed, none of the lights would light. If each bulb has a .05 probability of failing and failures are independent, what is the probability that a string of 12 bulbs will light properly? That a string of 24 bulbs will light properly?

3.89 A bank's 200,000 outstanding loans have been divided into the categories shown as follows. What is the probability that a randomly selected loan is **a.** for more than $1 million? **b.** for more than $1 million if it is high risk? **c.** high risk if it is for more than $1 million?

	Size of Loan	
Risk	$1 Million or Less	More Than $1 Million
Low	22,731	64,708
High	97,311	15,250

3.90 An analyst believes that there is a .05 probability than a certain ten-year junk bond will default during the first two years after it is issued, a .25 probability that it will default between the third and fifth year (if it hasn't defaulted during the first two years), and a .10 probability that it will default between the fifth and tenth year (if it hasn't defaulted before the fifth year). What is the analyst's probability that this bond will default sometime before maturity?

3.91 The chairman of the Los Angeles City/ County AIDS Task Force wrote:

> Several studies of sexual partners of people infected with the [AIDS] virus show that a single act of unprotected vaginal intercourse has a surprisingly low risk of infecting the uninfected partner—perhaps 1 in 100 to 1 in 1,000. For an average, consider the risk to be 1 in 500. . . . Statistically, 500 acts of intercourse with one infected partner or 100 acts with five different infected partners lead to a 100 percent probability of infection.[23]

What is wrong with his statistical calculation? What is the correct probability?

3.92 There are six boxes, five empty and one containing a check for $10,000. Two contestants will alternate choosing boxes until one selects the box with the $10,000 check. After a box had been opened, it is discarded. Is there any advantage to choosing first?

3.93 A fast-food chain operates 130 restaurants in Oregon and Washington, many of which are located near major highways:

	Near Highway	Not Near Highway
Oregon	42	22
Washington	48	18

If one of its restaurants is selected at random, what is the probability that this restaurant is **a.** In Washington? **b.** near a major highway? **c.** in Washington if it is near a major highway? **d.** in Washington or near a major highway?

3.94 Each week, three people with losing Montana state lottery tickets are allowed to spin a wheel with 100 slots, one of which pays a $1 million jackpot. After 222 tries in 1988 and 1989 in which no one won the $1 million jackpot, Montana state officials made a second slot worth $1 million. If each slot is equally likely to be chosen, what is the probability of 222 spins without a winner of 1 if 100 slots is a winner? If 2 of 100 slots are winners?

3.95 A marketing test of a new portable computer indicates that 10 percent of those who receive promotional material in the mail will attend a free demonstration of this computer and that of those who attend the demonstration 40 percent will purchase the computer. If so, what is the probability that a person receiving the promotional material will purchase the computer? What is the probability that of ten people who receive promotional material, none will make a purchase?

3.96 Medical tests are imperfect due not only to equipment limitations but also to errors in interpreting the test results. In one classic study an x-ray reader looking for signs of tuberculosis (TB) in 1,820 patients misread the x-rays for 26.7 percent of the patients with tuberculosis and for 2.8 percent of those without tuberculosis.[24] Using these error rates, if a patient has a .001 probability of TB before the x-ray, what is the revised probability of TB after the x-ray has been diagnosed as showing TB?

3.97 A 1968 story in a Denver newspaper argued that women are better drivers than men. Among the evidence cited: "Of 101 drivers involved in an accident while passing on a curve, 15 were women."[25] Explain why this evidence is not persuasive.

3.98 In Maleland, each couple has children until a son is born and then has no more children. What is probability that the number of children in a family will be 1, 2, 3, . . . , n? What fraction of the children can we expect to be male?

3.99 A family has two cars. Car A has a 0.002 probability of not starting (quitting); car B has a 0.004 probability of not starting. **a.** If the Bernoulli trial model applies, what is the probability that car A will start successfully 6,000 times in a row over a five-year period? **b.** What is the expected number of starts before car A quits? **c.** Car A and car B are started equally often. If a car quits, what is the probability that it is car A? **d.** Why might the Bernoulli trial assumptions be violated?

3.100 Smith is learning to play squash. Andrabi has set up an April 1 match in which Smith will be given a prize if he can win two consecutive games in a three-game match against Andrabi and Ernst alternately, either Andrabi-Ernst-Andrabi or Ernst-Andrabi-Ernst. Assume that Andrabi is a better player than Ernst and that Smith's chances of winning a game against either player are independent of the order in which the games are played and the outcomes of other games. Which sequence should Smith choose and why?

4

Probability Distributions

The Fairly Intelligent Fly
A large spider in an old house built a beautiful web in which to catch flies. Every time a fly landed on the web and was entangled in it the spider devoured him, so that when another fly came along he would think the web was a safe and quiet place in which to rest. One day a fairly intelligent fly buzzed around above the web so long without lighting that the spider appeared and said, "Come on down." But the fly was too clever for him and said, "I never light where I don't see other flies and I don't see any other flies in your house." So he flew away until he came to a place where there were a great many other flies. He was about to settle down among them when a bee buzzed up and said. "Hold it, stupid, that's flypaper. All those flies are trapped." "Don't be silly," said the fly, "they're dancing." So he settled down and became stuck to the flypaper with all the other flies. Moral: There is no safety in numbers, or in anything else.

James Thurber

TOPICS

4.1   THE IDEA OF A RANDOM VARIABLE

We have now been through the rules and tricks used to figure probabilities. You have learned how to calculate the probability of drawing four straight aces from a deck of cards, the probability of rolling at least one 6 with four dice, and the probabilities of even more complex outcomes. In many of these cases, numerical values are associated with the outcomes; for instance, the *number* of aces selected in four draws and the *number* of sixes obtained in four rolls. Sometimes a numerical monetary payoff may depend on uncertain events: the returns from buying stocks, drilling for oil, expanding a business, or cutting prices. Sometimes the numerical values are not monetary: the number of boys in a family of four, the life expectancy of a 25-year-old woman, the temperature in Dallas on January 1, and the proximity of a missile to its target. If, as in these many cases, there are numbers associated with the uncertain outcomes, we can call these assigned values a random variable.

> A **random variable** is a variable whose numerical value is determined by chance. More formally, it is a function that takes on unique numerical values determined by the outcome of an uncertain situation.

For example, a flipped coin could land heads or tails. We can define a random variable by assigning the value 1 if heads and 0 if tails. Alternatively, we could define a different random variable for the same sample space by assigning the value $+1$ if heads and -1 if tails.

Similarly, for the tossing of two coins simultaneously, we can label the number of heads a random variable X. This X is a function that can take on three different values: $X = 0$, $X = 1$, or $X = 2$, and X is random because we are not certain which particular value will occur in any particular flip of two coins. In this chapter and the next, we will only consider discrete random variables, which can take on a finite or countable number of distinct values; in Chapter 6, we will look at continuous random variables.

Probability Distributions

We may be able to assign probabilities to the possible outcomes. In our coin-flipping example, if heads and tails are equally likely, then

Outcome	Number of Heads X	Probability P
TT	0	.25
HT or TH	1	.50
HH	2	.25

Such a listing of the probabilities of the different possible values of X is called a probability distribution.

A **probability distribution** $p[x]$ for a discrete random variable X gives the probabilities of each possible value x.

Notice that the random variable itself is capital X, while the small x denotes a particular value of X. The phrase *probability distribution* indicates that we are telling how the probabilities (which add up to 1) are distributed among the possible outcomes. Here,

$$p[0] = .25$$

$$p[1] = .50$$

$$p[2] = .25$$

Another example can be found in Chapter 3, when we looked at all of the possible sums in the rolling of two dice. These outcomes and probabilities are reproduced in Table 4.1. The dice sum is a random variable, in that X can take on eleven different values and we do not know which particular value will occur. The listing in Table 4.1 of x and $p[x]$ for the roll of two dice is one way to show a probability distribution. A probability distribution also can be displayed visually, as shown in Figure 4.1.

The concept of a probability distribution is a valuable mathematical tool that can be used in theoretical work to define various concepts rigorously and to prove some important theorems. A graph of probability distribution can be useful in showing us at a glance the range and relative likelihoods of the possible outcomes. Figure 4.1 tells us very forcefully that 7 is the most likely outcome and that the probabilities decline symmetrically as X moves away from 7 toward 2 and 12.

Table 4.1 *Outcomes and Probabilities for the Roll of Two Dice*

Sum of Dice *x*	Probability $p[x]$
2	1/36
3	2/36
4	3/36
5	4/36
6	5/36
7	6/36
8	5/36
9	4/36
10	3/36
11	2/36
12	1/36

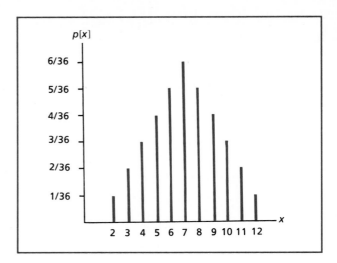

Figure 4.1 *Probability Distribution for the Roll of Two Dice*

Household Size

Now let's do a somewhat different example. The 1980 Census found that 217,482,000 U.S. citizens live in 79,108,000 households and that 8,958,000 U.S. citizens live in institutions and other group quarters.* Table 4.2 shows the distribution of households of different sizes.

Table 4.2   *Size of U.S. Households, 1980*

Size x	Number of Households	Fraction of Total Households p[x]
1 person	17,816,000	.225
2 persons	24,734,000	.313
3 persons	13,845,000	.175
4 persons	12,470,000	.158
5 persons	5,996,000	.076
6 persons	2,499,000	.032
7 or more	1,748,000	.022
Total	79,108,000	1.000

Source: U.S. Bureau of the Census, Current Population Survey, in *Statistical Abstract of the United States* (Washington, D.C.: U.S. Government Printing Office, 1981).

*The Census Bureau defines a *household* as persons who occupy a house, apartment, or other separate living quarters. One of the tests in determining a household is that the occupants do not live and eat with other persons in the same structure and that there are complete kitchen facilities for the exclusive use of the occupants. People who are not in households live in *group quarters* including rest homes, rooming houses, military barracks, jails, and college dormitories.

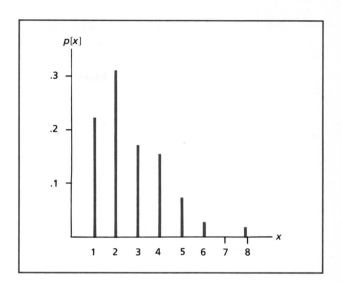

Figure 4.2 *Probability Distribution of U.S. Household Size, 1980*

These data show that 22.5 percent of all U.S. households contain just one person; 31.3 percent are two-person households, and so on. Now, if we were to pick a household at random, what is the probability that we would pick a one-person household? If each household, large or small, is equally likely to be picked, then there is a .225 probability of picking a one-person household.

Objectivists advise us to be careful in making such probability statements. They would wince if I said that there is a .225 probability that I live in a one-person household. I, in fact, live in a four-person household. There is zero probability that I am a one-person household. Instead, we should say that, if a household is selected at random, then there is a .225 probability of selecting a one-person household. With our words chosen carefully, we can then interpret household size as a random variable and the relative frequencies given in Table 4.2 as probabilities. Figure 4.2 shows this probability distribution.

We can readily see in this figure that most households contain from one to four persons, and that two is the most common size. We can also see at a glance that the probabilities steadily decline as we move away from a two-person household. Figure 4.2 also illustrates a commonplace aggregation problem. The category "7 or more" aggregates data for households where $X = 7, 8, 9, \ldots$, and so it would be misleading to plot this probability at the point $X = 7$. Instead, I've put this .022 probability at $X = 7.9$, which is the average size of households with seven or more persons.

EXERCISES

4.1 You are about to roll a normal six-sided die. Let the random variable X be the number that will come up. Make a table like Table 4.1 showing the possible values of X and their respective probabilities.

Now graph this probability distribution, making a figure comparable to Figure 4.1.

4.2 Mammoth Oil is considering an exploratory drilling in Cucamonga, California. The senior engineer in charge of this project says that the company is as likely to find significant oil and gas there as it is to draw the queen of spades out of a normal deck of cards. We can define the random variable X as follows: $X = 1$ if the drilling is successful and $X = 0$ if it isn't. What is the probability distribution of X?

4.2   EXPECTED VALUE, STANDARD DEVIATION, AND SKEWNESS

A complete probability distribution can be listed in a table or pictured in a graph. Sometimes, a few simple numbers can effectively summarize the most relevant characteristics of a probability distribution. The most popular are the expected value, standard deviation, and skewness coefficient. First, we will define these three measures and see how they describe a probability distribution. Then we'll look at several applications.

The Expected Value

You have seen how probabilities can be interpreted as long-run frequencies. For instance, if the probability of heads is .5, then heads should come up about half the time in a large number of flips. A similar logic applies to the example of the number of heads X obtained in two coin flips. The probabilities are $P[0] = .25$, $P[1] = .5$, and $P[2] = .25$; if we were to flip two coins many, many times, we could anticipate obtaining $X = 0$ about one-fourth of the time, $X = 1$ one-half of the time, and $X = 2$ one-fourth of the time. The average value of X would be 1. This predicted long-range average is called the *expected value*.

The **expected value** (or *mean*) of a discrete random variable X is

$$\mu = E[X] = \Sigma x p[x] \tag{4.1}$$

where the notation Σ means that we multiply each different possible value of x by its associated probability $p[x]$ and then add up these products $xp[x]$. The Greek symbol μ is pronounced "mu."

The expected value is a weighted average of the possible outcomes, with the probability weights reflecting how likely each outcome is. Thus, the expected value should be interpreted as what the long-run average value of X will be, if the frequency with which each outcome occurs is in accordance with its probability.

In our two-coin example, the expected value of X is

$$\mu = \Sigma x p[x]$$
$$= 0p[X = 0] + 1\,p[X = 1] + 2\,p[X = 2]$$
$$= 0(.25) + 1(.50) + 2(.25)$$
$$= 1$$

If two fair coins are flipped n times and the outcomes are in accordance with the probabilities, then "no heads" will occur $n/4$ times, "one head" will occur $n/2$ times, and "two heads" will occur $n/4$ times. The average value of X will be

$$\mu = \frac{0(n/4) + 1(n/2) + 2(n/4)}{n}$$

$$= 0(1/4) + 1(1/2) + 2(1/4)$$

$$= 1.0$$

To reiterate, because it is very important, this expected value calculation shows us that if we were to flip a pair of coins a great many times, carefully record the results, and calculate the average number of heads, then this average would be 1.0 if heads and tails come up as often as their probabilities say they should.

To make the example more sporting, imagine that, in return for a small participation fee, your statistics instructor offers to pay you a dollar for every head obtained when two coins are flipped. Imagine that you can play as many times as you want. In the long run, what is your predicted average payoff per game? This is the question that an expected value calculation answers. If heads and tails come up in accordance with their probabilities, you will average a $1 payoff per game. If your instructor charges you $1.25 per game to play, you can anticipate going broke in the long run.

For a second example, let's return to the roll of two dice, with the probabilities shown in Table 4.1. The possible outcomes are $X = 2, 3, 4, \ldots, 12$. We want to calculate an average of these possibilities, using the respective probabilities as weights.

$$\mu = 2(1/36) + 3(2/36) + 4(3/36) + \cdots + 12(1/36)$$

$$= 252/36$$

$$= 7$$

This calculation shows that if we roll a pair of dice many, many times, the average sum will be 7 if each number comes up as often as anticipated.

Imagine a casino game in which the bettor is paid a number of dollars equal to the numbers obtained when a pair of dice are rolled. In the long run, the average payoff per roll should be very close to $7. If it costs $7.50 to play this game, the bettor can anticipate losing a considerable amount of money in the long run. On any one play, the bettor may win up to $4.50 or lose up to $5.50. But, in the long run, the losses should average 50 cents per game.

Let's do one more example. In Table 4.2, we used the random variable X to represent the size of a U.S. household selected at random. What is the expected value of X? It is a calculation of the average size of X, using the respective probabilities as weights:

$$\mu = 1(.225) + 2(.313) + 3(.175) + \cdots + 7.9(.022)$$

$$= 2.75$$

The press reports this calculation as, "The average U.S. household contains 2.75 people." The casual reader then snickers that the typical U.S. household must contain some fractured people. This weak joke is a deliberate misinterpretation of expected value. The expected value of X is not the most likely or the most typical value of X. It is the long-run average value of X, if we repeatedly select households at random. Some households have fewer than 2.75 people; some have more. The average of these different household sizes is 2.75.

This 2.75 expected value is a summary figure for a complicated probability distribution that may be useful in comparing household sizes for different countries or for different time periods in the United States. For instance, the U.S. Census Bureau has reported the following average household sizes:

Census Year	Average Household Size μ
1950	3.37
1960	3.33
1970	3.14
1980	2.75

These figures chronicle a decline in the average size of U.S. households. To describe this decline more fully, we would want to look at the detailed probability distributions and some additional data. Are there now more single-person households? Are families now having fewer children? Are there now more older citizens who typically live in small households? The expected value does not answer these questions, but it does suggest that there are some interesting questions that can be asked.

The Standard Deviation

The expected value is a popular summary measure that may be useful in describing a probability distribution. Another favorite is the *standard deviation* σ (pronounced "sigma").

The **variance of a discrete random variable** X is

$$\sigma^2 = E[(X - \mu)^x] = \Sigma(x - \mu)^2 p[x] \tag{4.2}$$

The **standard deviation** σ is the square root of the variance.

An alternative computational formula for the variance is

$$\sigma^2 = E[X^2] - \mu^2$$

the expected value of X^2 minus the square of the expected value of X.

The interpretation of the variance is best understood by dissecting Equation 4.2. The mean is the expected value of X. The variance is the expected value of $(x - \mu)^2$; that is, the long-run average value of the squared deviations of the possible

Basketball Fouls

Basketball shots, called field goals, count for 2 or 3 points, depending on the player's distance from the basket. A basketball player who is fouled while attempting to make a field goal is allowed to shoot free-throws that count for 1 point each. If the field goal is made despite the foul, the player gets one shot; prior to the 1990–1991 basketball season, two free-throws were awarded if the fouled field goal shot missed. In 1990–1991 the NCAA changed this rule, allowing three free-throws if the missed field goal is a 3-point, rather than a 2-point, shot.

Three-point shots are generally taken by a team's best shooter. A 1990 study of 12 top NBA players found that they made 36.4 percent of their 3-point shots and 79.9 percent of their free-throws.[1] Suppose that time is running out and a team that is behind by 2 or 3 points passes the ball to its best shooter, who prepares to attempt a 3-point shot. Should the defense intentionally foul the shooter, preventing a 3-point shot but giving the player three free-throws? The probability of making all three free-throws is given by the multiplication rule, assuming independence: $.799^3 =$

.510 is considerably higher than the assumed .364 probability of making a 3-point shot.

The expected value of a 3-point shot is an average of the two possible outcomes (3 points or none) weighted by the respective probabilities (.364 and $1 - .364 = .636$):

3-point shot:
$$\mu = (3)(.364) + (0)(.636) = 1.092$$

Each free-throw has two possible outcomes, 1 point or none, with respective probabilities of .799 and $1 - .799 = .201$. The expected value of a free-throw attempt is

free-throw:
$$\mu = (1)(.799) + (0)(.201) = 0.799$$

The expected value of three free-throws is $3(0.799) = 2.397$.

Fouling a 3-point shooter increases the player's chances of making 3 points on the play, gives an expected value of 2.397 points, and creates a possibility of 4 points (if the player makes the 3-point shot plus a free-throw). This researcher's advice: Don't foul someone attempting a 3-point shot.

outcomes from the expected value μ. For each possible outcome, we calculate how far x is from the expected value μ, and then we square this deviation. By squaring, we do two things. First, we eliminate the distinction between positive and negative deviations. What matters to a variance calculation is how far the outcome is from μ, not whether it is above or below μ. Second, squaring gives primary importance to large deviations. One deviation of 10 squared is as large as four deviations of 5 squared.

After calculating the squared deviations from μ for each possible outcome, we then calculate the average squared deviation using the outcome probabilities as weights. This weighted average of the squared deviations from μ is the variance, and its square root is the standard deviation. The standard deviation measures the dispersion in the possible outcomes and gauges how "spread out" a probability distribution is. A distribution with a small standard deviation is very compact, while a distribution with a large standard deviation is dispersed.

Let's return again to the two-dice example in Table 4.1. The expected value is $\mu = 7$. Table 4.3 shows the squared deviations about this expected value. Notice that the squared deviation at $X = 7$ is zero. Notice also that the squared deviations are the same as we move in either direction away from the mean of 7, illustrating the point that the variance treats positive and negative deviations equally. Finally, notice how the squared deviations escalate as the outcomes get farther away from $X = 7$.

The variance is the probability-weighted average of these squared deviations:

$$\sigma^2 = 25(1/36) + 16(2/36) + 9(3/36) + \cdots + 25(1/36)$$
$$= 210/36$$
$$= 5.83$$

The standard deviation is the square root of the variance,

$$\sigma = \sqrt{5.83} = 2.415$$

To show how the standard deviation is related to the shape of a probability distribution, let's compare some distributions. First, imagine a pair of perfectly gimmicked dice that always come up 7. The probability distribution is shown in Figure 4.3. The expected value and variance are

$$\mu = 7(1) = 7$$
$$\sigma^2 = (7 - 7)^2(1) = 0$$

Table 4.3   *Squared Deviations about $\mu = 7$ for the Roll of Two Dice*

Outcome x	Squared Deviation $(x - \mu)^2$	Probability $p[x]$
2	25	1/36
3	16	2/36
4	9	3/36
5	4	4/36
6	1	5/36
7	0	6/36
8	1	5/36
9	4	4/36
10	9	3/36
11	16	2/36
12	25	1/36

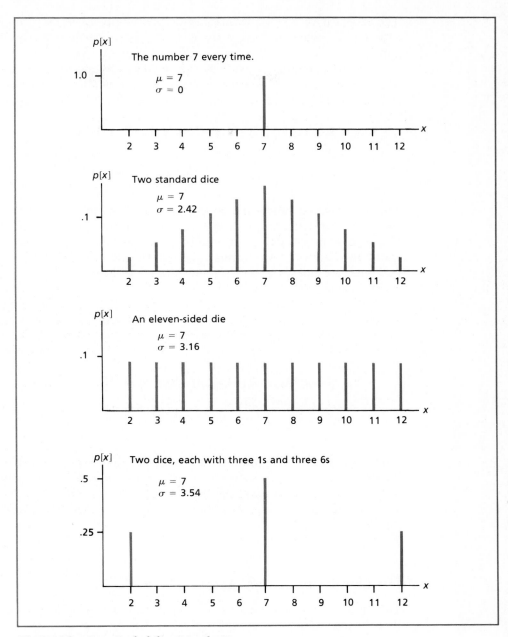

Figure 4.3 *Four Probability Distributions*

The expected value is 7 because this is the only possible outcome. The variance and standard deviation are zero because there is no uncertainty about the outcome.

Now let's imagine a single eleven-sided die, constructed so that the eleven numbers 2, 3, 4, . . . , 12 are all equally likely. This probability distribution is also shown in Figure 4.3. The expected value, variance, and standard deviation are

$$\mu = 2(1/11) + 3(1/11) + 4(1/11) + \cdots + 12(1/11) = 7$$

$$\sigma^2 = (2 - 7)^2(1/11) + (3 - 7)^2(1/11) + \cdots + (12 - 7)^2(1/11) = 10$$

$$\sigma = \sqrt{10} = 3.16$$

This eleven-sided die has the same expected value as the sum of two standard six-sided dice, but has a larger variance and standard deviation. Figure 4.3 shows that this larger standard deviation reflects the greater dispersion in the probability distribution. There is more uncertainty with the eleven-sided die because all eleven outcomes are equally likely. As compared to the sum of two standard dice, there is a higher probability of outcomes far from 7.

For a final example, consider two six-sided dice, each with three 1s and three 6s stamped on the faces. Now there are only three possible sums: 2, 7, and 12. This probability distribution is included in Figure 4.3, too. The expected value, variance, and standard deviation are

$$\mu = 2(9/36) + 7(18/36) + 12(9/36) = 7$$

$$\sigma^2 = (2 - 7)^2(9/36) + (7 - 7)^2(18/36) + (12 - 7)^2(9/36) = 12.5$$

$$\sigma = \sqrt{12.5} = 3.54$$

Again, the mean is 7. Notice that even though there is a .5 probability of rolling a 7, still the standard deviation is the highest of any of the distributions shown in Figure 4.3. The standard deviation is high because there is a .5 probability of a 2 or a 12, and the squaring of deviations emphasizes outcomes far from the mean.

The Coefficient of Skewness

The mean measures the center of a distribution and the standard deviation measures its dispersion.

The **coefficient of skewness** γ (pronounced "gamma") is a measure of the symmetry of the outcomes:

$$\gamma = \frac{E[(X - \mu)^3]}{\sigma^3} \tag{4.3}$$

Here, we calculate the average value of the cubed deviations from the expected value and, to provide a uniform scale, divide this average cubed deviation by the standard deviation cubed.

The cube of a positive deviation from the mean is positive, while the cube of a negative deviation from the mean is negative. Thus, the average cubed deviation could turn out to be either positive or negative. If the outcomes are symmetrical about the mean, so that the right half of the distribution is a mirror image of the left half, then the positive and negative deviations are equally balanced and the average cubed deviation will be zero. If, on the other hand, the outcomes are asymmetrical, then the coefficient of skewness may be positive or negative. The cubing makes large deviations in either direction very important in this calculation. If the probability distribution looks unbalanced, then the coefficient of skewness is probably not zero. Two simpler measures are Pearson's coefficient of skewness

$$\frac{(\text{mean} - \text{mode})}{\text{standard deviation}}$$

and, in a similar spirit,

$$\frac{3(\text{mean} - \text{median})}{\text{standard deviation}}$$

All of the dice examples in Figures 4.1 and 4.3 are, in fact, symmetrical with a zero coefficient of skewness. But look at the distribution of household sizes in Figure 4.2. This probability distribution is clearly not symmetrical about the mean μ = 2.75. Instead, household size has a high probability of being slightly less than 2.75 and a small probability of being much greater than 2.75. There is some chance of large positive deviations, and so we label this distribution *positively skewed*. The coefficient of skewness does, in fact, work out to be positive, $\gamma = 1.04$.

Figure 4.4 contrasts positively and negatively skewed distributions. The easiest

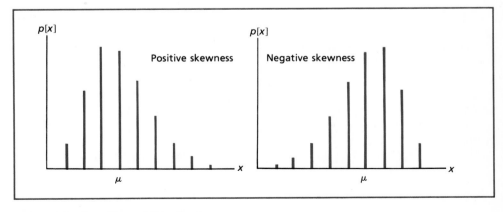

Figure 4.4 *Two Skewed Distributions*

way to remember the distinction is that a positively skewed distribution diminishes or "runs off" to the right (there is some chance of outcomes much higher than the mean), while a negatively skewed distribution "runs off" to the left.

Together, these three calculations—the mean, standard deviation, and skewness coefficient—often paint an accurate mental picture of a probability distribution. The mean tells us the location of the distribution; the standard deviation tells us how dispersed the distribution is; and the skewness coefficient tells us whether or not the distribution is symmetrical.

EXERCISES

4.3 In Exercise 4.1, you found the probability distribution for the number X obtained from the roll of a single die. **a.** What is the expected value of X? **b.** Is this expected value also the most likely value of X? **c.** What does this expected value tell us?

4.4 After the Internal Revenue Service (IRS) feeds a tax return into its main computer in West Virginia, a top secret formula estimates the probability that an audit will turn up an additional tax assessment. The local IRS decision to audit a return depends not only on this estimated probability, but also on estimates of the potential size of the additional taxes and the amount of time an agent will have to spend on the audit. Consider two equally time-consuming audits, one with a .80 probability of recovering an additional $5000 in taxes and one with a .60 probability of recovering $9000. Which has the higher expected value?

4.5 An investor is considering two mutual funds. Based on past performance, the following probabilities are assigned to the possible percentage rates of return:

First Fund		Second Fund	
Return	Probability	Return	Probability
−10	.10	−15	.20
0	.10	5	.20
10	.60	15	.20
20	.10	25	.20
30	.10	45	.20

a. Graph these two probability distributions, the first on the top half of a sheet of paper and the second on the bottom half. **b.** Which fund seems to have the higher expected return and which has the higher standard deviation? **c.** Calculate the means and standard deviations, and see if you are right.

4.6 Exercise 3.32 specified four inflation-unemployment scenarios and the assumed probabilities of each. Table 4.4 gives the anticipated revenues for each scenario. What is the expected value of revenue? Is the expected value also the most likely value? Is this probability distribution positively or negatively skewed?

• 4.7 The average income of 25,000 people who live in a mythical small country is nearly $50,000. There are 1,000 millionaires and then there are the 24,000 servants who earn $10,000 apiece. Given these data, the mean is an incomplete description of the income distribution. The standard deviation and skewness coefficient might help. Do you think that the standard deviation will be large or small? Do you think that the skewness coefficient will be positive, negative, or close to zero? Calculate their values and see if you are right.

Table 4.4 *Scenarios for Exercise 4.6*

Scenario	Inflation	Unemployment	Probability	Revenue (Millions $)
1	high	high	.16	2.0
2	high	low	.24	4.0
3	low	high	.36	1.0
4	low	low	.24	3.0

4.3 USING EXPECTED VALUES

The expected value undoubtedly is the most widely used summary of a probability distribution. The variety of applications can be appreciated best through several examples.

State Lotteries

Lotteries (or the casting of lots) were often used in ancient times to make equitable decisions. During a famine the Lydian king Atys had his people draw lots to decide who would stay with him and who would leave to look for food elsewhere. In Athens public officials were chosen by lottery to avoid political dynasties. The Romans used lots to select the person who would determine the sentence of criminals.

Lots were also used to predict the future, by allowing gods to communicate through the lottery. In the Bible, Jonah is thrown overboard when a casting of lots indicates that the storm threatening the ship is due to God's anger with him. In the popular short story "The Lottery," written by Shirley Jackson in 1948, a small New England town has a mysterious lottery; read the story and learn why.

In 1826 Thomas Jefferson proposed using a lottery to sell off much of his land. Since few people had enough cash to buy his property, a lottery gave everyone an opportunity to acquire the land and gave him a fair price. He argued that lotteries could be used as a voluntary tax, a relatively painless way for governments to raise revenue. Lotteries are now an important source of revenue for many governments throughout the world, though some criticize the official sanctioning of gambling and others argue that lotteries are a regressive tax, paid disproportionately by those with relatively low incomes.

To illustrate the use of probabilities, consider a stylized state lottery that costs $1 to enter and offers the chance of winning $50,000. To win, one need only correctly predict the five-digit number that the state will select. This scheme is a legal numbers game with a very large payoff and an even smaller chance of winning. Allowing for zeroes, there are 10 equally likely possibilities for each of the 5 digits. Therefore, the probability of winning is $(1/10)^5 = .00001$. The expected payoff from a $1 ticket is

$$\$50{,}000(.00001) + \$0(.99999) = \$0.50$$

This is, in fact, roughly the expected payoff from most state lotteries. The habitual player can, in the long run, anticipate getting 50 cents back for every dollar spent on tickets. The state treasury will keep, on average, 50 cents out of every dollar taken in, a whopping 50 percent take.

Many state lottery games have a variety of prizes, but most adhere, one way or another, to roughly a 50 percent payout target. For instance, a Massachusetts state lottery game is called Blackjack, the name of a popular card game in which the house take is very slight and experts can even beat the house by using probability theory and card counting. The Massachusetts lottery game is quite different. For a dollar, the player buys a ticket showing two concealed cards in the "dealer's hand" and three concealed cards in "your hand." The player wins if the sum of these three cards is larger than the sum of the dealer's two cards (with no penalty for exceeding 21). The name Blackjack and the three cards versus two are all intended to convey the impression that the game is easy to win. Of course, the "cards" are not randomly selected, but instead carefully chosen to make a lot of money for the state.

Beat the Dealer by	Prize x	Probability p	Product $(x)(p)$
Nine	$1000	.0000342	.0342
Eight	500	.0000684	.0342
Seven	100	.0004286	.0429
Six	50	.0012480	.0624
Five	20	.0033333	.0667
Four	10	.0066667	.0667
Three	5	.0166667	.0833
Two	2	.0700000	.1400
One	1	.1031000	.1031
			.6335

The expected payout of $0.6335 is better than most state lotteries, but still represents nearly a 40 percent take for the house. (The $1 prize here is actually just a free ticket that gives the player another chance to lose.)

One of California's lottery games includes the gimmick that 20 out of every 12,500 $100 winners are randomly selected to play the "Big Spin," a weekly televised event in which these 20 people spin a Grand Prize wheel with 100 equal slots, divided as follows:

Number of Slots	Prize
40	$10,000
30	$50,000
20	$100,000
10	$2,000,000

The Value of a Lottery Jackpot

In May 1984 the headlines read, "Thousands Seek Millions as Jackpot Fever Grips N.Y.," a fever that has broken out several times since.[2] The May 1984 story concerned a New York Lotto drawing with a $22.1 million jackpot that lured people into hour-long lines, beginning at 5:00 in the morning, betting more than a million dollars an hour—a total of $24.1 million.

In the New York Lotto game a machine randomly chooses 6 out of 44 Ping–Pong balls that are numbered 1 to 44. The jackpot is divided among those who correctly predict all six numbers (not necessarily in order). The probability of picking the winning number is given by the multiplication rule. The probability that the first ball will be one of the six numbers selected is 6/44. Given that the first number is a winner, the probability that the second ball (out of 43 remaining) will be one of the five other numbers selected is 5/43. Continuing, the probability of matching all six numbers is

$$\left(\frac{6}{44}\right)\left(\frac{5}{43}\right)\left(\frac{4}{42}\right)\left(\frac{3}{41}\right)\left(\frac{2}{40}\right)\left(\frac{1}{39}\right)$$
$$= .000000142$$

or about 1 in 7 million. The size of the jackpot depends on the amount wagered, with New York keeping about 60 percent of the amount wagered and paying out 40 percent in prizes. For the May 12, 1984, drawing, there had been no winner for three straight games and the state accumulated the prize money, expanding the jackpot to $22.1 million.

There turned out to be four winning tickets, with each winner receiving $22,100,000/4 = $5,525,000—not all at once, but in 21 annual installments of $5,525,000/21 = $263,095. A payment of $263,095 in 20 years is not as valuable as $263,095 now, since money received now can be invested for 20 years. At a 10 percent interest rate, $39,107 now will grow to $263,095 in 20 years. Thus $39,107 is the present value of $263,095 paid 20 years later.

The total present value of the 21 annual installments paid to these New York lottery winners is, at a 10 percent interest rate, $2.5 million—only 45 percent of the reported value of $5.5 million! The New York state lottery not only keeps 60 percent of the amount wagered but spreads the jackpots out so that the present value of the payoff is less than half its reported value. No wonder governments like lotteries.

The $2,000,000 prize is spread over 20 years and as explained in the accompanying box is, at a 10 percent interest rate, worth much less than $2,000,000—here only $936,492. The expected value of the prize from spinning the Grand Prize wheel is

$$\$10,000(.4) + \$50,000(.3) + \$100,000(.2) + \$936,492(.1) = \$122,649$$

Thus a person who has a $100 winning ticket receives $100 plus a 20/12,500 chance of spinning the wheel, which has an expected value of $122,649. The expected value of the $100 ticket is consequently

$$\$100 + \$122,649(20/12,500) = \$296.24$$

In 1985, a Californian holding a $100 lottery ticket placed an advertisement in the *Los Angeles Times* and sold his ticket for $3,500, explaining, "There's a fool born every minute, and some fools are bigger than others."[3]

Is a New Statistics Book a Good Investment?

A certain publisher is considering bringing out a new statistics book. This book will sell for $35, with the publisher receiving $25, the bookstore $6, and the author $4. Here are the demand and cost figures:*

Number of Copies Sold	Probability	Production Cost
5,000	.20	$275,000
10,000	.40	$300,000
20,000	.25	$350,000
30,000	.10	$400,000
40,000	.05	$450,000

What is expected value of profits (receipts minus production costs)?

The random variable, profits, is equal to the number of copies sold times $25, minus the production cost. If, for example, 5,000 copies are sold, then the revenue will be 5,000($25) = $125,000; the production cost will be $275,000; and there will be a net loss of $150,000. Here are all of the possible profit outcomes:

Sales	Profits x	Probability p	(x)(p)
5,000	−$150,000	.20	−$30,000
10,000	−$50,000	.40	−$20,000
20,000	$150,000	.25	$37,500
30,000	$350,000	.10	$35,000
40,000	$550,000	.05	$27,500
Total		1.00	$50,000

As shown, the expected value of profits is $\mu = \Sigma xp[x] = \$50,000$. The probabilities are such that this sort of book will lose money 60 percent of the time, but there is a small chance of considerable profits. The $50,000 expected value calculation tells us that if the publisher keeps bringing out books with comparable chances of success, then it can anticipate an average profit of $50,000 per book (with the occasional big success more than offsetting the frequent small failures).

---

*For the sake of simplicity, we'll assume there are only five possible outcomes. In a more practical sense, we can think of each outcome as a rough composite of similar but slightly different results.

Maxwell House's Price of Coffee

Maxwell House, a division of General Foods, is the nation's largest coffee producer. It acquires coffee beans from coffee-growing nations, grinds and roasts these beans, and sells them—either as ground or as instant coffee. For many years, ground coffee came in a can that had to be opened with a key, like a sardine can. In October 1962, Folger's began marketing a coffee can that could be opened quickly without a key. A year later, Maxwell House decided that it, too, should sell coffee in a keyless can.

The keyless can was slightly more expensive to manufacture, but consumers seemed willing to pay for this extra convenience. The extra cost to Maxwell House worked out to about 0.7 cent per pound of coffee. The company considered: (a) holding the price steady and absorbing this added cost or (b) raising the price per pound by 2 cents to cover the extra cost and then some. In Table 1 probabilities were assigned as to how profits would be affected by each strategy.[4]

If the price is held constant, the expected value of the effect of the keyless can on profits is

$$\mu = (\$4.1)(.20) \\ + (-\$0.6)(.50) + (-\$0.8)(.20) \\ + (-\$1.2)(.10) \\ = \$0.24 \text{ million}$$

The anticipated enthusiastic response by consumers more than makes up for the fact that the keyless cans cost more to produce.

A higher price would dampen consumer enthusiasm somewhat. On balance, their estimates imply an expected value with a 2-cents-per-pound price increase of

$$\mu = (\$11.9)(.25) + (\$6.5)(.25) \\ + (\$2.9)(.25) + (-\$1.1)(.25) \\ = \$5.05 \text{ million}$$

Not surprisingly, Maxwell House decided to introduce the keyless can and raise the price of its coffee by 2 cents a pound.

Table 1 *Effect of Keyless Can on Maxwell House Profits*

Hold Price Constant			Raise Price 2 Cents		
Market Share	Profits (Millions)	Probability	Market Share	Profits (Millions)	Probability
+2.8%	+$4.1	.20	+2.5%	+$11.9	.25
+1.0%	−$0.6	.50	+1.0%	+$6.5	.25
0.0%	−$0.8	.20	0.0%	+$2.9	.25
−0.6%	−$1.2	.10	−1.5%	−$1.1	.25

EXERCISES

4.8 Chisler Corporation provides a three-year warranty on the cars it sells. Suppose they anticipate the following typical repair record:

Warranty Repair	Cost to Chisler	Probability
None	$0	.25
Minor	$200	.50
Moderate	$500	.20
Major	$1000	.05

What is the expected value of these warranty repairs to Chisler? What use might they make of this expected value calculation?

4.9 A successful door-to-door vacuum cleaner salesman says that he works 12 to 14 hours a day, six days a week, talking to about 60 potential customers a day, of whom 2 allow him in to make a 90-minute sales pitch (in return for a hand-held vacuum that costs the salesman $20), and 1 buys a $600 vacuum (for which he earns $75 to $150, depending on the extras purchased).[5] If with each potential customer the probability of a sale is 1/60, what is the probability that he will have to try more than 60 potential customers before making a sale? More than 360? If the average commission on a sale is $100, what is the expected profit with each potential customer? The average profit per day? The average profit per week?

4.10 A standard slot machine has three reels. The first reel has 3 walnuts, 7 cherries, 3 oranges, 5 lemons, 1 bell, and 1 jackpot bar. The second reel has 7 cherries, 6 oranges, 1 lemon, 3 bells, and 3 bars. The third reel has 4 walnuts, 7 oranges, 5 lemons, 3 bells, and 1 bar. The payoffs are

Reel 1	Reel 2	Reel 3	Payoff in Coins
bar	bar	bar	85
bell	bell	bar	18
bell	bell	bell	18
lemon	lemon	bar	14
lemon	lemon	lemon	14
orange	orange	bar	10
orange	orange	orange	10

Reel 1	Reel 2	Reel 3	Payoff in Coins
cherry	cherry	bell	5
cherry	cherry	walnut	5
cherry	cherry	any other	3

What is the expected payoff if each picture on each reel is equally likely to appear? In the long run, what fraction of every dollar bet will this one-armed bandit keep?

4.11 The *Skimmer's Digest* sweepstakes mails 20,000,000 entry forms, of which

Winnings	Number
$3	12,000
$10	7,900
$500	90
$5,000	4
$10,000	3
$25,000	2
$100,000	1

If you enter this sweepstakes, what is the probability that you will win something? Is your expected return more than the cost of mailing a first class letter? If only 1 out of every 20 people actually enter the sweepstakes, what is the expected cost to *Skimmer's Digest* in prize money? What does this expected cost figure mean?

4.12 An April 28, 1986, article in the *Wall Street Journal* said that state-operated lotteries offer "such lousy odds and payouts that professional gamblers usually disdain them as sucker bets." A responding letter to the editor noted that 85 percent of the money spent on state lottery tickets is returned to the public: 50 percent in prizes and 35 percent for education and other programs. (The remaining 15 percent covers operating expenses.) In addition, the letter writer argued that a gambling casino's 95 percent payout applies only to a single wager. Fourteen wagers with a 95 percent payout return only 49 percent, less than a lottery's 50 percent payout in prizes. If you were an editor of the *Wall Street Journal*, how would you refute these arguments?

4.4 BETTING ODDS AND PROBABILITIES (OPTIONAL)

If we know both the possible payoffs and their respective probabilities, then we can calculate the expected payoff. Sometimes we can work the calculation in the other direction. We may know the payoffs and have an estimate of the expected payoff. If so, these data can be used to estimate some probabilities.

For example, in 1974 London bookies were offering 3 to 1 odds on a wager that Richard Nixon would finish his four-year term as president. We can use these quoted odds to estimate the bookies' subjective probability in 1974 that Nixon would finish his term.* Odds of 3 to 1 mean that, for every £1 wagered by the bettor that Nixon would finish, the bookie wagered £3 that he wouldn't, with the winner taking all. On a £1 wager, the bettor either loses £1 or is paid £4, for a £3 gain. If the probability of Nixon finishing his term is P, then the bettor's expected gain is

$$E[X] = (+£3)(P) + (-£1)(1 - P)$$
$$= 4P - 1$$

If the bets were "fair," so that the bookie just expects to break even in the long run, then the expected gain would be zero:

$$0 = E[X]$$
$$= 4P - 1$$

implying

$$P = 1/4$$

that in 1974 the probability of Nixon finishing his second term was perceived to be about 1/4.

Of course, bookies expect to do better than just break even. They are in business to make money. If they set the odds at 3 to 1, then they must have believed that Nixon had less than a 25 percent chance of finishing his term. How much less? We can reason that the odds offered are intended to give the bookie a profit large enough to cover expenses, and yet small enough to attract bettors. The bookies' average profit is generally in the range of 5 percent to 20 percent of the amount wagered, implying an expected return of $-.05$ to $-.20$ for the bettor. These profit margins give a probability range

5% profit: $-.05 = E[X] = 4P - 1$ implies $P = .24$

20% profit: $-.20 = E[X] = 4P - 1$ implies $P = .20$

---

*In quoting odds the bookie's stake is mentioned first. Thus odds of X to Y mean that the bettor can win X units by risking Y units.

Apparently the bookies felt that Nixon's chances of completing his term were in the range of .20 to .24.

Nixon's probability of survival was, of course, a very subjective probability that varied considerably from person to person. It is this subjective variation that fuels betting. If one person believes that there is a .2 probability of some event occurring and someone else believes that the probability is .3, then there is room for a bet in which each side believes its expected return is positive.* It is into this void that the bookie steps, offering bets to either or both parties. We can think of the bookies' odds as reflecting an expert or consensus opinion of the disputed probability. Those who disagree with this consensus will be tempted to back their belief with a wager.

Pari-Mutuel Bets

When a bookie offers 3 to 1 odds that Nixon will complete his term, the bookie is offering to bet against people who are more optimistic about Nixon's chances. The bookie is gambling, with no guarantee of a profit either on this bet or in the long run. The bookie counts on some expertise in assessing probabilities, and then sets the odds so as to leave some margin for error and still attract customers.

Horse race odds offer an interesting contrast in that they are explicitly set to reflect a consensus probability and, at the same time, guarantee a fixed percentage return to the track. The vehicle is the *pari-mutuel system* thought up by a French chemist, Pierre Oller, in 1872. The basic pari-mutuel ("wager-among-us-all") idea is to put the wagered money into a pot, take out a percentage for the track and the government, and divide up what's left among those who correctly picked the winner.

Table 4.5 gives a detailed example. A total of $117,647 is wagered on the winner of this race. Horse 1 is the favorite, and horse 6 has the fewest dollars bet on it. The track takes 15 percent of $117,647 (which comes to $17,647) out of the pool, conveniently leaving $100,000 to be paid out to the winning bettors.

If horse 1 wins, then $100,000 will be returned to the people who bet $32,937 on this horse. This payoff is $3.08 for every $1 bet. The track rounds this $3.08 down to $3.00—a practice known, in an interesting choice of words, as "breakage." A payoff of $3 on a $1 bet is a net gain of $2 and represents 2 to 1 odds. Similarly, if horse 6 surprises most bettors and wins, $100,000 will be paid to the people who bet $5,824 on this longshot. This payoff is $17.17 on every $1 bet, rounded down to $17.10, and reported as odds of 16.1 to 1. The payoff odds for each of the six horses is shown in Table 4.5.

Table 4.5 also shows, from the bettor's viewpoint, how high the subjective probability of each horse winning must be to make a wager on that horse a fair bet

---

*The mutually agreeable bet might be at 3 to 1 odds. The person with the .3 probability will put up the pound note, betting that the event will occur. The person with the .2 probability will bet £3 that the event doesn't occur. Each has a positive expected return:

$$P = .3: \quad E[X] = (\pounds 3)(.3) + (-\pounds 1)(.7) = \pounds 0.2$$
$$P = .2: \quad E[X] = (\pounds 1)(.8) + (-\pounds 3)(.2) = \pounds 0.2$$

Table 4.5 *Some Horse Race Odds*

Horse	Dollars Wagered	Odds	Probability for $E[X] = 0$
1	$32,468	2.0–1	.333
2	$31,915	2.1–1	.323
3	$21,052	3.7–1	.213
4	$14,022	6.1–1	.141
5	$12,366	7.0–1	.125
6	$5,824	16.1–1	.058
Total	$117,647		1.193

($E[X] = 0$). These probabilities sum to more than 1.0 because these are not fair bets; the track extracts 15 percent plus breakage. Nonetheless, each horse attracts the wagers of some optimistic bettors. As in the Nixon example, we can consider the consensus probabilities to be somewhat less than the fair-game probabilities implied by the betting odds. Optimists with subjective probabilities that are higher than the implied fair-game probabilities may be tempted to bet.

There is another way to look at this column of fair-game probabilities in Table 4.5. These are the fractions of a dollar one would have to wager on each horse to ensure a $1 payoff, no matter which horse wins. In total, one would have to bet $1.193 to be sure to get back $1. The track would make a percentage profit of .193/ 1.193 = .162, which is 15 percent plus breakage.

More Sports Bets

The pari-mutuel system provides the sure profit that bookies covet. Bookies do not like to gamble! They much prefer to be intermediaries, reaping a steady profit as brokers between gamblers with dissimilar subjective probabilities. However, for many bets, it is awkward to set up a formal pari-mutuel system.

For most sporting events, the bookie offers odds, often days or even weeks in advance, and then accepts bets as they trickle in. The conservative bookie would like to offer the odds that would be determined by a pari-mutuel system but is forced to guess what those odds would be. If the bookie guesses wrong, too much money will come in on one team and the bookie will either have to gamble, pass the bets on to another bookie, or adjust the odds. Unfortunately for the bookie, the traditional rules of the game are that the adjusted odds only apply to new bets and not to wagers made at odds offered earlier.

For many sports bets, bookies offer standard odds of 10 to 11 or 5 to 6, while using a point spread to try to elicit equal dollar bets on each team. For example, in the January 1987 Super Bowl, the New York Giants were offered as 6 1/2 point favorites over the Denver Broncos, meaning that 6 1/2 points would be added to Denver's score for betting purposes. Bettors could pick either team at, depending

The Rich Get Richer

Is there any advantage, in a fair game, to being rich or being poor? If the game is fair, then both sides have an expected gain of zero, but the player who is poor is much more likely to be the one who goes home broke. Let's say that two women, R and P, decide to play poker until one is bankrupt. These two women are equally skilled, but R has $90,000 while P has only $10,000. What is the probability that R will win?

This is a tricky problem, of a type that intrigued many of the greatest minds in the history of probability analysis. We could try to figure out all the possible hands and bets and the various twists and turns that this poker showdown might take, but there is a much simpler route. If they are equally skilled, then this is a fair game with a zero expected gain to each woman. When it is all over, R will either have won $10,000 or lost $90,000. A zero expected gain implies that

her probability of winning must be .9:

$$(\$10,000)(.9) + (-\$90,000)(.1) = 0$$

Similarly, with a .1 chance of winning, P has a zero expected gain, too:

$$(\$90,000)(.1) + (-\$10,000)(.9) = 0$$

The rich woman is much more likely to win but, of course, she will lose much more money if she doesn't win.

In general, in any fair game, a person with a certain percent of the total capital at stake has that same percent chance of winning. Here, R puts up 90 percent of the capital and has a 90 percent chance of winning. Although this precise rule only applies to a fair game, the moral can be extended to all sorts of situations, including a bettor trying to break the bank at Monte Carlo and two competitive firms trying to drive each other out of business.

on the bookie, 10 to 11 or 5 to 6 odds. Those who bet on New York win their bets if New York wins the game by at least 7 points. Those who bet on Denver win if Denver wins or New York wins by less than 7 points. (Thus the lament, "It's not whether your team wins or loses, but whether it beats the point spread.")

Let's say that, as an academic probability exercise, it is one day before the Super Bowl and you are considering betting on one of the teams that, you believe, has a probability P of winning (taking into account the point spread). With 10 to 11 odds, your expected return,

$$(\$10)(P) + (-\$11)(1 - P) = -\$11 + \$21P$$

will be positive only if

$$P > 11/21 = .524$$

If, taking into account the point spread, you believe that one team's probability of winning is greater than .524, then you might be tempted to bet. With 5 to 6 odds, a positive expected return requires one team to have a .545 probability of winning.

Because the oddsmakers want to equalize the number of dollars bet on each team, the point spread is set so that most people believe that the game is essentially a toss-up. There will be some people who believe that one team has more than a .524 (or .545) probability of winning. There will be other people who think that the second team has more than a .524 (or .545) probability of winning. If the point spread works as intended, an equal number of dollars will be bet on each team. With 10 to 11 odds, the contented bookie will take in $11 in bets on each team ($22 in all) and pay out $21 no matter who wins, for a safe 4.5 percent profit on every dollar wagered. With 5 to 6 odds, equally divided bets give the bookie a $1 profit on every $12 bet, an 8.3 percent take.

EXERCISES

4.13 In the game Hazard, the player rolls three standard dice, trying to get three of a kind. The bettor can wager on a particular three of a kind, such as three 6s, or can just bet that some three of a kind will be rolled. The payoffs are 180 to 1 for the first type of bet and 30 to 1 for the second type. What are the bettor's expected percentage returns for these two bets?

4.14 On three different occasions, the betting odds on the great racehorse Man O'War were 1 to 100. For such a bet to have a positive expected return,

how high would your probability of Man O'War's winning have to be?

4.15 With 5 to 6 betting odds on a football game, what probability of winning is necessary for a bettor to have a positive expected return?

4.16 (Gambler's Ruin) Moe and Joe agree to bet a dollar on a series of coin flips until one of them is bankrupt. Moe has $1,000 and Joe has $4,000. What is Moe's probability of winning?

4.5   INSURANCE (OPTIONAL)

Insurance is another area in which probabilities and expected values are very important. The premiums that companies charge for life, house, car, boat, business, and other insurance policies are naturally influenced by the probabilities of having to make a payoff. Companies want to charge more to insure the sickly than to insure the healthy. They want to charge more to insure a rundown wooden house than to insure a new brick house. They want to charge accident-prone drivers more than safe drivers.

Interestingly, this wasn't always so. Early life insurance policies seem to have largely ignored the life-expectancy data collected by statisticians. For example, at least as long ago as the 1500s, the British government sold life annuities, which paid a fixed annual amount over the life of the purchaser. But, up until 1789, the price of life annuities did not depend on the age of the purchaser! These life annuities were unattractive to the elderly and a bargain for the young. It is not surprising that life annuities were purchased mainly by the young and that the British government paid dearly for its neglect of life-expectancy probabilities.

Modern insurance companies pay close attention to probabilities. Within their legal constraints, they try to charge their high-risk customers more and charge everyone enough to make a generous profit.

Let's look at a specific example. In 1983, a $100,000 one-year, life insurance policy for a 25-year-old woman cost about $120. In 1980, the U.S. Census Bureau calculated that there were 2 million 25-year-old women in the country and that 1,300 died that year. If we use this observed frequency as an estimate of the probability P of a 25-year-old woman dying within a year, then

$$P = \frac{1,300}{2,000,000} = .00065$$

The expected value of the policy's payoff is consequently

$$(\$100,000)(.00065) = \$65$$

as compared to a cost of $120.

On any single policy, the insurance company will collect $120 and pay out either nothing or $100,000. On the average, if a fraction .00065 of the insured 25-year-old women die, the company will collect $120 and pay out $65 per policy. Roughly half of the insurance premiums collected will be paid back out to beneficiaries and half will be retained by the company to cover salaries, office buildings, and profits. This 50 percent benefit-to-premium ratio is representative of many insurance policies.

Table 4.6 shows some comparable insurance premium data for women and men of various ages. The data in this table are the typical costs in 1983, by age and sex, of a one-year $100,000 life insurance policy for someone who passes a mandatory medical exam. If we assume a 50 percent benefit-to-premium ratio, then we can calculate the implied probability P of death within a year for someone who passes such a test. Notice that men pay more than women and that the old pay much more than the young. There are actually even more detailed risk distinctions in insurance rates; for example, most companies have lower rates for nonsmokers than for smokers. Modern insurance companies do indeed use probabilities (and implicitly, expected values) to set their premiums.

Table 4.6   *The Structure of Life Insurance Premiums*

Age	Female		Male	
	Premium	*Implicit P*	*Premium*	*Implicit P*
25	$120	.0006	$120	.0006
35	$120	.0006	$120	.0006
45	$132	.0007	$165	.0009
55	$205	.0010	$315	.0016
65	$475	.0024	$870	.0044

Of course, insurance companies cannot sort out high-risk and low-risk people perfectly. They consequently must offer the same premiums to people of varying risks. In general, those who are above-average risks will be tempted to buy lots of insurance, while low-risk people will buy less or else buy no insurance at all. Economists call this sorting "adverse selection." Insurance companies try to protect themselves from adverse selection by considering such risk indicators as smoking habits, by requiring health examinations, and by leaving a generous profit cushion in their premium calculations.

EXERCISES

4.17 Here are some recently observed Swedish mortality rates (percent of those of the specified age dying during the year):[6]

Age	Male	Female
20	0.095%	0.036%
40	0.221	0.117
60	1.427	0.681
80	9.774	6.282
100	51.999	46.756

What prices would you have to charge for $100,000 one-year life insurance policies so that the expected value of each payout is equal to half of the price?

4.18 Under legislative and judicial pressures, life insurance companies have been moving toward "unisex" rates, in which males and females pay equal premiums. Who gains from a unisex premium structure: men or women? Why?

4.19 A land developer mails out 100,000 "Urgent Insta-Grams," announcing that the recipient has won a prize that can be claimed by visiting the developer's property and listening to a sales talk about "an exciting year-round vacation idea." The 100,000 awards consist of one check for $1,000, 2 color televisions costing $500 apiece, 4 black-and-white televisions costing $100 apiece, 4 ten-speed bicycles costing $100 apiece, and 99,989 35-millimeter cameras (cardboard, with fixed focus) costing $1 apiece. What is the expected value of claiming a prize? If 99 percent of the recipients throw these Urgent Insta-Grams into the trash and 1 percent claim their prizes, what is the developer's expected cost for the prizes?

• **4.20** In a laboratory experiment, a group of fish were offered a choice between A and B. The experiment was set up so that option A would give a reward 70 percent of the time and no reward 30 percent of the time. Option B was constructed to give a reward 30 percent of the time and no reward 70 percent of the time. The fish soon began choosing option A 70 percent of the time and option B 30 percent of the time. When a group of rats was offered the same sort of options, the rats soon began choosing option A every time.[7]

Did the fish or the rats behave more intelligently?

4.6 EXPECTED VALUE MAXIMIZATION (OPTIONAL)

The costs and benefits of most of the important decisions we make are not known with complete certainty. How, then, can we rationally make decisions? The expected values of the costs and benefits seem like a good place to start. If we can assign probabilities to the possible costs and benefits, then we can calculate their expected values. If the expected value of the benefits exceeds the expected value of the costs, then we should, averaged over many decisions, come out ahead.

An expected value calculation was, in fact, the earliest proposed criterion for rational decision making under uncertainty. According to the expected value rule, you should always choose the course of action with the highest expected net gain. An expected value criterion certainly seems appropriate when the uncertain situation will be repeated over and over again. It makes good sense to look at the long-run average when there is a long run to average over. Gambling casinos, state lotteries, and insurance companies are good examples. They will try to avoid offering gambles that give them a negative expected return because these would almost certainly lose money in the long run.

It is equally clear, however, that the expected return criterion is sometimes inappropriate. State lotteries are designed to have a positive expected return for the state, and they must therefore have a negative expected return for the people who buy the lottery tickets. Their gain is your loss. The people who buy lottery tickets are not maximizing their expected returns. Insurance is another example. Insurance rates are set to give insurance companies a positive expected return and, hence, to give insurance buyers a negative expected return. People who buy insurance must not be maximizing expected returns either. Stock selection provides yet another example. If you wanted to maximize your expected return, you would invest all of your funds in the stock that offers the highest expected return. Why, then, do investors instead buy dozens of securities?

We can also construct hypothetical situations to show that expected return maximization is not always appealing. Consider, for example, this pleasant scenario. A messenger from a rich recluse stops you after class and announces that you have been chosen to receive a gift of $1 million. He pulls out a thick wad of crisp $100 bills to show you that he is serious. As you blink and reach for them, the messenger asks if you would prefer a coin flip to make things more sporting: heads, you receive $2.1 million; tails, you get nothing but a good story. Think about it. If, like most people, you would take the sure million, then you are not maximizing expected return.

The St. Petersburg Paradox

Another counterexample is the famous *St. Petersburg Paradox* that was concocted by the Swiss mathematician Nicolas Bernoulli in the eighteenth century:

> Peter repeatedly tosses a coin until it comes up heads. He agrees to give you $1 if he gets a head on the very first throw, $2 if he gets the first head on the second throw, $4 if he gets it on the third throw, $8 if on the fourth throw, and so on, so that the number of dollars won doubles with each additional throw.

How much would you pay for an opportunity to play this game? Think about it for a minute, before we work out the arithmetic for the expected return.

Your possible returns are $1, $2, $4, $8, . . . with respective probabilities 1/2, 1/4, 1/8, 1/16, ... Your expected return is

$$E[X] = (\$1)(1/2) + (\$2)(1/4) + (\$4)(1/8) + (\$8)(1/16) + \cdots$$
$$= 1/2 + 1/2 + 1/2 + 1/2 + \cdots$$
$$= \infty$$

Therefore, if you are a dedicated expected-return maximizer, you should be willing to pay an unlimited amount for an opportunity to play this game. As few people are willing to pay more than a few dollars to play the St. Petersburg game, most people must be thinking about something other than a simple expected return maximization.

Risk Aversion

The primary inadequacy of expected return maximization is that it neglects **risk**—how certain or uncertain a situation is. An expected-return maximizer is indifferent given the choice between $1 million, a 50 percent chance at $2 million, or a 1 percent chance at $100 million because all three options have the same expected return. In an expected return calculation, it doesn't matter that in the first case there is a 100 percent chance of receiving $1 million and in the third case there is a 99 percent chance of getting nothing. But these differences do matter to many people. Diverse individuals have different attitudes toward risk, which they reveal by the decisions they make.

> **Risk-neutral:** Indifferent toward a fair bet. Five dollars with certainty and a 50 percent chance at $10 are equally attractive. Tries to maximize expected return. Does not buy insurance or lottery tickets because both have negative expected returns.
>
> **Risk-averse:** Turns down fair bets. Prefers $5 to a 50 percent chance at $10. Will sacrifice expected return to reduce risk. Buys insurance but not lottery tickets.
>
> **Risk-seeking:** Accepts fair bets. Prefers a 50 percent chance at $10 to a sure $5. Will sacrifice expected return to increase risk. Buys lottery tickets but not insurance.

This acknowledged variety in preferences explains why some people accept gambles that others shun and why some people buy insurance that others avoid. An individual's risk preferences may also depend on the particular situation. Milton Friedman, a famous economist, and Leonard Savage, a celebrated statistician, argued that the following tendencies are commonplace:[8]

1. Risk-averse toward substantial losses. Will buy insurance to protect against large reductions in living standards.
2. Risk-neutral toward small gambles.
3. Risk-averse toward moderate gains. Prefers a sure $5,000 to a 50 percent chance at $10,000.
4. Risk-seeking toward large, long-shot gains that would dramatically improve living standards. Will buy lottery tickets.
5. Risk-averse toward astounding gains. Prefers $1 million to a 50 percent chance at $2 million. Prefers two tickets on a $1 million lottery to one ticket on a $2 million lottery.

EXERCISES

4.21 On a football betting card, the bettor picks the winners of n selected games and is paid a specified amount if all n picks win. A point spread is used to equalize the games. We'll assume here that the point spread exactly equalizes each game, so that the bettor must choose between two equally likely winners. You can spend $\$x$ on any of these three cards and be paid the indicated amount: **a.** $\$6x$ for correctly picking 3 out of 3 games **b.** $\$11x$ for correctly picking 4 out of 4 games **c.** $\$16x$ for correctly picking 5 out of 5 games. What is the expected payoff on each of these three cards?

4.22 *(continuation)* Does this structure of payoffs on football betting cards suggest something about bettors' preferences?

4.23 Ken Harrelson, a professional baseball player, once offered the Cleveland Indians General Manager, Gabe Paul, a St. Petersburg deal. Harrelson offered to play an entire baseball season without a salary, except for 50 cents doubled for every home run he hit; his salary would be 50 cents if he only hit 1 home run, $1 if he hit 2 home runs, $2 if he hit 3 home runs, and so on. Gabe Paul turned him down and Harrelson then went out and hit 30 home runs that season. If Harrelson's offer had been accepted, how much would Cleveland have owed him?

• 4.24 Peter has offered to let you play the St. Petersburg paradox game, but the maximum payoff is some amount 2^m. What is your expected payoff?

4.7   THE STANDARD DEVIATION AS A MEASURE OF UNCERTAINTY (OPTIONAL)

Drawing on their statistical backgrounds, Harry Markowitz and James Tobin developed **mean-variance analysis** in the 1950s. Their objective was to find a useful and reasonable improvement on expected return maximization. Markowitz and Tobin proposed, as a plausible approximation, that people base their decisions under uncertainty on just two factors—the mean and the variance. The *mean* is just another word for the expected value and, as we have seen, it was long ago proposed that expected values influence decisions. The problem with an expected value criterion, as we have also seen, is that it ignores risk. Markowitz and Tobin suggested extending the expected value criterion by using the variance to measure risk.

The variance (or standard deviation) measures the amount of uncertainty because it gauges the probability of outcomes far from the expected return. If widely varying outcomes are probable, then the variance will be high. If there is little difference among the possible outcomes, then the variance will be low.

Four Examples

Consider the following four alternatives:

1. $5 with certainty
2. $10 with probability .5
   $0 with probability .5
3. $5 with probability .5
   $10 with probability .25
   $0 with probability .25

4. $ 5 with probability .5
   $105 with probability .25
   −$95 with probability .25

In mean-variance analysis, these four alternatives are evaluated by comparing their means and variances. To illustrate the calculations, let me show you that alternative 3 has an expected return

$$\mu = \$5(.5) + \$10(.25) + \$0(.25) = \$5$$

and a variance

$$\sigma^2 = (\$5 - \$5)^2(.5) + (\$10 - \$5)^2(.25) + (\$0 - \$5)^2(.25) = 12.50$$

The standard deviation is just the square root of the variance,

$$\sigma = \$3.54$$

In a similar manner, the expected returns, variances, and standard deviations for all four of the alternatives have been calculated and are found in Table 4.7.

The four alternatives were constructed to have the same $5 expected return. On the average, they should all do equally well in the long run, but as one-shot deals, they have widely varying degrees of risk or uncertainty. This risk could be gauged by either the variance or the standard deviation; the former is just the square of the latter. The alternative with the larger standard deviation also has the larger variance. It is just a matter of taste whether one measures risk by the standard deviation or the variance. The standard deviation easily wins this taste test. One reason is that its units are dollars, as are the units of the expected return, while the units for the variance are dollars squared. Another reason is that, as we will see in later chapters, many statistical measures and tests are based on the standard devia-

Table 4.7   *Four Expected Returns*

Alternative	Expected Return μ	Variance σ^2	Standard Deviation σ
1. $P[\$5] = 1.0$	$5	0	$0
2. $P[\$0] = .5$			
$P[\$10] = .5$	$5	25	$5
3. $P[\$5] = .5$			
$P[\$0] = .25$	$5	12.50	$3.54
$P[\$10] = .25$			
4. $P[\$5] = .5$			
$P[\$105] = .25$	$5	5000	$70.71
$P[-\$95] = .25$			

tion. By the time this course is over, you will be thinking in terms of "two standard deviations from the mean," but we're getting ahead of ourselves. The point here is just that we could use either the standard deviation or the variance to measure risk and uncertainty, and the standard deviation is the preferred choice. In fact, despite its name, the graphical and numerical analytics of mean-variance analysis are actually done in terms of the mean and the standard deviation!

Let's go back to Table 4.7 and look at the standard deviation to gauge the riskiness of these four alternatives. The first alternative has a standard deviation of zero, reflecting the fact that there is no uncertainty about its return. The payoff is sure to be $5. The second alternative has an uncertain return and, consequently, a nonzero standard deviation. Alternative 3 is also uncertain, but has a somewhat smaller standard deviation than alternative 2. Alternative 3 has a larger chance of returns that differ from the mean and this is what the standard deviation and variance measure. Alternative 4 has the largest standard deviation because there is a substantial chance of returns that are far from the mean.

According to mean-variance analysis, people will take both the expected return and the standard deviation into account when making decisions. The four alternatives shown in Table 4.7 have the same expected return, but quite different standard deviations. A risk-averse person will choose 1; a risk-seeker will select 4; and a risk-neutral person would be happy with any of these four alternatives.

Long Shots and Skewness

People seem to be mostly risk averse. Risk aversion is consistent with the personal introspection of cowardly professors and the observed behavior of decision makers. The most important qualification is that long shots have an undeniable appeal. For example, many people buy stock in young companies, even when the expected return is very low, hoping to catch the start of another IBM or Xerox.

One inadequacy of the standard deviation or variance as a measure of risk is the symmetrical treatment of gains and losses. The variance simply squares the deviations, positive or negative, about the mean. It does not matter if these deviations are pleasant or unpleasant surprises. Of course, they can't all be one or the other. By the definition of the expected return, if there is some chance of a pleasant surprise, then there is also some chance of an unpleasant one. These surprises need not be evenly distributed, however. There might be a large probability of a slightly above-average return and a small chance of a catastrophic loss, or there might be a large probability of a small loss and a small chance of a bonanza. The variance and standard deviation cannot distinguish between such differences. Consider these two assets:

Asset	Probability	Return	Expected Return	Standard Deviation
1	.001	$9,990	$0	$316.07
	.999	−$10		
2	.001	−$9,990	$0	$316.07
	.999	$10		

Risk Has Its Rewards

If investors are risk-averse, then they will shun risky investments—unless the price is right, that is, unless they are compensated for that risk with a relatively high anticipated return. A high return cannot, of course, be guaranteed for a risky investment because, by its definition, a risky investment has a very uncertain return. However, a return that has a high standard deviation can also be priced low enough to have a high expected value. And if investors are predominantly risk averse, high standard deviations should be associated with high expected returns.

A study of annual returns over the period 1926 through 1988 obtained the results shown in Table 1.[9] These are historical realized returns, how people did rather than how well they thought they would do. Nonetheless, the results are an intriguing confirmation of our common sense. The safest asset, short-term government bonds, has also been the least rewarding, while the riskiest, common stock, has done the best on average. Stocks went up 54% one year and down 43% another year. They have been a risky, but also a rewarding, investment.

Table 1

| | Returns | |
Asset	Mean	Standard Deviation
Common stocks	12.1%	21.0%
Long-term corporate bonds	4.8%	8.5%
Long-term government bonds	5.4%	8.5%
Short-term government bonds	3.5%	3.4%

By mean-variance criteria, there is no difference between these assets and investors should be indifferent to a choice between them. In practice, many people prefer Asset 1 to Asset 2.

Both of these assets are very asymmetrically *skewed* gambles. Asset 1 is a long shot in that there is a small chance of a large gain. Asset 2 is a "negative" long shot with a small chance of a large loss. We can gauge this asymmetry by calculating the coefficient of skewness:

$$\gamma = [E(X - \mu)^3]/\sigma^3$$

Asset 1, with its large positive deviation, has a positive skewness coefficient:

$$\gamma = \frac{(\$9,990 - \$0)^3(.001) + (-\$10 - \$0)^3(.999)}{(\$316.07)^3} = 31.6$$

Asset 2, with its large negative deviation, has a negative skewness coefficient:

$$\gamma = \frac{(-\$9,990 - \$0)^3(.001) + (\$10 - \$0)^3(.999)}{(\$316.07)^3} = -31.6$$

The purchase of a lottery ticket is a positively skewed gamble, like Asset 1, but it has a negative expected return. A decision not to purchase fire insurance is a negatively skewed gamble, like Asset 2, but it has a positive expected return. According to mean-variance analysis, a risk-seeker might buy a lottery ticket and spurn insurance. A risk-averter might avoid lottery tickets and buy insurance. But mean-variance analysis alone cannot explain the simultaneous purchase of both lottery tickets and fire insurance without resorting to such "irrational" explanations as a zest for the suspense of gambling or a mistaken assessment of the probabilities (what Adam Smith called "absurd presumptions in their own good fortune"[10]).

Skewness may be the rational explanation. The popularity of lottery tickets and insurance suggests that positive skewness is in fact attractive, so that people like Asset 1 but not Asset 2. As one ticket buyer said, "My chances of winning a million are better than my chances of earning a million."[11] This preference is also suggested by the overbetting on long shots at horse races. (Long shots would yield a negative average return even if the track did not take out 15 percent of the pool.) There have also been a number of empirical studies that provide more formal evidence.[12]

These apparent preferences imply that mean-variance analysis should be used cautiously in situations with highly skewed returns such as lottery tickets and fire insurance. In other situations, such as the purchase of financial assets, the returns may be sufficiently symmetrical for mean-variance analysis to be valuable.

There is a commonplace conflict here between simplicity and generality. The widespread popularity of mean-variance analysis is due to the fact that it is very simple and yet quite powerful. Most of the theoretical and empirical work on portfolio selection and the pricing of financial assets is an outgrowth of mean-variance analysis. If you take a course on security valuation and portfolio theory, you will see just how easy and useful this analysis is. These attractive features are brought out most vividly by a comparison with the alternatives. A simpler approach, such as expected return maximization, cannot handle such important questions as risk aversion. On the other hand, more general approaches are quite cumbersome and yield relatively few interesting conclusions.

EXERCISES

4.25 The standard deviation of the annual returns from 1926 through 1985 was 21.2 percent for all stocks and 36.0 percent for the stocks of small companies.[13] Which would you guess had the higher average return?

4.26 ABC stock has a .5 probability of yielding a 15 percent return and a .5 probability of yielding a 5 percent return. What are its expected return and standard deviation? CAB stock has a .3 probability of yielding a 20 percent return, a .5 probability of yielding 10 percent, and .2 probability of yielding 0 percent. What are its expected return and standard deviation?

4.27 *(continuation)* Which is riskier: ABC stock or CAB stock? Which stock would a risk-neutral

investor buy? Which stock would a risk-seeker buy? Which would a risk-averse investor buy?

● **4.28** *(continuation)* Are the returns from ABC and CAB stock positively skewed, negatively skewed, or symmetrical?

4.8   THE MEAN, STANDARD DEVIATION, AND SKEWNESS OF LINEAR TRANSFORMATIONS (OPTIONAL)

Throughout this course, we will find it useful to describe uncertain situations by referring to the mean, standard deviation, and skewness of a random variable. The mean tells us the long-run average value of the random variable. The standard deviation tells us how likely it is that the outcome will be close to its mean value. The skewness tells us whether or not the outcomes are symmetrically distributed about the mean.

It turns out, though, that our analyses will sometimes require us to look at random variables that are linear transformations of other random variables. For example, we might be interested in the temperature in Dallas on this coming January 1, at kickoff time for the Cotton Bowl. If X is the temperature in degrees Celsius, then $32 + 1.8X$ is the temperature in degrees Fahrenheit. The Fahrenheit temperature is a linear transformation of the Celsius temperature. As another example, we might spend \$5,000 on advertising, trying to sell a computer game for \$40 that we can manufacture for \$5. If the random variable X measures the number of games that we will sell, then $- \$5,000 + \$35X$ will be our profit. As a final example, we might buy 10 shares of MBI stock at \$100 apiece. If the random variable X is the price that we will receive when we sell this stock, then $10(X - \$100) = -\$1,000 + 10X$ is our capital gain.

In each of these examples, we have a pair of random variables that are linearly related to each other. If we know the mean, standard deviation, and skewness of one member of the pair, then we can find the mean, standard deviation, and skewness of the other. Here are the formulas (the derivations are in the study guide by Roy Erickson; see the Preface in this book). If X is a random variable with mean μ, standard deviation σ, and skewness coefficient γ, then the mean, standard deviation, and skewness of a linear transformation $a + bX$ are given by

$$\text{the expected value of } a + bX = a + b\mu \tag{4.4}$$

$$\text{the standard deviation of } a + bX = |b|\sigma \tag{4.5}$$

$$\text{the skewness coefficient of } a + bX = \begin{array}{l} -\gamma \text{ if } b < 0 \\ +\gamma \text{ if } b > 0 \end{array} \tag{4.6}$$

MBI Stock

These formulas make good sense, as a short example will show. Let's use the 10 shares of MBI stock that were purchased for \$100 apiece and assume that the selling price X has a .5 probability of being $120 and a .5 probability of being $100. The expected value of the selling price is

$$\mu = (\$120)(.5) + (\$100)(.5) = \$110$$

The variance is

$$\sigma^2 = (\$120 - \$110)^2(.5) + (\$100 - \$110)^2(.5) = 100$$

and the standard deviation is

$$\sigma = \sqrt{100} = \$10$$

The skewness coefficient is

$$\gamma = [(\$120 - \$110)^3(.5) + (\$100 - \$110)^3(.5)]/\sigma^3 = 0$$

If this situation were somehow to repeat itself over and over again, we would anticipate selling the stock for $120 half of the time and for $100 the remaining half. The mean $\mu = \$110$ gives the anticipated average selling price. The particular situation is, however, undoubtedly unique and will never be repeated exactly. The standard deviation $\sigma = \$10$ measures how likely it is that this unique outcome will be close to the mean of $110. A small standard deviation indicates that the outcome is almost certain to be very close to $110. A large standard deviation indicates that the price is likely to be far from $110. In this simple case here, there are only two possible outcomes, $120 and $100, and each is a 100 squared deviation from $110. Thus, the variance (the average squared deviation) is 100 and the standard deviation (the square root of the variance) is $10. The skewness coefficient is $\gamma = 0$ because the possible outcomes are symmetrical about $110.

Now what about our prospective capital gain $10(X - \$100) = -\$1,000 + 10X$? We could calculate the capital gain for each of the possible selling prices and then compute the mean, standard deviation, and skewness of these possible capital gains. With only two possible outcomes, these calculations would be no big deal. But, in practice, there are usually a great many possible selling prices and such capital gains calculations would be a big deal. The linear transformations, Equations 4.4–4.6, give us an easy alternative. Using $a = -\$1,000$ and $b = 10$, we immediately have

the expected value of $-\$1,000 + 10X = -\$1,000 + 10(\$110) = \$100$

the standard deviation of $-\$1,000 + 10X = 10(\$10) = \$100$

the skewness coefficient of $-\$1,000 + 10X = 0$

As we have 10 shares and the $110 expected value of the selling price is $10 more than we paid, the expected capital gain is $100. This gain is not certain; the average squared deviation is 100^2 and the standard deviation is $100. There is no skewness because the symmetrical selling prices make the possible capital gains symmetrical about $100.

EXERCISES

4.29 Let C be a random variable giving the Celsius temperature in Nowhere, North Dakota at noon on this coming February 8. Exhaustive National Weather Service records indicate that the mean, standard deviation, and skewness coefficient of C are $\mu = -10$, $\sigma = 10$, and $\gamma = 1$. The Fahrenheit temperature at noon on February 8 is given by $F = 32 + 1.8C$. What are the mean, standard deviation, and skewness coefficient of the random variable F?

4.30 I am going to spend $5,000 on advertising, trying to sell a computer game for $40 that I can manufacture for $5. If X is the number of games that I will sell, then $-\$5,000 + \$35X$ is my profit. My subjective probabilities are

x	p
0	.10
50	.40
100	.30
500	.10
1,000	.10

What are the means, standard deviations, and skewness coefficients of sales and profits?

● 4.31 Show that the formulas for the mean, standard deviation, and skewness coefficient of $a + bX$ are as given in Equations 4.4, 4.5, and 4.6.

4.9 SUMMARY

A random variable X can be used to tabulate the possible outcomes and associated probabilities in an uncertain situation. These probabilities $p[x]$ comprise a probability distribution. A graph of a probability distribution can clarify a complicated collection of probabilities.

Together, three calculations—the mean, standard deviation, and skewness coefficient—often paint an accurate picture of a probability distribution. The mean gives the location of the distribution, the standard deviation gauges its dispersion, and the skewness coefficient tells us whether or not the distribution is symmetrical. The formulas are

$$\text{mean:} \quad \mu = E[X] = \Sigma x p[x]$$

$$\text{variance:} \quad \sigma^2 = E[(X - \mu)^2] = \Sigma(x - \mu)^2 p[x]$$

$$\text{standard deviation:} \quad \sigma = \text{square root of variance}$$

$$\text{skewness:} \quad \gamma = \frac{E[(X - \mu)^3]}{\sigma^3}$$

The expected value is used in practice to measure the long-run average value of X; an example is the long-run payoff from a state lottery or the long-run return from a certain investment strategy. The standard deviation often is used to gauge the riskiness of a situation. A large standard deviation indicates a great deal of uncertainty, in that the outcomes are likely to be far from the expected value. The skewness coefficient tells us whether or not the possible outcomes are symmetrical about the mean. A positive skewness coefficient indicates a long-shot situation in which there is a small chance that X will be much larger than its expected value and a large chance that X will be slightly below its expected value.

Sometimes, we want to calculate the mean, standard deviation, or skewness coefficient for a linear transformation of X. If X is a random variable with mean μ, standard deviation σ, and skewness coefficient γ, then the mean, standard deviation, and skewness of $a + bX$ are given by

$$\text{the expected value of } a + bX = a + b\mu$$

$$\text{the standard deviation of } a + bX = |b|\sigma$$

$$\text{the skewness coefficient of } a + bX = +\gamma \text{ if } b > 0$$

$$-\gamma \text{ if } b < 0$$

APPENDIX: BIVARIATE PROBABILITY DISTRIBUTIONS

With two random variables X and Y, the probability that $X = x$ and $Y = y$ is given by the joint (or bivariate) probability distribution $p[x, y] = p[x]p[y|x]$. Suppose, for instance, that X has two possible values, $X = 0$ and $X = 1$, and that Y has two possible values, $Y = 0$ and $Y = 2$. The joint probabilities can be shown in a probability tree:

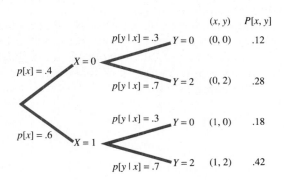

These particular random variables are independent. The probability that $Y = 0$ is equal to .3, regardless of the value of X; the probability that $Y = 2$ is .7, regardless of whether $X = 0$ or $X = 1$.

The joint probabilities $p[x, y]$ can be shown more concisely in the cells of a table:

	$Y = 0$	$Y = 2$	
$X = 0$	.12	.28	$P[X = 0] = .4$
$X = 1$	.18	.42	$P[X = 1] = .6$

$$P[Y = 0] = .3 \quad P[Y = 2] = .7$$

There are two important rules for bivariate random variables:

$$E[a + bX + cY] = a + bE[X] + cE[Y] \tag{4.7}$$

$$\mathrm{Var}[a + bX + cY] = b^2\,\mathrm{Var}[X] + c^2\,\mathrm{Var}[Y] + 2\,\mathrm{Cov}[X, Y] \tag{4.8}$$

where the covariance between two random variables X and Y is defined as

$$\mathrm{Cov}[X, Y] = E[(X - E[X])(Y - E[Y])]$$

The covariance is a measure of the association between X and Y. If the value of X tends to be above its mean when the value of Y is above its mean and to be below its mean when the value of Y is below its mean, then the product $(X - E[X])(Y - E[Y])$ is generally positive, and so is the covariance. If, on the other hand, $X - E[X]$ is likely to be negative when $Y - E[Y]$ is positive and vice versa, then the covariance is negative. If X and Y are independent, then their covariance is zero. (However, the reverse is not true; two variables may have a covariance of zero even if they are not independent.)

Since the covariance of independent random variables is zero, Equation 4.8 implies that the variance of the sum of independent random variables is equal to the sum of their variances: $\mathrm{Var}[X + Y] = \mathrm{Var}[X] + \mathrm{Var}[Y]$. This can be generalized to n independent random variables:

$$\mathrm{Var}\left[\sum_{i=1}^{n} X_i\right] = \sum_{i=1}^{n} \mathrm{Var}[X_i]$$

If all of the X_i have the same variance σ^2, then

$$\mathrm{Var}\left[\sum_{i=1}^{n} X_i\right] = n\sigma^2 \tag{4.9}$$

Equations 4.7, 4.8, and 4.9 can be confirmed by our specific numerical example. The expected values of X and Y are

$$E[X] = \Sigma xp[x] = (0)(.4) + (1)(.6) = 0.6$$

$$E[Y] = \Sigma yp[y] = (0)(.3) + (2)(.7) = 1.4$$

The variances of X and Y are

$$\text{Var}[X] = \Sigma(x - 0.6)^2 p[x] = (0 - 0.6)^2(.4) + (1 - 0.6)^2(.6) = 0.24$$

$$\text{Var}[Y] = \Sigma(y - 1.4)^2 p[y] = (0 - 1.4)^2(.3) + (2 - 1.4)^2(.7) = 0.84$$

The expected value of $X + Y$ is

$$E[X + Y] = \Sigma\Sigma(x + y)p[x, y]$$

$$= (0 + 0)(.12) + (0 + 2)(.28) + (1 + 0)(.18) + (1 + 2)(.42)$$

$$= 2.0$$

$$= E[X] + E[Y]$$

The variance of $X + Y$ is

$$\text{Var}[X + Y] = \Sigma\Sigma\{(x + y) - 2.0\}^2 p[x, y]$$

$$= (0 + 0 - 2)^2(.12) + (0 + 2 - 2)^2(.28) + (1 + 0 - 2)^2(.18)$$

$$\quad + (1 + 2 - 2)^2(.42)$$

$$= 1.08$$

$$= \text{Var}[X] + \text{Var}[Y]$$

REVIEW EXERCISES

4.32 Massive Insurance Liability Company (MILC) is considering insuring Howard Hardsell's voice for $1,000,000. They figure that there is only a .001 probability that they will have to pay off. If they charge $2,000 for this policy, will it have a positive expected value for them?

What is the expected value of this policy to Howard Hardsell? If the expected value is negative, why would he even consider buying such a policy?

4.33 In bridge, the four players are dealt 13-card hands from a standard 52-card deck. A "Yarborough" is a hand containing no honors; not a single jack, queen, king, or ace. What is the probability of being dealt a Yarborough?

4.34 The Yarborough hand is said to be named after Lord Yarborough, who offered 1,000-to-1 odds against a whist opponent being dealt a 13-card hand that did not contain a single ten, jack, queen, king, or ace. What is the probability of being dealt this original Yarborough hand? What is an opponent's expected percentage return from such a wager?

4.35 In May 1987, Kidder, Peabody estimated that Prime Computer, then selling for $27 a share, had a "potential price range" over the next twelve months of $18 to $36. What would be the percentage rate of return to someone buying at $27 if the price goes to $18? If the price goes to $36? What is the expected rate of return if each of the nineteen prices from $18 to

$36 is considered equally likely? (Assume that all prices are rounded off to the nearest dollar.)

• **4.36** Use the data in Table 4.2 to answer the following question: "If we were to select randomly one of the 217,482,000 U.S. citizens living in households, what is the probability that we would select a person who lives in a single-person household?" What is the probability that we would select a person who lives in a two-person household? Graph the complete probability distribution for this random variable (the number of people in the selected person's household).

• **4.37** Label the batting order on a baseball team 1, 2, 3, . . . , 9. If each position is equally likely to be the last batter in any given game, then over the course of a 180-game season, how many more expected times at bat will the #1 hitter have in comparison with #2? In comparison with #3? In comparison with #9?

•• **4.38** A certain weather bureau must forecast "rain" or "no rain" each day. They receive five times as many complaints when they incorrectly forecast "no rain" as they receive when they incorrectly forecast "rain." Nobody complains when their forecast is correct. Each day, they estimate a new probability P of rain. For which values of P should they forecast "rain" if they want to minimize the expected number of complaints?

4.39 A new restaurant will either be successful (and sold after one year for a $2 million profit) or unsuccessful (and liquidated after one year for a $400,000 loss). How high must the probability of success be for this restaurant to have a positive expected value?

4.40 A business is considering building either a one-story or two-story apartment complex. The value of the completed building will depend on the strength of the rental market when construction is finished. Here are the subjective probability distributions for the net profit with each option:

		Profit (Millions $)	
Rental Market	Probability	One-Story	Two-Story
Strong	.2	4	10
Medium	.5	2	3
Weak	.3	−1	−5

Use the mean and standard deviation to compare these two alternatives.

• **4.41** The Italian national lottery, "Lotto," dates back to the 1500s. Every week, five numbers are randomly drawn from 90 tickets bearing the numbers 1, 2, . . . , 90. The bettor may make any of the following wagers. **a.** Specify a number that will be among the five selected: payoff 10.5 to 1. **b.** Specify a number and the order in which it will be drawn: payoff 52.5 to 1. **c.** Specify two numbers that will be among the five selected: payoff 250 to 1. **d.** Specify three numbers that will be among the five selected: payoff 4,250 to 1. **e.** Specify four numbers that will be among the five selected: payoff 80,000 to 1. **f.** Specify all five selected numbers in order: payoff 1,000,000 to 1. What are the expected percentage returns from each of these bets? What fraction, on the average, of every lira wagered will the Italian government keep?

• **4.42** Sean Connery, the British actor who played James Bond so often, won a considerable amount of money at roulette in 1963 when he bet on the number 17 and it came up three times in a row. Joe College, an admirer of James Bond, has an extra $10 and has decided to try to duplicate this feat. He wagers $10 on the number 17. If he wins, he will have $360, which he will then bet on 17 again. If he wins again, he will leave his money on 17 for a third try. He will quit after his third win or if, unexpectedly, he loses at some point. With a 38-number wheel, what is Mr. College's chances of winning three times in a row? How much money will he have if he does win three times in a row? What is the expected value of his net gain?

4.43 A Las Vegas hotel used to advertise a $15 holiday, consisting of one night at their hotel and $10 in free play chips: "Your NET cost is $5." Each person taking advantage of this offer was given 10 special chips that could be bet once. If the person won, the chip was exchanged for $1. The hotel subsequently lost a $1,250,000 lawsuit over this ad. What, do you suppose, was the basis for this suit?

• **4.44** In August 1973, Jimmy the Greek quoted the following odds:

14 to 1 on Henry Aaron hitting a home run in his next time at bat.
4 to 1 on Henry Aaron hitting a home run in the next game he starts.

Simultaneously, Seymour Siwolff calculated that Henry Aaron, on the average, bats 3.24 times per game when he starts. Are Jimmy the Greek's odds consistent with Siwolff's calculation?

• **4.45** At the start of the 1981 NBA playoffs, the following odds of winning the championship were offered:

Celtics	3 to 1
Lakers	4 to 1
Suns	4 to 1
Sixers	5 to 1
Bucks	5 to 1
Spurs	20 to 1
Knicks	35 to 1
Bulls	40 to 1
Blazers	40 to 1
Rockets	100 to 1
Pacers	100 to 1
Kings	100 to 1

For each team, what would the probability of winning have to have been to give a bet a zero expected net return? What is the sum of these twelve implied probabilities? Why is this probability sum not equal to 1.0? To ensure a $100 payoff no matter who wins, how much would you have to had bet on each team? What would be your net percentage gain or loss from such a blanket bet?

4.46 "Program traders" simultaneously buy and sell stocks, stock options, and/or stock futures. When the options or futures expire, these traders unwind their position by liquidating their stock holdings. One study looked at the volume of trading on the New York Stock Exchange during the last hour on Friday afternoon, comparing those days when stock market index options and futures both expired with days when neither expired:[14]

	Expiration Day	No Expiration
Mean volume	31,156	15,959
Standard deviation	11,612	5,011

Based on these data, write a brief paragraph comparing the last hour of trading on expiration and nonexpiration Fridays.

• **4.47** Four quarrelsome brothers have inherited a prosperous farm. Convinced that they cannot work together, they decide to play "survival of the fittest," a vicious game that they have been playing since childhood. Each brother is given $25 at the start of the game. The two youngest brothers then flip coins, betting $1 on each flip, until one has lost his $25 stake. The winner than takes his $50 and flips coins with the next oldest brother until one is eliminated. The winner then has $75 to go up against the oldest brother. Whoever survives this final round gets the farm. Is there any advantage to being the youngest or the oldest brother in this ruthless game?

4.48 An example on page 157 discusses a Maxwell House pricing decision. Calculate the mean, standard deviation, and skewness of the change in profits for each of the two options considered. How would you characterize the choice?

• **4.49** Standardized tests, like the SAT and GMAT, convert the raw scores based on the number of right and wrong answers into scaled scores, typically with a mean of 500 and standard deviation of 100. For instance, each raw score X might be multiplied by 4.5 and then added to 180: $4.5 X + 180$. In this way a raw score of 80 is converted into a scaled score of $4.5 (80) + 180 = 540$. Where do you think the scaling numbers 4.5 and 180 come from?

• **4.50** On the roulette wheel used in the U.S. there are 38 numbers (00, 0, 1, 2, . . . , 36), each apparently equally likely to catch the spun ball. One bet is that one of the 12 numbers 1, 2, . . . , 12 will come up. If $1 is wagered on this bet, then the bettor will either win $2 or lose $1. Another bet is that a particular number will come up. If $1 is wagered on this bet, then the bettor will either win $35 or lose $1. What are the expected returns, standard deviations, and skewness coefficients for these two bets? Explain, in words, the different natures of the two bets.

4.51 Sketch the following probability distribution and calculate the expected value of X:

x	$p[x]$
2	.3
3	.4
4	.3

4.52 *(continuation)* What happens to the graph of the probability distribution and the expected value of X if the probability distribution is changed to

x	p[x]
2	.2
3	.6
4	.2

4.53 *(continuation)* What happens to the graph of the probability distribution and the expected value of X if the probability distribution is changed to

x	p[x]
1	.3
3	.4
5	.3

4.54 Sketch the following probability distribution and calculate the expected value of X:

x	p[x]
0	.4
1	.3
2	.2
3	.1

4.55 A developer borrowed $500,000 in order to purchase 10 acres of land and divide this land into 20 buildable lots. She has assigned the following probabilities to the number of lots that she will sell within the next 12 months. Sketch this probability distribution. What is the probability that more than 5 lots will be sold in the first 12 months? What is the expected value of the number of lots that will be sold during the first 12 months?

Number of Lots Sold in First 12 Months	Probability
0	.05
1	.10
2	.15
3	.20
4	.20
5	.15
6	.10
7	.05

4.56 A buyer for a large department store must decide how many bathing suits to order for the summer season. Based on past experience, the sales probabilities are as follows. Sketch this probability distribution. What is the expected value of sales. If she orders 50 dozen suits, what is the probability that they will all be sold?

Demand for Bathing Suits (Dozens)	Probability
46	.10
47	.15
48	.20
49	.25
50	.15
51	.10
52	.05

4.57 Exercise 2.3 in Chapter 2 describes the card game Between the Sheets, in which a player is dealt two cards and can either fold or wager that the value of a third card will be between the values of the first two cards. Money that is lost accumulates in a pot in the center until someone wins a between-the-sheets wager. Suppose that you have been dealt the two cards listed next. In each case, how much money must be in the pot for a $1 wager on a third card to have a positive expected value? **a.** A 4 and a 7? **b.** A 3 and a 10? **c.** A 6 and an 8?

4.58 Suppose that a random variable X has only two possible outcomes, $X = a$ and $X = b$, with $b > a$. Must the expected value of X be between a and b: $a \leq \mu \leq b$? Either prove that it must or give a counterexample.

4.59 Major League baseball players can file for salary arbitration. In arbitration, the player submits a contract, the owner submits a contract, and a three-person panel chooses one of these two contracts. Neither side knows the details of the other side's contract until both contracts have been submitted. In February 1990 California Angels pitcher Chuck Finley filed for arbitration, asking for a one-year contract for $810,000; the Angels filed a one-year contract for $600,000. After learning of each other's contract, Finley and the Angels canceled the arbitration hearing by agreeing to a one-year contract for $725,000. If P is the probability that the arbitration panel would have chosen Finley's $810,000 contract, for what value of P is the expected value of the arbitration ruling equal to the agreed-upon $725,000 contract?

4.60 Juan believes that the probability that a Republican will win the next U.S. presidential election is .7; Linda believes that this probability is .5. Specify a possible wager for which both have a positive expected return.

4.61 A business is considering building either a one-story or two-story apartment complex. The value of the completed building depends on the strength of the rental market when construction is finished. The subjective probability distributions for the net profit with each option are given in Table 4.8. Use the mean and standard deviation to compare these two alternatives.

4.62 A promoter has scheduled an outdoor rock concert for April 15 in New Jersey. If the weather turns out to be rain or worse, he will lose $300,000. If the weather is tolerable, he will make a profit of $600,000. What is the expected value of his profit if there is a .6 probability of good weather and a .4 probability of bad weather? For what probability of good weather is the expected value of his profit equal to zero?

4.63 A Boule wheel is similar to a roulette wheel but contains only 18 slots. Two of the slots are numbered 1, two are numbered 2, and so on through 9. Calculate the expected percentage returns from each of the following bets. In each case, the payoff is $1 for every $1 wagered: **a.** The ball lands on a low number (1, 2, 3, or 4).   **b.** The ball lands on a high number (6, 7, 8, or 9).   **c.** The ball lands on an odd number other than 5 (1, 3, 7, or 9).   **d.** The ball lands on an even number (2, 4, 6, or 8).

4.64 *(continuation)* At a Boule wheel, one can also bet that the ball will land on one selected number, 1 through 9. The payoff for this bet is $7 for every $1 wagered. What is the expected percentage return?

4.65 In 1987, with gold selling for $460 an ounce, a mining company offered investors a chance to buy gold at $250 an ounce—but they would have

to pay the money immediately and not receive their gold for 15 months. The company would purportedly use the investor's money to obtain the gold from a mine with uncertain reserves.[15] Say that there is a probability P that you will get the gold and a probability $1 - P$ that you won't. If you are risk-neutral and determined to buy gold (and to hold it for at least 15 months), what is the lowest value of P that will persuade you to accept this company's offer?

4.66 A stock analyst has found the following probability distribution "to be very useful":[16]

Stock Price Change During Next Six Months (%)	Probability
+40	0.1
+20	0.2
0	0.4
−20	0.2
−40	0.1

What is the expected value and standard deviation?

4.67 JRJ stock has a .25 probability of a 15 percent return, a .50 probability of a 10 percent return, and a .25 probability of a 5 percent return. What are the expected value and standard deviation of its return? An LL junk bond has a .90 probability of a 20 percent return and a .10 probability of a −100 percent return. What are its expected return and standard deviation? Which investment is riskier? Which investment is preferred by a risk-neutral investor? Are the returns from these two investments positively skewed, negatively skewed, or symmetrical?

4.68 A break in a barrier sandbar is threatening to wash away shorefront property. One of these threatened homes was appraised at $2,000,000 shortly before the break and is now offered for sale for $500,000. If the house and land underneath are washed away (an event with probability P), the prop-

Table 4.8   *Data for Exercise 4.61*

| Rental Market | Probability | Profit (Millions) | |
		One-Story	Two-Story
Strong	0.4	$5	$10
Weak	0.6	0	−5

erty will be worthless. If the break is closed (probability $1 - P$), the house will be worth $2,000,000. For what values of P will an expected return maximizer buy the house for $500,000?

4.69 A bank is going to open either 20 or 30 branches, with the aggregate value of these branches five years from now dependent on the strength of the economy. The bank's chief executive officer (CEO) has assigned the subjective probabilities given in Table 4.9 for the net increase in the market value of the bank for each option. Use the mean and standard deviation to compare these two alternatives. Which would a risk-neutral CEO choose? Which would a risk-averse CEO choose?

4.70 In the gambling game Chuck-A-Luck, you may bet a chip on any number from 1 to 6. Three dice are thrown and chips are returned to you depending on the number of times your number appears.

Number of Dice	Payoff
0	0
1	2
2	3
3	4

What is the expected value of your payoff?

4.71 An investment journal examined a Canadian Interprovincial Lottery, in which a Canadian couple bought nine tickets for $9 and won $13.89 million (Canadian dollars in each case): "Press accounts said that the odds against winning are 14 million to one."[17] This journal calculated the expected return from such an investment as follows:

$$-\$9(13,999,991/14,000,000)$$
$$+ (\$13,890,000 - \$9)(9/14,000,000)$$

This journal then calculated the standard deviation from this "investment" from the following steps:

> Subtract the individual returns from the expected return
> Square the difference
> Add the squared differences
> Divide this number by one less than the number of returns
> Take the square root and VOILA!

What error did they make in this calculation?

4.72 In this stylized gambling scheme, bets are placed on the outcome of a flip of a fair coin, with winnings equal to the amount wagered; for example, if you bet $1 and win, you get back $1 plus $1 of the gambling house's money. You have decided to bet over and over again until you either lose $100 or win all of the house's money ($100,000). What is your probability of going broke? Is this a fair game?

• **4.73** Here are the annual incomes Y (in $1,000s) of a population consisting of five households: $10, $20, $20, and $50. What are the values of the mean and standard deviation of Y? Now consider two alternative tax systems:

$$T = .3Y$$
$$T = .5(Y - 10)$$

Which system raises the most tax revenue? Which has the higher marginal tax rate (more taxes paid on an extra dollar of income)? Which is the more egalitarian in that it has the smaller standard deviation of after-tax income, $Y - T$?

4.74 A pound weighs .4539 kilogram. If the weights of hamburger patties at a certain fast-food restaurant have an expected value of .25 pound, a standard deviation of .05 pound, and a skewness coeffi-

Table 4.9 *Data for Exercise 4.69*

Economic Conditions	Probability	Profit (Millions) 20-Story	30-Story
Strong	0.2	$100	$150
Medium	0.5	50	50
Weak	0.3	−50	−100

cient of −.10, what are the expect value, standard deviation, and skewness coefficient of the weight in kilograms?

4.75 Merton, Scholes, and Gladstein compared the historical returns from two investment strategies, which are given in Table 4.10.[18] Based on these data, write a brief paragraph comparing the results from these two strategies.

4.76 In the Massachusetts daily lottery, the bettor pays $1 and can choose to make any one of the following bets by filling in from one to four blank boxes, which is then compared to the four-digit number drawn that night. To win, the numbers filled in must match the numbers drawn, in exact order. For instance, if you leave the first two boxes blank and write a 4 in the third box and a 7 in the fourth box, then you win if the third number drawn is a 4 and the fourth number drawn is a 7. The payoffs vary somewhat with the total amount bet but are approximately as follow. What is the expected value of the payoff for each of these bets?

Numbers Selected	Payoff
All four	$5,000
First three	700
Last three	700
Any two	60
Any one	6

4.77 An entrepreneur is considering spending $200,000 for this season's food concession at an NFL team's home games. She estimates net revenue (not including the cost of the concession) of $500,000 for a whole season, but only $100,000 if the season is shortened by a player's strike. If she estimates the probability of a strike to be $P = .50$, does the expected value of net revenue exceed the $200,000 cost? How large can the probability of a strike be before the

expected value of net revenue falls below the cost of the concession?

4.78 In the board game Careers, a player can pay $3,000 to play the stock market by rolling a single die and receiving $1,000 times the number rolled. In the long run, how much can a habitual stock market player expect to make or lose from this wager?

4.79 In 1987 an Amplicon Financial marketing manager had to assign a customer to one of three account executives. Executive A would receive a $1,000 commission on this account, while B or C would receive $500. The marketing manager intended to draw straws, but A proposed taking the account and splitting his commission three ways. C wouldn't agree to A's proposal and they drew straws. Was C maximizing expected return?

4.80 Jason, Jennifer, and Joshua decide to play poker until one player has won everything. Jason has $500, Jennifer $500, and Joshua $1,000. If the three players are equally skilled, so that each has a zero expected gain, what are each player's chances of winning?

4.81 A standard Wheel of Fortune is divided into 54 sections, each apparently equally likely to be selected by a stationary pointer after the wheel is spun fairly. Wagers are made by putting a dollar on one of six numbers, 1, 2, 3, 4, 5, or 6. Each of the 54 sections contains three numbers—either three different numbers, a pair and a different number, or three of a kind. The payoffs are $2 for matching a single number, $3 for matching a pair, and $4 for matching three of a kind. The wheel is arranged so that, out of 54 sections, each number has 4 chances for a $4 payoff, 4 chances for a $3 payoff, 7 chances for a $2 payoff, and 39 chances for no payoff. In the long run, what fraction of every dollar wagered can bettors expect to lose?

Table 4.10   *Data for Exercise 4.75*

Strategy	Mean	Standard Deviation	Coefficient of Skewness
Writing call options	7.7%	10.5%	.87
Writing put options	4.1%	5.6%	.44

4.82 *Donoghue's Moneyletter* developed a test to differentiate speculators from the risk-averse.[19] One of the questions is:

> You are on a TV game show and can choose one of the following. Which would you take?
> a. $1,000 cash.
> b. A 50 percent chance at winning $4,000
> c. A 20 percent chance at winning $10,000
> d. A 5 percent chance at winning $100,000

Which answer indicates risk-averse behavior? Risk-neutral? Risk-seeking?

4.83 Which would you prefer: *A,* $200,000; or *B,* a lottery with a 20 percent chance of winning $1,200,000 and an 80 percent chance of losing $50,000? How would you characterize a person who prefers *A* to *B*? Prefers *B* to *A*? Is indifferent between *A* and *B?*

4.84 In 1987 United Airlines reported that for every 10,000 passengers an average of 36.3 persons were denied boarding (bumped) in exchange for tickets and cash, of which 31.2 were bumped voluntarily and 5.1 involuntarily.[20] Use these data to estimate the probability that a randomly selected passenger will be bumped voluntarily and bumped involuntarily. If the tickets and cash have an average value of $200 for a voluntary bumping and $500 for an involuntary bumping, what is the expected value of the cost per passenger?

4.85 *(continuation)* American Airlines reported 17.3 bumpings per 10,000 passengers, of which 17.2 were voluntary and 0.1 involuntary, while Pan Am reported 18.9 bumpings per 10,000 passenger miles, of which 1.7 were voluntary and 17.2 involuntary. What is the expected value of the bumping cost per passenger for American Airlines and Pan Am?

4.86 You can invest $100,000 in a certain venture that will be worth either $1,000,000 or zero. For what probability of success is the expected value of the venture equal to your cost? Instead of investing $100,000 in this venture, you can spend $20,000 for an option that will allow you invest in the venture (again at a cost of $100,000) after you have gathered more information. You currently believe that there is a .075 probability that the venture will be a success. After gathering more information, this probability will be revised either upward to 0.15 or downward to 0.00. If you believe that the venture's probability of success is equally likely to be revised upward or downward, is the expected value of the option larger than its cost?

4.87 For $100,000, you purchased a 25 percent stake in a venture that you estimated had a .20 probability of being worth $4,000,000 (giving your 25 percent stake a value of $1,000,000) and an .80 probability of turning out to be worthless. Unexpected difficulties have caused you to revise these estimates. Now you believe that there is a .10 probability that the venture will be worth $3,000,000 and a .90 probability that it will be worthless. One of the other partners has offered to sell you her 25 percent stake for $80,000. If you are risk-neutral, should you accept her offer?

● **4.88** A lottery offers a 10 percent chance of winning $5, an 80 percent chance of winning nothing, and a 10 percent chance of winning another lottery ticket (with the same prizes and probabilities as the first ticket). What is the expected value of this ticket?

● **4.89** A farmer wants to estimate the area of a square field. When he measures the length of the field, his measurement X of the observed length is normally distributed with an expected value equal to the true length μ and a standard deviation σ. Concerned about his measurement error, he makes two independent measurements X_1 and X_2. He isn't sure whether he should average these values and then square,

$$Y = \left(\frac{X_1 + X_2}{2}\right)^2$$

or square first and then average,

$$Z = \frac{X_1^2 + X_2^2}{2}$$

For which procedure is the expected value of the estimate of the area closest to the true area μ^2? (*Hint:* $E(X^2) = \mu^2 + \sigma^2$ and, for independent variables, $E[X_1X_2] = E[X_1]E[X_2]$.)

4.90 Ms. Garcia is considering investing $10,000 in a mutual fund that buys German stocks. She believes that the most important uncertainty regarding the performance of this mutual fund over the next five years is the extent to which German companies will be able to sell products throughout Eastern Europe. Her probabilities follow. What is the expected value and standard deviation of her five-year return from this investment? If she makes more than 0 percent, what is the probability that her return is 200 percent?

Possible Event	Probability	Five-Year Return from Mutual Fund
Extensive sales	.50	200%
Moderate sales	.30	50%
Minor sales	.20	0%

4.91 You are playing a game in which you will either win or lose $100. Your opponent now offers to double the stakes to $200. According to the rules of the game, you can either quit the game and pay $100 or accept the double and play for $200. If you are risk-neutral, for what probability P of winning the game should you accept the double?

4.92 Most gambling systems are variants of the Martingale system: doubling your bets until you win. Suppose that you are playing a game with a probability P of winning $1 and a probability $1 - P$ of losing $1. In the simple Martingale system, you begin by betting $1. If you lose, you then bet $2. If you lose again, bet $4. And so on until, when you finally win, your net gain is $1.

To understand the principle, assume that your probability of winning each wager is 0.5. You will play until you win $1 and then stop. What is the probability that you will stop after one bet? After two bets? After three bets? Your limited funds and/or house betting limits prevent you from doubling your bet indefinitely. How high does the house limit have to be to allow you an eleventh bet? How much money do you need to make eleven doubled bets? What is the probability that you will lose eleven straight bets? If you do lose eleven straight bets, how much will you have lost? What is the expected value of your gain from this doubling system if you must stop after eleven straight losses?

4.93 (continuation) Now consider an American roulette wheel with 38 slots, of which 18 are red, 18 are black, and 2 green. If you bet $1 on red, you win $1 if red comes up and lose your $1 if black or green appears. What is your expected gain if you bet just $1? What is your expected gain if you try the Martingale system, stopping after one win or eleven straight losses? How does the Margingale system affect your chances?

4.94 Prior to their summer 1990 release, an independent movie industry analyst estimated the costs and projected revenue under two scenarios—hit or dud—for the five movies shown in Table 4.11.[21] (All data are in millions.) Assuming, unrealistically, that these are the only two possible outcomes, for each movie calculate the hit probability for which the expected value of projected profits is zero. Which of these movies has the highest required hit probability for the expected value of projected profits to be zero?

4.95 After the initial season of Twin Peaks (1989–1990), a television critic cited the following odds on who killed Laura Palmer:[22]

Dr. Lawrence Jacoby	5 to 2
Ben Horne	3 to 1
Leo Johnson	3 to 1
Leland Palmer	3 to 1
Dr. William Hayward	4 to 1
Catherine Martell	5 to 1
Josie Packard	5 to 1

Table 4.11   *Data for Exercise 4.94*

Movie	Production Cost	Releasing Cost	Total Cost	Projected Profit If a Hit	If a Dud
Another 48 Hours	$38	$24	$62	$100	−$47
Days of Thunder	45	30	75	101	−56
Dick Tracy	32	30	62	111	−47
Die Hard II	62	30	92	83	−69
Total Recall	60	27	87	102	−16

Three other persons were given odds of 10 to 1, two at 20 to 1, two at 25 to 1, two at 30 to 1 (including my pick, Laura Palmer), four at 50 to 1, and seven at 100 to 1. For each of the seven listed persons, calculate the implied probability of guilt P such that a bet at the quoted odds would have an expected value of 0. Is the sum of these seven implied probabilities approximately equal to 1?

• **4.96** Portfolio managers decide how to allocate the funds at their disposal among competing assets. The following simplified problem is intended to illustrate a general rule. Asset 1 has a .5 probability of yielding a return of 20 percent and a .5 probability of yielding 0 percent. If Asset 1 yields 20 percent, then Asset 2 has a .5 probability of yielding 30 percent and a .5 probability of yielding 10 percent. If Asset 1 yields 0 percent, then Asset 2 has a .5 probability of yielding −10 percent and a .5 probability of yielding 20 percent. **a.** Would you characterize the returns from Assets 1 and 2 as positively correlated, negatively correlated, or independent? **b.** What is the expected return from Asset 1? **c.** What is the expected return from Asset 2?

• **4.97** (continuation) What are the possible percentage returns on a $1,000,000 portfolio if $400,000 is invested in Asset 1 and $600,000 in Asset 2? What is the expected value of this portfolio percentage return? How is the expected value of the portfolio return related to the expected values of the returns from Assets 1 and 2? Calculate the standard devia-

tions of the percentage returns from Asset 1, Asset 2, and the specified portfolio.

• **4.98** Redo Exercises 4.96 and 4.97, this time assuming that, no matter what the return on Asset 1 turns out to be, Asset 2 has four equally likely possible returns: 30 percent, 20 percent, 10 percent, and −10 percent. Explain any changes in your answers.

• **4.99** In college football, after scoring a touchdown (which is worth 6 points), a team can try to score an additional 1 point with a kick or 2 points with a run or pass. A kick is considerably more likely to be successful. Suppose that a team is behind by 14 points and then scores two touchdowns. It can play for a tie by attempting kicks after both touchdowns. After Notre Dame was widely criticized for playing for a tie against Michigan State, most teams now attempt a kick after the first touchdown and then try for 2 points after the second touchdown. An alternative strategy is to try for 2 points after the first touchdown and after the second touchdown, attempt a kick if the 2-point attempt was successful and go for 2 points again if it wasn't. Use probabilities to compare these two go-for-the-win strategies.

• **4.100** You are offered a choice of two envelopes and told that one envelope has twice as much as the other and that you can change your mind after opening one of the envelopes. You choose Envelope B and find $100. If you now take the other envelope instead, is the expected value of your gain positive? By this reasoning, will you always switch envelopes?

5

Three Useful Distributions

In the long run we are all dead. Economists set themselves too easy, too useless a task, if in tempestuous seasons they can only tell us that when the storm is long past the ocean is flat again.

J.M. Keynes

TOPICS

You have now learned what probabilities are, where they come from, how they are calculated, and what they can be used for. In this chapter, I will introduce you to three special probability distributions. The first, the binomial distribution, was central to the development of probability and statistics, and will be used often in this course. If we flip a fair coin, in the long run we expect heads to come up about half the time. But what about in the short run? In ten flips, what is the probability of obtaining exactly five heads? What are the chances of six heads, seven heads, or even ten straight heads? And what about the hypothetical long run? Is there any guarantee that we will get 50 percent heads in the long run? Is there any chance at all of 60 percent, 70 percent, or even 100 percent heads in the long run? These interesting questions, and many more, can be answered by the remarkable binomial distribution.

5.1 BERNOULLI TRIALS

Before explaining and using this distribution, I want to convince you of its widespread relevance. There are many uncertain situations that have the same characteristics as the classic coin-flipping example. The relevant features are:

1. There are two possible outcomes, not necessarily equally likely, that are labeled generically "success" and "failure."
2. If the uncertain situation is repeated, the probabilities of success and failure are unchanged. The probability of success is not affected by the successes or failures that already have been experienced.

An uncertain situation that satisfies these two conditions is called a **binomial trial** or a **Bernoulli trial,** after James Bernoulli (1654–1705), one of an eminent family of Swiss mathematicians.

There are many situations that qualify as Bernoulli trials. A series of coin flips satisfies the two conditions. So does the roll of dice or the spin of a roulette wheel, if we think of certain numbers as "successes" and the remaining numbers as "failures." Perhaps some fraction of the shirts, toasters, or cars produced by a certain company are defective. If each item, as it is produced or purchased, has a constant probability of being defective that does not depend on whether or not other items turn out to be defective, then the selection of these shirts, toasters, or cars qualifies as Bernoulli trials. If someone is given independent medical tests by several different doctors and the probability of a particular diagnosis is the same with each doctor, then these tests, too, are Bernoulli trials. If we conduct a political poll and, for each independent interview, have the same constant probability of selecting someone who favors a particular candidate, then these interviews are Bernoulli trials. If, each day, the probability of stock prices rising is constant, no matter how the market did on other days, then the path of stock prices is also a series of Bernoulli trials.

In this book, we will encounter a multitude of situations that can be thought of as Bernoulli trials. There are several examples discussed in this chapter and many more in the exercises. Other examples will appear in later chapters, and there will be some on your exams. By the time this course is over, you may be seeing Bernoulli trials all around you, and indeed they are.

Sampling Without Replacement

If we draw cards, one after another, from a standard deck, these are not Bernoulli trials because the probabilities change as cards are removed from the deck. They are Bernoulli trials, though, if the cards are replaced and the deck is reshuffled after each draw. Similarly, if we are randomly drawing households from the U.S. population, a strict Bernoulli trial requires that each selected household be put back in the deck, possibly to be drawn again. The hypergeometric distribution, discussed later in this chapter, handles the messy details of sampling without replacement. But sometimes the "deck" is so large that we assume Bernoulli trials even though the "cards" are not replaced after each draw. Imagine that we actually have a million decks of cards shuffled together to form one enormous deck with 52 million cards. Perhaps success is drawing a heart. As there are 13 million hearts scattered among the 52 million cards in all, the initial probability of a heart is exactly .25. If the first card drawn is a heart and it is not replaced, then the probability of a heart on the second draw falls to

$$\frac{12,999,999}{51,999,999} = .249999986$$

If the first card withdrawn is not a heart, then the probability of a heart on the second try rises to

$$\frac{13,000,000}{51,999,999} = .250000005$$

For most purposes, it would be perfectly acceptable to simply assume that the probability of a heart stays at .25, regardless of the outcome of the first draw. There is undoubtedly more error introduced by inadequately shuffling 52,000,000 cards than by neglecting to replace a few cards withdrawn from the deck.

In the same spirit, we commonly apply the Bernoulli trial model to the random selection of people, to see if they are male or female, married or unmarried, young or old, rich or poor, Republican or Democrat, or whatever. There is a slight distortion because we do not put people back into the deck, possibly to be selected again, but usually this distortion can be safely neglected. The more worrisome danger is that the deck is not adequately shuffled, and that everyone does not have the same probability of being selected and interviewed.

EXERCISE

5.1 Identify which of the following are Bernoulli trials and explain your reasoning: **a.** head or tail when a coin is flipped. **b.** safe or out when a baseball player bats. **c.** black or not when a bird lands on a telephone wire. **d.** Republican or not as the winner of an election in Oregon. **e.** boy or girl when a baby is born in Cleveland.

5.2   BINOMIAL PROBABILITIES

With this introduction, let's now figure out the probability distribution for Bernoulli trials. We'll start with a specific example and then deduce the general formula. Consider some game of chance in which your probability of winning ("success") is π ("pi," a variable, not the scientific number 3.14 . . .) and your probability of losing ("failure") is $1 - \pi$. This game could involve coins, dice, a roulette wheel, or whatever. If you play the game five times, what is the probability of winning exactly three times?

The first thing to recognize is that there are several different routes to three successes and two failures. The first three games could be successes, followed by two failures: SSSFF. You could start with two failures and then finish strong with three straight successes: FFSSS. Or you could alternate SFSFS. And there are other three-success sequences. To figure the probability of three successes, we need to know how many three-success sequences are possible and how probable each sequence is.

Fortunately, because the games are independent, each three-success sequence has exactly the same probability of occurring. If the probability of a success is π, then

$$P[\text{SSSFF}] = \pi\pi\pi(1 - \pi)(1 - \pi) = \pi^3(1 - \pi)^2$$

$$P[\text{FFSSS}] = (1 - \pi)(1 - \pi)\pi\pi\pi = \pi^3(1 - \pi)^2$$

$$P[\text{SFSFS}] = \pi(1 - \pi)\pi(1 - \pi)\pi = \pi^3(1 - \pi)^2$$

Each sequence containing three successes and two failures has the probability $\pi^3(1 - \pi)^2$.

All that remains is to count the number of possible sequences carefully. What we have here is three successes and two failures, and we want to know the number of unique ways that these can be arranged. This is our old friend, the combination problem from Chapter 2! Think of the five trials as being distinct slots, numbered 1, 2, 3, 4, and 5. We want to pick three of these numbered slots to label successes. How many different three-number combinations can be chosen? The combination formula tells us that the number of three-item combinations that can be made out of five distinct items is

$$\binom{5}{3} = \frac{5!}{3!\ 2!} = 10$$

Let's check our logic by writing out all the possible combinations:

Sequence	Success Slots
SSSFF	1, 2, 3
SSFSF	1, 2, 4
SSFFS	1, 2, 5
SFSSF	1, 3, 4
SFSFS	1, 3, 5
SFFSS	1, 4, 5
FSSSF	2, 3, 4
FSSFS	2, 3, 5
FSFSS	2, 4, 5
FFSSS	3, 4, 5

There are ten possible sequences confirming our reasoning.

These, then, are the two parts to our answer. The probability of any particular three-success–two-failure sequence is $\pi^3(1 - \pi)^2$ and there are ten possible sequences. Therefore the probability of exactly three successes and two failures is

$$\binom{5}{3} \pi^3(1 - \pi)^2 = 10\pi^3(1 - \pi)^2$$

This logic is easily generalized:

The **binomial distribution** states that, in n Bernoulli trials, with probability π of success on each trial, the probability of exactly x successes is

$$P[x \text{ successes}] = \binom{n}{x} \pi^x(1 - \pi)^{n-x} \tag{5.1}$$

A complete binomial distribution is found by computing the probabilities for all of the possible values of X, from 0 to n. In practice, instead of actually calculating the probabilities, we usually find it easier and more accurate to simply look them up in a convenient table, such as Table 1 at the end of this book.[1]

If you do calculate an occasional binomial probability by yourself, be aware of the special cases $x = 0$ and $x = n$. If $x = 0$, the binomial formula is

$$P[0 \text{ successes}] = \binom{n}{0} \pi^0(1 - \pi)^n = (1 - \pi)^n$$

because $\binom{n}{0} = 1$ and $\pi^0 = 1$. The answer $P[0 \text{ successes}] = (1 - \pi)^n$ is quite logical because the probability of zero successes is the probability of n straight failures, which, by the multiplication rule, is $(1 - \pi)^n$. Similarly, for $x = n$,

$$P[n \text{ successes}] = \binom{n}{n} \pi^n (1 - \pi)^0 = \pi^n$$

A Few Examples

Let's try a few examples. First, what about those five plays of a game of chance? If the probability of winning a game is $\pi = .45$, then the probability of winning exactly three games is

$$\binom{5}{3} (.45)^3 (.55)^2 = .2757$$

This one is not in Table 1, so you will have to run it through a calculator if you want to confirm my arithmetic.

Here's another example. Mr. and Mrs. Steinem decide to have four children. What is the probability of two boys and two girls? In Chapter 2, we came up with .513 as the probability of a male baby. If these children's sexes are independent, then we can view Mrs. Steinem's childbirths as Bernoulli trials. The probability of exactly two boys is

$$\binom{4}{2} (.513)^2 (.487)^2 = .3745$$

which is far from a sure thing.

Now, let's try a considerably more complicated example. A wildcat driller believes that there is a .5 probability of oil at a certain site. A lease is obtained and five independent tests are conducted. Each test has a .8 probability of a positive reading when there is oil and a .8 probability of a negative reading when there is no oil. It turns out that four of the tests are positive and one is negative. What is the revised probability of oil, taking this evidence into account?

This is a Bayes' rule problem. If we let S and F denote positive and negative test results, then the formula is

$$P[\text{oil} \mid 4\text{S\&1F}] = \frac{P[\text{oil}]P[4\text{S\&1F} \mid \text{oil}]}{P[\text{oil}] \, P[4\text{S\&1F} \mid \text{oil}] + P[\text{no oil}] \, P[4\text{S\&1F} \mid \text{no oil}]}$$

We have $P[\text{oil}] = P[\text{no oil}] = .5$. The remaining probabilities can be computed from the binomial distribution. If there is oil, then the probability of a positive

Unrepresentative Representatives

In the 1960s, there was an increasing awareness of discrimination against women, blacks, Chicanos, and other groups. One of the arguments made was that these groups were deliberately underrepresented on juries, convention delegations, legislatures, and other bodies that are supposed to represent all segments of society. If we interpret "representative" as meaning randomly selected from the population as a whole, then these claims can be evaluated statistically.*

Let's say that we are randomly selecting 12 people, perhaps for a jury or as delegates to a political convention. Assume that the population is divided into two groups of equal size, say male and female. If the population is large and the selection is random, then each selection has a .5 probability of being male and a .5 probability of being female. What is the probability that all 12 selected people will be male? This is a binomial probability with $\pi = .5$, $n = 12$, and $x = 12$, so that

$$P[12 \text{ males}] = \binom{12}{12} (.5)^{12}(.5)^{0}$$
$$= (.5)^{12}$$
$$= 1/4096$$
$$= .00024$$

If there is reason to suspect the deliberate exclusion of females and it turns out that all 12 of the selected people are male, then the improbability of an all-male selection certainly fuels this suspicion. Similarly, the binomial distribution tells us that random selection is unlikely to result in 10 or 11 of the selected people being male:

$$P[11 \text{ males}] = \binom{12}{11} (.5)^{11}(.5)^{1} = .00293$$

$$P[10 \text{ males}] = \binom{12}{10} (.5)^{10}(.5)^{2} = .01611$$

If the selection of these 12 people is random with males and females equally likely to be

reading is $\pi = .8$, so that

$$P[4S\&1F \mid \text{oil}] = \binom{5}{4} (.8)^{4}(.2)^{1} = .4096$$

If there is no oil, then the probability of a positive reading is $\pi = .2$, so that

$$P[4S\&1F \mid \text{no oil}] = \binom{5}{4} (.2)^{4}(.8)^{1} = .0064$$

Putting these binomial probabilities into Bayes' formula, we have

$$P[\text{oil} \mid 4S\&1F] = \frac{(.5)(.4096)}{(.5)(.4096) + (.5)(.0064)}$$
$$= .9846$$

selected, then the probability that there will be fewer than 3 females selected is .00024 + .00293 + .01611 = .01928. If the selected group turns out to be so disproportionately male, then there is bound to be suspicion that the selection was not really random after all.

Now we do have to be careful here in that, if we examine many twelve-person groups, we are bound to find a few that are disproportionately male or female even if the selection is truly random. On the average, with random selection, we expect to find 2 out of every 100 twelve-person groups having fewer than 3 females (and a like number having fewer than 3 males). It is not correct to scrutinize 100 groups, find a few that are disproportionately male or female, and then claim that these specific groups were not randomly selected. If we look at data for 100 twelve-person groups, then we should report

the male-female split for all 1200 selected people, not just for some carefully chosen subsets.

A similar, but more subtle error, occurs when one set of data is examined many different ways. We might decide to carefully scrutinize a particular twelve-person group. First, we see if there is a disproportionate number of males or females. Then, we see if there is a disproportionate number of blacks or whites. Next, we turn to Chicanos, American Indians, and Jews. Then, on to the young, the poor, the rural, the blue-collar workers, the high-school dropouts, Republicans, NFL tight ends, and anything else we can think of. These categories are not independent but, still, we would be making the same kind of error in looking at many characteristics that we made in looking at many twelve-person groups.

*An alternative interpretation is that "representative" means a deliberate attempt to select people who represent all segments of society. In Chapter 8 we will see that such deliberate efforts may instead merely represent the biases of the person making the selections.

Surprisingly, four positive readings on five imperfect tests makes oil almost a certainty. This strong conclusion follows because the chances of four positive readings when there is oil are so much greater than the chances of four positive readings when there is no oil.

EXERCISES

5.2 On your first day at college, the dean tells each student to, "Look at the student on your left and the student on your right; one of you three won't graduate from this college." Assuming that one-third of the new students do not graduate and that each student is equally likely to be one of the nongraduates, what are the probabilities that, out of three new students (you and the ones on your left and right),

a. all three will graduate? **b.** one will not graduate? **c.** two will not graduate? **d.** all three will not graduate?

5.3 The Supreme Court once compared six- and twelve-person juries and declared that "there is no discernible difference between the results reached by the two different sized juries." If we assume that jurors are a random sample from a population of which a fraction $\pi = .95$ will vote for conviction and $1 - \pi$ against, compare the probabilities that all members of a six- and twelve-person jury will vote for conviction. What if $\pi = .8$?

5.4 An experienced door-to-door salesman feels that the probability of a successful sale to a potential customer is 1/50 for a good product and 1/100 for a bad one. In 300 attempts with a new product, he makes 5 sales. What is the probability of this result if it is a good product? If it is bad? In the absence of other information, would you tentatively classify this product as good or bad?

5.5 A prominent professor claims that no one can beat coin flips when predicting which stocks will do better than average ("winners") and which will do below average ("losers"). To refute this claim, a well-paid securities analyst selects six stocks to do better

than average over the coming year. If the professor is right, what is the probability that all six stocks will be winners? That five will be winners? That four will be winners?

• **5.6** An organization is having a national convention of ten-person delegations from each of the fifty states. In California, 30 percent of the members of this organization are black. If the delegates are selected randomly, what is the probability that none of the ten California delegates will be black? What is the probability that fewer than two delegates will be black?

Now assume, for the sake of simplicity, that in every state 30 percent of this organization's members are black. If the delegates are selected randomly, what is the probability that at least one state will have no black delegates? What is the probability that at least one state will have fewer than two black delegates?

5.3   THE SHAPE OF A BINOMIAL DISTRIBUTION

So far, we have focused on specific binomial probabilities. Now, for a full binomial distribution, let's go back to the five plays of a game of chance. You could win anywhere from zero to five games. If the probability of winning each game is, say, $\pi = .2$, then the probabilities of winning various numbers of games are

Number of Wins X	Probability $\binom{5}{x}(.2)^x(.8)^{5-x}$
0	.3277
1	.4096
2	.2048
3	.0512
4	.0064
5	.0003

You should check these probabilities in Table 1 in the back of the book to be certain that you know how to use this table.

This particular binomial distribution is graphed in Figure 5.1. If the probability of winning a game was some fraction other than .20, then the resulting binomial distribution would, of course, be different. Figure 5.1 shows two of these other distributions, one for $\pi = .5$ and another for $\pi = .8$.

Notice that the binomial distribution for $\pi = .5$ is symmetrical, while the other two distributions are asymmetrical. Notice also that, in each case, the largest probabilities seem to be near $X = \pi n$, which makes sense. If $\pi = .2$, then 1 out of 5 is

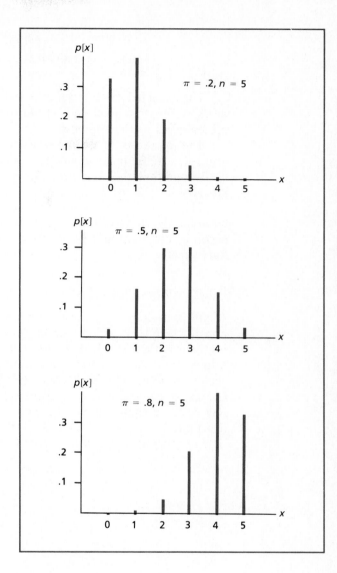

Figure 5.1   *Three Binomial Distributions*

the most likely result. If $\pi = .8$, then 4 out of 5 is most likely. For $\pi = .5$, we expect to win about half of the time. Half of 5 is 2.5, which of course is not feasible, but 2 and 3 wins are equally likely and the most likely outcomes.

It can also be shown (with some difficulty)[2] that the expected value, variance, standard deviation, and skewness coefficient for a binomial distribution are

$$E[X] = \pi n \qquad (5.2)$$

$$\text{variance of } X = \pi(1 - \pi)n \qquad (5.3)$$

$$\text{standard deviation of } X = \sqrt{\pi(1 - \pi)n} \tag{5.4}$$

$$\text{skewness coefficient of } X = (1 - 2\pi)/\sqrt{\pi(1 - \pi)n} \tag{5.5}$$

The expected value is quite logical. If we conduct a sequence of n trials many times, we expect a fraction π of our trials to be successes, that is, an average of πn successes per n-trial sequence.

The equations for the variance and standard deviation show that there is always dispersion in the possible values of X unless success or failure is certain ($\pi = 1$ or 0). Beyond that, the exact formulas are difficult to explain intuitively, so we should just accept them as being correct.

The skewness coefficient shows us that, with the exception of $\pi = 1$ or 0, the binomial distribution is only symmetrical if $\pi = .5$. As Figure 5.1 illustrates, the binomial distribution is positively skewed for π less than .5 and negatively skewed for π greater than .5.

Guessing on Multiple-Choice Tests

A typical multiple-choice question has four plausible answers (A, B, C, or D), only one of which is correct. An uninformed guesser has a probability $\pi = .25$ of selecting the right answer. If there are $n = 20$ questions on the test, then the expected number of correct answers for a pure guesser is $\pi n = (.25)(20) = 5$. Over many such tests, a guesser will almost certainly average about five correct answers per test. On any single test, of course, a guesser will probably get either more than five correct or fewer than five correct. The exact probabilities are given by the binomial distribution with $\pi = .25$ and $n = 20$. Table 1 reveals that there is, in fact, only a .2024 probability of getting exactly five correct answers:

Number Correct	Probability
<3	.0913
3	.1339
4	.1896
5	.2024
6	.1686
7	.1124
>7	.1018

Many multiple-choice tests are graded so that a guesser has an expected score of zero, the same as someone who simply doesn't bother to answer the questions. Because a guesser may get the right answer, points must be subtracted for a wrong answer to keep the expected score at zero. On a 20-question exam, the expected number of correct answers for a guesser is 5 and the expected number of wrong answers is 15. Therefore, if one point is given for a right answer, 1/3 of a point must be subtracted for a wrong answer to get a guesser's expected score down to zero.

The Expected Wait

In the St. Petersburg Paradox that was discussed in Chapter 4, a coin is flipped repeatedly until a head is obtained. What is the expected value of the number of tosses—the expected wait—until a head occurs? Table 1 lists four of the possible sequences that might occur and the probability of each sequence for the general case, where the probability of a heads is π, and for $\pi = .5$.

The expected value of the wait until obtaining a head is

$$1\left(\frac{1}{2}\right) + 2\left(\frac{1}{4}\right) + 3\left(\frac{1}{8}\right) + 4\left(\frac{1}{16}\right) + \cdots$$

Table 1

Wait Until Heads	Necessary Sequence	Probability	
		In General	$\pi = .5$
1	H	π	1/2
2	TH	$(1 - \pi)(\pi)$	1/4
3	TTH	$(1 - \pi)^2(\pi)$	1/8
4	TTTH	$(1 - \pi)^3(\pi)$	1/16
.			
.			
.			

A calculation of this sum shows that it approaches 2. A head can be expected to occur, on average, on one out of every two flips, and the expected wait until a head is two flips. Stated somewhat differently, the mean time between heads is two flips.

More generally, if the probability of success is π, then the expected wait until a success (the mean time between successes) is $1/\pi$. This rule is remarkably simple and useful. If the probability of rolling a 3 with a single die is 1/6, then the expected number of rolls until a 3 is obtained is 6. If a light blub has a .001 probability of failing each time that it is turned on, then the expected number of usages before failure is $1/.001 = 1,000$. If there is a .01 annual probability of a flood of a specified severity, then the expected wait until such a flood is $1/.01 = 100$ years. A 100-year flood should happen, on average, once every 100 years, and the expected wait until such a flood is 100 years.

In the United States the nationwide probability that a home will suffer a major fire in any given year is about .01. This probability implies that the expected wait until a major fire is approximately 100 years, considerably longer than the life expectancy of many of the materials used to construct houses. Most houses will not be destroyed by fire but by bulldozers.

More generally, if X is a random variable measuring the number of correct answers out of n questions; if 1 point is given for a correct answer; and if w points are subtracted for a wrong answer, then (using the fact that $E[a + bX] = a + bE[X]$), the expected score is

$$E[(X)(1) + (n - X)(-w)] = E[-wn + (1 + w)X]$$
$$= -wn + (1 + w)E[X]$$
$$= -wn + (1 + w)\pi n$$

The expected score equals zero if the penalty for a wrong answer is

$$w = \pi/(1 - \pi)$$

With four possible answers to each question, $\pi = .25$ and the penalty should be $w = .25/.75 = 1/3$. With five possible answers, $\pi = .2$ and the penalty should be $w = .2/.8 = 1/4$. These are, in fact, the penalties used on the SAT tests, GMAT tests, and other widely used multiple-choice exams.

Is there, then, any advantage to guessing? Well, informed guessing is clearly a good strategy. If you can eliminate even one of the possible answers, guessing will give a positive expected score—which is an improvement over the zero score received when a question is left blank. For instance, let's say that there is a tough question with four possible answers (A, B, C, and D). You can eliminate one of the answers, say A, but you have no idea which of the three remaining answers is correct. If you guess, you have a 1/3 chance of being right (and getting 1 point) and a 2/3 chance of being wrong (and being penalized 1/3 of a point). Your expected score on this question is positive:

$$1(1/3) + (-1/3)(2/3) = 1/9$$

It's not much, but every little bit helps. On the average, in the long run, this sort of barely informed guessing should raise your score by 1/9 of a point per question.

What about totally uninformed guessing? If you are running out of time, is there any advantage to simply marking answers willy-nilly without regard for the questions, let alone the right answers? The test booklets say, "random guessing will most probably yield a net raw score of zero, so it is not a useful strategy." It is true that guessing has an expected net score of zero and almost certainly will not be a

© 1968 United Feature Syndicate, Inc.

useful strategy, on the average, in the long run. But the test taker is interested in the single test at hand, not some hypothetical long-run average score. On a single test, guessing the answers to some questions may well raise or lower your score. We've seen that, if you guess the answers to twenty questions, there is only about a .2 probability that you will get exactly five right. There is a .8 probability that you will get either more or fewer than the expected number of correct answers. On a single test, guessing the answers to some questions will probably raise or lower your overall score. Your decision to try your luck depends on your situation.

Imagine that you somehow find yourself taking a test with no idea of the correct answers, and yet you know that a score of zero means failure. Should you leave the test blank and passively accept a zero or should you give yourself at least a slim chance by guessing? A slim chance is better than no chance, so guess away.

Fortunately, we seldom find ourselves in such a totally desperate position, but you may well encounter situations where analogous logic applies. Perhaps you are about to graduate from college and are torn between taking a job and going to business school. You decide that you will go to business school if you are accepted by one of the very best schools; otherwise, you will work for a few years and maybe go to business school later. With your grades, you figure that you need to score at least at the ninetieth percentile on the Graduate Management Admission Test (GMAT) to be accepted by a top business school. The GMAT test is 3 1/2-hours, consisting of approximately 200 multiple-choice questions covering reading recall, verbal aptitude, mathematics, data sufficiency, and business judgment. Each question has five possible answers, with one point given for the correct answer and one-quarter point deducted for a wrong answer. The GMAT test is used by many business schools to help evaluate applicants.

You dutifully take the GMAT test and find that, with 15 seconds remaining, you still have 20 questions left to answer. Should you leave them blank or should you hurriedly guess?

In the long run, if you were to take the test over and over again, it almost certainly wouldn't make any difference to your average GMAT score, but on this particular test, guessing probably will raise or lower your score a few points. The problem, of course, is that you don't know which will happen. Your decision to take this gamble should depend on how well you think you have done on the rest of the test. If you believe that you have done well enough to hit the magic ninetieth percentile, then don't jeopardize a good score by guessing. However, if you think that you have done poorly and have little hope of hitting the all-important ninetieth percentile, then you may as well guess and take your chances.

EXERCISES

5.7 A fair six-sided die is rolled 30 times. What is the expected number of times that the number 1 will come up? What is the probability that the number 1 will come up its expected number of times?

5.8 Len Deighton, a novelist, once wrote that a World War II pilot with a 2 percent chance of being shot down on a mission is "mathematically certain" to be shot down in 50 missions. What is the correct probability of being shot down in 50 missions, assuming Bernoulli trials with a 2 percent probability on each mission? What is the expected number of downed pilots per 50 missions, averaged over all pilots and missions? Why might the Bernoulli trial assumptions not apply here?

5.9 In 1984, a sports columnist for the Dallas Morning News had a particularly bad week picking the winners of NFL football games—he got 1 right and 12 wrong, with 1 tie. Afterwards, he wrote that, "Theoretically, a baboon at the Dallas Zoo can look at a schedule of 14 NFL teams, point to 1 team for each game and come out with at least 7 winners." The next week, Kanda the Great, a gorilla at the Dallas Zoo, made his predictions by selecting pieces of paper from his trainer. Kanda got 9 right and 4 wrong, better than all six Morning News sportswriters. **a.** Assuming no ties, what is the probability that a baboon will get at least 7 winners in 14 picks? **b.** What is the probability that a baboon will get no more than 1 out of 13 right? **c.** What is the probability that a baboon will pick as many as 9 out of 13 correctly?

5.10 For the thirteen months ending March 1987, only 65.9 percent of USAir's flights landed within 15 minutes of the scheduled arrival time.[3] (Of the five other airlines that released these data, two did slightly better and three did somewhat worse.) In the first three months of 1987, USAir Flight 42 from Pittsburgh to Boston was more than fifteen minutes late on 76 out of 81 flights. If the binomial distribution applies with $\pi = .659$ and $n = 81$, what is the probability that there will be five or fewer successes?

• 5.11 You are taking a multiple-choice exam where each question has five possible answers: A, B, C, D, and E. This test is scored so that a pure guesser has an expected score of zero. If one point is given for a correct answer, how many points are subtracted for a wrong answer? What is your expected score on a question if you can rule out **a.** one answer as definitely wrong? **b.** two answers as definitely wrong? **c.** three answers as definitely wrong? **d.** four answers as definitely wrong?

5.12 Explain why you agree or disagree with this test-taking advice: "Guessing is penalized on these examinations, since a percentage of the number of wrong answers is subtracted from the number of right answers. The applicant who feels that it would be a pity to leave answers blank and therefore chooses the answer that "looks good" will probably lower his score. The pupil who is not certain of the correct answer but has some knowledge of the question and can eliminate one or more choices as being unreasonable or contradictory, or entirely incorrect has a fairly good chance of choosing the correct answer.[4]

5.4   THE BINOMIAL PROPORTION

Often questions are phrased, not in terms of the absolute number of successes, but rather in terms of the proportion of the outcomes that are successes. Instead of asking the probability of winning three out of five games, we could ask the probability of winning 60 percent of five games. The probability is, of course, the same. We are just using different words to say the same thing.

Our formulas for the expected value, variance, standard deviation, and skewness coefficient do change because we are speaking of the proportion x/n rather than the absolute number x. Using the linear transformation Equations 4.4, 4.5, and 4.6, we have

$$P[\text{a fraction } x/n \text{ are successful}] = \binom{n}{x} \pi^x (1 - \pi)^{n-x} \tag{5.6}$$

$$E[x/n] = \pi \tag{5.7}$$

$$\text{variance of } x/n = \pi(1 - \pi)/n \tag{5.8}$$

$$\text{standard deviation of } x/n = \sqrt{\pi(1 - \pi)/n} \qquad (5.9)$$

$$\text{skewness coefficient of } x/n = (1 - 2\pi)/\sqrt{\pi(1 - \pi)n} \qquad (5.10)$$

Acceptance Sampling

When you, a business, or the government buy things, there is always a chance that your purchases will turn out to be defective, and yet, it is often uneconomical to inspect every purchase beforehand. When you buy a dozen oranges or a dozen avocados, it doesn't make sense to cut them all open before you buy them to see if they are all sweet and ripe, for then you will be buying rapidly rotting produce. When a business purchases a million transistors, it would be prohibitively expensive to test every single transistor before accepting delivery. When the Army purchases bombs, it cannot test each bomb by exploding it, for then it would have lots of test results, but no bombs.

What people do instead is to test a small sample and infer that comprehensive tests of all of the items would probably show similar results. Statisticians call this commonsense procedure **acceptance sampling** and use the binomial distribution to make the probability inference explicit. To use the binomial distribution, we need only assume that the test samples are Bernoulli trials and that the probability of selecting a defective item is constant and independent of the other selections.

Let's use the example of the Army and its bombs. Perhaps the Army has based its military planning and weapons deployment on the assumption that at least 90 percent of its bombs will explode properly. It may decide to test each new shipment of bombs by detonating some of them and seeing how many duds turn up. It would be natural to reject a shipment if more than 10 percent of those tested are defective.

If the Army does follow this procedure, what is the probability that a shipment will slip through that is 20 percent or even 30 percent defective? With the reasonable assumption that the tests are Bernoulli trials, we can let π be the true fraction of the bombs that are defective* and let x be the number of defective bombs found in a test sample of size n. The ratio x/n is the fraction of the bombs tested that prove to be defective. Equation 5.7 tells us that the expected value of x/n is π. If 20 percent of the bombs are defective, then, in the long run, the fraction found defective in the tests should average 20 percent. The actual results will, of course, vary from test to test.

Equation 5.9 says that the

$$\text{standard deviation of } x/n = \sqrt{\pi(1 - \pi)/n}$$

which is inversely related to the size of the sample. The larger the sample, the more certain it is that the fraction x/n found defective in the tests will be close to the

---

*For a Bayesian, π is the probability that a particular bomb is defective. An objectivist would point out that a bomb is not defective some of the time and explosive at others. It is either okay or it is a dud. For an objectivist, π is the probability that we will select a defective bomb to test.

proportion π of the entire shipment that is defective. If $\pi = .2$ and we only test five bombs, almost anything can happen, and so we won't be able to draw any firm conclusions from the results, whatever they turn out to be; if we instead test 1000 bombs, then the fraction x/n found defective is very likely to be close to .2. A 1000-bomb test is much more conclusive but, of course, it is also much more expensive. The whole point of testing a sample is to gauge inexpensively the quality of the product, but it must be recognized that there is a trade-off. The fewer the items tested, the less reliable the results will be.

Let's start our example by having the Army test 10 bombs, rejecting the shipment if there is more than 1 dud. The **acceptance number** is 1, in that the lot will be accepted if the number of defective bombs is less than or equal to 1. The binomial distribution tells us, for various values of π, the probabilities of accepting and rejecting shipments based on such a 10-bomb test. Table 5.1 shows some of these acceptance probabilities. If, for example, exactly 10 percent of the bombs are defective, then there is a .7361 probability that one or none of the 10 bombs selected will be defective, so most of the shipments that are 10 percent defective will pass this test. But notice that Table 5.1 also shows that there is an uncomfortably high .3758 probability of accepting shipments that are 20 percent defective. Even shipments that are 30 percent defective have a .1493 probability of having fewer than 2 duds in a 10-bomb test.

It appears that a 2-dud rule in a 10-bomb test does not catch unsatisfactory shipments often enough. What are the alternatives? Well, we don't want to make the test any easier by raising the rejection criterion to, say, 3 duds. If we want to weed out more of the unsatisfactory batches, then we must make the test tougher. We could decide to accept shipments only if all 10 bombs detonate properly. Table 5.1 shows that, as intended, the probability of accepting a 20 percent defective batch falls to .1074, and there is now less than a 3 percent chance of accepting a shipment that is 30 percent defective. The problem is that we will also frequently reject satisfactory shipments. If a batch is 10 percent defective, there is only .3487 probability that every single bomb will explode in a 10-bomb test. Sixty-five percent of the time, these acceptable shipments will be rejected, and even if a batch is only 5 percent defective, there is a 40 percent chance that it will be rejected by a 1-dud rule.

What are we to do? If we make the test easy, then a lot of unsatisfactory shipments will slip through. If we make the test tough, then we will reject a lot of sat-

Table 5.1   *The Probabilities of Accepting Shipments, Based on a 10-Bomb Test*

π	$x/n = 0$	$x/n \leq .1$	$x/n \leq .2$
.05	.5987	.9139	.9885
.10	.3487	.7361	.9298
.20	.1074	.3758	.6778
.30	.0282	.1493	.3828

Quality Control in the Soviet Union

In the Soviet Union a business's primary objective is to meet the production quotas set by the central planning agency Gosplan.[5] Because the emphasis is on quantity rather than quality, and there are no competing products available to Soviet citizens, consumer goods are generally of very low quality. For instance, televisions made in the Soviet Union have a reputation for often overheating and catching fire. In the military sector, in contrast, reliable equipment is considered a necessity and quality inspectors have long been used to monitor production.

One of the objectives of economic reform under Mikhail Gorbachev's program of *perestroika* is to improve the quality of con-

sumer goods. Military personnel were consequently reassigned to the consumer goods sector to inspect televisions and other products before they were shipped from factories. However, there was no change in the factory's objectives (to meet the production quotas set by Gosplan) and still no incentive to produce quality goods. The tough-minded military inspectors consequently created yet another problem—a shortage of televisions and other consumer goods, since they rejected a large fraction of what was produced. Instead of shipping 100,000 televisions, of which 10,000 catch fire, a factory might ship 30,000 safe televisions and throw out 70,000 potentially unsafe ones. To ease the shortage of consumer goods, the quality inspectors were sent back to the military.

isfactory shipments. This is, in fact, one of the basic trade-offs in acceptance sampling. The only real way to make this trade-off more appealing is to pay the additional costs of testing more bombs.

Table 5.2 shows some acceptance probabilities for a 100-bomb test. Look first at the column for the 10 percent rule, $x/n \leq .10$. A shipment that has only 5 percent defectives will pass this test almost every time. A shipment with 10 percent duds has a .58 probability of passing. Shipments that are 20 percent or 30 percent defective almost never pass this test. Increasing the size of the test from 10 bombs to 100 bombs has certainly increased the test's effectiveness in discriminating

Table 5.2 *The Probabilities of Accepting Shipments, Based on a 100-Bomb Test*

π	$x/n \leq .10$	$x/n \leq .13$	$x/n \leq .15$
.05	.9885	.9995	1.000
.10	.5832	.8761	.9601
.20	.0057	.0469	.1285
.30	.0000	.0001	.0004

What Are the Chances of a Run on the Bank?

Bank investment decisions are naturally influenced by the chances of depositor withdrawals. No bank wants to be caught short, without enough funds near at hand to satisfy withdrawals. The fraction of deposits withdrawn on a given day is a random variable that depends on the calendar and on business cycles, credit crunches, and other economic events. For illustrative purposes, let's assume that every depositor deposits a dollar and that the bank estimates that each dollar of deposits has a .1 probability of being withdrawn. If the bank has only $1 in deposits, then caution suggests that no loans should be made because there is a 10 percent chance that the lone depositor will want to withdraw the $1.

What if there are $2 in deposits? Then $2 will be withdrawn, or only $1, or none at all. Assuming independence, the binomial probabilities are:

Dollars Withdrawn	Probability
0	.81
1	.18
2	.01

Now the bank may be willing to loan out $1 because the probability of $2 being claimed is only .01. With $10 in deposits, the binomial probabilities are

Dollars Withdrawn	Probability
0	.3486784401
1	.3874204890
2	.1937102445
3	.0573956280
4	.0111602610
5	.0014880348
6	.0001377810
7	.0000087480
8	.0000003645
9	.0000000090
10	.0000000001

The odds of all deposits being withdrawn at once are now minuscule, at 1 in 10 billion. Surely some funds can be safely loaned out. Even if half of the deposits are loaned out and half are kept on reserve, there is only a 1-in-10,000 chance of deposit withdrawals exceeding reserves.

The most likely outcome is that one-tenth of the deposits will be withdrawn. As the quantity of deposits increases, it becomes

between satisfactory and unsatisfactory shipments. The price is the additional expense in testing (and destroying) 90 more bombs. If we're talking about sophisticated missiles costing millions of dollars apiece, then maybe the improved test isn't worth the additional expense, but if we're talking about inexpensive bullets, then a large test is certainly attractive. The final decision must hinge on a weighing of the need for a reliable test and the cost of a large test.

For any given test size, we still have to decide if we should use an acceptance rule that is easy or tough. The 10 percent rule shown in Table 5.2 seems a bit too tough because there is a 42 percent chance of rejecting shipments that actually have only 10 percent duds. Table 5.2 shows that, as we make the test easier, we are less more and more certain that withdrawals will be extremely close to one-tenth of deposits. With $100 deposited, the odds are only 1 in 1,000 that more than 20 percent of deposits will be withdrawn. With $1 million deposited, the odds are 1 in 1,000 that more than 10.09 percent will be withdrawn. Almost 90 percent can now be safely lent out.

This is an example of the general principle of *reducing risk through diversification.* In a large number of draws, independent risks will be diversified away. There will be some bad luck and good, but they will roughly balance each other out.

Notice, though, that in this numerical example I blithely assumed the independence required for applying the binomial distribution. Whether or not someone made a withdrawal had no effect on the likelihood of others making withdrawals. This is the individual risk that can be diversified away, and it concerns withdrawals made for private reasons that do not matter to others, such as the purchase of a new television, wedding expenses, or a sudden illness.

There are also public events that affect aggregate withdrawals. Some of these, like Christmas spending in December and tax payments in April, are easily anticipated. To the extent that they can be predicted, these variations are not risky. The bank simply adjusts its estimates of withdrawal probabilities and needed reserves. Other macroeconomic events may be more difficult to predict: recessions, credit crunches, and regulatory changes that allow other institutions to lure away depositors. The most dangerous and upredictable event is a panicky run on the bank. If depositors fear that they may not be able to withdraw their deposits, they are all sure to try. Because such macroeconomic risks affect all depositors, they cannot be diversified away simply by having a large number of depositors. With macroeconomic risks, there is no safety in numbers. A bank's only protection is to keep funds near at hand that can be mobilized in an emergency.

likely to reject satisfactory shipments, but are more likely to accept unsatisfactory shipments. A reasonable compromise might be $x/n \leq .13$.

Notice that we started out with a "natural" rule of allowing 10 percent duds in a 10-bomb test. Our binomial probability calculations suggested that a 10-bomb test wasn't large enough and that, in a larger test, a 10-percent-duds rule was too tough. The test that we finally select, perhaps a 13-percent-duds rule in a 100-bomb test, should not be chosen on the basis of its "naturalness." Instead, we should base our choice on the binomial probabilities and on a comparison of the various costs—of conducting the tests, of accepting unsatisfactory shipments, and of rejecting satisfactory shipments.

EXERCISES

5.13 Tweedle Dee and Tweedle Dum are running for president of the United States. Half of the voters prefer Dee and half prefer Dum. If *Newstime* magazine randomly polls 10 voters, what is the probability that at least 60 percent of those polled will prefer Mr. Dee? What if they poll 20 voters? What if they poll 100 voters? If they poll 100 voters, what is the probability that more than 60 percent will prefer one of the candidates, either Dee or Dum?

5.14 In any given year, about 5 percent of all mortgages are delinquent, with payments at least one month late. Assuming independence and that each mortgage has a 5 percent probability of being delinquent, what is the probability that at least 10 percent of a bank's mortgages are delinquent if the bank has **a.** twenty \$100,000 mortgages (\$2 million in mortgage assets)? **b.** fifty \$100,000 mortgages (\$5 million in mortgage assets)? Is there any reason for caution about your assumptions?

5.15 Sunspot Packing Company is inspecting the grapefruit at Littleman's farm in Porterville, California. The inspector cuts open 10 grapefruit selected at random. Sunspot will accept all of Littleman's crop if 9 of these 10 grapefruit are of satisfactory quality. If, in fact, 90 percent of all of Littleman's grapefruit are of satisfactory quality, what is the probability that his crop will be accepted?

• **5.16** *(continuation)* If 95 percent of all of Littleman's grapefruit are of satisfactory quality, what is the probability that his crop will nonetheless be rejected? What is the probability that his crop will be accepted if only 80 percent of his grapefruit are satisfactory? If only 50 percent are satisfactory? What would be some of the costs and benefits of the inspector testing more than 10 grapefruit?

• **5.17** The Quiet Night Motel has 35 rooms. Its experience has been that 20 percent of the people who make reservations never show up. How many no-shows can be expected if this motel accepts reservations for 35 rooms? For 40 rooms? For 50 rooms? What is the probability that there will not be enough rooms for those who do show up if this motel accepts reservations for 35 rooms? For 40 rooms? For 50 rooms?

5.5 THE LAW OF AVERAGES

The binomial distribution was developed by Bernoulli and other mathematicians to clarify the very meaning of probabilities. The *Bernoulli trial* label recognizes Bernoulli's work and, for similar reasons, the binomial coefficients ($\binom{n}{x}$) are often called *Pascal's Triangle.* These early probability mathematicians used the binomial distribution to argue that, in the long run, the proportion of successes x/n will converge to the probability of success π. Unfortunately, far too many have misinterpreted their argument as implying a fallacious "law of averages." Let's run through the mathematical logic and then see where intuition has so often been led astray.

The Law of Large Numbers

Consider some event that has a probability π of occurring. Let's label its occurrence a "success" and its nonoccurrence a "failure." Now, in n independent situations, how frequently will the event occur? This scenario qualifies as a sequence of Bernoulli trials and, therefore, the probabilities of various frequencies of occurrences x/n are given by the binomial probability distribution. The exact distribution depends, of course, on π and n.

Figure 5.2 shows four distributions, for $\pi = .5$ and $n = 10, 20, 50,$ and 100.

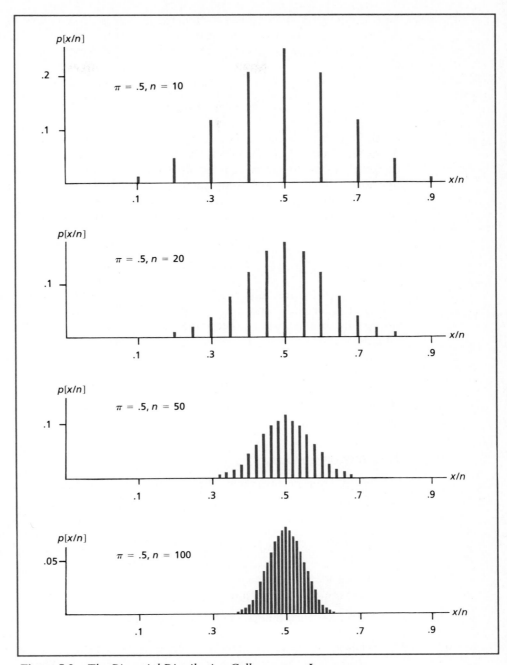

Figure 5.2   *The Binomial Distribution Collapses as n Increases*

Notice how each distribution is centered at $x/n = .5$ and how the distributions collapse on $x/n = .5$ as n increases. We can confirm this collapse more formally by looking back at Equations 5.8 and 5.9, which give the

$$\text{variance of } x/n = \pi(1 - \pi)/n$$
$$\text{standard deviation of } x/n = \sqrt{\pi(1 - \pi)/n}$$

As the number of trials n increases, the variance and standard deviation of x/n decrease. These declines mean that the dispersion in the possible outcomes decreases. In the long run, as n becomes infinitely large, the variance and standard deviation go to zero, meaning that there is very little uncertainty about what the value of x/n will be. As n approaches infinity, we are virtually certain that x/n will be extremely close to π. This is **Bernoulli's law of large numbers.***

> For large values of n, the probability is very close to 1 that x/n will be very close to π, its expected value.

Equivalently, for large n the probability is small that x/n will be far from π. Consider a fair coin, with a $\pi = .5$ probability of heads. For $n = 1$, there is a .5 probability that x/n will equal 0 (no heads) and a .5 probability that x/n will equal 1 (all heads). For a large n, the chances of no heads or all heads drop precipitously. The probability of no heads in 20 flips is .000000954. The probability of no heads in 100 flips is 7.89×10^{-31}. For 100 flips, in fact, there is only a .0176 probability that there will be fewer than 40 heads ($x/n \leq .4$). If n is very large, there is little chance that x/n will be far from π and, therefore, a high probability that x/n will be close to π.

A Common Fallacy

The law of large numbers is mathematically correct and intuitively obvious, but it is frequently misinterpreted as a "law of averages," which says that, in the long run, x/n must be exactly equal to π. For example, some people think that in 1,000 coin flips the number of heads must be exactly 500. It follows from this logic that if tails happen to come up more often than heads in the first 10, 50, or 100 flips, then heads will have to start coming up more frequently than tails to "average" things out. This argument is wrong, but it is all around us.

For example, one gambler said

> Flip a coin 1000 times and it'll come up heads 500 times, or mighty close to it. However, during that 1000 flips of the coin there will be frequent periods when heads will fail to show and other periods when you'll flip nothing but heads. Math-

---

*After its discoverer, James Bernoulli. His proof was published in his posthumous book *Ars Conjectandi* (1713).

ematical probability is going to give you roughly 500 heads in 1000 flips, so that if you get ten tails in a row, there's going to be a heavy preponderance of heads somewhere along the line.

Regardless of mathematics and the theory of probability, if I'm in a crap game and the shooter makes ten consecutive passes [wins], I'm going to bet against him on his eleventh throw. And if he wins, I'll bet against him again on the twelfth. Maybe it's true that the mathematical odds on every throw of the dice remain the same, regardless of what's gone before—but how often do you see a crapshooter make 13 or 14 straight passes?[6]

We don't see 13 or 14 straight wins very often, but we don't see 12 straight wins very often either. We see 13 straight wins about half as often as we see 12 straight wins. The probability of 12 heads in a row is $(1/2)^{12}$. The probability of 13 heads in a row is $(1/2)^{13}$, a rare event indeed. But it is a very different question to ask the probability of 13 straight heads, *given that you have already tossed 12 heads in a row.* This conditional probability is

$P[13$ straight heads $| 12$ heads already$]$

$$= \frac{P[13 \text{ straight heads and } 12 \text{ heads already}]}{P[12 \text{ straight heads}]}$$

$$= \frac{P[13 \text{ straight heads}]}{P[12 \text{ straight heads}]}$$

$$= \frac{(1/2)^{13}}{(1/2)^{12}}$$

$$= 1/2$$

The coin doesn't remember what happened on the last 12 flips. It is not interested in whether you win or lose, and it does not care about upholding some mythical law of averages. It is simply an inanimate object that gets repeatedly tossed into the air and examined by curious humans. If the coin is straight and fairly tossed, then heads and tails are equally likely to come up, no matter what happened on the last flip or the last 999 flips.

The Mathematical Error

The basis of the fallacious law of averages is the mistaken belief that heads and tails must come up equally often; a run of tails must consequently be balanced out by a run of heads. The law of large numbers does not say that the number of heads must equal the number of tails. Instead it says that, if we now begin to make a large number of throws, the probability is close to 1 that the fraction of these forthcoming throws that are heads will be very close to .5.

First, notice that the law of large numbers says nothing about the last 12 throws or the last 12 million throws. It is directed at the next n throws. The last 12 throws may have been all heads, all tails, or somewhere in between. The law of large numbers doesn't care. It says, looking ahead to predict the outcomes of the next 1,000 flips, that the probability is very high that heads will come up close to half the time.

Second, a relative frequency that is close to .5 is not inconsistent with there being many more heads than tails or vice versa. In fact, while we are almost certain that, in the long run, the heads proportion will be very close to .5, we are also equally certain that the number of heads will not equal the number of tails!

Table 5.3 shows that, as the number of trials increases, the relative frequency of heads can converge to .5 even while the absolute difference between the number of heads and tails grows larger. There are 57 heads in the first 100 flips, 7 more than an exact 50 percent, yet we don't need 7 extra tails to average things out. The extra 7 heads will not be offset by 7 tails, but instead diluted by thousands of tosses, in that 7 extra heads are unimportant in a long run of 1 million or 1 trillion flips. In fact, as shown in Table 5.3, heads can continue to come up more often than tails even while the relative frequency of heads converges to .5.

We can make an even stronger statement. Not only do we not need the number of heads to equal the number of tails in the long run, but we can, in fact, be almost certain that they will not be equal. Look at Table 1 at the end of this book. With 2 flips, the probability of 1 head and 1 tail is .5. With 4 flips, the probability of 2 heads and 2 tails drops to .375. The probability that the numbers of heads and tails will be equal keeps falling as n increases: to .3125 for $n = 6$, to .2734 for $n = 8$, and to .2461 for $n = 10$. This trend continues. Every increase in n makes it less likely that the number of heads will be exactly equal to the number of tails. For $n = 1,000$, there is only a .025 chance of exactly 500 heads.

More confirmation comes from a comparison of Equations 5.9 and 5.4. Equation 5.9 says that the standard deviation of the relative frequency x/n goes to zero as n becomes infinite. This fact is the valid law of large numbers. But Equation 5.4 says that the standard deviation of the absolute number of successes x increases as n increases. This fact denies the fallacious law of averages. In the long run, we are very confident that heads will come up close to 50 percent of the time, but we are very uncertain about exactly how many heads will come up.

Here are two final bits of evidence against the popular interpretation of the law

Table 5.3   *The Numbers of Heads and Tails Need Not Be Equal*

Number of Tosses n	Number of Heads x	Relative Frequency x/n	Heads Minus Tails x − (n − x)
100	57	.5700	14
1,000	520	.5200	40
10,000	5,040	.5040	80
100,000	50,060	.5006	120
1,000,000	500,100	.5001	200
1,000,000,000	500,001,000	.500001	2,000

of averages. We have already used the binomial distribution to calculate the probability that the number of heads will exactly equal the number of tails. We can also use the binomial distribution to calculate the probability that x/n will differ from .5 by less than some amount, such as .001, and to calculate the probability that the number of heads differs from the number of tails by some amount, such as 100. Table 5.4 shows the results of such calculations.

With 100 flips, a heads-tails difference of 100 is very unlikely; we would need all heads or all tails for it to occur. But, as n increases, we become more and more confident that there will be at least 100 more heads than tails or at least 100 more tails than heads. With 1 billion flips, we are almost certain that the numbers of heads and tails will differ by 100. At the same time, we are almost certain that the relative frequency of heads will be very close to .5. The law of large numbers does not require the numbers of heads and tails to be equal!

Some More Examples

The fallacious law of averages is all around us.[7] It is commonly appealed to by gamblers, hoping somehow to find a profitable pattern in the chaos created by random chance. When the roulette wheel turns up a red number several times in a row, there are always people eager to bet on black, counting on the law of averages. Of course, there are other people who will rush to bet red, trying to catch the "hot streak." The casino cheerfully accepts wagers from bettors of both faiths, confident that red and black numbers are still equally likely to come up.

One of the most dramatic roulette runs ever occurred at Monte Carlo on August 18, 1913. At one of the tables, black suddenly started coming up again and again. After about 10 blacks in a row, the table was surrounded by excited people, betting heavily on red and counting on the law of averages to reward them. But black came up again and again and again. After 15 straight blacks, there was a near panic as people tried to reach the table and place even heavier bets on red, and still black came up. By the 20th black, desperate bettors were wagering every chip they

Table 5.4 *We Are Almost Sure That, for a Large n, x/n Will Be Very Close to .5 and That There Will Be a Large Difference Between the Numbers of Heads and Tails*

Number of Tosses n	Probability That $.499 \leq x/n \leq .501$	Probability That $[x - (n - x)] \geq 100$
100	.08	1.6×10^{-30}
1,000	.08	.0015
10,000	.17	.32
100,000	.48	.75
1,000,000	.95	.92
1,000,000,000	≈ 1.00	.997

had left on red, hoping to recover a fraction of their losses. When this memorable run ended, black had come up 26 times in a row and the casino had won millions of francs.*

In honest games, at Monte Carlo and elsewhere, "balancing systems" based on the law of averages don't work, except by luck. In fact, the few gambling systems that have worked have been based on the opposite principle—that physical defects in a roulette wheel or other apparatus cause some events to occur more often than others. If the number 6 comes up an abnormally large number of times (not twice in a row, but maybe 35 times in 1,000 spins), then perhaps some irregularity is responsible. Instead of betting against 6, expecting Nature to even things out, you might bet on 6, hoping that there is some physical imperfection that will cause 6 to continue to come up 35 times in every 1,000 spins.

Another remarkable run occurred at a Chicago hospital in November 1949. Eighteen babies in a row turned out to be boys. The November 10, 1949 *Chicago Daily News* reported that the doctors and nurses expected a run of girl babies next. Yet 18 of the next 24 babies turned out to be boys. A few years later, the *Chicago Tribune* (February 3, 1953) reported that Mrs. Henry Drabik had given birth to 6 girls in a row, named Marybeth, Marykay, Marysue, Marylynn, Maryjan, and Marypat. When she became pregnant for the seventh time, "The odds against it were almost astronomical. Practically all Chicago was betting against the chance that [she] would have a seventh daughter. . . . Relatives offered to bet the Drabiks 10 to 1 the next child would be a boy," but it turned out to be another girl, named Maryrose.

The sports pages are another fertile source of law-of-averages fallacies. After the West Virginia football team lost to Penn State 13 years in a row, newspapers cheerfully reported that West Virginia at least had the law of averages on its side. West Virginia had last beaten Penn State in 1955; they played to a tie in 1958; and then Penn State won the next 25 in a row—a long wait for those counting on the law of averages to even things out. On October 27, 1984 West Virginia ended 28 years of frustration by finally beating Penn State's football team, 17–14. But it wasn't the law of averages that won the game for West Virginia. Going into their 1984 game, West Virginia was 6–1 and ranked 14th and 18th in AP and UPI polls; Penn State was 5–2 and ranked 19th and 20th. The West Virginia coach was quoted after the game as saying, "This is the greatest win I've been associated with since I've been here. Beating Oklahoma and Florida and Pitt twice is great, but to beat a team with the class and talent of Penn State ranks above them all."

Now, admittedly, sports do differ from games of chance in at least two ways. First we don't know the true probabilities. We don't really know how good a player or a team is until we gather some evidence. If Penn State and West Virginia had never played each other or common opponents, we might be very unsure of the outcome, but when Penn State beats West Virginia year after year, we become convinced that Penn State has the stronger football program. As the victories pile up,

---

*In 1931, 28 consecutive even numbers came up. Someone estimated that this should happen about once every 100 years, and it happened in Monte Carlo's 68th year.

it does not become more probable that Penn State will lose the next game. If any-thing, we revise Penn State's chances upward.

The second difference between sports and games of chance is that players do have memories and emotions. They do care about wins and losses and they do remember the last game. Sometimes a team will be beaten badly and play harder the next time, seeking revenge. But often, teams that lose a lot of games also seem to lose their self-confidence and grow accustomed to losing.

Many sports events require an almost mindless repetition of perfected physical motions. The successful baseball pitcher, golfer, bowler, tennis player, and so many others all seem to settle into "grooves" in which they perform well, without really consciously thinking about what they are doing. When a professional basketball player gets "hot," each basket he sinks makes him more confident and relaxed, allowing his body to do what it has been trained to do.[8]

When the basketball player misses six shots in a row, the self-doubts begin to build. He starts to worry about what he is doing wrong, and the more he worries, the harder it is for his body to perform well. These destructive self-doubts are seen in almost all sports. Athletes have turned to sports psychologists and even hypno-tists to help them relax and forget past failures.

If West Virginia loses to Penn State year after year, West Virginia may get fired up and Penn State may get overconfident, but it is also possible that Penn State will play with confidence while West Virginia is plagued by self-doubts. Similarly, when Eric Davis goes hitless in 12 at-bats, some sports commentator usually announces that, "Davis has the law of averages on his side" or that "he is due for a hit." But the probability of getting a hit doesn't go up just because he hasn't had one lately. It may be that his 12 outs in a row were just bad-luck line drives hit right at fielders. If so, his bad luck hasn't made him any more likely to get a hit in his next time at bat. If he is a .300 hitter, then his chances of a hit are still around 30 percent.

If it's not just bad luck, then 12 outs in a row suggests that a physical or mental problem may be causing him to play poorly. If I were the manager of a baseball player who was 0 for 12, I would tend to be pessimistic. By the same reasoning, if someone has had 12 hits in his last 24 at-bats, I would not bench him because he is "due for an out." I'd prefer batting someone who was 12 for 24 to an equally talented player who was 0 for 12.

On another occasion, I watched a Penn State placekicker miss three field goals and an extra point in an early season game against the Air Force. The television commentator said that Joe Paterno, the Penn State coach, should be happy about those misses as he looks forward to some tough games in coming weeks. This com-mentator explained that every kicker is going to miss some over the course of a season and it is good to get the misses "out of the way" early in the year. This optimism is the law-of-averages fallacy again. Those misses don't have to be bal-anced by successes. If I had been coaching Penn State, I would have been very worried by this poor performance. It suggests that this placekicker really isn't very good, and even if the kicker does have some talent, he is certainly going to be very nervous on his next attempts.

Mary Sue Younger, a statistics professor at the University of Tennessee, has written of another misuse of the law of averages: "I was told by my insurance salesman that some insurance companies will raise your premiums after so many years without a claim, figuring you are coming due for an accident!"

EXERCISES

5.18 The Steinems are planning their family and both want an equal number of boys and girls. Mrs. Steinem says that their chances are best if they plan on having two children. Mr. Steinem says that they have a better chance of having an equal number of boys and girls if they plan on having four children. Who is right? (Assume that boy and girl babies are equally likely.)

5.19 Explain why you think that this logic is sound or unsound:

> A famous baseball pitcher, Ted Lyons, was once removed for a pinch hitter on his fifth time at bat after having made a base hit on each of his first four times at bat. The reason given was that it was virtually impossible for a pitcher to get five hits in a row![9]

5.20 In World War I, soldiers were advised to sit in shell holes because it was considered extremely improbable that two shells would hit the same spot. Is this probability logic valid?

5.21 Exercise 2.1 involved the professor who reads only 1 of the 10 essays that each student writes on the final exam. Let's assume now that 3 of your essays would be graded A, 4 would be graded B, and 3 would be graded C. Would you prefer that this professor read all of your essays or just one of them? Explain your reasoning.

• 5.22 Two baseball teams, the Stags and the Sagehens, will play each other for the league championship. The Stags are, in fact, the better team. (But the Sagehens have the better yell, "That's all right, that's okay. You'll be working for us someday.") If these two teams were to play each other many, many times, the Stags would win 60 percent of the time. Assuming baseball games are Bernoulli trials, what is the probability that the Stags will win the championship if these two teams play just 1 game? If they play a 3-game series? If they play a 5-game series? If they play a 7-game series? What is the moral?

Can you think of any reason why a series of baseball games is not, strictly speaking, a sequence of Bernoulli trials?

• 5.23 Which of the following data is more convincing evidence that male and female babies are not equally likely: (*a*) of 3 babies, 2 are male; (*b*) of 100 babies, 52 are male; or (*c*) of 1000 babies, 520 are male?

5.6   THE HYPERGEOMETRIC DISTRIBUTION (OPTIONAL)

The binomial distribution assumes that the probability of success does not change as the trials proceed. If, however, we are sampling without replacement from a finite population, then the probabilities do change as observations are withdrawn from the population.

Consider, for instance, the drawing of cards from a 52-card deck, with an ace being a success and any other card a failure. If we replace the card and shuffle the deck after each draw, then the binomial distribution applies, since the draws are independent and the probability of success holds firm at 4/52. If, however, we pick 5 cards without replacement, then only the first card has a 4/52 chance of being an ace. The probability that the second card is an ace will either be 3/51 (if the first card turns out to be an ace) or 4/51 (if the first card is not an ace). The probability that the third card drawn is an ace is 4/50, 3/50, or 2/50.

Although the binomial distribution does not apply, we can figure out the probability distribution in a similar manner. The probability of no aces is

$$P[X = 0] = \left(\frac{48}{52}\right)\left(\frac{47}{51}\right)\left(\frac{46}{50}\right)\left(\frac{45}{49}\right)\left(\frac{44}{48}\right)$$

There are five ways to get a single ace,

$$\text{first card:} \quad \left(\frac{4}{52}\right)\left(\frac{48}{51}\right)\left(\frac{47}{50}\right)\left(\frac{46}{49}\right)\left(\frac{45}{48}\right)$$

$$\text{second card:} \quad \left(\frac{48}{52}\right)\left(\frac{4}{51}\right)\left(\frac{47}{50}\right)\left(\frac{46}{49}\right)\left(\frac{45}{48}\right)$$

$$\text{third card:} \quad \left(\frac{48}{52}\right)\left(\frac{47}{51}\right)\left(\frac{4}{50}\right)\left(\frac{46}{49}\right)\left(\frac{45}{48}\right)$$

$$\text{fourth card:} \quad \left(\frac{48}{52}\right)\left(\frac{47}{51}\right)\left(\frac{46}{50}\right)\left(\frac{4}{49}\right)\left(\frac{45}{48}\right)$$

$$\text{fifth card:} \quad \left(\frac{48}{52}\right)\left(\frac{47}{51}\right)\left(\frac{46}{50}\right)\left(\frac{45}{49}\right)\left(\frac{4}{48}\right)$$

Notice that each sequence has the very same probability! Thus the sum of these five probabilities is simply the number of possible sequences (5) multiplied by the probability of any particular one-ace sequence,

$$P[x = 1] = 5\left(\frac{4}{52}\right)\left(\frac{48}{51}\right)\left(\frac{47}{50}\right)\left(\frac{46}{49}\right)\left(\frac{45}{48}\right)$$

Similarly, there are

$$\binom{5}{2} = 10$$

two-ace sequences and

$$P[x = 2] = 10\left(\frac{4}{52}\right)\left(\frac{3}{51}\right)\left(\frac{48}{50}\right)\left(\frac{47}{49}\right)\left(\frac{46}{48}\right)$$

Obviously, the logic parallels that for the binomial distribution, the only difference being that the probability of, say, a particular sequence of two successes is more complicated than

$$\left(\frac{4}{52}\right)^2\left(\frac{48}{52}\right)^3$$

A General Formula

The easiest way to find the general probability formula is to count the possible outcomes, and divide the number of outcomes with the requisite number of successes and failures by the total number of outcomes. Letting

$$N = \text{size of population}$$
$$S = \text{number of successes in population}$$
$$n = \text{size of sample}$$
$$x = \text{number of successes in sample}$$

$$P[x \text{ successes}] = \frac{\left(\begin{array}{c}\text{number of ways } x \\ \text{successes can be} \\ \text{selected from } S \\ \text{possible successes}\end{array}\right)\left(\begin{array}{c}\text{number of ways } n - x \\ \text{failures can be} \\ \text{selected from } N - S \\ \text{possible failures}\end{array}\right)}{\begin{array}{c}\text{number of ways a sample} \\ \text{of size } n \text{ can be selected} \\ \text{from a population of size } N\end{array}}$$

The first part of the numerator counts the number of ways to get x successes, the second part the ways to get $n - x$ failures; the product gives the number of ways to get x successes and $n - x$ failures.

If we substitute the combination formulas into this general expression, we get the *hypergeometric* probability distribution:

$$P[x \text{ successes}] = \frac{\binom{S}{x}\binom{N - S}{n - x}}{\binom{N}{n}}$$

In the aces example,

$$P[x = 2] = \frac{\binom{4}{2}\binom{48}{3}}{\binom{52}{5}}$$

$$= \frac{\binom{4\ 3}{2\ 1}\binom{48\ 47\ 46}{3\ 2\ 1}}{\dfrac{52\ 51\ 50\ 49\ 48}{5\ 4\ 3\ 2\ 1}}$$

$$= 10\,\frac{4}{52}\,\frac{3}{51}\,\frac{48}{50}\,\frac{47}{49}\,\frac{46}{48}$$

we get the same answer as derived above by another route. The virtue of the first approach is its close analogy to the binomial distribution; the second approach yields a general formula that can be applied readily.

Mathematicians have figured out that the mean and variance for a hypergeometric distribution are

$$\mu = \left(\frac{S}{N}\right)n$$

$$\sigma^2 = \left(\frac{S}{N}\right)\left[1 - \frac{S}{N}\right]n\left(\frac{N-n}{N-1}\right)$$

Notice the correspondence to the binomial distribution when the population N is sufficiently large so that the success probability is effectively constant, $\pi = S/N$, and the "finite population correction factor," $(N - n)/(N - 1)$, is very close to one: $\mu = \pi n$ and $\sigma^2 = \pi(1 - \pi)n$.

Three Brief Examples

For another example of the hypergeometric distribution, consider a shipment of $N = 15$ missiles, of which $S = 10$ are satisfactory and $N - S = 5$ defective. If $n = 3$ are tested, what is the probability that all will be satisfactory? That two will be satisfactory and one defective? And so on. The hypergeometric distribution is

$$P[x \text{ successes}] = \frac{\binom{10}{x}\binom{5}{3-x}}{\binom{15}{3}}$$

In particular,

$$P[3 \text{ successes}] = \frac{\binom{10}{3}\binom{5}{3-3}}{\binom{15}{3}}$$

reduces to

$$= \frac{10}{15} \frac{9}{14} \frac{8}{13}$$

$$= .2637$$

Similarly,

$$P[2 \text{ successes}] = \frac{\binom{10}{2}\binom{5}{3-2}}{\binom{15}{3}}$$

$$= 3 \frac{10}{15} \frac{9}{14} \frac{5}{13}$$

$$= .4945$$

and

$$P[1 \text{ success}] = \frac{\binom{10}{1}\binom{5}{3-1}}{\binom{15}{3}}$$

$$= 3 \frac{10}{15} \frac{5}{14} \frac{4}{13}$$

$$= .2197$$

Finally,

$$P[0 \text{ successes}] = \frac{\binom{10}{0}\binom{5}{3-0}}{\binom{15}{3}}$$

$$= \frac{5}{15} \frac{4}{14} \frac{3}{13}$$

$$= .0220$$

Since we have found the complete distribution, we can check our calculations by seeing if the probabilities add to one. They do, but for .0001 of rounding error.

Here is another quick example. A certain technical analyst claims that she has an uncanny ability to predict which stocks will go up and which will go down. She

The Case of the Misplaced Ballots

In the 1984 election of Guam's (nonvoting) delegate to the U.S. House of Representatives, the incumbent, Antonio Won Pat, appeared to have won reelection over his challenger Ben Blaz, when a power failure disrupted the computerized tally. When the power was restored two hours later and the votes counted and recounted on three subsequent occasions, Ben Blaz was declared the winner by amounts ranging from 328 to 368 votes out of more than 31,000 cast. Won Pat challenged the results, charging a variety of "substantial irregularities," including the mysterious appearance the day after the election of a box containing 197 valid ballots. Of these 197 found ballots, 137 were from the Santa Rita precinct, with the rest scattered among 72 other precincts. Curiously, while the initial count from the Santa Rita precinct showed 181 votes for Blaz and 179 for Won Pat, the found ballots were divided 83 for Blaz and only 54 for Won Pat, giving Blaz a total vote in this precinct of 264 to Won Pat's 233.

The hypergeometric distribution can be used to compute the probability that randomly misplaced ballots would differ so markedly from those remaining. Imagine 497 slips of paper of which 264 are red and 233 blue. If 137 of these slips of paper are dropped randomly into a box, what is the probability that as few as 54 will be blue? Here

$$N = 497$$
$$S = 233$$
$$n = 137$$
$$P[x = 54] = \frac{\binom{233}{54}\binom{264}{83}}{\binom{497}{137}}$$
$$= .00969$$

Similar calculations for 53 blue, 52 blue, and so on, show that the overall probability of fewer than 55 blue slips is .0249. Thus the contention that 137 randomly misplaced ballots would show such a preponderance for Blaz may, to some skeptics, seem too improbable to be credible. Nonetheless, the U.S. House of Representatives, after spending months in acrimonious debate over an Indiana congressional election and finally deciding that the Democrat had won by four votes, decided that the grounds for Won Pat's challenge were, in the words of one committee member, "not significant enough for Congress to get involved."

is given price charts for fifteen stocks, ending on December 31, 1983. Ten of these stocks went up in 1984 and five went down. Looking only at the pre-1984 data, she picks three stocks that she is confident went up in 1984. If her techniques are worthless, what is the probability that her random selection will yield three stocks that went up? two? one? none?

This is a different context, but uses the very same calculations used in the previous example: $N = 15$, $S = 10$, and $n = 3$. By random guessing, she has a .2637 chance of 3 successes, .4945 chance of 2 successes, .2197 chance of one success, and .0220 chance of no successful picks. Since even a guesser has a .26 chance of picking three stocks that went up, her predictions are destined to be unconvincing. For the evidence to be persuasive, she needs to do something that cannot be easily explained by chance alone—either by picking more than three winners, or by picking from a group that does not have so many winners.

EXERCISES

5.24 The text discusses a technical analyst's selection of three "winners" from a group of fifteen stocks, ten of which went up. What if she tried to pick three winners from a group of fifteen, ten of which went down? What is the probability that by guessing alone she will select three stocks that went up? two? one? none? If she does pick three winners, can this be explained easily as lucky guesses?

5.25 A week before the final exam, Professor Sailors gives his students ten questions. At the beginning of the final examination, a student will reach into a wastebasket and blindly select three of these ten questions, which then become the final exam. You are well prepared to answer six of the ten questions and poorly prepared to answer the other four. What are the probabilities that the three selected questions **a.** will all be ones that you are well prepared to answer? **b.** will include two that you are well prepared for and one that you are poorly prepared for? **c.** will include one that you are well prepared for and two that you are poorly prepared for? **d.** will all be ones that you are poorly prepared to answer?

5.26 In Keno, there are 80 ping pong balls, numbered 1, 2, . . . , 80. Bettors are given cards containing these 80 numbers and can select from 1 to 15 numbers. If the bettor decides to pick 10 numbers, for example, then this is called a 10-spot card. After the bettors have filled out their cards, a "Keno goose" picks 20 balls at random. Payoffs are based on the type of card and the number of correct picks. On a 60-cent, 8-spot card, for instance, the payoffs are $5 for five correct picks, $50 for six correct, $1,100 for seven correct, and $12,500 for all eight correct. What is the expected percentage payoff?

5.27 There are four players in bridge and each is dealt 13 cards. You have 5 spades, your partner has 4 spades, and your two opponents have a total of 4 spades between them. What is the probability that these 4 spades are evenly divided, 2 to each opponent? Is it more likely that these 4 spades divide 2–2 or 3–1?

5.28 On a recent English test, students were given the names of four authors and four novels, and asked to match each novel with the correct author. If a student just guesses randomly, what is the probability of getting all four correct? Three correct? Two correct? One correct? None correct? The teacher gives one point for a correct answer and subtracts one point for a wrong answer. Is a guesser more likely to have a positive score or a negative score?

5.7   THE POISSON DISTRIBUTION (OPTIONAL)

The Poisson distribution was discovered in the early 1800s by a French mathematician, Simeon Denis Poisson, who was trying to apply probability theory to legal issues. The binomial distribution is concerned with the number of successes in n Bernoulli trials, each with success probability π; the expected number of successes is $\pi = \pi n$. Consider, however, an attempt to apply the binomial distribution to the number of failures of certain parts on a submarine during a 60-day mission. We could divide the mission into 60 days and think of each day as a trial, during which a part may or may not fail. But the part may fail twice on a single day, and the binomial distribution allows only a single success or failure during a trial. Even if we divide the 60-day mission into shorter intervals, say hours, and think of a mission as 24(60) = 1440 hourly trials, there might still be two or more failures during an hour. We must make the interval of time so short that only one failure can possibly occur during an interval. But since a failure can occur at any instant, we must divide the 60-day mission into an infinite number of infinitesimally short intervals.

The Formula

As the number of trials n becomes infinitely large, the probability π of success during any particular trial becomes very small. If the expected number of successes for any given interval remains constant, at $\mu = \pi n$, then the binomial distribution turns into the Poisson distribution,

$$P[x \text{ successes}] = e^{-\mu} \left(\frac{\mu^x}{x!} \right)$$

where $e = 2.71828$ is the base of the natural logarithms. Since π is assumed to be very small, the Poisson distribution is sometimes called the *probability of rare events.*

Notice that the Poisson probability distribution has only one parameter, μ, the expected value of x. We do not specify π, which is very close to zero, nor n, which is very large, but rather their product, $\mu = \pi n$, the expected number of successes during the total period being analyzed. (As it turns out, with a Poisson distribution, the variance of x is also equal to πn.) Also, unlike the binomial distribution, where the number of successes is limited by the number of trials, there is no limit to the number of possible successes with a Poisson distribution, since the number of trials is considered infinite; of course, while any number of successes is possible, the probability of an exceptionally large number of successes is quite small.

A Variety of Applications

Sometimes the Poisson distribution is simply used to approximate the binomial distribution for small π; for example, the Los Angeles court case with the alleged 1/12,000,000 probability. More often it is applied to situations where, instead of a finite number of trials, it is more appropriate to think of a continuum of time or space in which an unlimited number of trials takes place. Part failures is one example and the U.S. Department of Defense did, in fact, use the Poisson distribution to decide how many spare parts to carry on Polaris submarine missions and on inventory ships that restock the submarines at the end of a mission.[10]

The Defense Department used data from 61 submarine missions to calculate the average number of breakdowns per mission, which was then used as an estimate of $\mu = \pi n$, the expected value of the number of breakdowns during a 60-day mission. Notice again that the number of trials n is not specified, nor is the probability of breakdown during one of these infinitesimally brief trials; instead, we need only an estimate of their product, πn, the expected number of breakdowns during the mission as a whole.

The Defense Department's estimates of $\mu = \pi n$ ranged from 0 to 5 depending on the particular type of part. For a value of, say, $\mu = 0.5$, the Poisson distribution implies that

$$P[0] = e^{-.5}\left(\frac{.5^0}{0!}\right) = .6065$$

$$P[1] = e^{-.5}\left(\frac{.5^1}{1!}\right) = .3032$$

$$P[2] = e^{-.5}\left(\frac{.5^2}{2!}\right) = .0758$$

$$P[3] = e^{-.5}\left(\frac{.5^3}{3!}\right) = .0126$$

$$P[4] = e^{-.5}\left(\frac{.5^4}{4!}\right) = .0016$$

The probability of more than 4 failures during a mission is

$$P[X > 4] = 1 - .6065 - .3032 - .0758 - .0126 - .0016 = .0003$$

that is, a 3 in 10,000 chance. The Defense Department can use such information, together with an assessment of the consequences of running out of spare parts, to decide how many spare parts to take on each mission and to carry on the inventory ships.

The Poisson distribution has been applied successfully to a wide variety of phenomena, including the number of

phone calls received during a period of time
cars arriving at a toll booth during a time period
customer demand during a time period
accidents occurring during a time period
runs per inning in a baseball game
defects in a line of wire
defects in a bolt of cloth
blemishes on a car's exterior
weed seeds in a bag of grass seeds
fires caused by lightning in a forest
bombs landing in a town
fish in a patch of ocean
bacteria in a drop of water
stars in space

In each case, we are interested in the number of successes or failures, and assume that Bernoulli trial assumptions hold (that the success probability is constant and the results independent) but are dealing with a continuum of time or space in which, logically, the number of Bernoulli trials is infinite.

EXERCISES

5.29 Here you will apply the Poisson distribution to the Los Angeles court case discussed in Chapter 3. Assume that the probability that a randomly selected couple has the specified characteristics is $\pi = 1/12{,}000{,}000$ and that there were n couples in the general area of the robbery. The California Supreme Court assumed that $n = 12{,}000{,}000$; calculate the following probabilities for their assumption and for $n = 1{,}000{,}000$: **a.** What is the probability that none of these n couples has the relevant characteristics? **b.** What is the probability that at least one couple does? **c.** What is the probability that exactly one couple has these characteristics? **d.** What is the probability that more than one couple have the characteristics? **e.** What is the probability that more than one couple have the characteristics, given that at least one couple (Collins) does?

5.30 During the period 1:00 PM to 3:00 PM, cars arrive at a certain toll booth at a rate of 2 cars a minute. If the Poisson distribution applies, what is the probability of 5 cars during a minute? Why might the

highway department be interested in such information?

5.31 The Winning Edge Company makes personal computers and related equipment, and its technical support office has found that it receives an average of 30 calls an hour. Assuming that the Poisson distribution applies, what is the probability that it will receive 2 calls during a one-minute interval? that it will receive 4 calls during a two-minute interval? Why might the office be interested in such information?

5.32 A fishing company estimates that there is one school of tuna per 50,000 square miles of sea. Use the Poisson distribution to find the probability that a search of a 10,000-square-mile area will turn up empty.

5.33 A store sells an average of 2 jars of Joe's Special Salsa a week, and brings its inventory up to 4 jars every Saturday, when the delivery truck stops by. Assuming the Poisson distribution applies, what is the probability that more than 4 jars will be sold in a week's time?

5.8 SUMMARY

A series of n uncertain outcomes, in which the probabilities of "success" and "failure" are constant, irrespective of the other outcomes, is called a sequence of Bernoulli trials. This model is applied to a wide variety of situations, including games of chance, sampling tests of defective items, and public opinion polls.

If the constant success probability is π, then the probability of exactly x successes is given by the binomial distribution,

$$P[x \text{ successes}] = \binom{n}{x} \pi^x (1 - \pi)^{n-x}$$

Many of these binomial probabilities are in Table 1 at the end of this book.

The expected value, variance, standard deviation, and skewness coefficient for a binomial distribution are

$$E[X] = \pi n$$

$$\text{variance of } X = \pi(1 - \pi)n$$

$$\text{standard deviation of } X = \sqrt{\pi(1 - \pi)n}$$

$$\text{skewness coefficient of } X = (1 - 2\pi)/\sqrt{\pi(1 - \pi)n}$$

Sometimes we are interested in the fraction x/n of the trials that are successes:

$$P[\text{a fraction } x|n \text{ are successful}] = \binom{n}{x} \pi^x(1 - \pi)^{n-x}$$

$$E[x/n] = \pi$$

$$\text{variance of } x/n = \pi(1 - \pi)/n$$

$$\text{standard deviation of } x/n = \sqrt{\pi(1 - \pi)/n}$$

$$\text{skewness coefficient of } x/n = (1 - 2\pi)/\sqrt{\pi(1 - \pi)n}$$

The law of large numbers says that if n is very large, then the probability is very high that the proportion of successes x/n will be very close to π. The fallacious law of averages says that an unusual run of successes must be "balanced out" by a run of failures, so that x/n will equal π exactly.

The hypergeometric distribution is used for sampling without replacement and the Poisson distribution is used when, instead of a finite number of trials, a continuum of time or space with an unlimited number of trials is more appropriate.

REVIEW EXERCISES

5.34 In 1987 an Iraqi missile hit the U.S. frigate *Stark*. One informed observer wrote that

It's quite possible that the Stark's radar did not see the weapons separate from the launching aircraft. I have seen radar miss an aircraft carrier sitting in plain view on the horizon. The same radar will "paint" the blip nine times out of 10, but the law of statistics mean that occasionally it will miss on two of the three consecutive passes.[11]

Assuming that the binomial distribution applies, what is the probability of two or more misses in three trials?

5.35 A newly formed corporation has raised enough money to build ten restaurants, each in a different city. If the probability of a restaurant being successful is .1 and the outcomes are independent, what

is the probability that **a.** exactly one will be successful? **b.** at least one will be successful?

5.36 Each player on the Sagehens baseball team has a .300 on-base average; that is, each gets on base, on the average, 30 times per 100 times at bat. If we assume that a baseball game is a sequence of Bernoulli trials, what is the probability that the Sagehens will be the victims of a perfect game (27 straight outs) the next time they play? Can you think of any plausible objections to the assumption that a baseball game is a sequence of Bernoulli trials?

There are about 2150 games (each involving two teams, of course) in a major league baseball season. At this rate, on average, how often should we expect to see a perfect game if your earlier probability calculation is reasonably accurate?

5.37 A basketball article in the January 16, 1983 *Los Angeles Times* was titled "Wright Defies the Percentages, and USC Holds Off Oregon, 62–54." The article explained that, "Oregon's strategy was obvious in the closing minutes—foul Gerry Wright, a 37.5 percent free throw shooter. But Wright, a reserve center, didn't fold under the pressure. He made 3 of his 6 foul shots in the final two minutes."

Assuming the binomial distribution to be appropriate, what is the probability of Wright making so many free throws (at least 3 of 6)? Is the fact that he was so successful an unlikely defiance of the percentages? What assumptions are needed for the binomial distribution to be appropriate?

• **5.38** Long ago, the astragali (heel bones) of animals were used as dice. An astragalus of a hooved animal has four sides. Experiments have shown that the probabilities of each of these four sides appearing are .39, .37, .12, and .12. In ancient Greece, one of the games was to roll four astragali simultaneously, with the best outcome being "Venus," in which four different sides appear. What is the probability of throwing a Venus?

• **5.39** Let's investigate the often-heard claim that if enough monkeys were given enough typewriters and enough time, they could generate every book that was ever written. If we neglect the distinction between uppercase and lowercase letters, there are 50 keys on my typewriter. What is the probability that one monkey who randomly hits 11 keys on one typewriter will correctly type the word "probability?"

Now imagine that we have a billion monkeys and a billion typewriters. Each monkey hits 11 keys every 12 seconds (a key a second and then a one-second rest before trying again). If these billion monkeys type at this rate 24 hours a day, 365 days a year for 1985 years, what is the expected number of times that "probability" will be typed correctly?

5.40 Find a computer program that will simulate 1,000 flips of a fair coin. Repeat this experiment ten times and write a paragraph summarizing your results. Be sure to mention whether there were exactly 500 heads each time and whether the percentage and absolute number of heads were more nearly constant after 100, 500, or 1,000 flips.

• **5.41** The Jolly Roger computer quiz invites contestants to complete ten sentences by selecting the more appropriate of two words. For example, in the February 6, 1973 quiz, the contestant could complete sentence number 5, "Teachers sometimes have to __ their emotions," with either "suppress" or "control." There is a $2 entry fee, but the contestant wins $1,000 if all ten sentences are completed correctly. If we assume that each of the two answers is equally likely to be right, what is a contestant's expected percentage return?

5.42 Mutual funds promise superior management; critics allege that funds do no better than coin flippers, but charge fancy fees. Exercise 1.11 in Chapter 1 gives the annual percentage returns for twenty-four randomly selected mutual funds in 1984, 1985, and 1986. Here are the median returns for these twenty-four funds compared to the S&P 500 index:

	1984 (%)	1985 (%)	1986 (%)
Mutual Funds	−10.3	26.0	30.0
S&P 500	− 4.8	30.9	35.8

Although most funds did worse than the market, perhaps a few selected funds were consistent winners. To see if there is any consistency to performance, identify the eight funds in Exercise 1.11 that were above the median in both 1984 and 1985. If performance were due to luck rather than skill, then, on average, what fraction of the funds that did well in the past would you expect to be above the median in the future? How many of these eight funds were in fact above the median in 1986?

• **5.43** Professor Lay Z. Mann requires a twenty-page paper in his course. The paper is graded pass-fail. The rumor is that Professor Mann grades each paper by reading a few pages selected at random. Each year, a few risk-seeking students hand in papers with some pages containing acceptable material and other pages containing nonsense straight out of *Alice in Wonderland*.

If half of the pages in a twenty-page paper are nonsense, what are the chances of being caught if Professor Mann reads just one page? Two different pages? If Professor Mann is known to read just one page and a student wants at least a 90 percent chance of passing, what is the maximum number of nonsense pages to sneak into a twenty-page paper?

•• **5.44** In major league baseball, the World Series is won by the first team to win four games. If the games are independent and the teams are evenly matched (each with a .5 probability of winning any game), what is the probability that the Series will end after four games? After five games? After six games? After seven games?

5.45 *(continuation)* In 1976, it was observed that of 31 World Series since World War II (1945–1975), 18 had lasted seven games.[12] When this statistic was reported nationwide, many people wrote to the statistician, offering a variety of conspiratorial theories ranging from gamblers to television executives. If the teams are evenly matched and the binomial model is applicable, what is the probability that 18 or more World Series out of 31 will last seven games?

5.46 *(continuation)* Look in an almanac or other source of sports information and find out how many World Series since 1975 have lasted seven games.

• **5.47** A certain textbook has 500 pages with approximately 1,000 printed characters per page. The publisher claims that there will typically be only 50 misprinted characters randomly scattered through a book of this size. To check this claim, we are going to pick a page at random and carefully count the typographical errors. Answer the following questions, assuming that the selected page has 1,000 characters and the publisher's claim is correct. **a.** What is the expected number of misprinted characters? What, exactly, does this "expected number" calculation mean? **b.** What is the probability of no error? **c.** What is the probability of one error? **d.** What is the probability of two errors? **e.** What is the probability of more than two errors? **f.** If you find more than two errors on this selected page, will you be skeptical of the publisher's claim? Why?

5.48 During the January 1984 NFC championship game between Washington and San Francisco, it looked as if the game might be decided by a field goal at the end. One of the CBS radio commentators, Jack Buck, pointed out ominously that Mark Mosely has already "missed four field goals today." Hank Stram responded, "The percentages are in his favor." Do you agree?

5.49 Critically evaluate the following:

This man has kept records of play on Blackjack, craps, and Roulette, and the house percentage on all three games works out inexorably, within a fraction of a percentage point—but there are times shown by his records when there are lengthy streaks of steady house wins along with other streaks of house losses.

His records show that such streaks exist. But no matter how long he studies his figures, he hasn't been able to arrive at any pattern in them. He can see that the streaks are there, but he hasn't anything even resembling an explanation of why they occur when they do or why they last for a certain period.

"Nevertheless," he insists, "if I'm scoring red and black on a Roulette game and red has come up 500 times in 900 spins of the wheel, I'm going to put my money on black on the next 100 spins. What's more, I'll bet you that I come out ahead of the game."[13]

5.50 Explain why you either agree or disagree with Edgar Allan Poe:

Nothing . . . is more difficult than to convince the merely general reader that the fact of sixes having been thrown twice in succession by a player at dice, is sufficient cause for betting the largest odds that sixes will not be thrown in the third attempt. . . . It does not appear that the two throws which have been completed, and which now lie absolutely in the Past, can have influence upon the throw which exists only in the Future. . . . this is a reflection which appears so exceedingly obvious that attempts to controvert it are received more frequently with a derisive smile than with anything like respectful attention. The error here involved—a gross error redolent of mischief—I cannot pretend to expose within the limits assigned me at present.[14]

•• **5.51** What is (*a*) the probability of *x* heads in *n* = 2*x* flips of a fair coin, and (*b*) the probability of *x* + 1 heads in *n* = 2*x* + 2 flips of a fair coin? Is the ratio of (*a*) to (*b*) greater than, equal to, or less than 1? What is the moral?

•• **5.52** This exercise is intended to illustrate the detection of a manufacturing breakdown that causes an unacceptably large number of defective products. Imagine a box that is supposed to contain 95 white balls and 5 black ones—the probability is, in fact, .999 that it will contain these numbers of balls. But, occasionally (with .001) probability, someone makes an error and puts in 90 white balls and 10 black ones. What is the probability that this error has been committed if you observe the following drawings (with replacement): **a.** 1 black ball? **b.** 10 black balls out of 100 drawn? **c.** 50 black balls out of 500 drawn?

5.53 Ten college students are asked to choose between two different brands of cola. These colas are, in fact, so similar that each student is equally likely to choose either brand. What is the probability that five students choose each brand? That at least seven students choose one brand or the other?

5.54 Jean and Choi agree to flip a coin nine times, with Jean winning if at least five heads appear and Choi winning if at least five tails occur. After four flips, there have been three tails and one head. What is Jean's probability of winning?

5.5 The *Wall Street Journal* runs a monthly stock picking contest in which four investment professionals are matched against four stocks selected by tossing darts at the financial pages of the *Wall Street Journal*. During the first eight months of the contest, the professionals beat the darts six times.[15] If the professionals had only a 50 percent chance each month of beating the dartboard portfolio, what is the probability that they will win six or more times in an eight-month experiment?

5.56 A certain very tired student is asleep half the time in math class. What is the probability that this student will be caught sleeping if the professor looks up from the lecture notes three times? What is the probability that the student will be sleeping all three times that this professor glances up from the notes?

5.57 Mary bets Irene that a coin flipped eight times will give four heads and four tails. Irene bets that it won't. What is Mary's probability of winning this bet? If Mary bets $100, how much must Irene wager to give this bet an expected value of zero?

5.58 Based on his knowledge of the computer market, an executive believes that the 60 percent of computer users will react favorably to a new software program. A marketing survey of 20 computer users finds that only 8 react favorably to the software. If the executive is right, what is the probability that 8 or fewer would react favorably in this survey?

5.59 Professor Smith decides to give a short and sweet multiple-choice exam consisting of 10 questions, each with three possible answers: A, B, or C. If she gives 1 point for each right answer, how much should she subtract for a wrong answer to give a pure guesser an expected score of zero?

5.60 In the California Lotto game, 6 numbered balls are randomly selected out of 49 balls numbered 1 to 49. To win the grand prize, an entrant's six selected numbers must match the six winning numbers, not necessarily in the order in which they are drawn. The Thursday, February 21, 1990, Lotto game had a $68.56 million grand prize, to be divided among those who correctly picked all six numbers. **a.** What is the probability of matching the six winning numbers? **b.** A total of 50.5 million tickets were sold. What was the expected number of grand prize winners? **c.** What is the probability that none of these tickets will win the grand prize? **d.** As it turned out, there were four grand prize winners, who each received $17.1 million. Why was each of their prizes worth less than $17.1 million?

5.61 Big Mo and Little Mo are two basketball players. Big Mo makes 80 percent of his free-throws, while Little Mo makes 90 percent. They have a contest in which they each shoot 10 free-throws. What are the probabilities that **a.** Big Mo will make all 10 shots? **b.** Little Mo will make all 10 shots? **c.** Both will make all 10 shots? **d.** Big Mo will make 9 shots, but Little Mo will make all 10? What assumptions did you need to make in order to calculate these probabilities?

• **5.62** *(continuation)* What is the probability that this free-throw shooting contest will end in a tie? What is the probability that Big Mo will win? What is the probability that Little Mo will win?

5.63 A marketing test of a new automobile indicates that 5 percent of a target group who receive a promotional video in the mail will go to a dealer and test drive the car, and that of those who test drive the car, 25 percent will purchase it. If so, what is the probability that a person in the target group who receives the promotional video will purchase the car? Ten thousand videos are mailed to households in the target group. What is the expected number of car purchases? What is the probability that fewer than 100 of these 10,000 people will purchase the car?

5.64 A company is introducing $1,000,000 one-year insurance policies for college students. Public Health Service statistics indicate that a 20-year-old U.S. male has a .0020 probability of dying during a one-year period, while a 20-year-old U.S female has a .0006 probability of dying during a one-year period. If the company assumes that these probabilities are appropriate, how much must it charge males and females for these policies for the expected value of its payout to equal half of the amount charged?

5.65 Twenty men of age 40 apply for a 10-year life insurance policy. Past data indicate that each has a .05 probability of dying by age 50. What is the expected value of the number of deaths for this group of 20? What is the probability that none will die by age 50? That all will? Assume that the Bernoulli trial model is appropriate.

5.66 *(continuation)* Redo the computation in the preceding exercise, this time analyzing 20 women of age 40, each of whom has a .03 probability of dying by age 50.

5.67 The *Wall Street Journal* reported that the San Francisco Giants scored "runs by the bushel" in winning the 1989 National League baseball playoffs but scored only 1 run in the first two games of the World Series that year. Their explanation: "Baseball, that game of the long haul, is the quintessential sport of the mean, and the mean ol' law caught up with the San Francisco Giants."[16] What is the "mean ol' law" that the *Journal* alludes to? Do you agree with this use of the law?

5.68 A certain manufacturing process involves 20 distinct tasks performed by 20 different people. If each person has a 2 percent chance of making an error in his or her task, what is the probability that a product will be made with no errors at all? That exactly one of the tasks will have an error? That more than one task will have an error? What is the expected number of tasks with errors? What if each person has a 4 percent chance of making an error? Assume that the Bernoulli trial model applies.

5.69 A state lottery ticket has five randomly selected numbers lined up with five concealed numbers. The buyer wins if some of the concealed numbers match the corresponding open numbers. the ticket costs $1 and pays $1,000 for five matched numbers, $10 for four matches, $3 for three, $2 for two, and $1 for one. What is the expected value of the payoff? What fraction of every dollar collected, on the average, does the state expect to pay out in prizes?

5.70 Explain the error in this argument: "I prefer pitching on the day after we've been shut out because I figure we're going to score. I guess the odds are really in my favor now" (Orel Hershiser, in 1989, before pitching a baseball game for the Los Angeles Dodgers, who had not scored a run for 24 consecutive innings).[17]

5.71 In 1980 the United States tried a daring rescue of American hostages that were being held by Iranians in the American Embassy in Teheran. The American plan was to bring in soldiers in eight helicopters traveling 800 miles across the desert at low altitudes during the night. The military commanders believed that at least six of the eight helicopters would have to reach Teheran for the mission to have a chance of succeeding. As it turns out, the rescue attempt was canceled when three of the helicopters became disabled. Use the binomial distribution to calculate the probability that at least six of eight helicopters would reach Teheran if each has a .75 probability of success. If each has a .90 probability of success.

• **5.72** *(continuation)* If each helicopter has a .75 probability of succeeding, how many helicopters are needed for there to be a .90 probability that at least six will reach Teheran?

5.73 A life insurance company recently purchased a package of 100 home mortgages issued by a savings and loan association. If the Bernoulli trial model applies, with each mortgage having a 2 percent chance of default, what is the probability that none of these mortgages will default? That more than five will default?

5.74 *(continuation)* Do you think that the probabilities you calculated in the preceding exercise would increase or decrease if the probability of default were 4 percent instead of 2 percent? Calculate these probabilities and see if you are right.

5.75 Some newspapers print a "Lotto Analysis" column each week, in which they report which lottery numbers have been selected the most frequently in the past 52 weeks. Would a believer in the law of averages bet on numbers that have been selected frequently or infrequently? If you were to buy a lottery ticket, would you find a "Lotto Analysis" useful?

5.76 During the 1989 World Series, the batting coach for the San Francisco Giants said: "I believe in the law of averages. I'd rather see a so-so hitter who's hot come up for the other side than a good hitter who's cold."[18] What do you say?

5.77 What is the probability that the roll of a pair of dice will give snake-eyes (a pair of 1s)? What is the expected number of rolls until snake-eyes occurs?

5.78 Mary claims that when she cuts a standard deck of cards, the queen of hearts often turns out to be the top card. If the deck is thoroughly shuffled and randomly cut, what is the probability that the queen of hearts will be the top card? If Mary has no special skills and this experiment is repeated over and over, with the cards thoroughly shuffled each time, what is the expected wait until the queen of hearts is the top card?

5.79 The chances of winning the New York Lotto game are 1 in 7 million. A determined player has vowed to purchase one ticket every week until he wins. What is the expected value of the number of tickets that he will have to purchase?

5.80 A coin will be flipped until two heads in a row occur. What is the expected wait until this happens?

5.81 The Garcias want to have a son and a daughter. What is the expected value of the number of children that they will need to achieve their objective? Assume that male and female children are equally likely.

5.82 Jimmy the Greek has been quoted as saying that

> The smart professional gambler, when heads comes up four times in a row, will bet that it comes up again.... The amateur bettor will figure that heads can't come up again, that tails is "due."

Do you agree with the professional or the amateur?

• 5.83 In the binomial distribution, what is the variance of X, the number of successes in n trials? For what value of π is this variance the largest?

5.84 Explain and evaluate this sportswriter's use of the law of averages:

> Thursday night's game was another chapter in the Raiders' stunning success in prime time. Put the Raiders on Monday night, Thursday night, Sunday night football and the only thing in the opponent's favor is the law of averages. Thursday night's win gave the Raiders a 22-2-1 record in prime time.[19]

5.85 *(Hypergeometric)* A firm has 100 accounts receivable, of which 10 are delinquent. If an outside auditor samples 10 of these accounts, what is the probability that it will find exactly 1 to be delinquent?

If it samples 20 accounts, what is the probability that it will find 2 to be delinquent?

5.86 *(Hypergeometric)* A sociologist claims that tall persons have an advantage in the MBA job market. To test this theory, the sociologist examined the job offers of the 20 male students with the highest grades from a prestigious business school in 1990. (A similar comparison was made for 20 female students.) The students were divided into two groups of 10 according to their heights (short and tall) and into two groups of 10 according to their starting salaries (rich and poor). Seven of the 10 students with the highest salaries were tall; only 3 were short. What is the probability that 7 or more of the students with high salaries would be tall if height were unimportant?

5.87 *(Hypergeometric)* In the game Joker Seven, a standard deck of 52 cards and one joker is shuffled and 7 cards are dealt. The payoffs for five of the $1 bets that can be made are as follow. Calculate the expected value of the payoff for each of these five bets.

A particular card will be red	$2
There will be exactly 4 red cards	3
There will be exactly 5 red cards	6
There will be 6 or 7 red cards	10
The joker will appear	6

5.88 *(Hypergeometric)* The President is preparing for a news conference at which he will answer questions from 5 randomly selected journalists. His aides have compiled a list of 25 questions, each equally likely to be asked, and sketches of knowledgeable responses. If he concentrates on 10 of these questions and neglects the other 15, what is the probability that he will be well prepared for all 5 questions that are asked? For at least 4 of 5? (Assume that no journalist asks a question not on this list.)

5.89 *(Hypergeometric)* A bag contains 4 red balls, 5 white balls, and 6 blue balls. What is the probability that 3 balls drawn out of the bag will be red, white, and blue (not necessarily in that order) if the drawing is with replacement? Without replacement?

5.90 *(Poisson)* In the manufacture of a certain cloth, there is, on average, one defect per square yard of cloth. What is the probability that a square yard of this cloth will have no defects? Exactly one defect? Exactly two defects? More than two defects?

5.91 *(Poisson)* Here are data on the number of fatal automobile accidents in the United States in 1983, per 100 million vehicle miles:

Total	2.30
urban	1.78
rural	3.01
Interstate	1.07
urban	0.90
rural	1.29
Noninterstate	2.62
urban	2.00
rural	3.46

Consider a 20-mile stretch of road, carrying 5 million vehicles a year. Use the Poisson distribution to determine the probability of 3 or more fatal accidents if this road is an urban interstate. If the road is a rural noninterstate.

5.92 *(Poisson)* One of the most famous applications of the Poisson distribution was to the number of German calvarymen killed by horse kicks each year during the period 1875 to 1894. There were a total of 196 deaths for 14 corps over this 20-year period. If we let our unit of measurement be a single corps during a year's time, then the average number of deaths was $\mu = 196/280 = .7$. What is the probability that a corps will have no deaths from horsekicks during a year's time? One death? Two deaths? Three deaths? Four deaths? More than four deaths?

5.93 A stock market adviser guarantees that the portfolio of New York Stock Exchange (NYSE) stocks that he recommends at the beginning of each year will not decrease in value. If his portfolios, in fact, have a 2/3 probability of increasing in value and a 1/3 probability of decreasing (and the returns are independent from year to year), what is the expected number of years until his pledge is broken?

5.94 Explain the error in this reasoning: "My next marriage is sure to work; the odds against five divorces are astronomical."

5.95 Mr. Wrists has a .3 probability of getting a hit each time he bats for the Sagehens. If the outcomes of each time at bat are independent, what are Mr. Wrist's probabilities of getting 4 hits in 4 times at bat? Of getting 3 hits in 4 at-bats? What would you predict Mr. Wrist's lifetime batting average to be? Does the confidence with which you make this prediction depend on how many times he bats during his career?

5.96 "Five out of 10 people who have this disease die from it. It is lucky you came to me; my last 5 patients all died." Would you be comforted by such advice? If there is a .5 probability of dying from this disease, what is the probability that 5 out of 5 patients will die from it? That 9 out of 10 patients will die from it?

• **5.97** A baseball fan wants to collect a complete set of 100 baseball cards. The only way to obtain such cards is to buy one card at a time, for $1 apiece. Each card is equally likely to be obtained and the identity of the card is not revealed until after the card has been purchased and unwrapped. What is the expected value of the cost of obtaining a complete set?

5.98 Pete Rose and Wee Willie Keeler both hit safely in 44 consecutive baseball games. What is the probability that a .300 hitter will have at least one hit in each of the next 44 baseball games? In 1941 Joe DiMaggio hit safely in 56 games. What is the probability that a .325 hitter will have at least one hit in each of the next 56 baseball games? Assume that the Bernoulli trial model applies and that a player has four times at bat in each game.

•• **5.99** An airline finds that, on average, 20 percent of the reservations made on a certain 20-seat commuter flight turn out to be no-shows. Use the binomial tables to find the probability distribution for the number of empty seats and the number of overbooked passengers who show up and are bumped, if this airline accepts 20 reservations. If it accepts 25 reservations. (Neglect probabilities less than .001.) If the cost of an empty seat is $25 and the cost of an irate bumped passenger is $50, is the expected cost lower if this airline accepts 20 or 25 reservations?

5.100 Imagine that John and Mary are each going to repeat the following experiment 10 times: flip a coin n times and record the number of heads and tails. John uses $n = 1,000$ and Mary uses $n = 100$. Do you think John is more likely to get 500 heads on his first try or that Mary is more likely to get 50 heads on her first try? Do you think that John is more likely to get 500 or 501 heads on his first experiment? Looking at the 10 experiments, do you think that John or Mary will show greater variation in the number of heads? In the percentage of heads?

6

The Normal Distribution

I know of scarcely anything so apt to impress the imagination as the wonderful form of cosmic order expressed by the "Law of Frequency of Error." The Law would have been personified by the Greeks and deified, if they had known of it. . . . Whenever a large sample of chaotic elements are taken in hand and marshaled in the order of their magnitude, an unsuspected and most beautiful form of regularity proves to have been latent all along. The tops of the marshaled row form a flowing curve of invariable proportions; and each element, as it is sorted into place, finds, as it were, a preordained niche accurately adapted to it. . . .

Sir Francis Galton

TOPICS

We examined the binomial distribution in great detail because of its wide applicability and its important role in the intellectual history of probability analysis. In this chapter we will study a distribution that was even more important in the development of probability and statistics and has wider applications—a distribution so pervasive that it is called the *normal distribution.* It is, by far, the most widely used probability distribution. It is so popular that, when a statistician is unsure of the appropriate probabilities, a normal distribution will be usually assumed. In this chapter, you will learn what the normal distribution is and why it is so alluring. But, first, I will explain the difference between discrete and continuous probability distributions, because we have so far only used the former, and the normal distribution is an example of the latter.

6.1   CONTINUOUS PROBABILITY DISTRIBUTIONS

All of the examples analyzed in earlier chapters involved *discrete* random variables, in which there were a countable number of possible outcomes. The coin could land heads or tails, the die could come up 1 through 6, and the student could answer 0 to 20 questions right.

A Spinner

In theory, however, some random variables can take on a continuum of values. Figure 6.1 shows one example. I've drawn a circle similar to the face of a clock, but instead of clock hands, there is a spinner that will randomly select some point on the circle. We can imagine that this is a clean, well-balanced device in which each point is equally likely to be picked, but how many points are there on this circle? In theory, there are an infinite number.

Let's say that the perimeter of the circle measures 12 inches. If our markings were so crude that we could only divide the circle into 12 equal 1-inch intervals, then the probability of stopping in any particular interval would be 1/12. If we used

Figure 6.1   *Pick a Number, Any Number*

finer markings to divide the circle into 120 intervals of length 0.1 inch, then the probability of stopping in any particular interval would be 1/120. If we could accurately measure intervals to five places, there would be 1,200,000 intervals of length 0.00001 inch; the probability for each interval would be 1/1,200,000. As we measure the stopping place more precisely, we get a larger number of possible outcomes and a smaller probability of observing any particular outcome. With infinite precision, we could identify an infinite number of points, and the probability of hitting any one point would be zero.

This argument suggests how mathematicians handle the vexing situation of an uncountable number of possible outcomes. There is a nonzero probability for a measurable interval of outcomes, but the probability of any specific outcome must be zero. The probability that the spinner will stop at some point between 1.0 and 2.0 is 1/12. The probability that the spinner will precisely hit the point 1.33 . . . is zero.

We can display these probabilities by using the continuous probability distribution shown in Figure 6.2. This distribution is constructed in such a way that the probability that the outcome is within a particular interval is shown by the corresponding area under the probability distribution. The shaded area, for example, shows the probability that the spinner will stop between 1 and 2. This area is (base)(height) = (1)(1/12) = 1/12. Similarly, what is the probability that the spinner will stop between 0 and 12? This probability is the entire area under the probability distribution. We have (base)(height) = (12)(1/12) = 1, as it should be.

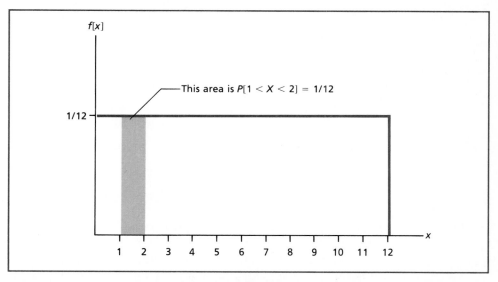

Figure 6.2 *A Continuous Probability Distribution for the Spinner Shown in Figure 6.1*

The Formal Distinction

The formal distinction between the discrete and continuous probability distributions can be summarized as follows:

> In a **discrete probability distribution,** the possible outcomes are countable. We use a discrete random variable X and a discrete probability distribution $p[x]$. Each of the possible outcomes can have a nonzero probability.

A graph of a discrete probability distribution will (as in Figure 4.1) consist of probability spikes showing the probability of each outcome.

> In a **continuous probability distribution,** the possible outcomes are not countable. We use a continuous random variable X and a continuous probability density function $f[x]$. Each possible outcome has zero probability, while an interval of possible outcomes has a nonzero probability.

A graph of a continuous probability density function will (as in Figure 6.2) show the probability of an interval of outcomes as the area under the density function.

A Few Details

Notice that we use the different symbols $p[x]$ and $f[x]$ to distinguish discrete and continuous probability distributions. The special symbol $f[x]$ is intended to remind us that $f[x] = 1/12$ is a continuous probability density function, not a probability. The density function $f[x]$ gives the height of the probability distribution at the point x. At $X = 1.33\ldots, f[x] = 1/12$, while $p[x] = 0$. Therefore, for a continuous random variable, nonzero probabilities only apply to intervals of outcomes, such as $P[1.0 < X < 2.0] = 1/12$.

Another fine detail is that, for continuous random variables, it doesn't really matter whether the endpoints are included in an interval. All of the following probabilities are the same:

$$P[1.0 < X < 2.0] = 1/12$$

$$P[1.0 \leq X < 2.0] = 1/12$$

$$P[1.0 < X \leq 2.0] = 1/12$$

$$P[1.0 \leq X \leq 2.0] = 1/12$$

because $P[X = 1.0] = P[X = 2.0] = 0$.

A final point is that, in practice, our data are never really continuous. We cannot measure anything to an infinite number of places. Our measurement limitations force us to aggregate all points within a certain interval. Perhaps, in our spinner example, we can make accurate measurements to three decimal places. The point 1.333 can be distinguished from 1.334, but we cannot separate 1.3331 from

1.3332 (so, we call them both 1.333). Our measurement limitations then force us to divide the circle into 12,000 intervals of length 0.001. The probability of hitting any one of these intervals is 1/12,000, not zero.

We can recognize that a circle contains an infinite number of points, even while acknowledging our inability to identify precisely which point is hit. The advantage of a theoretical image of a continuum of points is that it allows us to readily deduce the probability of landing in an interval of any size, and it is certainly easier to draw a continuous probability distribution than to draw 12,000 probability spikes.

EXERCISES

6.1 Which of the following variables are discrete and which are continuous?   **a.** male births   **b.** height   **c.** weight   **d.** rainfall   **e.** income   **f.** number of cars sold   **g.** votes received by a candidate   **h.** time

6.2 Figure 6.1 shows a spinner on a numbered circle. What is the probability that the spinner stops **a.** between 6 and 12?   **b.** between 9 and 12?   **c.** between 3 and 9?   **d.** between 1 and 2?

6.3 My computer has a random-number generator that will select a digit from 0 to 9. Each of these ten digits is equally likely to be picked. I am going to use this generator to randomly select a number between 0 and 1.   **a.** I let the random-number generator pick a single digit, which I put to the right of a decimal point. If, for example, the selected digit is 3, then my answer is 0.3. Graph the probability distribution for my answer.   **b.** I let the random-number generator pick a digit, which I put to the right of a decimal point. Then, I let the random number generator pick a second digit, which I put to the right of the first. If, for example, the computer selects a 3 and

then a 7, then my answer is 0.37. I then let the computer pick a third digit, which I put to the right of the first two. I can continue this process indefinitely, to obtain as precise a fraction as I want. Graph the probability distribution for my final answer.

6.4 A certain continuous random variable X has this continuous probability density function:

$$f[x] = 1, \quad 0 < x < 1$$
$$f[x] = 0, \quad \text{otherwise}$$

Graph this density function and determine the following probabilities: **a.** $P[0 \leq X \leq 0.5]$ **b.** $P[0.25 \leq X \leq 0.75]$ **c.** $P[X \leq 0]$ **d.** $P[X \geq 0.5]$ **e.** $P[X = 0.5]$

6.5 A continuous random variable X has this probability density function

$$f[x] = 2 - 2X, \quad 0 \leq x \leq 1$$
$$f[x] = 0, \quad \text{otherwise}$$

Graph this density function and determine the following probabilities: **a.** $P[0 \leq X \leq 1]$ **b.** $P[X = 0]$ **c.** $P[0.5 \leq X]$ **d.** $P[X \leq 0.5]$ **e.** $P[X \leq 0.25]$

6.2   STANDARDIZING VARIABLES

Random variables can take on a wide variety of units: dollars, inches, pounds, percentages, or whatever. Usually it is easiest to interpret our models and results if we measure variables in these natural units. But, in the eighteenth and nineteenth centuries, scientists discovered that when variables are *standardized*, in a way that I'll soon show you, the probability distributions often look remarkably similar.

Many random variables are the *cumulative* result of a sequence of random events. For instance, a random variable giving the *sum* of the numbers when eight dice are rolled can be viewed as the cumulative result of eight separate random events. A second example is the percentage change in stock prices over the next six months. We can view this six-month change as the *cumulative* result of a great many daily, hourly, or even more frequent random price changes. A third example is the height of 11-year-old males. At birth, each person's future height is unknown and, therefore, is a random variable. Again, we can think of the eventual height as the *cumulative* result of a great many events, some hereditary and some having to do with diet, exercise, and so on.

These three examples involve very different units of measurement. In addition, the values of each random variable depend on our horizon—whether we are looking at the roll of 8 dice or 800 dice, stock prices six months hence or six years hence, the heights of 11-year-olds or 21-year-olds. In each case, as we lengthen the horizon we might anticipate that both the expected value and the standard deviation of the variable will change. If we were to graph the probability distributions for varying horizons of these quite different variables, they would look very dissimilar, but researchers have found that, when the variables are standardized, the probability distributions are strikingly similar! This remarkable similarity is the most important discovery in the long history of probability and statistics.

A Mean of Zero and a Standard Deviation of One

We have seen that two of the most important tools for describing differences among probability distributions are the expected value and the standard deviation. A natural standardization would be to transform the units of our random variables somehow, so that they would all have the same expected value and the same standard deviation. This reshaping is easily done in the statistical beauty parlor.

To standardize a random variable X, we subtract its expected value μ and then divide by its standard deviation σ:

$$Z = \frac{X - \mu}{\sigma} \tag{6.1}$$

If we recall the linear transformation Equations 4.4 and 4.5, we can easily see that, no matter what the initial units of X are, the **standardized random variable** Z has an expected value of zero and a standard deviation of 1:*

$$E\left[\frac{X - \mu}{\sigma}\right] = \frac{1}{\sigma}(E[X] - \mu) = \frac{1}{\sigma}(\mu - \mu) = 0$$

$$\text{std. dev. of } \frac{(X - \mu)}{\sigma} = \frac{1}{\sigma}(\text{std. dev. of } X) = \frac{\sigma}{\sigma} = 1$$

The standardized variable Z measures how many standard deviations X is above or below its mean. If X is equal to its mean, then Z is equal to zero. If X is one

---

*The standardized variable $(X - \mu)/\sigma$ can be rearranged as $(-\mu/\sigma)X$, which has a standard deviation $(1/\sigma)\sigma = 1$ because Equation 4.5 says that the standard deviation of $a + bX$ is $b\sigma$.

standard deviation above its mean, then Z is equal to 1. If X is two standard deviations below its mean, then Z is equal to -2. Suppose for instance, that X has a mean of 20 and a standard deviation of 10. Then, here are some corresponding values of X and Z:

X	Z
0	-2
10	-1
20	0
30	$+1$
40	$+2$

Some Standardized Distributions

Let's look at some examples. Any random variable can be standardized to have a zero mean and a standard deviation of 1. Here we are especially interested in random variables that can be viewed as the *cumulative* result of a sequence of random events. First, consider the roll of a single six-sided die, letting the random variable X be the number that appears. We can readily calculate the expected value and standard deviation to be $\mu = 3.5$ and $\sigma = 1.708$.* Table 6.1 shows how these numbers can be used to construct the standardized variable Z. A bit of arithmetic confirms that, as intended, Z does have an expected value of zero and a standard deviation of 1.

In the same fashion, I have constructed standardized variables for the *sum* of the numbers when $n = 2, 4,$ or 8 dice are rolled. These probability distributions are shown in Figure 6.3. Notice how each distribution is centered about zero. Notice also, that as n increases, the probability distribution loses its "block" shape and becomes increasingly "bell" shaped.

Table 6.1 *Standardizing the Roll of a Die*

P	X	$Z = (X - 3.5)/1.7$
1/6	1	-1.464
1/6	2	-0.878
1/6	3	-0.293
1/6	4	0.293
1/6	5	0.878
1/6	6	1.464

*Here are the details:

$$\mu = 1(1/6) + 2(1/6) + 3(1/6) + 4(1/6) + 5(1/6) + 6(1/6) = 21/6 = 3.5$$
$$\sigma^2 = (1 - 3.5)^2(1/6) + (2 - 3.5)^2(1/6) + (3 - 3.5)^2(1/6)$$
$$+ (4 - 3.5)^2(1/6) + (5 - 3.5)^2(1/6) + (6 - 3.5)^2(1/6) = 17.5/6 = 2.917$$
$$\sigma = \sqrt{2.917} = 1.708$$

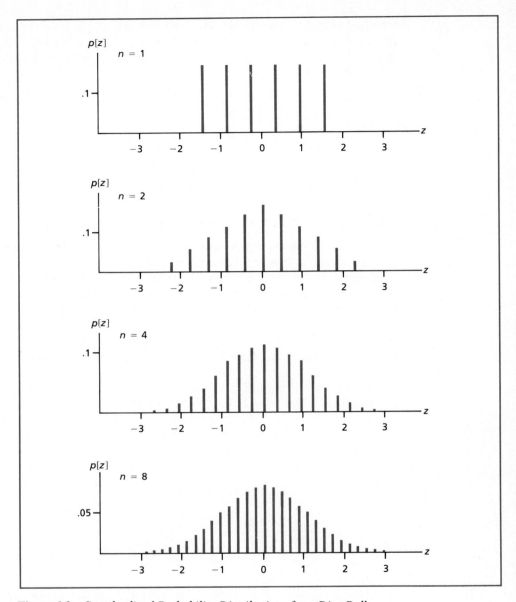

Figure 6.3 *Standardized Probability Distributions for n Dice Rolls*

For a second example, let X be the number of heads that come up when n coins are tossed. This is a binomial distribution with $\pi = .5$. The random variable X is the *sum* of the successes in n flips. Figure 6.4 shows the standardized probability distributions for $n = 1, 2, 10, 20$, and 50. Again, a distribution that is very unextraordinary for $n = 1$ turns into a pleasing bell shape as n increases.

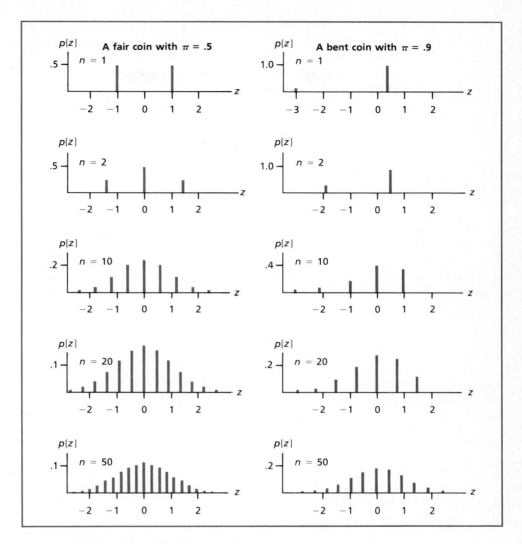

Figure 6.4   *Standardized Probability Distributions for n Coin Flips*

Our first two examples involve simple symmetrical distributions. Now let's do a strongly skewed distribution. Imagine a sharply bent coin with a .9 probability of landing heads. This bent coin is a proxy for any sequence of events, each with a .9 probability of "success." Again, the random variable X is the *sum* of the successes. Figure 6.4 shows that, as n increases, even a very skewed distribution is gradually transformed into the now-familiar bell shape.

Look again at Figures 6.3 and 6.4. We began with three very different probability distributions. But as we *cumulated* a sequence of these varied uncertain events, the same bell-shaped probability distribution kept emerging. You can imagine the excitement that scientists must have felt when they first discovered this regularity. Here they were modeling uncertainty and, amidst the chaos, a regular pattern kept reappearing. Even more unexpectedly, they were eventually able to prove that this same bell-shaped distribution almost always appears when the outcomes are cumulated, no matter what the shape of the initial probability distribution! No wonder Sir Francis Galton called it a "wonderful form of cosmic order" deserving deification.

EXERCISES

6.6 Use the data in Table 6.1 on the roll of a single die to calculate the mean and standard deviation of Z.

6.7 Let the random variable X be the number 1, 2, 3, or 4 that comes up when a four-sided die is rolled. Graph the probability distribution of X. Calculate the expected value μ and standard deviation σ of X. Now compute the possible values of the standardized variable $Z = (X - \mu)/\sigma$. What are the expected value and standard deviation of Z?

6.8 Elizabeth is just starting her career, teaching English as a Second Language (ESL). In one of her textbooks she read about grading by the curve, so she decides to give an F for $Z < -1.5$, a D for $-1.5 \leq Z < -0.5$, a C for $-0.5 \leq Z < 0.5$, a B for $0.5 \leq Z < 1.5$, and an A for $Z \geq 1.5$. On her first test,

everyone gets 100. What grades should she give her students?

6.9 *(continuation)* On her second test, Elizabeth decides to use a standardized test that has been given nationally with a mean score of 75 and a standard deviation of 10. Elizabeth's ten students receive scores of 94, 89, 83, 81, 78, 76, 74, 71, 67, and 67. If she uses the national mean and standard deviation, what will her ten students' standardized scores be? If she then uses the same grading scale given in the preceding question, how many As, Bs, Cs, Ds, and Fs will she give on this test?

6.10 *(continuation)* If she uses the national mean and standard deviation, what unstandardized scores would a student need to receive an A, B, C, D, and F on this test?

6.3   THE NORMAL CURVE AND THE CENTRAL LIMIT THEOREM

An English mathematician, Abraham De Moivre (1667–1754), was the first to deduce the equation for the bell-shaped curve that emerged in our Figures 6.3 and 6.4.

The probability density function of a standardized continuous random variable Z with a **normal distribution** is

$$f[z] = \left(\frac{1}{\sqrt{2\pi}}\right) e^{-z^2/2} \tag{6.2}$$

This π in the equation for the normal distribution is the celebrated number 3.14159 . . . , and must not be confused with the use of the symbol π in Bernoulli trials to designate the success probability. The symbol e in the normal distribution is the base of natural logarithms, $e = 2.71828. \ldots$

The graph of Equation 6.2 is shown in Figure 6.5. Notice that the normal distribution is continuous. The probabilities are not shown by discrete spikes, but are instead given by areas under the curve. The normal curve is a continuous approximation to the spikes shown in Figures 6.3 and 6.4. In 1733, De Moivre concluded that, for a binomial distribution with a .5 probability of success, this approximation becomes virtually perfect as n increases.

The Central Limit Theorem

Nearly a century later, a French mathematician, Pierre Laplace (1749–1827), proved that this convergence is true for all success probabilities (except 0 and 1, of course). Laplace believed that his remarkable theorem could be used to explain most of the uncertainties that fill our lives, and he consequently applied the normal distribution to a wide variety of phenomena in both the natural and social sciences. Karl Gauss (1777–1855), the German "prince of mathematicians," applied the normal distribution to measurements of the shape of the earth and the movements of planets. His work was so extensive and influential that the normal distribution is now often called the *Gaussian distribution*—except in France, where it is called *Laplacean.*

Others, following in the footsteps of Laplace and Gauss, applied the normal distribution to all sorts of physical and social data. They found that these data often conform to a normal distribution, and they proved that many specific theoretical probability distributions converge to a normal distribution when they are cumu-

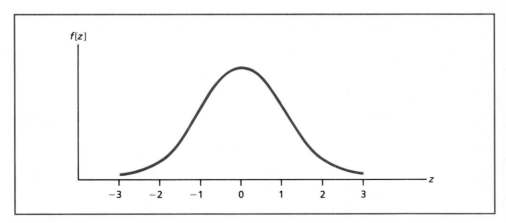

Figure 6.5 *The Normal Distribution*

lated. In the 1930s, mathematicians proved that this convergence is true of virtually all probability distributions.* This theorem, which is the culmination of 200 years of investigation, is one of the most famous mathematical theorems.

> The **central limit theorem** states that if Z is a standardized sum of any n independent, identically distributed (discrete or continuous) random variables, then the probability distribution of Z tends to a normal distribution as n increases.

The requirements are the same as for a sequence of Bernoulli trials. The probability distributions must not vary from one trial to the next and must not be affected by other outcomes in the sequence. If these assumptions are satisfied, then the sum of the outcomes will tend to the normal distribution.

As remarkable as it is, the central limit theorem would be of little practical value if the normal curve did not emerge until n was very, very large. The most important and intriguing thing about the normal distribution is that it so often appears even when n is quite small. Figures 6.3 and 6.4 showed three examples. A common rule of thumb is that the distribution of Z is usually acceptably close to the normal distribution as n gets up to 20 or 30. If the distribution for $n = 1$ is reasonably smooth and symmetrical (as with the dice roll or coin flip), then the approach to normality is very rapid. If, as with the $\pi = .9$ bent coin, the distribution is quite skewed or otherwise exotic, then the path to normality may take a bit longer.

Similarly, the central limit theorem would be just a mathematical curiosity if the Bernoulli-trial assumptions had to be strictly satisfied. The Bernoulli-trial assumptions do apply to games of chance, but in practical affairs, they are seldom if ever exactly true. Probabilities may vary slightly from one trial to the next as conditions change, and in many cases, the probabilities are affected by the outcomes of earlier trials. But, as you will see in the next section, the central limit theorem is fairly robust, in that even if the assumptions aren't quite true, the normal distribution is still a pretty good approximation. This is why the normal distribution is so popular and the central limit theorem is so celebrated.

Physical Identification

Gauss was led to the normal distribution by his study of astronomical measurements. When a distance or shape is measured repeatedly, perhaps by the same person, or by a different astronomer using different equipment, the measurements vary somewhat, due to imperfections in the equipment and its usage. But Gauss noticed that repeated measurements conform to a normal distribution—about half above average and half below, with most clustered in the middle and then tapering away in less frequent large deviations, giving a familiar bell-shaped curve. In accord with the central limit theorem, these measurement variations, called *measurement*

---

*The important restriction is that the standard deviation exists and is not equal to zero.[1]

errors, are apparently the cumulative result of a large number of small imperfections.

Even the most carefully repeated measurements with the most advanced scientific equipment vary slightly. Repeated measurements give an approximate normal distribution, and scientists use the center of this distribution (the mean) as an estimate of the "true" distance, length, weight, or other physical characteristic. The normal distribution of measurement errors also underlies what has been labeled *biometrics*—modern computerized physical identification equipment.

When we see a friend, how do we recognize the identity of that person? Somewhere in our minds we have stored a record of the person's physical characteristics (height, weight, hair color, shape of face, and so on) and when the characteristics of the person we see match our mental record, we recognize who it is. Of course, we sometimes make mistakes—when poor lighting impairs our vision or when we see only the back of the head of someone who superficially resembles a friend.

In theory, computers equipped with video cameras could be programmed to recognize people in much the same way—by storing a record of each person's physical characteristics and then matching those to the person the computer sees. Several companies have, in fact, developed computerized identification systems for use in nuclear plants, military buildings, and wherever access is to be restricted to a few authorized individuals.[2]

For instance, the Hand Scan Identimat requires the individual requesting access to insert four fingers into slots and then compares the measured length, thickness, curvature, and other physical characteristics with the values stored in the computer. The measurements for any single individual vary slightly with each use, depending on the way in which the fingers are inserted, the condition of the machine, and even the temperature that day. These measurement errors do conform to a normal distribution, as Gauss observed centuries ago with very different data. Figure 6.6 shows the distributions of measured index finger length that might be obtained for two different individuals. As indicated by the assumed cutoff in this figure, there is some chance that finger measurement of the authorized individual

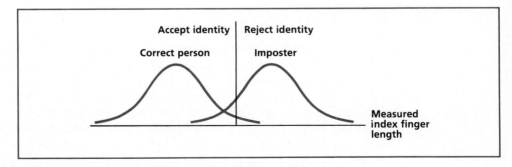

Figure 6.6 *Separating the Imposter from the Valid*

is so large that the machine incorrectly denies access and some chance that the imposter's measurement will happen to be so low that it is accepted.

Taking into account the several measured characteristics of the four fingers, the makers of the Identimat claim that there is only a 1 percent chance of rejecting an authorized person and a 1.5 percent chance of accepting an imposter. Other systems are based on fingerprints, thumbprints, palmprints, and even voiceprints. The most accurate equipment now available is Eyedentify (and most expensive at $12,000 per machine), which uses an infrared camera to examine the pattern of retinal blood vessels and reports error rates of .1 percent and .0001 percent.

EXERCISES

6.11 Explain the central limit theorem in your own words. Use some examples to show when the underlying assumptions are and are not satisfied.

6.12 Professor I.B. Gessen has a .5 probability of picking a stock whose price goes up and a .5 probability of picking a stock whose price goes down. If he picks two stocks, what is the probability that both will go up? That both will go down? That one will go up and the other down? If he picks ten stocks, give the probabilities that all ten will go up, that nine will go up and one down, and so on.

6.13 *(continuation)* Let the number of stocks that Professor Gessen selects be n and let the number that go up be X. For $n = 1, 2,$ and 10, **a.** Calculate the expected value μ of X. **b.** Calculate the standard deviation σ of X. **c.** Calculate the values of the standardized variable $Z = (X - \mu)/\sigma$. **d.** Graph the probability distribution of Z.

What do you think will happen to the probability distribution of Z as n becomes infinitely large?

6.14 There are 25 workers on an assembly line. Each has a .2 chance of making a mistake, irrespective of whether or not the other workers have made mistakes. The total number of mistakes X in assembling the product can range from 0 to 25. What are the expected value and standard deviation of X? Calculate the possible values of the standardized variable $Z = (X - \mu)/\sigma$ and graph its probability distribution. What would happen to the probability distribution of Z as the number of workers is increased indefinitely?

• **6.15** Find a computer program with a random-number generator that allows you to simulate the roll of a four-sided die. Let n be the number of dice rolled and let X be total number obtained (the sum of the numbers on the n dice). Conduct the following experiments for $n = 1, 2,$ and 8. **a.** Make the n rolls 1,000 times. **b.** Calculate the relative frequencies with which each total X appears. **c.** Graph this observed frequency distribution.

6.4   CLOSE ENOUGH? (OPTIONAL)

In practice, researchers commonly appeal to the central limit theorem when n is large even if the Bernoulli-trial assumptions are not strictly met. We may have a random variable, like the height of an 11-year-old boy, which logic tells us will be the cumulative result of a large number of uncertain events. We imagine that a boy's height depends on such factors as heredity, diet, health, and exercise, but we really have little, if any, information about the underlying probability distributions and their independence. Nonetheless, we might hope tentatively that the number of specific events that influence height is so very large that an approximately normal distribution may still emerge.

The boxed example on page 248 shows how an early study of heights did, in fact, find a striking correspondence to the normal distribution. The normal distribution also appears when we examine weights, the breadths of human skulls, IQ scores, the number of kernels on ears of corn, the number of ridges on scallop shells, the number of hairs on dogs, the heights of trees, and even the number of leaves on trees. If some phenomenon appears to be the cumulative result of a great many separate influences, then the normal distribution is quite often a very useful approximation.

Varying Probabilities

To demonstrate the robustness of the central limit theorem more formally, let's consider first a sequence of independent events that have varying probability distributions. Each event can be a success or a failure, but the probability of success is not constant. For example, a baseball player's chances of a base hit depend on the opposing pitcher, the ballpark, the month, and whether it is a day game or a night game. What would a probability distribution for the *total* base hits by this player over the coming season look like? We cannot formally invoke the central limit theorem because we are violating the assumption of constant probabilities. Yet, maybe the theorem is robust enough to tolerate some deviations.

This scenario of varying probabilities also applies to a student's *total* test score if the questions vary in difficulty and, similarly, to the *total* number of defective products made by workers of varying skills. It would also apply to the *total* amount of dollars withdrawn from a bank because withdrawal probabilities vary from account to account and from day to day. Clearly, there are many situations where varying probabilities are more realistic than constant probabilities. Will normality still emerge if we have a large number of trials?

Here's one experiment that offers some hope. Let's consider 19 independent events with success probabilities .05, .10, .15, . . . , .95. That is, on one occasion, the probability of success is .05; another time, the probability of success is .10; and yet another time, it is .15. It does not matter whether these probabilities appear in any particular order, only that the 19 events have these 19 different (in fact, very different) probabilities. Now, as per the central limit theorem, let's look at the total number of successes, a random variable that can be as low as zero or as high as 19. With considerable effort, I computed the standardized probability distribution and graphed it in Figure 6.7. As you probably anticipated, it has that familiar bell shape again.

There is a rough, intuitive explanation. First, the extremes are almost certainly going to be very improbable. There is only one way to get $X = 19$: to have 19 successes in a row, a most improbable event. Nineteen failures in a row ($X = 0$) is also quite unlikely. The results $X = 18$ and $X = 1$ are more likely, because there are more ways in which they could occur. For $X = 18$, we could have 18 successes and then a failure, 17 successes followed by a failure and then a success, and so on—19 different sequences in all, as compared to only 1 possible sequence for $X = 19$. There are similarly 19 different ways to get $X = 1$, as compared to only 1

The Heights of 11-Year-Old Boys

In the late 1800s, H.P. Bowditch measured the heights of several thousand Massachusetts schoolchildren.[3] His data for 1293 11-year-old boys are shown in Figure 1. These 1293 boys had an average height of 52.9 inches, with a 2.5-inch standard deviation. We can standardize his height data by subtracting the mean and dividing by the standard deviation. The spikes in Figure 1 show the distribution of these standardized heights among the 1293 boys he measured. We can think of each spike as the probability that a boy who is randomly selected from these 1293 children will be a certain height. Alternatively, we can view these 1293 heights as relative frequencies that can be used to estimate the distribution of heights for all 11-year-old boys at the time that this study was conducted. In either case, the normal distribution appears to be a good approximation.

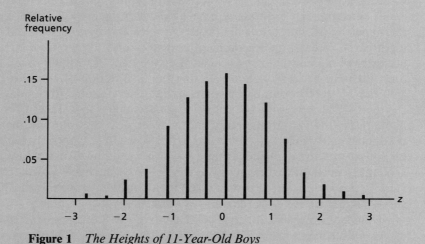

Figure 1   *The Heights of 11-Year-Old Boys*

way to get $X = 0$. There are $\binom{19}{2} = 171$ possible sequences for $X = 17$ and for $X = 2$, and $\binom{19}{3}$ possible sequences for $X = 16$ and $X = 3$. This escalation of possible sequences peaks in the middle at $\binom{19}{10} = \binom{19}{9} = 92{,}378$ for $X = 10$ and $X = 9$. There are many, many ways to get a middling number of successes and very few ways to get an extreme result. This strong sequence-counting pattern overwhelms the varying probabilities and produces a bell-shaped probability distribution.*

*This argument does not depend on the probabilities varying evenly from .05 to .95. They could instead vary from .2 to .4 (as with batting averages) or even from .01 to .05 (as with bank withdrawals or defective products). We saw in the bent-coin example how the cumulation of n outcomes makes skewed distributions become symmetrical as n increases.

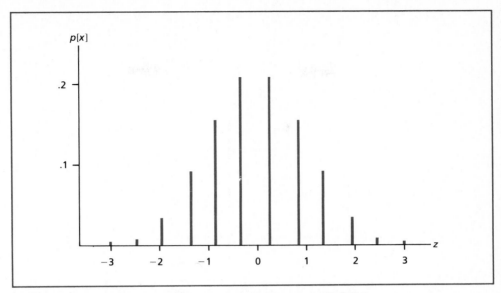

Figure 6.7   *Nineteen Independent Events with Varying Probabilities*

Events That Are Not Always Independent

For a second example, let's consider a hypothetical sequence of *n* nonindependent events. The probability of success is .5 on the odd-numbered trials. On the even-numbered trials, the probability of success is .7 if the preceding trial was a success and .3 if it was a failure. This scheme is a stylized model of a situation in which some of the events are dependent and some are independent. One illustration would be an athlete who has periodic hot and cold spells as physical and mental conditions fluctuate. Another illustration is deposit withdrawals or stock prices that are dependent during the course of a particular day but independent between days. A final illustration is heights. Some influences, such as lifetime dietary habits, are not independent. A person with an advantageous diet at age 13 will probably also have an advantageous diet at age 14. Other influences, such as genetics, are largely independent of diet.

   Well, it turns out that some independence is enough to satisfy the central limit theorem. The probability distribution for the number of successes on the first and the (dependent) second trial is shown in Figure 6.8. This very same distribution applies to any odd-even pair of trials: 1 and 2, 3 and 4, and so on. As these paired trials are independent of each other, we can think of, say, 16 trials as really being 8 independent paired trials, each with the same probability distribution for *n* = 2 that is shown in Figure 6.8. It follows that the central limit theorem applies. Figure 6.8 shows that the familiar bell shape is, indeed, apparent by the sixteenth trial.

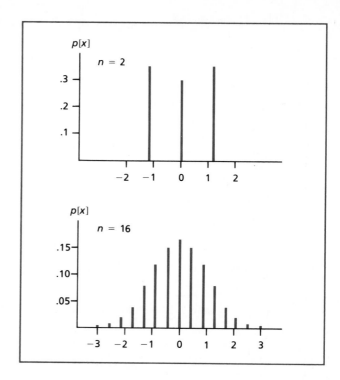

Figure 6.8 *A Mixture of Dependent and Independent Events*

Events That Are Never Independent

For one final example, consider a scenario in which events are never really independent. For the first event, there is a .5 probability of success and a .5 probability of failure. Thereafter, whenever a success occurs, the probability is .7 that the next outcome will be a success and .3 that it will be a failure. Whenever a failure occurs, the probability that the next event will also be a failure is .7, while the probability of a success is .3.

This is a stylized representation of a situation in which events tend to repeat themselves. The persistence might be due to psychological factors, such as the confident athlete doing well and the discouraged athlete doing poorly, or, in a Bayesian fashion, we might be repeatedly revising our probabilities in light of recent experience. Suppose we are making a string of preliminary tests for oil, water, or some mineral. After each test, we move to another nearby location and make another test. At each location, our subjective probability of a positive test result is undoubtedly influenced by the outcome of the preceding test. We will be optimistic if the last test was positive and pessimistic if the last test was negative. Similarly, when a business introduces a new product, there is uncertainty about how well it will sell. If the product starts selling well, then the firm becomes more optimistic. If sales slump, then the firm reduces its sales projections.

This persistence model also applies to situations in which momentum is apparent. A nation's economy has considerable momentum because people's spending depends on their incomes, which in turn depend on the spending of others. The chances of an increase in GNP are greater if GNP went up last month than if GNP declined. Birth rates tend to follow cycles, too. The probability that a nation will have a high birth rate this year is larger if this nation had a high birth rate last year. There is also momentum in the life of cities. A city that is improving is more likely to improve again next year. A decaying city is likely to continue to decay. You can undoubtedly think of many more examples.

This model doesn't conform strictly to the assumptions of the central limit theorem, but let's see what happens to the probability distribution for the number of successes in n trials. For $n = 2$, the probability distribution is the same as in the top half of Figure 6.8. The built-in momentum makes two straight successes or two straight failures more likely than one success and one failure. This distribution contrasts sharply with two independent coin flips, where the most likely outcome is one success and one failure.

As n increases, however, the probability of an extreme outcome inexorably declines. Even with momentum, the chances of 19 or 20 successes in 20 trials is very slight. With a large number of trials, it becomes more and more certain that the results will be a mixture of successes and failures. Figure 6.9 shows the standardized probability distribution for the number of successes in 25 trials. You guessed it, the bell shape again!

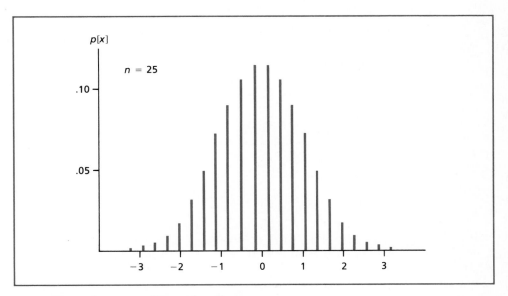

Figure 6.9   *A Sequence of Dependent Events*

Baseball Batting Averages

Baseball is a wonderful game for figuring probabilities and a great source of statistical data—the most common being batting averages. A player with a .300 batting average gets a base hit, on the average, 3 out of every 10 times at bat. (Walks don't count as official times at bat.) Each player's chances of a base hit depend on the player's mental and physical condition, the opposing pitcher, the shape of the ballpark, and even whether the game is played during the day or at night. Dedicated students of the game will collect all sorts of data on a player's performance under these varying conditions; for example, how well Robin Yount bats in August, against a certain lefthanded pitcher in night games.

Although batting is not a strict Bernoulli trial, if we look at the cumulative result of a large number of times at bat, we may find a normal distribution. To see if this is so, I looked at the annual batting averages of nine players with lifetime batting averages of about .300: Hank Aaron, Lou Brock, Al Kaline, Mickey Mantle, Willie Mays, Minnie Minoso, Joe Morgan, Duke Snider, and Carl Yastrzemski. I only looked at years in which the players had 400 times at bat. In each of these years, the season's batting average is the cumulative result of at least 400 times at bat, although admittedly not with constant and independent probabilities. Table 1 shows the data I found, and Figure 1 shows the approximate bell shape.

Apparently, a player's performance each season does conform roughly to the normal distribution. A .300-lifetime hitter averages above .300 some years and averages below .300 other years. Most years, the player's average is near .300, but there are some years in which the player's batting average is much higher or far lower than .300. The popular press attributes these "good" years and "bad" years to fluctuating skills. A statistician might find them consistent with a chance model in which the player comes to the plate over and over again, each time with about a .3 probability of getting a hit. The difference between a .320 year and a .280 year may be due to changes in the player's fortunes, rather than skills.

Table 1   *The Annual Batting Averages of Nine Lifetime .300 Hitters*

Batting Average	Number of Years	Batting Average	Number of Years
.225–.234	1	.295–.304	23
.235–.244	3	.305–.314	21
.245–.254	2	.315–.324	13
.255–.264	7	.325–.334	12
.265–.274	11	.335–.344	3
.275–.284	14	.345–.354	3
.285–.294	20	.355–.364	2
		Total	135

Figure 1   *The Distribution of Annual Batting Averages for Nine Lifetime .300 Hitters*

A Robust Theorem

The central limit theorem tells us that the standardized sum of n independent, identically distributed random variables converges to the normal distribution. In practice, though, there are many interesting situations in which the central limit theorem's assumption of independent, identical distributions is not strictly valid. Fortunately, mathematicians have proven several other limit theorems that show that the normal distribution still emerges even if these assumptions are relaxed somewhat.[4] Unfortunately, the exact conditions that are sufficient to ensure normality are too complex to be very useful to us here.

Instead, I've used several examples to illustrate the robustness of the central limit theorem. It can tolerate some dependence and some variation in each trial's distribution and still yield an approximate normal distribution for the sum of the outcomes. This is why the normal distribution is so very important. However, don't be lulled into thinking that probabilities always follow the normal curve; this is only true if the outcomes are the cumulative result of a large number of Bernoulli trials, or some facsimile thereof. It is also not true that the sum of the outcomes must conform perfectly to a normal distribution; the bell shapes that have appeared in our examples are approximately, but not perfectly, normal.

EXERCISES

6.16 Bowditch also recorded the heights of 1253 12-year-old boys:

Height (Inches)	Number	Relative Frequency
46	1	.0008
47	3	.0023
48	7	.0055
49	13	.0103
50	31	.0247
51	73	.0582
52	111	.0886
53	176	.1404
54	189	.1508
55	198	.1580
56	162	.1293
57	106	.0846
58	77	.0615
59	49	.0391
60	31	.0247
61	10	.0080
62	9	.0072
63	4	.0032
64	1	.0008
65	1	.0008
66	1	.0008
Total	1253	1.0

a. What are the mean and the standard deviation of these heights? How do these values compare with the mean and standard deviation for 11-year-old boys that are given in the text? **b.** Use your computed mean and standard deviation to calculate the 21 standardized heights. **c.** Plot the relative frequencies for these standardized heights.

6.17 I asked a local Boy Scout troop to measure the length of a school soccer field. Each of the 20 boys made five measurements, which are given in Table 6.2.

Calculate the number of observations that fall in each of the intervals 68.50–68.99, 69.00–69.49, . . . , 71.00–71.49, and then graph a histogram of these data.

Table 6.2   *Data for Exercise 6.17*

Boy	Measurement (Meters)				
	1	*2*	*3*	*4*	*5*
1	70.50	70.30	70.10	69.85	69.90
2	68.70	69.80	68.60	69.50	69.80
3	69.15	69.00	69.20	69.25	69.80
4	70.15	70.00	70.50	71.00	70.05
5	70.05	70.65	69.80	69.95	69.65
6	69.70	69.40	69.75	70.40	70.30
7	70.40	70.10	69.85	69.80	70.60
8	69.95	69.15	69.50	70.00	70.10
9	70.80	70.50	70.15	70.45	69.85
10	69.75	70.10	69.90	70.55	69.60
11	70.55	70.80	70.20	70.50	70.55
12	70.10	70.45	70.60	70.85	69.95
13	70.00	69.95	69.95	70.25	70.00
14	70.05	70.25	70.05	70.20	69.75
15	70.50	69.95	69.30	71.45	69.40
16	70.90	70.75	69.75	69.45	70.65
17	69.40	69.50	70.20	69.70	69.90
18	70.60	70.10	70.35	70.40	70.15
19	70.35	70.70	70.25	70.20	70.05
20	70.60	69.80	70.65	70.25	69.85

6.18 A 1-acre wheat field was divided into 500 plots, each measuring 0.002 of an acre, and the grain yields from each of these 500 plots were recorded. Why might the yields vary from one plot to the next? Is there any hope that the yields might be approximately normally distributed? Here is the actual yield distribution:[5]

Yield (Pounds)	Number of Plots
2.7–2.9	4
2.9–3.1	15
3.1–3.3	20
3.3–3.5	47
3.5–3.7	63
3.7–3.9	78
3.9–4.1	88
4.1–4.3	69
4.3–4.5	59
4.5–4.7	35
4.7–4.9	10
4.9–5.1	8
5.1–5.3	4
Total	500

Calculate and graph the relative frequencies.

• **6.19** The seasonal batting averages shown in Table 1 and in Figure 1 on pages 233 seem to have somewhat more extreme observations than a bell-shaped curve; for instance, examine the three years in the .235–.245 range. Can you think of any reason for this?

• **6.20** A multiple-choice test has 20 questions, each with 4 possible answers. An uninformed guesser answers these 20 questions randomly. To penalize guessers, the teacher gives 1 point for a right answer and subtracts 1/3 of a point for a wrong answer. Construct a table showing the probability distribution for the guesser's score on this test. Now calculate the mean and standard deviation of this score and use these to calculate the possible standardized scores. List these standardized scores in your table and graph the standardized probability distribution. Does this distribution look bell-shaped? Carefully explain why you might or might not expect the distribution to be bell-shaped.

6.5   THE LAW OF DISORDER (OPTIONAL)

Modern physics uses the normal distribution to describe the movements of molecules. The motion of each individual molecule is quite disordered, and yet their overall behavior is very predictable. This disordered movement is known as a *random walk*. The idea of a random walk was actually used by Laplace and others to analyze a gambler's chances of wandering into bankruptcy. Today, the random walk model is applied to many phenomena, including the stock market.

The Movement of Molecules

At the temperature $-273°C$ ($-459°F$), molecules are motionless. This completely frozen state is known as the *absolute zero* temperature. Near absolute zero, molecules have so little energy that they are locked together into a rigid solid. At higher temperatures, molecules have more energy and begin bouncing against each other. At a sufficiently high temperature, this bouncing about is violent enough to allow molecules to break their locked pattern and move around, changing the solid into a fluid. The temperature at which this melting occurs varies from substance to substance, depending on the strength of each substance's cohesive forces. Frozen hydrogen melts at $-259°C$; water (ice) melts at $0°C$; and iron melts at $1535°C$. At still higher temperatures, the molecules become so agitated that they break completely free of each other and disperse. At this point, the liquid boils and turns into a gas. Liquid hydrogen vaporizes at $-253°C$, water boils at $100°C$, and liquid iron turns into a gas at $3000°C$. At even higher temperatures (on our sun, for instance), the thermal agitation is so violent that the molecules break up into atoms and then into bare nuclei and free electrons.

A Drunkard's Walk

At absolute zero, molecules are frozen into a rigid pattern, but at higher temperatures, the molecules move about, randomly bouncing off each other. As the temperature increases, this random movement becomes more agitated and the molecules become more disordered. It becomes increasingly difficult to predict where a particular molecule will be after a few seconds of chance collisions, but we can give a probabilistic prediction, which is known as the **law of disorder.**

In the early 1800s, Robert Brown, an English botanist, watched tiny plant spores through his microscope and noticed how these spores were being randomly pushed about by the surrounding molecules. The movement of each spore was the random consequence of chance collisions with molecules that were themselves randomly bouncing off of each other. This haphazard movement is now known as *Brownian motion.* The popular analogy is a drunkard leaving a lamp post in the middle of an empty city square. Each wobbly step could be in any direction. Can we say anything about the result of this drunkard's walk?

We can. Let's pick a direction, say west-east, in which to measure the drunkard's progress. A west-east line is shown in Figure 6.10. Perhaps home is to the west and a favorite bar is to the east. After an hour, the drunkard is equally likely to be 20 feet to the west or 20 feet to the east. The expected value of the west-east position is zero. There is even more to be said; we can think of each step to the west or east as a Bernoulli trial with $\pi = .5$. If there are many such steps, then the central limit theorem implies that the sum of these steps, measuring the west-east distance from the starting point, will follow a normal distribution! The drunkard is most likely to end up slightly west or east of the lamp post. The chances of being farther west or east are given by the normal distribution.

Galton's Apparatus

The same conclusion applies to the spores that Brown watched and, indeed, to any object moving randomly. In a classroom or science exhibit (or even on the television show, *The Price Is Right*), you may have seen the apparatus (originally devised by Galton) that is depicted in Figure 6.11. Many identical balls are dropped, one

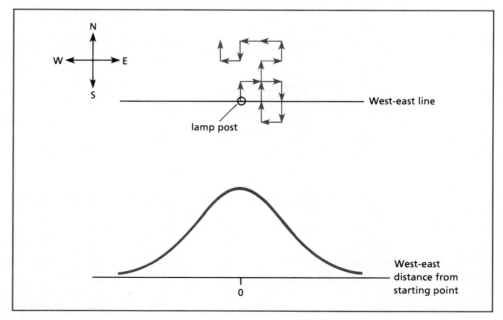

Figure 6.10   *A Drunkard's Walk*

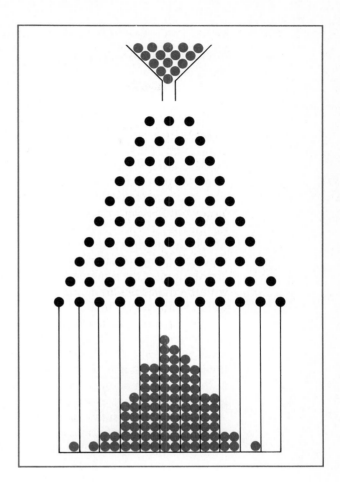

Figure 6.11 *An Apparatus for Demonstrating Brownian Motion*

at a time, onto a symmetrical array of posts. Each ball hits the first post, bounces to the left or right, and then bounces left or right again off a post in the next row. After randomly working its way through these posts, each ball drops into a bin that measures its horizontal distance from the starting point. As with the drunkard's walk, this horizontal distance is the summation of many independent left-right bounces, and as shown, the horizontal dispersion of the balls follows the normal distribution.

This apparatus is an ingenious depiction of the Brownian movement of plant spores or small particles suspended in a liquid. Einstein showed that the probability distribution for the distance that a particle will move from its initial position during a time period t is normally distributed with a mean zero and a variance $2D^2t$ where D is the liquid's *diffusion constant*. The same logic holds for molecules of gas colliding with other molecules and for small stars that are buffeted by the gravitational fields of other stars.

Some Improbable Events

We have seen how the movement of a molecule is random and yet can be probabilistically described by the normal distribution. A substance contains billions of molecules that are repeatedly involved in billions of collisions. The average outcome of these collisions is also governed by the central limit theorem. Think of the individual air molecules in the room that you are now in. Where will they be 60 seconds from now? Each air molecule will be involved in about 60 million collisions during the next minute. Some molecules will end up a bit to the right of where they are now; some will end up a bit to the left of their current position. A few will move far to the right, while others move far to the left. On the average, the dispersion of the molecules should be about the same as it is now, and that's a good thing! You would be very uncomfortable if all of the air molecules suddenly migrated to the far right corner of the room.

Such an unpleasant event is not impossible, but it is very improbable. It would require billions of air molecules to experience the far-right tails of their normal distributions. One scientist has estimated that we should expect to observe a room with the molecules all in one-half about once every $10^{299,999,999,999,999,999,999,999,999,991}$ years,[6] which is not very often.

Other scientists have figured out the probabilities of other improbable molecular arrangements. There is an extremely small chance that many more air molecules will strike this book from below than above, causing the book to leap upwards. This startling event is more likely to occur than all of the air molecules in this room bouncing to one side, but it is still not very probable. It should happen less often than once every 10^{100} years.[7]

There is also a small chance that most of the molecules in a substance will have velocities much higher or lower than the expected velocity for a given temperature.* There is a very slight probability that a teakettle of water on a hot stove will freeze. In fact, there is even a chance for the proverbial snowball in hell. One scientist estimated that the probability that a piece of celluloid placed in a temperature 2.8×10^{12} absolute (the assumed temperature of hell) would survive unburned for a week.[8] This slim probability is $1/10^{10^{50}}$, which he estimated is as likely as reproducing the manuscript of Hamlet 10^{30} times in a row without error by randomly spattering ink on paper.

EXERCISES

6.21 Two evenly matched people play backgammon for $1 a game. If they play 100 games, what is each player's expected gain? What else can you say about the possible outcomes? Explain your reasoning carefully.

6.22 A gambler bets $1 on red over and over again at a Las Vegas roulette wheel. On each bet, she has a 18/38 probability of winning $1 and a 20/38 probability of losing $1. If she plays 1,000 times, what is her expected gain? What else can you say about her

*The relevant probabilities come from Maxwell's law of velocities: the velocity in any fixed direction of a molecule with mass M in a gas at absolute temperature T is normally distributed with mean 0 and variance M/kT where k is Boltzmann's constant.

possible gains and losses after 1,000 plays?

6.23 A certain drunkard is trying to get from here to there. Every 10 seconds, this drunk takes a step, which is equally likely to be one foot to the left or right. After an hour, this drunk may be 360 feet to the left, 360 feet to the right, or somewhere in between. What is the exact probability distribution? Why would the normal distribution be a good approximation?

•• **6.24** Figure 6.11 shows an apparatus in which dropped balls form a normal distribution. This is, of course, not an exact normal distribution because the balls and the bins are discrete instead of continuous. Try to figure out the discrete probability distribution that gives the probabilities of a ball falling into the various bins. Assume that there are 10 rows of pegs and 11 bins, and that at each post the ball is equally likely to fall to the left or the right.

6.6 STOCK RETURNS (OPTIONAL)

In his 1900 doctoral dissertation, Louis Bachelier assumed that the prices of stocks, bonds, and other securities follow a random walk and that future price changes cannot be predicted from past price changes. Bachelier worked out several interesting implications of his model and he found a striking correspondence between his theories and the actual behavior of French security prices.

Bachelier's work went largely unnoticed and unappreciated until the late 1950s, when there began a flurry of research on the predictability of stock market prices. Here, too, researchers concluded that prices follow a random walk and that past movements of stock prices are of no value in predicting future price movements.[9] As with Brownian motion, there is no way to tell in advance in which direction prices will be pushed. The stock market is apparently very much like a large casino, where a playful god rolls dice to decide whether prices are going up or down.

The Efficient Market Hypothesis

What is the source of this randomness? Is it really a god rolling dice? Or is it, as Keynes suggested, the vagaries of unpredictable mass psychology? In the 1960s, a Nobel-prize-winning economist, Paul Samuelson, offered another possibility in a short mathematical paper entitled "Proof that Properly Anticipated Prices Fluctuate Randomly."[10] Samuelson's idea is now called the **efficient market hypothesis** and is used by many economists to explain the apparent randomness in stock prices.

In an efficient market, all traders have access to the same information, and therefore current prices take into account all available information. It is new information that causes price changes, with positive information pushing prices up and negative information pulling prices down. But, because new information is, by definition, unpredictable, we cannot predict in advance whether prices will go up or down.[11]

As you can imagine, this is a very provocative argument because it implies that security analysis is a waste of time. There is no need to pay lavish fees to experts who pick securities for you. You can do just as well by throwing darts at the financial pages. In later chapters, we will test this interesting theory in a variety of ways.

For now, the important point is simply that, if the returns are independent, then the normal distribution may be appropriate.

A Normal Distribution Again

If stock prices do follow a random walk, then we again have a *law of disorder,* based on the application of the central limit theorem to Brownian motion. The cumulative price change over any period of time is the sum of all of the price changes during the period. Suppose, for instance, that the price is initially $20 a share and then follows this path:

Price Change	New Price
+1/8	20 1/8
+1/4	20 3/8
−3/8	20
+1/4	20 1/4
−1/8	20 1/8
−1/4	19 7/8
+1/8	20
−1/4	19 3/4
+1/4	20
−1/8	19 7/8
−3/8	19 1/2
−1/4	19 1/4
+1/8	19 3/8
+1/8	19 1/2
+1/4	19 3/4
Total −1/4	

The net price change, from $20 a share to $19.75, is the sum of the 15 price changes during this period of time. Now, if these price changes follow a random walk (in which they are independent and identically distributed), then the central limit theorem implies that their sum, the net price change over the period, is a random variable whose probability distribution converges to the normal distribution!

This argument applies to any period of time sufficient for the net price change to be the cumulation of many price changes. We could look at annual, monthly, weekly, or daily returns. As long as there are many independent and identically distributed price changes during the interval, the net price change over some longer interval becomes normally distributed.

There is, in fact, pretty convincing evidence that stock price changes are independent. But it strains credulity to assert that probability distributions are fixed forevermore. Expected returns change with financial markets and with the anticipated returns on other assets. Standard deviations vary, too, as the future of the economy is sometimes more and sometimes less certain. Nonetheless, the robustness of the central limit theorem suggests that stock returns may still be described approximately by a normal distribution.

One Year with IBM

I collected data on the daily returns on IBM stock during 1983. We can think of each daily return as the sum of several price changes during the day as the price of IBM stock fluctuated. If these individual price changes are independent and identically distributed, then the central limit theorem implies that the daily returns will come from a normal distribution. My data are a fairly small sample of observed returns during one year. Figure 6.12 shows the frequency distribution of these daily returns and they, apparently, may have come from a normal distribution.

Sixty-Five Years with the Market

Roger Ibbotson and Rex Sinquefield[12] have assembled data on the overall monthly returns since January 1926 for the 500 stocks included in the Standard and Poor's Composite Index. If each month's return is the cumulative result of independent and identically distributed returns over shorter intervals, then the monthly returns should come from a normal distribution. Figure 6.13 supports this hypothesis.

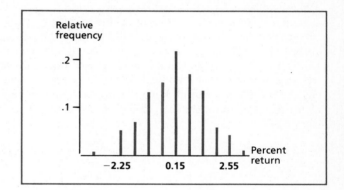

Figure 6.12 *Daily Changes in the Price of IBM Stock, 1983*

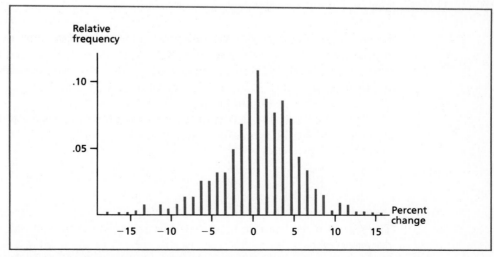

Figure 6.13 *The Distribution of Monthly Changes in Stock Prices, 1926–1980*

EXERCISES

6.25 Empirical investigations of the distribution of price changes have generally focused on speculative prices, such as the prices of stock and pork-belly futures. What assumptions are needed to justify the hypothesis that these price changes come from a normal distribution?

• **6.26** *(continuation)* Why might the daily change in the price of automobiles or the annual births in a country not come from a normal distribution?

6.27 I have sometimes used the following experiment to generate stock price patterns, which I then give to my technical-analyst friends to analyze, without explaining that the data are fictitious. (The moral is that they can't tell the difference between real and fictitious data.) The stock's initial price is set at $20. A silver dollar is flipped, with heads signifying a 25-cent increase in the stock's price and tails signifying a 25-cent decline. The net result of ten flips gives the change in the stock's price that day; for example, 6 heads and 4 tails implies that this imaginary stock went up 50 cents. One thousand flips gives a price chart over a 100-day period. What is the (exact) probability distribution for the daily dollar change in this stock's price?

• **6.28** *(continuation)* What is the probability distribution for the dollar change in the stock's price after 10 days? After 20 days? Explain why the distribution of the dollar change in this stock's price does or does not converge to a normal distribution as the horizon lengthens.

•• **6.29** Pick a stock and compute the daily percentage changes in its price over a one-year period. These data can be computed from the financial pages of newspapers at your college library. Separate these returns into ten equal intervals and make a relative frequency graph. Does it look bell-shaped?

•• **6.30** Use the financial pages of newspapers at your college library to compute the percentage change in the Dow-Jones industrial average each day the stock market was open in 1987. These percentage changes are an estimate of the daily stock market returns. (They are only estimates because they neglect dividends and because the Dow-Jones average only includes 30 stocks.) Separate these returns into 10 equal intervals and make a relative frequency graph. Does it look bell-shaped?

6.7   SUMMARY

In earlier chapters, we looked at discrete random variables such as the number of successes in a sequence of Bernoulli trials. This chapter introduced the bell-shaped normal distribution, which is continuous. With a continuous random variable X, the probability that X falls in the interval $a < X < b$ is shown by the area under the probability distribution between the points a and b.

A random variable X is standardized by subtracting its expected value μ and then dividing by the standard deviation σ:

$$Z = \frac{X - \mu}{\sigma}$$

The standardized random variable Z has an expected value of zero and a standard deviation of 1.

The central limit theorem says that if Z is the standardized sum of any n independent, identically distributed random variables, then the probability distribution of Z tends to the normal distribution. Many things are approximately normally distributed, even if: n is not very large; there is not complete independence; the probabilities are not constant from one trial to the next. As a consequence, the normal distribution is widely used by virtually all of the physical and social sciences.

REVIEW EXERCISES

6.31 A continuous random variable X has this probability density function:

$$f[x] = 1 + x, \quad -1 \leq x \leq 0$$
$$f[x] = 1 - x, \quad 0 \leq x \leq 1$$
$$f[x] = 0, \quad \text{otherwise}$$

Graph this probability distribution and determine the following probabilities: **a.** $P[-1 \leq X \leq 1]$ **b.** $P[-1 \leq X \leq 0]$ **c.** $P[0 \leq X]$ **d.** $P[0 \leq X \leq .5]$ **e.** $P[-.5 \leq X \leq .5]$

6.32 Use the data in Table 4.1 on the roll of two dice to calculate the mean and standard deviation of X. Then calculate the 11 standardized values of Z. Check your work by calculating the mean and standard deviation of Z. Graph the probability distribution of Z.

• 6.33 Find a book of baseball statistics at the library and choose 20 baseball players with lifetime batting averages of about .280 (calculated over a period of at least 10 years). This will give you annual batting averages for at least 200 seasons. (Ignore seasons with fewer than 400 times at bat.) Divide these season averages into the intervals shown in Table 1 on page 252, calculate the relative frequencies for each interval, and make a relative frequency graph such as Figure 1 on page 252. Is an approximate bell shape apparent? If not, what differences do you detect?

• 6.34 Repeat Exercise 6.15, this time using 100 rather than 1,000 rolls. Which results look more bell shaped? Do you think this is a fluke, or should we expect systematic differences in the results of 100 and 1,000 rolls?

•• 6.35 *(continuation)* Does the central limit theorem relate to the number of dice (1, 2, or 8) or the number of rolls (100 or 1,000)? Why do both seem to matter here?

6.36 Which of the following variables are discrete and which are continuous? **a.** age **b.** body temperature **c.** number of children **d.** length of right index finger **e.** number of hairs on head

6.37 Graph this probability density function for the continuous random variable X,

$$f[x] = .5, \quad 2 < x < 4$$
$$f[x] = 0, \quad \text{otherwise}$$

and find the following probabilities: **a.** $P[X > 0]$ **b.** $P[2 < X < 4]$ **c.** $P[3 < X < 4]$ **d.** $P[2.5 < X < 3.5]$ **e.** $P[X = 3]$

6.38 Graph this probability density function for the continuous random variable X,

$$f[x] = 2x, \quad 0 < x < 1$$
$$f[x] =, \quad \text{otherwise}$$

and find the following probabilities: **a.** $P[0 < X < 1]$ **b.** $P[X < .5]$ **c.** $P[X > .5]$ **d.** $P[X > .8]$

• 6.39 *(Requires calculus)* Graph this probability density function for the continuous random variable X,

$$f[x] = 1.5x - 0.75x^2, \quad 0 < x < 2$$

and then use integration to find the following probabilities: **a.** $P[0 < X < 2]$ **b.** $P[0 < X < 1]$ **c.** $P[.5 < X]$

6.40 Here is the probability density function for the continuous random variable X:

$$f[x] = ax, \qquad 0 < x < 1$$
$$f[x] = 2a - ax, \qquad 1 < x < 2$$

What is the value of a?

6.41 Which of the following events is better described by the normal distribution?

The sum of the two numbers obtained when two dice are rolled

The sum of the ten numbers obtained when ten dice are rolled

Explain your reasoning.

6.42 Is there any logical explanation for the observation that the numbers of leaves on trees (of the same species, in similar locations, and of similar age) are approximately normally distributed?

6.43 Karl Pearson and Alice Lee tabulated the heights of 1,052 mothers in 1903. Their data are shown in the following table.[13] (The 0.5 frequencies occur because those with borderline heights were counted half in one category and half in the other.) Plot these data and draw a freehand smooth curve that fits these data points. Does your curve look like a normal distribution?

Height (Inches)	Number	Height (Inches)	Number
52–53	1.5	61–62	163.0
53–54	0.5	62–63	183.0
54–55	1.0	63–64	163.0
55–56	2.0	64–65	114.5
56–57	6.5	65–66	78.5
57–58	18.0	66–67	41.0
58–59	34.5	67–68	16.0
59–60	79.5	68–69	7.5
60–61	135.5	69–70	4.5
		70–71	2.0

6.44 *(continuation)* Calculate the mean and standard deviation of the height data in Exercise 6.43. Now use this computed mean and standard deviation to calculate the values of the standardized variable Z for $X = 65$, $X = 67$, and $X = 55$.

6.45 The per-share price of a certain stock has a 0.5 probability of going up $10 and an equal probability of going down $10. (Strangely enough, no other outcomes are possible.) What are the expected value and the standard deviation of the price change? If the random variable X (with possible values $+10$ or -10) represents the price change, calculate the possible values of the random variable Z that is standardized to have an expected value of zero and standard deviation of one.

• 6.46 *(Requires calculus)* The exponential probability distribution has this density function:

$$f[x] = \lambda e^{-\lambda x}, \quad x > 0, \quad \lambda > 0, \quad \text{and } e = 2.71828$$

Graph this density function and use integration to find the following probabilities: **a.** $P[0 < X < \infty$ **b.** $P[0 < X < 1/\lambda]$

•• 6.47 *(continuation)* Determine the expected value of an exponentially distributed random variable. (*Hint:* $\int xe^{ax}\, dx = (e^{ax})/(a^2)\,(ax - 1)$ and $\mathrm{Lim}(x/e^{ax}) \to 0$ as $x \to \infty$.)

• 6.48 *(Exponential)* If the probability of y successes in a specified interval is described by the Poisson distribution,

$$P[y] = e^{-\mu}\frac{\mu^y}{y!}$$

then the probability density function for the amount of time or space x between consecutive successes is given by the exponential distribution, as in Exercise 6.46, with $\lambda = \mu$. Chapter 6 discusses the example of a submarine mission with μ the expected number of part breakdowns during a mission. If $\mu = 0.5$ for a certain type of part, use the Poisson distribution to determine the probability of zero breakdowns on a mission. Now use the exponential distribution to determine the probability that the length of time between breakdowns will be less than one mission. Do these probability calculations seem consistent or contradictory?

• 6.49 *(Exponential)* The arrival of customers at a certain store is described by the Poisson distribution with an average rate of 30 customers an hour. Show how the Poisson distribution can be used to determine the probability of 20 or fewer customers in an hour. (Just set up.) Then use the exponential distribution to determine the probability that the length of time between customers will be more than six minutes. What is the probability that the length of time between customers will be less than one minute?

• **6.50** *(Exponential)* The number of phone calls received by a certain firm during its business hours can be described by a Poisson distribution with an average rate of 10 calls an hour. Use the exponential distribution to determine the probability that the length of time between calls will be less than six minutes. What is the probability that the length of time between calls will be less than three minutes?

7

Using the Normal Curve

When you can measure what you are speaking about and express it in numbers you know something about it; but when you cannot measure it, when you cannot express it in numbers, your knowledge is of a meager and unsatisfactory kind.

Lord Kelvin

TOPICS

By now, you should be semiconvinced that the normal distribution is worth knowing. So, let's look at the details of this pervasive bell shape. The actual equation for the normal distribution,

$$f[z] = \left(\frac{1}{\sqrt{2\pi}}\right) e^{-z^2/2} \tag{7.1}$$

is pretty complicated. We could, if we had to, try to calculate probabilities by using this equation and some calculus to estimate areas under the normal curve. But fortunately, others have already calculated a great many of these probabilities for us and I've put these probabilities in Table 2 at the back of this book. It is far easier to glance at this table than to do an area computation yourself.

7.1 THE NORMAL PROBABILITY TABLE

Table 2 has been set up so that it shows the area in the right-hand tail of the standardized normal distribution, as demonstrated in Figure 7.1. The left-hand column of Table 2 shows values of Z in 0.1 intervals, such as 1.1, 1.2, and 1.3. For finer gradations, the top row shows 0.01 intervals for Z. By matching the row and column, we can, for instance, find the probability that Z is greater than 1.25. The logic of this calculation is shown in Figure 7.1. For even finer gradations, we can interpolate the probabilities shown in Table 2; for instance,

$$P[Z > 1.257] = .1056 + (7/10)(.1038 - .1056)$$

$$= .1056 - .13$$

$$= .1043$$

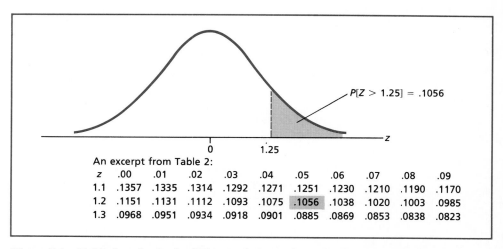

Figure 7.1 *Table 2 at the Back of This Book Gives the Right-Hand Tail Probabilities*

For probabilities other than the right-hand tails, we use a bit of logic plus the fact that the normal distribution is symmetrical and about the point $Z = 0$. One immediate implication, for instance, is that the area in a left-hand tail is the same as the area in the corresponding right-hand tail. The probability that Z is less than -1.25 is equal to the probability that Z is greater than $+1.25$. Figure 7.2 shows this equivalence.

The symmetry about $Z = 0$ also implies that there is a .50 probability that Z is greater than zero and a .50 probability that Z is less than zero. This 50-50 split allows us to find the probability that Z is between zero and some number. For instance,

$$P[0 < Z < 1.25] = .5 - P[Z \geq 1.25]$$
$$= .5 - .1056$$
$$= .3944$$

Figure 7.3 shows the logic of such a calculation.

Yet another case is the probability that Z lies between two positive numbers; for example, 0.75 and 1.25 as shown in Figure 7.4. Table 2 tells us that the probability of Z exceeding 0.75 is .2266, and we already know that the probability of Z exceeding 1.25 is .1056. Therefore, the probability that Z will be between 0.75 and 1.25 must equal the difference between these two probabilities, $.2266 - .1056 = .1210$.

For a final illustration, consider the probability that Z is between -0.5 and 1.25. Figure 7.5 shows this situation. The desired probability can be found by subtracting the excluded tail probabilities from 1.0. The probability that Z is greater than 1.25 is still .1056. Table 2 tells us that the probability that Z is less than -0.5 is .3085. Therefore, the probability that Z will be between -0.5 and 1.25 is $1.0 - .1056 - .3085 = .5859$.

With a bit of thought, Table 2 gives us the information we need to figure out all sorts of probabilities for the normal distribution. It is usually safest to make a

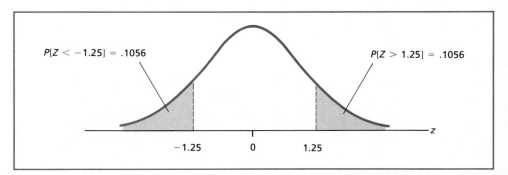

Figure 7.2   *The Probabilities in the Right-Hand and Left-Hand Tails Are Equal*

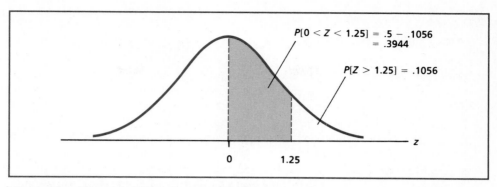

Figure 7.3 *The Probability That Z Is Between 0 and 1.25*

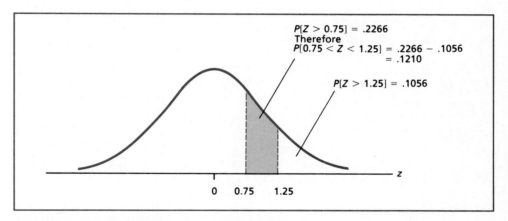

Figure 7.4 *The Probability That Z Is Between 0.75 and 1.25*

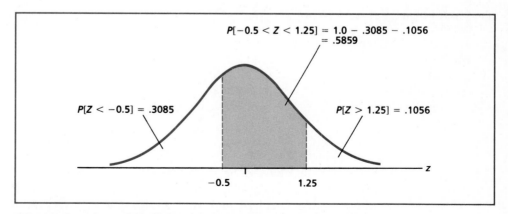

Figure 7.5 *The Probability That Z Is Between −0.5 and 1.25*

Grading by the Curve

Some professors grade by absolute standards; for example, 90 percent correct is an A score, 80 percent a B, 70 percent a C, 60 percent a D, and below 60 percent is an F. On essay questions, the professor can decide beforehand what sorts of answers merit an A grade, a B, and so on. With absolute grading standards such as these, it is possible for every student in the class to get an A or for every student to get an F.

A very different approach is to judge each student's performance relative to the others in the class, for example, "grading by the normal curve." Each student is given a numerical test score, perhaps the number of questions right or the number right minus a fraction of the number wrong. The professor then uses the mean and standard deviation of these scores to calculate a standardized score, Z, for each student. A value $Z = 0$ is an average score. Those who did above average have positive Zs, while those who scored below average have negative Zs. If the scores are approximately normally distributed, they will be symmetrically scattered about $Z = 0$, in the bell shape shown in Figure 1. This particular grading scale gives 7 percent As, 24 percent Bs, 38 percent Cs, 24 percent Ds, and 7 percent Fs.

Proponents argue that grading by the curve is an objective way of determining how difficult a test really is. If the average score is 40 out of 100, the teacher should realize that the test was very difficult and grade accordingly. Many students, though, abhor grading by the curve. They want to be judged by whether or not they have mastered the material, and not by what others have done. Why, they ask, if everyone does well, do some have to be given Ds and Fs? Is it fair that the grade distribution should be the same in a class full of good students as in a class with many weak students? (And of course, they are always in the class with all the good students.) They rightly point out that grading by the curve discourages student cooperation and camaraderie. When others do well, it makes your own performance look that much worse. Grading by the curve also encourages such unseemly practices as bribing weak students to show up for tests.

Most professors agree and use some combination of absolute and relative grading standards. They have some gut feeling about what a good score is and what a bad score is; and they will adjust their beliefs if the class does unexpectedly well or surprisingly poorly.

Figure 1   *Grading by the Curve*

quick sketch, as I have done in Figures 7.1–7.5. It is then readily apparent what numbers we have to add or subtract to arrive at the final answer. We will do several examples in this chapter—and there are, of course, exercises for you to practice on.

EXERCISES

7.1 Use Table 2 to find these five probabilities; in each case sketch the probability distribution of Z and shade in the area showing the appropriate probability: **a.** $P[Z > 0.0]$ **b.** $P[Z \geq 1.0]$ **c.** $P[Z > 2.0]$ **d.** $P[Z \geq 3.0]$ **e.** $P[Z < -1.0]$

7.2 Determine each of these five probabilities; sketch the probability distribution of Z and shade in the area showing the appropriate probability: **a.** $[-1.0 < Z < 1.0]$ **b.** $P[-2.0 < Z < 2.0]$ **c.** $P[Z > 2.0 \text{ or } Z < -2.0]$ **d.** $P[Z \geq 1.65]$ **e.** $P[0.0 < Z < 1.65]$

7.3 In each of these three cases, determine the value of z: **a.** $P[Z > z] = .500$ **b.** $P[Z > z] = .025$ **c.** $P[Z < z] = .025$

7.4 A certain statistics instructor gave a class a standardized test that had been administered throughout the United States. Nationwide, scores on this test are normally distributed with a mean of 0 and a standard deviation of 1. What fraction of the students nationwide **a.** got negative scores? **b.** got scores between 0 and 1? **c.** got scores above 3.0?

7.5 *(continuation)* This instructor wants to convert her students' scores into percentiles. That is, a score "in the 70th percentile" would be one that was better than 70 percent of the students taking the test nationwide. Find the percentiles for the following scores: **a.** Ralph, 0.0 **b.** Mike, -1.0 **c.** Connie, 2.0

7.2 NONSTANDARDIZED VARIABLES

So far, to emphasize the pervasive bell shape, we have only dealt with variables that have been standardized by subtracting the mean and dividing by the standard deviation. But what about the probability distributions of the underlying, nonstandarized variables? We can determine these probabilities via an important mathematical theorem:

If a random variable X is normally distributed, then any linear function $a + bX$ is also normally distributed.

We can use this theorem to go from a nonstandardized variable to a standardized variable, or vice versa. Let's say that X is some nonstandardized, normally distributed variable with a mean μ and standard deviation σ. We can summarize these facts with the shorthand notation

$$X \sim N[\mu, \sigma]$$

The symbol $\sim$ stands for "is distributed." The letter N signifies a normal distribution, with the first number inside the brackets giving the mean and the second giving the standard deviation. If X is normally distributed with a mean of 100 and a standard deviation of 15, we write $X \sim N[100, 15]$. For a standardized variable, we write $Z \sim N[0, 1]$. The theorem implies that there is a close relationship between standardized and nonstandardized variables.

$$X \sim N[\mu, \sigma] \text{ implies } Z = \frac{X - \mu}{\sigma} \sim N[0, 1] \qquad (7.2)$$

$$Z \sim N[0, 1] \text{ implies } X = \mu + \sigma Z \sim N[\mu, \sigma] \qquad (7.3)$$

In practice, we typically begin with a variable X in its natural units—dollars, feet, pounds, or whatever. Then we standardize this variable by subtracting its mean μ and dividing this difference by its standard deviation σ,

$$Z = \frac{X - \mu}{\sigma}$$

If this standardized variable Z has the $N[0, 1]$ distribution that is shown in Figures 7.1–7.5 and spelled out in Table 2, then X must have a normal distribution, too. The only differences are that the mean of X is μ instead of 0 and the standard deviation of X is σ instead of 1. But these are mere cosmetic differences. No matter what the values of μ and σ, all normal distributions are closely related in that they all have the very same probability of being a certain number of standard deviations away from their respective means. Stated somewhat differently, the only thing that distinguishes one normal distribution from another is that they have different means and standard deviations. Or, as the son of the great language mangler, Yogi Berra, explained, "Our similarities are different."

Other Means, Other Standard Deviations

The mean and standard deviation of a normal probability distribution tell us two things: location and dispersion. The mean tells us where the center of the distribution is, while the standard deviation tells us how spread out the distribution is. Together, the mean and standard deviation describe a normal distribution.

Figure 7.6 shows two normal distributions with the same standard deviation, but with different means. Perhaps these random variables are the percentage returns on two different investments. One investment has a 10 percent expected

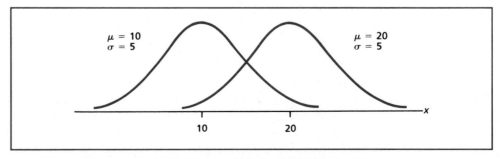

Figure 7.6   *The Mean Locates the Center of a Normal Distribution*

return and the other has a 20 percent expected return. In the long run, it is almost certain that the second investment will have the higher average return. In the short run, the equal standard deviations suggest that these two investments are equally risky, in that their returns are equally uncertain. For instance, the probability that the return will turn out to be 5 to 10 percentage points less than the expected return is the same for both investments. It is natural to describe these two investments by saying that they have expected returns of 10 percent and 20 percent and that we are equally confident of how close the returns will be to their respective expected values. If offered such a choice, most investors would choose the second investment.

Figure 7.7 shows a different situation. Here, the expected values are the same, but the standard deviations differ. If these are percentage returns, then all three investments should have nearly equal average returns in the long run. But, for the short run of a single outcome, the unequal standard deviations imply that these alternative investments are not equally risky. Look again at Figure 7.7. For the investment with a standard deviation of 2.5 percent, we are almost certain that the return will turn out to be between 5 and 15 percent. With $\sigma = 5$ percent, however, there is a good chance that the return will be outside this 5 to 15 percent range. For the investment with $\sigma = 10$ percent, it is even more likely that the actual return will be far from its expected value. Risk-averse investors will prefer the investment with the low standard deviation, because it is expected to do as well, on average, as the others and its return is more certain.

Figure 7.8 shows two normal distributions with different means and standard deviations. The distribution with $\mu = 8$ percent is centered to the right of the one with $\mu = 5$ percent. The first distribution is also more spread out because it has a

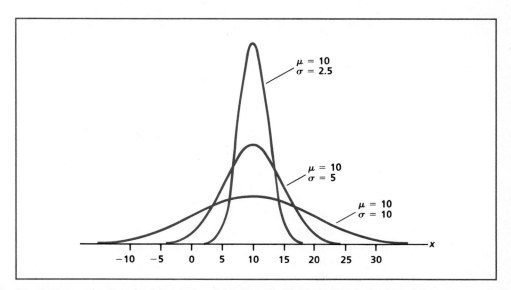

Figure 7.7 *The Standard Deviation Describes the Spread of a Normal Distribution*

The Weights of Coins

The graph in Figure 1 shows the distribution of the weights of quarters when minted and after 5 and 20 years in circulation.[1] Each distribution is approximately normal, for reasons explained by the central limit theorem: The initial weight is the cumulative result of many steps in the minting process and, as time passes, each coin's weight is altered by the cumulative effects of repeated handling.

Notice that the average weight declines over time, shifting the distribution to the left, as wear and tear takes its toll. The standard deviation also increases, widening the distribution, as the experiences of the coins diverge. Some happen to be handled very gently and retain almost all of their initial weight; others are nicked and bruised almost beyond recognition.

Figure 1   *The Distribution of the Weights of Quarters*

higher standard deviation. If these are investment returns, then the first investment has the higher long-run average return; however, in the short run, its return is more uncertain. Gamblers might go for $\mu = 8$ percent and $\sigma = 10$ percent, while the cautious might be more comfortable with $\mu = 5$ percent and $\sigma = 5$ percent.

Standardizing to Find Probabilities

We have used Figures 7.6–7.8 to help us visualize how the mean and the standard deviation describe the location and spread of a nonstandardized normal distribution. Now, let's see how the mean and standard deviation can be used to make specific probability calculations. How, then, do we find probabilities for a nonstandardized variable X? It's easy. We simply standardize the variable and look up the probability in Table 2.

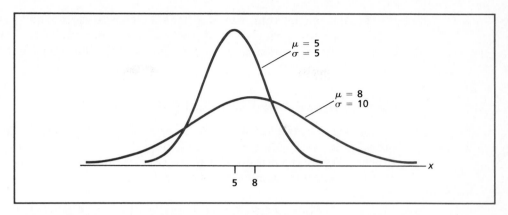

Figure 7.8   *Two Normal Distributions with Different Means and Standard Deviations*

Let's say, for instance, that X, the percentage return on an investment, is normally distributed with a mean of 10 percent and a standard deviation of 10 percent. What is the probability that the return will be greater than 20 percent, that is $X > 20$? We translate to a standardized Z variable to see how many standard deviations 20 is from the mean of X:

$$Z = \frac{X - \mu}{\sigma}$$

$$= \frac{20 - 10}{10}$$

$$= 1$$

With $\mu = 10$ percent and $\sigma = 10$ percent, a 20 percent return is one standard deviation above the mean. Therefore the probability that $X > 20$ equals the probability that $Z > 1$, which Table 2 tells us is .1587.

Let's do another example. If X still has a mean of 10 and a standard deviation of 10, what are the chances of losing money; that is, of $X < 0$? The Z value

$$Z = \frac{X - \mu}{\sigma}$$

$$= \frac{0 - 10}{10}$$

$$= -1$$

tells us that if X is less than 0, then it is one standard deviation below its mean, which again has a .1587 chance of occurring.

Try this one on your own. If X has a mean of 10 percent and a standard deviation of 5 percent, what is the probability of $X < 0$ percent? If your answer is .0228, then you have learned how to find probabilities for nonstandardized normal variables.

One, Two, and Three Standard Deviations

In practice, it is often convenient to use some simple rules of thumb for normal probabilities, particularly if you don't carry Table 2 around with you. Three good benchmarks are the probabilities that a normally distributed variable will be within one, two, and three standard deviations of its mean:

$$P[\mu - 1\sigma < X < \mu + 1\sigma] = P[-1 < Z < 1] = .6826$$

$$P[\mu - 2\sigma < X < \mu + 2\sigma] = P[-2 < Z < 2] = .9544$$

$$P[\mu - 3\sigma < X < \mu + 3\sigma] = P[-3 < Z < 3] = .9973$$

These useful rules are displayed in Figure 7.9. A normally distributed random variable has

a 68 percent (roughly about a 2/3) chance of being within one standard deviation of its mean,

a 95 percent chance of being within two standard deviations of its mean, and

a better than 99 percent chance of being within three standard deviations of its mean.

Equivalently, there is less than a 1 percent chance that a normally distributed variable will be more than three standard deviations away from its mean. There is a 5 percent chance of being more than two standard deviations from the mean and a 32 percent chance of being more than one standard deviation away.

Let's say, for instance, that we are trying to explain, as I tried earlier, why an investment with a mean return of 10 percent and a standard deviation of 5 percent is safer than an investment with a mean of 10 percent and a standard deviation of 10 percent. Our benchmarks tell us immediately that there is only a .05 probability that the first investment will have a return outside the range 0–20 percent (10 percent, plus or minus two 5 percent standard deviations). With the second investment, there is a .32 probability of being outside this 0–20 percent range (10 percent, plus or minus one 10 percent standard deviation). Clearly, the return on the first investment is the more certain.

Mathematical and Other Skills

An individual's skills in playing baseball, singing songs, and solving math problems depend on the cumulation of a great many hereditary and environmental factors.

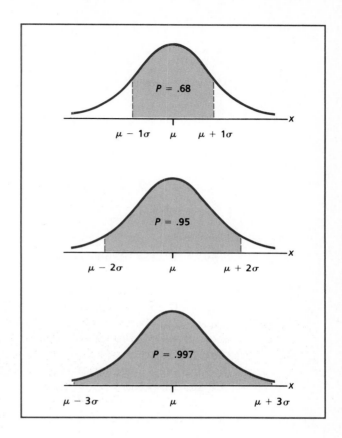

Figure 7.9 *Three Normal Benchmarks*

The central limit theorem suggests that measures of such skills may follow a normal distribution. And, in fact, many skill tests do yield normally distributed scores.

For example, the Scholastic Aptitude Test (SAT), Graduate Record Examination (GRE), and Graduate Management Aptitude Test (GMAT) all measure verbal, mathematical, and analytical abilities, and the scores on each of these various tests are approximately normally distributed. The raw scores (number right minus a fraction of the number wrong) are translated into scaled scores with a mean of about 500 and a standard deviation of approximately 100 for some reference group of test takers. For instance, the GMAT tests are scaled so that those who took the original test in 1955 had an average score of 500 and a standard deviation of 100. The precise means and standard deviations vary from year to year with the students taking the tests, but a rough normal distribution persists. Figure 7.10 shows the distribution of a half million GMAT scores during the years 1980–1983. The overall mean works out to be 469 and the standard deviation 104. The figure shows a reasonably close correspondence between a theoretical normal distribution and the actual distribution of these test scores.

The shape of a normal distribution can be used as a rough gauge of a student's performance. If, for instance, the scores on one of these standardized tests are nor-

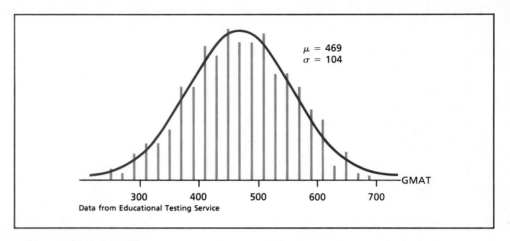

Figure 7.10   *The Distribution of GMAT Scores, 1980–1983*

mally distributed with a mean of 500 and standard deviation of 100, then how good is a score of 700? A 700 score is two standard deviations above the mean of 500, and our rule of thumb tells us that about 5 percent of the scores should be two standard deviations away from the mean, in either direction. Because the normal distribution is symmetrical, approximately 2.5 percent of the scores will be above 700 and 2.5 percent below 300.

Let's do one more. How good is a score of 638? This is almost 1.5 standard deviations above the mean; not as good as 700, of course, but still very respectable. The exact Z value is

$$Z = \frac{X - \mu}{\sigma}$$

$$= \frac{638 - 500}{100}$$

$$= 1.38$$

and Table 2 tells us that the corresponding probability is $P[Z > 1.38] = .0838$; thus only about 8 percent of the people who take this test score higher than 638.

Scores on most of the general aptitude tests that you have taken or will take follow a rough normal distribution. However, we could expect a nonnormal distribution, especially for adults, whenever there has been a focused, nonrandom, nonindependent cultivation of certain skills by some of the people who take a skill test. For example, Figure 7.11 shows a hypothetical distribution of athletic skills for 10-year-old-boys. At this age, most of the differences in athletic ability are due to random hereditary and environmental influences, and a normal distribution emerges. But the sweaty pursuit of fame and fortune will persuade many of the more skilled

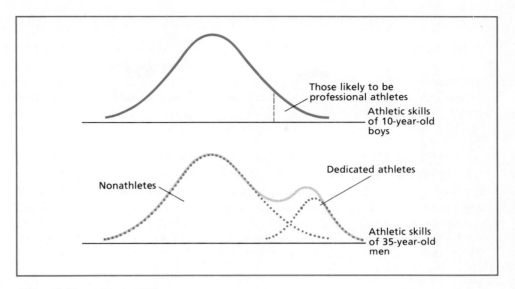

Figure 7.11 *Athletic Skills*

athletes to devote a great deal of time to perfecting their athletic skills. Other, less athletically skilled males will work equally hard at earning their livelihoods in other ways. If we measure the athletic skills of 35-year-old men, it would not be too surprising to find the sort of twin-peaked distribution that is shown in the bottom half of Figure 7.11. The bulk of the population is relatively unskilled, while a small cadre is highly skilled. The overall, twin-peaked distribution might reflect two separate normal distributions: one for highly trained, dedicated athletes and one for the rest of the population.

Similar, twin-peaked distributions could be expected if we measured people's fluency in reading Spanish, skill in playing the piano, or any other trait where the population separates into those who practice a skill and those who don't. If we gave a test to either group alone, however, we could expect an approximately normal distribution. For example, the Educational Testing Service prepares a Graduate Record Exam (GRE) Advanced Economics Test that is used for admissions and fellowships for graduate study in economics. If everyone took this test, we might expect the scores to split into two groups: those who have studied economics and those who haven't. But, in practice, only the former take the GRE economics test and their scores are fairly close to a normal distribution (see Exercise 7.11).

Hurricanes: To Seed or Not to Seed?

A classic study by Howard, Matheson, and North considered the question of whether the federal government should seed hurricanes with silver iodide.[2] It was thought that seeding was likely to reduce the destructive wind velocity of a hurri-

IQ Tests

There are a number of tests designed to measure a person's IQ (Intelligence Quotient). These tests are descended from 54 mental "stunts" that two psychologists (Alfred Binet and Theodore Simon) conducted in 1904 to weed dull students out of the Paris school system. The intention is to test a person's general intelligence, including an accurate memory and the ability to reason clearly and logically.

Scores on IQ tests are approximately normally distributed for people of the same age. The standard scaling gives a mean IQ of 100 with a standard deviation of 16. Thus, a score of 100 indicates that a person's intelligence is average for people of that age. About half of the people tested will score above 100, while half will score below 100. Our two-standard-deviation rule implies that about 5 percent of the population will score more than 32 points away from 100: about 2.5 percent will score above 132, while another 2.5 percent will score below 68. Our three-standard-deviation rule similarly implies that less than one-half of 1 percent of the population have IQs above 148. Figure 1 shows the complete distribution of IQ scores.

cane; but seeding does cost money and the involvement of the federal government might expose it to additional cleanup costs, legal expenses, and political reverberations. Historical data suggested that, for a "typical" 100 mph hurricane, the percentage change in the maximum sustained surface wind speed within 12 hours before landfall was normally distributed with a mean of 0 and standard deviation of 15.6. These researchers divided the normal curve into the five regions shown in Figure 7.12 and estimated the damage for each of these wind speed ranges.

The normal distribution was then used to calculate the probability of each of these wind speed ranges. For instance,

$$P[X > +25] = P\left[\frac{X - \mu}{\sigma} > \frac{25 - 0}{15.6}\right]$$
$$= P[Z > 1.603]$$
$$= .054$$

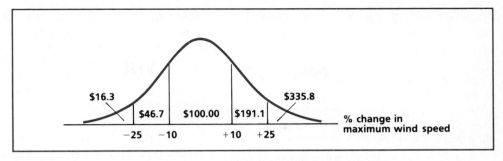

Figure 7.12   *The Cost of a Normal Hurricane, in Millions of Dollars*

The five probabilities and costs are shown below, along with the estimates for a seeded hurricane:

Change in Wind Speed	If Unseeded Probability	Cost	If Seeded Probability	Cost
> +25%	.054	335.8	.038	504.0
+10% to +25%	.206	191.1	.143	248.7
−10% to +10%	.480	100.0	.392	105.3
−25% to −10%	.206	46.7	.255	47.0
< −25%	.054	16.3	.172	16.6

Seeding reduces the chances of an increase in wind speed, but raises the cost dramatically if seeding fails. The expected value of the cost works out to be

$$\$335.8(.054) + \$191.1(.206) + \cdots + \$16.3(.054) = \$116.0$$

without seeding, and

$$\$504.0(.038) + \$248.7(.143) + \cdots + \$16.6(.172) = \$110.9$$

with seeding. On balance, seeding reduces the expected cost by about 4 percent.

EXERCISES

7.6 On average, approximately three defects are discovered in new U.S. cars within the first month of service. If the number of defects is roughly normally distributed with a mean of 3 and a standard deviation of 2, what percentage of cars have more than 5 defects?

7.7 The monthly returns in the stock market over the years 1904–1974 averaged 0.68 percent with a standard deviation of 6.85 percent.[3] If the monthly returns are, in fact, normally distributed with a mean of 0.68 percent and a standard deviation of 6.85 percent, then how frequently is the monthly return

a. greater than 10 percent? **b.** negative? **c.** less than −10 percent? **d.** greater than 20 percent?

7.8. *(continuation)* A security analyst predicts that the market will go up next month. How unusual is such a month? How impressed would you be if this prediction is right? Another analyst predicts that the market will go up at least 30 percent next month. How often does that happen? Would you be impressed if this prediction turns out to be correct?

7.9 Scores on an IQ test are normally distributed with a mean of 100 and a standard deviation of 16. What is the probability that a randomly selected person has an IQ above 120? Above 130? Above 140? Above 150? One psychologist has estimated that Galileo had an IQ of 185 and that Goethe had an IQ of 210. How often can we expect to find people with such high IQs?

7.10 The Scholastic Aptitude Test (SAT) scores were originally intended to be normally distributed with a mean of 500 and a standard deviation of 100. If this is the actual distribution, then what fraction of the students taking this test will be eligible for admission to a college that does not accept students who score **a.** below 400 on the English SAT test? **b.** below 500 on the English SAT test? **c.** below 600 on the English SAT test? **d.** below 700 on the Math SAT test?

7.11 The following scores were recorded on 10,720 GRE Advanced Economics Tests given between October 1978 and September 1981:

Scaled Score	Percent of Tests
>860	1
820–860	2
780–820	4
740–780	5
700–740	9
660–700	10
620–660	14
580–620	13
540–580	14
500–540	11
460–500	9
420–460	6
<420	3

The mean and standard deviation of these 10,720 scores are 606 and 108, respectively. If the scores were normally distributed, with this mean and standard deviation, what fraction of the scores would you expect to be above 860; to be between 820 and 860; and so on? Construct a table comparing these 13 theoretical percentages with the actual distribution of test scores.

7.12 Joanna is a new professor who wants to grade strictly by the curve. But, to keep up with grade inflation, she doesn't want to give as many Fs as As and as many Ds as Bs. Instead, she wants to give, on average, 20 percent As, 30 percent Bs, 35 percent Cs, 10 percent Ds, and 5 percent Fs. If her numerical scores are normally distributed, which standardized values should be graded A, B, C, D, and F? Draw a rough sketch of the normal curve, as in Figure 1 on page 251, showing Joanna's grade distribution.

7.13 At Big State University (BSU), the weights of the 20,000 male students are normally distributed with a mean of 160 and a standard deviation of 25. How many of these 20,000 males weigh more than 200 pounds? How many weigh between 150 and 170? How many weigh less than 100 pounds?

7.14 Auto manufacturers design their cars for an average-size person, with some allowance for individual variation consistent with their desires for an attractive and fuel-efficient design. For each year's new car models, *Consumer Reports* estimates the front-seat head room for a person 69 inches tall. Here are some of its 1986 estimates:

Nissan 300ZX	2.0 inches
Pontiac Fiero	3.5 inches
Ford Taurus	4.5 inches

The heights of U.S. males are approximately normally distributed with a mean of 69 inches and a standard deviation of 2.4 inches, while the heights of U.S. females are approximately normally distributed with a mean of 64 inches and a standard deviation of 2.3 inches. What fraction of each sex would bump their heads in each of these three cars? (Assume that every additional inch of height requires an extra one-half inch of head room.)

7.3 CONTINUITY CORRECTION

The central limit theorem says that the sum of independent, identically distributed (discrete or continuous) random variables converges to a normal distribution. So far, we have focused on discrete variables because these are the easiest to visualize. But the central limit theorem holds for continuous variables, too. Figure 7.13 shows two examples.

The convergence is, in some sense, cleaner with a continuous variable in that the distribution is visibly transformed into a normal distribution. With a discrete variable, in contrast, a normal distribution is a continuous approximation of a discrete distribution. We saw with the spinner in Chapter 6 how a continuous distribution is, in fact, the theoretical limit of a discrete distribution as the size of the interval goes to zero. But we have also seen that, in practice, a bell shape often emerges very quickly. In Figure 6.3, for instance, the probability distribution for two dice rolls is already starting to look bell-shaped. Yet, eleven possible sums are a very far cry from the infinite number of possible outcomes allowed by a continuous normal distribution.

It turns out that, if we make a *continuity correction,* then the normal distribution may be a remarkably accurate approximation even if the number of discrete outcomes is quite small. The logic behind this continuity correction is as follows. Suppose we want to use a normal distribution to approximate the probability of obtaining a 7 when two dice are rolled. We could hardly use the probability that a normally distributed variable is exactly equal to 7, because there is zero probability that a continuous variable is precisely equal to a specific number. Let's instead imagine that the outcomes are somehow continuous and that we can group them into discrete intervals such as 6, 7, or 8. We put everything from 6.5 to 7.4999 . . . into the interval labeled "7." We put the outcomes from 5.5 to 6.4999 . . . into the interval labeled "6" and we put the outcomes from 7.5 to 8.4999 . . . into the interval "8." Figure 7.14 illustrates the logic. This argument suggests that the continuous interval 6.5 to 7.4999 . . . corresponds to the discrete outcome 7. And this is, in fact, the recommended continuity correction.

> The **continuity correction** for a given discrete interval consists of a continuous interval that is a half-unit wider at each end than the corresponding discrete interval.

Thus, for the discrete number 7, the corresponding continuous interval is 6.5 to 7.5.* For the discrete interval 6 to 8, the corresponding continuous interval is 5.5 to 8.5.

---

*I've conveniently replaced 7.4999 . . . with 7.5 because the probabilty that a continuous variable X is less than 7.5 is the same as the probability that X is less than or equal to 7.5.

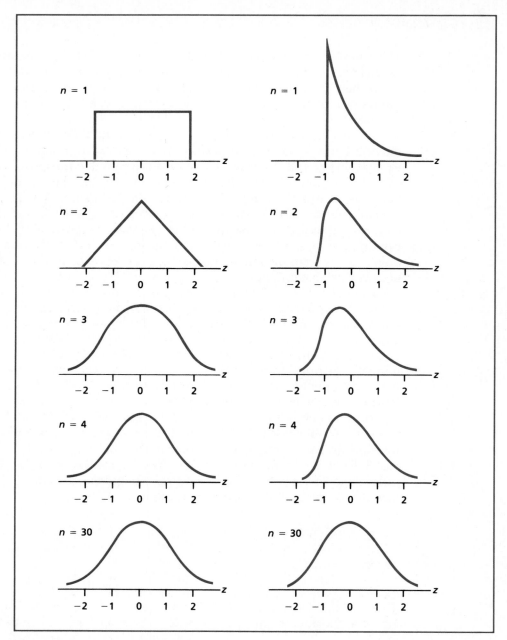

Figure 7.13   *The Central Limit Theorem*

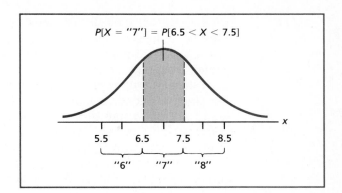

Figure 7.14 *The Discrete Grouping of a Continuous Random Variable*

It Works!

Does it work? Well, the exact probability of rolling a 7 with two dice is 6/36 = .167. For a normal approximation, we can use the mean $\mu = 7$ and standard deviation $\sigma = 2.415$ calculated in Chapter 4 to obtain the corresponding Z values: $Z = (6.5 - 7)/2.415 = -0.207$ and $Z = (7.5 - 7)/2.415 = +0.207$. Table 2 tells us that

$$P[-0.207 < Z < +0.207] = .164$$

which, for most purposes, is close enough.

Let's try another. With two dice rolls, what is the probability of a number from 8 to 10? Table 4.1 reminds us that the correct answer is 5/36 + 4/36 + 3/36 = .333. The normal approximation uses these Z values: $Z = (7.5 - 7)/2.415 = +.207$ and $Z = (10.5 - 7)/2.415 = 1.449$. The probability

$$P[0.207 < Z < 1.449] = .344$$

is close, but certainly not perfect.

Remember, though, that we are looking at the case $n = 2$, which is a long way from the theoretical $n = \infty$ assumed by the central limit theorem. As n increases, the normal approximation becomes even more accurate. For example, with just eight dice here are the chances of rolling numbers that sum to 28 and of rolling either a sum of 25, 26, 27, or 28:

	$X = 28$	$X = 25, 26, 27,$ or 28
Exact probability	.081	.307
Normal approximation	.082	.303

The normal approximation won't always be this close for small values of n. It depends on how quickly the probability distribution becomes bell-shaped. In Figure

6.3, we saw that the bell shape emerges very quickly for dice rolls. In some of the other examples shown in Chapter 6, it takes a while longer. The usual rule of thumb is that a normal approximation will be satisfactory if the number of trials, n, is greater than 20 or 30. If you look back at the figures in Chapter 6, you will see that a bell shape is usually pretty apparent by $n = 20$ or $n = 30$. You might use a more conservative rule, such as $n > 50$, if you demand a very accurate probability or if you're not quite sure that the central limit theorem's assumptions are satisfied. Conversely, if you have 100 trials, you can be pretty confident in using a normal approximation; with 15 trials, you might be a bit cautious.

One final point is that the continuity correction becomes less important as the number of trials increases. With a large number of trials, there will be a very large number of possible outcomes. There is, consequently, little chance that the outcome will be in an interval that is only a unit wide, and thus it doesn't matter whether or not we add a half-unit on either side to make up an interval.*

For a normal approximation of the probability that X is larger than some number a, we would (with a continuity correction) use

$$Z = \frac{a + 0.5 - \mu}{\sigma}$$

$$= \frac{a - \mu}{\sigma} + \frac{0.5}{\sigma}$$

If we ignored the continuity correction, instead we would use

$$Z = \frac{a - \mu}{\sigma}$$

The $0.5/\sigma$ difference between these two numbers becomes less important as σ increases with n.

I'll use the dice example to illustrate this point. With one die, the standard deviation of X is 1.708. With n dice, it can be shown that the standard deviation is $1.708\sqrt{n}$. As n increases, σ increases, too, and the continuity correction $0.5/\sigma$ declines. The first column of Table 7.1 gives some specific numbers. Using this column, I've calculated some probability differences that would occur if we didn't make a continuity correction. The maximum difference is always right at the center of the normal distribution, where $X = \mu$ and $Z = 0$. As one moves away from this center, the continuity correction is even smaller. You can see in Table 7.1 that a continuity correction can be very important for small values of n, but it becomes less important as n increases. Perhaps $n = 100$ is a reasonable place to stop bothering with continuity corrections.

*To obtain a nonzero probability of obtaining a single number, such as $X = 28$, we still must add 0.5 to each side.

Table 7.1 *A Continuity Correction Becomes Less Important as n Increases*

		Probability Difference Without Correction	
n	*0.5/σ*	*At Z = 0*	*At Z = 1*
2	.207	.082	.045
8	.104	.041	.024
20	.065	.026	.015
50	.041	.016	.010
100	.029	.012	.007
500	.013	.005	.003
1000	.009	.004	.002

For Example, Approximating the Binomial

The various rules of thumb for using normal approximations must necessarily be very rough because they are intended to apply to a great many different probability distributions. We can be more concrete when we restrict our attention to a particular distribution, such as dice rolls. In this section, I want to focus on normal approximations to the binomial distribution.

The Bernoulli scheme involves n independent trials, each with probability of success π. If we let $X = 1$ when there is a success and $X = 0$ when there is a failure, then the sum of the outcomes is the number of successes. The binomial distribution gives the corresponding probabilities. Because Bernoulli trials satisfy the conditions for the central limit theorem, it follows that the binomial distribution turns into a normal distribution as the number of trials increases. Indeed, it was De Moivre and Laplace's discovery of this metamorphosis that led to the more general central limit theorem.

Look back at Figure 6.4, where I have graphed some binomial distributions for $\pi = .5$ and $\pi = .9$. With $\pi = .5$, the binomial distribution starts out symmetrical at $n = 1$ and quickly becomes bell-shaped. With $\pi = .9$, however, the distribution is very skewed for low values of n and a bell shape takes longer to emerge.

In Chapter 5, we saw that the mean and standard deviation of the number of binomial successes are

$$\mu = \pi n$$
$$\sigma = \sqrt{\pi(1 - \pi)n}$$

A normal approximation will run into trouble if the mean $\mu = \pi n$ is close to the extreme values of X, either 0 or n. Figure 7.15 shows the example of $\pi = .9$ and n

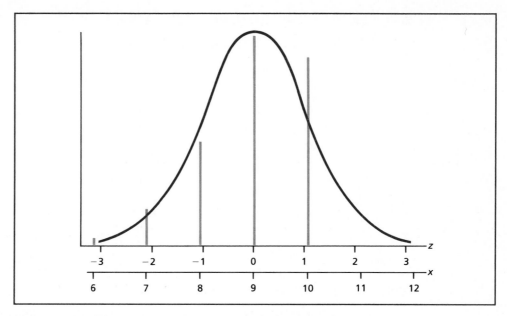

Figure 7.15   *Normal Approximation to the Binomial for n = 10, π = .9*

= 10. In this figure, I've put the unstandardized values of X on the axis so that you can see the endpoint problem clearly. The number of successes cannot be greater than 10, but with a mean of 9, the normal distribution has some substantial probability all the way out to $X = 12$. With this erroneous probability of X being greater than 10, the normal approximation is bound to be off in its probability estimates for when X is less than 10, too.

The other problem is that when, as here, the mean is close to an extreme boundary, the binomial distribution has no room for a symmetrical bell shape. Look again in Figure 7.15 at the mean $\mu = 9$. There is only one possible outcome, $X = 10$, to the right of the mean. One outcome can hardly show a bell pattern, and sure enough, if we look at the probabilities for $X = 9$ and $X = 10$, there is no hint of a bell.

A satisfactory normal approximation requires a large enough n to get the mean πn safely away from the extremities $X = 0$ and $X = n$. If there is a comfortable distance between the mean and the extremes, then there will be room for the binomial distribution to show a bell shape and the normal distribution won't go slopping over the boundaries into values of X that are impossible. How large an n is required? It depends on π. For $\pi = .5$, an $n = 10$ gives a mean of 5 that is safely distant from 0 and 10. For $\pi = .01$, even $n = 100$ gives a mean of only 1, which is uncomfortably close to 0. A reasonable rule of thumb is that a normal approximation to a binomial distribution will be satisfactory if the mean πn is at least 5 units away from both 0 and n, or if

$$\pi n - 5 \geq 0 \quad \text{and} \quad \pi n + 5 \leq n$$

Let's do a few quick computations to see if it really works. First, let's try $\pi = .5$ and $n = 10$. The mean $\pi n = 5$ is barely 5 away from 0 and n. Table 1 gives .451 as the binomial probability that X is equal to 5 or 6. Using the standard deviation $\sigma = \sqrt{\pi(1 - \pi)n} = 1.58$, the Z values for a normal approximation (with a continuity correction) are $Z = (4.5 - 5)/1.58 = -0.316$ and $Z = (6.5 - 5)/1.58 = 0.949$, and

$$P[-0.316 < Z < 0.949] = .452$$

which is pretty accurate.

Second, let's try $\pi = .9$ and $n = 50$, so that the mean $\pi n = 45$ is barely 5 away from n. The probability that $X = 45$ is .185. Using $\sigma = \sqrt{\pi(1 - \pi)n} = 2.12$, we have $Z = (44.5 - 55)/2.12 = -0.236$ and $Z = (45.5 - 45)/2.12 = 0.236$, so that

$$P[-0.236 < Z < +0.236] = .186$$

which, again, is strikingly accurate.

So, the normal distribution seems to be a useful approximation to the binomial distribution, as long as the mean πn is at least 5 away from the extremes 0 and n.

The Statistical Reason for Long Multiple-Choice Tests

Here's a very practical example. A statistics professor gives a test consisting of 48 multiple-choice questions. Each question has 4 possible answers, so that a pure guesser has a .25 probability of picking the correct answer. What is the probability that a guesser will get 16 or more correct answers?

This is a binomial probability with $\pi = .25$ and $n = 48$ that isn't included in Table 1 at the end of this book. But, because the mean $\mu = \pi n = 12$ is more than 5 units away from 0 and n, we can use a normal approximation. The standard deviation is $\sigma = \sqrt{(.25)(.75)(48)} = 3$ and, with a continuity correction, $Z = (15.5 - 12)/3 = 1.167$ and

$$P[Z > 1.167] = .122$$

so that a guesser has about a 12 percent chance of getting at least 16 out of 48 questions right.

If the professor wants a guesser to have only a .001 chance of passing, how many correct answers should be required? In the first calculation, we had a score $X = 16$ and needed to determine a probability. Now, we have a probability and need to determine the corresponding score. Table 2 at the end of this book states that .001 $= P[Z > 3.08]$. This fact can be manipulated as follows:

$$\frac{X - \mu}{\sigma} = 3.08$$

implies

$$X = \mu + 3.08\sigma$$

Notice this characteristic format. There is a .001 probability that Z will be greater than 3.08, which translates into a .001 probability that X will be more than 3.08 standard deviations above its mean. Because, here, the mean is 12 and the standard deviation is 3, we have

$$.001 = P[X > 12 + 3.08(3)]$$
$$= P[X > 21.24]$$

If the professor fails those who get fewer than 20 questions right, then a guesser will have less than a .001 chance of passing.

Now let's consider the plight of a reasonably good student who can correctly answer half of all of the multiple-choice questions that this professor might conceivably ask. If the professor randomly selects 48 questions to put on the test, what is the probability that this reasonable student will get fewer than 20 of these 48 questions right? Now, the mean is $\mu = \pi n = .5(48) = 24$ and the standard deviation is $\sigma = \sqrt{\pi(1 - \pi)n} = \sqrt{(.5)(.5)(48)} = 3.364$. With a continuity correction, $Z = (19.5 - 24)/3.464 = -1.3$ and

$$P[Z < -1.3] = .097$$

By using a minimum of 20 correct for a passing grade, the professor will make it very difficult for guessers to pass, but will also fail quite a few reasonable students.

Is there any way that the professor can increase the reasonable student's probability of passing, while still giving the guesser only a .001 chance? The answer, I'm sorry to relate, is that a more reliable separation of the good student from the guesser requires a longer exam. Just as in the case of acceptance sampling of bombs, more data improves the identification of the duds. With 96 questions, for example, the guesser has less than a .001 chance of getting 38 correct answers, while our reasonable student has better than a .98 probability of getting at least 38 correct. (The details are left as Exercise 7.15.) So, remember, when a professor gives a long exam, it is really for the student's sake!

Election Polls

A very similar analysis applies to opinion polls on such matters as the better-tasting cola and an upcoming election. Let's do the election example. Suppose that 40 percent of the voters prefer candidate A. If we pick 10 voters at random, what is the probability that at least half will be supporters of A? This is a binomial probability with $\pi = .4$ and $n = 10$. Because the mean $\mu = \pi n = 4$ is not 5 units from 0, we should resist the temptation to use a normal approximation. Table 1 gives the binomial probability as .3669 that at least half of the 10 people polled prefer A.

To make the poll more reliable, we need to make it larger. Consider a poll of 100 people. Now $\mu = \pi n = .4(100) = 40$, $\sigma = \sqrt{\pi(1 - \pi)n} = \sqrt{(.4)(.6)(100)} = 4.9$, and a normal approximation should be satisfactory. With a continuity correction, $Z = (49.5 - 40)/4.9 = 1.94$ and the probability that at least half of the 100 voters polled prefer A is given by $P[Z > 1.94] = .026$. As the sample gets larger, it becomes increasingly certain that the fraction of those polled who prefer A will be very close to .4 and increasingly improbable that the poll will erroneously predict that A will win the election.

EXERCISES

7.15 The multiple-choice example in this chapter involved two students: a reasonable student with a .5 probability of correctly answering a selected question and a guesser with only a .25 chance. If this professor asks 96 questions and requires 38 correct answers for a passing grade, what are these two students' chances of passing?

7.16 The random walk hypothesis says that the probability that stock prices will increase today is independent of past performance. If, on any given day, there is a .52 probability that the Dow Jones Industrial Average will go up and a .48 probability that it will fall, then (assuming independence), what is the probability that during the course of a year the average will have more up days than down days? Assume 250 trading days in a year.

7.17 A manufacturer uses an acceptance sampling rule that a batch of products is rejected if 10 percent of a sample is found to be defective. What is the probability that a batch that is 25 percent defective will be rejected if the size of the sample is 10? 20? 200?

• **7.18** Mendel postulated that the self-fertilization of hybrid yellow-seeded sweet peas would yield offspring that were three-quarters yellow seeded and one-quarter green seeded. In 1865 he reported that 8,023 such experiments yielded 6,021/8,023 = .7505 yellow-seeded plants and 2,002/8,023 = .2495 green-seeded plants. It has been suggested that these results are too good to be true, and that Mendel must have reported what he wanted rather than what he found. What is the probability of this close a correspondence between the theory and the data? Use a normal approximation.

• **7.19** Karl Pearson reported flipping a coin 24,000 times and obtaining 12,012 heads and 11,988 tails. If heads and tails are, in fact, equally likely, what is the probability that the numbers of heads and tails would come out this close to 12,000? Use a normal approximation.

7.20 In the election poll example in this section, I stated that a normal approximation is risky for $\pi = .4$ and $n = 10$. Compare a normal approximation with the true .3669 binomial probability that X is greater than or equal to 5.

I did use a normal approximation for $\pi = .4$ and $n = 100$ to obtain .026 as an estimate of the probability that X is greater than or equal to 50. Compare this approximate calculation with the true binomial probability.

7.21 Mr. Noxin is running for dogcatcher, but only 45 percent of the voters favor him. What is the probability that more than half of the voters polled will favor Noxin if 100 voters are surveyed? If 1,000 voters are surveyed?

• **7.22** (continuation) There are 1,000,000 voters eligible to vote in this special election to replace the old, badly mauled dogcatcher. Only 45 percent of these eligible voters favor Noxin. But, because the weather is terrible and the race is boring, only 100 people bother to vote. If these 100 voters are a random selection from all eligible voters, what are Noxin's chances of winning? What if 1,000 people vote?

From a purely statistical point of view, does a large turnout favor the underdog or the favorite? What reasons can you think of why voter turnout is not entirely random?

7.23 A nation should have enough hospital beds to meet variations in demand and, yet, doesn't want to waste resources by having too many beds. In 1985, the United States had 1,250,000 hospital beds of which, on average, 912,500 (73 percent) were occu-

pied; experts estimated that 1.0 to 1.1 million would be sufficient.[4] If the population of the United States is 250 million, then the fraction that are hospitalized on any given day is

$$\pi = 912,500/250,000,000 = .00365$$

Use the binomial distribution to determine the expected number of hospital beds (μ) needed on any given day, the standard deviation (σ), and the value of $\mu + 3\sigma$. Can you think of any reasons why this latter value may be an underestimate of the number of hospital beds needed to meet variations in demand?

7.4 SUMMARY

Table 2 at the back of this book shows the probabilities in the right-hand tail of the standardized normal distribution. This table can be used to find the probability that the standardized variable Z is greater than some number, less than some number, between two numbers, and so on.

In practice, we usually work with nonstandardized variables X, but Table 2 also can be used to find these probabilities. If a normally distributed variable X has a mean μ and standard deviation σ, then $(X - \mu)/\sigma$ is normally distributed with a mean of 0 and a standard deviation of 1. That is,

$$X \sim N[\mu, \sigma] \text{ implies } Z = \frac{(X - \mu)}{\sigma} \sim N[0, 1]$$

For instance, let's say that X is a normally distributed random variable with mean $\mu = 20$ and standard deviation $\sigma = 10$. To find the probability that X is greater than 40, we calculate the Z value $Z = (40 - 20)/10 = 2$ and use Table 2 to determine that $P[Z > 2] = .0228$.

A rough rule of thumb is that a normally distributed variable has a 68 percent (or about 2/3) chance of being within one standard deviation of its mean, a 95 percent chance of being within two standard deviations, and a better than 99 percent chance of being within three standard deviations.

The central limit theorem says that the sum of n independent, identically distributed random variables becomes normally distributed as n increases. In many cases, this convergence is quite rapid, in that the sum is approximately normally distributed for very small values of n. In general, the approximation is quite good for when n is approximately equal to 20 or 30. For the binomial distribution, we have the more specific rule of thumb: A normal approximation to the binomial distribution with success probability π is satisfactory if the mean πn is at least 5 units away from both 0 and n.

With the binomial and other discrete distributions, the approximation can be made more accurate by a continuity correction: For a given discrete interval, the corresponding continuous interval encompasses the discrete interval plus an additional half-unit at each end. For instance, to find the probability that the sum of ten dice will be either 35 or 36, we would look at the probability that the normally distributed variable X is between 34.5 and 36.5.

REVIEW EXERCISES

7.24 The text says, "Together, the mean and standard deviation describe a normal distribution." Why didn't we also look at the skewness coefficient?

7.25 The heights of 17-year-old U.S. males are approximately normally distributed with a mean of 175.3 centimeters and a standard deviation of 6.8 centimeters. About how many of these young men are **a.** more than 176 centimeters tall? **b.** more than 182 centimeters tall? **c.** more than 188 centimeters tall? **d.** more than 194 centimeters tall? **e.** more than 200 centimeters tall?

7.26 *(continuation)* A centimeter is equal to 0.3937 inch. If we looked at the heights of 17-year-old U.S. males, measured in inches, would these be normally distributed? What would be the mean and standard deviation? About how many of these young men would be **a.** shorter than 5 feet, 5 inches? **b.** taller than 6 feet? **c.** taller than 6 feet, 2 inches?

7.27 An 1883 study estimated the heights of adult males born in the British Isles to have a mean of 67.46 inches and a standard deviation of 2.57 inches.[5] If the heights were normally distributed with this mean and standard deviation, then what fraction of the adult males were **a.** shorter than 5 feet, 5 inches? **b.** taller than 6 feet? **c.** taller than 6 feet, 2 inches? Using the data given in Exercises 7.25–7.27, give a brief comparison of the heights of English adults in 1883 and modern U.S. 17-year-old males.

7.28 Inevitably, there are variations in manufacturing processes due to variations in the quality of materials, condition of the machinery, and skill of the operators. Industrial buyers usually set specifications ("specs"), stating the permissible variation in size or performance of the product. For example, the manufacturing process may produce a steel rod with lengths that are normally distributed with a mean of 20 cm. and a standard deviation of .1 cm, while the buyer sets specs of 20 cm plus or minus .2 cm. If so, what percentage of the rods will violate the specs?

7.29 Assume that the age at death of 20-year-old females is normally distributed with a mean of 76.4 years and a standard deviation of 8.1 years. If so, what fraction of these females will live past 80? What fraction will live past 90? What fraction will live past 100?

7.30 A woman wrote to Dear Abby, saying that she had been pregnant for 310 days before giving birth. Completed pregnancies are normally distributed with a mean of 266 days and a standard deviation of 16 days.[6] What are the probabilities that a pregnancy lasts **a.** longer than 270 days? **b.** longer than 300 days? **c.** longer than 310 days?

7.31 The Scholastic Aptitude Test (SAT) scores were designed to have a mean of about 500 and a standard deviation of about 100. After 1963, though, the average scores steadily declined throughout the 1960s and 1970s. In 1980, the average scores were 470 on the Math SAT and 430 on the verbal SAT. The scores are approximately normally distributed and the standard deviations have stayed at about 100. What fraction of the students score above 500 if the mean is **a.** 500? **b.** 470? **c.** 430?

7.32 Males and females do about equally well on the verbal SAT test, but on the average, males do better than females on the math SAT test. In 1980, for instance, the mean male score on the Math SAT test was 490, while the mean female score was 450. Both male and female scores are approximately normally distributed, with standard deviations of about 100. On the 1980 math SAT test, what fractions of the males and females scored **a.** above 500? **b.** above 600? **c.** above 700?

7.33 A study of the adequacy of New York City's snow-fighting capability estimated that the number of snowstorms in a season was approximately normally distributed with a mean of 6.5 and a standard deviation of 2.25.[7] How often does New York City have more than ten snowstorms in a season?

7.34 On average, 70 percent of the passengers on a flight from Los Angeles to Boston prefer chicken to fish. Assume that these choices can be modeled as Bernoulli trials with $\pi = .70$. If there are 200 passengers on a flight and the airline carries 140 chicken and 60 fish dinners, what is the probability that no one will be disappointed? That more than 5 passengers will be disappointed? (Use a normal approximation.)

7.35 The weights of the grapefruit grown on Harvey's farm in Porterville, California are normally distributed with a mean of 15 ounces and a standard deviation of 1.5 ounces. On David's farm, the grapefruit weights are also normally distributed, but with a mean of 14 ounces and a standard deviation of 1 ounce. What fractions of Harvey's and David's grapefruit can be classified as "Jumbo," i.e., greater than 16 ounces?

7.36 *(continuation)* The Mammoth packing plant will only accept grapefruit that weigh at least 12 ounces. What fraction of Harvey's grapefruit and David's grapefruit are unacceptable?

7.37 Whirlaway claims that the useful lifetime of its dishwashers is normally distributed with a mean of three years and a standard deviation of one year. If they guarantee their machines for one year, what fraction will break down within the warranty period? What fraction will break down within two years? What fraction will last more than five years?

If it costs Whirlaway $250 to replace a guaranteed dishwasher, how much do they have to add to the price to cover these costs?

• **7.38** *(continuation)* Consumer Rejects magazine tested two of Whirlaway's machines for two years and found that both broke down. If Whirlaway's claims are correct, what is the probability that two tested machines will both break down in less than two years? Does this probability calculation cast some doubt on Whirlaway's claims?

7.39 In Exercise 5.13 in Chapter 5, the binomial distribution was used to find the probable results of a Newstime magazine poll of voters' preferences for Tweedle Dee or Tweedle Dum. If the voters are evenly divided between these two candidates, what is the probability that a poll of 100 voters will find that more than 60 percent prefer one of the candidates, either Dee or Dum? Use a normal approximation and then compare your answer with the binomial probability obtained in Exercise 5.13. If 400 voters were polled, what is the probability that 60 percent would prefer one of the candidates?

• **7.40** *(continuation)* If Newstime wants there to be less than a .005 chance that more than 60 percent of those polled will prefer one of the candidates when the voters are, in fact, evenly divided, then how many people must they poll? What if they want less than a .001 chance of such a misleading poll result?

7.41 Gamma Airlines has found that 20 percent of the people who make reservations cancel at the last minute. If a plane holds 200 passengers, how many cancellations can they expect if they accept 200 reservations? If they accept 220 reservations? If they accept 240 reservations? How many reservations should they accept if they want the expected number of people who show up to equal 200?

7.42 *(continuation)* What is the probability that more than 200 people will show up if Gamma Airlines accepts 200 reservations? If it accepts 220 reservations? 240 reservations? 250 reservations? Use a normal approximation.

• **7.43** *(continuation)* How many reservations should Gamma Airlines accept if they want the probability of more than 200 passengers showing up to be .50? To be .05?

•• **7.44** *(continuation)* If more than 200 people show up, Gamma's policy is to give these overbooked passengers twice the value of their tickets. Gamma executives have consequently decided to accept reservations from a number of passengers n, such that the probability of empty seats is twice as high as the probability of insufficient seats. What is this value of n? With this booking policy, how frequently can they anticipate having empty seats on Gamma flights? How frequently can they anticipate having insufficient seats on Gamma flights?

7.45 A certain college has found that half of the students they accept for admission decide to go elsewhere. How many students should they admit if they want the expected size of their freshman class to be

400? If they follow this admission policy, what is the (approximate) probability that the entering freshman class will be larger than 400? What is the probability that it will be smaller than 400?

• **7.46** *(continuation)* By doubling up in some rooms and putting some students in a limited amount of off-campus housing, the college can accommodate a maximum of 450 new students. How many students should they accept if they want the probability of more than 450 new students coming to be less than .10? Less than .01? (*Hint:* Use trial and error or the quadratic formula.)

• **7.47** *(continuation)* If too few students show up, then a severe financial strain is put on the college. How many students should they accept if they want the probability of fewer than 350 students coming to be less than .10? Less than .01?

•• **7.48** *(continuation)* The board of trustees has decided that, all things considered, inconvenient housing for the students is not as serious as a tuition shortfall. They have consequently instructed the admissions office that the number of students admitted should be such that the probability of fewer than 400 students coming is only 1/4 the probability of more than 400 students coming. To implement this guideline, how many students should they admit?

•• **7.49** Figure 6.10 pictured a drunkard's walk. Assume that this drunkard is equally likely to step 12 inches to the north, south, east, or west. Let X_i be the distance (in feet) that this drunkard moves eastward on step i. This distance X_i can be $+1$, -1, or 0. What are the expected value and standard deviation of X_i? Now, let $X = X_1 + X_2 + \cdots + X_n$ be the total eastward distance covered by this drunkard after n steps. What are the possible values of X? What are the mean and standard deviation of X? (*Hint:* Because the X_i are independent, the variance of X is the sum of the n variances of the X_i.) If $n = 50$, what is the probability that the drunkard will be east of the lamp post? More than 5 feet eastward? More than 10 feet eastward? More than 10 feet westward?

•• **7.50** *(continuation)* Now let's apply the drunkard's walk analogy to stock prices. Consider some stock that is selling for $20 a share. Let X_i be the dollar change in the price of this stock during week i and let X be the total price change after $n = 50$ weeks. We can apply the drunkard's walk by assuming that a step

east is a $1 increase in the stock's price, a step west is a $1 decline, and a step north or south is no change. What is the probability that, at the end of 50 weeks, the price of this stock will be above $20? Above $25? Above $30? Below $15?

7.51 If two probability distributions have the same mean and standard deviation, are they identical?

7.52 Over the period 1926 through 1988, the annual returns from common stocks averaged 12.1 percent with a standard deviation of 21.0 percent. If the returns followed a normal distribution, with this mean and standard deviation, what is the probability that an annual return will be **a.** greater than 30 percent? **b.** negative? **c.** less than -20 percent?

7.53 Over the period 1926 through 1988, the annual returns from long-term corporate bonds averaged 4.8 percent with a standard deviation of 8.5 percent. If the returns followed a normal distribution, with this mean and standard deviation, what is the probability that an annual return will be **a.** greater than 30 percent? **b.** negative? **c.** less than -20 percent?

7.54 The heights of the 20,000 men at BSU are normally distributed with a mean of 70 inches and a standard deviation of 2.5 inches. How many of these men are at least 6 feet, 5 inches tall? How many are between 5 and 6 feet tall?

7.55 Assuming an annual batting average is the random result of 500 times at bat, each with a .3 probability of hitting safely, use a normal approximation to determine the probability of an annual batting average **a.** above .350. **b.** below .250. **c.** between .280 and .320.

7.56 Most products are designed for the dimensions of the average person, with a margin that accommodates most but not all people. For instance, the torsos of adult U.S. males are approximately normally distributed with a mean of 32 inches and a standard deviation of 2 inches. If a certain automobile is built with a seat-to-ceiling distance that accommodates a maximum torso length of 35 inches, what percentage of U.S. adult males will not fit comfortably in this car?

7.57 The useful life of a certain battery is a random variable from a normal distribution with a mean of 150 hours and a standard deviation of 15 hours. What is the probability that a battery will last more than 150 hours? More than 160 hours? More than 170 hours?

7.58 The breaking strength of a commonly used support beam is designed to be 500 pounds with a standard deviation of 25 pounds. If these breaking strengths are normally distributed, what is the probability that a randomly selected beam will have a breaking strength of less than 500 pounds? Less than 450 pounds? Less than 400 pounds?

7.59 In 1989 *Sports Illustrated* reported that Kansas City's Pat Tabler, a career .290 hitter, had 38 hits in 66 times at bat with the bases loaded, an astonishing .576 batting average in these key situations.[8] Use a normal approximation to the binomial with a continuity correction to calculate the probability that an event with a .290 probability of success would happen 38 or more times in 66 trials.

7.60 The U.S. Postal Service has on-time standards for its mail delivery; for instance, letters from Washington, D.C. to Manhattan should arrive within two days. The Postal Service claims that it delivers 90 percent of the mail on time in Manhattan. In the summer of 1988, *New York* magazine tested this claim by mailing 144 letters from around the country to Manhattan zip codes and found that 50 percent of these letters did not arrive on time. If the Postal Service's claims are correct, what is the probability that 72 of more of these 144 letters would not arrive on time? Assume that the Bernoulli trial model applies, and use a normal approximation with a continuity correction.

7.61 If a professor asks 60 multiple-choice questions and requires 20 correct answers for a passing grade, what is the probability of a passing grade for a student who has a .5 probability of answering a question correctly? Use a normal approximation.

7.62 A *Wall Street Journal* poll in December 1986 asked 35 economic forecasters to predict the interest rate on 30-year Treasury bonds in June 1987. The mean of these 35 predictions was 7.05 percent and the standard deviation was 0.53 percent. If the predictions of all forecasters are normally distributed with a mean of 7.05 percent and a standard deviation of 0.53 percent, what fraction of the forecasters predicted an interest rate below the actual value of 8.50 percent in June 1987?

7.63 The life of one brand of light bulbs is approximately normally distributed with a mean of 1,000 hours and a standard deviation of 100 hours. What is the probability that a randomly selected bulb will last more than 1,100 hours? Will last less than 900 hours?

7.64 *(continuation)* What percentage of these bulbs last between 900 and 1,100 hours? Between 800 and 1,200 hours?

7.65 Since 1962 both the American and National Leagues have used a 162-game baseball schedule. Since the number of teams has increased from 20 in 1962 to 26 in 1989, let X be the total number of home runs hit during a season divided by the number of teams playing that season. During the years 1962–1989 (excluding 1981, when there was a baseball strike), X averaged 127.3 with a standard deviation of 16.6.[9] If $X \sim N[127.3, 16.6]$, what is the probability that the value of X will be as large as 170.5, its value in 1987? What is the probability that at least 1 of 27 seasons will have this large a value of X?

7.66 The temperature on March 15 in City 1 is normally distributed with a mean of 40° and a standard deviation of 8°; the temperature in City 2 is normally distributed with a mean of 50° and a standard deviation of 18°. Which city is more likely to have temperature below freezing, 32°?

7.67 The enrollment in an introductory college statistics class is approximately normally distributed with a mean of 250 students and a standard deviation of 30 students. If this class is scheduled for a lecture hall with a maximum capacity of 300 students, what is the probability that the enrollment will exceed the capacity of this room?

7.68 What is the relationship between the mean and the median of a normal distribution?

7.69 A famous major league baseball scout rates prospects on a scale of 50 to 70. If his ratings are normally distributed with a mean of 60 and a standard deviation of 2.5, what is the probability that a randomly selected prospect will be rated 65 or higher? If he scouts 400 prospects this year, what is the probability that he will rate at least ten of them 65 or higher?

7.70 Give a logical explanation of why many physical measurements—including height, hip width, and shoulder width—are approximately normally distributed.

7.71 An influential handbook gives the following guidelines for designing products that accommodate the physical dimensions of the population: "In civilian work, it is desirable to include at least 95 percent of the people. . . . Work for the armed forces requires the inclusion of 90 percent."[10] If the relevant physical dimension, such as foot width or torso length, is normally distributed, what simple rule involving the mean and standard deviation meets the recommended 95 percent civilian guideline? What comparable rule satisfies the military's 90 percent requirement?

7.72 The heights of adult U.S. males are approximately normally distributed with a mean of 68.8 inches and a standard deviation of 2.65 inches. What fraction of the U.S. adult male population is able to stand barefoot without stooping in a 76-inch doorway? In an 80-inch doorway?

7.73 The heights of adult U.S. females are approximately normally distributed with a mean of 63.6 inches and a standard deviation of 2.50 inches. What fraction of the adult U.S. female population is taller than 68.8 inches, the height of the average adult of U.S. male?

7.74 *(continuation)* An inch is 2.54 centimeters. If we measure the heights of U.S. adult females in centimeters, will the measurements be approximately normally distributed? What are the mean and standard deviation, in centimeters?

7.75 The shoulder widths of adult U.S. males are approximately normally distributed with a mean of 17.7 inches and a standard deviation of 0.85 inch. If an airline seat is designed to be comfortable for someone with a shoulder width of 18 inches or less, what fraction of adult U.S. males will be uncomfortable?

7.76 *(continuation)* The shoulder widths of adult U.S. females are approximately normally distributed with a mean of 16.0 inches and a standard deviation of 0.85 inch. What fraction of adult U.S. females will be uncomfortable in an airline seat designed for people with shoulder widths of 18 inches or less?

7.77 A table height that can be adjusted in the range 26.8 to 30.0 inches allows 95 percent of all adult males to position the table at an optimal height. For adult women, the corresponding range is 25 to 28.2 inches. If the distribution of optimal table heights for each sex is described by a normal distribution, what are the apparent means and standard deviations?

7.78 *(continuation)* If a table is designed so that its height can be adjusted in the range 25.9 to 29.1 inches, what fraction of adult males and what fraction of adult females will not be able to set the height at an optimal level? Draw a graph that shows the male and female distributions of optimal heights and identifies those who will not be able to get an optimal table height.

• **7.79** The median sitting hip width of adult U.S. males is 13.9 inches. About 2.5 percent have sitting hip widths of 12.4 inches or less, and 97.5 percent have sitting hip widths of 15.8 inches or less. How can you tell from the information given that hip widths are not normally distributed? Are hip widths positively or negatively skewed? Draw a rough sketch, identifying 12.4, 13.9, and 15.8 inches.

7.80 Here is the distribution of 512,811 Graduate Record Examination (GRE) Quantitative Ability scores for the period October 1, 1981, through June 30, 1983:

Score	Percent of Scores
800–840	1
760–800	3
720–760	7
680–720	7
640–680	8
600–640	10
560–600	11
520–560	11
480–520	10
440–480	8
400–440	7
360–400	6
320–360	5
280–320	2
240–280	2
200–240	2

The average score was approximately 540 and the standard deviation 140. If the scores were normally distributed, what fraction of the scores would lie within one standard deviation of the mean? Within two standard deviations? Approximately what fractions of the actual scores lie within one and two standard deviations of the mean?

7.81 *(continuation)* Regroup the data into these intervals: 120–240, 240–360, 360–480, 480–600, 600–720, and 720–840. Compare the actual fraction in each interval with the theoretical fraction. Were the scores normally distributed with a mean of 540 and standard deviation of 140?

7.82 Each week the Federal Reserve (Fed) estimates various monetary aggregates, such as M1, which includes currency outside banks, traveler's checks, balances in checking accounts, and other checkable deposits. Because these data are estimates, they are subject to error, and thus the Fed often revises its monetary aggregate data months or even years later. Suppose that the correct value of M1 is μ and that the Fed's initial estimate X is a normally distributed random variable with an expected value equal to μ and a standard deviation equal to $.01\mu$. How often will the Fed's initial estimate of M1 differ from the correct value by more than 1 percent? By more than 2 percent?

7.83 Exercise 5.58 in Chapter 5 used the binomial distribution to calculate the probability that 8 or fewer computer users out of 20 surveyed will react favorably to a marketing survey. Compare this binomial probability to the probability given by a normal approximation with a continuity correction.

7.84 Exercise 5.73 in Chapter 5 calculates the probability of zero mortgage defaults and of more than five defaults, when each mortgage has a 2 percent chance of default. Calculate these probabilities, using a normal approximation to the binomial with a continuity correction.

7.85 Twenty thousand men of age 40 apply for a 10-year life insurance policy. Past data indicate that each has a .05 probability of dying by age 50. What is the expected value of the number of deaths by age 50 for this group of 20,000? What is the probability that more than 1,000 will die by age 50? That more than 1,050 will? Assume that the Bernoulli trial model is appropriate, and use a normal approximation without a continuity correction.

7.86 *(continuation)* For a woman of age 40, the probability of dying by age 50 is approximately .03. What is the expected value of the number of deaths by age 50 of a group of 20,000 women of age 40? What is the probability that more than 1,000 will die by age 50? Assume that the Bernoulli trial model is appropriate, and use a normal approximation without a continuity correction.

7.87 Consider a school with 2,000 undergraduate students, each with a 0.05 probability of requiring access to a computer on a given evening. What is the probability that more than 120 (6 percent) of the students will require a computer? Assume that the Bernoulli trial model applies, and use a normal approximation. Why might the Bernoulli trial model be inappropriate?

7.88 *(continuation)* Now consider a school with 2,000 undergraduate students, each with a 0.05 probability of requiring access to a computer on a given evening. What is the probability that more than 1,200 (6 percent) of the students will require a computer? Assume that the Bernoulli trial model applies, and use a normal approximation. Explain any difference between this probability and the probability calculated in the preceding exercise.

7.89 *(continuation)* Assume that the conditions in the previous exercise apply. If computer access is provided at this college by computer terminals, how many terminals are needed so that there is only a .01 probability that there will be insufficient terminals to satisfy student demand on any given evening?

7.90 Exercise 6.43 gives the heights of 1,052 mothers in 1903. The mean height is 62.484 inches and the standard deviation is 2.3904 inches. If these heights were from a normal distribution with this mean and standard deviation, what is the probability of a height greater than 65 inches? Less than 58 inches? How close are the observed frequencies to these theoretical probabilities?

7.91 A machine produces bolts with a diameter that is a normally distributed random variable with a mean of 1.000 centimeter and a standard deviation of 0.004 centimeter. The company that uses these bolts specifies that the diameter must be 1.000 $\pm$ .010 centimeter, that is, between 0.990 and 1.010 centimeters. What fraction of the bolts produced by this machine violate these specifications?

7.92 *(continuation)* What if the machine becomes misaligned, so that the bolt diameters are normally distributed with a mean of 1.001 centimeters and a standard deviation of 0.004 centimeter?

7.93 Exercise 5.55 describes the *Wall Street Journal*'s monthly stock picking contest in which four investment professionals are matched against four stocks selected by tossing darts at the financial pages. If the professionals have only a 50 percent chance each month of beating the dartboard portfolio, what is the probability that they will win six or more times in an eight-month experiment? Use a normal approximation to a binomial distribution with a continuity correction and compare this answer to the exact binomial probability.

7.94 Since a normal approximation to a binomial distribution is just an approximation, why would a statistician ever use a normal approximation in place of the exact binomial probability?

7.95 A business experiences financial distress when its cash flow does not cover its fixed obligations—including scheduled debt repayment and leasing charges. If it defaults on its financial obligations, it may be forced into bankruptcy. Suppose that a company's annual cash flow X (before meeting its fixed obligations) is normally distributed with an expected value of $150 million and a standard deviation of $25 million. What is the probability that its cash flow will be inadequate to cover fixed obligations of $a = \$100$ million?

7.96 *(continuation)* During the five years 1984–1988, U.S. corporations issued $400 billion in additional debt in order to repurchase $400 billion of their outstanding stock. This increased their leverage, magnifying the shareholders' percentage return when the firm does well and magnifying the losses when it does not. The issuance of additional debt increases a firm's fixed obligations and increases the chances that a given cash flow will be inadequate to cover these obligations. What is the probability that the cash flow of the firm described in the preceding exercise will be inadequate to cover fixed obligations of $a = \$125$ million?

• **7.97** *(continuation)* What are the expected value and standard deviation of $X - a$ if $a = \$100$ million? If $a = \$125$ million?

7.98 Some portfolio managers follow a contrarian strategy by deliberately doing the opposite of what others are doing—buying unpopular securities and avoiding popular ones. One gauge of a stock's popularity is its price-earnings (P/E) ratio, the ratio of its price per share to its annual earnings per share. Presumably, stocks with low P/Es are out of favor among investors and might offer bargains to contrarians. If P/Es are approximately normally distributed and a portfolio manager wants to screen 1,000 stocks, how many stocks should she expect to find with P/Es that are two standard deviations below the average P/E ratio?

7.99 *(continuation)* If the portfolio manager wants to identify the 5 percent of stocks with the lowest P/Es, how many standard deviations below the mean should she look?

7.100 *(continuation)* Some contrarians also look at companies with small market values (small-cap stocks). This portfolio manager decides to screen 1,000 stocks, identifying the 50 stocks with the lowest P/E ratios and the 50 stocks with the smallest market values. She finds that 12 of the small-cap stocks are also low P/E stocks. Are small-cap and low P/E independent? How can you tell? How many of the low P/E stocks are also small-cap stocks?

PART

II

Statistics

8

Sampling

"Data! data! data!" he cried impatiently. *"I can't make bricks without clay!"*

Sherlock Holmes
The Adventure of the Copper Beeches

TOPICS

8.1 Introduction
8.2 Random Sampling
8.3 Opinion Polls
8.4 Summary

8.1   INTRODUCTION

This chapter begins the statistics part of the course. You have learned now to think about the world as being uncertain and to use probabilities to quantify these uncertainties. You have also learned how to manipulate probabilities and when to use various distributions. Probability analysis is like most mathematical reasoning: We make certain assumptions and then deduce the logical implications. If 5 cards are fairly dealt from a standard, well-shuffled 52-card deck, then what is the probability of obtaining 4 aces? If 30 percent of the bombs in a large shipment are duds, then what is the probability that none of 10 tested bombs will be duds? If an asset return is normally distributed with a mean of 10 percent and a standard deviation of 10 percent, then what is the probability that the return will be negative? These are well-defined probability questions that you have learned how to answer. Few, if any, of you knew how to answer these probability questions when you began this course, but now you do, and that's a substantial accomplishment.

Probability is, however, only half of probability and statistics. Except in games of chance, we seldom know the correct assumptions needed to figure probabilities. In an apparently honest game of poker, we can reasonably assume that there are 52 cards in the deck, of which 4 are aces. But how many duds should we assume are in a particular bomb shipment? And how do we know that a certain asset return has a mean of 10 percent and a standard deviation of 10 percent? This is where statistics come in. Probability tells us, under certain assumptions, how often we can anticipate observing certain data—an ace, a dud, or a negative return. Statistical analysis is concerned with the reverse question. Having observed certain data, what can we infer about the number of aces in the deck, the number of duds in the shipment, or the mean and standard deviation of an asset return?

Another way of seeing the difference between probability and statistics is to distinguish between a population and a sample.

> A **population** consists of all the data that we might observe. A **sample** is the data that we actually do observe. **Probability analysis** is used to deduce the samples that are likely to come from an assumed population. **Statistical analysis** is used to infer the characteristics of the population that generated the sample at hand.

In the poker example, the population is the deck from which cards will be dealt. A sample is the particular cards that are dealt. Probabilities tell us how likely various poker hands are, assuming the game is fair. A statistical analysis could involve using the hands that you are dealt to infer that the game is, in fact, either fair or unfair. In the bomb example, the population is the entire shipment, while a sample is the bombs selected for testing. Probability analysis can be used to figure the likelihood of finding a certain number of duds in a sample, assuming that a certain percentage of all of the bombs are duds. Statistical analysis can be used to make inferences, based on the sample results, about the number of duds in the entire shipment. In the asset return example, the population is the returns that might occur, while a sample is the returns that do occur. Probability analysis tells us how likely various returns are, assuming a normal distribution with a particular mean and standard deviation. Statistical analysis can be applied to the returns we actually observe to help us estimate the population mean and standard deviation for the asset return.

Statisticians use samples to draw inferences about the population because it is often impractical to scrutinize the entire population. If we detonate every bomb to see how many are duds, we will have none left to use. If we cut open every grapefruit to check for quality, we will have none left to sell. If we burn every lightbulb to see how long it lasts, we will have only burnt-out bulbs left. Many tests are not this destructive, but are simply too expensive to apply to the entire population. Companies simply cannot afford to test every single item they manufacture. Instead, they test a sample. Editors sample a few pages of unsolicited manuscripts, because they haven't the time or patience to read every page of every draft that crosses their desk promising to be the great American novel. Opinion polls are given to a sample of people, because it is too expensive to ask everyone. Even the national census is done only once every 10 years. And even then, only relatively few people are given the long questionnaire that is the basis for most of the census data.

Social scientists are often compelled to use small samples, simply because usable data are scarce. In chemistry and physics, great amounts of data can be collected in laboratories with carefully controlled, repetitive experiments. In the social sciences, however, most data come from "natural experiments" that are beyond their control. Economists would like to know the effects of income fluctuations on consumer spending, but thankfully, they don't have the power to raise and lower the national income to test their theories. Instead, they stand back and record income and consumption data as the economy is tossed and turned by a variety of factors. They gather only a small amount of data because they cannot use the economy as a laboratory and because time is limited. They don't use data from the distant past, because the economy was very different then. And they don't wait until the distant future, because the economy will be different then, too. Also, they don't have the patience to wait for lots of data to accumulate. They want to test their theories soon, to make useful recommendations (and to publish, rather than perish). The same is true of many interesting social issues, ranging from school busing to capital punishment. Samples are small simply because nature's experiments provide few useful data.

EXERCISES

8.1 Pollsters interview samples of registered voters to predict the outcomes of presidential elections. What is their intended population? What is the costs and benefits of interviewing only a sample rather than the entire population?

8.2 Congressional committees often have pollsters testify about the public's views on policy questions. If you were a member of Congress, why might you do the same thing? Can you think of any reasons why your vote might not hinge on the results of a national poll, even if you believed the survey was completely accurate?

8.3 When a new product is designed, businesses test several prototypes to see if they do what they are designed to do. What is the intended population? Why do they test a sample rather than the population?

8.4 Businesses also make marketing tests of new products to gauge consumer enthusiasm. What are the benefits of sampling consumer reaction? Why do businesses use only a sample rather than the entire population?

8.5 Manufacturing processes inevitably fluctuate as there is wear and tear on people and machines. To keep the products tolerably close to specifications, companies routinely test production samples. If the output begins violating specifications, then adjustments are made in the manufacturing process. What are the costs and benefits of this monitoring?

8.6 Some managers occasionally drop in unexpectedly on their subordinates to see whether they are working or playing. These surprise visits are intended to be a sample from what population? What are the costs and benefits of using a sample rather than the entire population?

8.2   RANDOM SAMPLING

We are liable to draw invalid inferences from a sample if it isn't representative of the population. For instance, we might be interested in the physical strength of the students at the University of Nebraska. The relevant population is the entire student body, while a sample is the students who we actually test. If we use test results for the varsity football team, perhaps because these data are readily available, this would not be a representative sample. The average Nebraska football player may be able to benchpress 400 pounds, but the average student at Nebraska cannot. For measuring physical strength, the football team is a **biased sample,** in that it is clearly atypical. A biased sample such as this will give a distorted picture of the population.

We could try to put together a representative sample by wandering around the Nebraska campus and carefully selecting students who appear to be "typical," but if we did, we would surely assemble a sample that understates the diversity of students at Nebraska. We would pick what we think are "average" students and end up with a sample that has far less variance than the population. It would probably also be biased, because those we exclude for being "above average" and those we exclude for being "below average" are unlikely to just balance each other out. Worst of all, the biases would depend, in unknowable ways, on the researcher's mistaken perception of the "typical" Nebraska student.

These subjective biases are reminiscent of the story of the Chinese researcher who was trying to estimate the height of the emperor. The researcher was not allowed to look at the emperor. Instead, he estimated the emperor's height by asking the opinions of many people, none of whom had ever seen the emperor. Each person had his or her own, uninformed opinion of how tall the emperor was. It's the same here. None of us knows how strong the typical Nebraska student is. There may not even be a typical Nebraska student, just as there's no average U.S. household with 2.75 people in it. If we try to deliberately pick a sample that is representative, all we will get is a sample that shows how strong we think Nebraska students are!

To avoid the subjective biases of the researcher, statisticians advise that, paradoxically, the "fairest" sample is one in which the researcher exercises no control over who is selected! Similarly, a "fair" hand in a card game is not one in which the dealer turns the deck face up and carefully selects representative cards. Instead, a fair hand is whatever results from a blind deal from a well-shuffled deck. What card players call a fair deal, statisticians call a random sample.

> In a **random sample,** each member of the population is equally likely to be selected for the sample, and the chances of being selected do not depend on the previous selections.

There is, admittedly, a possibility that a sample chosen by chance methods will, by the luck of the draw, be unrepresentative. But that slight risk is needed to avoid samples that tell us little more than the researcher's subjective opinions of what's typical.

One of the subtle advantages of a random sample is that probabilities can be calculated and used to make statistical inferences about the population; we will be able to make statements, such as, "We are 95 percent confident that Candidate A will receive 54 percent of the vote, plus or minus 2 percent." With a subjectively picked sample, we are helpless. We may be able to say something about the person who made the selections, but there is no basis for drawing inferences about the population that the sample is intended to represent.

When analyzing the theoretical properties of a random sample, it is assumed that the drawing is with replacement, so that the probability of being picked remains constant. Otherwise, the mathematics are overly complicated. In practice, however, most sampling is done without replacement, because it seems silly to pick the same item twice. This divergence between theory and practice is justified by the fact that samples are usually taken only when the population is so large that the probabilities are barely affected by sampling without replacement.

Notice also that the word *random* is used differently in the probability phrase *random variable* and in the statistical phrase *random sample.* A random variable is just a variable whose value is uncertain. In this sense, random simply means uncertain. A random sample means much more, in that it is a series of outcomes conforming to the strict Bernoulli-trial assumptions of constant probabilities and independent outcomes. The strengths of Nebraska football players are random variables, but they do not constitute a random sample from the student population.

What if we were to test the physical strength of every entering freshman at Nebraska? This is a large sample, but it is not a random sample from the Nebraska student body—most students would have no chance of being included in this sample. This exclusion causes an obvious bias because there are systematic physical differences between freshmen and more mature students. Similarly, graduating seniors, the students in an advanced math class, or the females in a physical education class are not a random sample. In each case, there are easily identifiable reasons why the group is not representative of the student body as a whole.

We would ideally like to put each student's name into a box and pick names

Statistics in World War II

Many United States and British statisticians used statistical analyses to help their countries during World War II. In the process, they refined their techniques and proved the value of statistical analysis. There are many examples, but I'll only mention a few here.

With war rationing, the U.S. Office of Price Administration (OPA) needed estimates of the inventories held by tire dealers. They first tried to make a complete tabulation by mailing questionnaires to every single dealer in the country, but many dealers didn't mail the questionnaires back. The OPA suspected that the inventories of the dealers who didn't respond were quite different from those who did, so they put together a small sample of dealers and used follow-up phone calls and visits to ensure responses from almost every dealer in the sample. The small but reliable sample proved to be less expensive and more accurate than the large tabulation.

A similar thing happened with estimates of German output. The Germans tried to make a complete tabulation of how much they were producing, but the reports from individual factories were often late and not always reliable. The British and U.S. statisticians were keenly interested in estimating German war production too, but they could hardly ask the German factories to send them reports. Instead, they were forced to base their estimates on the manufacturing serial numbers on captured equipment. These serial numbers provided a very small sample, but it was random. Special studies made after the war discovered that the British and U.S. estimates of German production were more accurate and timely than Germany's own estimates!

out randomly. Each student, whether weak, strong, or somewhere in between, would then have an equal chance of inclusion in our sample. In practice, somewhat different methods are often used, but these are satisfactory as long as there is no systematic relationship between the method and what we are measuring. For example, we may have access to Nebraska student identification numbers. If so, then we can construct a random sample by picking each student whose identification number ends with, say, the number 3. Our sample will then include about 10 percent of the student body, selected in a manner that is apparently unrelated to their physical strength. If a smaller sample is desired, then we could pick those students whose last two digits are, say, 24. In each case, we want to use a selection method that seems unrelated to what we are measuring—in this case, it is physical strength.

The 1954 Nationwide Tests of a Polio Vaccine

Poliomyelitis (polio) is an acute viral disease that causes paralysis, muscular atrophy, permanent deformities, and even death. It is an especially cruel disease in that most of its victims are children. The United States had its first polio epidemic in 1916 and, over the next 40 years, hundreds of thousands of Americans were

What Makes a Book a Best-Seller?

The Sunday *New York Times* book review section contains widely followed lists of best-selling fiction and nonfiction books.[1] Many people consult these lists before they buy books, reasoning that they should read what other people are reading. Bookstores prominently display the *New York Times* best-selling list; some even have a special section reserved for these best-selling books. Thus success breeds success, in that a book that makes the *Times* best-seller list will sell many more copies, while a book that doesn't make the list is apt to vanish without a trace.

How does the *New York Times* compile this influential list of best-selling books? They can't rely on publishers to provide sales data because, recognizing the power of the list, publishers would exaggerate the sales of books they want to promote. Instead, the *Times* surveys bookstores—a reasonable course but one that, for many years, it pursued in a fairly haphazard manner. Each week, *Times* staffers telephoned several bookstore managers and asked them to name, in no particular order, books that were selling well. The managers' answers may have been based on actual sales data or memory, perhaps influenced by a recollection of titles on the current *New York Times* best-seller list.

The bookstores surveyed by the *Times* are not a random sample and may not be representative of books with special audiences. There is also too much opportunity for subjective judgment and manipulation. For example, although a biography of former Alabama football coach Bear Bryant sold more copies than many books on this best-seller list, it never made the list because the *New York Times* did not survey enough—if any—bookstores in southern states. When a novel written by Chuck Barris, a television personality, did not make the best-seller list, he personally purchased thousands of copies from bookstores that he knew were surveyed by the *Times,* enough copies to put his book on the list and to generate many more sales to people whose book purchases are influenced by the best-seller list.

afflicted. In the early 1950s, more than 30,000 cases of acute poliomyelitis were reported each year.

At the same time, researchers discovered that most adults had experienced a polio infection during their lives. In most cases, this infection had been mild and had caused the person's body, in trying to ward off the infection, to produce antibodies that made the body immune to another attack. Similarly, they found that polio is rarest in societies with the poorest hygiene. The explanation is that all of the children in these societies are exposed to this contagious virus while still young enough to be protected by their mother's antibodies, and so they develop their own antibodies without ever suffering from the disease.

Scientists hoped to find a safe vaccine that would provoke the body to develop antibodies, without causing paralysis or worse. A few vaccines had been tried in the 1930s and then abandoned after apparently causing the disease that they had been designed to prevent. By the 1950s, extensive laboratory work had turned up several promising vaccines that seemed to safely produce antibodies against polio.

In 1954, the Public Health Service organized a nationwide test of Jonas Salk's polio vaccine, involving 2 million schoolchildren.[2] Polio epidemics varied greatly from place to place and from year to year throughout the 1940s and early 1950s. For this reason, the Public Health Service decided not to offer the vaccine to all children, either nationwide or in a particular city. Otherwise, they would not have been able to tell if variations in polio incidence were due to the vaccine or to the vagaries of epidemics. Instead, it was proposed that the vaccination be offered to all of the second graders at selected schools. The experience of these children (the *treatment group*) could then be compared to first and third graders (the *control group*) at these schools, who were not offered the vaccine. Ideally, they would have liked the treatment group and the control group to be alike in all respects, but for the fact that the treatment group received the vaccine and the control didn't.

There were two major problems with this proposed test of the Salk vaccine. First, participation was voluntary and those who agreed to be vaccinated tended to have higher incomes and better hygiene and, as explained earlier, to be more susceptible to polio. Thus, there was selection bias in the sample. Second, school doctors were instructed to look for both acute and mild cases of polio, and the mild cases were not easily diagnosed. If the doctor knew that some second graders were vaccinated, while first and third graders were not, this knowledge might influence the diagnosis. A doctor who hoped the vaccine would be successful would subconsciously be more apt to see polio symptoms in the unvaccinated than in the vaccinated.

A second proposal was to run a *double-blind* test, in which only half of the volunteer children would be given the Salk vaccine, and neither the children nor the doctors would know whether a child received the vaccine or a placebo solution of salt and water. Each child's injection fluid would be chosen randomly from a box and the serial number recorded. Only after the incidence of polio had been diagnosed, would it be revealed whether the child received vaccine or placebo. The primary objection to this second proposal was the awkwardness of asking parents to support a program in which there was only a 50 percent chance that their child would be vaccinated.

As it turned out, about half of the schools decided to inoculate all second-grade volunteers and use the first and third graders as a control group. The remaining half insisted on a double-blind test using the placebo children as a control group. The results are shown in Table 8.1. In both approaches, the Salk vaccine reduced the incidence of polio, but the gains are more dramatic in the double-blind experiment. The number of diagnosed polio cases fell by about 54 percent in the first approach and by 61 percent with the placebo control group. The effectiveness of the Salk vaccine is partially masked in the first experiment because selection bias makes the second-grade volunteers different from the first and third graders. If the double-blind experiment had not been conducted, the case for the Salk vaccine would have been less convincing because the observed differences were smaller and a skeptic could have attributed these differences to a subconscious desire by doctors to have the vaccine work.

Table 8.1 *The Results of a Nationwide Test of a Polio Vaccine in 1954*

	First and Third Graders as Control Group		Double-Blind Test with Placebo Control Group	
	Number of Children	Polio Incidence per 100,000	Number of Children	Polio Incidence per 100,000
Treatment group	221,998	25	200,745	28
Control group	725,173	54	201,229	71
No consent	123,605	44	338,778	46

Source: Holden-Day, Inc., Oakland, CA 94609.

The Salk vaccine was eventually replaced by the even safer and more effective Sabin preparation, but the 1954 tests stand as a landmark national public health experiment that provided convincing evidence of the value of a polio vaccine. Nowadays, 90 percent of all U.S. children have received three or more oral doses of polio immunization, and the nationwide incidence of polio has been reduced to about a dozen cases a year.

The Nielsen Ratings

Arthur Charles Nielsen was trained as an electrical engineer, earning the highest grade-point average in the history of the University of Wisconsin's engineering school. During the 1930s he started a business, the A.C. Nielsen Company, that collected and sold data on product sales in drugstores. He compiled these data by having his employees interview drugstore customers, but he soon found that his data for two soaps, Lux and Lifebuoy, were not consistent with actual sales data. Nielsen's son later explained, "Dad realized that a lot of women said they bought Lux because of its image as a prestige soap for movie stars while Lifebuoy was only for people with b.o."[3]

Nielsen learned that data on what people do are sometimes more reliable than data on what they say they do. He stopped interviewing customers and began paying druggists to allow his employees to examine their sales records. Today, the A.C. Nielsen Company makes most of its money from its Retail Index, which keeps track of store sales of food and drug products.[4] Companies may know how much they are producing, but they don't have up-to-the-minute data on whether their products are selling or gathering dust on store shelves, so they pay the Nielsen Company to monitor a sample of stores for them.

Nielsen is best known, though, for monitoring our television viewing habits. The primary tool is an electronic gadget, called a people meter, that works like a remote control switch and keeps track of which show is on and which family members are watching. Before 1987, when people meters were introduced, Nielsen relied on viewer diaries and on audimeters, electronic devices that automatically recorded the channel being watched, but did not count the number of people watching. People meters require less effort than a written diary and keep a minute-by-minute record of viewing. The people meter even has the ability to determine if people fast forward through commercials when watching a show recorded on a video cassette recorder.

Nielsen pays 2,000 households throughout the country a small fee for allowing them to attach monitors to their TV sets and phone lines. Each day, a Nielsen computer telephones the people meter and automatically processes the viewing habits that have been recorded.

Nielsen then computes a "rating" and an "audience share." The Nielsen rating is an estimate of what fraction of all of the homes in the nation watched a particular program. On a typical evening in "prime time," about two-thirds of the nation's homes are watching television. If they all watch the same show, that lucky show will get a Nielsen rating of 66.7 percent and the other networks will get big fat zeros. In practice, the three major networks divide up the 67 percent pretty evenly, with local stations, UHF stations, and cable television taking what's left over. Nielsen also reports audience-share figures, which estimate what fraction of the televisions that are turned on are tuned to that show. A network show is usually in trouble if it gets a rating of less than 20 percent and a share of less than 30 percent.

Because this sample of 2,000 households is used to estimate the behavior of 80 million homes, each represents 40,000 households—an awesome responsibility! If 20 of these chosen households switch channels, that will cost a show a rating point and perhaps its life. With all of this power, each of Nielsen's 2,000 households must sign an agreement promising not to let anyone connected with the communications industry know that they are among the chosen few.

The nice thing about Nielsen's people meters is that they really do show what channels the television is turned to. With personal interviews or viewing diaries, there is a great temptation to vote for shows you like, even if you didn't find time to watch them that week. Of course, the people meter only shows that buttons were pushed, not that people watched. A *Los Angeles Times* reporter broke through Nielsen's code of secrecy and located one of the 2,000 dubbed "Deep Eyes." Deep Eyes claimed that he had his television on at least 50 hours a week, winning points for favored shows, even though he didn't really spend nearly that much time watching television. After the story was printed in the *Times,* other Nielsen households contacted the reporter to say that they were honest watchers and to express outrage at Deep Eyes's abuse of the system. Nielsen is working on a way to combat this— an electronic gadget that uses sound waves or heat sensors to determine how many people are in the room.

The 2,000 people meters are too thinly spread across the country to give reliable ratings of local stations or local viewing of the network stations. For these data, Nielsen uses viewer diaries. There are four "sweep weeks" each year, when Nielsen mails out a few hundred thousand viewer diaries to gather detailed data on local viewing habits and the demographic and economic characteristics of viewers. Advertisers want to know not only how many sets are on, but how many big spenders are watching. They also want to avoid advertising shaving cream on shows watched only by women and children. The television stations respond to sweep weeks in a predictable fashion, showing their blockbuster movies and other specials during these four important rating weeks. The next time you notice that all three networks are showing great movies at the same time, you can reasonably conclude that it is a sweep week again.

Random-Number Generators

Instead of putting names into a box, random sampling is usually done through some sort of numerical identification combined with the random selection of numbers. First, a numbering is decided on that encompasses every member of the target population. If we're sampling people, we might use social security numbers, telephone numbers, or street addresses. If we're conducting a scientific experiment, then we could number each batch, beaker, or bunny as it is received. Sometimes, events can be labeled by a time of day or by the day of the week, month, or year.

The second step is to randomly select numbers that determine the items to be included in the sample. You or I could easily design a brief experiment for randomly picking these numbers. For instance, we could play a scientific numbers game by using the digits in the attendance figures at a local race track. Or we could set up a small game of chance: To pick a digit from 0 to 9, we could shuffle ten cards, ace through 10, and select one card. For a two-digit number, we could do this twice.

To illustrate some other approaches, let's say that we have 64 patients that we want to divide into two groups, one group to receive a new medication and the other group to receive a placebo. For each patient, we could simply flip a coin, with heads being the new medication and tails the placebo; we could separate the patients alphabetically, according to their last names; we could number the patients 1 to 64, and pick 32 numbers out of a hat; or we could use ten cards numbered 0 to 9 to randomly pick 32 two-digit numbers between 1 and 64, the number 0 and numbers greater than 64 being simply ignored.

A statistician who doesn't want to go to the trouble of flipping coins or drawing cards can use a table of *random numbers* constructed by someone else. Random numbers are generated by some physical process in which each digit from 0 to 9 has a one-tenth chance of being selected. The most famous of these tables contains a million random digits obtained by the RAND Corporation from an electronic roulette wheel using random frequency pulses.[5] I have put some of these digits, which I picked at random, in Table 3 at the end of this book. These random numbers are arranged in 25 columns of 50 two-digit numbers, and are separated by a space after every fifth column and row for easier reading.

I'll illustrate the use of this table by the selection of 32 out of 64 numbered patients. First, we see that we're interested in two-digit numbers, ranging from 01 to 64. Next, we have to find a place to begin in Table 3. We could begin at the beginning, with the very first number 10. But statisticians don't like to always begin at the beginning. There is a slight danger that the researcher will eventually memorize the numbers at the beginning of the table and thereby destroy their randomness. If I know that 10 is always the first number picked, then, when I am numbering patients, I may (consciously or subconsciously) give the number 10 to someone who I want to receive the medication. Or, later, when I am evaluating the patient's condition, I may remember that 10 always gets the medication and consequently I will be more inclined to see an improvement in patient 10's condition.

Statisticians advise randomly picking a place to begin in a table of random numbers. You could simply close your eyes and point, or you could make up some more formal procedure. I used the stopwatch function on my digital wrist watch. I started the stopwatch, stopped it after several seconds, and looked at the reading in the hundredths-of-a-second place; every two-digit number from 00 to 99 is equally likely to be selected by this procedure. First, I went for the row number and got 34. My first try at the column number was 60; but there are only 25 columns in Table 3. So I tried again and got 16. The number in the thirty-fourth column and sixteenth row of Table 3 is 13.

Now I can quickly move up, down, sideways, or even diagonally to find additional numbers. I used the stopwatch again with 00 to 24 being north, 25 to 49 east, and so on, and got 28 (east). So, my first four numbers are 13, 02, 12, and 48. The next two, 92 and 78, are ignored (as are any repeats), since there are only 64 patients, and I move on to 56, 52, 01, and 06. Having hit the end of this row, I go down to the next and continue with 46, 05, and so on. Thus the patients numbered 13, 02, 12, 48, 56, . . . receive the new medication.

Random numbers also can be plucked out of computers. Most computers, large and small, have built-in procedures for randomly generating numbers. Sometimes, two large numbers are multiplied and then divided by a third large number. Other programs take a large number and raise it to a power. In either case, some of the digits in the answer are used as the random number. Modern computers can make many such calculations extremely quickly. Many years ago, Karl Pearson spent weeks flipping a coin 24,000 times. Now, with a computer program for generating random numbers, 24,000 coin flips can be simulated in a few minutes.

Some theorists object that the digits appearing in a table or programmed into a computer are not strictly random. They call these digits "pseudo-random numbers." This is a variant of the argument that, once a coin has been flipped, the outcome is no longer a random variable, even if no one yet knows what that outcome was. By the same logic, the numbers in a table cannot be random because those numbers have already been determined. (Indeed, the RAND table was constructed in 1947.) For the objectivist, randomness comes from picking a random path through the table. Call them what you will, random-number tables can be used to construct random samples successfully, as long as they give each member of the population an equal chance of being selected.

Dr. Spock's Overlooked Women

In 1969, Dr. Benjamin Spock, the author of a best-selling book on child care, was tried in a Massachusetts federal court for conspiracy to violate the Military Service Act of 1967.[6] The potential jurors were allegedly selected at random from lists of adult residents of the district: The court clerk testified that he put his finger randomly on a page and made a mark next to the name selected. The potential jurors were sent questionnaires and after those disqualified by statute were eliminated, batches of 300 were called to a central jury box, from which a panel of 100 (called a "venire") was selected. As it turned out, even though more than half of the residents in the district were female, only 9 of the 100 on Spock's panel were. Spock was convicted, but appealed.

The jury selection process may not have been truly random. A finger placed blindly on a list of residents will almost certainly be close to several names, and the court clerk may favor males subconsciously when deciding which name to mark. The court records do not reveal how the panel of 100 was selected from the central jury box, but this, too, may have been influenced by subjective, nonrandom factors.

Several months after the end of the trial, the defense obtained data showing that out of 598 jurors used recently by Spock's trial judge, only 14.6 percent were female, while 29 percent of the 2,378 jurors used by the other six judges in this district were female. To suggest how this bias may have prejudiced Spock's chances of acquittal, the appeal cited an April 1968 Gallup poll in which 50 percent of the males labeled themselves hawks and 33 percent doves on Vietnam, as compared to 32 percent hawks and 49 percent doves among females. It is also conceivable that women who had raised their children "according to Dr. Spock" might have been sympathetic to his antiwar activities. The appeal argued that the probability that a randomly selected jury would be so disproportionately male was infinitesimally small* and that, "The conclusion, therefore, is virtually inescapable that the clerk must have drawn the venires for the trial judge from the central jury box in a fashion that somehow systematically reduced the proportion of women jurors."

Spock was eventually acquitted, though on First Amendment grounds rather than because of a flawed jury selection. However, a law was passed in 1969 mandating the random selection of juries in federal courts using mechanical, statistically accepted techniques. Subsequently, males and females have been equally represented on venires in federal courts.

EXERCISES

8.7 Answer this student's question:

First, statisticians tell us that a sample should be representative of the population. Then, they tell us to pick the sample completely randomly, with no guarantee that it will be representative. Why, if a representative sample is so important, don't we make some effort to ensure that the sample is representative?

8.8 A vaccine called BCG was tested to see if it could reduce deaths from tuberculosis.[7] A group of doctors gave the vaccine to children from some tubercular families and not to others, apparently depending on whether or not parental consent could be obtained easily. The results over the subsequent six-year period (1927–1932) were

	Vaccinated	Not Vaccinated
Number	445	545
TB deaths	3	18
Percent	0.67	3.30

A second study was then conducted in which doctors vaccinated only half of the children with parental con-

---

*As in the case of Guam's misplaced ballots, we can apply the hypergeometric distribution with $N = 2{,}976$, $S = 777$, and $n = 598$; the probability of 87 or fewer females works out to be 3.6×10^{-14}.

sent. The results over an eleven-year period were

	Vaccinated	*Not Vaccinated*
Number	556	528
TB deaths	8	8
Percent	1.42	1.52

Is there any logical explanation for these contradictory findings?

8.9 One week, 400 of Nielsen's 2,000 households watched a certain television show on CBS, 360 watched NBC, 440 watched ABC, 140 watched other stations, and 660 had their sets off. What are Nielsen's ratings and audience shares for the three networks?

8.10 Before people meters, three ambiguities for the Nielsen computations were that most households have more than one person in them, many have more than one television set, and some may watch only part of a show. The Nielsen ratings assumed that everyone in a household watched every set that was on and watched all of any program if they watched at least six minutes of it. How did these assumptions bias Nielsen's estimate that two-thirds of us each watch television at 9:00 at night? Explain your reasoning carefully, with numerical examples if you wish.

8.11 The Chesapeake and Ohio Railroad Company once studied the economic advantages of allocating interline passenger revenue on the basis of a sample of tickets rather than by an exhaustive tabulation. A complete tabulation determined that their share of the revenue over a four-month period should come to $212,164 from Railroad A and $79,710 from Railroad B. A 5 percent sample yielded the respective estimates $212,063 and $80,057. **a.** How much revenue would the Chesapeake and Ohio have lost if it had used these samples? **b.** If sampling cost $1,000 while a complete tabulation cost $5,000, what would have been the net gain to the Chesapeake and Ohio of using this sample instead of a complete tabulation? **c.** If the samples are randomly chosen, how much can the Chesapeake and Ohio anticipate saving, on average, each four-month period by sampling these tickets?

8.12 Pepsi-Cola had a "blind taste test" in which people tasted Pepsi from a glass marked *M* and Coca-Cola from a glass marked *Q*.[8] More than half of the people preferred Pepsi. Coca-Cola then ran its own test, letting people drink Coke from a glass marked *M* and Coke from a glass marked *Q*. They found that most people preferred Coke from the glass labeled *M*! This prompted their advertising headline: "The day Coca-Cola beat Coca-Cola." Is there any rational explanation for these strange results?

Pepsi then ran another test, serving Pepsi in glasses marked *L* and serving Coke in glasses marked *S*. Again, most people preferred Pepsi. As a statistician, how would you have designed a "blind taste test"?

8.13 Critically evaluate the following passage:

> . . . it is the farmer, not the importuning salesman, who leads in the consumption of alcohol. Results of a survey of drinkers classed by occupation, which was published last week by the Keeley Institute of Dwight, Illinois . . . show that of 13,471 patients treated in this well-known rehabilitation center from 1930 through 1948, a total of 1,553 (11.5%) were farmers. Next in line came salesmen, merchants, mechanics, clerks, lawyers, foremen and managers, railroad men, doctors and manufacturers.
>
> Since Keeley is located in the heart of the farm belt, it might be expected that its proportion of farmer patients would be unusually large. This is not the case, according to James H. Oughton, institute director. The patients in the survey were drawn from all over the world.[9]

8.14 A researcher wants to estimate the average weight of students at Boston College. A convenient sample is obtained by asking the weights of the first 10 students in line when the dining hall opens for lunch at noon. What systematic biases might be introduced by this researcher's procedure?

8.15 Estimate the proportion of students at your college who are left-handed by asking 25 students. Tell how you selected these students and explain why you think they are a good sample.

8.16 Estimate the proportion of students at your college who come from out-of-state by asking 25 students. Tell how you selected these 25 students and explain why you think they are a good sample. Can you think of any reason why more care should be exercised in selecting this sample than in selecting a sample to estimate the number of left-handed people?

8.17 The Educational Testing Service wanted a representative sample of college students[10] so they first divided all schools into groups, such as large public universities, small private colleges, and so on. Then they chose a "representative" school from each group. Finally, they wrote to the administrations of these selected schools and asked them to choose some students for the study. In what ways might their sample not be representative of the college student population?

8.3   OPINION POLLS

Public opinion polls are undoubtedly the most difficult random sample to construct, but they also can be among the most interesting. In an election, for instance, the statistical population consists of every person who votes. The winner will not be known for certain until the votes are actually cast and counted, but statisticians can use a sample of a few thousand voters to predict the election outcome ahead of time. That is, they will use the expressed preferences of a few voters to estimate the preferences of all voters. For these estimates to be accurate, statisticians need a sample that is fairly chosen and representative of the population as a whole. Sometimes samples are not representative and the results are very misleading. The 1936 *Literary Digest* fiasco is a famous example of an embarrassingly inaccurate poll.

The 1936 *Literary Digest* Poll

In 1932, the Great Depression drove Herbert Hoover out of the White House and brought in Franklin Roosevelt. Roosevelt's efforts to end the Depression were largely unsuccessful, but he retained the public's faith that he was trying at least and that he was doing as well as any president could. In 1936, he ran for reelection against Alf Landon, the rather unexciting governor of Kansas. Most veteran political observers thought that Roosevelt would be reelected easily. But the *Literary Digest* attracted considerable attention with its prediction of a landslide victory for Landon, 57 percent to 43 percent. The *Digest* polled 2.4 million people, the largest political poll ever conducted, and claimed that its prediction would be "within a fraction of 1 percent" of the actual vote. The *Digest* had, in fact, successfully forecast the five presidential elections. But this time they were woefully wrong. It was Roosevelt who won in a landslide, 62 percent to 38 percent, and the *Literary Digest* soon folded, with its reputation permanently tarnished.

The *Digest* poll was incredibly ambitious. They mailed questionnaires to 10 million people, a full quarter of the voting population. But, in sampling, quality is much more important than quantity. Two thousand randomly sampled voters is much better than 2 million unrepresentative voters. The *LiteraryDigest* got the bulk of its 10 million names from an examination of every single telephone book in the United States. Nowadays, this sort of examination would encompass virtually every household—except those with unlisted numbers, new numbers, or those few households without phones. Indeed, one present-day pollster, Albert Sindlinger, conducts polls successfully by having interviewers dial randomly selected telephone numbers.

In 1936, however, telephones were far from universal. Phone service was still relatively new and the Depression had made the telephone a luxury for many households. There were only 11 million residential phones in 1936 and these homes were disproportionately well-to-do and in favor of Republican candidate Landon. Those people without phones were relatively poor and overwhelmingly for the Democrat, Roosevelt. The telephone books in 1936 yielded a biased sample that was unrepresentative of the voter population. The *Literary Digest* also used some names and addresses collected from car registrations, club memberships, and its own subscriber lists, but these sources were even more biased than the telephone directories.

To get a random sample, the *Literary Digest* should have focused on lists of registered voters. They were, in fact, one of the first to use voter registration lists at all, but they only used a few lists and these were swamped by the names from telephone directories and other atypical sources. Thus, the 1936 *Literary Digest* poll had a great deal of selection bias.

Selection bias is a systematic exclusion of certain groups from consideration for the sample.

Mail Polls

The *Literary Digest* poll had another problem that is common to polls conducted by mail. Most people do not waste time carefully reading and responding to their junk mail. Only those who care a great deal about an issue or an election are going to bother to read a poll's instructions, fill out the form, and mail it in. Those who do take the time and trouble may be atypical of the population as a whole. This sampling problem is called nonresponse bias.

Nonresponse bias is systematic refusal of some groups to respond to a poll.

For example, the *Literary Digest* mailed voting questionnaires to one-third of the registered voters in Chicago, but only 20 percent of these people bothered to fill out the questionnaires and mail them back to the *Digest.* Landon was favored by more than half of the people who responded to this mail poll, but, in the election, two-thirds of all Chicago voters voted for Roosevelt. The *Literary Digest* poll apparently had nonresponse bias as well as selection bias.[13]

Statisticians have discovered that one pattern in nonresponses is that mail questionnaires tend to be discarded by lower-income and upper-income households. Perhaps the lower-income groups don't trust polls and the upper-income groups have more valuable things to do with their time, or maybe these two income groups don't bother reading any junk mail. In any case, those who do respond to mail polls tend to be disproportionately middle-income households. So, if a statis-

New Coke Versus Old Coke

In 1985, after 99 years with essentially the same taste, Coca-Cola decided to switch to a new, high-fructose corn syrup, to make Coke sweeter and smoother—more like its arch rival, Pepsi. This historic decision was preceded by an enormous, top-secret survey of 190,000 people, in which the new formula beat the old by 55 percent to 45 percent. What Coca-Cola apparently neglected to take into account was that many of the minority who preferred old Coke did so passionately. While the 55 percent who voted for new Coke may have been able to live with the old formula, many on the other side swore that they could not stomach new Coke. Coca-Cola's announced change provoked outraged protests and panic stockpiling by old-Coke fans. Soon, Coca-Cola backed down and brought back old Coke as "Coke Classic." A few cynics suggested that this was Coca-Cola's game plan all along, and that the anticipated protests were just a clever way of getting some free publicity and causing, in the words of a Coke senior vice-president for marketing, "a tremendous rebonding with our public." For 1985, new Coke captured 15.0 percent of the entire soft-drink market and Coke Classic 5.9 percent with Pepsi at 18.6 percent. In 1986, new Coke collapsed to 2.3 percent, Coke Classic surged to 18.9 percent, and Pepsi held firm at 18.5 percent.

In 1987 the *Wall Street Journal* commissioned an interesting test of 100 randomly selected cola drinkers, of whom 52 declared themselves beforehand to be Pepsi partisans, 46 to be Coke Classic loyalists, and 2 to be new Coke drinkers.[11] In the *Journal*'s test, new Coke was the winner with 41 votes, followed by Pepsi with 39 and Coke Classic with 20. Seventy of the 100 people who participated mistakenly thought they had chosen their favorite brand. Some were very indignant. A Coke Classic drinker who chose Pepsi said: "I won't lower myself to drink Pepsi. It is too preppy. Too yup. The New Generation—it sounds like Nazi breeding. Coke is more laid back." A Pepsi enthusiast who chose Coke said: "I relate Coke with people who just go along with the status quo. I think Pepsi is a little more rebellious and I have a little bit of rebellion in me."

In 1990 Coca-Cola relaunched new Coke with a new name—Coke II—and a new can with some blue color, Pepsi's traditional color. Coca-Cola executives and many others in the soft-drink industry remain convinced that cola drinkers prefer the taste of new Coke[12] even while they remain fiercely loyal to old Coke and Pepsi—a loyalty due perhaps more to advertising campaigns than taste.

tician wants all income groups to be fairly represented, then a mailed questionnaire is not a good approach.

The main distinguishing feature between respondents and nonrespondents is, of course, that the respondents feel strongly enough about the questions asked to take the time to complete the poll. Who is going to bother to mail in an answer "I don't know," "no opinion," or "I haven't made up my mind"? It follows that those who do respond will be more sharply divided than the population as a whole. Let's say, for example, that questionnaires are mailed to 1,000 people, of whom 200 strongly favor a certain policy, 200 strongly oppose the policy, and 600 are largely indifferent. The responses might come from 100 of those in favor and 100 of those opposed, indicating that half the people are in favor and half opposed. This survey result correctly shows that opinion is evenly divided, but it completely misses the important fact that most people haven't made up their mind on the issue.

Now let's change the numbers slightly. It's eighteen months before a presidential election and four candidates are beginning their long run for the Democratic nomination. One thousand questionnaires are mailed out: to 200 people who favor candidate A, 100 people who favor B, 50 people who favor C, 50 people who favor D, and 600 people who don't really know any of these candidates, let alone which one they favor. Responses might be received from 100 people who favor A, 50 who favor B, 25 C, and 25 D. This mail poll shows that A is favored by a 2-to-1 margin over B and that C and D have almost no support. The fragility of these numbers is completely hidden. The mail poll doesn't tell us that 60 percent of the people polled haven't made up their minds yet. When the undecided do make up their minds, any of these candidates could turn out to be the people's choice.

Let's change the numbers one more time. Some time has passed and now there are just two candidates, A and B, left in the race for the Democratic nomination. One thousand questionnaires are mailed out: to 200 people who strongly favor A; to 400 people who would vote for B, but don't feel very strongly about it; and to 400 people who still haven't made up their minds. It turns out that 150 of the A supporters promptly mail in their choice, while only 50 of the B supporters bother to respond. Even though B is favored 2-to-1 by those who have made up their minds, the mail poll shows a 3-to-1 lead for A. Again, the mail poll is distorted by nonresponse bias.

In each case, we can't tell how many people haven't made up their minds, because the undecided don't bother to express this indecision. In addition, those with strong opinions are more apt to respond than those with more balanced opinions. Unfortunately, the nature of the U.S. politics is that candidates who do not do well in the early polls do not get the financial support they need to run a competitive campaign. Thus, special-interest candidates who are strongly favored by a small number of voters have a big advantage over candidates whose appeal is more broadly based, but not as fervent. Mail polls exaggerate this distortion and this is one reason why impartial political pollsters no longer use mail questionnaires.

A related example occurred in Massachusetts, where pollsters found that a majority of voters approved of mandatory busing to help integrate the schools, but the polls overlooked the strength of the voters' feelings. Most of those who favored busing did not feel very strongly about it, certainly not enough for this issue to be decisive in an election campaign. But those who were opposed to busing were very opposed, enough so that they would automatically vote against any candidate favoring busing. Politicians who looked at the polls and thought that they could win votes by endorsing busing soon learned otherwise.

A mail poll is naturally more suspect, the fewer the people who bother to respond. In the 1940s, the makers of Ipana Tooth Paste boasted that a national survey had found that "Twice as many dentists personally use Ipana Tooth Paste as any other dentifrice preparation. In a recent nationwide survey, more dentists said they recommended Ipana for their patients' daily use than the next two dentifrices combined." The Federal Trade Commission banned this ad after it learned that less than 1 percent of the dentists surveyed had named the brand of toothpaste they used and that even fewer had named a brand recommended to their patients.[14]

An even more extreme example occurred in 1983, when the *Winnipeg Sun* asked its readers to vote on whether the city should build a new hockey arena for the Winnipeg Jets.[15] A total of 430 readers mailed in ballots clipped from the newspaper, with 280 in favor of the arena and 150 against it. Despite the very small returns, the *Sun* ran a headline proclaiming: *Winnipeg Wants a Downtown Arena.* City officials nonetheless rejected the proposal.

A year later a rival newspaper reported that employees of the Winnipeg Jets had mailed in 200 of the 280 "yes" ballots in a carefully orchestrated plan: buying newspapers in small batches, filling out the ballots with different color pens and in different handwriting, and mailing the ballots in envelopes of different sizes, with stamps of different denominations, from neighborhood mailboxes throughout the city. Very few people cared enough to share their opinion with the *Sun;* those who did were not a random sample.

Quota Sampling

Let's go a bit farther with the example of a presidential election poll. In theory, we would like to put every voter's name into a large box, mix the pieces of paper thoroughly, and then pull out some names for our sample. But, in practice, we don't really know the names of all voters. It would be a major effort to collect accurate, up-to-date voter registration lists from every precinct in the nation. Also, many registered voters will not actually vote on election day. The correct population is those who will vote, not those who could vote, and there is no way to identify these voters ahead of time. Even if we could somehow get the right names in the box, it would be very difficult to track down every person whose name is drawn. Some will be in very out-of-the way places. Does it make sense to spend two days of a pollster's time finding a recluse in Nowhere, North Dakota? Some people will have moved. Many will not be home when the pollster calls. We may get a badly biased sample if we ignore people in isolated places, people who have moved, and people who aren't home at convenient times for the pollster.

Because of these problems, pollsters don't follow the theoretical random sampling model of collecting every name and then drawing a sample. One alternative is quota sampling.

> In **quota sampling,** a sample is constructed by filling quotas of certain characteristics that are thought to be true of the population as a whole.

Based on surveys of past elections, pollsters can estimate the fractions of voters who are in various age, sex, geographic, and economic categories. Interviewers can then poll people who are in these same categories. For example, in the 1948 presidential election, one interviewer for the Gallup poll was told to interview fifteen people in St. Louis, of whom seven were to be male and eight female.[16] In addition, the interviewer was told that nine of the people polled should live downtown while six should live in the suburbs. Six of the men were to be white and one black. Three of the men were to be under 40 years of age and four were to be over 40. Two of the white men should pay less than $18 a month for rent, three should pay between $18 and $44, and one should pay more than $44. There were analogous quotas for the eight women. Beyond these quotas and the requisite mental gymnastics, the interviewer was free to select the persons to be polled.

Quota sampling avoids the painstaking drudgery of mechanical random sampling, but the danger is that it leaves too much discretion in the hands of the interviewer. One simple mechanical problem is to verify the categories. Where do you draw the dividing line between city and suburb? Also, people may well misstate (or not reveal) their ages or rents. To "play it safe," the interviewer may decide to avoid people near the dividing points. If you only poll 20-year-olds and 60-year-olds, then the subjects definitely will be under or over 40, as required. But playing it safe means that people near 40 years of age will get slighted in the survey, and 40-year-olds may well have different opinions from those who are 20 or 60 years old.

In addition, most interviewers meet their quotas in the easiest possible ways, which means that people in risky, unfamiliar, or inconvenient places will be ignored. Interviewers do not want to waste their time or risk their lives. If it's easiest to interview people during the day, then those with 9-to-5 jobs will be slighted. If door-to-door interviews in the evening are easier for the interviewer, then those who go out at night will be missed.

If you take a part-time polling job to help pay your way through college, you will probably interview a lot of your fellow students. You may even misstate their ages, addresses, and so on, because that's a lot easier than finding people who do fit the categories. Interviewers have even been known to fill out the questionnaires themselves to avoid venturing into unfamiliar parts of town. Even if a conscientious interviewer does try to locate people to fill every category, there will still be a subconscious tendency to avoid those who look disreputable or possibly dangerous. In theory, quotas ensure a more balanced sample than mailing lists, but in practice, quota sampling tends to be "convenience" sampling rather than genuine random sampling. In his provocative book, *How to Lie With Statistics,* Darrell Huff put it this way,

> You have pretty fair evidence to go on if you suspect that polls in general are biased in one specific direction, the direction of the *Literary Digest* error.
>   This bias is toward the person with more money, more education, more information and alertness, better appearance, more conventional behavior, and more settled habits than the average of the population he is chosen to represent.[17]

After the 1936 *Literary Digest* fiasco, political pollsters turned from mail questionnaires to quota sampling as an inexpensive way of assuring a representative sample. But, for the reasons just discussed, quota sampling is biased, too. The bias may be less severe than with mail surveys, but it is still uncomfortably large. Given the freedom to select their poll subjects, interviewers tend to choose people who dress well, live in nice neighborhoods, and are also disproportionately Republican.

The Gallup poll, for example, was just starting at the time of the 1936 presidential election. Using quota sampling, the Gallup poll did better than the *Literary Digest*. Still, the Gallup poll overestimated the vote the Republican candidate would receive in each of the first three elections it covered:

Year	Gallup Prediction of Republican %	Actual Republican %	Prediction Minus Actual
1936	44%	38%	+6%
1940	48%	45%	+3%
1944	48%	46%	+2%

The selection bias in quota sampling caused Gallup to overestimate the vote for the Republican candidate, but luckily, in each case the bias was not large enough to cause an embarrassing, erroneous prediction of Republican victory. In the next election, though, the vote was close and Gallup was one of several pollsters who incorrectly predicted that the Republican candidate would be elected.

Franklin Roosevelt died in 1945 and the presidency passed to his vice-president, Harry Truman, a scrappy, but largely unknown, small-time politician from Kansas City. In 1948, Truman had to seek election on his own against Thomas Dewey, who achieved national fame as a crime-fighting New York City district attorney and then gained political prominence as the governor of New York. Gallup predicted that Dewey would get 50 percent of the vote and Truman only 44 percent, with 6 percent going to two minor party candidates. The other major pollsters, using quota sampling too, agreed that Dewey would win easily. The October issue of *Fortune* stated that

> Barring a major political miracle, Governor Thomas E. Dewey will be elected the thirty-fourth president of the United States in November. So decisive are the figures given here this month that *Fortune,* and Mr. Roper, plan no further detailed reports. . . . Dewey will pile up a popular majority only slightly less than that accorded Mr. Roosevelt in 1936 when he swept by the boards against Alf Landon.[18]

The editors of the *Chicago Tribune* were so certain of the outcome that they printed up the Dewey victory headlines even before the votes were counted. But there was a Republican bias in the polls' quota sampling: Truman got 50 percent of the vote and Dewey only 45 percent. Truman went to bed thinking he would lose and woke up to find that he had won. He triumphantly posed for photos, waving a newspaper that had prematurely announced Dewey's victory. Truman went on to become a successful president, while the pollsters went looking for ways to improve their samples.

Cluster Sampling

The mistaken pollsters were an easy target. One oft-repeated quip was that, "Everyone believes in public opinion polls—from the man in the street right on up to

President Dewey." Even the former editor of the *Literary Digest* managed a few digs: "Nothing malicious, mind you, but I get a very good chuckle out of this."[19] The embarrassed political pollsters concluded that they had made two big mistakes. The first was that they had stopped polling too early, and a lot of people didn't decide to vote for Truman until the last two weeks before the election. The second was that their samples had not been really representative of the voter population.

They wanted to find a way to make their samples more random, without incurring the staggering expenses associated with a complete random sample. What they settled on is a sequential sampling procedure called **cluster sampling,** in which they first sample a map and then sample people. The nation is divided into geographic regions and then into states, counties, and cities of different sizes. A random-number generator is used to pick specific cities, with each city's chance of being selected made proportionate to its size. Then, in each of these selected cities, a random-number generator is used to choose some voting precincts. Voter registration lists are obtained for the selected precincts and a random-number generator is used again, this time to choose households. Interviewers (who, incidentally, are overwhelmingly female) are then sent to the selected households to conduct the survey. To maintain randomness, interviewers are told to interview a specific person, not necessarily the person who answers the door. Otherwise, a systematic bias may be introduced by the fact that age, sex, and work habits may influence both who opens the door and their political preferences.

The sequential nature of the sampling process makes it unnecessary to list every single household in the country. Yet the random selection at each stage means that, in theory, every household has an equal chance of being included in the sample, no matter what its physical appearance or the condition of the neighborhood. Inconvenient, out-of-the-way precincts may well be chosen. This possibility is needed for a true random sample. Yet clustering also makes it more economical to sample out-of-the-way places because several households are interviewed within each precinct. Remember, one of the problems with a one-stage random sample is that the selected names will probably be scattered all over the map. Interviewers would spend more time traveling than interviewing. Cluster sampling is more convenient for interviewers, without giving them the discretion of choosing their own sample.

In practice, the pollsters' procedures are not quite this pure. Most don't consider excessively out-of-the-way sampling places, such as Alaska. Many allow their interviewers to choose the particular houses to visit and people to interview within a given sampling area. Many scrutinize the age, sex, race, and income of the respondents, and will either adjust the results or throw out the poll if the sample seems unrepresentative.

Thus, pollsters' practices are a common-sense blend of random and quota sampling, a hybrid with most of the economic advantages of quota sampling and few of the disadvantages. It is a logical and economically feasible approximation of a simple one-stage random sample from the entire population. And it works! Here is Gallup's presidential forecasting record since 1948:

Year	Gallup Prediction of Republican %	Actual Republican %	Prediction Minus Actual
1952	51%	55.4%	−4.4%
1956	59.5%	57.8%	+1.7%
1960	49%	49.9%	−0.9%
1964	36%	38.7%	−2.7%
1968	43%	43.4%	−0.4%
1972	62%	61.7%	+0.3%
1976	49%	48.0%	+1.0%
1980*	47%	50.7%	−3.7%
1984	59%	58.8%	+0.2%
1988	56%	53.9%	+2.1%

*The 1980 election included an independent candidate who received, as predicted, 7 percent of the vote. The Democrat, Jimmy Carter, was predicted to get 44 percent, but only got 41 percent. A *New York Times*/CBS postelection poll found that 20 percent of the voters changed their minds on who to vote for or whether to vote at all during the last four days of the campaign, and 60 percent of this group decided not to vote for Carter.[20]

The Republican bias has been eliminated by cluster sampling in that the Republican vote is now underestimated as often as it is overestimated. In addition, the size of the prediction error has dropped dramatically. With quota sampling, Gallup's average prediction error in the 1936 through 1948 elections was 4 percent. Using cluster sampling, the average prediction error has been less than 2 percent. What is most impressive is that Gallup and the other political pollsters have become more accurate even while sharply reducing the size of their samples. Gallup's embarrassing 1948 prediction was based on a poll of 50,000 registered voters. Nowadays, using cluster sampling, Gallup obtains much more accurate predictions from interviews with less than 4,000 people! To be reliable, a sample needs to be random much more than it needs to be large.

More Polling Pitfalls

There are, of course, still some pitfalls in opinion polling that cannot be avoided just by sampling procedure. Many registered voters do not vote, and the pollsters' objective is to predict the preferences of those who do vote, not the preferences of those who could vote. Good pollsters try to identify nonvoters by asking such questions as "Did you vote in the last election?" and "Do you feel strongly about the upcoming election?"

Another response problem is that some people in the sample may not be home at the right time, even if the interviewer tries again and again. And even if people are home, they may have things to do that seem more important than answering some nosy questions. Pollsters usually "piggyback" their polls, having their interviewers ask hundreds of questions for many different clients. After an hour or two of this grilling, many bored or irritated subjects will understandably become more careless in their answers.

Another problem is that interviewers are human and some may be a bit lazy or even dishonest. For example, two weeks before the 1968 presidential election, *U.S. News & World Report* reported that a poll by a Wisconsin editor found that George Wallace was favored over Richard Nixon by a 5-to-3 margin.[21] The Wisconsin editor conducted this poll by quizzing eight men in a bar! Additionally, it is always easier to give a questionnaire to your friends or fill it out yourself than to track down the people who are supposed to be interviewed, particularly if they live in an unfamiliar or seemingly dangerous neighborhood. Good pollsters discourage such dishonesty by making random followup calls to see if the people who were supposed to be interviewed really did get interviewed.

Appearances can work the other way, too. The answers people give may be influenced by the demeanor of the interviewer. We have a natural tendency to suppress feelings that might be provocative or embarrassing and, instead, say what we think people want us to say. With the occasional exception of a Barry Goldwater or George McGovern, our presidential candidates are usually so bland that this is not a problem for political polls. It doesn't take much courage to admit who you are going to vote for. In addition, the interviewers sometimes give each person a ballot that can be marked secretly and dropped unsigned into a box that the interviewer carries. (Of course, there is then no guarantee that the person answers the questions at all!)

An interesting postelection phenomenon is that, often, more people claim they voted for the winner than actually did. An extreme example is John F. Kennedy, who got barely half of the votes cast in 1960. After his assassination and elevation to hero status, pollsters found it almost impossible to find anyone who would say he voted against Kennedy in 1960. This particular halo effect is undoubtedly exaggerated by the fact that Kennedy's 1960 opponent was Richard Nixon, who has fallen from grace. Nowadays, most people say they didn't vote for Nixon in 1972 either, even though he won that election by a landslide over George McGovern.

In other, even more controversial opinion polls, some people either are evasive or misstate their position. Darrell Huff[22] tells the example of the National Opinion Research Center, which used two teams of interviewers, one white and one black, to poll 500 southern blacks during World War II. One question asked was, "Would Negroes be treated better or worse here if the Japanese conquered the United States?" The black interviewers reported that 9 percent of those polled thought the Japanese would treat blacks better and 25 percent thought blacks would be treated worse. The white interviewers, however, found that only 3 percent of the blacks said that blacks would be treated better while 45 percent said they would be treated worse. The most reasonable explanation for this disparity is that blacks were less candid with white interviewers. Another possibility is that the white interviewers were more reluctant to poll surly looking blacks.

Another controversial opinion poll involved the 1990 Nicaraguan presidential election, in which Daniel Ortega of the ruling Sandinista National Liberation Front campaigned against an outsider, Violeta Barrios de Chamorro.[23] Three independent opinion polls conducted during the month preceding the election showed Ortega winning a landslide victory over Chamorro by 16 to 24 percentage points. Two days before the election, Ortega told a reporter that defeat was "inconceivable." Yet Chamorro won the election with a decisive 55 percent of the vote. One pollster later said that voter surveys had underestimated the citizens' fears of the ruling Sandinistas. Many voters apparently thought that the people conducting the polls were Sandinistas. They publicly voiced support for Ortega but, in the privacy of their secret ballots, voted against him.

Sometimes, it's the question itself, rather than the questioner, that biases the answers. A 1964 issue of *Fact* magazine had the attention-grabbing headline,

1,189 Psychiatrists say Goldwater Psychologically Unfit to be President

The story was based on a mail poll of 12,000 doctors, asking such loaded questions as

"Does he seem prone to aggressive behavior and destructiveness?"
"Can you offer any explanation for his public temper tantrums?"
"Do you believe that Goldwater is psychologically fit? No or yes?"

About 20 percent of the doctors responded, with the provocative headline based on the fact that about half of these answered "no" to the vague question of Goldwater's psychological fitness. Goldwater sued for libel and was eventually awarded $50,000 in damages.

For a less extreme example, consider this question that the Gallup poll once asked:

The U.S. Supreme Court has ruled that a woman may go to a doctor to end pregnancy at any time during the first three months of pregnancy. Do you favor or oppose this ruling?

Gallup found that 47 percent were in favor, while 44 percent were opposed. These answers were undoubtedly influenced by the careful avoidance of the word "abortion" and by the inclusion of the Supreme Court decision. A pro-abortion group substituted the phrase "for an abortion" for "to end pregnancy" and found that only 41 percent approved while 48 percent were opposed. They then omitted the Supreme Court reference by asking this question: "As far as you yourself are concerned would you say that you are for or against abortion, or what do you think?" Then they found only 36 percent in favor and 59 percent opposed.[24]

There is no best way to ask a controversial question. The point is just that the phrasing sometimes has a big effect on the answers. One pollster, Peter Hart, claims that by deliberately wording the questions to favor your position, "You can come up with any result you want." Of course, most pollsters don't have favored positions. They are honest professionals, trying to obtain an accurate gauge of public opinion. But many controversial issues are not as simple as "Who are you going to vote for?" They are complex questions that are difficult to word fairly in a way that will still yield simple "yes" or "no" answers. Here, the art of polling is not in choosing a random sample, but in asking the right questions. And, as in all fields, there

Randomizing Answers to Sensitive Questions

Many interesting questions are pretty personal and perhaps embarrassing:

"What is your annual income?"
"Do you frequently steal from your employer?"
"Have you ever had an abortion?"
"Are you a homosexual?"

The most honest answers would be given anonymously, but because of nonresponse bias we don't want to use mail polls to provide this anonymity. We could instead use a questionnaire that is filled in while the interviewer's back is turned and then sealed in a plain envelope. The Gallup presidential poll, for example, uses a secret ballot. However, some respondents' answers may be influenced by a suspicion that the curious pollster will promptly rip open the envelope and snoop at the answers.

An ingenious alternative is to use randomized questioning, so that the pollster doesn't know which question is being answered.[25] For example, the person can be handed a questionnaire that says:

To protect your privacy, you have been given a bag containing two balls: one red and the other blue. Please close your eyes and draw one ball out of this bag. If the ball is red, write "yes" below. If the ball is blue, answer the question "Have you ever cheated on an exam?" Your answer is completely confidential since no one knows which color ball you drew.

If it turns out that 60 percent of the answers are "yes," what is our estimate of the fraction

π of the population who cheat on exams? On average, we expect that

1/2 draw a red ball and answer yes
$(1/2)\,\pi$ draw a blue ball and answer yes
$(1/2)\,(1 - \pi)$ draw a blue ball and answer no

Overall,

$$\begin{aligned} \text{yes answers:} &\quad (1/2) + (1/2)\pi \\ \text{no answers:} &\quad \underline{(1/2)\,(1 - \pi)} \\ \text{total:} &\quad 1 \end{aligned}$$

If, in fact, there are 60 percent yes answers, then we can estimate π by

$$(1/2) + (1/2)\pi = .60$$

implies that

$$1 + \pi = 1.2$$

so that

$$\pi = .2$$

Say, for instance, that 200 people are polled. We expect 100 yes answers from people who draw red balls and, if $\pi = .2$, 20 yes answers from the 100 people who draw blue balls. Altogether, there would be $120/200 = .6$ yes answers.

With randomized responses fewer people answer the question of interest. But giving anonymity to those who answer sensitive questions encourages truthful answers. And, again, a small, unbiased sample is much better than a large, biased one.

are some unscrupulous pollsters whose primary objective is getting the answers that the client wants to hear. The founder of the Roper poll put it this way,

> While we are probably the newest profession in existence, we have managed in a few short years to take on many of the characteristics of the world's oldest profession.[26]

Some polls are biased by the fact that they are mailed out with material arguing one side of a particular issue. For example, I used to get letters from one congressman boasting how hard he was fighting to control wasteful government spending. Included with these letters were questionnaires that asked whether we thought government spending should be reduced or increased.

Darrell Huff provides another example.[27] The *New York Times* once ran a story "DOCTORS REPORTED FOR SECURITY PLAN" that began, "A recent poll in New Jersey indicates that most of the country's doctors would like to come under the Social Security old age and survivors insurance program, according to Representative Robert W. Kean, Republican of New Jersey." It turns out that this story was based on a postcard questionnaire that was included in an issue of the *Bulletin of the Essex County Medical Society* that contained an article by Representative Kean explaining why he thought doctors should be included in the social security program. Can you spot the major reasons for giving little credence to this poll?

First, there is a potential sampling bias, because Essex County doctors may not be representative of "most of the country's doctors." Second, there is the nonresponse bias inherent in a postcard questionnaire (which 80 percent of the readers did not answer). Third, those who did respond may have been temporarily biased by having just finished reading arguments on one side of the issue. After their emotional momentum subsided and after they heard the other side, they might have felt differently.

EXERCISES

8.18 A Los Angeles television station includes a poll as part of its nightly local news show. A provocative "question of the day" flashes on the screen and viewers are asked to call in their "yes" or "no" votes. Can you think of any reason why these poll results might not be representative of community opinion?

8.19 Congresswoman Louise Day Hicks mailed 200,000 questionnaires to the people in her district, asking whether or not they favored using mandatory busing to integrate the public schools. The *Boston Globe* reported that 80 percent of the 23,000 people who responded were opposed to busing, which "proved decisively that the 'silent majority' does care about the country and its problems and, given the opportunity, will speak out on the issues."[28]

Is there any reason to use caution in interpreting these poll results? Who was the real "silent majority" here?

8.20 In July 1983, Congressmen Gerry E. Studds (a Massachusetts Democrat) and Daniel Crane (an Illinois Republican) were censured by the House of Representatives for having had sex with teenage pages. A questionnaire in the Sunday (July 17) *Cape Cod Times* asked readers whether Studds should "resign immediately," "serve out his present term, but not run for reelection," or "serve out his present term, and run for reelection." Out of a paid Sunday circulation of 51,000 people, 2,770 returned this questionnaire. Of these, 45.5 percent (1,259) wanted Studds to resign immediately, while 46 percent (1,273) wanted him to serve out his term and run for reelection. Another 7.5 percent (211) wanted him to serve out his term but not seek reelection and 1 percent (27) were undecided. Why might these survey results be misleading?

8.21 Estimate the average grade point average (GPA) of students at your college by asking 10 students to tell you their GPAs. Tell how you selected these 10 students and explain why you think they are a good sample. Did any students refuse to tell you their GPAs? What sort of nonresponse bias might arise in a GPA survey? Are there any other reasons why your estimate might be used cautiously?

8.22 Ask 25 people this question:

> Karl Marx said, "Whenever any form of government becomes destructive of these ends, it is the right of the people to alter or to abolish it." Do you agree?

Do you think you would get substantially different results if you identified the correct source as the U.S. Declaration of Independence?

8.23 A poll taken by the National Opinion Research Center asked blacks if the Army is unfair to blacks. Only 11 percent said yes to white interviewers, while 35 percent said yes to black interviewers.[29] What could account for this disparity?

8.24 A variety of data were collected on the boys at a summer Boy Scout camp. One scout leader noticed an interesting relationship between foot size and speed:

Foot Length	Average Time for 50-Yard Dash
< 9 inches	9.5 seconds
≥ 9 inches	8.6 seconds

Do big feet help people run faster? Why do you suppose that the boys with big feet were faster than the boys with small feet?

This scout leader also discovered that boys with big feet had more merit badges than boys with small feet. Do big feet make people smarter, too?

8.25 A newspaper selected 100 names at random out of the city telephone book and sent a reporter out to ask these 100 people whether they felt social security benefits are "too high," "too low," or "about right." Only 60 of these people were home when the reporter tried to interview them. Rather than use a sample with such a small response rate, the newspaper selected another 100 names at random and sent the reporter out again to try names in this second batch until 40 more people were interviewed successfully. Does the newspaper's procedure eliminate the small response rate? Is their poll a random sample or could it be biased in some way?

8.26 In 1952, the Chicago Police Department banned the public showing of the Italian film, "The Miracle." The American Civil Liberties Union (ACLU) subsequently showed this film at several of its private meetings and reported that,

> Of those filling out questionnaires after seeing the film, less than 1 percent felt it should be banned. "It thus seems," said Sanford I. Wolff, Chairman of the Chicago Division's Censorship Committee and Edward H. Meyerding, the Chicago ACLU's Executive Director, "that the five members of the [police] Censorship Board do not represent the thinking of the majority of Chicago citizens."[30]

Why is the ACLU's poll not convincing statistical evidence?

8.27 A randomized response poll was given to 30 students in a Tuesday statistics class:

> Is your birthdate an even or odd number? If it is even, please answer question (a); if it is odd, please answer question (b).
> a. Is today Tuesday?
> b. Have you ever handed in a course paper that was substantially written by someone else?

Eighteen "yes" answers were handed in along with twelve "no" answers. If we assume that even and odd birthdates are equally likely, what is your estimate of the fraction of the class that has handed in papers that were written by others?

8.4   SUMMARY

Samples are used in a variety of situations because a complete enumeration is too expensive or difficult. We sample food when we take a taste: "You don't have to eat the whole thing to know its rotten." Many of the courses you take at college are a sample of a discipline, to help you make career decisions. Your dates can be a sample of married life. Soon, an employer will take a sample of your intelligence, training, and personality by using a job interview to learn something about you. Government agencies sample products to see if they are safe. The military samples weapons to see if they will work. In business, samples are used to help design new products, test product reliability, gauge the quality of raw materials, monitor production methods, measure worker productivity, and conduct advertising campaigns.

Samples are intended to be representative of an entire population. If they are, then they can provide accurate estimates of population characteristics. Deliberate attempts to construct representative samples are unwise, though, because "representative" is in the eye of the beholder. If a researcher carefully assembles what is thought to be a representative sample, then we will learn something about this researcher's opinion, but little more. Instead, statisticians recommend that a random sample be used, in which each member of the population is equally likely to be selected. This is analogous to a fair deal from a deck of cards. In practice, a random sample can be constructed by enumerating the members of the population and then using a table of random numbers to select the sample.

Polls are used to gauge public opinion on all sorts of things, including the President's performance, political candidates, controversial legislation, economic conditions, new products, the effectiveness of advertising campaigns, and even the appeal of prospective movie plots. These can be very interesting and even useful, but require considerable care in construction and interpretation. A selection bias may inadvertently exclude some members of the population from consideration for the poll. Nonresponse bias involves the refusal by some groups to participate in an opinion poll. In addition, it is difficult to phrase questions in a balanced way that invites simple answers. Slight changes in the wording of an opinion poll may well affect many people's answers. Also, many questions are too complex to be answered easily with a simple "yes" or "no." The statistics of opinion polling is to design a random sample; the art is to ask good questions.

REVIEW EXERCISES

8.28 Estimate the average parental income of students at your college by asking 10 students their parents' income. Tell how you selected these 10 students and explain why you think that they are a good sample. Do you have any reason to be cautious about interpreting these income declarations? If you suspect a bias, explain your reasoning. Do you think that the average declared income is more likely to be too high or too low?

8.29 In the United States, more people die from polio vaccine than from polio. Does this show that it is safer not to be vaccinated?

8.30 A Chinese census for famine relief showed a population of 105 million people. Yet a census taken a few years earlier for military and tax purposes had shown a population of only 28 million. Why do you suppose these census figures were so far apart?

8.31 *Consumer Report's* annual automobile issue includes frequency-of-repair records, based on letters received from their readers. Is this a random sample? If not, what biases might appear?

8.32 The Harris Poll doesn't include people in Alaska or Hawaii, because it is too expensive to send interviewers to these two states. How might this economization bias the poll results?

8.33 In 1964, pollsters found that presidential candidate Barry Goldwater did better in their polls when interviewers used secret ballots. What could explain this curious phenomenon?

• **8.34** A certain advertising agency is looking for magazines in which to advertise inexpensive imitations of designer clothing. Their target audience is households earning less than $15,000 a year. A survey finds that the readers of one magazine have an average household income of $17,500. Is the agency correct in concluding that the people in their target audience do not read this magazine?

8.35 The Census Bureau has estimated the lifetime earnings of people by age, sex, and education. The *Cape Cod Times* (July 7, 1983) reported that

> 18-year-old men who receive graduate education will be "worth" more than twice as much as peers who don't complete high school. On the average, that first 18-year-old can expect to earn just over $1.3 million while the average high-school dropout can expect to earn about $601,000 between the ages of 18 and 64.

Mike O'Donnell was getting poor grades in high school and was thinking of dropping out and going to work for his dad's construction firm. But his father, an avid reader of the *Cape Cod Times,* saw this article and told Mike that dropping out would be a $700,000 mistake. Is Mike's dad right?

8.36 A newspaper columnist wrote that women's goal of

> an equal chance at a good job . . . has happened. . . . Consider the changes from census to census—from 1970 to 1980. Briefly, there were only 13,000 female "lawyers and judges" back then. Today there are 74,000. Stop the presses: There were 61,000 "editors and reporters" then—and 103,000 now. Another X-ray scan: There were 45,000 "physicians and dentists" then—and 76,000 now.[31]

Do these statistics prove that women now have an equal chance at a good job? What other data might be of interest?

8.37 During World War II, 408,000 U.S. citizens were killed in military duty and 375,000 U.S. citizens were killed by accidents in the United States. Does this show that it was almost as dangerous to stay home as to fight in the war?

8.38 In 1984, a congressman asked his constituents "Which statement is closest to your view with respect to the federal budget deficit?" Of three options, 69 percent of those who responded chose the statement "We should reduce the deficit by raising taxes and by reducing spending in many areas of the federal budget, including the military."[32]

Explain any reservations you might have about this congressman's conclusion that "69 percent . . . oppose the current buildup in military spending . . . confirm[ing] my belief that the people of Southeastern Massachusetts are appalled to see defense spending soar while the rest of the federal budget is suffering severe cutbacks."

8.39 Ann Landers once asked her readers "If you had it to do over again, would you have children?" and received 10,000 responses, 70 percent saying no they wouldn't. Why should we be cautious in interpreting this poll?

• **8.40** The useful life of a certain car is normally distributed with an expected value of 70,000 miles and a standard deviation of 10,000 miles. What is the probability that one of these cars will last longer than 100,000 miles? The advertising agency has suggested an ad featuring an owner whose car has lasted 100,000 miles. The agency telephones 100 people who have owned this car sufficiently long so that the car either has 100,000 miles or has died. What is the probability that at least one of these 100 cars will have lived to see 100,000 miles? Why are the owners cited in this ad a biased sample?

8.41 Males who have never smoked average 14.8 days of restricted activity per year, while present smokers average 22.5 days and former smokers average 23.5 days.[33] Do these data show that it is healthiest to never smoke, but once you start, it's better not to stop?

8.42 A polio vaccine was tested in a town with 1,000 children. Half of the children were vaccinated and half weren't. The study found that none of the children who were vaccinated got polio. Does this prove that the vaccine worked? What other data would you like to see? The children who were not vaccinated did not get polio either. What was wrong with this sample?

8.43 In a 1986 television commercial for pain killers, an announcer says:

We asked 1000 doctors if stranded on an island, which would they want: Brand A, Extra Strength Brand A, or Brand B? More doctors chose Brand B, nearly 2 to 1 over Extra Strength Brand A.

The actual percentages were shown in small print:

> Brand B 44%
> Extra Strength Brand A 23%
> Brand A 20%
> Brand C 5%

Why might the makers of Brand A claim that the announcer is misleading?

8.44 Explain any possible flaws in this conclusion:

> A drinker consumes more than twice as much beer if it comes in a pitcher than in a glass or bottle, and banning pitchers in bars could make a dent in the drunken-driving problem, a researcher said yesterday.
>
> E. Scott Geller, a psychology professor at Virginia Polytechnic Institute and State University in Blacksburg, Va., studied drinking in three bars near campus. . . .
>
> Observers found that, on average, bar patrons drank 35 ounces of beer per person when it came from a pitcher, but only 15 ounces from a bottle and 12 ounces from a glass.[34]

8.45 During the 1976 presidential primaries, Lou Harris asked Democrats and Independents to choose from a variety of candidates. When a full list of 25 possibilities was offered, Edward Kennedy got the most votes and George Wallace finished second with 17 percent of the votes. When Kennedy's name was removed, Hubert Humphrey finished first and Wallace second, this time with 19 percent. Jimmy Carter, the eventual nominee, got only 2 percent. Yet when Harris asked this sample to choose between Carter and Wallace, Carter got 43 percent and Wallace 38 percent, with 19 percent undecided.[35]

If you were Wallace's campaign manager, how would you interpret these poll results?

8.46 A *Newsday* (July 13, 1983) story was titled "Poll Shows Strong Work Ethic." The story told how a poll had counted workers as having a strong work ethic if they agreed with the statement, "I have an inner need to do the best I can regardless of pay." The poll found that the percentage of U.S. workers expressing a strong work ethic was higher than in West Germany or Japan. Yet, U.S. worker productivity has been lagging behind these other countries. The report accompanying the poll results concluded that, "Studies of work behavior show that slackening effort may be making a significant impact on the U.S. economy. . . . The challenge, therefore, is to find means to harness the latent but untapped power of the U.S. work ethic." One of the researchers associated with the poll offered the following theory:

> From an economy dominated by assembly line jobs, we have moved to a situation where white-collar and service workers have a great deal more control over their work. The decision to work harder is more than ever in the hands of the individual worker. Many managers don't yet realize how you go about motivating a worker who has a lot of freedom in his job.[36]

This seems to be a farfetched explanation of the disparity between the poll's finding of a strong work ethic and the evidence of slackening worker effort. If workers have a strong work ethic and more freedom to work as hard as they want, then I would expect them to work harder. If, as the report concludes, people are not working harder, then perhaps the strength of their work ethic is exaggerated. Is there any reason to suspect that the results of this sort of poll might be biased?

8.47 *Dial* magazine published a long article on how different people are using home computers, based on a survey of its readers. One of their conclusions was that "the huge majority (82 percent) feel 'very favorable' about the computers in their homes. Another 12 percent feel 'somewhat favorable,' and not one of the 1,000 in the random sample tabulated for the survey indicated feeling 'very unfavorable.'" Comment on the following description of their random sample:

> *Dial* asked its readers who use home computers or who have experience with computers at work to respond to a two-page questionnaire in the magazine. We received nearly 5,200 responses; more than 600 of them included letters of up to 15 pages in length. Most of these letters were written on computers.
>
> A random sample of 1,000 responses was pulled from the group. The statistics in the article are based on a tabulation of that sample. . . .[37]

• **8.48** Critically evaluate:[38]

> The more children you have, the less likelihood there is that you will ever be divorced.
>
> This is an idea that many have suspected to be true. But it remained for the Metropolitan Life Insurance Company to prove it with the statistics printed

below. The possibility that a childless couple will be divorced is about twice as great as it would be if they had two or more dependent children. . . .

Number of Children	Divorces per 1,000 Marriages
0	15.3
1	11.6
2	7.6
3	6.5
4 or more	4.6

8.49 A randomized response poll included this question:

> Please flip a coin. If it is heads, write down "yes." If it is tails, please answer this question truthfully, "Have you ever shoplifted?"

The answers turned out to be 62 "yes" and 38 "no." Estimate the fraction that has shoplifted. The researcher who administered this poll suspects that some people who were proud of never having shoplifted may have answered "no" even though their coin flip turned out to be a head. If so, is your estimate of the fraction of the population who has shoplifted too high or too low?

• **8.50** A randomized response poll of 300 people included this question:

> Please flip this coin and then answer question (a) if it lands heads and (b) if it lands tails.
> (a) Please roll this die and write down the number that comes up.
> (b) Please write down the number (from 1 to 6) that is closest to your monthly income (in 1,000s of dollars).

The answers turned out to be

Number:	1	2	3	4	5	6
Responses:	85	61	52	40	34	28

Use these data to estimate the income distribution.

8.51 The *Cape Cod Times* conducts a weekly poll in the summer, asking a question of a half dozen or so people at a public beach. The answers, together with each person's name, occupation, and usually a picture, appear in the Saturday edition of the paper. Why wouldn't a statistician consider this poll to be scientific?

8.52 The *Atlanta Journal–Constitution* conducted a poll in February 1985, asking its 571,646 readers whether the U.S. football League (USFL) should switch from a spring to a fall season.[39] There were 12 responses, 4 wanting the USFL to switch to the fall so that it would be wiped out by the NFL, 4 wanting the USFL to stay alive either in the fall or spring, and 4 not caring one way or the other. One of the newspaper editors said that it was probably the first poll in history in which every response was printed. Even though every response was printed, why might this poll be a misleading gauge of public opinion?

8.53 "The most dangerous place to be is in bed, since more people die in bed than anywhere else." Do you agree?

8.54 Since 1985 students at the University of Delaware who are not in the Honors Program or in remedial sections may choose a freshman composition course that uses IBM computers or one that uses Macintosh computers. In 1990 the assistant director of the writing program at the University of Delaware wrote an article arguing that "there is some sort of effect on students' writing when they use a Mac, different from when they use an IBM."[40] In particular, essays written on the Mac tended to be on frivolous topics, to use childish prose, and to have more misspelled words. How, as a statistics student and Mac enthusiast, could you criticize her student samples as possibly biased?

8.55 During the 1985 All-star game, NBC asked viewers to telephone a special 900 number (at 50 cents a call) to vote for or against the designated hitter rule. There were 234,832 calls, 41 percent voting yes and 59 percent voting no. Why was this not a scientific random sample of the U.S. population?

8.56 An Associated Press story titled "College Broadens Not Just Mind, Study Says" reported that students at Penn State gained an average of 9.1 pounds their freshman year, 7.3 pounds sophomore year, 7.8 pounds junior year, and 6.5 pounds senior year.[41] What other data would you like to see before jumping to the conclusion that college broadens the stomach?

8.57 An industrial psychologist routinely told supervisors, "If you've got 14 people working for you, at least one of them has a drinking problem. If you don't believe it, then you're kidding yourself."[42] Why is such advice misleading, even if 1/14 = 7 percent of all workers have drinking problems?

8.58 To research a magazine article on what makes a perfect husband, Sam Blum reported, "I interviewed, at random and with relish, secretaries, college girls, school teachers, a young woman standing in line to see a movie, two dancers, a salesgirl, several women temporarily between marriages, the wives of friends."[43] Do you think this sample qualifies as a random sample? Why or why not?

8.59 A student tried to determine whether there is a relationship between a student's social life and academic major. Since all of the telephones at his college are hall phones, he did not call individual students but, instead, randomly selected the numbers of the hall phones and surveyed the person who answered. Why might his results be biased?

8.60 During the last six weeks of the 1976 presidential election, General Cinema conducted a "straw vote" by offering soft drink customers at its 645 movie theaters the choice of a straw marked with either Jimmy Carter's or Gerald Ford's name. Nearly 600,000 straws were taken, with Carter leading Ford by 5 percentage points. Why might this poll be biased?

8.61 Shere Hite sent detailed questionnaires to 100,000 women (which, according to Hite, took an average of 4.4 hours to fill out) and received 4,500 replies, 98 percent saying that they were unhappy in their relationships with men. A *Washington Post-ABC News* poll telephoned 1,505 men and women and found that 93 percent of the women considered their relationships with men to be good or excellent. How would you explain this difference?

8.62 A cigarette advertisement boasted that "Triumph Beats Merit," explaining that "An amazing 60% said that 3-milligram Triumph tastes as good or better than 80-milligram Merit." The actual sample results follow:

Much better tasting than Merit	14%
Somewhat better tasting than Merit	22%
About the same in taste as Merit	24%
Somewhat worse tasting than Merit	29%
Much worse tasting than Merit	11%

If you were an executive for Philip Morris, the producer of Merit, how might you object to this advertisement?

8.63 A *Wall Street Journal*–NBC poll on President Ronald Reagan's performance included the question "Do you think Mr. Reagan will change his management style, or do you think he is too set in his ways to change?"[44] Nationwide, 56 percent of those sampled answered that Reagan would not change his management style, leading to a *Wall Street Journal* article headlined "President Is Viewed as Too Set in His Ways to Change His Governing Style, Poll Shows." If you were a spokesperson for Reagan, how might you have objected to the wording of this question?

8.64 The 7-Eleven stores in New York City let their customers answer a weekly question by choosing to have their coffee in a cup labeled "yes" or "no." A January 1990 survey asked if Congress deserved a pay raise, illustrating the question with a picture of a fat congressman lighting a cigar with a $100 bill; 71 percent of the 682,285 persons served chose a no cup. Why would a statistician be skeptical of these survey results?

8.65 Many college applicants take the Scholastic Aptitude Test more than once, and most college admissions offices only consider the applicant's best scores. Why would a statistician be skeptical of this procedure?

8.66 A Volvo advertisement said that the average American drives for 50 years and buys a new car, on average, every 3 1/4 years—a total of 15.4 cars during one's lifetime. Since Volvos last an average of 11 years in Sweden, Volvo owners need to buy only 4.5 cars in 50 years. Carefully explain why these data are very misleading.

8.67 Explain why you agree or disagree:

If a random sampling of a universe is taken, that sampling will accurately represent the entire universe.

Edward Lazarus, a partner in the Washington polling firm of Information Associates, . . . says that "a randomly selected sample would be in every way representative of the population at large because every individual in the universe has an equal probability of being picked. If that holds true, your attitudes, demographics, geography will be represented proportionately in the sample."[45]

8.68 During the May 1987 sweeps week, Los Angeles's ABC station ran a widely advertised eight-part special report on its 11 P.M. news show about the Nielsen households that are surveyed for television ratings data. The competing stations complained loudly (even threatening lawsuits). Why?

8.69 Market research firms often use focus groups to test advertisements, movies, and television shows, either to gauge their potential or for fine tuning—perhaps changing the nature or prominence of characters or adjusting the plot. A focus group consists of a dozen or so people who watch the film or video and afterward discuss what they like and dislike about it. A discussion leader solicits answers to specific questions and also allows some free-ranging discussion. Imagine that you work for an advertising firm that has produced a new television commercial. What do you consider to be the advantages and disadvantages of a focus group?

Focus group participants are gathered by bounty hunters, firms that dial random telephone numbers looking for volunteers. The market research firms that run focus groups pay each participant from $20 to $40, exclude people who know each other or have participated in a focus group before, and do not allow parents to bring their children. Give a plausible reason for each of these rules. In what way does a focus group differ from a scientific random sample?

8.70 *(continuation)* Why do you suppose that market research firms often have members of a focus group fill out a questionnaire after watching the video or film and before the discussion begins? Why do you suppose that focus groups are usually run by a single individual while other executives watch through a one-way mirror?

8.71 *(continuation)* In 1990 some research firms introduced special video hookups that allow executives in a corporation's headquarters to ask and answer questions of focus group members. If you were working for a market research firm, what advantages and disadvantages could you identify for this procedure?

8.72 A market research firm in Burbank, California, used to test movies and television shows at a movie theater that it ran, called the Preview House. Although in a somewhat disreputable part of town, no admission was charged at the Preview House, making it a popular cheap date for some people. In what ways was the Preview House not a scientific random sample?

8.73 *(continuation)* Members of the Preview House audience recorded their reactions by adjusting

individual dials as they watched, moving the dial toward 10 for parts they liked and toward 0 for parts they disliked. What advantages and disadvantages do you see for such a dial system in contrast to focus groups?

8.74 New television commericals are often tested by purchasing a half-hour of time on a cable television station and asking randomly selected people to watch during this time period. The volunteers are told beforehand that they are previewing a pilot for a new television series. Afterward, they are asked questions that gauge their recall of the commercials. Why do you suppose market research firms mislead volunteers about their intentions? Why do they ask the volunteers beforehand if they have ever participated in such marketing research?

8.75 *(continuation)* Sometimes, in place of recall tests about the commercials, a market research firm will ask the volunteers their opinion about only the phony pilot. A few weeks later, the volunteers are sent a list from which they can choose several free products as compensation for their help in evaluating the pilot show. One of the products offered will be the advertised one. What advantages, if any, do you see to this testing procedure?

8.76 Some firms, including many Japanese automakers, use observational research. Instead of asking people what they think about a product, they monitor their reactions, either in dealer showrooms or at home, after they have purchased the product. In designing the Mazda Miata and developing its advertising campaign, observational researchers made thousands of hours of videotapes of the life-styles of people in Orange County, California; some researchers moved in with an Orange County family for a month. What advantages and disadvantages do you see for this procedure, in contrast to a traditional focus group?

8.76 Before A.C. Nielsen introduced the audimeter, radio networks gauged the popularity of their shows by weighing sacks of fan mail. Explain why this procedure was unreliable.

8.77 In the 1950s A.C. Nielsen's researchers invented what came to be called the "Nielsen whoopee cushion." The idea was to monitor who watched television by putting special pressure-sensitive electronic cushions where each family member sat while watching television: a father cushion in the father's chair, a mother cushion in the mother's chair, and child cushions in the children's chairs. What advantages and disadvantages did these cushions have in comparison with audimeters?

8.78 Preliminary tests of Nielsen's people meters found that ratings of Saturday morning children's television shows and of the most popular nighttime shows were significantly lower than ratings based on diaries. How would you explain these declines?

8.79 The Nielsen Company has been working on a passive system for measuring television audiences with infrared scanners that are sensitive to body heat. Imagine that you work for a large advertising agency. What advantages and disadvantages do you see in infrared scanners as compared to people meters?

8.80 Television advertisers are concerned that a substantial number of viewers record shows on VCRs and watch them later, fast-forwarding through the commercials—a practice called zapping. A network-sponsored study of zapping asked randomly selected people to watch a pilot television show on a cable television station. These people were told that they would be telephoned the next day and asked their opinion of the pilot. When the researchers called, they asked if the volunteers had recorded the show and zapped the commercials—and found that very few had. Imagine that you work for an advertising firm. What faults would you find in this study?

8.81 Exercise 8.80 describes zapping. In another study,[46] volunteers were told that they had been selected to give their reactions to the content of music videos. Half watched the videotape with no opportunity to zap, and the other half were given remote control devices and told to fast-forward through the commercials. In each case, the video was viewed by one person at a time. Afterward, the viewers were tested individually on their recall of the commercials. Why do you suppose that this researcher disguised his true intentions? This study was conducted in California. Why did the researcher use some commercials for product brands that are available only in Texas or Massachusetts?

8.82 In 1990 a movie market research firm reported that its surveys show that only 4 percent of U.S. movie audiences react negatively to commercials shown before the beginning of a movie. It turned out that this firm didn't actually poll anyone. The basis for the 4 percent statistic was that its researchers heard

audible boos at only 4 percent of the movies that they viewed. As a statistician, why do you find their survey results potentially misleading?

8.83 *Consumer Reports* invites readers to judge movies on a scale of 1 to 5: "If you attend movies and would like your votes recorded here, write to . . ." Each month, *Consumer Reports* lists all current movies and the average of the votes received since the movie was first released. In its May 1990 issue it printed ratings for 90 movies, only three of which had been rated by more than 300 readers: *Parenthood* (343 votes and an average rating of 3), *Lethal Weapon II* (325 votes and an average rating of 3), and *Driving Miss Daisy* (304 votes and an average rating of 5). Twenty-seven of the movies rated had fewer than 25 votes. Explain why their poll is not a random sample of reader opinion, and give some biases that might be expected.

8.84 *(continuation)* Do you suppose that the average movie rating is above or below 3? Go to a library, find the most recent issue of *Consumer Reports,* and calculate the average rating. (Ignore the fact that some movies have more votes than others.)

● **8.85** *(continuation)* Redo the calculation of the average rating, this time weighting each movie by the number of votes received; that is, a movie with 25 votes and an average rating of 3 is counted as 25 votes of 3.

8.86 A razor company used to run a television commercial showing a number of barbers shaving, one after the other, with the same blade. The 12th, 13th, 15th, and 17th barbers were interviewed, and all said the shave was satisfactory—the implication being that this blade was good for at least 17 shaves. Why might this conclusion be unjustified?

8.87 Most polls today are conducted by telephone because this method is much cheaper than sending people out for face-to-face interviews. Each telephone interview costs clients from $25 to $50 and may run 15 minutes to an hour, as pollsters piggyback questions for a variety of clients. Explain why a lengthy telephone poll may be biased. Give specific examples.

8.88 Sometimes the most economical way to do sampling is in batches. For example, the military takes a blood sample from each new soldier to test for syphilis. Instead of testing each individual's blood sample separately, it is less expensive to mix several blood samples and then test this mixture. If the batch shows no evidence of syphilis, then one test did the work of several separate tests. If the batch shows evidence of syphisis, then each person in that group is tested separately. This retesting results in one extra test, the batch test, that could have been avoided by testing everyone separately in the first place. Batch tests save time and money when the disease is sufficiently rare and the batches sufficiently small so that often, no disease shows up in the batch test.

During World War II the Selective Service Administration estimated that a fraction, $\pi = 0.0453$, of the men aged 21–35 had syphilis. Compare the expected number of tests per 100 soldiers if each soldier is tested separately with the expected number of tests if soldiers are tested in batches of five, as was actually done during World War II.

8.89 When a shipment goes over several railroad lines, each railroad is entitled to part of the total shipping charge; a waybill is a document that accompanies a railroad car, recording its weight and the route traveled. Because the necessary clerical deciphering of the waybill and the arithmetic computations are tedious and time consuming, a railroad company once conducted a study to see how much difference there would be between a sample and a comprehensive tabulation of 23,000 waybills.[47] (They found that there was only an $83 difference between the results of a sample and a complete tabulation and that a complete tabulation cost an additional $5,000.) For its sample, the waybills were divided into five strata, according to the size of the total shipping charge, as shown in the following table. Why do you suppose their statisticians advised them to take relatively larger samples from the population of expensive waybills?

Waybill Charge	Fraction Sampled (%)
$5.00 or less	1
$5.01 to $10.00	10
$10.01 to $20.00	20
$20.01 to $40.00	50
More than $40.00	100

8.90 A school estimated the average number of school-aged children per family by questioning each child in the school. Why was their estimate much too high?

8.91 Explain the error in this newspaper article:

U.S. Statistics Prove it . . . Marry and Live longer: There's no question about it: Married people live longer than single people; and people who were once married live longer than people who were never married.

The Public Health Service's National Office of Vital Statistics has proved this to be a fact. . . . Deaths among bachelors were almost two-thirds greater than among husbands. Among divorced men and widowers, the rate was half again more.[48]

8.92 Two students wrote term papers on the relationship between a student's grade-point average (GPA) and extracurricular activities. One student surveyed only freshmen and the other only seniors. The average GPAs that they obtained were markedly different. Which do you suppose was higher? Explain your reasoning. If you were writing a term paper comparing student GPAs and extracurricular activities, would you survey freshmen or seniors?

8.93 Most murder victims are relatives, friends, or former friends of their assailants. Does this mean that you are safer with strangers in unfamiliar places?

8.94 Your college food service has commissioned you to survey students regarding their preferences for water, milk, coffee, tea, and different brands of soft drinks. Explain the inadequacies of each of the following procedures and identify some potential biases: **a.** A student recreation building has a soft-drink vending machine. Ask the company managing the machine for sales records. **b.** The library has a vending machine that dispenses coffee, tea, and fruit juices. Go to the library for seven evenings and tabulate the drinks that people buy. **c.** The college dining hall serves water, milk, coffee, tea, and several soft drinks. Spend breakfast, lunch, and dinner on Friday of this week recording student choices.

8.95 (continuation) Explain, in detail, how you would accomplish the task commissioned by your college food service.

8.96 (continuation) Implement this procedure and report your results.

8.97 A newspaper advertisement was headlined "Half of Our Furniture Is Free!" and went on to explain, "More than half of the items are marked down 50%."[49] Explain why, as a consumer, you would prefer either half of the furniture to be free or the furniture to be half-price. In politics, explain why there is a difference between half of the people preferring A and everyone indifferent between A and B.

8.98 *(continuation)* Explain why there is a difference between surveying the food expenditures of one household over a 30-week period and of 30 households during a one-week period.

8.99 *Time* magazine once surveyed college graduates by obtaining lists from colleges throughout the country and mailing questionnaires to all graduates whose last name began with the letters *Fa*. What

advantage is there to this procedure? Can you see any perils?

• **8.100** Red Lion Hotels & Inns ran full-page advertisements in 1989 claiming: "For every 50 business travelers who try Red Lion, a certain number don't come back. But forty-nine out of 50 do." The basis for this claim was a survey of Red Lion customers, of whom 49 of 50 agreed that "they would usually stay in a Red Lion Hotel or Inn when they travel to one of our cities." Carefully explain why the survey results do not prove the advertising claim. You may find a numerical example helpful.

9

Estimation

In solving a problem of this sort, the grand thing is to be able to reason backward. That is a very useful accomplishment, and a very easy one, but people do not practice it much. . . . Most people, if you describe a train of events to them, will tell you what the result would be. . . . There are few people, however, who, if you told them a result, would be able to evolve from their own inner consciousness what the steps were which led up to that result. This power is what I mean when I talk of reasoning backward.

Sherlock Holmes
A Study in Scarlet

TOPICS

Sampling is used as an economical way to estimate characteristics of the underlying population. Election polls estimate the fraction of the population who prefer a candidate; opinion polls estimate the fraction of the population who favor certain positions; bomb tests estimate the fraction of a shipment that are duds; and product tests estimate the fraction of manufactured items that are defective. In each of these four examples, a fraction is estimated. Absolute numbers and averages can be estimated, too. The Nielsen ratings can be used to estimate the number of people who watched a particular television program. Marketing samples are used to estimate the demand for a new product. The Federal Reserve uses samples to estimate the nation's money supply. Samples are used to estimate the amount of cholesterol in a person's body, the number of unmarried senior citizens in Florida, and the number of fish in a lake. Weather stations use samples at different locations to estimate the average temperature and rainfall. Samples are used to estimate the average breaking strength of fishing lines, the average acidity of a pond, and the average rate of increase of prices.

In each case, sample data are used to estimate a population value. But exactly how should the data be used to make these estimates? And how much confidence can we have in estimates that are based on a small sample? We will answer these questions by focusing on one particular example. Then we will apply these procedures to other situations.

9.1   USING A SAMPLE MEAN TO ESTIMATE A POPULATION MEAN

Consider a certain brand of crackers sold in a box that is marked "net weight 16 ounces." There may, on average, be 16 ounces of crackers in a box, but imperfections in the manufacturing and packaging process cause variations in the actual weight from box to box. Some boxes contain slightly more than 16 ounces, others somewhat less.

Because the imperfections are many and varied, the central limit theorem suggests that net weights may be normally distributed. For the sake of argument, suppose that the net weight is, in fact, a random variable from a normal distribution with a mean of $\mu = 16$ ounces and a standard deviation of $\sigma = 0.1$ ounce. Thus, when a person buys a box of these crackers, there is a 50 percent chance that the chosen box will turn out to have a net weight of more than 16 ounces and a 50 percent chance that the net weight will be less than 16 ounces. Ninety-five percent of the time, the net weight will be within two standard deviations of the mean—between 15.8 and 16.2 ounces.

Of course, no one can really know the mean weight μ of the complete population of cracker boxes unless every single box is opened and the contents weighed. This is where statistics comes in. The company, the government, or some other interested party can use a sample to estimate the population mean μ. A sample of

10 boxes might yield the following net weights (in ounces):

| 16.10 | 16.01 | 15.82 | 15.93 | 16.05 |
| 16.05 | 15.86 | 15.90 | 16.05 | 16.13 |

How could we use these data to estimate the mean net weight of all of the boxes of crackers sold by this company? It is natural to use the sample mean $\overline{X}$ to estimate the population mean μ. The sample mean works out to be

$$\overline{X} = \frac{16.13 + 16.01 + \cdots + 16.10}{10} = 15.99$$

This sample mean, $\overline{X} = 15.99$ ounces, happens to be slightly less than the population mean $\mu = 16$ ounces. Another sample would almost surely yield a somewhat different result. Someone else conducting such a test would select 10 different boxes and might obtain these net weights:

| 16.08 | 16.01 | 16.06 | 15.93 | 16.02 |
| 16.00 | 15.97 | 16.06 | 16.23 | 15.94 |

The mean of this sample is 16.03. One sample mean was too low, while another was too high, which is just what we should expect to happen.

Let's retrace our steps and, this time, affix statistical labels as we go. The net weight of a cracker box is a random variable with a population mean and standard deviation that are, in practice, unknown. A sample can be used to estimate the population mean (and the standard deviation, too).

> The population mean is a **parameter** that is unknown, but can be estimated by a sample statistic that is labeled an **estimator**. The specific value of the estimator obtained in a particular sample is an **estimate**.

Sampling Error

The sample mean is a natural estimator of the population mean. In our cracker example, one sample yielded an estimate of 15.99, while another sample gave an estimate of 16.03. Estimates vary from sample to sample because of the luck of the draw. By chance alone, a particular sample may contain mostly boxes with below-average weights or mostly boxes with above-average weights.

> The variation of estimates from sample to sample is called **sampling error**.

Sampling error is not due to mistakes or sloppy procedures. It is the inevitable result of the fact that we are looking at a small sample whose members are determined by chance.

In practice, we take just a single sample and use this sample mean to estimate the population mean. What we must recognize is that our particular sample is just one of many samples that might have been obtained. If we select 10 cracker boxes, open them, weigh the contents, and calculate the average weight to be 15.99 ounces, there is a temptation to treat this result as definitive. This temptation should be resisted. A sample mean of 15.99 ounces is just one of many sample means that could have been obtained. Another sample might yield a mean of 16.03 ounces. A third sample would yield yet another mean.

The **sample mean is a random variable** that is determined by the particular items picked for the sample.

We cannot say whether a particular sample mean, such as 15.99, is above or below, close to, or far from the population mean, because we don't know the value of the population mean. We could get a 15.99 sample mean from a population with a mean of 16.0, 15.99, or 43.7, and we wouldn't know for sure which is the case. However, we can use probability theory to deduce how likely it is that a sample will be picked whose mean happens to be far from the population mean.

The Distribution of the Sample Mean

The net weight of a cracker box is a random variable X, which, we assumed, is normally distributed with a mean of 16 ounces and a standard deviation of 0.1 ounce.

$$X \backsim N[16, 0.1]$$

Our random sample consists of 10 independent draws from this population. Let's label the net weight of the first selected box X_1. This net weight X_1 is a random variable with

$$X_1 \backsim N[16, 0.1]$$

The net weight of the second selected box, X_2, is another random variable from the same normal distribution.

$$X_2 \backsim N[16, 0.1]$$

The same is true of the net weights of the other boxes $X_3, X_4, \ldots, X_{10}$. Each of the 10 net weights is a random variable that is normally distributed with a mean of 16 and a standard deviation of 0.1 ounce.

The sample mean is the sum of these 10 random variables, divided by 10.

$$\overline{X} = \frac{X_1 + X_2 + \cdots + X_{10}}{10}$$

Least Squares Estimation

The German scientist Karl Gauss analyzed a great deal of data that had been collected in the fields of astronomy and geodesy (measurements of the shape of the earth and the location of places on it). Several observers had made many measurements at varying locations and times. These measurements inevitably had errors, due to imperfections in their methods and the imprecision of their equipment. Gauss wanted to combine these different measurements somehow so that "the accidental errors might as far as possible mutually destroy each other,"[1] thereby yielding a single, "best" estimate.

Gauss' basic model is as follows. In attempting to measure some distance μ, the measurements $X_1, X_2, \ldots, X_n$ are obtained. The difference between each measurement X_i and the true distance μ is measurement error ϵ_i:

$$X_i = \mu + \epsilon_i \qquad (9.1)$$

Now, let's say that we want to use some estimator $\hat{\mu}$ (pronounced "mu hat") of μ to predict the measurements X_i that people obtain. Our prediction errors will be $X_i - \hat{\mu}$. Gauss argued that we should use the estimator $\hat{\mu}$ that minimizes the sum of squared prediction errors.*

$$(X_1 - \hat{\mu})^2 + (X_2 - \hat{\mu})^2 + \cdots + (X_n - \hat{\mu})^2 \quad (9.2)$$

Gauss showed that this **least squares** estimator turns out to be the sample mean of the measurements.

$$\overline{X} = \frac{X_1 + X_2 + \cdots + X_n}{n}$$

This is Gauss' novel justification for using the sample mean to estimate the population mean. He then went on to develop least squares estimators for more complicated models. This work laid the basis for what we now call *regression analysis,* which is one of the most useful and powerful statistical techniques. We will look at regression analysis in later chapters.

Gauss also showed that when measurement errors are independent and identically distributed, then the sample mean tends to be normally distributed with an expected value equal to the population mean. This important finding, *Gauss' Law of Errors,* is an implication of the central limit theorem and is closely related to the primary subject of this chapter.

*The least squares idea apparently was introduced around 1805 by a French mathematician, Adrien M. Legendre (1752–1833).

In general, if n items are sampled, then the sample mean is the sum of n random variables, divided by n:

$$\overline{X} = \frac{X_1 + X_2 + \cdots + X_n}{n}$$

The sum of independent, normally distributed random variables is normally distributed, too, with a mean equal to the sum of the individual means and a variance equal to the sum of the individual variances.* The sample mean is normally distributed with an expected value μ and a standard deviation $\sigma/\sqrt{n}$:

$$\overline{X} \sim N[\mu, \sigma/\sqrt{n}]$$

In practice, we cannot be certain that we are sampling from a normal distribution. Here, we blithely assumed that the errors in the manufacturing and packaging process are such that the net weights turn out to be normally distributed. What if we are a little more cautious in our assumptions? Or, even more drastically, what if we know that what we are measuring is not normally distributed? It doesn't matter, as long as the sample is reasonably large.

The central limit theorem implies that, no matter what the underlying distribution, the sample mean will tend to a normal distribution.

$$\overline{X} \sim N[\mu, \sigma/\sqrt{n}] \tag{9.3}$$

A random sample is a set of independent and identically distributed measurements; these are the very assumptions that are used in the central limit theorem. This theorem states that the sum of n independent, identically distributed random variables approaches a normal distribution as n increases. Therefore, the sample mean, which is the sum of n independent, identically distributed random variables divided by n, must approach a normal distribution as n increases.

The central limit theorem is not only a very important part of probability, it is also an absolutely crucial part of statistics. In probability analysis, the central limit theorem tells us that many of the uncertainties about us, ranging from IQ scores to the movements of molecules, conform to a normal distribution. In statistical analysis, the central limit theorem tells us that the mean of a random sample conforms to a normal distribution. The only cautions we need to exercise are to ensure that it really is a random sample and that it is large enough to justify an appeal to the central limit theorem. With something like cracker weights, which are themselves approximately normally distributed, a sample of size 10 is large enough. With an unknown distribution, the usual rule of thumb is that a sample of size 20 or 30 will be satisfactory. If the underlying distribution is known to be highly skewed, then 50, 100, or more observations may be needed. Figure 9.1 shows four examples.

If a random sample is drawn from a normal distribution, then Equation 9.3 always holds. Even if the underlying distribution is not normal, a sufficiently large sample will ensure that Equation 9.3 still holds. This is undoubtedly the most important formula in statistics. It is absolutely crucial that you remember it and understand it.

---

*Equations 4.7 and 4.9 in Chapter 4's appendix imply that the sum of the Xs has an expected value $n\mu$ and a variance $n\sigma^2$. Equation 9.3 then follows from the linear transformation Equations 4.4 and 4.5.

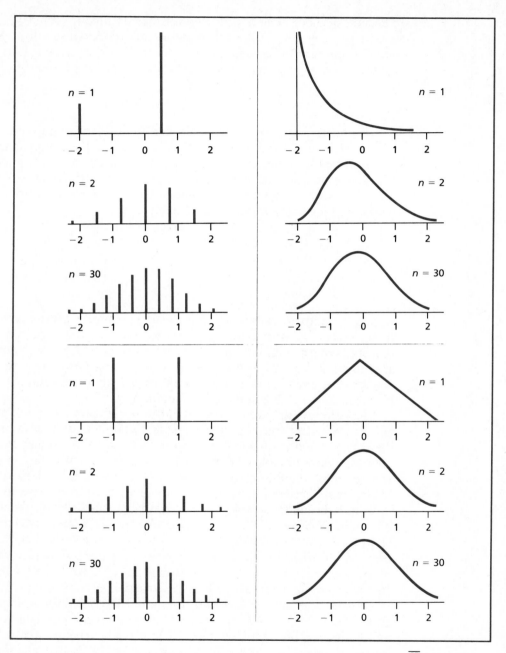

Figure 9.1   *The Distributions of Some Standardized Sample Means, $Z = \dfrac{\overline{X} - \mu}{\sigma/\sqrt{n}}$*

The Expected Value of the Sample Mean

The sample mean is a random variable that is determined by the luck of the draw in selecting the sample. Equation 9.3 implies that the probability distribution for the sample mean is symmetrical about the population mean, μ. There is a .5 probability that the sample mean will be above the population mean and a .5 probability that it will be below it. The expected value of the sample mean is equal to the population mean. The sample mean is consequently said to be an **unbiased estimator** of the population mean.

> An estimator is **unbiased** if its expected value is equal to the parameter being estimated.

Unbiased estimators have a considerable appeal. It would be discomforting to use estimates that one knows are systematically too high or too low. A statistician who uses unbiased estimators can anticipate estimates with errors that, over a lifetime, average very close to zero. But, of course, average performance is not the only thing that counts. There was the lawyer who summarized a career by saying that, "When I was young, I lost many cases that I should have won. And when I was old, I won many cases that I should have lost. So, on average, justice was done." The conscientious statistician should be concerned with not only how good the estimates are on average but also how accurate they are in particular cases.

The Standard Deviation of the Sample Mean

Equation 9.3 states that the standard deviation of the sample mean is $\sigma/\sqrt{n}$. If the sample has only one observation ($n = 1$), then its standard deviation is σ. As n increases, the standard deviation of the sample mean declines. To understand this, remember that the standard deviation is a measure of the uncertainty of the outcome. With a small sample, we are not at all sure what the sample mean will turn out to be. With a large sample, however, it is extremely unlikely that all of the observations will be far above μ and equally unlikely that all of the observations will be far below μ. Instead, we are almost certain to have a sample whose mean is very close to μ. As the sample size n becomes infinitely large, the standard deviation of the sample mean goes to zero, reflecting our common-sense judgment that, as the sample encompasses the population, the sample mean is certain to be the population mean. As Sherlock Holmes observed, "While the individual man is an insolvable puzzle, in the aggregate he becomes a mathematical certainty. You can, for example, never foretell what any one man will do, but you can say with precision what an average number will be up to. . . . So says the statistician."[2]

Let's apply Equation 9.3 to the cracker example, again assuming that the manufacturing and packaging process yields boxes with net weights that are normally distributed with a mean $\mu = 16$ ounces and a standard deviation $\sigma = 0.1$ ounce. If we weigh a random sample of 10 boxes, the mean of this sample is a random variable that is normally distributed with a mean of 16 ounces and a standard deviation of $0.1/\sqrt{10} = 0.032$ ounce. Figure 9.2 shows this probability distribution. Using our two-standard-deviation rule, there is a .95 probability that the average weight of 10 boxes will turn out to be in the range $16 \pm 2(0.032)$ ounce, that is, between 15.936 and 16.064 ounces.

These calculations tell us that there is only a 5 percent chance that the sample mean will be more than 0.064 ounce away from the population mean. If we use the sample mean to estimate the population mean, then there is only a 5 percent chance that our estimate of the population mean will be off by more than 0.064 ounce. If we use a larger sample, we can be even more confident that our estimate will be very close to μ. Figure 9.2 shows how the probability distribution for the sample mean collapses as the size of the sample increases. We can use the two-standard-deviation rule to quantify this convergence. If we base our estimate of the population mean on the average weight of 25 boxes, there is only a 5 percent chance that our estimate will be more than $2(0.1/\sqrt{25}) = 0.04$ ounce away from μ. With a sample of 100 boxes, there is only a 5 percent chance that our estimate will be more than $2(0.1/\sqrt{100}) = 0.02$ ounce away from μ.

Confidence Intervals

We can rephrase these probability computations to show the confidence we have in using the sample mean $\overline{X}$ as an estimate of the population mean μ. If $\sigma = 0.1$ and $n = 10$, there is a .95 probability that the sample mean will be within 0.064 of μ. Figure 9.3 depicts this argument. Ninety-five percent of the time, the sample mean (our estimate of μ) will be between 15.936 and 16.064 ounces. In each of these cases, our estimate is within 0.064 ounce of μ. Five percent of the time, the sample mean will be either less than 15.936 or greater than 16.064. In each of these cases, the estimate is more than 0.064 ounce away from μ.

Although we don't know the true value of μ, we can quantify our uncertainty—our sampling error—by saying that there is a .95 probability that the difference between μ and $\overline{X}$ will be less than 0.064 ounce. This interval is called a **confidence interval** and the .95 probability is the interval's **confidence level**. If the sample mean does turn out to be 15.99 ounces, then we report our estimate of μ as 15.99 ounces and state that our 95 percent confidence interval is 15.99 ± 0.064 (i.e., 15.926 to 16.054). A standard, shorthand reporting form is

$$15.99 \pm 0.064 \text{ ounce}$$

These are quick-and-dirty computations using the two-standard-deviation rule of thumb. A more formal calculation would proceed as follows. First, decide the confidence level for your interval. If you want to be 99 percent certain that the interval will contain μ, then you must make the interval wide. If you are satisfied with a 50 percent chance of containing μ, then a narrow interval will do. Ninety-five percent confidence levels are standard, but there is no reason why you can't use others.

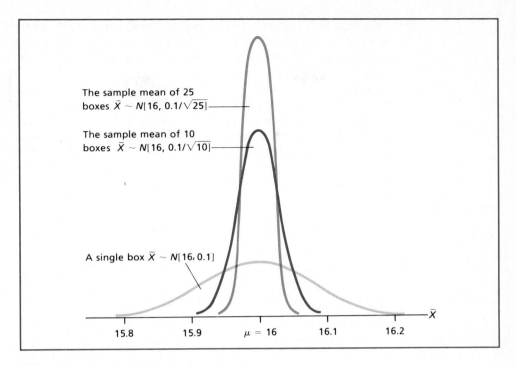

The sample mean of 25
boxes $\bar{X} \sim N[16, 0.1/\sqrt{25}]$

The sample mean of 10
boxes $\bar{X} \sim N[16, 0.1/\sqrt{10}]$

A single box $\bar{X} \sim N[16, 0.1]$

$\bar{X}$

15.8   15.9 $\mu = 16$ 16.1   16.2

Figure 9.2 *The Distribution of Net Weights*

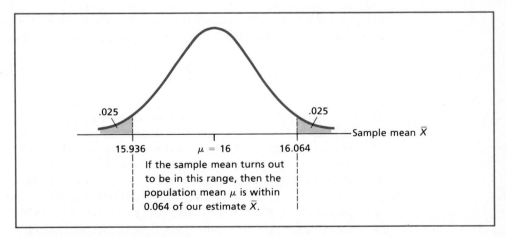

.025   .025

Sample mean $\bar{X}$

15.936 $\mu = 16$ 16.064

If the sample mean turns out
to be in this range, then the
population mean μ is within
0.064 of our estimate $\bar{X}$.

Figure 9.3 *Confidence Interval Logic*

I'll work out the standard 95 percent confidence interval here and leave the others for you to do as exercises. We've taken the first step, which is to decide how much confidence to attach to the interval. The second step is to consult the normal distribution in Table 2 at the end of this book to see how many standard deviations are appropriate. Our rule of thumb is that about 95 percent of the observations should be within two standard deviations of the mean. But let's look in Table 2 to get a more precise number. Table 2 shows that, in fact, there is a .95 probability that a standardized normal variable Z will be between -1.96 and $+1.96$. The last step is to translate this probability statement about Z into a probability statement about the sample mean. To standardize the sample mean, we subtract its expected value μ and divide by its standard deviation $\sigma/\sqrt{n}$:

$$Z = \frac{\overline{X} - \mu}{\sigma/\sqrt{n}}$$

Therefore,

$$.95 = P[-1.96 < Z < +1.96]$$

$$= P[-1.96 < \frac{\overline{X} - \mu}{\sigma/\sqrt{n}} < +1.96]$$

which can be rearranged to show that

$$.95 = P[\overline{X} - 1.96(\sigma/\sqrt{n}) < \mu < \overline{X} + 1.96(\sigma/\sqrt{n})] \qquad (9.4)$$

This is the general equation for a 95 percent confidence interval. The logic is apparent: Because there is a .95 probability that Z will be between -1.96 and $+1.96$, it follows that there is a .95 probability that the sample mean will be within 1.96 standard deviations $(\sigma/\sqrt{n})$ of its expected value (μ). The shorthand formula is

$$\mu = \overline{X} \pm 1.96\,\sigma/\sqrt{n}$$

In the cracker example, with $\overline{X} = 15.99$, $n = 10$, and $\sigma = 0.1$, $1.96\,\sigma/\sqrt{n} = 1.96(0.1)/\sqrt{10} = 0.062$ and the 95 percent confidence interval differs slightly from our two-standard-deviations rule of thumb answer, 15.99 ± 0.064.

A slip that students sometimes make is to forget that we are concerned with the standard deviation of the sample mean. They don't divide σ by the square root of the sample size, and consequently, they get the wrong answer. What must be remembered is that we are using the sample mean to estimate μ and that the accuracy of this estimate depends critically on the sample size. With a large sample, there is little chance that the sample mean will be far from its expected value μ and, therefore, we can be more confident that the sample mean is an accurate estimate of μ. This intuition shows up in the arithmetic when we calculate the standard deviation of the sample mean $\sigma/\sqrt{n}$ and see that it declines as n increases.

Remember, too, that sampling error is the reason for the uncertainty concerning our estimate. We assume that the boxes are randomly selected and that the weighing scales are unbiased. There is uncertainty only because the net weights vary from box to box and the sample mean depends on the luck of the draw. Our estimate may be too high or too low, and we gauge this sampling error by computing a confidence interval. With a sample of 10 boxes, we are 95 percent confident that the sample mean will not be more than 0.062 ounce away from μ. This 0.062-ounce leeway is a margin for sampling error.

> The spread in a confidence interval is a **margin for sampling error.** For a 95 percent confidence interval, the margin for sampling error is $\pm 1.96(\sigma/\sqrt{n})$, or about ± 2 standard deviations of the sample mean.

Notice also that it is the sample mean and the confidence interval that vary, and not the population mean. The equation

$$.95 = P[\overline{X} - 1.96(\sigma/\sqrt{n}) < \mu < \overline{X} + 1.96(\sigma/\sqrt{n})]$$

is interpreted as meaning, "There is a .95 probability that the sample mean will turn out to be sufficiently close to μ that the confidence interval I calculate includes the true value of μ. There is a .05 probability that the sample mean will turn out to be so far from μ that my confidence interval does not include μ." The probabilities refer to the confidence intervals that result from repeated sampling; they do not refer to μ. In the long run, it can be anticipated that 95 percent of these confidence intervals will include μ and 5 percent won't. A clever analogy is to a game of horseshoes,

> A ringer occurs when the shoe rings or encloses the peg. What is the interpretation of the statement that the probability is .95 that a ringer will be thrown? Does it mean that 95% of the shoes thrown will ring the peg? Or does it mean that 95% of the time the peg will come up through the horseshoe? The correct statement is fairly obvious. It is a particularly good example, because the wider the spread of the shoe, the more chance there is of making a ringer, other things being equal.[3]

There is a subtle distinction, though, between Bayesian and objectivist interpretations of specific confidence intervals, such as 15.99 ± 0.062. A Bayesian is willing to put probabilities on different possible values for μ, but an objectivist isn't. Thus, a Bayesian will say that there is a .95 probability that a particular confidence interval, such as 15.99 ± 0.062, includes μ. The interpretation is that the Bayesian is indifferent between a bet that μ is somewhere in this particular interval, and a bet that a red card will be selected from a box containing 19 red cards and 1 black card.

An objectivist resists putting probabilities on μ, because μ is a fixed parameter.

The population mean μ either is 16 ounces or it isn't 16 ounces. It is not sometimes 16 ounces and sometimes 15 ounces. There is no relative frequency interpretation of a .95 probability for μ. Similarly, an objectivist will not say that there is a .95 probability that a particular interval includes μ, because that would place a probability on something that has already occurred. What the strict objectivist will say is that, "There is a .95 probability that the interval I am about to calculate contains μ" or "My interval is 15.99 $\pm$ 0.062; you should know that I chose my procedure anticipating that 95 percent of such intervals will include μ."

Choosing a Sample Size

We took a sample of 10 boxes and calculated the average net weight to be 15.99 ounces. Recognizing that this is just a sample, we allow for sampling error by calculating the width of a 95 percent confidence interval, based on the fact that 95 percent of the time, a normally distributed variable will be within 2 (actually, 1.96) standard deviations of its expected value. For a sample mean, the expected value is μ and the standard deviation is $\sigma/\sqrt{n}$. With $\sigma = 0.1$ and $n = 10$, the standard deviation of the sample mean works out to be 0.032 ounce. Thus, the 1.96 standard deviation margin for sampling error is 0.062 ounce, and we report our estimate as 15.99 $\pm$ 0.062 ounces.

How could we obtain a more precise estimate, that is, one with a smaller margin for sampling error? Well, the margin for sampling error is

$$\pm 1.96(\sigma/\sqrt{n})$$

The 1.96 Z value comes from the normal table and there's nothing that can be done about that. The underlying standard deviation, $\sigma = 0.1$ ounce, comes from the manufacturing and packaging process, which the statistician has no control over. The only lever the statistician has is the sample size n. A larger sample will give a smaller margin for sampling error, as shown by the following illustrative calculations:

Sample Size n	Margin for Sampling Error $\pm 1.96(\sigma/\sqrt{n})$
10	± 0.062
25	± 0.039
100	± 0.020
400	± 0.010
1,000	± 0.006
10,000	± 0.002

If we sample 100 boxes, our 95 percent confidence interval will be $\overline{X} \pm 0.02$ ounces, whatever the sample mean $\overline{X}$ turns out to be. With 10,000 boxes, the confidence interval will be $\overline{X} \pm 0.002$ ounces, again using whatever $\overline{X}$ turns out to be.

We can use these sorts of calculations to decide in advance how many boxes we want to sample. If we will settle for an estimate with a 0.062-ounce margin for sampling error, then 10 boxes is enough. If we need an estimate with only a 0.02-ounce margin for error, then 100 boxes are needed. If we need an extremely precise estimate, one that we are 95 percent sure will be within 0.002 ounce of μ, then we will have to open and weigh 10,000 boxes. The tradeoff, of course, is the time and energy spent on a larger sample plus the larger number of boxes that will be tested instead of sold.

In making this kind of choice, it is usually best to lay out a table of the alternatives, as I have done. You can then see at a glance exactly how the margin for sampling error varies with the size of the sample. Such a table isn't needed, of course, if you happen to know in advance exactly what margin of error you want. Perhaps you've decided that your estimate of net weights should be accurate to within 0.05 ounce. If so, then you can directly calculate the requisite sample size by setting the margin for sampling error, $1.96(\sigma/\sqrt{n})$, equal to 0.05. This equality,

$$1.96(\sigma/\sqrt{n}) = 0.05$$

implies

$$n = (1.96(\sigma)/0.05)^2$$
$$= (1.96(0.1)/0.05)^2$$
$$= 15.4$$

or 16 boxes, to be on the safe side. (If you are going to weigh four-tenths of a box, you may as well weigh the whole thing!)

Notice that one very interesting characteristic of this requisite sample size is that it does not depend on the size of the population. Regardless of whether the population consists of 100,000 boxes or 100,000,000 boxes, a 0.05 margin for sampling error requires a sample of 16 boxes. At first glance, this conclusion may seem surprising. If we're trying to estimate a characteristic of a large population, then there is a natural tendency to believe that a large sample is needed. If there are only 100 boxes in the population, then a sample of 16 uses 1 out of every 6 boxes. But, if there are 100,000 boxes in the population, then a sample of size 16 only looks at 1 out of every 6,000 boxes. And if there are 100 million in the population, then a sample of 16 uses only 1 out of every 6 million! If there are 100 million in the population, how can we possibly obtain a reliable estimate with a sample of size 16, looking at just 1 out of every 6 million boxes?

A moment's reflection shows why the size of the population doesn't matter. Sampling error is present because the luck of the draw yields samples whose means differ from the population mean. What matters to these chances is the distribution of the population, and not its absolute size. In particular, sampling error depends on the number of boxes that are far from the population mean *relative* to the number near the mean, and not how many boxes there are altogether. Thus, the requisite sample size depends on the standard deviation of the population distribution, but not the size of the population.

In fact, our calculations implicitly assume that the population is infinite, because we don't adjust the probabilities even though we are sampling without replacement. The exact formula for the standard deviation of the sample mean when the sampling is without replacement from a finite population of size N is

$$\frac{\sigma}{\sqrt{n}} \ \sqrt{\frac{N-n}{N-1}} \tag{9.5}$$

The new term

$$\sqrt{\frac{N-n}{N-1}}$$

is known as the **finite population correction.** It scarcely matters whether a sample of 16 is drawn from a population of 100,000 or 100,000,000. The sample size n has to be pretty close to the population size N for our probability calculations to change significantly. The effect would be to shrink the confidence intervals and reduce the requisite sample sizes. If the population had only 16 boxes, for an extreme example, then a sample (without replacement) of 16 boxes would have no sampling error, because the sample mean is sure to be the population mean. In practice, the finite population correction is usually ignored unless the sample includes more than 10 percent of the population, that is, $n > .1N$.

Once More

We have just been through the essential features of most statistical analyses, so let's recall the highlights. We want to estimate a population mean, here the average net weight of all of the boxes produced by a certain company. A random sample is selected and measurements are taken; in this case they are the net weights of the boxes. The average of these weights (the sample mean) is an estimate of the population mean. This estimate is subject to sampling error, because the particular sample selected depends on the luck of the draw. The sample mean may turn out to be above or below the population mean. This risk can be gauged by calculating a confidence interval, which is equal to the sample mean, plus or minus the margin for sampling error. With a 95 percent confidence interval, the margin for sampling error is $1.96(\sigma/\sqrt{n})$, or about two standard deviations of the sample mean. The larger the sample, the smaller the sampling error will be. The choice of a desired margin for sampling error determines the requisite size of the sample.

[handwritten annotations at top of page: "9.1 a) p = x−μ / 6   b) P(x>4000)=.0228   c   x−μ / σ/√n   b) P(x>4000)= .00003"]

EXERCISES

9.1 The life of a Rolling Rock tire is normally distributed with a mean of 30,000 miles and a standard deviation of 5,000 miles. What are the probabilities that **a.** a tire will last more than 30,000 miles? **b.** a tire will last more than 40,000 miles? **c.** the average life of four tires will be more than 30,000 miles? **d.** the average life of four tires will be more than 40,000 miles?

9.2 A certain grade of oranges has a weight that is normally distributed with a mean of 9 ounces and a standard deviation of 1 ounce. What percent of these oranges weighs less than 8 ounces? These oranges are sold by the bag, with 25 oranges in each bag. What percent of these bags has a net weight of less than 200 ounces?

9.3 A package of Cow Country butter says "net weight = 16 ounces." The net weight is, in fact, normally distributed with a mean of 16.05 ounces and a standard deviation of 0.05 ounce. What proportion of the packages weigh less than 16 ounces? How high would the company have to raise the mean weight so that only 1 percent of the packages will weigh less than 16 ounces?

A government agency has decided to weigh a sample of Cow Country butter packages to see if the mean net weight really is at least 16 ounces. What is the probability that the average weight of n packages will be less than 16 ounces if $n = 1$? If $n = 4$? If $n = 16$? Why does this probability change as n increases?

9.4 A manufacturer wants to estimate the speed of its new printer. A random sample yielded the following measured speeds (characters per second):

112	120	102	107	118	131	101	97
108	114	121	116	119	99	107	110
124	113	122	104	112	103	115	110

Assuming that the standard deviation is known to be 8 characters per second, **a.** construct a 95 percent confidence interval for the population mean. **b.** construct a 99 percent confidence interval for the population mean. **c.** is your 99 percent confidence interval wider or narrower than your 95 percent confidence interval? Explain the reason carefully.

9.5 A statistics textbook gives this example:

Suppose a downtown department store questions forty-nine downtown shoppers concerning their age. . . . The sample mean and standard deviation are found to be 40.1 and 8.6, respectively. The store could then estimate μ, the mean age of all downtown shoppers, via a 95% confidence interval as follows:

$$\bar{x} \pm 1.96(s/\sqrt{n}) = 40.1 \pm 1.96(8.6/\sqrt{49})$$
$$= 40.1 \pm 2.4$$

Thus, the department store should gear its sales to the segment of consumers with average age between 37.7 and 42.5.[4]

Explain why you agree or disagree with the following interpretations: **a.** There is a .95 probability that the average age of downtown shoppers is between 37.7 and 42.5. **b.** Ninety-five percent of downtown shoppers are between the age of 37.7 and 42.5.

9.6 Assume that Scholastic Aptitude Test (SAT) scores are normally distributed with a mean of 500 and a standard deviation of 100. If student SAT scores are picked at random: **a.** What is the probability of picking a student whose math SAT score is above 650? **b.** What is the probability of picking 100 students whose average math SAT score is above 650? A certain college has an entering freshman class of 100 students whose average math SAT score is 650. Do you think that these freshmen are a random sample of the students who take the SAT test?

9.7 Explain why you agree or disagree with this reasoning:

. . . the household [unemployment] survey is hardly flawless. Its 60,000 families constitute less than 0.1 percent of the work force.[5]

9.8 What is the probability that a statistician who estimates 20 independent 95 percent confidence intervals will mistakenly miss a population mean at least once? Only once? Not at all?

9.2 A FEW EXAMPLES

Fishing Line Strength

The strength of a fishing line depends on its composition and thickness. However, even lines of the same material and thickness are not identical, because inevitably there are imperfections in the raw materials and in the manufacturing process. Because there are many such imperfections, let's assume that their effects on breaking strength are normally distributed. A certain company may make 30-weight nylon Proline fishing line with a breaking strength that is normally distributed, with a mean $\mu = 30$ pounds and a standard deviation $\sigma = 3$ pounds. A person who buys 30-weight nylon Proline is drawing fishing line from this normal population. Half the time, the line will turn out to have a breaking strength greater than 30 pounds, and half the time, the line purchased will have a smaller breaking strength. Ninety-five percent of the time, the line will have a breaking strength within two standard deviations of the mean, or between 24 and 36 pounds.

Now, let's say that a consumer research group wants to estimate the mean breaking strength of this fishing line. They take a sample of 25 lines and find that the average breaking strength of these lines is 30.5 pounds. To take into account sampling error, they note that when 25 items are drawn from a population with a standard deviation of 3 pounds, the mean of such a sample has a standard deviation of $\sigma/\sqrt{n} = 3/\sqrt{25} = 0.60$ pound. For a 95 percent confidence interval, a two-standard-deviation margin for sampling error is ± 1.2 pounds. Thus, the consumer group's conclusion is "Our estimate of the average breaking strength of 30-weight nylon Proline fishing line is 30.5 pounds ± 1.2 pounds."

Insurance Risks

Insurance companies take advantage of the statistical fact that there is less uncertainty about an average outcome than about an individual outcome. Automobile accidents, fires, and serious illnesses are events that are very unlikely, but financially crippling if they do occur. There is a great risk for the individual, because we cannot predict who will be hit by catastrophe. However, there is little risk for a large group, to the extent that we can predict the number of automobile accidents, fires, and serious illnesses that will occur.

I'll use life insurance as an example. Assume that the age of death of men who are now 20 years old has a mean of 65 years and a standard deviation of 10 years. The individual does not know how long he will live, and the early death of a breadwinner can be financially crippling to a family. Consider an insurance company that writes life insurance policies for 10,000 20-year-old men. If we can assume that the deaths are independent (which insurance companies try to ensure by writing their policies to exclude war and other national catastrophes), then we can invoke the central limit theorem. The average age at death of these 10,000 men is normally distributed with a mean of 65 years and a standard deviation of $10/\sqrt{10,000} = 0.1$ year. Because there is only a .01 probability that a standardized

Does the Enzyme Work or Doesn't It?

A pharmaceutical company tested a new enzyme that they hoped would increase the yield from a certain manufacturing process.[6] The data were calculated as the ratio of the actual yield with the enzyme to an assumed yield with the old process (based on a lot of past experience). A value of 110.2 meant that, in this particular test, the yield was 10.2 percent higher with the enzyme than with the old process. If the enzyme had no effect on the expected yield, the measured yields should average about 100. A sample of 41 batches found an average yield of 125.2 with a standard deviation of 20.1. A member of the company's research department argued that

There is no good theoretical reason for believing that the mean yield should be higher than 100 with the enzyme. Moreover, in the 41 batches studied, the standard deviation of individual yields was 20.1. In my opinion, 20 points represents a large fraction of the difference between 125 and 100.

There is no real evidence that the enzyme increases yield.

This logic is wrong! It is the standard deviation of $\overline{X}$, and not the standard deviation of X, that is of interest. If the standard deviation of X is 20.1, then the standard deviation of $\overline{X}$ is only

$$\sigma/\sqrt{n} = 20.1/\sqrt{41} = 3.14$$

If the expected value is 100 and the standard deviation 20.1, then there is, indeed, a good chance that a single observation might be as high as 125, but there is not much chance that the average of 41 observations would be so high. A 95 percent confidence interval for the expected value of the yield with the enzyme is

$$\overline{X} \pm 2(\sigma/\sqrt{n})$$
$$125.2 \pm 2(3.14)$$
$$125.2 \pm 6.28$$

which is far from 100.

normal variable Z will be less than -2.33, there is only a 1 percent chance that the average age at death of these 10,000 men will be 0.233 year (about 3 months) less than 65 years. As long as those insured are a random sample from the assumed population, there is little chance that the insurance company will be financially crippled.

Control Charts

Walter Shewart, an employee of Bell Telephone Laboratories, developed the foundations of statistical control in the 1930s. He recognized that production processes are inevitably affected by fluctuations in the performance of humans and machines. For instance, cut lumber can be slightly longer or shorter than intended. Too much or too little sugar can be put in soft drinks. Inventories can accumulate or be depleted. Small fluctuations may be tolerable, but sometimes there are unacceptably large deviations between the actual and intended characteristics of a product.

Lumber that is within 1/8 inch of the stated measurement may be acceptable, while 1-inch errors are unacceptable.

Shewart demonstrated that statistical sampling procedures can provide an economical means of distinguishing between typical run-of-the-mill fluctuations and a breakdown in the production process—a process that is out of control. If there is some minimal standard—a well-defined separation between acceptable and defective—such as an ignition that starts or an explosive that detonates, the firm can use acceptance sampling, as discussed in Chapter 5. At periodic intervals, a product sample is tested to see whether the number of defective items exceeds a specified threshold. If so, then the production process has broken down and the firm should determine what is causing this unacceptably large number of defective products.

In other cases, product characteristics are quantitative, for example, the dimensions of lumber, the sugar content of soft drinks. Shewart developed **control charts,** as illustrated in Figure 9.4, to monitor the variation over time in specific aspects of a production process, for example a product's length, weight, or thickness. Each point X on a control chart is a single observation or, more commonly, the average of several observations. The mean line on a control chart shows the intended value or the historical average value. The upper and lower control limits give the boundaries within which observed fluctuations are considered to be typical and acceptable. Observations outside the control limits signal that the production process has broken down and that corrective action should be taken.

U.S. and Canadian industries commonly set the upper and lower control limits at three standard deviations from the mean line. If the typical fluctuations are the cumulative result of a large number of factors, the central limit theorem suggests that these variations may conform to a normal distribution. If $X \sim N[\mu, \sigma]$, then

Figure 9.4   *Two Control Charts*

the probability that X will be more than three standard deviations from μ is .0013 + .0013 = .0026. (When a normal distribution is inappropriate, the mean X can be replaced by a box plot, which is discussed in an optional section of Chapter 1.)

An abrupt breakdown of equipment will be noticed immediately by workers, But a gradual wearing out or slippage of machine parts may be detected only by the observation that a measured characteristic has wandered away from its target value and violated a control limit. The problem that the firm faces is to distinguish between normal, acceptable fluctuations and assignable, correctable variation. If the firm stops the production process every time there is a slight variation, almost nothing will be produced. If it never stops the production process, it may produce a large number of defective products. Control charts are intended to provide a timely detection of the presence of assignable variation before a large number of defective items are produced.

A process might be monitored every few minutes, hours, days, or even less frequently. Figure 9.4 depicts two examples. The first, the lumber width control chart, shows a process in control. The second chart shows that the sugar content hasn't yet violated its lower control limit, but the steady decline is worrisome and suggests a gradual deterioration of the production process that perhaps should be monitored more closely. One advantage of a control chart is that it may reveal such trends. Another advantage is that the identification of the time that a process goes out of control may help identify its cause—perhaps a change in personnel, a replacement of raw materials, or an adjustment of equipment.

These examples focus on industrial production, but control charts can be used to monitor virtually any measurable activity, including sales, accidents, and expenses. In some cases, one of the control limits may reflect an improvement, for instance, a large increase in sales or a decrease in reported accidents. The firm may nonetheless want to identify these large deviations. The recorded improvement might not be real but merely reflect inattention or exaggeration. If the improvement is real, the firm will want to learn why it happened and encourage its continuation.

W. Edwards Deming, a U.S. government statistician and friend of Shewart, gave hundreds of lectures to U.S. and Japanese managers in the 1940s and 1950s, arguing that comprehensive statistical process control at every stage of production is the key to making high-quality products. When a process is under control, a business not only achieves consistency but can discern how quality is affected by changes in the production process. Without process control, it is virtually impossible to determine whether most changes are beneficial or detrimental.

Deming has long criticized U.S. managers for paying too little attention to the importance of quality and for focusing their quality control efforts on identifying defective items after they have been produced. To Deming, process control should be used at every stage to build quality into the product in the first place. Japanese managers were far more receptive to Deming's message, and since 1951, the Union of Japanese Scientists and Engineers (JUSE) has awarded highly coveted annual Deming prizes to individuals and businesses for quality improvements. In 1960 the Japanese government awarded Deming one of its highest awards, the Second Order of the Sacred Treasure.

EXERCISES

9.9 Whirltag claims that the useful lifetime of its dishwashers is normally distributed with a mean of 3 years and a standard deviation of 1 year. Consumer Rejects magazine tested 25 of these dishwashers and found that the average useful life was only 2.5 years. If Whirltag's claims are correct, what is the probability that the average life of 25 machines will turn out to be no greater than 2.5 years? Using Whirltag's reported 1-year standard deviation, what is Consumer Rejects' 95 percent confidence interval for the average lifetime of Whirltag dishwashers?

9.10 The manufacturer of a television guarantees the picture tube for 2 years. The lifetimes of these picture tubes are normally distributed with a mean of 4 years and a standard deviation of 2 years. **a.** What is the probability that one of these picture tubes will need to be replaced before the guarantee expires? **b.** If it costs the manufacturer $200 to replace a picture tube, how much must be added to the price to cover the expected value of the cost of the guarantee? **c.** A consumer research group tested four of these televisions and found that none of the tubes failed within 2 years. What is the probability of this occurring? **d.** The average lifetime of these four picture tubes turned out to be 5.1 years. Using a standard deviation of 2 years, what is a 95 percent confidence interval for the life of this manufacturer's picture tubes?

9.11 Over the period 1929–1933, scientists measured the speed of light 2,500 times and obtained an average value of 186,270 miles per second, with a standard deviation of 8.7 miles per second. Assuming that these are independent, unbiased measurements, find a 99 percent confidence interval for the speed of light.

9.12 A company wants to estimate the average Friday lunch break taken by its salaried executives. One Friday, 25 executives are monitored and it is found that their average lunch break is 94.6 minutes. If the standard deviation is assumed to be 25 minutes, calculate a 95 percent confidence interval for the average Friday lunch break for all this company's executives.

• **9.13** *(continuation)* Sometimes statisticians report 50 percent confidence intervals, with the sampling error described as the *probable error*. The estimate of the mean lunch break, 94.6 minutes, is said to have a probable error of E minutes if there is a .5 probability that the interval $\overline{X} \pm E$ includes the population mean. What would the value of E be in this example?

•• **9.14** An elevator is designed to carry a maximum of 2,000 pounds. The weights of the people who will use this elevator are normally distributed with a mean of 175 pounds and a standard deviation of 25 pounds. Government regulations specify that a sign must be posted saying, "Maximum capacity is n people" where n is chosen so that if n randomly chosen people are on the elevator, there is less than a 1 percent chance that their weight will exceed the maximum weight that the elevator can carry. What is the appropriate n for this elevator?

9.3　ESTIMATES OF THE STANDARD DEVIATION

So far, to simplify matters, I have assumed that the underlying standard deviation of cracker weights is known to be $\sigma = 0.1$ ounce. However, the standard deviation will seldom, if ever, be known by the statistician. It, too, is a parameter that must be estimated. The most natural course is to use the sample standard deviation as an estimate of the population standard deviation. The sample standard deviation s is the square root of the sample variance, which in turn is the average squared deviation of the data about the sample mean:

$$s^2 = \frac{(X_1 - \overline{X})^2 + (X_2 - \overline{X})^2 + \cdots + (X_n - \overline{X})^2}{n - 1} \qquad (9.6)$$

The relationship between the population variance and the sample variance is analogous to the relationship between the population mean and the sample mean. The *population mean* μ is the expected value of the random variable X, and is determined by the probabilities of observing the various values of X. The *sample mean* is an average of the observed values of X, and is determined by how frequently each value of X happens to occur. The population mean and sample mean will be essentially equal in a very large sample in which the relative frequencies are very close to the probabilities. In the same way, the *population variance* σ^2 is the expected value of the squared deviation of X about the population mean μ. This expected value of the squared deviations is determined by the probabilities of observing the various values of X. The *sample variance* s^2 is an average of the observed squared deviations of X about its observed sample mean. This average value is determined by how frequently various values of X happen to occur. The population variance and sample variance will be essentially equal in a very large sample in which the relative frequencies are quite close to the probabilities.

There is one little hitch in this analogy, however. The sample variance is calculated from the squared deviations about the sample mean, instead of the true population mean. A brief appendix at the end of this chapter shows that, if the sample mean doesn't equal the population mean, then the squared deviations about the sample mean are always less than the squared deviations about the population mean. Thus, the sample variance is a biased estimator of the population variance. To correct this bias, the sample variance divides the sum of the sample squared deviations by $n - 1$ rather than n.

One inadequacy of this adjustment is that it relates to the bias of variance estimates, whereas it is usually the standard deviation that we are interested in. Our calculations of sampling error, confidence intervals, and the requisite sample sizes all use the standard deviation, and not the variance. It can be shown that s is, in fact, a biased estimator of the standard deviation! Unfortunately, statisticians don't know how to ensure an unbiased estimate of the standard deviation, because the expected value of a square root is very complicated and depends on the particular probability distribution being analyzed. Of course, in large samples, the denominator doesn't make much difference. If $n = 100$, for instance, then division by 100 or 99 will give almost the same answer.

Student's *t* Distribution

The need to estimate the standard deviation creates another source of uncertainty in gauging the reliability of the estimated mean. The sample mean is still a good estimator of the population mean, but the calculation of a 95 percent confidence interval, using Equation 9.4,

$$.95 = P[\overline{X} - 1.96(\sigma/\sqrt{n}) < \mu < \overline{X} + 1.96(\sigma/\sqrt{n})]$$

requires a value for the standard deviation σ. If we must estimate σ, then we cannot be quite so confident that our estimated interval will encompass μ. Sometimes, the confidence interval will miss μ, not because the sample mean is off, but rather because our estimate of the width of the confidence interval is off. This added uncertainty implies that we have to make the calculated confidence interval a little bit wider to allow a little more margin for error.

In 1908, W.S. Gosset figured out the exact widening that is needed when the sampling is from a normal distribution. He was a statistician for the Irish brewery, Guinness, which encouraged statistical research but not publication. Gosset persuaded Guinness to allow his work to be published under the pseudonym "Student," and these calculations became known as **Student's _t_ distribution.**

When the sample mean is standardized by subtracting its expected value and dividing by its known standard deviation:

$$Z = \frac{\overline{X} - \mu}{\sigma/\sqrt{n}}$$

this Z variable has a standard $N[0, 1]$ distribution. Gosset determined the distribution when the sample mean is standardized, using an estimate of the standard deviation:

$$t = \frac{\overline{X} - \mu}{s/\sqrt{n}} \tag{9.7}$$

The distribution of t depends on the sample size n, because the reliability of the standard deviation estimate is greater when the sample size is larger. For an infinite sample, s will equal σ and the distributions of t and Z coincide. With a small sample, s may be either larger or smaller than σ and the distribution of t is consequently more dispersed than Z.

This dispersion depends not only on the size of the sample, but also on how many parameters must be estimated with the sample. The more data we have, the more confidence we can have in our results; the more parameters we have to estimate, the less confidence we have. Statisticians keep track of these two factors by calculating the **degrees of freedom:**

$$\begin{array}{c} \text{degrees of} \\ \text{freedom} \end{array} = \begin{array}{c} \text{number of} \\ \text{observations} \end{array} - \begin{array}{c} \text{number of parameters that} \\ \text{must be estimated beforehand} \end{array}$$

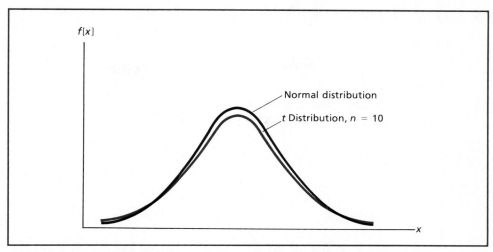

Figure 9.5   *A t Distribution, in Comparison with the Normal Distribution*

Here, to construct a confidence interval, we calculated s by using n observations and estimating one parameter (the mean). Thus there are $n - 1$ degrees of freedom here.*

The distribution of t depends on the degrees of freedom, ranging from a normal distribution when $n - 1 = \infty$ to much more dispersed distributions for small $n - 1$. Figure 9.5 compares the distributions of Z and t for $n = 10$ (9 degrees of freedom); they are very similar, with the t distribution slightly more spread out. One way of gauging the greater dispersion of the t distribution is to see how wide an interval will encompass 95 percent of the observations. The probability is .95 that Z will be between -1.96 and $+1.96$. With $n = 10$, there is a .95 probability that t will be between -2.26 and $+2.26$. This implies that if we must estimate σ on the basis of 10 observations, then our 95 percent confidence interval should be

$$.95 = P[\overline{X} - 2.26(s/\sqrt{n}) < \mu < \overline{X} + 2.26(s/\sqrt{n})] \qquad (9.8)$$

The confidence interval would be even wider for a smaller sample size. As the sample size increases, we are more confident of our estimate of σ and the margin for sampling error approaches ± 1.96 standard deviations, as when σ is known with

*There is another way to think of degrees of freedom that is more closely related to the name itself. We calculate s from n squared deviations about a mean, but this mean is calculated so that the sum of these n deviations is zero. Thus if we know $n - 1$ deviations, then we know the last deviation, too. Only $n - 1$ deviations are freely determined by the sample.

Using Confidence Intervals to Protect the Environment

The U.S. Environmental Protection Agency (EPA) typically sets industrial effluent discharge standards by measuring the discharge by companies that use the "best practical control technology currently available [BPCTCA]." These sample data (on the daily effluent of soap, for instance) are used to calculate an average $\overline{X}$ and standard deviation s and, from these, a 99 percent confidence interval for the mean effluent discharge μ from companies using BPCTCA technology. Since the standard deviation must be estimated, a t distribution is appropriate for determining the upper and lower limits for this confidence interval. With 100 or more observations, Table 3, at the end of this book, tells us that a 99 percent confidence interval is

$$\overline{X} \pm 2.6 \left(\frac{s}{\sqrt{n}} \right)$$

The effluent discharge from an exemplary plant using BPCTCA can be expected to fall within the EPA's 99 percent confidence interval 99 percent of the time. In those 1 percent of the cases when the effluent is outside this interval, half will be on the low end and half on the high end. Thus there is a $.01/2 = .005$ probability that a plant using BPCTCA will exceed the upper limit of the EPA's 99 percent confidence interval. Firms using less effective technology will exceed this limit more often and are consequently more likely to be cited by the EPA for violating effluent discharge standards.

certainty. The following table shows this pattern:

Sample Size n	Degrees of Freedom $n - 1$	95 Percent Confidence Interval for μ
2	1	$\overline{X} \pm 12.71(s/\sqrt{n})$
5	4	$\overline{X} \pm 2.78(s/\sqrt{n})$
10	9	$\overline{X} \pm 2.26(s/\sqrt{n})$
15	14	$\overline{X} \pm 2.15(s/\sqrt{n})$
20	19	$\overline{X} \pm 2.09(s/\sqrt{n})$
25	24	$\overline{X} \pm 2.06(s/\sqrt{n})$
30	29	$\overline{X} \pm 2.05(s/\sqrt{n})$
∞	∞	$\overline{X} \pm 1.96(s/\sqrt{n})$

Frankly, the difference is usually pretty slight. If you have the 20 or 30 observations recommended for invoking the central limit theorem, then you also have a large enough sample to ignore the distinction between the normal distribution and Student's t distribution. A two-standard-deviations rule of thumb is close enough. It is more important to ensure the randomness of the sample than to worry about whether the margin for sampling error is precisely 2.05 or 1.96 standard deviations.

But for those occasional very small samples from what we are willing to assume is a normal distribution, I've put several t distributions in Table 4 at the end of the book.

EXERCISES

9.15 *The Official Preppy Handbook*[7] claims that the students at Hamilton College are unusually short. Let's assume that the heights of all male college students are normally distributed with a mean of 70 inches and a standard deviation of 3 inches. What is the probability that 900 male students, who are selected randomly with regard to height, will have an average height of less than 69.5 inches?

Now assume that 16 students at Hamilton College are selected at random and their heights measured. It turns out that the sample average is 66 inches and the sample standard deviation is 4 inches. Construct a 95 percent confidence interval for the average height of the students at Hamilton College.

9.16 An economist wants to estimate the average annual income of widows living in a certain city. A random sample of 25 widows gives a sample mean of $11,700 with a sample standard deviation of $2,400. Calculate a 90 percent confidence interval for the average income of the population.

9.17 A farmer wants to estimate the average wheat yield of some land that has been used to grow corn. Ten plots are set aside and planted with wheat. The yields (in bushels per acre) are

14.8	12.0	6.8	16.8	15.6
16.4	17.2	12.0	14.2	16.0

Use these data to construct a 95 percent confidence interval for the mean wheat yield of this farmland.

9.18 An investment counselor wants to estimate the average return from a certain investment, based on past experience. During the past 10 years, the percent returns have been

12	9	24	26	9
29	10	8	0	7

Use these data to construct a 95 percent confidence interval for the mean return from this investment.

9.19 A sample of 10 cracker boxes yielded the following net weights (in ounces):

16.10	16.01	15.82	15.93	16.05
16.05	15.86	15.90	16.05	16.13

a. What is the value of the sample mean? **b.** What is the value of the sample standard deviation s? **c.** Use the t distribution to construct a 95 percent confidence interval for the population mean μ.

9.20 The 50,000 students at Enormous University have an average age of 18.8 years, with a standard deviation of 1.7 years. The 25 students on an entertainment committee are selected at random from the student body. What is the probability that the average age of these 25 students will be under 18 years? To answer this question, do you have to assume that student ages are normally distributed? What if this 1.7 standard deviation had been estimated from this sample of 25 students?

9.4 ESTIMATES OF A POPULATION PROPORTION OR SUCCESS PROBABILITY

In many sampling situations, we want to estimate a *population proportion* or *success probability*. Political polls, for instance, are used to estimate the fraction of the population who will vote for a candidate. Similarly, opinion polls estimate the proportion of the population who behave in certain ways or have certain opinions. Television polls estimate the proportions of the population who watch various television shows. Samples of manufactured products are used to estimate the fraction that are defective.

In probability analysis, our logic runs from assumed probabilities to predicted outcomes. We use the binomial distribution to tell us how likely we are to observe x successes in n trials, when the probability of success is assumed to be π. In statistical analysis, our logic runs in the other direction, from outcomes to probabilities; we observe x successes in n tests and want to use these data to estimate the probability of success π.

The statistical analysis of such tests follows exactly the same lines as when estimating a population mean. It is easy to see why this should be so. Think of each item sampled as being either a "success" or a "failure." These designations could correspond to "yes" or "no," "candidate A" or "not candidate A," or whatever is appropriate. Now let X be a random variable such that X equals 1 if the ith sampled item is a success and X equals 0 if it is a failure. The sample mean $\overline{X}$ is then the number of successes divided by the number of items tested, that is, the fraction that are successes. Thus, the fraction that are successes is really just the sum of the observed values of X divided by n—a sample mean that can be analyzed by the very same procedures that you have already learned.

The Mean and Standard Deviation

The only quirk is in the relationship between the mean and the standard deviation. When X is normally distributed, the mean and standard deviation are specified separately, but when X can only equal 1 or 0, then there is an intimate relationship between the mean and standard deviation. If we take a sample of size n from a population in which the probability of a success is π, then the probability of x successes is given by the binomial distribution:

$$P[x \text{ successes in } n \text{ trials}] = \binom{n}{x} \pi^x (1 - \pi)^{n-x}$$

In Chapter 5, Equations 5.7 and 5.9 gave the expected value and standard deviation of the success proportion $p = x/n$ as

$$\mu = \pi \tag{9.9}$$
$$\sigma = \sqrt{\pi(1 - \pi)/n} \tag{9.10}$$

The mean and standard deviation are both determined by the success probability π. Therefore, instead of separately specifying the mean and standard deviation, we specify π, and this jointly determines the mean and standard deviation.

A Political Poll

A political poll will illustrate the details of the statistical analysis. Two candidates, Mary Jones and John Martin, are running for governor. It so happens that 55 percent of the voting population favor Mary and 45 percent favor John. These population proportions are, of course, not known to anyone until after the votes are counted, but we can use a sample of voters to estimate them. If the poll is a genuine random sample, then each voter has an equal chance of being interviewed. As each participant is chosen, there is a $\pi = .55$ chance of picking a Mary supporter and a $1 - \pi = .45$ chance of picking a John partisan. A sample of 100 voters is taken and the results turn out to be

54 favor Mary

46 favor John

We want to use these data to estimate π and $1 - \pi$. The most natural estimator of the population proportion π is the sample proportion p:

$$p = 54/100 = .54$$

Indeed, as explained earlier, this corresponds precisely to using a sample mean to estimate the population mean.

Based on the outcome of this poll of 100 voters, we would estimate that 54 percent of the voters favor Mary and 46 percent favor John. We must realize though that samples are subject to sampling error. We only interviewed 100 voters and, even in a well-designed poll, the luck of the draw may turn up 54 people who favor an unpopular candidate. We can gauge the reliability of our poll result by estimating a margin for sampling error and a corresponding confidence interval for our estimate.

The central limit theorem tells us that the sample proportion p is approximately normally distributed because it is a sample mean. It immediately follows that there is a .95 probability that the sample proportion p will be within two (actually 1.96) standard deviations of its expected value π. Equation 9.10 gives the standard deviation of p.

> For a 95 percent confidence interval for a population proportion π, the margin for sampling error is
>
> $$\pm 1.96 \sqrt{\pi(1 - \pi)/n} \qquad (9.11)$$
>
> or about ± 2 standard deviations of the sample proportion p. The shorthand formula for the confidence interval is
>
> $$p \pm 1.96 \sqrt{\frac{\pi(1 - \pi)}{n}} \qquad (9.12)$$

Notice, though, that the calculation of a confidence interval requires a value for π, which is the very thing we are trying to estimate. One logical solution is to use our estimate p. A more conservative approach is to use $\pi = .5$, which gives the theoretical maximum value for the standard deviation of p. (You can quickly convince yourself of this mathematical fact by trying a few different values of π.) The value

$\pi = .5$ also greatly simplifies the approximate margin for sampling error, in that $2\sqrt{\pi(1 - \pi)/n} = 2\sqrt{.5(.5)/n} = 1/\sqrt{n}$ implies a 95 percent confidence interval $p \pm 1/\sqrt{n}$.

In practice, if p is close to .5, then it won't make much difference whether we use p or .5. But if p is close to 0 or 1, then the conservative confidence interval will be the much wider of the two. The following chart provides some illustrative calculations for $n = 100$:

π	$\pm 2\sqrt{\pi(1 - \pi)/n}$
.5	$\pm .100$
.4 or .6	$\pm .098$
.3 or .7	$\pm .092$
.2 or .8	$\pm .080$
.1 or .9	$\pm .060$
.01 or .99	$\pm .020$

In our political poll example, we have $p = .54$. If we use this estimate of π to calculate the standard deviation of p, we get

$$2\sqrt{\pi(1 - \pi)/n} = .0997$$

which is very close to the .10 margin for sampling error obtained when the conservative value $\pi = .5$ is used, so let's just use this latter value.

A margin of $\pm .10$ for sampling error implies that the 95 percent confidence interval that goes with our estimate of Mary's vote is

$$.54 \pm .10$$

or

$$.44 < \pi < .64$$

Based on our poll, we estimate that Mary will receive 54 percent of the votes, but because our sample only encompasses 100 voters, there is considerable uncertainty attached to that prediction. Our ± 10 percent margin for sampling error is too large to rule out the possibility that John is actually the people's choice.

Choosing a Sample Size

Elections are almost always sufficiently close so that a ± 10 percent margin for sampling error renders a poll inconclusive. The sampling error Equation 9.11 confirms our intuition that a more useful poll requires a larger sample. The following chart shows some illustrative calculations using $\pi = .5$ (and hence, $2\sqrt{\pi(1-\pi)/n} = 1/\sqrt{n}$)

Sample Size n	Margin for Sampling Error $1/\sqrt{n}$
10	$\pm.316$
100	$\pm.100$
400	$\pm.050$
1,000	$\pm.032$
1,600	$\pm.025$
2,500	$\pm.020$
4,000	$\pm.016$
10,000	$\pm.010$

In general, if you use $\pi = .5$ and can select a desired margin for sampling error, say, .07, then you can solve for n:

$$1/\sqrt{n} = .07$$

implies

$$n = 1/.07^2$$
$$= 204.1$$

In practice, it is usually best to base one's choice on the previous table of options and a comparison of the need for reliability with the costs of interviewing people. Sampling error falls very rapidly at first as the sample size increases. However, beyond $n = 1,000$, significant reductions of sampling error require substantial increases in sample size. Modern political pollsters base their predictions on a sample of 1,500 to 4,000 people, giving a sampling error of around ± 3 to ± 1.5 percent.

Table 9.1 shows some data for the Gallup presidential polls. Several of the questions asked are used to predict whether or not the person will actually vote in the election. Only about 60 percent of those polled are identified as likely voters and used to predict the outcome. If we assume that this 60 percent is a true random sample from the population of actual voters, then we can calculate the standard deviation and sampling errors as I have done. In theory, there is a .95 probability that the prediction will be within about two standard deviations of the actual election outcome. In practice, Gallup has missed by more than two standard deviations in four of these ten elections (1952, 1956, 1964, and 1980). However, two of these large misses were back in the 1950s shortly after Gallup had switched from quota sampling to cluster sampling, for the reasons explained in Chapter 8. It is tempting to believe that the polls have become more accurate over the years as pollsters have refined and improved their sampling procedures.

Table 9.1 *The Gallup Presidential Poll's Theoretical Sampling Errors and Actual Forecast Errors**

Year	Number Polled	Number of Likely Voters	Standard Deviation	Sampling Error Assuming a Simple Random Sample	Actual Error
1952	5,385	3,350	0.9%	± 1.9%	4.4%
1956	8,144	4,950	0.7%	± 1.4%	1.7%
1960	8,015	5,100	0.7%	± 1.4%	0.9%
1964	6,625	4,100	0.8%	± 1.6%	2.7%
1968	4,414	2,700	1.0%	± 2.0%	0.4%
1972	3,689	2,100	1.1%	± 2.2%	0.3%
1976	3,439	2,000	1.1%	± 2.2%	1.0%
1980	3,500	1,950	1.1%	± 2.2%	3.7%
1984	3,700	2,100	1.1%	± 2.2%	0.2%
1988	3,800	1,970	1.1%	± 2.2%	2.1%

*Gallup's estimates of the sampling error are somewhat larger since it is not a simple random sample.
Source: The Gallup Poll (American Institute of Public Opinion), and personal communications.

It must be remembered, too, that the Gallup poll is not a simple random sample. First, it is an approximation obtained from a sequence of selecting cities, precincts, and then people. Second, some of those selected may refuse to be interviewed, give misleading answers, or change their minds before the actual election. Third, there is inevitably a substantial chance for error in predicting which of the interviewees will actually vote. Public opinion polls present a most difficult sampling problem. There are very good reasons why the theoretical sampling error, which assumes a pure random sample, may substantially understate the margin for error in opinion polls. And yet, in the last several presidential elections, the pollsters' accuracy has been impressive.

EXERCISES

9.21 A 1986 sample of 118 California voters by Pitzer College students found that 66 (56 percent) planned to vote for Alan Cranston for senator while 52 (44 percent) favored his opponent, Ed Zschau. What is the probability of finding this strong a sample preference for Cranston if, in fact,   **a.** 50 percent of all voters favor Zschau?   **b.** 55 percent of all voters favor Zschau?   **c.** 60 percent of all voters favor Zschau? The early election returns, representing nearly 200,000 voters (3 percent of the total) gave Zschau nearly a 2-to-1 lead, but the final returns showed Cranston to be the winner by 50.8 percent to 49.2 percent. How could a poll of 118 voters predict the total vote more accurately than did the first 200,000 votes counted?

9.22 A *New York Times* columnist called the 1986 campaign to remove California Supreme Court Chief Justice Rose Bird the "most significant political event in the country."[8] On the eve of the election, a Pitzer College poll of 124 voters found that 53 planned to vote for Bird and 71 against. Calculate a 95 percent confidence interval for the fraction of all Californians planning to vote for Bird.

9.23 A marketing survey will be conducted to estimate the fraction of the population that prefer a new spaghetti sauce to the old formula. The pollsters advertise that 95 percent of the time, their sample estimate is within 2 percentage points of the actual fraction of the population. How large a sample is needed to justify this claim?

9.24 Charlie Hustle is a professional baseball player. Each time at bat, he has a .3 probability of getting a hit and a .7 probability of making an out. Batting averages are calculated by dividing the number of hits by the number of times at bat. A person who gets 50 hits in 200 times at bat is batting .250 and is said to be hitting "two-fifty." Assume that Charlie Hustle has 500 times at bat each baseball season. What are the probabilities that he will have a batting average of **a.** at least .300 next season? **b.** at least .325 next season? **c.** at least .350 next season? **d.** at least .400 next season? **e.** at least .325 averaged over the next five seasons?

9.25 A pollster asked people if they thought a poll of only 2,000 people could be trusted. The answers were 47 percent "yes," 45 percent "no," and 8 percent "undecided." How would you answer? Explain your reasoning.

• 9.26 In political polls, the margin for sampling error depends on the standard deviation of the sample proportion p: $\sqrt{\pi(1 - \pi)/n}$. Use calculus to show that this standard deviation is maximized by $\pi = .5$.

• 9.27 Probability theory claims that if the probability of some event is 1/2, then in a very large number of trials the relative frequency with which this event occurs is "almost certain" to be "very close" to 1/2. To make this claim more precise, consider the flip of a fair coin with the probability of obtaining heads apparently equal to 1/2. Let p be the fraction of n flips in which heads appear. Calculate $P[.5 - .001 < p < .5 + .001]$ for **a.** $n = 10,000$ **b.** $n = 1,000,000$ **c.** $n = 100,000,000$

9.5 SUMMARY

Samples are used to estimate characteristics of a population; for instance, a sample mean is used to estimate a population mean and a sample proportion is used to estimate a success probability. Such estimates are subject to sampling error, in that the luck of the draw determines the particular sample that will be obtained.

If a random variable X is normally distributed with mean μ and standard deviation σ, then the mean $\overline{X}$ of a sample of size n will also be normally distributed,

$$\overline{X} \sim N[\mu, \sigma/\sqrt{n}]$$

Notice that the expected value of the sample mean is equal to the expected value of X. Thus, $\overline{X}$ is said to be an unbiased estimator of μ. The standard deviation of the sample mean gauges the variation to be expected in $\overline{X}$ from one sample to the next. The larger the sample size n, the smaller the standard deviation.

Even if X itself is not normally distributed, the central limit theorem tells us that the mean of a random sample will tend to be normally distributed as n increases. In practice, a common rule of thumb is that a normal approximation is satisfactory for a sample size of around 20 or 30.

Confidence intervals are a standard way of measuring the degree of sampling error. There is a .95 probability that a sample mean will be within approximately two of its standard deviations of the expected value μ. Thus estimates are commonly written as

$$\overline{X} \pm 2(\sigma/\sqrt{n})$$

The interpretation is that 95 percent of the intervals so constructed should encompass the true population mean μ.

The confidence interval formula is usually used to compute a margin for sampling error to accompany our estimates. If we know our desired margin for sampling error in advance, then we can use the confidence interval formula to determine the requisite sample size.

If X is normally distributed and, as is usually the case, the standard deviation σ is not known but must be estimated, then Student's t distribution can be used in place of the normal distribution for figuring confidence intervals. The effect is to make the confidence interval more than two estimated standard deviations wide, and thus take into account our possible error in estimating the standard deviation. In practice, if the sample is large enough to invoke the central limit theorem, then there is not much difference between a normal distribution and Student's t distribution.

A very similar logic applies when a success probability π is estimated by the sample proportion p, giving the fraction of the observed outcomes that are successes. If we now think of the random variable X as equaling 1 if the outcome is a success and 0 if a failure, then the sample proportion p is just a sample mean and can be analyzed like any other sample mean. The only new wrinkle is that both the expected value of p and its standard deviation are determined by π:

$$\mu = \pi$$
$$\sigma = \sqrt{\pi(1 - \pi)/n}$$

APPENDIX: SOME STANDARD DEVIATION ARITHMETIC

The sum of the squared deviations about the sample mean can be written as

$$
\begin{aligned}
(X_1 - \overline{X})^2 + &\cdots + (X_n - \overline{X})^2 \\
= {} & (X_1 - \mu + \mu - \overline{X})^2 + \cdots + (X_n - \mu + \mu - \overline{X})^2 \\
= {} & (X_1 - \mu)^2 + \cdots + (X_n - \mu)^2 \\
& + (\overline{X} - \mu)^2 + \cdots + (\overline{X} - \mu)^2 \\
& - 2(\overline{X} - \mu)\{(X_1 - \mu) + \cdots + (X_n - \mu)\} \\
= {} & (X_1 - \mu)^2 + \cdots + (X_n - \mu)^2 + n(\overline{X} - \mu)^2 \\
& - 2(\overline{X} - \mu)\{(X_1 - \overline{X} + \overline{X} - \mu) + \cdots + (X_n - \overline{X} + \overline{X} - \mu)\} \\
= {} & (X_1 - \mu)^2 + \cdots + (X_n - \mu)^2 - n(\overline{X} - \mu)^2
\end{aligned}
$$

because the sum of the deviations about the sample mean is zero. Thus, the sum of the squared deviations about the sample mean is less than the sum of the squared deviations about the population mean by an amount equal to $n(\overline{X} - \mu)^2$.

Now, if we take the expected values of both sides of this equation (and use the fact that the variance of the sample mean is σ^2/n), we get

$$E[(X_1 - \overline{X})^2 + \cdots + (X_n - \overline{X})^2] = n\sigma^2 - \sigma^2$$
$$= (n - 1)\sigma^2$$

It follows that dividing the sum of squared deviations about the sample mean by n gives a biased estimator of σ^2, while dividing by $n - 1$ gives an unbiased estimator.

REVIEW EXERCISES

9.28 Estimate the number of words in this book. Explain how you made your estimate.

9.29 Use a random sample to estimate the fraction of the students at your school who are female. Explain how you constructed your sample and give a 95 percent confidence interval.

9.30 The gasoline mileage for A cars is normally distributed with a mean of 26 miles per gallon (mpg) and a standard deviation of 4 mpg. If an impartial group computes the average mpg for a sample of these cars, what is the probability that their average will be greater than 30 mpg if only 1 car is tested? If 25 cars are tested? Is the assumption that mileage is normally distributed crucial to your calculation? Explain your reasoning.

9.31 *(continuation)* Why does gasoline mileage have a probability distribution in the first place? If the average mpg of 25 cars turns out to be 30, then doesn't that show that the mean is, in fact, 30 and not 26? Finally, why is there a probability that the average mpg of 25 cars will be greater than 30, less than 20, or some other value?

9.32 The American Coinage Act of 1792 specified that the gold ten-dollar eagle would contain 247.5 grains of pure gold. An 1837 Act of Congress permitted a .25 grain deviation in the weight of a single eagle and a .048 gain deviation in the average weight of 1,000 eagles. If the minted coins were a random sample from a normal distribution with a mean of 247.5 grains and an (unknown) standard deviation σ, would the Mint be more likely to fail the test of a single coin or the test of the average weight of 1,000 coins?

9.33 A consumer research group believes that the lives of television picture tubes are normally distributed with a standard deviation of 2 years. How many sets must they test if they want to be 95 percent sure that their estimate of the mean lifetime will not be off by more than **a.** 2 years? **b.** 1 year? **c.** 1 month?

• **9.34** If we want to halve the size of the confidence interval, then must we double the size of the sample?

•• **9.35** In a boxed example on page 344, I asserted that the sample mean of the measured distances is the least squares estimate of the true distance. Use calculus to show that this assertion is true.

9.36 The Gallup poll interviews fewer than 4,000 people to predict who almost 100,000,000 voters will elect president of the United States. How can they make accurate predictions when they only interview 1 out of every 25,000 voters?

9.37 Chisler Corporation is considering issuing a three-year warranty for major repairs on its cars. How many cars must it test to be 95 percent sure that its estimate of the probability of a major repair is within .01 of the actual probability?

9.38 There are 10,000 people and one hospital in Smalltown. On any given day, there is a .001 probability that a person will require a hospital bed. How many beds must the hospital have to be 99.9 percent sure that, on any given day, there will be enough beds for everyone who requires one?

• **9.39** A random sample of 1,000 homes finds that 999 have television sets. Find a 95 percent confidence interval for the percentage of all homes that have television sets.

9.40 The Nielson television ratings are based on a sample of 2,000 homes out of a population of 80,000,000 homes. If they estimate that 30 percent of these homes were watching a football game, what is their margin for sampling error?

9.41 A critic of a 1952 government survey wrote

Only 25,000 families were interviewed.... The main fault of the report is the limited scope of its available figures. 25,000 families are not a sound representation for the 46,000,000 families in the U.S.[9]

Is this logic sound?

9.42 The manager of a nationwide discount shopping club is concerned about employee turnover. A random sample of the files of 100 employees yielded the following data for males under 21 and for all employees, including males under 21. Calculate two 99 percent confidence intervals for the population means, one for males under 21 and one for all employees. Do these confidence intervals overlap?

	Males Under 21	All Employees
Number	30	100
Employment (days)		
Average stay	147.0	315.5
Standard deviation	38.1	105.9

•• 9.43 *(continuation)* Do you think that the mean and standard deviation for the 70 employees other than males under 21 are larger or smaller than the mean and standard deviation for males under 21? Use the formulas for the sample mean and standard deviation to calculate these values and see if you are right.

9.44 Find a computer program that will simulate the repeated sampling of 100 voters from a population in which 56 percent favor candidate A and 44 percent favor B. Draw 5,000 such samples and draw a frequency distribution showing the number of samples in which x percent favored A. What are the mean and standard deviation of this empirical distribution? How do you think the mean and standard deviation would have differed had you drawn 5,000,000 instead of 5,000 samples? If each sample had contained 400 instead of 100 voters?

• 9.45 A government agency wants to estimate the number of fish in a lake. They catch 100 fish, tag them, and put them back in the lake. A short while later, they catch another 100 fish and find that 10 have been tagged. Think of this second batch as a random sample of all of the fish in the lake. Based on this sample, what is your estimate of the fraction of the fish in the lake that are tagged? Because you know that 100 fish are tagged, what is your estimate of the total number of fish in the lake?

• 9.46 Engineers are trying to estimate the probability π that a certain system will work. So far, it has been successful in 8 out of 10 trials. The chances of succeeding 8 times in 10 trials depends of course on the probability of success π. What is the exact formula for the probability of 8 successes in 10 trials? For what value of π is this probability highest?

• 9.47 A circle has a radius ρ and an area $\pi\rho^2$. It is difficult to measure the radius of a circle accurately. We can view a measurement as a random variable X that has an expected value equal to the true radius ρ. Two statisticians have measured the radius of a circle and obtained measurements X_1 and X_2. **a.** Is the average of their measurements, $r = (X_1 + X_2)/2$, an unbiased estimator of the radius ρ? **b.** Is πr^2 an unbiased estimator of the area $\pi\rho^2$? **c.** If they each estimate the area and then average, $(\pi X_1^2 + \pi X_2^2)/2$, will they have an unbiased estimator of the area? (*Hint:* The variance of X can be written as $\sigma^2 = E[X^2] - \mu^2$.)

•• 9.48 Statistical theory assumes, for convenience, that sampling is done with replacement, while the usual statistical practice is to sample without replacement. This exercise is intended to explore the implications of this divergence between theory and practice. Consider a population of 10 machines, of which 5 are acceptable and 5 are defective. A prospective buyer does not know that half of these machines are defective, but has decided to test 4 of them. Let p be the fraction of these sampled machines that turns out to be defective. **a.** What are the possible values of p? **b.** What are the probabilities of these different values of p if the sampling is done with replacement? **c.** What are the probabilities if the sampling is done without replacement? **d.** Is sampling with replacement or without replacement more likely to give a value of p that is close to the population value .5?

e. Why? **f.** Does this conclusion suggest that the true margin for sampling error is larger or smaller than the margin that we calculate by assuming sampling with replacement?

•• **9.49** Two statisticians are trying to estimate the mean shelf-life of a new drug. The first statistician has tested a sample of 25 packages and obtained an average shelf-life of $\overline{X}_1 = 48$ months. The second statistician has tested 100 packages and found an average shelf-life of $\overline{X}_2 = 40$ months. The government wants to combine these two independent samples into a single estimate. Show that the standard deviation of $\overline{X}_1$ is twice as large as the standard deviation of $\overline{X}_2$. The following formulas have been suggested for obtaining a combined estimate: **a.** $(1/2)\overline{X}_1 + (1/2)\overline{X}_2$ **b.** $(1/3)\overline{X}_1 + (2/3)\overline{X}_2$ **c.** $(1/4)\overline{X}_1 + (3/4)\overline{X}_2$ **d.** $(1/5)\overline{X}_1 + (4/5)\overline{X}_2$ **e.** $\overline{X}_2$ Which, if any, of these estimates are unbiased? Which has the smallest standard deviation? (*Note:* The variance of $a\overline{X}_1 + b\overline{X}_2$ is a^2(variance of $\overline{X}_1$) + b^2(variance of $\overline{X}_2$).)

•• **9.50** A random sample $X_1, X_2, \ldots, X_n$ can be used in many different ways to estimate μ, the expected value of X. Some possible estimators are X_1, $(X_1 + X_2)/2$, and $(3X_1 + 2X_2 + X_3)/4$. Each of these estimators is an example of a linear combination of the data:

$$\hat{\mu} = \alpha_1 X_1 + \alpha_2 X_2 + \cdots + \alpha_n X_n$$

a. What constraint must be placed on the α_i to ensure that the estimator is unbiased? **b.** Which unbiased linear estimator has the smallest variance? (Use calculus.)

9.51 The life expectancy of a certain television picture tube is normally distributed with a mean of 6 years and a standard deviation of 2 years. **a.** What is the probability that a randomly selected tube will last more than 10 years? **b.** What is the probability that the average life of a random sample of four tubes will be more than 8 years?

9.52 *(continuation)* A consumer group tests four of these picture tubes and finds that the average life is 8 years. If they assume a standard deviation of 2 years, what is their 95 percent confidence interval for the average life? If they want to halve the width of their confidence interval, what size sample should they use?

9.53 In a 1975 lawsuit concerning the discharge of industrial wastes into Lake Superior, the judge observed that a court-appointed witness had found the fiber concentration to be "0.0626 fibers per cc, with a 95 percent confidence interval of from 0.0350 to 0.900 fibers per cc."[10] Explain the apparent statistical or typographical error in this statement.

9.54 Exercise 1.86 reported the results of 100 drives of a newly designed golf ball. Use these sample data to calculate a 90 percent confidence interval for the expected value of the distance of a drive by this amateur.

9.55 A *Wall Street Journal* poll conducted on November 30 and December 1, 1987, asked a nationwide random sample of 1,599 adults a number of questions about an upcoming summit meeting between U.S. President Reagan and Soviet leader Gorbachev.[11] The *Journal* said that "households were selected by a method that gave all telephone numbers, listed and unlisted, a proportionate chance of being included." If you were the *Journal*'s pollster, how would you have carried out this instruction?

9.56 *(continuation)* Explain this observation by the *Journal:*

Chances are 19 of 20 that if all households with telephones in the U.S. had been surveyed using the same questionnaire, the findings would differ from the poll results of all respondents by no more than three percentage points in either direction.

9.57 *(continuation)* One of the questions asked was whether Reagan or Gorbachev had a better understanding of international problems. Overall, half said Reagan and a third said Gorbachev. But among those with a postgraduate education, 48 percent picked Gorbachev and 23 percent Reagan. Assuming 10 percent of those surveyed have a postgraduate education, use a normal approximation to give a 95 percent confidence interval for the fraction of the population with a postgraduate education who think Gorbachev has the better understanding of international problems. (Use the sample frequencies when estimating the standard deviation.)

9.58 In the 1989 and 1990 NCAA basketball tournaments, 1,891 3-point shots were attempted, of which 840 were taken by favored teams and 1,051 by underdog teams (using NCAA seedings to identify the favorite and underdog in each game).[12] Assuming that these data are a random sample from NCAA tournament games, calculate a 95 percent confidence interval for the probability that a randomly selected 3-point shot was taken by an underdog team. (Use a normal approximation and use a conservative $\pi = .5$ when estimating the standard deviation.)

9.59 In 1983 *Sports Illustrated* reported that,

> The Atlanta Braves aren't the least bit impressed with Rod Carew's pursuit of a .400 season. After all, they have Terry Forster. The left-handed reliever got his seventh save of the year in an 8–1 win over St. Louis last week and also slapped a single through the infield to maintain his season average at 1.000 (3 for 3) and raise his career mark to .424 (28 for 66).[13]

Assuming Bernoulli trials, use a normal approximation to calculate a 95 percent confidence interval for Forster's probability of a base hit. (Use a conservative $\pi = .5$ when estimating the standard deviation.)

9.60 *Consumer Reports* characterizes its typical reader as "well educated, married, suburban, a homeowner."[14] In 1990 this magazine mailed a nutrition survey to a sample of 1,400 subscribers and received 928 responses, of whom 15 percent were smokers. Use a normal approximation to calculate 95 percent and 99 percent confidence intervals for the fraction of all *Consumer Reports* subscribers who are smokers. (Use the sample frequencies when estimating the standard deviation.)

9.61 In the game Roshambo (or rock-scissors-paper), two players simultaneously move their fists up and down three times and then show a fist (rock), two fingers (scissors), or an open hand (paper). Rock beats scissors, scissors beats paper, and paper beats rock. If there is a tie, the players repeat the game. A student played 50 matches against randomly selected opponents and recorded the frequency of the initial moves.[15] Considering these data to be a random sample, estimate the student's probability of playing rock on the initial move and use a normal approximation to give a 99 percent confidence interval. Do the same for the opponents' probability of playing rock on the initial move. (Use $\pi = .5$ for estimating the standard deviations.)

	Student	Opponents
Rock	26	16
Scissors	9	17
Paper	15	17

9.62 *(continuation)* If this student wanted to be 99 percent confident that his estimate of the opponents' probability of playing rock on the initial move was within .05 of the actual probability, how large a sample should he have taken?

9.63 A 37-car train carrying 2,500 kegs of beer was derailed during a snowstorm.[16] The shipper's insurance policy covered the complete cost of this accident if more than 10 percent of the kegs were damaged. The shipper examined a sample of 50 kegs and found that 12 (24 percent) were damaged. The insurance company disputed the use of a sample in place of an examination of all 2,500 kegs; however, a court awarded damages based on the calculation of a 95 percent confidence interval. Determine a 95 percent confidence interval for π, the fraction of kegs damaged, using a normal approximation and assuming an infinite population. (Estimate the standard deviation with $\pi = .5$) Repeat your calculation using a finite population correction factor. Do these confidence intervals include $\pi = .10$?

9.64 *(continuation)* If 10 percent of an infinite population of kegs are damaged, what is the probability that 24 or more kegs will be found damaged in a random sample of 50 kegs? (Use a normal approximation with a continuity correction.)

9.65 A *Wall Street Journal* (July 6, 1987) poll asked 35 economic forecasters to predict the interest rate on three-month Treasury bills in June 1988. These 35 forecasts had a mean of 6.19 and a variance of 0.47. Give a 95 percent confidence interval for the mean prediction of all economic forecasters and then explain why each of these interpretations is or is not correct: **a.** There is a .95 probability that the actual Treasury bill rate on June 1988 will be in this interval. **b.** Approximately 95 percent of the forecasts are in this interval. **c.** If the *Journal* took another random sample, there is a .95 probability that the new confidence interval would include 6.19.

9.66 A study of a sample of 286 NBA basketball games during the 1989–1990 season found that 6,841 personal fouls were called against home teams and 6,574 against visiting teams.[17] Use a normal approximation (and $\pi = .5$ for estimating the standard deviation) to calculate a 95 percent confidence interval for the probability that a randomly selected personal foul is against the visiting team.

9.67 *(continuation)* How large a sample would the pollster have needed for a 95 percent confidence interval with a width of .010? For a 99 percent confidence interval with a width of .010?

9.68 A 1990 survey of 1,000 *Macworld* readers found that 486 were extremely or very likely to update to Apple's forthcoming System 7.0.[18] Assuming that these readers are a random sample of all Macintosh users, use a normal approximation to estimate a 99 percent confidence interval for the fraction of all Mac users in 1990 who were extremely or very likely to update to System 7.0. (Use $\pi = .5$ to estimate the standard deviation.)

9.69 A study of a sample of electronic mail messages found 64 typographical errors, of which 7 involved the letter *e*.[19] Calculate a 95 percent confidence interval for the probability that a typographical error involves the letter *e*. Use a normal approximation and use the observed frequency x/n when calculating the standard deviation.

In ordinary written English,[20] the letter *e* is used with a frequency of .13. Is this .13 frequency inside your 95 percent confidence interval for typographical errors? What conclusion would you draw if a letter of the alphabet had a disproportionately large number of typographical errors?

9.70 Some firms place the upper and lower limits in control charts at two, rather than three, standard deviations from the mean. How often will a process that is in control violate these limits? Would you characterize a two-standard-deviations rule as more or less conservative than a three-standard deviations rule?

9.71 A manufacturing process is designed to produce machine bolts with a diameter that is normally distributed with a mean of 1.000 centimeter and a standard deviation of 0.050 centimeter. A quality control engineer proposes sampling four bolts every hour. If the manufacturing process is functioning as intended, what is the expected value and standard deviation of the mean of a four-bolt sample?

What upper and lower control limits should the quality control engineer set for the mean of a four-bolt sample for there to be a .01 probability of violating the control limit when the process is functioning normally?

9.72 *(continuation)* What are the appropriate upper and lower control limits if nine bolts are sam-

pled instead of four? Explain any difference from your answer to the preceding exercise.

9.73 When the temperature control system for a large office building is set at T, the temperature at a certain point in the building is normally distributed with an expected value of T and a standard deviation of 3° Fahrenheit. If the management wants only a .01 probability that the temperature at this point will exceed 75°, what value of T should it choose? If it uses this value of T, what is the probability that the temperature at this point will be less than 65°? Less than 60°?

9.74 The management of an office building monitors the temperature control system by taking four independent readings of the temperature at four locations where, if the system is functioning properly, the temperature is normally distributed with an expected value of 68° and a standard deviation of 3°. What is the expected value and standard deviation of the average of these four readings if the system is functioning properly? What upper and lower control limits should the building's management set for the average of these four temperature readings for there to be a .01 probability of violating a control limit when the temperature control system is functioning normally? What are the appropriate control limits if it wants a .005 probability of violating a control limit when the temperature control system is functioning normally?

• **9.75** *(continuation)* How many locations should it sample if it wants there to be a .01 probability that the average temperature at these loctions will be outside the range 65° to 71° if the temperature control system is functioning normally?

9.76 An examination of the historical records of an oil-drilling company indicates that its daily use of a well lubricant is approximately normally distributed with a mean of 30,000 gallons and a standard deviation of 20,000 gallons. If this company wants only a .01 probability that its day-end inventory will be less than 20,000 gallons, how many gallons should it have at the beginning of the day?

9.77 *(continuation)* This company works seven days a week and monitors its lubricant use on a daily basis. If the average daily use during a seven-day week is less than 20,000 gallons or more than 40,000 gallons, it conducts a thorough review of its drilling operations. If the seven daily readings are a random sample and there has been no change in the expected value or standard deviation of its lubricant use, what is the probability that it will violate these control limits?

9.78 *(continuation)* What upper and lower control limits are appropriate if this company wants a .01 probability of violating the upper control limit and a .05 probability of violating the lower control limit when there has been no change in the expected value or standard deviation of its lubricant use?

9.79 On average, 5 percent of the gadgets manufactured by Handcraft Industries are defective. If a random sample of 20 gadgets is selected, what is the probability that at least one will be defective?

9.80 *(continuation)* If a random sample of 200 gadgets is selected, what is the probability that at least 10 will be defective?

9.81 Explain why an objectivist would disagree with the following interpretation of confidence intervals:

> We might say that a 95 percent confidence interval for the percentage of voters favoring candidate X is 45 percent plus or minus 6 percent. This means that we can be 95 percent certain that the population percentage is within 6 percent of the sample percentage; in this case, between 39 percent and 51 percent of the population favor candidate X.[21]

9.82 A sample of 100 voters finds that 54 prefer Mary while 46 prefer Mario. Use a normal approximation to determine the probability of finding this strong a sample preference for Mary if, in fact, 50 percent of all voters prefer Mario. If 60 percent of all voters prefer Mario.

9.83 It is two weeks before the election and an underdog politician is favored by only 45 percent of the voters. Badly needing some last-minute financial support, this desperate politician commissions three independent pollsters to each poll 100 people. What is the probability that at least one of these polls will find this politician to be the favorite? If the politician releases only the most favorable of the three poll results, why is this sample biased? If you were a reporter trying to predict the winner, how would you use the three polls?

9.84 The net weights of cereal boxes are known to be normally distributed with a standard deviation of 0.1 ounce. A sample of 10 boxes yields a mean of 25.99 ounces. Construct a 90 percent and then a 99

percent confidence interval for the population mean. Which is wider? Why?

9.85 In 1988 the PGA Tour made a special study of the putting success of professional golfers.[22] At 15 tournaments they selected one green with a relatively flat and smooth surface and measured and recorded the result of each putt on that green. Here are some of their results:

Length of Putt (Feet)	Number of Putts Measured	Fraction of Putts Made
5	353	.589
15	167	.168

For each of these two distances, calculate a 95 percent confidence interval for the probability of making a successful putt, using a normal approximation and assuming that the binomial model applies. (Use the sample frequencies when estimating the standard deviation.)

9.86 A researcher tested 72 birthday candles by lighting each and calculating the burning time, in seconds.[23] Use the following data to estimate the population mean burning time and use the t distribution to make a 99 percent confidence interval for your estimate:

700	695	660	765	620	650	715	715	660
790	635	645	690	705	695	815	715	615
720	655	690	675	790	725	755	660	630
730	800	620	670	645	740	730	720	630
710	660	650	730	785	690	730	615	580
735	690	765	800	725	840	675	645	680
820	775	745	810	780	780	715	790	765
785	720	760	785	770	755	710	735	740

9.87 Following are some data on Michael Jordan's shooting accuracy in the regular season during his first five years in the National Basketball Association.[24] In each of these three categories, calculate a 95 percent confidence interval for his probability of making a shot, assuming these shots to be Bernoulli trials. Use a normal approximation and use $\pi = .5$ when calculating the standard deviation. Why are his basketball shots not Bernoulli trials?

Category	Attempted	Made
3-point shots	287	58
2-point shots	8,025	4,120
Free-throws	3,496	2,965

9.88 Each of twelve college intramural basketball players was asked to take 30 shots 15 feet from the basket—10 shots from the free-throw line, 10 shots from a 45° angle to the basket, and 10 shots from the baseline.[25] Assuming that the Bernoulli trial model is appropriate, use the following results to give a 95 percent confidence interval for the probability of making each of these three shots. (Use the sample frequencies for estimating the standard deviations.)

Shot	Attempts	Successes
Free-throw line	120	84
45° angle	120	62
Baseline	120	53

9.89 During the 1989 Major League baseball season, home teams won 1,157 of 2,106 games. Assume that these are a random sample of all professional baseball games and use these data to estimate a 99 percent confidence interval for the probability that the winner of a randomly selected game is the home team. Use $\pi = .5$ when calculating the standard deviation.

9.90 A researcher used a random number generator to select 25 stock funds from *Forbes* annual mutual fund issue, which tabulates returns over a 12-month period that does not coincide with the calendar year.[26] The returns from four such issues are shown in Table 9.2. For each of these four years, use the mean and standard deviation of these 25 returns to estimate a 95 percent confidence interval for the average return of all mutual funds. Since the standard deviation must be estimated, use a *t* distribution. The returns on the S&P 500 stock index during these four years were, respectively, 35.8 percent, 25.1 percent, −6.9 percent, and 20.4 percent. In which of these years was the aver-

Table 9.2 *Data for Exercise 9.90*

Fund	1986	1987	1988	1989
Seligman Common Stock Fund	41.3%	17.5%	−8.0%	11.4%
Federated Stock Trust	29.6	22.0	−6.2	11.6
United Services–Growth	30.9	3.1	−11.0	8.0
Sigma Venture Shares	32.7	0.9	−19.9	4.1
Ivy Growth Fund	26.4	21.5	−8.8	17.4
Prudential-Bache Option Growth	24.6	19.1	−4.1	16.2
IDS Strategy Fund–Equity	26.5	17.0	10.2	19.4
Kidder Peabody Special Growth	18.0	3.8	−9.8	21.5
National Growth Fund	39.6	6.2	−10.3	−1.9
Vanguard Explorer Fund II	16.8	5.9	−6.4	7.4
Country Capital Growth Fund	28.9	14.8	−6.7	19.9
Mass Financial Special Fund	35.4	20.9	−6.1	21.2
United Vanguard Fund	31.7	20.3	2.8	11.9
Merrill Lynch Capital Fund	30.7	18.6	0.1	16.2
Amcap Fund	29.7	25.0	−3.3	17.2
Sigma Investment Shares	39.2	17.2	−5.1	17.1
American Capital Pace Fund	23.7	26.3	−11.8	17.3
Oppenheimer Time Fund	46.6	16.5	−2.6	19.4
Nationwide Growth Fund	36.1	18.9	−1.7	16.5
Dreyfus Growth Fund	37.2	24.5	−2.1	12.4
American Investor Growth Fund	22.6	6.6	−7.1	24.6
Commonwealth Indenture A&B	27.2	12.5	−0.7	14.2
Surveyor Fund	45.2	7.9	0.0	24.4
Quasar Associates	54.2	1.8	7.8	26.5
ISI Trust Fund	15.5	10.5	3.8	13.4

age sample fund return better than the S&P 500? Is the S&P 500 return within the funds' 95 percent confidence interval each year?

• **9.91** Suppose that the damage award favored by individual potential jurors in a personal injury case can be described by a normal distribution with a mean of $5,000,000 and a standard deviation of $2,000,000. What percentage of potential jurors favor an award of more than $5,500,000? Of less than $4,500,000? Now assume that a jury is a random sample from this distribution and that the jury's damage award is the average of the awards favored by those who serve on the jury. Write a paragraph comparing a six-person jury system with a twelve-person system.

9.92 Each semester, Professor Smith gives students a copy of the final exam from the preceding semester to help them study for their final exam. In the spring of 1990 he passed out the exam from the fall of 1989 and warned that "there is inevitably variation in the difficulty of tests. This exam was 0.6 standard deviations above the mean." As a student, how would you interpret this remark?

9.93 Explain how you, as a statistician, would evaluate this advice: "Look for a sample of at least 600 interviews. If fewer than 400 people were polled, you can skip the survey."[27]

9.94 Mr. Young is running against Mr. Old for governor. How large a random sample is needed if a pollster wants to be 95 percent confident that the percentage of the sample who prefer Young will be within 1 percentage point of the actual percentage of all voters who prefer Young?

9.95 A student used a table of random numbers to select 80 books in her library's card catalog.[28] Of these, only 57 could be found in the library. The other 23 were either checked out, stolen, or shelved incorrectly. Assume an infinite population and use a normal approximation to calculate a 95 percent confidence interval for the proportion of all of the library's books that cannot be found in the library. Use the

lower boundary of this interval and the library's estimate that it has 1.4 million books and 6,000 users to estimate a minimum for the average number of missing books per user.

9.96 "I know that, on average, women in this country are paid more than men. I've seen the department paychecks, and the women's annual salaries are $44,000, $39,000, and $28,000, while the men's salaries are $42,000 and $30,000." Write a paragraph explaining why this evidence is unpersuasive.

9.97 A student mailed a short questionnaire to the 495 professors at her school.[29] One of the questions was, "Overall, would you consider your views more consistently in agreement with Democrats or Republicans?" Of the 248 professors who responded, 193 answered Democrats and 55 Republicans. Assuming (unrealistically) the population to be infinite, calculate a 95 percent confidence interval for the fraction of all the professors at her school who would answer Democrat. (Use the sample frequencies when estimating the standard deviation.) If we were to take into account the finite population, would this shrink or widen our confidence interval?

• **9.98** *(continuation)* Recalculate your 95 percent confidence interval, this time using the finite population correction factor.

9.99 The federal government wants to estimate the fraction of the population living below the poverty level. In a random sample of 2,500 people, 13.8 percent are found to have incomes below the poverty level. Use a normal approximation to calculate a 99 percent confidence interval for the fraction of the entire population with incomes below the poverty level. (Use the sample frequencies when estimating the standard deviation.) Must you assume that incomes are normally distributed?

9.100 How many times must a fair coin be randomly flipped to be 90 percent certain that between 40 percent and 60 percent of the flips will turn up heads? Use a normal approximation.

10

Hypothesis Tests

It is a capital mistake to theorize before you have all the evidence. It biases the judgements.

Sherlock Holmes

TOPICS

10.1   INTRODUCTION

Sample data are often used to confirm or refute theories. Chapter 8 related how nearly two million school children participated in a nationwide test of the effectiveness of the Salk polio vaccine. Sample data have been collected to test whether vitamin C wards off colds, aspirin reduces the risk of heart attack, cigarette smoking is dangerous to your health, and the death penalty deters murder. Data have been used to support the contention that minorities have been systematically excluded from certain schools, occupations, and juries. National income data were used to test Keynes' theories that consumer spending depends on income and that investment spending is influenced by stock prices.

It would be nice if data such as these could prove conclusively that a theory is either true or false, but few things are 100 percent certain. The data are almost always a sample and therefore subject to sampling error. A new sample might turn out to be sufficiently different from earlier ones to discredit a previously confirmed theory. For example, if 52 percent of a random sample of 400 voters prefer a certain presidential candidate, these data support the theory that this candidate is favored by a majority of all voters. But a second survey of another 400 voters may find that 53 percent favor the other candidate. We cannot be certain who will win until all of the votes have been counted.

Most theories are in a permanent state of uncertainty because there is no final counting of the votes. New experiments and fresh observations of the world around us continually provide new data. Social scientists have so few data that old results can easily be undone by a few more observations of people's behavior. Keynes' consumption function was confirmed in the 1940s and found inadequate in the 1950s. Even in the physical sciences, where innumerable laboratory experiments can generate mountains of data, theories remain fragile. Edward Leamer, a very talented UCLA statistician, wrote that,

> All knowledge is human belief, more accurately human opinion. What often happens in the physical sciences is that there is a high degree of conformity of opinion. When this occurs, the opinion held by most is asserted to be an objective fact, and those who doubt it are labeled "nuts." But history is replete with examples of opinions losing majority status, with once objective "truths" shrinking into the dark corners of social intercourse.[1]

For centuries before Copernicus, people believed that the sun revolved around the earth. Experts taught that heavy objects fall faster than light ones, until Galileo's experiments at the Leaning Tower of Pisa. Newtonian physics gave way to Einstein. Scientists think of new ways to test old theories or fresh ways to interpret old data, and thus the weaknesses of old theories may become exposed.

If we can never be sure that a theory is true, then the next best thing would be to make probability statements such as, "Based on the available data, we are 90 percent certain that this theory is true." A Bayesian wouldn't hesitate to say this, but the classical, objectivist interpretation of probabilities precludes such statements because a theory is either true or it isn't. A theory can't be true 90 percent of the time, with its truth or falsity determined each day by a playful nature's random-number generator.

Statistics has followed another route. Instead of this forbidden conditional probability,

$$P[\text{theory is true} \mid \text{observed data}]$$

statisticians calculate the reverse conditional probability,

$$P[\text{observed data} \mid \text{theory is true}]$$

The first statement says, given the available data, this is the probability that the theory is true. The second statement says, if this theory were true, this is the probability that the data would look like they do.

These are two very different statements, and a recognition of the difference is crucial to an understanding of the meaning and limitations of classical hypothesis tests. Hypothesis tests are an attempted proof by *statistical contradiction.* For any particular theory, statisticians can deduce that, if the theory is true, then a sample of data is likely to look like this and unlikely to look like that. Then they go out and gather some data. If the sample data are in the likely category, then the data are consistent with the theory and, so, confirm the theory. If, however, the data are improbable, were the theory true, then the data are not consistent with the theory, and the theory is refuted; this is what Thomas Huxley called "the great tragedy of science—the slaying of a beautiful hypothesis by an ugly fact."

A theory is rejected if it can be shown statistically that the data we observe would be very unlikely to occur if the theory were in fact true. In this sense, it is a proof by statistical contradiction. Because the data are unlikely if the theory were true, we conclude that the theory is not true. Of course, we are never 100 percent certain, because the unlikely may well have happened. Notice, too, that for a theory to be accepted, it need only not be rejected by the data. This is a relatively weak conclusion, because there are invariably many other theories that are consistent with the data, too. In the next section we'll look at a specific example to make these ideas more concrete.

EXERCISES

10.1 There are two candidates for governor, Wiseman and Fullhead, and on the eve of the election each predicts that he will get 53 percent of the vote. A random sample of 100 voters finds 49 percent for Wiseman and 51 percent for Fullhead. Explain what each of these probability statements means: **a.** $P[51$ percent of those polled favor Fullhead $\mid$ 53 percent of the voters favor Fullhead] **b.** $P[53$ percent of the voters favor Fullhead $\mid$ 51 percent of those polled favor Fullhead]. Which probability statement would an objectivist feel most comfortable with?

10.2 *(continuation)* Which of these probabilities do you think is larger: **a.** $P[51$ percent of those polled favor Fullhead $\mid$ 53 percent of the voters favor Wiseman] **b.** $P[51$ percent of those polled favor

Fullhead | 53 percent of the voters favor Fullhead]? Check your reasoning by computing these probabilities.

• **10.3** *(continuation)* If we were to use Bayes' formula to compute P[53 percent of the voters favor Fullhead | 51 percent of those polled favor Fullhead], what information would we need? How could we interpret a Bayesian probability that Wiseman is the people's choice?

• **10.4** *(continuation)* Assuming that, prior to the poll, we believe that each candidate's claim is equally likely to be correct, P[Wiseman will get 53 percent of the vote] = P[Fullhead will get 53 percent of the vote] = .5. (For simplicity, assume that no other outcomes are possible!) Compute the revised probabilities, taking the poll into account: P[53 percent of the voters favor Wiseman | 51 percent of those polled favor Fullhead] and P[53 percent of the voters favor Fullhead | 51 percent of those polled favor Fullhead].

10.2   CRAMMING FOR THE MATH SAT TEST

Let's imagine that we work for a consumer advisory group that wants to test a two-week cram course that is intended to increase scores on the math Scholastic Achievement Test (SAT). The necessary proof is by statistical contradiction of the "null hypothesis" that the course is worthless; this terminology comes from the fact that the hypothesis being tested is often, as here, that there is no, or *null,* effect.

Average SAT scores for the nation as a whole vary slightly from year to year, but we'll assume here that math SAT scores are normally distributed with a mean of 500 and a standard deviation of 100. A natural test procedure is to enroll a random sample of 25 students in the course and then see how well they do on the math SAT test.* If the students' average score turns out to be 500, these data provide no evidence against the null hypothesis that the course is worthless.

However, even if the course is worthless, we have data for only a small random sample, and the sample mean most likely will not turn out to be exactly 500. There is inevitably sampling error because the 25 students selected by chance may be either above average or below average. Whether the course is worthwhile or worthless, the sample mean may be 452, 521, 554, or some other number.

Taking sampling error into account, how well do the 25 students have to score to convince us that the course is worthwhile? An average score of 500 is certainly unconvincing, but what about 520? Or 530? Or 550? Where do we draw the line and say, "Now, those are sufficiently high scores to persuade me that this course is worthwhile"?

The Chances of Incorrectly Approving a Useless Course

Let's think a bit more carefully about the likely size of the sampling error. If the course is worthless, then our sample of 25 students are drawn from a population

---

*Another possible design is to have 25 students take the math SAT, take the cram course, and then take the SAT again to see if there is any improvement in their scores. The problem with this procedure is that their scores may improve simply from the experience gained by taking the test.

whose scores X are normally distributed with a mean $\mu = 500$ and a standard deviation $\sigma = 100$.

$$X \frown N[500, 100]$$

The average score $\overline{X}$ of a sample of size 25 is normally distributed, too, with a mean $\mu = 500$ and a standard deviation $\sigma/\sqrt{n} = 100/\sqrt{25} = 20$:

$$\overline{X} \frown N[500, 20]$$

This distribution of the sample mean, under the null hypothesis that the course is worthless, is shown in Figure 10.1.

What is the probability that, by the luck of the draw, the average score of these 25 students will be above, say, 510? We standardize $\overline{X}$ to be a $Z \frown N[0, 1]$ variable that can be looked up in Table 2 at the end of this book:

$$Z = \frac{X - \mu}{\sigma/\sqrt{n}}$$
$$= \frac{510 - 500}{100/\sqrt{25}}$$
$$= 0.5$$

and

$$P[Z > 0.5] = .31$$

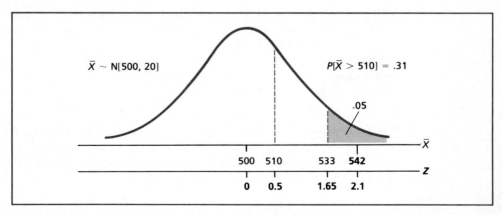

Figure 10.1 *The Distribution of the Sample Mean $\overline{X}$ of 25 Math SAT Scores When a Preparation Course Is Worthless*

An average score of 510 is half a standard deviation above the expected value of 500 and there is a .31 probability that the sample mean will be at least this high.

A sample mean of 510 is not impressive evidence in favor of the course because, even if the course is worthless, there is a .31 probability that, by the luck of the draw, we will select 25 students who do at least this well. An average score of 510 is too easy a hurdle, in that there is too high a probability that a worthless course will clear this hurdle. We want a criterion that is sufficiently difficult that there is only a small chance that a worthless course will appear effective. We are trying to distinguish between merit and luck here and, to do so, we need to set a cutoff point that is unlikely to be achieved by luck alone.

A standard rule is to set a cutoff point so that there is only a 5 percent chance of rejecting the null hypothesis that the test is worthless. In the next chapter, we'll scrutinize the logic behind this rule; for now, let's just work out the details of its application. How much higher do we have to raise the hurdle to reduce that probability from .31 to .05? Table 2 tells us that there is a .05 probability that Z will exceed 1.65. Our standardized variable is $Z = (\overline{X} - \mu)/(\sigma/\sqrt{n})$. We can combine these two facts to obtain the answer:

$$\frac{\overline{X} - \mu}{\sigma/\sqrt{n}} = 1.65$$

$$\overline{X} - \mu = 1.65(\sigma/\sqrt{n})$$

$$\overline{X} = \mu + 1.65(\sigma/\sqrt{n})$$

$$= 500 + 1.65(100/\sqrt{25})$$

$$= 533$$

Notice the characteristic form. There is a .05 probability that Z will exceed 1.65. Therefore, there is a .05 probability that the sample mean will be 1.65 standard deviations above its expected value. Here, there is a .05 probability that the sample mean will exceed $500 + 1.65(20) = 533$.

Z-Statistics and P-Values

A sample mean of, say, 542 is persuasive evidence that the course matters because it is unlikely that the luck of the draw alone would give such a high sample mean. What, in fact, is the probability that a random sample of 25 students from a population with a mean of 500 and standard deviation of 100 would average as high as 542? Translating again to a standardized Z variable,

$$Z = \frac{(\overline{X} - \mu)}{(\sigma/\sqrt{n})}$$

$$= \frac{(542 - 500)}{(100/\sqrt{25})}$$

$$= 2.1$$

and

$$P[Z > 2.1] = .0179$$

from Table 2. If the null hypothesis is true and the course is worthless, the students are drawn from a population with a mean of 500. A sample mean of 542 is 2.1 standard deviations above its expected value according to the null hypothesis, and there is less than a 2 percent chance that a sample mean will be so far from its expected value. Because this probability is so slight, the luck of the draw (sampling error) is an unconvincing explanation. A difference of 2.1 standard deviations between the sample mean and its presumed expected value is too large to be explained plausibly by chance alone.

Let's generalize the argument by retracing our steps. First, we look at the sample mean, whatever it turns out to be. Then we standardize it by subtracting the expected value of the sample mean under the null hypothesis and dividing by the standard deviation of the sample mean:

$$\textbf{Z-statistic} = \frac{\text{observed sample mean} - \text{expected value under null hypothesis}}{\text{standard deviation of the sample mean}} \tag{10.1}$$

$$= \frac{(\overline{X} - \mu)}{\sigma/\sqrt{n}}$$

Next, we use Table 2 to find the probability of observing such an extreme value. This probability is called a **P-value.** If the P-value is less than 5 percent, then the null hypothesis is rejected. A glance at Figure 10.1 confirms that this procedure is fully equivalent to our first method. If the null hypothesis is true, there is less than a 5 percent chance of observing sample mean higher than 533. If the sample mean does turn out to be larger than 533, then the value of the Z-statistic is greater than 1.65, the P-value is less than 5 percent, and the null hypothesis is rejected.

The Chances of Incorrectly Slighting a Useful Course

Sampling error can, of course, work in the other direction, too. Perhaps the firm administering the two-week cram course claims that it adds 50 points to SAT scores, and it does. Yet, we may happen to pick 25 below-average students who would score only 450 on their own. Even though the course raises their scores by

50 points, these below-average students will still only average 500 and we will erroneously conclude that the course is worthless. Or we might pick 25 students who would average 480 on their own and 530 with the course. Because 530 is below 533, we again mistakenly conclude that the course didn't work as advertised.

Figure 10.2 shows how to figure the probability of making this kind of mistake. If the course raises math SAT scores by 50 points, then our sample is drawn from a population that, after taking the course, would average $\mu = 550$ with a standard deviation $\sigma = 100$. The average score $\overline{X}$ for the 25 students in our sample comes from a normal distribution with an expected value $\mu = 550$ and a standard deviation $\sigma/\sqrt{n} = 100/\sqrt{25} = 20$. The probability that the sample mean will fall below the 533 cutoff point is determined by calculating the Z-value, $Z = (\overline{X} - \mu)/(\sigma/\sqrt{n}) = (533 - 550)/(100/\sqrt{25}) = -0.85$. Table 2 tells us that

$$P[Z < -0.85] = .1977$$

or, roughly 20 percent.

Sampling error inevitably implies that there are two kinds of erroneous statistical conclusions that can be drawn. We may conclude that a worthless course is useful or we may conclude that a useful course is worthless. With a 533 cutoff point, there is a 5 percent chance that a bad course will appear to be successful and a 20 percent chance that a course that adds 50 points will appear to be a failure.

Notice that we did not try to calculate the probability that the course is useful. We resisted putting a probability on the value of the course, because objectivists insist that it isn't a fit subject for a probability: either the course works or it doesn't. Instead, we calculated the probability of an average score above 533 if the course is useless and the probability of an average score below 533 if the course is useful.

This logic, plus the inevitable sampling error, is why we never "prove" theories to be true or false. Instead, we say that the data "reject" a theory in the sense that, were the theory true, such data would be very unlikely. A theory that is not rejected by the data is one that is consistent with the data in the sense that the deviation

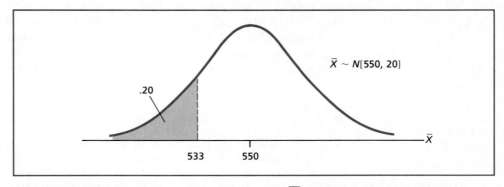

Figure 10.2   *The Distribution of the Sample Mean $\overline{X}$ of 25 Math SAT Scores When a Preparation Course Raises Scores by 50 Points*

The Importance of a Control Group

Medical researchers tested a vaccine for the common cold, using 548 college students who believed themselves especially susceptible to colds.[2] Half were given the experimental vaccine and half, the control group, were given plain water. The group taking the vaccine reported an astonishing 73 percent decline in colds in comparison with the previous year—astonishing, except for the fact that the control group reported a 63 percent decline! Apparently, this particular season was relatively cold-free, there were psychological benefits from the presumed vaccine, or the students reported a decline in colds because they thought that this was what the researchers wanted to hear.

If there had been no control group, the researchers would have considered the vaccine to be a great success. Having a control group for comparison, they concluded that the vaccine had little if any effect on colds. The importance of a control group is made even more striking by the researchers' report that afterward they received letters from doctors stating, "I have a patient who took your cold vaccine and got such splendid results that he wants to continue it. Will you be good enough to tell me what vaccine you are using?"[3] When they checked the records, the researchers often found that the patient with such splendid results had received plain water.

between theory and experience can plausibly be explained by sampling error. An average score of 520 by 25 students doesn't prove that the course is worthless. It is just that a 520 average could well be due to the luck of the draw (sampling error) and, so, is not sufficiently improbable to rule out a population mean of 500. For this reason, statisticians say "the data do not reject this theory" rather than "the data prove this theory." Nonetheless, the statement "The evidence does not show that cramming pays off" is often misinterpreted as meaning "The evidence shows that cramming does not pay off."

EXERCISES

10.5 A Texas A&M tradition is for members of the student body to stand during football games, indicating that each is a "12th Man," ready to play if needed. In 1983, the coach decided to let some of the student fans play:

They all laughed when Texas A&M Coach Jackie Sherrill came up with his version of the 12th Man, a kickoff return team made up entirely of students.

It's no longer a joke. Going into today's game against Texas, the 12th-man unit has allowed an average of only 13.6 yards per return.

"These guys cover kickoffs better than anybody who plays against us," Sherrill said.

And A&M's road kickoff-return unit, made up of regular players since the 12th-Man team doesn't travel, is allowing more than 18 yards per return.

Even the White House has been impressed. After

a presidential press aide attended an Aggies game, President Reagan sent each 12th Man member an autographed picture along with a personal message.[4]

What other information would you like to see before concluding that the students cover kickoffs better than the regular football players?

10.6 The text uses the example of gauging the effects of a cram course on math SAT scores. In this example, 25 students are selected at random, enrolled in the course, and then given the math SAT. Why do you suppose the students were picked at random and then enrolled in the course, rather than picked at random from those already enrolled in the course?

10.7 The earlobe test was introduced in a letter to the prestigious *New England Journal of Medicine,* in which Dr. Sanders Frank reported that 20 of his patients with creases in their earlobes had many of the risk factors (such as high cholesterol levels, high blood pressure, and heavy cigarette usage) associated with heart disease. For instance, the average cholesterol levels for his patients with noticeable earlobe creases was 257 (mg per 100 ml), compared to an average of 215 with a standard deviation of 10 for normal middle-aged men. If these 20 patients were a random sample from a population with a mean of 215 and a standard deviation of 10, what is the probability that their average cholesterol level would be as high as 257? Can you think of any reason why these 20 patients may not be a random sample from the population of middle-aged men?

10.8 A seminal study in the 1930s investigated whether blacks who had recently moved to New York City from the South had lower IQ scores than blacks who had been born and raised in New York.[5] Among 12-year-old black males born in New York, IQ scores were known to be normally distributed with a mean of 87 and a standard deviation of 29. IQ tests were administered to a sample of 56 black 12-year-olds who had lived in New York for two years or less. If these 56 recent migrants were a random sample from a normal distribution with a mean of 87 and a standard deviation of 29, from what distribution was the sample mean drawn? What cutoff point would be appropriate if we wanted only a 5 percent chance that sampling error would cause such a low value for the sample mean?

The sample mean turned out to be 64. Does this suggest that something more than sampling error is involved? What are the Z- and P-values? Can you think of any logical explanation, other than sampling error, why recent migrants to New York might have lower IQ scores than those born in New York?

10.9 A professor claims that the college's physical education majors have an average IQ of 90, while the tennis coach claims that their average IQ is 100. To test these competing claims a random sample of 16 P.E. majors will be given IQ tests. If the standard deviation of IQ scores is 16, what is the probability that the sample mean will be below 100 if the professor is right? If the tennis coach is right? Where should we set the cutoff if we want only a 5 percent chance of rejecting the tennis coach's claim if it is true? What is the probability that the sample mean will be above this cutoff if the professor is correct?

10.3   A GENERAL FRAMEWORK

The SAT example illustrates the logic that underlies statistical tests of all sorts of hypotheses. This logic has been codified into a standard framework that will be described here and then applied to several situations.

The Null and Alternative Hypotheses

Statistical tests are designed to choose between two competing theories. One theory is labeled the **null hypothesis, H_0,** while the other is labeled the **alternative hypothesis, H_1.** In our example, the null hypothesis is that the course has no effect. We will assume temporarily (and unrealistically) that the alternative hypothesis is a single value, $\mu = 550$, perhaps based on a claim by the firm offering the course that it raises math SAT scores by 50 points:

$$H_0: \mu = 500$$

$$H_1: \mu = 550$$

Typically, we gather data that we hope represent a random sample and then, based on these data, decide whether or not to reject the null hypothesis. In the example, we pick 25 students, have them take the cram course, and then calculate their average SAT score. Our *decision rule* (cutoff point) is to reject H_0 if the sample mean turns out to be larger than 533.

Type I and Type II Errors

Whichever hypothesis is correct, there is a chance that sampling error will lead us to a false conclusion. We may reject a null hypothesis that is true or we may not reject a null hypothesis that is false. These two types of mistakes are imaginatively called Type I and Type II errors.

A **Type I error** is the rejection of a null hypothesis that is true. A **Type II error** is a failure to reject a null hypothesis that is false.

The probabilities of committing these errors are labeled α and β:

$$\alpha = P[\text{Type I error}] = P[\text{reject } H_0 \mid H_0 \text{ is true}]$$
$$\beta = P[\text{Type II error}] = P[\text{do not reject } H_0 \mid H_0 \text{ is false}]$$

The actual probabilities depend, of course, on where the cutoff point is drawn for choosing between the two hypotheses. In the example, the 533 cutoff gives an α of .05 and a β of .20. A comparison of Figures 10.1 and 10.2 shows that raising the cutoff above 533 would reduce α, but raise β. Lowering the cutoff below 533 would reduce β, but raise α. This is the fundamental trade-off in hypothesis testing. In practice, researchers traditionally choose the cutoff so that α equals .05 and then let β be whatever it turns out to be.

If the observed data reject the null hypothesis, then the results are said to be **statistically significant** at the 5 percent level. What this cryptic phrase means is that there is less than a .05 probability of observing such a large difference between the sample mean and its expected value under the null hypothesis. An observed difference that is too large to be attributed plausibly to chance alone is statistically significant in the sense that its statistical improbability persuades us to reject the null hypothesis.

An Appropriate Sample Size

In the previous example, we had a sample of size 25, chose a cutoff point of 533 to set α at .05, and then computed β to be .20. Often the researcher has some freedom in choosing the sample size and can exercise that freedom to set α and β at prescribed levels.

By increasing the sample size, the researcher can reduce α, β, or both. The exact formula is shown in the footnote below,* but the idea is more important than the specific formula for its implementation. A larger sample gives more information about the population and thus provides a more reliable identification of correct and incorrect hypotheses. With only a few observations, we cannot make a very informed judgment; if we observe the entire population, we'll be certain of the answer. Because a large sample can be expensive, a reasonable strategy is to list the possible values of α and β for different sample sizes, and then make a choice.

In the math SAT example, with α fixed at .05, the sample sizes needed for a variety of values of β are

Desired β	Needed n
.20	25
.10	34
.05	43
.01	63
.001	89

EXERCISES

10.10 Look again at the Hand Scan Identimat example in Chapter 6. If the null hypothesis is that the person's identity is valid, explain clearly the meaning of the claim, "The probability of Type I error is 2 percent." Assume now that the test for Jimmy Bond is based solely on a measurement of the length of his index finger and that his measurements are normally distributed with a mean of 7.8 cm. with a standard deviation of .1 cm. If an imposter's measurements are distributed $N[8.2, .1]$ and the cutoff is 8.1 cm., what is the probability of Type II error?

10.11 Let's modify the math SAT example so that, because of budget limitations, only 9 students are

*The values of α and β are determined by

$$\overline{X}_c = \mu_0 + Z_\alpha \sigma / \sqrt{n}$$
$$\overline{X}_c = \mu_1 - Z_\beta \sigma / \sqrt{n}$$

where $\overline{X}_c$ is the cutoff point, μ_0 and μ_1 are the values of μ under the null and alternative hypotheses, and Z_α and Z_β are the Z values corresponding to the selected values of α and β. The subtraction of the second equation from the first gives

$$0 = (\mu_0 - \mu_1) + (Z_\alpha + Z_\beta)\sigma / \sqrt{n}$$

Solving for the sample size,

$$n = \left\{ \frac{(Z_\alpha + Z_\beta)\sigma}{(\mu_1 - \mu_0)} \right\}^2 \tag{10.2}$$

tested. The null and alternative hypotheses are

$$H_0: \mu = 500$$

$$H_1: \mu = 550$$

and the standard deviation is assumed to be 100. What cutoff sets the probability of Type I error at $\alpha = .05$? What is the probability of Type II error? What are the Z- and P-values if the sample mean turns out to be 525?

10.12 In the math SAT example, a cutoff of 533 for the average score of a random sample of 25 students implied a .05 probability of Type I error and a .20 probability of Type II error. For a given sample size, how can the probability of Type II error be diminished to .10? Is there any reason why a researcher might choose not to so reduce β? Alternatively, how might the sample be altered to reduce β to .10, while keeping α at .05?

10.13 A large corporation's secretaries have typing speeds that are normally distributed with a mean of 70 words per minute and a standard deviation of 15 words per minute. A manufacturer of word processors claims that its revolutionary keyboard can increase average typing speeds by 10 percent, with no

change in the standard deviation. The corporation agrees to let 25 randomly selected secretaries try these machines to test this claim. Set up a test of the competing hypotheses

$$H_0: \mu = 70$$

$$H_1: \mu = 77$$

based on the average typing speed $\overline{X}$ of these 25 secretaries. Find the cutoff point so that there is only a 5 percent chance of concluding that there has been a 10 percent increase in typing speed if there has, in fact, been none at all. If the word processing firm's claim is correct, what is the probability that your test will incorrectly reject their claim? Calculate the Z- and P-values if the data turn out as follows:

63	70	91	75	64	84	77	95	84	99
66	85	67	76	105	84	68	88	72	73
53	45	57	78	79					

10.14 *(continuation)* If 100 secretaries are tested, where should the cutoff be set to keep the probability of Type I error at 5 percent? Now what is the probability of Type II error?

10.4 COMPOSITE HYPOTHESES

We are seldom confronted with just two possible values for a population parameter; instead, there is usually a continuum. In our math SAT example, we constructed a test for choosing between two possibilities for the expected value: 500 or 550. But there is no good reason for ruling out an expected value of 549, 573, or 484 and, consequently, we need to restructure our test to allow for these many possibilities. The details depend on the precise nature of the hypotheses, but the principles are the same in each case.

A One-Sided Alternative Hypothesis

The firm may claim that its cram course significantly raises math SAT scores, without putting a precise number on the magnitude of the improvement. To test this claim, the null and alternative hypotheses are

$$H_0: \mu = 500$$

$$H_1: \mu > 500$$

The null hypothesis is that the course has no effect, and the alternative hypothesis is that it raises scores by an unspecified amount. A natural test of these competing hypotheses would be to see if the average score of 25 students who take this course is much above 500. The exact cutoff for choosing between the two hypotheses is determined by specifying the probability of Type I error. If we set $\alpha = .05$, then a sample mean of 533 (with a Z-value of 1.65) is again the cutoff. If the null hypothesis is true, and the course has no effect on student scores, there is only a 5 percent chance that we will pick 25 students whose average score is above 533. Thus, with a 533 cutoff, there is only a 5 percent chance that we will be persuaded that a useless course is useful.

The one change in our analytics is that, with many possible alternative values for μ, there are now many possible values for β (the probability of Type II error), depending on how much the course actually does raise scores. Figure 10.3 shows the β values for $\mu = 525$, 550, and 575. The Type II errors for all values of μ between 500 and 575 are shown in Figure 10.4. If the course has only a small effect on student scores (raising them by just 25 points, for example), then we are very likely to conclude erroneously that it has no effect at all. If the course has a very large effect on scores (raising them by 75 points, for example), then there is almost no chance that sampling error will hide this positive effect.

A Bayesian could calculate an overall probability of committing a Type II error as an average of these βs, weighted by the probabilities that the course raises scores by 25 points, 50 points, and so on. If the Bayesian thinks that the course probably has only a small effect on student performance, then the probability of Type II error will be high. If this Bayesian thinks that the course is likely to have a large effect, then the probability of Type II error is small. An objectivist avoids such calculations, because they require that subjective probabilities be placed on the population parameter μ.

Composite Null and Alternative Hypotheses

So far, we have ruled out the possibility that the firm's course worsens math SAT scores. To allow for a negative effect, we can let our competing hypotheses be

$$H_0: \mu \leq 500$$

$$H_1: \mu > 500$$

The null hypothesis is now that the course does not have a positive effect on math SAT scores: it either worsens them or leaves them unchanged. The alternative hypothesis is that the effect is positive. Again, it is natural to choose between these two hypotheses based on the scores of a random sample of students who take the course. If these students' average score is high enough, we will be persuaded to reject the null hypothesis of "no benefits."

The added complication here is that, because the null hypothesis includes many different values of μ, there are many different probabilities of Type I error.

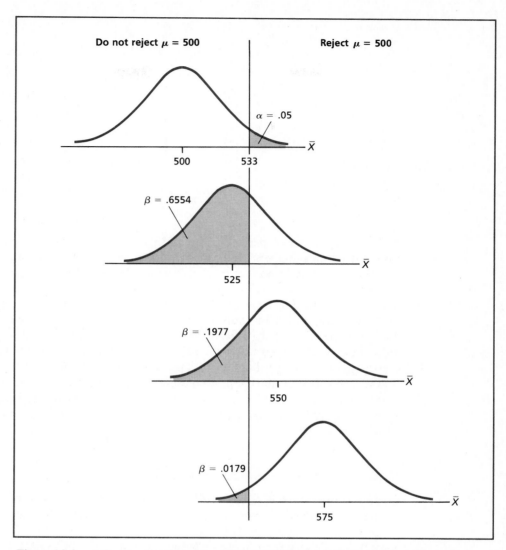

Figure 10.3 *A Composite Alternative Hypothesis Implies a Variety of Possible Type II Errors*

Suppose that we keep 533 as the cutoff point. Figure 10.5 shows that there are now many possible values of α and β. Again, a Bayesian could obtain an overall probability of Type 1 error by assigning subjective probabilities to the different possible values of $\mu \leq 500$. Objectivists will not make such a weighting; instead they look at a single α, to be conservative, the maximum value of α. This maximum α occurs at the borderline value of the mean, $\mu = 500$. In Figure 10.5 you can see that if μ is less than 500, then the probability distribution is shifted to the left and the prob-

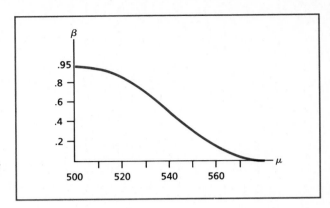

Figure 10.4   *A Graph of Type*
II Errors for μ Between 500
and 575

ability of Type I error declines. Thus, 533 is still the appropriate cutoff point, but now it is because it gives a maximum value of α equal to 5 percent.

We have now looked at three hypothesis-testing situations and all three have given the same decision rule: An average score of 533 or higher (a Z-value of at least 1.65, reflecting a sample mean that is 1.65 standard deviations above its expected value under the null hypothesis) indicates that the course does improve student scores. The only differences are in the implications of using this decision rule. If there are only two possible values of μ, then there is one probability of Type I error (α) and one probability of Type II error (β). If the alternative hypothesis allows a variety of values of μ, then there are a variety of βs. If the null hypothesis allows a variety of values of μ, then there are a variety of αs. In each case, the standard procedure for choosing the cutoff point is to set α (or the maximum value of α) equal to .05.

Two-Sided Composite Alternative Hypotheses

So far, we have looked at *one-sided* tests, in that the alternative hypothesis has been on just one side of the null hypothesis (above it). There has consequently been a single cutoff for choosing between the competing hypotheses. We reject the null hypothesis if the sample mean is above 533 and we don't reject it if the mean is below 533. Often, researchers use *two-sided* tests in which the alternative hypothesis is on both sides of the null hypothesis. The null hypothesis may be that a certain action has no effect on the population mean, while the alternative hypothesis is that the action has some effect, either positive or negative, but we don't know which.

In the cram course example, the competing hypotheses would be

$$H_0: \mu = 500$$

$$H_1: \mu \neq 500$$

The null hypothesis is that a two-week cram course has no effect on math SAT scores. These scores are determined by a student's intelligence and education, and

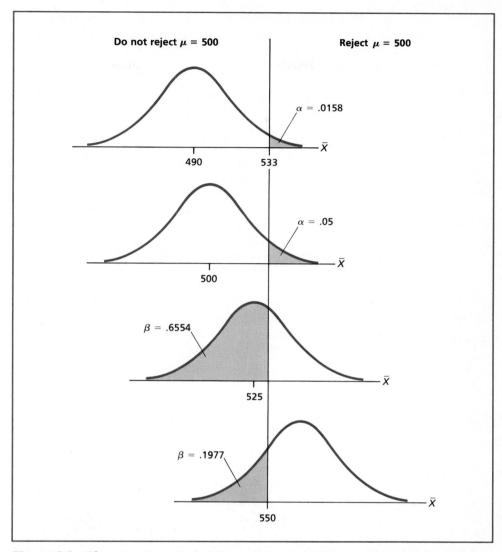

Figure 10.5 *There Are Many Probabilities of Type I and Type II Errors When There Are Composite Null and Alternative Hypotheses*

two weeks of studying isn't going to make any difference one way or the other. The alternative hypothesis is that the course does have some effect, either positive or negative. The course may stuff in a few useful facts and may give students some useful test-taking strategy, or the crash course may leave students so exhausted and irritable that they do poorly. We don't know which effect will dominate, so we leave them both open as alternatives.

Now what sort of scores would be decisive evidence against the null hypothesis of no effect? If the students score either much above 500 or much below 500, then we will be persuaded that the course does indeed affect their performance; a two-sided alternative hypothesis leads to a two-sided test.

The exact cutoff points are set by fixing the probabilities of Type I error. Remember that a Type I error occurs when we incorrectly reject the null hypothesis. Here, the luck of the draw might give us either 25 above-average students or 25 below-average students. In the former case, the above-average students do so well that we are persuaded that the course has positive effects. In the latter case, the below-average students do so poorly that we conclude the course has negative effects. We want to set our cutoff points sufficiently far from 500 to keep these error probabilities low. The standard practice is to allow for a 2.5 percent chance of incorrectly concluding that there are positive effects and a 2.5 percent chance of incorrectly concluding that there are negative effects, giving an overall 5 percent chance of committing a Type I error.

Figure 10.6 illustrates this logic. When you first construct hypothesis tests, it is a good idea to draw such a sketch to crystalize your own reasoning. Because the normal distribution is symmetrical, the cutoff points are equidistant from 500. How far? Our two-standard-deviations rule of thumb tells us that there is about a 5 percent chance that a normally distributed variable will be more than two standard deviations away from its mean. Here, if the null hypothesis is correct, individual scores are normally distributed with a mean of 500 and a standard deviation of 100. The sample mean of 25 scores is then normally distributed with a mean of 500 and a standard deviation of $100/\sqrt{25} = 20$. Thus, there is approximately a 5

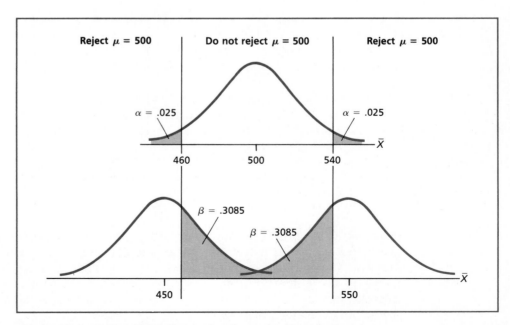

Figure 10.6   *A Two-Sided Composite Alternative Hypothesis*

percent chance that the sample mean will be more than 40 points above or below 500, and our cutoff points are consequently 540 and 460.*

Equivalently, if the sample mean turns out to be above 540 or below 460, the Z-value will be larger than 2 or less than -2, showing that the sample mean is more than two standard deviations away from its expected value under the null hypothesis. Since the probability of this large a disparity, on one side or the other, is less than .05, we reject the null hypothesis at the 5 percent significance level. For instance, if the sample mean turns out to be 560, the Z-value is

$$Z = \frac{(\overline{X} - \mu)}{(\sigma/\sqrt{n})}$$

$$= \frac{(560 - 500)}{(100/\sqrt{25})}$$

$$= 3$$

and the P-value is .0013, less than the .025 required for rejection with a two-tailed test. If the sample mean instead turns out to be 530, the Z-value is

$$Z = \frac{(530 - 500)}{(100/\sqrt{25})}$$

$$= 1.5$$

and the P-value of .0668 is not sufficiently small to reject the null hypothesis.

Figure 10.6 shows that the probabilities of Type II error are also two-sided. These probabilities are displayed in Figure 10.7. Again, we are most likely to err by concluding that the course has no effect if the effect is small. If the course has a large effect, either positive or negative, on student scores, then we are almost certain to see that effect reflected in their scores.

Let's briefly review the logic and implications of one-sided and two-sided tests. For a one-sided test at the 5 percent level, we see whether the observed sample mean is 1.65 standard deviations away from its expected value under the null hypothesis. The cutoff point in our example works out to be 533. For a two-sided test at the 5 percent level, we see whether the observed sample mean is two standard deviations above or below its expected value under the null hypothesis. This works out to be 40 points in either direction, either above 540 or below 460.

Notice that with a two-sided test, the cutoff scores must be higher than with a one-sided test to convince us that the course has a positive effect. This means that a two-sided test has the greater chance of making a Type II error by concluding that

*A more formal derivation would proceed as follows. Table 2 says that there is a .025 probability that a standardized normal variable Z is greater than 1.96. The sample mean is standardized by subtracting its mean $\mu = 500$ and dividing by its standard deviation $\sigma/\sqrt{n} = 20$. Therefore, $(\overline{X} - \mu)/(\sigma/\sqrt{n}) = 1.96$ implies $\overline{X} = \mu + 1.96(\sigma/\sqrt{n}) = 500 + 1.96(20) = 539.2$.

Process Control at the American Stove Company

Manufacturing processes are not unfailingly perfect. They are influenced by variations in temperature, humidity, materials, and human performance. These inevitable fluctuations cause variations in the size, shape, and performance of finished products. In addition, wear and tear on the equipment and operators may cause the process to deteriorate. Manufacturers would like to separate process deterioration from the inevitable variation, because process deterioration can usually be corrected but, if left uncorrected, can result in unacceptable product defects. Naturally, they would like to catch a process deterioration as soon as possible, but there are at least two constraints on their actions. First, it is often impractical to inspect every single item closely. Second, it doesn't make sense to stop the process for repairs every time a slight deviation is noticed. Most such deviations are normal, inevitable variations and do not indicate a breakdown in the manufacturing process.

As explained in Chapter 9, statistical **process control** is a reasonable procedure for balancing these competing concerns. A mean and standard deviation for some attribute of interest (size, weight, or performance) can be determined from historical records or by the company's target objectives (its "specifications"). The standard deviation gauges the normal variation that can be expected when the manufacturing process is operating properly. A periodic sample of manufactured items can then be compared to this normal range of variation. If the observed deviation between a sample and the historical mean is too large to be explained by chance variation, then the process is apparently deteriorating and requires some corrective action.

To reinterpret process control as a periodic hypothesis test, we'll look at a procedure used by the American Stove Company.[6] This firm produced a certain metal piece whose height, under normal manufacturing conditions, was approximately normally distributed with a mean of 0.8312 inch and a standard deviation of 0.006169 inch. If the

a beneficial course has no effect. You can see these higher βs by comparing Figures 10.4 and 10.7 for $\mu = 550$. A two-sided test has to be more conservative because it has to worry about erring in two directions. If we can rule out one of these directions ahead of time, then we should do so. We can reduce the probability of making a Type II error if we can legitimately justify using a one-sided test rather than a two-sided test.

Two Quick and Easy Extensions

In the math SAT example, I assumed that student scores were known to be normally distributed with a standard deviation of 100. What if the random sample is not drawn from a normal distribution? It makes little difference, as long as the sample is large enough to invoke the central limit theorem. With a sufficiently large manufacturing process deteriorated, these items might become so much longer or shorter than usual that they would be unusable.

To guard against this, the American Stove Company carefully measured a random sample of five items from each day's production to see whether or not the mean height of this sample was within three standard deviations of its expected value, assuming no deterioration in the process. If the observed disparity was greater than three standard deviations, then they concluded that something more than chance variation was at work—production was stopped and corrective action taken.

Implicitly, they were making a two-tailed test of the null hypothesis that the expected height is 0.8312 inch.

$$H_0: \mu = 0.8312$$
$$H_1: \mu \neq 0.8312$$

If the standard deviation of an individual item is 0.006169, then the standard deviation of the mean of a random sample of 5 items is $0.006169/\sqrt{5} = 0.00276$. Their decision rule was to not reject H_0 as long as the sample mean was within three standard deviations of its expected value under the null hypothesis:

$$0.8312 \pm 3(0.00276)$$

that is, if

$$0.8229 < \overline{X} < 0.8395$$

The probability of Type I error is the chance that a sample mean will be outside this range even though the null hypothesis is correct. Table 2 at the end of this book tells us that there is a .0013 chance that a normally distributed variable will be more than three standard deviations above its expected value and a similar chance of being three standard deviations below its expected value. Thus, the probability of Type I error is .0013 + .0013 = .0026.

sample, the central limit theorem implies that the sample mean will be normally distributed and we can consequently construct our test just as we have done before.

What if we don't know the standard deviation? Then we can estimate it. With a large sample, we can assume that our estimate of the standard deviation is correct and proceed as shown earlier. With a small sample, though, the estimate may be off by a bit. If the underlying distribution is approximately normal, then the extra uncertainty can be taken into account by using the t distribution rather than the normal distribution to calculate the cutoff points. The effect will be to make the cutoff points a bit farther from the expected value under the null hypothesis. To keep the probability of incorrectly rejecting the null hypothesis at 5 percent while acknowledging that we may be off in our estimate of the standard deviation, we require the sample mean to be even farther removed from what would be expected if the null hypothesis were true.

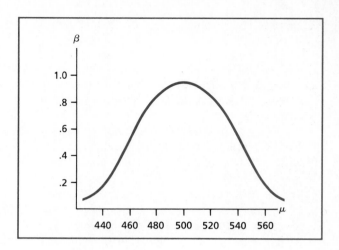

Figure 10.7 *Type II Errors for a Two-Sided Composite Alternative Hypothesis*

Instead of a Z-value, we calculate the value of the t-statistic:

$$t\text{-statistic} = \frac{\text{observed sample mean} - \text{expected value under null hypothesis}}{\text{estimated standard deviation of the sample mean}} \quad (10.3)$$

$$= \frac{(\overline{X} - \mu)}{s/\sqrt{n}}$$

Notice that the only difference between the Z-statistic and t-statistic is that the former uses the known value of the standard deviation while the latter uses an estimate based on the sample at hand.

For instance, the American Stove Company based its process control test on a daily sample of just five items. If they had estimated the standard deviation from each day's sample, instead of using the historical value of 0.006169, this would have provided an extra reason for error, and hence, for caution. Table 4 at the back of this book shows the t values for $n - 1 = 4$ degrees of freedom to be 2.776 for $\alpha = .05$ and 4.604 for $\alpha = .01$. Thus, a test at the 5 percent significance level requires a t-value of 2.776, rather than the Z-value of 2 used when the standard deviation is assumed to be known ahead of time. A test with a 1 percent chance of Type I error requires a t-value of 4.604, nearly a five-standard-deviations rule.

EXERCISES

10.15 Proline fishing line has a breaking strength that is normally distributed with a mean $\mu = 30$ pounds and a standard deviation $\sigma = 3$ pounds. The engineers have developed a new, less expensive manufacturing process, but are worried that the mean breaking strength may be adversely affected. The competing hypotheses are

$$H_0: \mu = 30$$

$$H_1: \mu < 30$$

They assume that the breaking strength is unchanged, unless there is convincing evidence to the contrary. A sample of 100 of these new lines will be tested and their average breaking strength calculated. Assuming that the standard deviation is unchanged, find the cutoff point for the sample mean that gives a 5 percent probability of incorrectly rejecting the null hypothesis. The sample mean turns out to be 28.4 pounds. What are the Z- and P-values? Can this be explained by sampling error or is it compelling evidence that the new lines are weaker?

10.16 Exercise 1.11 gives data on the annual returns for 1984, 1985, and 1986 for 24 randomly selected mutual funds. The returns for the S&P 500 index were -4.8 percent in 1984, 30.9 percent in 1985, and 35.8 percent in 1986. For each of these three years, calculate the mean and standard deviation for the sample of 24 mutual funds and make a t-test at the 5 percent level of the null hypothesis that these returns are drawn from a normal distribution with an expected value equal to the return on the S&P index that year.

10.17 A tobacco company claims that its new brand of cigarette, Life, contains less than 1/2 mg of nicotine. Tests of 16 Life cigarettes yield the data shown below. Make a t-test at the 5 percent level of the company's claim.

.65   .45   .75   .80   .75   .60   .80   .60
.65   .40   .70   .55   .30   .50   .70   .75

10.18 A researcher examined the admissions to a mental health clinic's emergency room when the moon was full.[7] For the 12 full moons from August 1971 to July 1972, the number of people admitted was

5   13   14   12   6   9   13   16   25
13   14   20

Using a 5 percent probability of Type I error, test the null hypothesis that these are a random sample from a normal distribution with a mean equal to 11.2, the average number of admissions on other days.

• 10.19 When there are several possible values for μ, there are several βs. A Bayesian could calculate an overall probability of Type II error by putting subjective probabilities on the possible values for μ. Consider the math SAT cram course and assume, for the sake of simplicity, that if the null hypothesis is false, then there are only three possible values for μ:

$$P[\mu = 525 \mid \mu \neq 500] = .50$$

$$P[\mu = 550 \mid \mu \neq 500] = .30$$

$$P[\mu = 575 \mid \mu \neq 500] = .20$$

The overall probability of Type II error is then

$$\beta = P[\text{accept } \mu = 500 \mid \mu \neq 500]$$
$$= P[\text{accept } \mu = 500 \mid \mu = 525]P[\mu = 525 \mid \mu \neq 500]$$
$$\quad + P[\text{accept } \mu = 500 \mid \mu = 550]P[\mu = 550 \mid \mu \neq 500]$$
$$\quad + P[\text{accept } \mu = 500 \mid \mu = 575]P[\mu = 575 \mid \mu \neq 500]$$

If a sample of 25 is used with a cutoff of 533, find the probabilities of accepting $\mu = 500$ if, in fact, $\mu = 525$, 550, and 575, and then find the overall probability of Type II error.

10.5   BINOMIAL TESTS

Our tests also can be easily extended to the binomial distribution. R.A. Fisher, who was one of the pioneers in developing hypothesis tests, used the example of an English lady who claims that, when having tea with milk, she can usually tell whether the tea or the milk was poured into her cup first. To test this remarkable claim, 8 cups of tea are prepared. For each cup, a coin is flipped to determine whether to pour in the tea or the milk first. These cups are brought to the woman, who takes a sip from each and correctly identifies 6 of the 8 cups!

Is that performance good enough to satisfy the statistician in the crowd? Or might the statistician reason that, by chance alone, a person will frequently get 6 out of 8 correct? This is a hypothesis-testing situation, in which the binomial distribution is appropriate rather than the normal. We can let π be the probability that this woman will correctly identify whether the tea or the milk was poured first. The null and alternative hypotheses can be

$$H_0: \pi = .5$$

$$H_1: \pi > .5$$

The null hypothesis is that this woman has only a 50 percent chance of being correct, the same as if she simply guesses without tasting. The alternative hypothesis is her claim that she is right more often than wrong.

To be considered statistically significant, how good does her performance have to be? The relevant binomial probabilities are given in Table 1 at the back of this book using the null hypothesis assumption that $\pi = .5$. If she were merely guessing, there would be a .0039 probability of correctly identifying 8 out of 8 cups, a .0312 probability for 7 out of 8, a .1094 probability for 6 out of 8, and so on. We want to set our cutoff point so that there is less than a 5 percent chance of making a Type I error by incorrectly rejecting the null hypothesis that she is just guessing. The appropriate cutoff is thus 7 out of 8, because there is only a .0039 + .0312 = .0351 probability that a guesser will get at least 7 out of 8 right.* Because this lady only got 6 out of 8 right, we do not reject the null hypothesis that she is an uninformed guesser. Another way of looking at her record is that a pure guesser has a .0039 + .312 + .1094 = .1445 probability of getting at least 6 out of 8 right. Therefore, her performance is not remarkable enough to rule out lucky guesses as the explanation.

Acceptance Sampling at Dow Chemical

Chapter 5 included a discussion of *acceptance sampling,* which is the use of a sample to determine whether or not to accept a box of oranges, a batch of batteries, or a shipment of bombs. This kind of decision can be recast as a hypothesis test. The null hypothesis is that the population meets some minimal standards. The quality of a random sample is then tested and the null hypothesis is either accepted or rejected. The cutoff between acceptance and rejection can be determined by setting the probability of Type I error at, say, 5 percent.

An example is provided by Dow Chemical's auditing of invoices for bulk mailing and duplicating items.[8] Their null hypothesis was that no more than one-fourth

---

*For an exact 5 percent probability of making a Type I error, we can use a randomized hypothesis test in which a random-number generator is used to determine whether 6 out of 8 will be accepted as statistically significant. The details are left as Exercise 10.48. Frankly, such a procedure has an unscientific aura and is usually more trouble than it's worth.

Does the Dropped Bread Fall Jelly Side Down?

Murphy's law states that, "If anything can go wrong, it will." Most people can remember lots of personal examples of Murphy's law in action. One researcher was apparently so impressed (or devastated) by the law that he conducted a (tongue-in-cheek) laboratory experiment.[9] His vehicle was the familiar lament, "If I'm putting jelly on a slice of bread and drop it, why does it usually land jelly side down?" One of this researcher's experiments involved 1,712 slices of bread smeared with peanut butter and a variety of jellies. Each was dropped and the results dutifully recorded:

jelly-side-up: 206
jelly-side-down: 1,506

The null hypothesis is that the bread has an equal chance of landing jelly side up or down. Murphy's law suggests a preponderance of jelly-sides-down:

$$H_0: \pi = .5$$
$$H_1: \pi > .5$$

where π is the probability of landing jelly side down. The outcomes can be described by a binomial distribution. To use a normal approximation, we need only remember that the sample proportion p has an expected value π and a standard deviation $\sqrt{\pi(1 - \pi)/n}$. Under the null hypothesis, the expected value is .5 and the standard deviation is $\sqrt{.5(.5)/1,712} = .012$. The observed proportion 1,506/1,712, = .88 gives a Z-value of $Z = (.88 - .5)/.012 = 31.4$, showing decisively that dropped bread is not equally likely to land jelly side up or down. This researcher concluded that Murphy's law is 88 percent correct. An improved formulation is, "If something can go wrong, it probably will."

of one percent of the weekly entries on these invoices contained clerical errors:

$$H_0: \pi \leq .0025$$

$$H_1: \pi > .0025$$

where π is the fraction of the week's entries that are incorrect.

They decided to examine a random sample of 225 entries each week. This sample is not really large enough to use a normal approximation to the binomial distribution, because the expected number of errors, $\pi n = (.0025)225 = .5625$, is not larger than 5, the rule of thumb given in Chapter 5. If the null hypothesis is true, then the binomial probabilities are

$$P[\text{no errors}] = (.9975)^{225} = .569$$

Are Mendel's Data Too Good to Be True?

Gregor Mendel's probabilistic model of how genes pass traits on from one generation to the next is one of the most important and elegant scientific theories. Mendel reported many experimental results confirming his theory, and many subsequent supporting experiments have been conducted by others.

One of Mendel's original experiments involved the self-fertilization of hybrid yellow-seeded sweet peas. His theory said that the offspring have a .75 probability of being yellow-seeded and a .25 probability of being green-seeded. Mendel reported the following results:

Offspring	Number
Yellow-Seeded	6,021
Green-Seeded	2,002
Total	8,023

Do these numbers support or reject Mendel's theory?

We can use the binomial model with π the probability of a yellow-seeded offspring. The competing hypotheses are

$$H_0: \pi = .75$$
$$H_1: \pi \neq .75$$

The null hypothesis is that Mendel's theory is correct. The two-sided alternative hypothesis is that Mendel is wrong. The sample, $n = 8{,}023$, is sufficiently large so that the sample proportion p should be normally distributed. Under the null hypothesis,

$$p \sim N[\pi, \sqrt{\pi(1 - \pi)/n}]$$
$$\sim N[.75, \sqrt{(.75)(.25)/8023}]$$
$$\sim N[.75, .005]$$

There is about a .95 probability that a normally distributed variable will be within two standard deviations of its expected value, here $.75 \pm 2(.005) = .75 \pm .01$. Thus, allowing the usual 5 percent chance of Type I error, the

$$P[\text{one error}] = 225\,(.0025)(.9975)^{224} = .321$$

$$P[\text{two errors}] = \left(\frac{(225)(224)}{2}\right)(.0025)^2(.9975)^{223} = .090$$

$P[\text{more than two errors}]$

$$= 1 - P[\text{no errors}] - P[\text{one error}] - P[\text{two errors}] = .02$$

If Dow rejected the week's invoices based on the discovery of more than two errors in their sample, there would have been a .02 probability of Type I error. If, instead, they refused to accept invoices if more than one error was found, then their probability of Type I error would have been $.02 + .09 = .11$. The two-error cutoff is closer to $\alpha = .05$, and that is just what Dow did.

cutoffs for rejecting Mendel's theory are an observed value of p below .74 or above .76.

What do his data show? Mendel's reported value of p is 6,021/8,023 = .7505, which is almost exactly as predicted. Or, put somewhat differently, 75 percent of 8,023 is 6,017.25. Mendel's reported number, 6,021, is off by less than 4 plants in an experiment involving over 8,000 plants! Mendel's data support his theory strikingly, perhaps too strikingly. Some 70 years later, the eminent statistician, R.A. Fisher, argued that Mendel's data are literally too good to be true.

Fisher asked the reverse question: Assuming Mendel's theory is true, what is the probability of observing a value of p so close to .75? Mendel's reported .7505 is 1/10 of a standard deviation away from its expected value. Table 2 tells us that there is only about an 8 percent chance that a normally distributed variable will be within 0.1 standard deviation (in either direction) of its mean. Such a close correspondence between theory and data takes a bit of luck or something less innocent.

In fact, Mendel studied seven inherited characteristics of pea plants and, in every single case, there was this close a correspondence between his results and his predictions. The probability of such a phenomenal run of experimental luck is about .000004. It is very hard to believe that Mendel was that lucky. Even worse, in one of his studies, he miscalculated the theoretical probability and then reported data that fit this erroneous probability almost exactly and reject the correct probability. It is hard to disagree with Fisher's conclusion that Mendel's data are too good to be true. Fisher was, however, so impressed by Mendel's scientific contributions and by the fact that Mendel was a monk that he placed the blame elsewhere.

I have no doubt that Mendel was deceived by a gardening assistant, who knew only too well what his principal expected from each trial made.[10]

Racial Awareness in Children

The binomial test of the exceptional tea taster used a very small sample of just eight observations. If the sample were larger, we could use a normal approximation to the binomial distribution. As an illustration, consider a 1958 study of the racial awareness of black children.[11] The sample consisted of 253 black children, 134 from Arkansas and 119 from Massachusetts. The children's ages ranged from 3 to 7 years. Each child was offered four dolls, two white and two black, and told to, "Give me the doll that you would like to play with." It turned out that 83 children chose a black doll and 169 chose a white doll. (One child was apparently not interested in playing with dolls and refused to choose one.)

Overall, about two-thirds of these black children (169/252 = .671) preferred to play with a white doll. Can this result reasonably be explained by chance, either in the random selection of the children or in their haphazard choice of a doll? Or does it suggest that young black children are well aware of race and, indeed, have a negative image of their race and of themselves? Let π be the probability that a randomly selected black child will choose a white doll. The competing hypotheses are

$$H_0\colon \pi = .5$$

$$H_1\colon \pi \neq .5$$

The null hypothesis is that the color of the doll's skin is unimportant to black children as a whole; for every black child who would pick a white doll, there's another who would pick a black doll. The alternative, two-sided hypothesis is that the color of a doll's skin does matter to black children. (The test must be two-sided, because we cannot rule out a priori the possibility that black children prefer black dolls.)

The binomial distribution can be applied here, and the sample is large enough to allow a normal approximation to the binomial distribution. Under the null hypothesis, the sample proportion p preferring a white doll has a normal distribution with a .5 expected value and 0.0315 standard deviation:

$$p \sim N[\pi, \sqrt{\pi(1-\pi)/n}]$$
$$\sim N[.5, \sqrt{(.5)(.5)/252}]$$
$$\sim N[.5, 0.0315]$$

For a normal approximation to the binomial distribution, the Z-statistic is

$$Z\text{-statistic} = \frac{\text{observed sample mean} - \text{expected value under null hypothesis}}{\text{standard deviation of the sample mean}}$$

$$= \frac{\text{sample proportion} - \text{population probability}}{\text{standard deviation of proportion}}$$

$$= \frac{p - \pi}{\sqrt{\pi(1-\pi)/n}} \tag{10.4}$$

Equivalently, we get the same Z-value if we work with the number of successes, rather than the success proportion:

$$Z\text{-statistic} = \frac{\text{observed successes} - \text{expected number of successes under } H_0 0}{\text{standard deviation of sample successes}}$$

$$= \frac{x - \pi n}{\sqrt{\pi(1-\pi)n}} \tag{10.5}$$

Here, using the sample proportion, the Z-value is

$$Z = (.671 - .5)/.0315 = 5.43$$

(i.e., the observed .671 proportion is more than five standard deviations from the expected value under the null hypothesis), decisively rejecting the null hypothesis that black children are color blind.

EXERCISES

10.20 A small firm claims that its patent medicine can cure half of the people suffering from a certain disease. The Food and Drug Administration administers this medicine to 10 people suffering from this disease and finds that 2 are cured. Do these data support the firm's claim using a 5 percent significance level?

10.21 Seymour Siwoff, Steve Hirdt, and Peter Hirdt compiled 1975–1984 batting averages during late-inning pressure situations—seventh inning or later with the batter's team tied or trailing by three runs or less (four if bases are loaded).[12] Here are the data for two reputed clutch hitters:

	Pressure Situations		Overall Batting Average
	At Bat	Hits	
Reggie Jackson	649	174	.263
Steve Garvey	904	282	.302

For each batter, make a one-tailed test (at the 5 percent significance level) of the null hypothesis that his performance in pressure situations is a Bernoulli trial with success probability equal to his overall batting average.

10.22 Karl Pearson reported that when he flipped a coin 24,000 times, he obtained 12,012 heads and 11,988 tails. Are these data consistent with the null hypothesis that the probability of a head is .5? Are Pearson's data too good to be true?

10.23 In what way is the following a misinterpretation of the meaning of Fisher's probability calculation?

Gregor Johann Mendel, whose experiments with garden peas first revealed the basic laws of heredity, was such a hero of modern science that scientists in the thirties were shocked to hear that this pious monk probably doctored his data. R.A. Fisher, a famous British statistician, checked Mendel's reports carefully. The odds, he concluded, are about 10,000 to 1 that Mendel gave an inaccurate account of his experiments.[13]

10.24 A home scientist claims that people are weakened when they wear watches. To test his claim, 10 volunteers have their strength measured with and without watches. If the variations in these measurements are due just to imperfections in measuring strength (and not at all to the wearing of watches), then what is the probability that 7 of the 10 volunteers will show a decrease in strength when they wear a watch? If we decide to accept this person's watch-weakness theory if at least 7 volunteers do show a decrease in strength, then what is the probability of Type I error? (Be sure to reveal your null and alternative hypotheses.)

10.25 In 1977, the U.S. Supreme Court *(Castaneda v. Partida)* observed that persons with Spanish surnames constituted 79 percent of the population of Hidalgo County, Texas, but only 339 of 870 recent jurors in this county. The Court calculated a Z-value of 29 and commented that "as a general rule for such large samples, if the difference between the expected value and the observed number is greater than two or three standard deviations, then the hypothesis that the jury drawing was random would be suspect to a social scientist." **a.** What parameter value constitutes the null hypothesis? **b.** Show how the Court calculated this Z-value. **c.** Explain the reference to "large samples." **d.** Explain the reference to "two or three standard deviations."

Do Some NFL Teams Consistently Beat the Point Spread?

In football wagers, points are commonly added to the underdog team's score. If, for instance, the Dallas Cowboys are favored by 2 1/2 points over the Washington Redskins, then 2 1/2 points are added to Washington's score. Those who pick Dallas win their bets if the Cowboys beat the Redskins by more than 2 points. Those who bet on Washington win if the Redskins win or lose by less than 2 points. The bookmaker makes a profit by only putting up $10 against every $11 wagered by the bettor.

A natural null hypothesis is that the bookmaker's point spreads are set so that each team has an equal chance of winning. There are, of course, a variety of systems that are designed to bankrupt the bookies, and two researchers tested some of these theories statistically over the period 1969–1974.[14]

One theory is that some teams consistently beat the point spread while others repeatedly fall short. Thus, the bettor need only find teams that have paid off in the past and then begin putting wagers on these teams. These researchers applied this theory by following teams that had winning records against the point spread in the preceding year. The competing hypotheses are

$$H_1: \pi = .5$$
$$H_1: \pi > .5$$

where π is the probability that one of these selected teams will beat the point spread. They found that this strategy led to 616 bets, of which 312 were won and 304 lost. The observed proportion $312/616 = .506$ is not significantly different from .5. Under the null hypothesis, the expected proportion is .5 with a standard deviation of $\sqrt{.5(.5)/616} = .020$, giving a Z-value of $(.506 - .500)/.020 = .3$. That is, the observed .506 is only a third of a standard deviation away from .5, and not nearly far enough to persuasively reject the null hypothesis. They also looked at a similar strategy of betting against teams that had losing records against the point spread the previous year. This led to 693 bets, of which 343 were won and 350 lost—not statistically significant and the wrong direction to boot! We can only hope that this research interest was purely academic.

10.6   CONFIDENCE INTERVALS

You have learned the standard procedure for testing hypotheses. Now I will show an alternative method, which is really just a different way of saying the same thing. Because the logic of hypothesis tests is a bit complicated, it is helpful to look at other ways of explaining what it is we're doing.

Let's say that we had set out to estimate the expected value μ of the math SAT scores of students who take this cram course. We would use the sample mean as our estimate and use a two-standard-deviations rule of thumb to compute a 95 percent confidence interval. If the sample mean is, as shown earlier, 520, then our

95 percent confidence interval is

$$520 \pm 2(20)$$

$$520 \pm 40$$

Because 500 is well within this confidence interval, we cannot rule out the possibility that the course has no effect on student performance.

There is, in fact, a logical equivalence between hypothesis tests and confidence intervals, in that a hypothesis will be rejected only if it lies outside a corresponding confidence interval! This equivalence is easily seen. A two-sided hypothesis test asks if the sample mean is within two standard deviations of the null hypothesis.* A confidence interval includes all values for μ within two standard deviations of the sample mean. If the sample mean is more than two standard deviations from the value specified by the null hypothesis, then the null hypothesis is outside the confidence interval. Therefore, hypothesis tests can be conducted simply by using the sample to construct a confidence interval.

Even more fundamentally, we can think of hypothesis tests as being just a reinterpretation of estimation. When we say that we are 95 percent certain that a confidence interval will include the true value of μ, we are also saying that, if μ were equal to one of the excluded values, there would be no more than a 5 percent chance that the sample mean would be so far from this value. The improbability of this disparity rules out these excluded values.

EXERCISES

10.26 At some colleges, students write senior essays that are graded twice, once by an advisor and once by an anonymous reader. It is conceivable that the advisor, who knows the student and may have helped with the essay, will grade more leniently than the reader: "Anonymity encourages objectivity, or meanness at least." A sample of 26 senior essays in economics written at Yale in 1976 yielded the following data:

Reader's Grade	Number
Higher than advisor's	5
Equal to advisor's	7
Lower than advisor's	14

Let π be the probability that a grade that is different from the advisor's will be greater than the advisor's grade. Use a 95 percent confidence interval to test the null hypothesis that $\pi = .5$ against the alternative $\pi \neq .5$.

10.27 Researchers separately asked 153 married men and their wives to state the highest school grade completed by the wives.[15] The difference, $X =$ husband's answer − wife's answer, is a measure of the extent to which men exaggerate or belittle their wife's educational accomplishments. Whelpton and Kiser found that the average value of X was 0.32 with a standard deviation of 1.1. Use a 95 percent confidence interval to make a two-sided test of the null hypothesis

*A slight adjustment is needed with a one-sided test. Because a one-sided test asks whether the sample mean is within 1.65 standard deviations of the null hypothesis, the corresponding confidence interval should be $\pm$ 1.65 standard deviations, that is, a 90 percent confidence interval. You will, of course, only look at the appropriate side of the confidence interval.

that the population mean of X is zero. Can the positive sample mean plausibly be explained by sampling error?

10.28 The chemists at a major soft-drink company have developed a new, less expensive formula for their cola. These chemists claim that the new cola tastes better, too. However, the company's executives fear that their customers won't like this new concoction. A random sample of 400 people are given unlabeled glasses of the old and new colas: 170 prefer the new cola, while 230 prefer the old cola. Use both a Z-value and a confidence interval to see whether this difference can be explained plausibly by sampling error. Be precise in stating your statistical conclusion. What are the sources of sampling error in this problem?

10.29 In 1973, a federal judge ruled that statistics proved a prima facie case of discrimination in the Montgomery County, Alabama jury rolls. The Montgomery Jury Commission was ordered to reconstitute the jury rolls to include a cross-section of the community. The judge noted that blacks made up 30 percent of the jury-age population, but only 12 percent of the jury rolls, and that women made up 54 percent of the jury-age population, but only 16 percent of the jury rolls. If there are 100 people on a jury roll, what are the probabilities that random selection would yield **a.** no more than 12 percent blacks? **b.** no more than 16 percent females? Do these data support or reject the null hypotheses that jury rolls are chosen without regard to race or sex?

10.30 Courts have held that there is prima facie evidence of discrimination (under Title VII) if there is a statistically significant difference between the percentage of a firm's employees who are black and the percentage of people in the area who are black. Assume that 20 percent of the people in the area are black and that every firm has exactly 100 employees. What would be the cutoff for showing a statistically significant shortfall in the number of black employees? (Use a one-tailed test with a 5 percent probability of Type I error.) If a firm's hiring decisions are, in fact, color-blind, what are its chances of being found guilty of race discrimination by this criteria? If all firms are color-blind, what fraction of them can expect to be found guilty of race discrimination?

10.7   SUMMARY

A hypothesis test asks the question, "If the null hypothesis is true, what is the probability that a random sample will yield a statistic whose value is so far from its expected value?" If this probability is very small, then we reject the null hypothesis by concluding that the discrepancy is too large to be explained by chance alone. The observed difference is then said to be statistically significant. If, on the other hand, the probability is not small, then we do not reject the null hypothesis, because the observed discrepancy may well be due to chance, that is, sampling error. In this case, the observed data are said to be statistically consistent with the null hypothesis.

There are a variety of ways to conduct statistical tests about the population mean μ for a random variable X. Each approach uses the fact that the mean of a random sample is normally distributed if X is normally distributed or (according to the central limit theorem) if the sample is large. The expected value of the sample mean is μ and the standard deviation is $\sigma/\sqrt{n}$, where σ is the standard deviation of X and n is the size of the random sample. The standard deviation of the sample mean reflects sampling error—the fact that chance affects the mean of a randomly selected sample. This sampling error is why we cannot be 100 percent confident about the implications of statistical tests. Naturally, the smaller the random sample, the larger the margin for sampling error must be.

The usual testing procedure begins by specifying the null and alternative hypotheses. Second, we figure out which values of the sample mean have a small chance of occurring if the null hypothesis is true. For a two-tailed test at the 5 percent level, these are values that are approximately two standard deviations from the sample mean, that is, $2(\sigma/\sqrt{n})$, away from the expected value under the null hypothesis. For a one-tailed test at the 5 percent level, 1.645 standard deviations are used. Then, see whether or not the sample mean falls in the unlikely region.

The Z-statistic measures how many standard deviations the sample mean is from its expected value under the null hypothesis,

$$Z\text{-statistic} = \frac{\begin{array}{cc}\text{observed} & \text{expected value} \\ \text{sample mean} & - \quad \text{under null hypothesis}\end{array}}{\text{standard deviation of the sample mean}}$$

$$= \frac{(\overline{X} - \mu)}{\sigma/\sqrt{n}}$$

For a two-tailed test, there is about a 5 percent chance of being two standard deviations away from the expected value, and thus of having a Z-value greater than 2. For a one-tailed test, there is about a 5 percent chance of being 1.645 standard deviations to one side of the expected value. Z-values greater than these numbers cast doubt on the null hypothesis. For any particular observed Z-value, the normal table shows the probability of being so far from the expected value. This probability is called a P-value.

If the standard deviation must be estimated from the sample, then the t-statistic can be used if the sample is drawn from a normal distribution:

$$t = \frac{(\overline{X} - \mu)}{s/\sqrt{n}}$$

The P-value is then taken from Student's t distribution.

With Bernoulli trials, the P-values can be calculated directly from the binomial distribution or, if the sample is large, we can use a normal approximation and calculate the value of the Z-statistic using either the sample proportion p:

$$Z = \frac{p - \pi}{\sqrt{\pi(1 - \pi)/n}}$$

or the number of sample successes x:

$$Z = \frac{x - \pi n}{\sqrt{\pi(1 - \pi)n}}$$

Yet another approach is to use the sample mean and standard deviation to construct a confidence interval. A null hypothesis will be rejected at the 5 percent level if and only if it lies outside a corresponding 95 percent confidence interval. The reasoning is simple: If the sample mean is more than two standard deviations from its expected value under the null hypothesis, then it has a Z-value greater than 2, a P-value less than .05, and a 95 percent confidence interval that excludes the null hypothesis.

REVIEW EXERCISES

10.31 In a large high school in New York City, 60 randomly selected students tried a new math textbook, while the rest of the students used the old text. On an achievement test given at the end of the year, the students using the old book averaged 74, with a standard deviation of 9, while the students using the new book averaged 79, with a standard deviation of 9. Using a 5 percent significance level and assuming scores to be normally distributed with a mean of 74 and a standard deviation of 9, can the observed performance of the students using the new book be plausibly attributed to sampling error? Why might there be sampling error here? Is the normal distribution assumption important?

10.32 Critically evaluate this statistical argument:

Left-handers and right-handers continue to debate about who hits a baseball better, but the answer may lie in the eye of the beholder.

A University of North Carolina researcher gives the edge to left-handers. In a *New England Journal of Medicine* article, Arthur Padilla explains that the right eye is dominant in most right-handers. But most natural right-handers who bat from the right side are forced to watch the ball coming toward the plate with their left eye.

"But the pattern does not hold as frequently with left-handers," Padilla said, "because many who hit from the left side also have a dominant right eye, the one closest to the pitcher." He said only a small percentage of right-handers have a dominant left eye.

However, the ultimate authority—statistics—do not support Padilla's theory. Among the 10 leading hitters in the major leagues as of Wednesday, five bat right-handed, two left-handed, and three are switch hitters.

"Morning Briefing," *Los Angeles Times,* April 26, 1984

10.33 Consumer Inspects magazine has decided to test the claim of a peanut butter maker that its jars contain an average net weight of 18 ounces. From previous tests, it is known that the standard deviation of peanut butter weights is 0.2 ounce. If Consumer Inspects weighs the contents of 100 jars, what average weight should they use for a cutoff if they want there to be no more than a .01 probability of risking a lawsuit by falsely rejecting the company's claim?

10.34 The stopping distance (at 55 mph) for a company's premium tire is 54 meters with a standard deviation of 10 meters. Two consultants conduct independent tests of a new tread design. The first consultant tests 25 tires and gets an average stopping distance of 48 meters. The second consultant tests 100 tires and gets an average stopping distance of 50 meters. Which test result provides the more convincing statistical evidence that the new tread design does reduce the average stopping distance? Explain your reasoning.

10.35 The average weight of a certain breed of cattle is 1,800 pounds, with a standard deviation of 125 pounds. A new feed is tried on 25 cattle and it gives an average weight of 1,750. Is this difference statistically significant at the 5 percent level? Use a two-sided test and give the Z- and P-values. Give a 95 percent confidence interval for the expected weight of cattle using this new feed.

10.36 Ralph is an avid bowler. He has an expected score of 180 with a standard deviation of 40. Assume that his scores are normally distributed and independent. What are the probabilities that **a.** his score will be above 200? **b.** his score will be below 100? **c.** his average score for three games will be above 200? **d.** his average score for five games will be above 200?

10.37 *(continuation)* Ralph moves to a new city and wants to join a first-place bowling team, the Boomers, that happens to have an opening. Ralph tells the Boomers that he averages 195, and they ask him to bowl five games to verify this claim. The statistician on the Boomers views the situation as follows:

$$H_0: \mu = 195$$
$$H_1: \mu < 195$$

What cutoff five-game average should the Boomers' statistician set so that there will be only a 5 percent chance of incorrectly rejecting Ralph's claim? (Assume that the standard deviation for bowlers of this caliber is known to be 40.) What are Ralph's chances of successfully passing himself off as a 195-average bowler?

10.38 Gusto beer claims that its cans contain 12 ounces of beer. However, a chemistry fraternity carefully weighed the contents of 144 Gusto beer cans and found an average net weight of only 11.9 ounces, with a standard deviation of 0.9 ounce. Construct a 95 percent confidence interval for the net weight of Gusto beer cans. Does this fraternity's study refute Gusto's claim? (Use a two-sided test with a 5 percent significance level.)

10.39 A carnival wheel of fortune is divided into 10 sections, numbered 1, 2, . . . , 10. A prominent local citizen has complained to the sheriff that this wheel is rigged somehow. The sheriff suggests that the citizen may have been unlucky, but the citizen protests that too much money was lost for it to be luck alone. The sheriff enlists the help of a statistician from a nearby college. The statistician has several students take turns observing the operations of this wheel, and they report the following data:

Number of bets placed: 1542
Number of bets won: 91

What are the Z-value and P-values for the null hypothesis that there is a 10 percent chance of winning? Do the data reject the citizen's theory or the sheriff's theory?

• 10.40 An Ace delivery truck has a maximum capacity of one ton, in that if the net weight exceeds this amount, there is likely to be severe engine damage. This truck is to deliver cartons whose weights average 40 pounds, with a standard deviation of 4 pounds. What is the maximum number of cartons that can be loaded if the probability of exceeding one ton is to be kept below 1 percent?

10.41 The assistant manager of the Yale Cafe at Commons suspected that the supplier of hamburger patties was delivering hamburgers that weighed less, on average, than the advertised 16 ounces. A random sample of 48 patties was weighed very carefully on a balance scale and found to have an average weight of 15.968 ounces and a variance of 0.028. Do these data reject (at the 1 percent level) the null hypothesis that the population mean is 16 ounces? After the test was completed, the assistant manager learned that the balance scale has a negative bias and remembered that he had not taken into account the weight of the paper on each patty. Together, these errors meant that 0.010 ounces should be added to each of his recorded weights. How does this adjustment affect the results?

10.42 In the 1960s, only males who were 21 years old and older were eligible for jury duty in Alabama. About 26 percent of the 21-year-old males in Talladega County, Alabama at that time were black. The murder conviction of a black man in Talladega County was appealed to the Supreme Court on the grounds that there were no blacks on the jury, and indeed no black men had served on juries in Talladega County "within the memory of persons now living." The U.S. Supreme Court (*Swain* v. *Alabama,* 1965) denied the appeal, noting that there had been 8 blacks on the 100-person panel from which the final jury was selected. (The prosecution had ruled out these 8 blacks with constitutionally protected peremptory challenges.) What is the probability that a randomly selected panel of 100 citizens will have as few as 8 blacks? Do these data support or refute the hypothesis that the panel selection is color-blind?

10.43 An upcoming election was considered to be a toss-up until one of the candidates admitted that she didn't like apple pie. A new poll of 100 randomly selected voters found 46 percent favoring this candidate and 54 percent favoring the opponent. Are these data sufficient to reject the null hypothesis at the 5 percent level that the voters are evenly divided between the two candidates, or can it reasonably be attributed to sampling error? What is the nature of the sampling error here?

10.44 Consumer Retorts magazine tested the reliability of a new electronic table wiper. Three of these machines were used for three months. Consumer Retorts certifies an appliance as trouble-free if there are no problems with any of the three machines.

What is the probability that these electronic table wipers will be certified as trouble-free if a machine's probability of breaking down within three months is **a.** .10? **b.** .20? **c.** .50?

10.45 A certain professor gives true-false exams because they are so easy to grade. During his long career, this professor has accumulated nearly 8,000 true-false questions, which are conveniently stored on index cards. For each exam, the professor shuffles the cards and selects the exam questions. (The students also have accumulated a set of these questions over the years and pass them on from one class to the next.) The professor's intention is to pass students who have learned enough to answer 70 percent of these 8,000 questions correctly. If an exam consists of 100 questions, how many correct answers should this professor require for a passing grade in order to be 99 percent sure that a student who knows the answers to 70 percent of these 8,000 questions passes the course? What, then, is the probability of passing for a student who only knows the correct answers to 60 percent of these 8,000 questions? To 50 percent?

10.46 Two researchers looked at the performance of NFL teams that were big underdogs (given more than 15 points by bookmakers) over the period 1969–1974.[16] Including their point spreads, these underdogs won 32 of 50 games. Is this winning proportion statistically significantly different from .50? Use a two-tailed test with a 5 percent significance level.

10.47 Willard H. Longcor conducted some tests of inexpensive and precision-made dice.[17] Two million rolls were made with the precision-made dice, recording on each roll whether an even or odd number appeared. A new die was used after every 20,000 tosses, to guard against imperfections from the wear and tear of being rolled over and over. The same experiment was conducted with inexpensive dice, but Longcor stopped after 1,160,000 rolls:

Dice	Rolls	Fraction Even
Precision-made	2,000,000	.50045
Inexpensive	1,160,000	.50725

For each type of dice, test the null hypothesis at the 1 percent level that even and odd numbers are equally likely to be rolled.

• 10.48 The text discusses R.A. Fisher's example of a lady who claims that she can usually tell whether

milk or tea is poured into her cup first. The probability of a Type I error is .035 when we decide to accept her claim only if she correctly identifies at least 7 out of 8 randomly prepared cups. We can make this probability of Type I error exactly 5 percent by using a randomized binomial test: If she identifies 6 out of 8 cups, there is a probability P that we will accept her claim and a probability $1 - P$ that we will reject her claim. (We will still always accept her claim if she gets 7 or 8 right.) How high should we set P to make the probability of Type I error 5 percent? How could we implement this randomized binomial test?

•• 10.49 Eight cups are used to test a lady's claim that she can usually tell whether the tea or milk was poured first. If she correctly identifies 7 of the 8 cups, is this performance more convincing if she knows that **a.** with each cup, a coin is flipped to determine whether the tea or milk is poured first, or **b.** it is decided in advance to pour the tea first in 4 cups and the milk first in the other 4 cups?

•• 10.50 A chartist claims an ability to find patterns in stock prices that help tell whether a stock's price is going to go up or down. To test this claim, the chartist is shown graphs of stock prices for 10 unidentified companies over unidentified periods of time. For 5 of these companies, the stock's price rose sharply shortly after the end of the period shown in the chart. For the other 5 companies, the stock's price dropped sharply shortly after the period charted. The chartist is asked to identify which 5 stocks went up in price and which 5 went down. As it turns out, 3 of the 5 stocks identified as "winners" did go up in price, while 3 of the 5 put in the "losers" pile did go down. If the selections had been made randomly, what would be the chances of doing this well?

If we demand a performance that would have no more than a 5 percent chance of occurring by luck alone, how many must the chartist identify correctly to convince us?

10.51 A friend says that she can predict with remarkable accuracy which stocks will do better than average (above the median) in any given two-year period. Hearing of your statistical training, she asks you to test her ability. The null hypothesis is $\pi = .5$, where π is the probability of selecting a stock whose return will be above the median. If she makes predictions for 100 stocks, what is your cutoff for a one-tailed test at the 1 percent level? What is the probability that she will fail this test if she has a .6 probability of picking a stock whose return will be above the median? (In both calculations, assume that the outcomes are independent and use a normal approximation with a continuity correction.)

10.52 *(continuation)* Your friend says that the analytical techniques she uses can be applied to only a small number of stocks. Which of the following three sample results is more convincing evidence that she really can identify stocks whose returns will be above the median? **a.** Three successes in four selections. **b.** Seven successes in ten selections. **c.** Thirteen successes in twenty selections.

• **10.53** *(continuation)* How many predictions must she make, using a one-tailed test at the 1 percent level, for there to be only a 1 percent chance of not passing this test when her probability of picking winners is .6?

10.54 A builder has contracted for the delivery of 400,000 pounds of loam. The supplier proposes making delivery in 100 truckloads that he claims will average 4,000 pounds. The builder insists on weighing a random sample of 16 truckloads and obtains these data:

3820	3900	4000	3820	3780	3880
3700	3620	4380	3560	3840	3980
4040	4140	3780	3780		

Do these data reject the supplier's claim at the 5 percent level? (Assume normality and use a one-tailed t-test.)

10.55 In a 1979 court case, a military supplier argued that the U.S. government had tested too small a sample when it rejected a shipment of 20,000 nose fuse adaptors, a component of artillery shells.[18] Military procurement standards specified that a shipment of 10,001 to 35,000 adaptors would be tested by sampling 315 pieces, yet the government rejected this shipment of 20,000 pieces after testing 20 adapters and finding them all to be defective. What is the probability that every item in a random sample of 20 items from an infinite population will be defective if the probability of selecting a defective item is $\pi = .10$? $\pi = .50$? $\pi = .95$?

10.56 *(continuation)* Suppose that before inspecting the shipment, the government had the following prior probabilities $P[\pi = .10] = .98$, $P[\pi =$

.50] $= .01$, and $P[\pi = .95] = .01$. Calculate the revised probability that $\pi = .95$ after 20 of 20 tested items are found defective.

10.57 Explaining their reluctance to recommend approval of TPA, a genetically engineered drug intended to dissolve blood clots in the arteries of heart-attack victims, an FDA drug advisory committee cited the case of thalidomide, a drug widely prescribed as a tranquilizer in the late 1950s and early 1960s, until it was found to cause birth defects: "Early approval of a good drug, with minimal supporting data, looks wise in retrospect; early approval, with minimal supporting data, of a poor drug appears extremely unwise in retrospect."[19] If the null hypothesis is that a drug does not have the intended beneficial effects, what events would be described as Type I and Type II errors? If the null hypothesis is that a drug does not have harmful side effects, what events constitute Type I and Type II errors?

10.58 Exercise 9.58 reports that, of the 1,891 3-point shots attempted in the 1989 and 1990 NCAA basketball tournaments, 1,051 were taken by underdog teams. Treating these data as a random sample, test (at the 1 percent level) the null hypothesis that a randomly selected 3-point shot has a .5 probability of being taken by an underdog team. Use a normal approximation and calculate the Z-value and P-value.

10.59 The National Research Council, a private, nonprofit agency of the National Academy of Sciences, has been urged to increase its recommended dietary allowance (RDA) of vitamin C to as much as 1,000 milligrams, based on research indicating a variety of medical benefits. The current RDA of 60 milligrams of vitamin C for adult males is based on studies showing that the average daily utilization of vitamin C by healthy adult males is 21.5 milligrams with a standard deviation of 8 milligrams. If the daily utilization of vitamin C by adult males is normally distributed with a mean of 21.5 milligrams and a standard deviation of 8 milligrams, what is the probability that a member of this group utilizes more than 60 milligrams?

10.60 The mean systolic blood pressure of males in their 20s is 120 with a standard deviation of 9. Twenty-five randomly selected male soccer players in their 20s had an average blood pressure of 115. Is this 5-point difference statistically significant, using a .05

probability of Type I error? What, exactly, does statistically significant mean here?

10.61 IQ scores are normally distributed with a standard deviation of 16. Using a 5 percent chance of Type I error, find the cutoff points for using a random sample of 100 people to test the hypotheses

$$H_0: \mu = 100$$

$$H_1: \mu \neq 100$$

A random sample of 100 children who watch "Seesaw Street" regularly have an average IQ of 102.2. Is this sufficient evidence to show that the average IQ of children who watch this show is not 100?

10.62 During 1975–1979, there were 16,466,000 recorded births in the United States, of which 8,444,000 were male. Assume that these data are considered a random sample from a population that includes births from other times and/or places. Calculate the Z-value and P-value for a test of the null hypothesis that male and female babies are equally likely.

10.63 Camilla Benbow and Julian Stanley identified 300 U.S. children under the age of 14 who scored 700 or higher on the math SAT. Only about 5 percent of college freshmen and about 1 in 10,000 13-year-olds score that high. Benbow and Stanley reported the following four characteristics for these 300 mathematically gifted children. For each of these four characteristics, use a normal approximation to find the probability that 300 children selected at random would show the disparities reported.   **a.** 280 are boys, 20 girls.   **b.** 20 percent are left-handed, compared to 8 percent for the population as a whole.   **c.** 60 percent have allergies or asthma, as compared to 10 percent of the population as a whole.   **d.** 70 percent are nearsighted, as compared to 15 percent of the population.

10.64 James Bond insists that his martinis be shaken, not stirred. A cynical bartender tests Bond with 6 martinis (using a coin flip to determine whether to shake or stir). Bond correctly identifies 5 of the 6 before passing out. Is this evidence sufficient (at the 5 percent level) to reject the null hypothesis that Bond can't tell the difference and is merely guessing? What is the P-value? (Do not use a normal approximation.)

10.65 *(continuation)* If Bond had correctly identified 50 of 60, would this evidence be sufficient at the 5 percent level to reject the null hypothesis that Bond

is merely guessing? Use a normal approximation with a continuity correction to calculate the P-value. If Bond correctly identifies either 5 of 6 or 50 of 60, this is 83.3 percent accuracy. Is the difference between the P-values for 5 of 6 and 50 of 60 consequently due to the use of a normal approximation in place of the exact binomial probability?

10.66 Many professional football fans believe that the team that receives the ball first in a sudden-death playoff has an unfair advantage. In the first 109 NFL games decided this way, the team receiving the ball won 56 times and lost 53 times. If we interpret these data as a random sample, do they reject the null hypothesis (at the 5 percent level) that each team is equally likely to win? Use a normal approximation and calculate the Z-value and P-value.

10.67 The monthly sales of a certain product are normally distributed with a mean of 10,000 and a standard deviation of 2,000. A new MBA claims that sales can be increased by printing "new and improved" on the box. If the first month's sales with this new box are 12,000, is the increase statistically significant (at the 5 percent level)? Calculate the Z-value and the P-value. What if sales average 12,000 a month for 4 months? For 25 months? Why does the number of months matter?

10.68 High doses of anticancer drugs can be effective in killing malignant cells in women with breast cancer, but these drugs also destroy vital bone marrow. In May 1990 Professor Karen Antman of the Harvard Medical School reported that 150 of 259 patients with advanced breast cancer who received a combination of bone marrow transplants and high doses of anticancer drugs were free of cancer after 12 months.[20] Normally, 30 percent of the patients with advanced breast cancer who receive conventional treatment are free of cancer after 12 months. Is this difference statistically significant, using a one-tailed test at the 1 percent level?

10.69 Sixteen pregnant women were asked to read a children's story aloud three times a day during the last six and a half weeks of pregnancy.[21] These readings were tape recorded and, shortly after birth, the babies were allowed to choose (by varying the rhythm of their sucking on special bottles) between a tape recording of their mother reading the familiar story and reading an unfamiliar one. In 13 of 16 cases, the babies chose the familiar story. What is the P-value for a test of the null hypothesis that each baby has a .5 probability of choosing the familiar story? (Do not use a normal approximation.)

10.70 A high school basketball coach said that a missed free-throw by a right-handed shooter is more likely to bounce to the right, while the reverse is true of a left-hander. To test this theory, a student asked two college basketball players to shoot 50 free-throws apiece.[22] These players missed 21 of 100 free-throws, of which 15 landed on the same side as their shooting hand and 6 bounced to the opposite side. Show how you would use the binomial model to make a two-tailed test at the 5 percent level. (Do not use a normal approximation.)

10.71 Exercise 9.60 reports that a 1990 survey of 928 *Consumer Reports* subscribers found that 15 percent were smokers. At the time, 25 percent of all Americans were smokers. Use a *Z*-value and *P*-value to make a two-tailed test of the null hypothesis that 25 percent of all *Consumer Reports* subscribers are smokers. Is the difference between the survey results and the smoking habits of the U.S. population as a whole statistically significant at the 5 percent level? At the 1 percent level?

10.72 Historically, 17 percent of a company's tractors show rust damage after three years. A new paint is tried and, after three years, 90 of 600 tractors show rust damage. Is this reduction in the incidence of rust damage statistically significant at the 5 percent level? (Use a normal approximation without a continuity correction to calculate the *P*-value.)

10.73 A mail order firm fills orders in an average of 6 days with a standard deviation of 2.5 days. The mail room is now reorganized for a trial period of 400 orders. If the standard deviation is still 2.5 days, what results would be considered statistically significant, using a one-tailed test at the 1 percent level?

10.74 The U.S. Postal Service claims that 94 percent of all first-class letters and postcards mailed from Los Angeles County to Contra Costa County, California, are delivered within two days. To test this claim, five postcards were mailed in 1990 from each of ten locations on two separate days—100 postcards in all.[23] Eighty-nine of these cards were delivered within two days, 11 were not. Test the Postal Service's claim, assuming the Bernoulli trial assumptions are appropriate. Use a one-tailed test at the 5 percent level, and do not use a normal approximation.

10.75 To see if there was a relationship between a basketball's teams rebounding and winning, a student looked at data for the first 164 games of the 1989–1990 NBA season.[24] His null hypothesis was that the team that got the most rebounds in a game would have a 0.50 chance of winning the game. In 118 of these 164 games, the team with the most rebounds won. Assuming the binomial model is appropriate, use a normal approximation for a one-tailed test at the 5 percent level. Why did he use a one-tailed test? Why might the binomial model be inappropriate?

10.76 The number of loaves of bread sold daily by a small bakery is approximately normally distributed with a mean of 300 and a standard deviation of 30. What is the probability that this bakery will sell more than 320 breads in a single day? That their average daily sales for a 6-day week will average more than 320?

10.77 *(continuation)* After a slight change in the ingredients, the bakery observes that during a 6-day period, daily sales average only 290 loaves. Assuming that these 6 days constitute a random sample, is this change statistically significant using a two-tailed test at the 1 percent level? What if daily sales average 290 loaves over a 24-day period?

10.78 *(continuation)* For how many days must daily sales average 290 loaves for this change to be statistically significant using a two-tailed test at the 1 percent level?

• **10.79** *(continuation)* Redo Exercise 10.77, this time assuming that each 6-day week consists of 6 independent days, 4 days during which bread sales are normally distributed with an expected value of 200 loaves and a standard deviation of 30 loaves and 2 days during which bread sales are normally distributed with an expected value of 500 loaves and a standard deviation of 30 loaves. Assume that the two periods during which sales average 290 loaves consist of one 6-day week and four 6-day weeks.

10.80 Over a long period of time, Lil Phillips counted an average of 19 station wagons for every 100 cars that were not station wagons (a ratio of 19 out of every 119 cars).[25] However, on a trip in 1984 she counted 24 wagons to 100 nonwagons, then 22 to 100, and then 24 to 100 again. Assuming Bernoulli trials and using a normal approximation, do these 1984 data show a statistically significant difference (at the 5 percent level) from her historical ratio of 19/119?

10.81 A *Wall Street Journal* poll in December 1986 asked 35 economic forecasters to predict the interest rate on three-month Treasury bills in June 1987. The mean prediction was 4.98 and the standard deviation 0.47. Use these data to test (at the 5 percent level) the null hypothesis that the predictions were a random sample from a population with a mean of 5.73, the actual value of the Treasury bill rate in June 1987.

10.82 In a 1978 Pennsylvania lawsuit, evidence was presented that 11.5 percent of the local labor pool consisted of blacks and other minorities, but that only 336 of 6,974 members of an operating engineers union were minorities.[26] Use a normal approximation to the binomial distribution to calculate the probability that 6,974 persons selected at random from an infinite population that is 11.5 percent minority will have 336 or fewer minorities. Would an accounting for the finite population (of 7.7 million) increase or reduce this probability?

10.83 A pizza store guarantees delivery within a 10-mile radius within 30 minutes of receipt of a telephone order, with the price of the pizza reduced by \$2 if more than 30 minutes elapses. If delivery times are independent and normally distributed with a mean of 26 minutes and a standard deviation of 3 minutes, what is the probability that a delivery will take more than 30 minutes? What is the expected value per delivery of the cost of this guarantee? Use a normal approximation with a continuity correction to determine the probability that on a given day with 100 deliveries, 10 percent or more of the pizzas will be delivered late.

10.84 *(continuation)* The first day after three pizza deliverers are replaced, 8 out of 78 pizzas are delivered late. Are these data consistent with the null hypothesis that pizza deliveries are independent with a .05 probability of being late? Calculate the Z-value and the P-value, using a normal approximation with a continuity correction and a two-tailed test with a .05 probability of Type I error.

10.85 In 1988 a random sample of 118 U.S. executives and 79 Japanese executives were asked who had the lead in supercomputer technology, the United States or Japan.[27] Of the U.S. executives, 106 chose the United States and 12 chose Japan. Of the Japanese executives, 73 chose the United States and 6 chose Japan. In each case, test the null hypothesis that half of the population from which the sample was drawn would answer the United States and half would answer Japan. Calculate the P-value for each of these two tests, but do not use a normal approximation. Is either of the sample proportions statistically significantly different from .50 at the 1 percent level?

10.86 *(continuation)* Now calculate the P-value for each of these two samples using a normal approximation to the binomial distribution with a continuity correction. Does the normal approximation give a P-value that seems reasonably close to the exact binomial probability?

10.87 *(continuation)* When the survey cited in Exercise 10.85 asked who had the lead in computer-integrated manufacturing, 53 U.S. executives chose the United States and 65 chose Japan. Of the Japanese executives, 57 chose the United States and 22 chose Japan. For each of these samples, test the null hypothesis that half of the population from which the sample was drawn would answer the United States and half would answer Japan. Use the binomial distribution to calculate the P-value for each of these two tests. Is either sample proportion statistically significantly different from .50 at the 1 percent level?

10.88 *(continuation)* Use a normal approximation to the binomial distribution with a continuity correction to recalculate the P-value for each of the two samples in Exercise 10.87. Does the normal approximation seem reasonably accurate?

10.89 Hazelwood is a rural area in St. Louis County with relatively few black students. In *Hazelwood School District* v. *United States* (1977), the federal government argued that the Hazelwood School District had followed a "pattern or practice" of employment discrimination as evidenced by the fact that only 15 of 406 (3.7 percent) of the Hazelwood teachers hired during the preceding two years were black, while 15.4 percent of the teachers in St. Louis County, including the City of St. Louis, were black. The Supreme Court accepted the school district's argument that the 3.7 percent figure in Hazelwood should be compared with data showing that 5.7 percent of the teachers in St. Louis County, excluding the City of St. Louis, were black. Use a normal approximation with continuity correction to calculate the Z-values and P-values for these two different values of π

(.154 and .057) under the null hypothesis that, with regard to race, hiring is a binomial process with a success probability π.

10.90 The 1960 World Series matched the New York Yankees versus the Pittsburgh Pirates. During the regular season, the Yankee team batting average had been .260, while the Pirates had batted .276. During the World Series, the team batting averages were .338 for the Yankees (in 269 times at bat) and .256 for the Pirates (in 234 times at bat). Despite being outhit (and outscored, 55 runs to 27), the Pirates won the World Series, four games to three. Use these data to test two null hypothesis: first, that the Yankees' probability of getting a basehit during the 1960 World Series was a constant .260 and, second, that the Pirates' probability of getting a base hit during the World Series was a constant .276. Use a normal approximation to the binomial distribution to calculate these two Z-values and P-values.

10.91 Table 9.2 gives the S&P 500 return and the returns from a sample of 25 mutual funds for each of the years 1986, 1987, 1988, and 1989. For each of these four years, use a t-statistic to test (at the 5 percent level) the null hypothesis that these annual mutual fund returns are a random sample from a normal distribution with a mean equal to that year's S&P 500 return. What conclusion regarding mutual fund performance do you draw from this exercise?

10.92 At a tuna packing plant, five cans are weighed every hour and the production process is stopped if the average weight of these five cans is less than 6.425 ounces or more than 6.575 ounces. What is the probability that the average weight of five cans will fall outside these limits if the weight of each can is an independent draw from a normal distribution with a mean of 6.50 ounces and a standard deviation of 0.06 ounce? Why does this plant stop the production process every time this limit is violated?

10.93 An examination of a leasing firm's historical data on annual gross revenue per employee found an approximate normal distribution with a mean of $470,000 and a standard deviation of $180,000. A new employee, Carol, generates $650,000 in gross revenue her first year. What is the probability that a single observation from a normal distribution with a mean of $470,000 and a standard deviation of $180,000 will be as large as $650,000? Assuming independence, what is the probability that the average of five observations (corresponding to her average gross revenue over five years) will be $650,000 or higher? Why might the assumption of independence be inappropriate here?

10.94 Federal law requires that the painted lines on interstate highways have a brightness reading of 7.0 as measured by the average reading at 25 locations. If the brightness of a certain paint after three years of normal wear has an expected value of 7.20 and a standard deviation of 0.80, what is the probability that a random sample of 25 locations will give an average reading of less than 7.0?

10.95 *(continuation)* Compare this paint to a second paint whose brightness after three years of normal wear has an expected value of 7.25 and a standard deviation of 1.00.

10.96 A 1986 Gallup poll of 504 U.S. teenagers found that of those surveyed 52 percent believed in astrology and 19 percent believed in witchcraft.[28] Use a normal approximation to the binomial distribution to calculate 95 percent confidence intervals for each these figures. Use the astrology confidence intervals to test the null hypothesis that 50 percent of all U.S. teenagers believe in astrology. Do the same with the witchcraft confidence interval. What is the probability of Type I error when a 95 percent confidence interval is used to test the null hypothesis that $\pi = .5$?

10.97 Ozone is one measure of air pollution. Ozone levels at noon on August 8 in the Inland Empire are approximately normally distributed with a mean of 170 and a standard deviation of 40. How often does the August 8 noon ozone level exceed 200 (which represents a Stage 1 smog alert)? Noon readings on August 8 for ten consecutive years gave the following data. Use the sample mean to calculate the Z-value and the P-value for the null hypothesis that these ten observations are a random sample from a normal distribution with a mean of 170 and standard deviation of 40. Do these data reject this null hypothesis using a two-tailed test with a 5 percent probability of Type I error?

238	218	206	178	122
282	134	150	186	166

10.98 *(continuation)* Use the sample data in the preceding exercise to estimate both the mean and standard deviation. Calculate the t-value and P-value for a two-tailed test of the null hypothesis that these ten observations are a random sample from a normal distribution with a mean of 170 and an unknown standard deviation. Explain why this P-value is either larger or smaller than the P-value calculated in Exercise 10.97.

10.99 *(continuation)* Exercise 10.97 used a sample of ten annual noon August 8 observations to test whether there had been a change in the mean ozone reading. If there had been 310 observations with the same sample mean as these ten observations, would the P-value have been larger or smaller? Why? Why might daily or hourly observations violate a crucial assumption underlying the calculation of a Z-value and P-value?

10.100 Explain the error in this interpretation of a statistical test, at the 5 percent level, of the null hypothesis that employment decisions are made without regard to race: "If there is only a 5 percent chance that the null hypothesis is true, then it is rather unlikely that race has no relationship to employment. We can be about 95 percent certain that rejecting the null hypothesis is the right thing to do."[29]

11

The Use and Abuse of Statistical Tests

If you torture the data long enough, Nature will confess.

Ronald H. Coase

TOPICS

Hypothesis tests are undoubtedly one of the most commonplace applications of statistics. The professional journals are full of statistical tests of all sorts of interesting, if somewhat esoteric theories. Indeed, it is almost impossible to publish empirical work without doing the requisite statistical tests.

The more useful or outrageous theories find their ways into the popular press. Newspapers, magazines, and television news shows frequently report theories that have been "proven" statistically. For instance, I have before me newspaper clippings reporting that

> Vitamin C does help ward off colds. . . . there were statistically significant differences between treatment groups in the duration of morbidity.

> Statistics proved a "prima facie" case of discrimination against blacks and women [in Alabama jury rolls].

> Irrefutable Proof the History of United States is Shaped and Controlled by an Eternal, Non Human Mind and Hand!

In order to be intelligent consumers and/or producers of such tests, we must think a bit more carefully about just what it is that hypothesis tests do and don't do. First we'll examine the structure of the tests and then we'll look at their application.

11.1   SETTING UP TESTS

There are two troublesome, yet very practical questions in classical hypothesis testing. The first is choosing which theory to label the null hypothesis and which to label the alternative hypothesis. The second is why α, the probability of Type I error, should be set precisely equal to .05, with little or no attention paid to β, the probability of Type II error. These issues are closely related. Think back to Chapter 10 and these two hypotheses describing the effectiveness of a cram course intended to improve SAT scores:

$$H_0: \mu = 500$$

$$H_1: \mu = 550$$

With $\alpha = .05$, the cutoff for rejecting $\mu = 500$ is $\overline{X} > 533$. If we reverse the labels, so that

$$H_0: \mu = 550$$

$$H_1: \mu = 500$$

then $\alpha = .05$ implies that the cutoff is 33 points below the new null hypothesis 550, at 517. Similarly, the cutoff is 517 if we use the original labels, but set $\beta = .05$ rather than $\alpha = .05$. Figure 11.1 shows this logic.

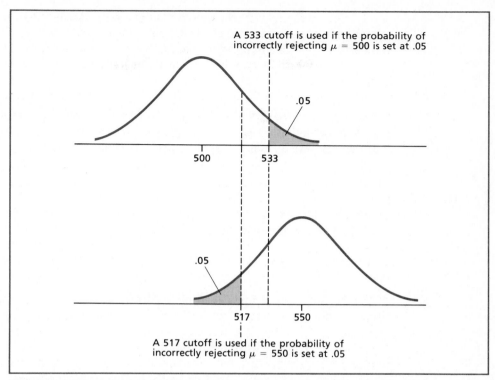

Figure 11.1   *The Cutoff Depends on Which Error Probability Is Set at .05*

Thus, there is really a dual question: On what basis do we choose a null hypothesis, with an accompanying 5 percent probability of Type I error? A formal rationale is complicated and not completely satisfying because it requires the placement of subjective probabilities on the competing hypotheses. A brief exposition of this rationale is in the appendix at the end of this chapter, but the general idea can be explained intuitively.

Choosing a Null Hypothesis

In most situations, there are a variety of competing possible values for the parameter of interest, say μ, and a multiplicity of αs and/or βs. The usual testing procedure is to set the cutoff point so as to fix $\alpha = .05$ for one particular value of μ. That is, we choose one particular hypothesis and set the test so that there is only a small probability that we will reject this hypothesis if it is, in fact, correct. Little or no attention is paid to the other possible values of μ, with their various αs and βs. This concentration of interest implies that the incorrect rejection of this null hypothesis is our paramount concern. What reasons, then, could justify focusing on the probability of incorrectly rejecting one particular hypothesis? Several ideas come to mind.

1. One of the hypotheses represents the conventional wisdom, based on theoretical logic and earlier studies. The weight of this wisdom argues that we should not hastily reject this hypothesis, based on our meager study, and so we make this favored hypothesis the null hypothesis to ensure a small probability of incorrectly abandoning it.

2. We don't have strong feelings favoring any of the competing hypotheses, but we do know that there will be serious consequences if we erroneously reject a particular hypothesis. So, we make this hypothesis the null hypothesis to ensure that there is only a small probability of incorrectly rejecting it.

3. The statistical rejection of a hypothesis is a stronger statement than is statistical acceptance. All of the values inside a confidence interval are acceptable hypotheses. Accepting a hypothesis just puts it on a list with other unrejected hypotheses. The rejection of a hypothesis, however, is a definitive conclusion. Thus, researchers may choose as the null hypothesis a *straw hypothesis* that they hope to reject. In fact, the terminology *null* hypothesis comes from this approach. To prove that there is an effect, we attempt to reject the hypothesis that there is no *(null)* effect.

Let's run through a few examples, starting with the cram course. The conventional wisdom is that such courses have little or no effect, arguing for "no effect" as the null hypothesis. It is not clear whether it is more dangerous to recommend a worthless course or to spurn a useful course. Those who want to show that the course is effective are virtually compelled to make "no effect" the null hypothesis, hoping to show its effectiveness by rejecting this straw hypothesis. So, there are at least two reasons for making "no effect" the null hypothesis.

Now consider the example of a new additive that gives food a more appetizing color, but may cause cancer. The consequences of mistakenly approving a carcinogenic additive are far worse than the consequences of incorrectly withholding a safe additive. Unless it is a substance that scientific knowledge tells us is almost surely safe, the null hypothesis should be that the additive is a carcinogen. The additive should be presumed dangerous until proven otherwise. The same logic applies to our justice system, although trials aren't usually decided by statistical analysis. The consequences of sending an innocent person to prison are thought to be far worse than the consequences of letting a guilty person go free. So, the appropriate null hypothesis is that a person is innocent until proven guilty.

What about a medicine that may be either beneficial or worthless? There may be some scientific knowledge about whether or not a particular substance is of any value. An independent researcher's feelings also may be influenced by the fact that there are always plenty of unscrupulous charlatans around who are eager to sell anything that people can be persuaded to buy. On the other hand, the consequences of withholding a miraculous drug may be more serious than the consequences of allowing the peddling of a worthless drug. This argument suggests letting the null hypothesis be that the drug is effective. Finally, the straw hypothesis approach argues for proving a drug's value by making "no effect" the null hypothesis. Here, we have a split decision that could go either way.

The Right Level of α

We can continue this line of reasoning by thinking more carefully about the appropriate α. The selected probability of Type I error, α, is called the **significance level of the test**. With the usual $\alpha = .05$, the researcher will report that the results either were or were not "significant at the 5 percent level." In the 1920s, the great British statistician, R.A. Fisher, endorsed a 5 percent cutoff:

> It is convenient to draw the line at about the level at which we can say: "Either there is something in the treatment, or a coincidence has occurred such as does not occur more than once in twenty trials . . . "
>
> If one in twenty does not seem high enough odds, we may, if we prefer it, draw the line at one in fifty (the 2 percent point), or one in a hundred (the 1 percent point). Personally, the writer prefers to set a low standard of significance at the 5 percent point, and ignore entirely all results which fail to reach that level.[1]

The 5 percent α is now so ingrained that some researchers simply say that their results "are statistically significant," with the reader understanding that the test was constructed to allow a 5 percent chance of Type I error. But why use 5 percent? Why not 1 percent or 0.0001 percent? Frankly, there is not much more than acquired habit behind the 5 percent rule. There is no single right level for α! It's like asking "How far is far?"

The logic of our tests is that we choose as the null hypothesis some hypothesis that we want to be cautious about incorrectly abandoning. If deviations between this hypothesis and the data can be plausibly explained by sampling error, then we'll explain them that way. If it strains credulity to appeal to sampling error, then we will reluctantly abandon the null hypothesis. So the α question is really, "How improvable does the sampling error explanation have to be before we abandon it?" Five percent seems like a reasonable, ballpark answer. It's pretty improbable, but it's not so incredible that we never let go of a null hypothesis.

In some situations, a little more caution may be called for. If we're testing a potentially dangerous drug, then we may well demand very strong evidence before approving it. A tough standard can be imposed by having the null hypothesis be that the drug is dangerous and setting α at 1 percent, 0.1 percent, or even lower. Or, in cases where the conventional wisdom is very strong, convincing evidence will have to be even stronger. If the null hypothesis is a widely accepted conventional wisdom, then α should be set very low, so as to rule out firmly the possibility of sampling error. An interesting example of this logic is the *Journal of Experimental Psychology,* which publishes research on extrasensory perception (ESP) and other controversial matters. The editor of this journal explained that

> In editing the *Journal* there has been a strong reluctance to accept and publish results . . . when those results were significant [only] at the .05 level, whether by one- or two-tailed test. This has not implied a slavish worship of the .01 levels, as some critics may have implied. Rather, it reflects a belief that it is the responsibility of the investigator in a science to reveal his effect in such a way that no reasonable man would be in a position to discredit the results by saying they were the product of the way the ball bounces.[2]

The best that can be said for using a standard α is that it provides a common vocabulary and benchmark, but there is no reason to pay slavish homage to $\alpha =$ .05, .01, or to any other particular number, and few researchers do. Everyone realizes that a P-value of 4.9 percent is "barely significant," that 5.1 percent is "almost significant," that 1 percent is "highly significant," and so on. As long as the P-value or Z-value is reported, readers can judge for themselves whether the sampling error explanation is sufficiently implausible to discredit the null hypothesis.

EXERCISES

11.1 A police department requires tires with an expected life of at least 30,000 miles. A random sample of 16 tires from one dealer yields these tire lives:

24,000	34,000	27,000	16,000
24,000	30,000	38,000	15,000
18,000	36,000	20,000	29,000
35,000	14,000	20,000	32,000

Use a t test with a 5 percent significance level to test the null hypothesis that the average tire life for the population is at least 30,000 miles.

11.2 *(continuation)* Does this test give the benefit of the doubt to the tire dealer? Assume that the population standard deviation is known to be 8,000 miles and that a sample of 16 tires is tested. Using a 5 percent probability of Type I error, determine the cutoffs for these two alternative criteria: **a.** Buy from the dealer if the sample does not reject the null hypothesis that the population mean is at least 30,000 miles. **b.** Buy from the dealer if the sample rejects the null hypothesis that the population mean is less than 30,000 miles. What are the chances of passing these tests if the tires are drawn from a population with a mean life of 27,500 miles?

11.3 The text has used the example of testing whether or not a cram course raises SAT scores by 50 points. Two competing hypotheses are that the average test scores of students taking this course are $\mu = 500$ and $\mu = 550$. If we conduct a standard statistical test at the 5 percent level, does it make any difference which value of μ we use as the null hypothesis and which we use as the alternative hypothesis? Why do you think $\mu = 500$ was used as the null hypothesis in the text?

11.4 A 1985 *Wall Street Journal* editorial said

We wrote recently about a proposed Environmental Protection Agency ban on daminozide, a chemical used to improve the growth and shelf life of apples. EPA's independent scientific advisory panel last week ruled against the proposed prohibition. "None of the present studies are considered suitable" to warrant banning daminozide, the panel said. It also criticized the "technical soundness of the agency's quantitative risks assessments" for the chemical. EPA can still reject the panel's advice, but we see some evidence that the agency is taking such evaluations more seriously. There is no better way to do that than to keep a firm grip on the scientific method, demanding plausible evidence when useful substances are claimed to be a danger to mankind.[3]

Does the *Wall Street Journal*'s interpretation of "the scientific method" implicitly take the null hypothesis to be that daminozide is safe or that it is dangerous?

11.5 Some chemists have invented a cola sweetener that is slightly less expensive to produce than the sweetener that a company now uses. Fearful of bad publicity and expensive lawsuits, the company's president orders some tests to determine whether or not this new sweetener is a carcinogen. Should the null hypothesis be that it does or does not cause cancer?

11.6 A computer software company has stumbled on a bizarre computer game concept unlike any other now on the market. The company's president thinks that it is equally likely that this new game will be a sensation or a flop. If it does flop, there will be small losses; but if it is a sensation, the gains will be enormous. The president orders a very small marketing survey to test consumer interest without letting the new concept fall into the hands of its competitors. Should the null hypothesis be that the game will be a sensation or that it will be a flop? What value of α would you recommend?

11.7 A doctor suspects that a patient may have a rare disease. This disease is seldom serious, but the treatment is less expensive the earlier the disease is discovered. In testing for the disease, should the null hypothesis be that the patient does or does not have the disease? What value of α would you recommend?

11.8 A renowned political forecaster conducts an election-day opinion poll, hoping to make an early prediction of a presidential race that has been considered too close to call. There will be considerable glory if the prediction is correct, but even larger embarrassment if the prediction is wrong. If poll's results are not decisive, the forecaster will just say, "It is too close to call." If the null hypothesis is that the incumbent will get exactly 50 percent of the votes, what value of α do you recommend? What if you were a novice pollster, eager for notoriety, feeling that the publicity accompanying a correct prediction far outweighs the neglect accompanying a wrong one?

11.2 BUT IS IT IMPORTANT?

In testing hypotheses, it is easy to be confused by the distinction between statistical significance and practical importance. A statistically significant result may be of little or no consequence. Conversely, a researcher may find a potentially very important result that does not happen to be statistically significant. To illustrate what may appear paradoxical, let's return again to the math SAT cram course.

A random sample of n students takes the cram course and then the SAT test. Their average score $\overline{X}$ is normally distributed,

$$\overline{X} \sim N[\mu,\ 100/\sqrt{n}]$$

The standard deviation is assumed to stay at 100, but it is not known whether their expected score will go up, go down, or be unchanged. The null hypothesis is that the average score of students who take the course is 500, the same as for students who don't take the course,

$$H_0\colon \mu = 500$$

One way of using the sample to test this null hypothesis is to compute a 95 percent confidence interval and see whether or not 500 is inside this interval. Using our two-standard-deviations rule of thumb, we have

$$.95 = P[\overline{X} - 2(100\sqrt{n}) < \mu < \overline{X} + 2(100/\sqrt{n})]$$

To exaggerate our impending conclusion, let's pretend that this SAT cram course garners as much national attention as the polio vaccine tests in the 1950s. A sample of 1,000,000 students take the cram course and obtain an average score of 500.3. The standard deviation is $100/\sqrt{1,000,000} = .1$ and our confidence interval is $500.3 \pm .2$. These data are sufficient to reject the null hypothesis. The average score of 500.3 is three standard deviations above 500, and so we conclude that cramming does have a statistically significant effect on math SAT scores. But this statistically significant effect is of little or no practical importance; no rational person would spend time and money to raise the score by less than half a point! The course should be abandoned, even though it has a statistically significant effect on SAT scores.

Now let's look at the reverse case. Our underfunded researcher can only afford to enroll and test four students. Their average score after taking the course turns out to be 598, which is better than 84 percent of the students taking the test. Our confidence interval is very wide, though, because we have so few data:

$$598 \pm 2(100/\sqrt{4}) = 598 \pm 100$$

The confidence interval is so large that we cannot rule out the possibility that the course has no effect. By the luck of the draw, we may have picked four 84th-percentile students and not raised their scores at all. Or we may have picked four 16th-percentile students and raised their scores 200 points. Our best guess is a 98-point improvement, but the small sample size gives a 100-point margin for error, ranging from a 2-point decrease to a 198-point increase. The result is not statistically significant, but it may be of great importance. A course that, at worst, lowers scores 2 points and, at best, raises scores 198 points would appeal to many students. This researcher has found a potential bonanza; the course will thrive and additional research will be done.

These two examples are deliberate caricatures that are designed to make a point. As the sample size increases, our confidence interval shrinks to zero. If our sample is the entire population, the sample mean will exactly equal the population mean and there will be no sampling error at all. Any null hypothesis would be rejected, except for the one that has exactly the right number for the population mean. And, yet, without knowing the whole population, there is no chance of picking the precise value of this number beforehand. It follows that, whatever null hypothesis we use, we are sure to reject it if we just gather enough data! Thus, a cynical way of looking at hypothesis tests is that accepted hypotheses are those that we don't yet have enough data to reject.

Remember, hypothesis tests only evaluate statistical significance, and not practical importance. Think of the expected value of math SAT scores after a cram course. If scads of data give a confidence interval 500.0003 ± 0.0002, then 500 is indeed rejected by the data. But, as a practical matter, the distinction between 500 and 500.0003 is of no importance whatsoever. The meaningful conclusion is that the course has little or no effect, and that conclusion should be drawn whether the confidence interval is 501 ± 2 (with 500 "accepted" by the data) or 500.0003 ± 0.0002 (with 500 "rejected" by the data).

Hypothesis tests can be a contrived and misleading exercise. When we select a specific hypothesis and then conclude that it is either accepted or rejected by the data, we are only giving a partial glimpse at the implications of the data. A confidence interval often paints a clearer picture of the data than does a hypothesis test. With a confidence interval we can see the estimate and use our common sense to judge whether, as a practical matter, this estimate is a large or small number. A confidence interval also shows the margin for sampling error, whose practical importance can also be judged. Finally, a confidence interval shows which numbers are rejected by the data and, again, we can judge whether particularly interesting numbers differ from the estimate in a practical as well as a statistical sense.

Unfortunately, hypothesis tests are the usual reporting style, perhaps because they sound so decisive. But they may be misleading. Consider, for example, the question of whether men and women are equally likely to be right-handed. A sample of 6,672 people found that 90 percent of the men and 92 percent of the women were right-handed. The rest were left-handed or ambidextrous. This difference is statistically significant, but of no real importance.

Darrell Huff gives the example of a *Reader's Digest* study in the 1950s that was designed to see if there were any real differences in the amounts of nicotine, tar, and so on in different brands of cigarettes. The conclusion was that the brands they studied were virtually identical. But, as Huff explains,

> In the lists of almost identical amounts of poisons, one cigarette had to be at the bottom, and that one was Old Gold. Out went the telegrams, and big advertisements appeared in newspapers at once in the biggest type at hand. The headlines and the copy simply said that of all the cigarettes tested by this great national magazine Old Gold had the least of these undesirable things in its smoke. Excluded were all the figures and any hint that the difference was negligible.[4]

On the other hand, an advertiser will claim "Laboratory tests show that no cleaner is more effective than Deep Klean," suggesting that Deep Klean is the most effective cleaner when what the laboratory really found is that there was no statistically significant difference among the cleaners tested. Or consider the conclusion of the two economists studying the effect of inflation on election outcomes. They estimated that, in the 1976 election, the inflation issue added 7 percentage points to the Republican vote. A standard deviation of 5 percentage points gave them a 95 percent confidence interval of 7 percent $\pm$ 10 percent. They concluded that "in fact, and contrary to widely held views, inflation has no impact on voting behavior."[5] That is, of course, not at all what their data say. The fact that they do not rule out 0 percent does not not show that 0 percent is the true value. The confidence interval does include 0 percent, but it also includes everything from -3 percent to $+17$ percent. Their best estimate is 7 percent $\pm$ 10 percent, and 7 percent is more than enough to swing most elections one way or the other.

Sometimes, it is a comparison of apples and oranges that leads to an erroneous conclusion. For instance, during the Spanish-American War, the death rate was 9 per 1000 in the U.S. Navy and 16 per 1000 in New York City. Navy recruiters claimed that it was safer to be fighting in the Navy than to be living in New York. A meaningful conclusion, however, would require a comparison of the death rates of the same group, here healthy young men.

EXERCISES

11.9 A junior high school student made a wooden die in wood shop. The instructor suggested that this die looked a bit lopsided, so that the number 6 might not come up exactly one-sixth of the time. In 18,000 careful rolls of this die, the number 6 came up 3,126 times. Is this difference from 3,000 **a.** of statistical significance at the 5 percent level?   **b.** of practical importance?

11.10 The president of a large state university boasts that the entering freshman class has significantly better English skills than the average high school student who is considering a college education. Nationwide scores on a standardized English test are normally distributed with a mean of 500 and a standard deviation of 100. Is this president's boast justified by the fact that the average score of this college's 10,000 freshmen is 503?

11.11 *(continuation)* The owner of a small bookstore boasts that four new employees have significantly better-than-average English skills. Is this owner's boast justified by the fact that the average score of these four new employees is 580?

11.12 A "new, improved" toothpaste claims that it significantly reduces cavities for children in their cavity-prone years. Cavities per year for this age group are normally distributed with a mean of 3 and a standard deviation of 1. A study of 2,500 children who used this toothpaste found an average of 2.95 cavities per child. Explain why you are or are not persuaded of this toothpaste's cavity-reducing ability.

11.13 Mercury and Venus are "inner" planets, in that they are closer to the sun than the earth. The other six planets in our solar system are "outer" planets. Relative to the earth, the two inner planets have an average mass of 0.43, while the six outer planets average 74. A statistics text proposes to test whether this difference is statistically significant.[6] Are these data a random sample?

11.14 An article in the *New York Times* business section suggested that the Chinese calendar might be helpful in predicting whether stock prices will go up or down:

> Trying to discern any sort of seasonality in the stock market may be a futile exercise. Perhaps the most ingenious "analysis" was done by the Rothschild firm, which noted that February begins the Year of the Dragon in the Chinese calendar.... During the 20th century ... there were twice as many up Dragon years [4] as down Dragon years [2].[7]

What other data do you need to determine whether Dragon years are especially good years to buy stocks? Why are you confident that the relationship is not statistically significant? Why, even if the relationship were statistically significant, would you be unpersuaded?

•• **11.15** Critically evaluate the following:

> Is it ever possible to have too large a sample? Some statisticians feel that it is. Suppose we are looking at two groups of students, each of which has been taught to read by a different system. We want to know if one system works better than the other. If the groups are small, then the amount of error to be expected will be large, so only a very large difference between the performance of the two groups will count as evidence for the superiority of one method over the other. Now as we increase the size of the groups, the amount of error will decrease. Perhaps there is a point at which a really insignificant difference in performance will count as evidence because small as it is, it is still larger than the expected amount of error. Although this situation is theoretically possible, it is hardly a common problem, since the time and money needed to do any research make it relatively unlikely that too large a sample will be studied.[8]

11.3   DATA MINING

A selective reporting of results is potentially the most serious abuse of hypothesis tests. To illustrate the pitfalls, I'll pretend that I'm a psychic and you pretend that you're a statistician. I come into your office, claiming that I have a psychic ability to predict the outcomes of coin flips. I don't claim to be perfect, but I do get more than 50 percent right. My "secret" is that I have discovered a pattern in coin flips: After a head occurs, the next flip is likely to be a tail, and vice versa. You pull out a pencil, paper, and coin and dutifully test my wild theory.

Because the conventional wisdom is that there are no patterns in coin flips, you let the null hypothesis be that my probability of a correct prediction is $\pi = .5$. The alternative hypothesis is my claim that $\pi > .5$. Pressed for time and scornful of my claim, you have the test consist of just 20 flips of a coin. How many do I have to get right to convince you that I am not just a guesser? Ten correct is unimpressive, while 20 out of 20 is astonishing. Where do we draw the line? If the null hypothesis is true, so that $\pi = .5$, then the binomial distribution (in Table 1 at the back of this book) tells us that there is a .058 probability of getting 14 or more right. So, let's make that the cutoff.

Well, the truth of the matter is that I am not a psychic and my theory is worthless. I make my 20 guesses and perhaps I only get 8 right. You rightly throw me out of your office and go back to work. As I go out the door, another alleged psychic walks in, with her theory that coin flips have momentum: a head is likely to be followed by another head and a tail is likely to be followed by another tail. In the interest of science, you flip your coin 20 more times and test her theory. She's a fraud, too, and only gets 12 out of 20 right. Not bad, but not good enough. Out she goes and in comes another, with yet another theory: a head tends to be followed by a head, but two or more heads in a row tend to be followed by a tail—another test and another fraud exposed.

Somewhere along the line, however, one of these frauds is going to get lucky and get 14 or more right. A 5 percent chance of Type I error implies that about 1 out of 20 worthless theories will be statistically significant. This reasoning is not limited to fraudulent psychics; it applies to all statistical tests of all theories. Even

Cartoon by Rob Pudin, courtesy *The Skeptical Inquirer*

if you never test anything but fraudulent theories, you are sure to give your statistical seal of approval to some, if only you test enough of them.

In practice, with powerful computers, mountains of data, and incredible pressures to produce statistically significant results, untold numbers of worthless theories get tested. The harried researcher will try 20 theories, write up the result that is most statistically significant (lowest P-value, for instance) and throw the rest in the trash. Twenty struggling graduate students may each try one theory; one will get a statistically significant result, a thesis, and a research grant, and the other 19 will transfer to law schools. The problem for science and society is that we only see the tip of the statistical iceberg. We see the result that is statistically significant, but not the other 19 that didn't work out. If we knew that this one was 1 out of 20 tries, we would surely be less impressed by its statistical significance.

This testing of many theories is known as **data mining** (also "data grubbing" or "fishing expeditions"). The eager researcher takes a body of data, tests countless theories, and reports the one that is the most statistically significant. The end result, which is that a statistically significant result has been obtained, shows little more than the energetic efforts of the researcher. We know beforehand that, even if only worthless theories are examined, the researcher will eventually stumble on a worthless theory that seems confirmed by the data, due to the inevitable sampling error. It is not surprising, then, when a data miner finds a statistically significant result, and it should not convince us that the researcher has found anything more than sampling error. We cannot tell whether the end product of a data mining expedition shows the veracity of a useful theory or the perseverance of a determined researcher.

The Sherlock Holmes Inference

Many theories get rejected out of hand without a formal statistical test. Think of the 20 coin flips again. In these tests, each quack walked into the office with a harebrained scheme for predicting coin flips, and a coin was flipped 20 times. In practice, researchers often have the data in hand before they concoct the theories to be tested. A practiced observer can look at the data and quickly see which theories are promising and which aren't going to work; maybe 5 tests are run through the computer and 50 are run through the mind.

For example, I flipped a coin 20 times and the results were

T H H T H T T T T T H H T H T T H H T T

In less than 30 seconds, I found a pattern, a "theory," to explain these coin flips. My theory is that a flipped coin tends to give a tail followed by two heads, over and over again. If we test my theory using these data, we find that it correctly predicts 15 out of 20 outcomes, a statistically significant result! These data reject your suspicion that I am just a hopeless guesser.

Yet, would you bet that my system will continue to hit 15 out of 20? Of course not, because I told you how I came up with my "theory," and my method is not convincing. I looked at the data, tried many possible theories in my mind, and found one that would have been right 15 out of 20 times, if only I had known the theory before the coin was flipped—but I didn't.

This is pure and simple data mining. When you see how it is done and then see it applied to something like coin flips, the results are not persuasive; your intuition resists such theories. But professional data mining is usually applied to situations where we don't have much intuition. And you don't get to see the data mining in action; you only see the final results.

If I came to you and said that, as a professional economist, I have been studying the stock market and have come up with a theory that works, you might pay attention. If I told you that in 15 of the last 20 years, this theory has correctly predicted whether the stock market will go up or down, you might get very interested. Yet, I may have found my stock market system in the very same way that I found my system for predicting coin flips. If H stands for an up year and T for a down year, then the 20 coin flips recorded earlier might well be a history of the stock market over some 20-year period. I could look at this history of ups and downs and, in less than 30 seconds, find a pattern that seems to explain the stock market: The market goes backward one step and then up two steps. And sure enough, this theory is right 15 out of 20 years. Subjected to statistical tests, we would find that this 75 percent accuracy decisively rejects the null hypothesis that I have only a 50 percent chance of guessing which way the market is going. The statistical evidence sounds pretty impressive, unless you know how I did it.

The theoretical logic behind hypothesis tests is quite straightforward. The researcher has some competing theories, gathers some data to test these theories, and reports the results. But, in a hurried effort to get publishable (or salable) results, some researchers work in the other direction; they study the data until they find theories that will be shown to be statistically significant. This is the Sherlock Holmes method of reasoning, as indicated by the quotations prefacing the last three chapters: examine the data to find theories that explain these data. It can be a very useful method of operation. Many important scientific theories, including Mendel's genetic theory, were discovered by attempts to find theories that explain the data. But Sherlock Holmes research has also been the source of thousands of quack theories.

How do we tell the difference? How can we separate the genius from the quack? There are two good antidotes to unrestrained data mining: common sense and fresh data. Unfortunately, there are short supplies of both. The first thing we should think about when confronted with statistical evidence in support of some hypothesis, is "Does it make sense?" If it is a ridiculous theory, then you shouldn't be persuaded by anything less than mountains of evidence. One of the reasons you laughed off my claims of predicting coin flips is that your common sense tells you that coin flips are unpredictable. Even if I walked in with data showing that I had successfully predicted 99 out of 100 flips, you would still be unconvinced. Instead of accepting my coin-calling prowess, you would suspect that somehow I had cheated or massaged the data. As Thomas Paine once wrote, "Is it more probable that nature should go out of her course, or that a man should tell a lie?"

Most situations are not as clear cut as these coin predictions, but nonetheless, you should still exercise common sense in evaluating statistical evidence. Unfortunately, common sense is an uncommon commodity, and many silly theories have been seriously tested by honest researchers. I'll give you some examples shortly.

The second antidote is fresh data. I've explained before why it is not a fair test of a theory to use the very data that were used to choose the theory. For an impartial test, we should specify the theory before we see the data that will be used to test the theory. If certain data have been used to concoct the theory, then look for fresh data that have not been contaminated by data mining. Think of the coin-flipping example again. When I announce that I have a system that was right 15 out of 20 times, your likely response is, "Oh, yeah. Let me get out a coin and we'll see how well you do." This will be fresh data that will most likely expose my fraudulent system. Researchers should do the same with any theory that is dug up by data mining.

In chemistry, physics, and the other natural sciences, this is standard procedure. Whenever someone announces an important new theory, a dozen people will rush to their laboratories to see if they replicate the results. Unfortunately, most social scientists don't have laboratories where they can conduct experiments and generate fresh data. Instead, they have to wait for the world to turn and then patiently record data as they occur. If they have a theory about the economy, presidential elections, or peace, they may have to wait several years or even decades to accumulate enough fresh data to test their theories.

When economists have done this further testing with fresh data, the results are usually disappointing. They often find that their theory doesn't explain fresh data nearly as well as it did the original data. Their usual explanation is that there have been "structural changes" in the economy that necessitate modifications of their original models. They then mine the fresh data for a new theory. In this way, a provocative first paper turns into a career. An alternative explanation of the disappointing results with fresh data is the data mining itself.

EXERCISES

11.16 An eager graduate student has found several worthless theories to test. This student hopes that for at least one of these theories, the data will reject (at the 5 percent significance level) the null hypothesis that the theory is worthless. If so, the student will have provocative conclusions and perhaps a thesis topic. Assuming that the tests are independent, what is the probability that this student will be successful if the number of theories tested is **a.** one? **b.** two? **c.** ten? **d.** twenty? **e.** ∞?

11.17 A psychic claims an ability to predict the outcome of coin flips. To test this claim, a coin is flipped 100 times, with the psychic predicting each flip in advance. It is decided to accept the psychic's claim if at least 60 outcomes are forecast correctly. What is the probability that this psychic's claim will be accepted if, in fact, the psychic is a fraud? What is the probability that the psychic's claim will be rejected if, in fact, the psychic's probability of making a correct prediction is 51 percent? 60 percent? 75 percent? 90 percent?

11.18 A professor conducts a coin-calling experiment with a class of 32 students. Half of the students predict heads for the first flip and half predict tails.

The coin is flipped: it's a head and the first half is right. These 16 students then predict a second flip, with half saying heads and the other half tails. It comes up tails. The 8 students who have been right twice in a row now try for a third time. Half predict heads and half tails, and it comes up tails. The 4 students who were right divide again on whether the next flip will be heads or tails. It's tails and now we're down to 2 students. One calls heads and the other tails. It's heads and we have our winner—the student who correctly predicted five in a row.   **a.** What is the probability of correctly calling five coin flips in a row?   **b.** Does five in a row satisfy the usual statistical test of the null hypothesis that a claimed psychic is just a guesser?   **c.** Would you bet $2 against my $1 that the winner of this contest will call the next flip correctly?

11.19 An unscrupulous investment adviser mails sample investment advice to 3,200 wealthy investors. Half of the letters predict the stock market will go up this month and half predict that it will go down. At the end of the month, it turns out that the market went down. This adviser crosses out the 1,600 investors who were given bad advice and mails another sample prediction to the 1,600 who were given correct advice. Half of these 1,600 people are told that the market will go up in the second month and half are told that it will go down. At the end of this month, 400 investors have been given two correct predictions in a row. These people are given a third month's predictions; again, half are told that the market will go up and half are told that it will go down. By the end of five months, there are 100 investors who have received five correct predictions in a row and they should be eager to subscribe to the adviser's expensive investment service.   **a.** If the stock market is equally likely to go up or down, what is the probability of five correct predictions in a row?   **b.** As an investor who received five sample predictions, all of which were correct, would you be impressed?   **c.** As someone who saw how it was done, would you be impressed by five correct predictions?   **d.** What is the statistical fallacy here?

11.20 While playing bridge recently, Ira Rate picked up his 13-card hand and found that it contained no spades (and very few honors, either). He promptly threw his cards down and declared that the deal had not been random. He read in a book on probability and statistics that (a) the probability of being dealt a hand with no spades is less than 5 percent and (b) this low a probability is sufficient evidence to reject the null hypothesis of a fair deal. Point out a fault or two in his reasoning.

11.21 A study of the relationship between socioeconomic status and juvenile delinquency tested 756 possible relationships and found 33 to be statistically significant at the 5 percent level.[9] What statistical reason is there for caution here?

11.22 A researcher wants to show that a subject has a statistically significant ability to predict the outcomes of coin flips. The conventional null hypothesis is that the probability of a correct prediction is $\pi = .5$. If a test consists of 10 coin flips, then how many must this subject predict correctly for the probability of Type I error to be about 5 percent?

This researcher has decided that if the subject does not make enough correct predictions to pass the first test, then these results will be discarded and a second 10-coin test will be given. What is the probability that a guesser   **a.** will pass the first test?   **b.** will fail the first test, but pass the second?   **c.** will fail two tests, but pass a third?   **d.** will eventually pass a test?

• **11.23** *(continuation)* This researcher has been criticized for discarding data and, in response, decides on a modified procedure. If the subject does not get enough correct answers to pass the first test, then 10 more flips will be predicted and the data combined to give a 20-flip test. How many correct predictions are required in a 20-flip test for about a .05 probability of Type I error?

If the subject does not pass this 20-flip test, then 10 more flips will be predicted and the researcher will use a 30-flip test. What is the probability that the subject will eventually pass one of these expanded tests?

11.4   SOME FLIMSY THEORIES

Let's now look at several examples of theories that are statistically significant, but dubious at best.

How Not to Win at Craps

A book was written (and sold, of course) explaining a system for winning at craps. The author had recorded the outcomes of 50,000 dice rolls at a Las Vegas casino. The numbers came up just about as often as would be expected with fair dice, but a careful scrutiny of the sequence in which the numbers appeared revealed some unusual patterns. For example, the sequence 4-4-11 could be expected about 20 times in 50,000 roles, but it happened 31 times! If you had bet $100 on 11 whenever 4 came up twice in a row, you would have made a $13,700 profit. Similarly, the author found that the sequence 7-12-7 occurred 38 times and that on 10 of these 38 occasions, the next number was either a 2, 3, or 12. If you had bet $100 on craps (2, 3, or 12) whenever the sequence 7-12-7 appeared, you would have won $4,200.

Are you convinced by this evidence? This is exactly analogous to my finding a pattern in 20 coin flips. Your common sense should reject this system strongly. I hope that common sense is enough and that you don't go to Las Vegas to gather fresh evidence.

Tall Presidents

A month before the 1972 presidential election pitting Richard Nixon against George McGovern, *Parade* magazine reported that sociologist Saul Feldman of Case Western Reserve University,

> . . . whose research interest is height, took a reading of Presidential candidates for the past 40 years. He reports that with only one exception (Wendell Wilkie in 1940), the taller of the two candidates has always won the electoral victory. In 1968, he points out, Nixon just barely defeated Humphrey who at 5'11" stands a half-inch shorter than the president. . . . "Heightism" is a prejudice just like sexism and racism, Feldman believes.[10]

Do Feldman's data show a statistically significant relationship between height and electoral success? If height is unimportant, then the winner is equally likely to be the shorter or taller candidate. Under this null hypothesis, the probability that the winner is the taller of the two candidates is $\pi = .50$. The binomial distribution tells us that if $\pi = .50$, then there is only a .01 probability of as many as 9 successes in 10 trials. Thus, the evidence seems to decisively reject the null hypothesis that height is unimportant and to confirm Feldman's "heightism" theory.

But let's think about this evidence a little more carefully. Presidential elections are not the Bernoulli trials assumed by a binomial distribution. Presidential candidates often run again and again, so that the outcomes are not independent. Dwight D. Eisenhower beat Adlai Stevenson in 1952 and he beat him again in 1956. The 1956 election did not provide fresh evidence about candidates of different heights; it just told us once again that Eisenhower was more popular than Stevenson. In fact, four of Feldman's ten elections involved Franklin Roosevelt winning again and again and again. So, the data are not quite as conclusive as they seem.

In addition, we can't help but wonder why Feldman started with the 1932 election. Why did he look back 40 years, instead of 20 years, 60 years, or some other number? Perhaps this 40-year period yielded the data that was most favorable to his theory, but it is not appropriate to discard data simply because it does not support one's theory. Whenever someone uses only part of the available data, we should wonder why the rest wasn't used.

What about common sense, our first antidote to data mining? My mind is not entirely made up, but my common sense rejects Feldman's simplistic theory. I cannot believe that a half-inch difference in height is responsible for Nixon's victory over Humphrey. Nor can I believe that height explains the four victories by Roosevelt, whose legs were paralyzed. If someone who is 6'4" runs against someone who is 5'4", then this height difference may well influence some voters. However, hardly anyone will notice an inch or two of difference, especially because the candidates hardly ever stand next to each other. Did anyone other than Feldman really know that Humphrey was a half-inch shorter than Nixon, let alone care?

What about the second antidote, fresh data? Well, one reason Feldman's theory was reported in *Parade* magazine was that he was virtually alone in predicting a McGovern victory in 1972. Because McGovern was 1 1/2 inches taller than Nixon, Feldman predicted that McGovern "should inch out Nixon this November." McGovern lost every state but Massachusetts. In 1976, Jimmy Carter beat Jerry Ford despite a 2 1/2-inch height disadvantage. In 1980, and again in 1984, the taller man, Ronald Reagan, did win. So did George Bush in 1988, a contest that some pundits called "the wimp versus the shrimp." In all, using fresh data, Feldman's tall-president theory has been right an unimpressive three out of five times.

Presidential Names

If we scrutinize the characteristics of successful presidential candidates, it is easy to find patterns that make even less sense than "heightism." One is that the major party candidate with the longest last name wins. From 1900 through 1960, this theory was right an astounding 14 out of 15 times. It missed 4 of the next 5 times, but after the 1988 election it was still 15 out of 21 correct. What about the first letter of the last name being later in the alphabet than the opponent's? This one was right 10 out of 13 times from 1900 through 1948. It has only been right 6 out of 10 times since then, but that's still a credible 16 out of 23.

The statistical success of both of these theories is enhanced by the fact that many candidates win or lose more than once. For instance, five of these twentieth-century winners were Roosevelts, which happens to be a long name that begins with a letter late in the alphabet. Once we've noticed all the Roosevelt victories, it is easy to find theories that work. Nominate a candidate named Roosevelt, for instance. Or if you want to be more adventurous, nominate a candidate whose last name has an S in the middle, but not on either end. There have been 11 such candidates and they all have won. If you believe this theory, that "S-ism" leads voters to favor candidates with an S in the middle of their name, then you don't yet understand the pitfalls of data mining.

Baseball and the President

The President of the United States has often thrown out the first pitch to start a new baseball season. What many people do not realize is that during presidential election years, the league that wins the World Series determines which party wins the White House. *The New York Times* carried an article in October 1976 explaining that the Democrats win the White House if the National League wins the World Series and the Republicans win if the American League does. As of 1976, there had been 18 presidential elections since the first World Series in 1903, and this theory worked 12 out of 18 times. In the modern baseball era, since 1940, the theory has been right 8 out of 9 times—the exception being 1948, when Truman surprised everybody in winning. So, with the Cincinnati Reds winning the 1976 World Series, this article concluded, "I am pleased to announce that the winner of the 1976 Presidential election will be Jimmy Carter." And what do you know, it worked again! (In 1980, though, the National League's Phillies beat Kansas City, while the Democrat Carter lost to Reagan; in 1984, the American League's Detroit Tigers won and so did Reagan; and in 1988 it was the National League Dodgers and the Republican Bush.)

Football and the Stock Market

On Super Bowl Sunday in January 1983, both the business and sports sections of the *Los Angeles Times* carried articles on the "Super Bowl Stock Market Predictor."[11] The theory is that the stock market goes up if the National Football Conference (or a former NFL team now in the American Football Conference) wins the Super Bowl and goes down if the AFC wins. This theory had proven correct for 15 out of 16 Super Bowls—the one exception was 1970, when Kansas City beat Minnesota and the market went up 0.1 percent. Thus the *Times* quoted a broker for Dean Witter Reynolds as saying, "Market observers will be glued to their TV screens. . . . Today it will be hard to ignore an S&P [market index] indicator with an accuracy quotient that's greater than 94 percent." And indeed, it was. An NFC team, the Washington Redskins, won, the market went up, and the Super Bowl system was back in the newspapers the next year, stronger than ever. In 1984, however, the AFC's L.A. Raiders won and the market went up anyway. From its discovery through 1989, the Super Bowl system has been correct six out of seven times.

The accuracy of the Super Bowl system is obviously just an amusing coincidence, since the stock market has nothing to do with the outcome of a football game. Over these 23 years, the stock market has generally gone up and the NFL has usually won the Super Bowl. The correlation is made more startling by the gimmick of including the Pittsburgh Steelers, an AFC team, in with the NFL. The excuse is that Pittsburgh once was in the NFL; the real reason is that Pittsburgh won the Super Bowl four times in years when the market went up.

Inspired by the success of the Super Bowl indicator, a *Los Angeles Times* staff writer did a little data mining in 1989 and discovered several comparable coincidences.[12] The "Yo, Adrian Theory" holds that if a Rocky or Rambo movie is released, the stock market will go up that year. On average, the Dow increased by 173 points in the four years when a Rocky movie appeared and 245 points in the years when a Rambo movie was released. The Geraldo Rivera indicator notes that the Dow has fallen an average of 13 points on the day after seven major Geraldo Rivera specials. The George Steinbrenner indicator monitors his firing of Yankee managers. On the five occasions that he fired Billy Martin, the Dow rose an average of 5 points the following day; the eight times Steinbrenner fired someone else, the Dow fell an average of 3 points the next day.

Patterns in the Stock Market

In 1982, *Fortune* magazine told of a company selling a computer program for forecasting the stock market 80 days ahead. The company selling this program claimed to have found a complex pattern in stock prices that would have enabled a user to realize a 72.1 percent profit over the preceding three years, while the stock market as a whole went down 6.1 percent. Just like my 20 coin flips and the 50,000 dice rolls, it should be easy to find such apparent patterns even if the stock market, like coin flips and dice rolls, is completely random.

A *Fortune* editor ordered one of these programs and used it for 80 days, from August 13 through December 6, 1982. The August 13th forecast was that the market would drop 7.5 percent over the next 80 days. Indeed, every single day during this 80-day period, it advised selling stocks. Well, it turned out that these three months were the heart of the great 1982 stock market rally. The Dow Jones Industrial Average went up 11 points on August 13, and 300 more points over the next few weeks. This was one of the strongest advances in stock market history and the computer program missed it completely. Back to the charts.

Extrasensory Perception

Professor J.B. Rhine and his associates at Duke University produced an enormous amount of evidence concerning extrasensory perception (ESP). For example, they have conducted several million trials involving persons attempting to identify cards that they cannot see. For a stylized illustration, consider an experiment with five cards numbered 1, 2, 3, 4, and 5. (In practice, they usually use five pictures instead of five numbers, but this distinction is of no statistical importance.) The five cards are shuffled and a card is selected at random. The subject states which of the five cards he or she believes to have been selected and the results are recorded. The experiment is repeated 25 times.

How many cards must the subject identify correctly to reject the null hypothesis of no ESP? Under the null hypothesis that mere chance is involved, the probability of a correct selection is $1/5 = .2$. Because these are Benoulli trials, with independent outcomes and constant probabilities, we can use the binomial distribution. The most likely result for a pure guesser is to be right 1/5 of the time: 5 right and 20 wrong. Some of the exact binomial probabilities are

Number of Successes ($\pi = .2$, $n = 25$)	Probability
0	.0038
1	.0236
2	.0708
3	.1358
4	.1867
5	.1960
6	.1633
7	.1108
8 ———— cutoff	.0623
9	.0294
10	.0118
11	.0040
12	.0012
13	.0003

For a 5 percent probability of Type I error, we can draw the cutoff line between 8 and 9 correct. A pure guesser has only a .047 chance of getting more than 8 out of 25 correct. Therefore, 9 or more correct is sufficiently improbable to enable us to reject the null hypothesis of no ESP.

A messy complication is that many people are tested. Even if only guessers are tested, about 1 out of every 20 guessers will get at least 9 correct. If 400 guessers are tested, we expect about 20 of them to do sufficiently well to lead to the conclusion that they have ESP. If millions of tests are conducted, some astoundingly remarkable performances should be found, even if nothing more than lucky guesses are involved.

This, then, can be seen as another form of data mining—run a large number of tests and report only the most remarkable results. A score with a one-in-a-thousand chance of occurring is not really so remarkable, if you take into account the fact that this is just one of out of a thousand tests that were conducted.

Our first antidote to data mining is common sense. Here, common sense is divided. Many people do not believe in ESP, but others do. Perhaps the safest thing to say is that if there is such a thing as ESP, the evidence is not yet strong enough to convince the skeptics. Even if ESP is of statistical significance, it has yet to be shown to have practical importance. It is of little or no importance to find three people whose probability of identifying a chosen card is .2001 instead of .2000. There is certainly no public evidence that people have such strong mental powers that they can knock planes out of the sky, have long distance conversations without a telephone, or even win in Las Vegas.

The second antidote to data mining is fresh data. Those people who score well should be retested to see whether it is ESP or data mining. Of course, with many,

Beating the Lottery

State lotteries, especially the multimillion dollar Lotto payoffs, offer people a chance to change their lives completely—and the popularity of Lotto games indicates that many are very eager to do so. The catch, of course, is that the chances of winning a Lotto game are minuscule. This chasm between dreams and reality gives entrepreneurs an opportunity to sell gimmicks that purportedly enhance one's chances of winning.

A mail order company sells a Millionaire Maker for $19.95, using this pitch:

Lotto players face a dilemma each time they buy a ticket. What numbers to pick? Studies have shown that most Lotto *winners* don't use any sort of special system to select their numbers. Instead, they tap the power of *random selection.*[13]

Their $19.95 product for tapping the power of random selection is a battery-powered sphere filled with numbered balls that "mixes the balls thoroughly" and chooses a "perfectly random set of numbers."

Gail Howard sells a very different system, a report on how to handicap lottery numbers, just like handicapping horses or doing a technical analysis of stock prices.[14] She has reportedly appeared on the "Good Morning America" television show, written an article in *Family Circle* magazine, and publishes a monthly report ("Lottery Buster")—"But you'd better hurry, while distribution of this controversial information is still permitted."

Howard's report says, "You don't have to worry about the state lottery games being crooked or rigged. . . . The winning numbers are selected through a completely random process." Nonetheless, she offers several tips for "greatly improving your chances of winning." For instance, don't bet on 6 consecutive numbers or 6 numbers that have won before, because these are highly unlikely to win (as are any 6 numbers, a skeptic would point out). Her main tip is to chart the winning numbers, look for trends, and bet hot numbers because "ABOUT HALF OF THE WINNING NUMBERS DRAWN HAVE HAD A HIT WITHIN THE PREVIOUS *THREE* GAMES!!" Not coincidentally, she sells an encyclopedia of winning numbers. (When 6 numbers are drawn from 44, the probability that a particular number will be picked is 6/44. The probability that it will be picked at least once in three games is one minus the probability that it won't be picked: $1 - (38/44)^3 = .36$.)

No doubt, just as careful study can uncover worthless patterns in coin flips or dice rolls, so too can tedious scrutiny discern worthless patterns in lottery numbers. The most persuasive reason for skepticism about lottery-beating systems is that anyone who really had a system that worked would become rich buying lottery tickets, rather than peddling books and gadgets.

many subjects, some will doubtless compile high scores through several rounds. When Rhine retested his high scorers, he reported that they almost always show a marked decline in ability after their initial success. He wrote that, "This fatigue is understandable . . . in view of the loss of original curiosity and initial enthusiasm." An alternative explanation is that the high scores on the early rounds were just lucky guesses.

The large number of subjects tested isn't the only difficulty in assessing the statistical significance of Rhine's results. He also looked for test periods where the subjects did well: at the start of a test, in the middle, or at the end. This is very similar to the statistical pitfall of testing several people and reporting only the highest scores. A remarkable performance by 1 of 20 subjects or by 1 of 5 subjects on one-fourth of a test isn't really that remarkable.

Rhine also looked for either "forward displacement" or "backward displacement," where the subjects' choices didn't match the contemporaneous cards, but did match the next card, the previous card, two cards hence, or two cards previous. In this way, one test can be multiplied into several tests, increasing the chances of finding "remarkable" coincidences. Rhine also considered it remarkable when there was "avoidance of target" or "negative ESP," in which the subject got an unusually low score. Rhine explained how a subject

> . . . may begin his run on one side of the mean and swing to the other side just as far by the time he ends the run; or he may go below in the middle of the run and above at both ends. The two trends of deviations may cancel each other and the series as a whole average close to "chance."

He described the results of one test as follows:

> . . . the displacement was both forward and backward when one of the senders was looking at the card, and only in the forward direction with another individual as sender; and whether the displacement shifted to the first or the second card away from the target depended upon the speed of the test.[15]

With so many subjects and so many possibilities, it should be easy to find patterns, even in guesses.

ESP research is very interesting and even provocative. But, unfortunately, the enthusiasm of the researchers has often led to a variety of forms of data mining. They would be less controversial and more convincing—one way or the other—if there were more uses and fewer abuses of statistical tests.

EXERCISES

11.24 The following letter to the editor was printed in *Sports Illustrated:*

> In regard to the observation made by *Sports Illustrated* correspondent Ted O'Leary concerning this year's bowl games (Scorecard, January 16), I believe I can offer an even greater constant. On Jan. 2 the team that won in all five games started the game going right-to-left on the television screen. I should know, I lost a total of 10 bets to my parents because I had left-to-right in our family wagers this year.
>
> John Doe
> Anytown, U.S.A.

Under the null hypothesis that the eventual winner is equally likely to have started the game from the left or right side of the television screen, what is the probability that all five winners would start on the right side? Are you convinced by this statistical proof that the right-to-left team has an advantage? Is there any logical explanation for this apparent advantage?

11.25 Scott Ostler, a *Los Angeles Times* reporter, noticed that in 1983 the San Francisco 49ers were 11–0 in games in which they flew in DC10s. Does this astounding fact persuade you that the 49ers would benefit from only flying DC10s? What hesitations, if any, might you have about jumping to this conclusion?

11.26 Critically evaluate the following statistical argument that a god must have shaped the history of the United States:

Q—What are the odds against accident that the 2nd and 3rd presidents should both die on the 50th anniversary of 7/4/1776, on the 50th Jubilee Day July 4, 1826?

A—The odds against a person taking any action on any particular day of the calendar is 365/1. So it is that any two persons dieing on the same day of the year is 365/1 $\times$ 365/1. That *two presidents* of the United States, should both die on the *day of independence of that nation* is then 365/1 $\times$ 365/1 $\times$ 365/1.

The odds that two individuals present at an event should both die on the 50th Jubilee Day of that event are then increased by 50/1 for each person, the odds then are ... 365 $\times$ 365 $\times$ 365 $\times$ 50 $\times$ 50 = odds against accident are 121,568,812,500/1....

Reason says—This can never be an accident. Some cause must be sought that can do this thing.[16]

11.27 Choose a subject and conduct an ESP test. Carefully record how you conducted your test and what your results were.

11.28 An ESP test has been set up with five different cards. These five cards will be shuffled, a prediction made, and one card selected. If this procedure is repeated 25 times, how many correct predictions are needed to reject the no-ESP hypothesis   **a.** at the 5 percent level?   **b.** at the 1 percent level?

• **11.29** *(continuation)* If everyone in your class conducts an ESP test, what is the probability that, by chance alone, at least one person will obtain scores that are statistically significant at the 5 percent level? at the 1 percent level?

• **11.30** *(continuation)* An ESP test has been calibrated so that there is a 5 percent chance that a pure guesser will make enough correct predictions to pass. If it turns out that there are not enough correct predictions, then the researcher will look for "negative ESP," a statistically significant (at the 5 percent level) small number of correct predictions. If this doesn't work, then the researcher looks for forward displacement by one card, forward displacement by two cards, backward displacement by one card, and finally backward displacement by two cards (in each case, at the 5 percent level). If we assume, for simplicity, that the results of each of these six tests are independent, then what is the approximate probability that a pure guesser will demonstrate a statistically significant amount of ESP on at least one of these tests?

11.5   SUMMARY

Hypothesis tests focus on the probability of incorrectly rejecting a particular hypothesis, the null hypothesis. By setting up the test to give a low probability of Type I error, the researcher ensures that sampling error is unlikely to lead to the incorrect rejection of this particular hypothesis. Which specific hypothesis should be chosen for this special attention? At least three considerations may influence the choice of the null hypothesis and the probability of Type I error.

1. Theoretical logic or previous empirical studies may indicate that one hypothesis is likely to be true. If so, we can guard against a hasty rejection based solely on this test, by making this "conventional wisdom" the null hypothesis.

2. If there are serious consequences of incorrectly rejecting a particular hypothesis, then we can keep the chances of making that error slight by making this hypothesis the null hypothesis.

3. A researcher may want to reject a particular hypothesis convincingly to demonstrate the correctness of the alternative. Often, the null hypothesis is a "no effect" hypothesis that the researcher hopes will be rejected decisively.

A 5 percent probability of Type I error is commonplace, but there is little more than acquired habit behind this particular number. The objectives just enumerated may argue for a lower or higher number. For instance, a very low α defuses the argument that the results may well be due to chance. One attractive approach is simply to report the P-value and let the reader decide whether or not it is convincingly low. Still, the question most readers will instinctively ask is whether or not the reported P-value is less than 5 percent.

Statistical significance is not at all the same as practical importance. A statistically significant result is one that cannot be reasonably explained by sampling error. Such a result may well be of little or no practical importance. Conversely, research may indicate a very important phenomenon, even though there are too few data to rule out sampling error. A confidence interval is an informative reporting format that allows readers to judge for themselves whether the result is both important and statistically significant.

Hypothesis tests reject a selected hypothesis by showing that, were the hypothesis true, the observed data would be very improbable. However, the testing of many hypotheses makes it very likely that the determined researcher will find at least one case in which the data, considered in isolation, seem improbable. It is not at all improbable that an examination of 20 situations will yield 1 situation with only a 5 percent chance of occurring. Because data mining will inevitably lead to statistically significant results, we cannot tell whether the results of such an expedition reflect the discovery of a useful theory or just perseverance. The only way to tell is by using common sense and fresh data.

APPENDIX: A BAYESIAN RATIONALE

If we adopt a Bayesian perspective for a few moments, an intuitive rationale can be provided for the structure of hypothesis tests. To keep the analytics simple, consider the choice between the two hypotheses

$$H_0: \mu = 500$$

$$H_1: \mu = 550$$

The probabilities of Type I and Type II errors are

$$\alpha = P[\text{reject } H_0 \mid H_0 \text{ true}]$$

$$\beta = P[\text{accept } H_0 \mid H_1 \text{ true}]$$

For a Bayesian, it is more natural to inquire of the reverse error probabilities,

$$P[H_0 \text{ true} | \text{reject } H_0]$$

$$P[H_1 \text{ true} | \text{accept } H_0]$$

The conversion from the first set of probabilities to the second is made via Bayes' theorem, using subjective, a priori probabilities for the two hypotheses, $P[H_0 \text{ true}]$ and $P[H_1 \text{ true}]$. For example,

$$P|H_0 \text{ true} | \text{reject } H_0] = \frac{P[H_0 \text{ true}]P[\text{reject } H_0 \mid H_0 \text{ true}]}{P[H_0]P[\text{reject } H_0 \mid H_0 \text{ true}] + P[H_1]P[\text{reject } H_0 \mid H_1 \text{ true}]}$$

$$= \frac{P[H_0 \text{ true}]\alpha}{P[H_0 \text{ true}]\alpha + P[H_1 \text{ true}](1 - \beta)}$$

Similarly,

$$P[H_1 \text{ true} | \text{accept } H_0] = \frac{P[H_1 \text{ true}]\beta}{P[H_1 \text{ true}]\beta + P[H_0 \text{ true}](1 - \alpha)}$$

If the two hypotheses are, a priori, judged to be equally likely,

$$P[H_0 \text{ true}] = P[H_1 \text{ true}]$$

then the Bayesian conditional error probabilities become

$$P[H_0 \text{ true} | \text{reject } H_0] = \frac{\alpha}{\alpha + 1 - \beta}$$

$$P[H_1 \text{ true} | \text{accept } H_0] = \frac{\beta}{\beta + 1 - \alpha}$$

The 533 cutoff analyzed at the beginning of the chapter gives $\alpha = .05$ and $\beta = .1977$, so that

$$P[H_0 \text{ true} | \text{reject } H_0] = .059$$

$$P[H_1 \text{ true} | \text{accept } H_0] = .172$$

Here it turns out that the reverse probabilities are not far from α and β, but this isn't necessarily so. For example, if the (equally likely) alternative hypothesis is $\mu = 525$, then $\beta = .6554$ and

$$P[H_0 \text{ true} | \text{reject } H_0] = .13$$

$$P[H_1 \text{ true} | \text{accept } H_0] = .41$$

If the null and alternative hypotheses are not perceived to be equally likely, then there is even more room for disparities.

A Bayesian would report these revised probabilities and let the user decide how to use them. A yes-no decision, such as accepting or rejecting hypotheses, requires more than probabilities. We have to know what the consequences of such decisions will be. We'll use the symbols

$$G_0 = \text{gain if accept } H_0 \text{ when } H_0 \text{ is true}$$

$$G_1 = \text{gain if accept } H_1 \text{ when } H_1 \text{ is true}$$

$$L_0 = \text{loss if reject } H_0 \text{ when } H_0 \text{ is true}$$

$$L_1 = \text{loss if reject } H_1 \text{ when } H_1 \text{ is true}$$

These gains and losses should be measured in dollars or whatever, so as to express the relative seriousness of the consequences. The net expected gain is then

$$\begin{aligned}
E[\text{gain}] &= P[H_0 \text{ true and accept } H_0](G_0) + P[H_0 \text{ true and reject } H_0](-L_0) \\
&\quad + P[H_1 \text{ true and accept } H_1](G_1) + P[H_1 \text{ true and reject } H_1](-L_1) \\
&= P[H_0 \text{ true}]\{(1 - \alpha)G_0 - \alpha L_0\} + P[H_1 \text{ true}]\{(1 - \beta)G_1 - \beta L_1\} \quad (10.1)
\end{aligned}$$

Let's take as a starting point the case where the competing hypotheses are equally likely and the costs are symmetrical,

$$P[H_0 \text{ true}] = P[H_1 \text{ true}], \quad G_0 = G_1, \text{ and } L_0 = L_1$$

In this central case, the expected gain is

$$E[\text{gain}] = .5\{G_0 - (\alpha - \beta)(G_0 + L_0)\}$$

The expected gain is maximized if $(\alpha + \beta)$ is minimized, and this occurs when the cutoff is put halfway between the two hypotheses, at $\overline{X} = 525$. You can see from the shapes of the curves (or from some quick calculations) that α rises more than β falls if the cutoff is reduced below this midway point, and β rises more than α falls if the cutoff is increased. So, the first conclusion (and it makes good sense) is:

> If the two hypotheses are equally likely and the potential gains and losses are symmetrical, then the cutoff should be midway between the hypotheses, at the point where $\alpha = \beta$.

Conversely, intuition or the application of a little arithmetic to Equation 10.1 shows that, if the hypotheses are not judged to be equally likely or if the gains or losses are not symmetrical, then the cutoff should be set so that α and β are not equal. In particular, because

lower α implies less likely to reject H_0 if H_0 is true

higher β implies more likely to accept H_0 if H_1 is true

it follows that α should be set lower than β if

1. $P[H_0 \text{ true}] > P[H_1 \text{ true}]$. Because H_0 is the more likely hypothesis, we want to reduce the chances of making an error if H_0 is true.

2. $L_0 > L_1$. If the losses from incorrectly rejecting H_0 are greater than the losses from incorrectly rejecting H_1, then we want to reduce the chances of incorrectly rejecting H_0.

3. $G_0 > G_1$. If the gains from correctly accepting H_0 are greater than the gains from correctly accepting H_1, then we want to reduce the chances of incorrectly rejecting H_0.

With more than two hypotheses, the analysis and the conclusions are considerably more complex, but the flavor lingers.

REVIEW EXERCISES

11.31 The 1980 census recorded the sizes of 79,108,000 U.S. households. The distribution (shown in Table 4.2) has a mean of 2.75 persons and a standard deviation of 1.56 persons. Why shouldn't you use these data to test the null hypothesis that the average size of U.S. households is 3.0?

11.32 Explain what is wrong with these conclusions: **a.** The data show that this person has a statistically significant ability to forecast whether the stock market is going up or down. Therefore, we are 95 percent sure that this person can predict which way the market is going. **b.** The data do not show a statistically significant ability to forecast in which direction the market is going. Therefore, we are 95 percent certain that this person cannot predict which way the market is headed. **c.** This person's forecasting record has a P-value of 8 percent. Because it is significant at the 10 percent level, but not the 5 percent level, the data show that this person has a weak forecasting ability.

11.33 It has been said that the null hypothesis is the "favored" hypothesis. Is it always true that the probability of accepting the null hypothesis when it is, in fact, correct is greater than the probability
a. of rejecting the null hypothesis when it is false?
b. of accepting the alternative hypothesis when it is true?

11.34 It has been claimed that pickles cause cancer, war, communism, and travel accidents:[17] 99.9 percent of all cancer victims have eaten pickles during their lifetimes, as have 100 percent of all soldiers, 96.8 percent of all communists, and 99.7 percent of those involved in auto and plane accidents. Do you agree that this evidence is persuasive?

11.35 What is the probability that 13 cards randomly dealt from a standard deck of cards will all be the same suit? Oswald Jacoby, a backgammon, bridge, and poker expert, has estimated that four or five times a year someone playing bridge in the United States should be dealt a hand in which all thirteen cards are the same suit. Yet two dozen or so such hands get reported each year, and the suit is usually spades. How would you explain this phenomenon?

11.36 On the eve of the 1987 Super Bowl, *Sports Illustrated* advised that

> there may be something to the J theory, as elucidated by San Francisco accountant Steve Carroll. Carroll expects the Redskins or 49ers to come out on top because they fit the trend of this decade. "All six 1980s Super Bowls were won by teams with starting quarterbacks whose first names began with the letter J and had three letters," says Carroll. The Raiders' Jim Plunkett was on the winning side in 1981 and '84; Joe Montana led the 49ers to titles in '82 and '85; Joe Theismann's Redskins won in '83, and Jim McMahon's Bears won last year.[18]

How would you, as a statistician, respond to this evidence in support of the J theory?

11.37 A certain tire has an estimated life of 45,000 miles with a standard deviation of 8,000 miles. A competitor tests 1,000 tires made of a new "space-age" material and finds that 20 of these tires last more than 60,000 miles. Why is it statistically misleading for this company to launch an advertising campaign showing these 20 tires and announcing that, "In actual road tests, these tires lasted more than 60,000 miles!"

11.38 Table 3 at the end of this book is supposed to contain random numbers. But look at the twenty-second through twenty-fourth numbers in the twenty-fifty row (20, 73, 17) and the eighth through fifth numbers in the thirtieth row—they are the same! The chances that three randomly selected numbers will match three other randomly selected numbers is 1/1,000,000. Does this discovered match show that these are not random numbers after all?

11.39 It has been argued that a football bettor should bet on the team that had the most lopsided win in the previous week. As evidence, it was reported that during the 1970 season, this strategy would have produced 11 winning bets and only 2 losses, taking the bookmaker's point spread into account. An investigation of the strategy over the six-year period 1969–1974 yielded these results:[19]

Season	Wins	Losses
1969	5	8
1970	11	2
1971	6	7
1972	2	11

Season	Wins	Losses
1973	8	5
1974	4	9
Total	36	42

The researchers concluded that the 1970 season was "distorted evidence," reflecting a chance aberration. Why is it a distortion to focus on the 1970 season and ignore the data from other years? Do the results from these six years as a whole, 36 wins and 42 losses, show a statistically significant deviation at the 5 percent level from the null hypothesis that the probability of a winning bet is $\pi = .5$?

11.40 Critically evaluate this news article:

BRIDGEPORT, Conn.—Christina and Timothy Heald beat "incredible" odds yesterday by having their third Independence Day baby.

Mrs. Heald, 31, of Milford delivered a healthy 8-pound 3-ounce boy at 11:09 a.m. in Park City Hospital, where her first two children, Jennifer and Brian, were born on Independence Day in 1978 and 1980, respectively.

Mrs. Heald's mother, Eva Cassidy of Trumbull, said a neighbor who is an accountant figured the odds are 1-in-484 million against one couple's having three children born on the same date in three different years.[20]

11.41 X is a normally distributed random variable with a known standard deviation of 100. We want to choose between the following two, equally likely, hypotheses about the expected value of X:

$$H_0: \mu = 100$$
$$H_1: \mu = 150$$

The mean of a random sample of size 25 is found to be 136. At the 5 percent level of significance, do we accept $\mu = 100$ or 150?

If we had set the hypotheses up as

$$H_0: \mu = 150$$
$$H_1: \mu = 100$$

which hypothesis would have been accepted? For which values of the sample mean does it make a difference whether the null hypothesis is $\mu = 100$ or $\mu = 150$?

• **11.42** *(continuation)* Now assume that the costs of incorrectly concluding that $\mu = 100$ are greater than the costs of incorrectly concluding that $\mu = 150$

(and that there are no costs associated with being correct). Will a test at the 5 percent level have lower expected costs if the null hypothesis is $\mu = 100$ or $\mu = 150$?

•• **11.43** *(continuation)* It is decided to let the null hypothesis be $\mu = 100$ and let the alternative hypothesis be $\mu = 150$. The cutoff point is set so that the probability of Type I error is 5 percent. If the two hypotheses are initially regarded as being equally likely to be correct, what is the probability that H_0 is true if we reject H_0? What is the probability that H_0 is false if we accept H_0?

•• **11.44** It is known that the random variable X has this continuous density function

$$f[x] = 1/\theta \qquad \text{for } 0 \le x \le \theta$$
$$f[x] = 0 \qquad \text{otherwise}$$

but it is not known whether the value of θ is 1 or 2. You will have just a single observation of X on which to base your choice between

$$H_0: \theta = 1$$
$$H_1: \theta = 2$$

a. What are the probabilities of Type I and Type II errors if you decide to accept $\theta = 1$ if $X \le 0.9$ and to accept $\theta = 2$ if $X > 0.9$? **b.** What are the probabilities of Type I and Type II errors if you decide to accept $\theta = 1$ if $X \le .5$ and to accept $\theta = 2$ if $X > .5$? **c.** Which of these two rules is more appropriate if the consequences of a Type II error are more serious than the consequences of a Type I error? **d.** Which of these two rules is more appropriate if, *a priori*, you think that $\theta = 1$ is the more likely value?

• **11.45** "As January goes, so goes the year" is an old stock market adage. The idea is that if the market goes up in January, then it will also go up for the entire year; if the market drops in January, then it will be a bad year. *Newsweek* cited this theory in its February 4, 1974 issue, and reported that "this rule of thumb has proved correct in twenty of the past 24 years."[21] Let's assume that the probability that the stock market will go up in any given month is .5 and that the probability that the market will go up in any given year is .5. If the market's performance in January and for the entire year are independent, then what is the probability that the market will go the same direction in January and during the entire year? Use the given data to test the null hypothesis that how the market does in January is independent of how it does during the entire year.

• **11.46** *(continuation)* Check data on the Dow Jones Industrial Average for the years 1974 to the present to see how well this theory has done since the *Newsweek* article. Use these data to test the null hypothesis that how the market does in January is independent of how it does during the entire year.

•• **11.47** *(continuation)* Still assume that how the market does each month and how it does for the entire year are completely independent. Also assume that the probabilities that the market will go up in any given month and for the entire year are both .5. If we examine 24 years of data, what is the probability that we will find some month (January, February, or some other month) for which, in at least 20 out of 24 cases, the market has gone the same direction in that month and for the entire year?

•• **11.48** *(continuation)* There are at least two problems with the calculations so far. First, January and the entire year are not really independent because January is part of the year. What we should be looking at (both as statisticians and as investors) is January and the remainder of the year (after January). Second, we shouldn't be using the same probabilities for a month and for a year (unless P really is .5, which it isn't).

To remedy these deficiencies, go to the library and collect some actual data for the years 1950 to the present. For each year, see whether the stock market (as measured by the Dow Jones Industrial Average) went up or down in January and during the remainder of the year. Use these data to test the null hypothesis that the market's performance in January is unrelated to its performance during the remainder of the year.

•• **11.49** The Usual Foods Company claims that there are, on average, 100 raisins in a box of its cereal. To test this claim,

$$H_0: \mu \ge 100$$

a consumer group considers the following possible test designs:

1. $n = 25$, $\alpha = .05$
2. $n = 25$, $\alpha = .01$
3. $n = 100$, $\alpha = .05$

a. Which design has the smallest probability of falsely rejecting the company's claim? **b.** Does design 1 or

2 have the smaller probability of incorrectly accepting the company's claim? Why? **c.** Does design 1 or 3 have the smaller probability of incorrectly accepting the company's claim? Why?

•• **11.50** A bag contains 4 balls, of which n are black and $4 - n$ are white. The null hypothesis is that there are an equal number of black and white balls. The alternative hypothesis is that the numbers are unequal:

$$H_0: n = 2$$

$$H_1: n \neq 2$$

We will draw 2 balls from the bag and accept the null hypothesis if 1 ball is black and the other white. We will reject the null hypothesis if the 2 balls are the same color. Find the probabilities of Type I and Type II errors if **a.** the sampling is without replacement. **b.** the sampling is with replacement. Which sampling procedure is superior?

11.51 A doctor's credo is "first do no harm." Does this suggest that the null hypothesis for an untested drug should be that it is safe or that it is dangerous?

11.52 A company has developed a new product that it will market in either red or blue packages. A one-month marketing test is made of the null hypotheses

$$H_0: \pi = 0.5$$

$$H_1: \pi \neq 0.5$$

where π is the probability that a purchased package is red and the null hypothesis is that the customer is as likely to buy a blue package as a red one. It turns out that 160,000 of these items are sold, of which 80,432 are red and 79,568 blue. Is this difference statistically significant at the 5 percent level? (Use a two-tailed test and a normal approximation.) Does the difference seem substantial to you? Can there ever be a conflict between statistical significance and practical importance?

11.53 A machine produces 180,000 pieces a year, of which 2 percent are defective and must be replaced at a cost of $5 apiece. A salesman claims that using teflon gaskets in the machine will reduce the number of defectives significantly. In a one-month trial, 15,000 pieces are produced, of which 250 are defective. Is this difference statistically significant at

the 5 percent level? (Use a one-tailed test and a normal approximation.) If the teflon gaskets must be replaced each year at an additional cost of $5,000, would you recommend a switch?

11.54 In December 1984 *Sports Illustrated* reported that

> Mindy, 6, a dolphin at the Minnesota Zoo, predicts the outcomes of NFL games by choosing among pieces of Plexiglas, each bearing a different team's name, that are dropped into its pool. This season, Mindy has a 32–21 record.

Is this record sufficient to reject (at the 5 percent level) the null hypothesis that Mindy has a .5 probability of picking the winner? Use a two-tailed test and the exact binomial P-value. Explain why this report might be considered "data mining."

11.55 In 1984 the *Boston Globe* reported

> Reader Ismael Dambolena, Associate Professor of Quantitative Analysis at Babson College, points out that the Lakers won all the Sunday games against the Celtics, while Boston won on the weeknights. "Had each team had an even likelihood of winning each game, the probability of this pattern occurring by chance alone would have been less than 2 percent," writes Dambolena.[22]

Why aren't you persuaded that the Lakers have a better chance of winning on Sunday and that Boston should be considered the favorite on weeknights?

11.56 After a family Thanksgiving dinner, a statistics student conducted a simple ESP experiment.[23] He shuffled 20 cards, each labeled with a single number from 1 to 10, drew one card and concentrated on it while each of his 23 family members present wrote down a guess. He reshuffled the cards and repeated the process 20 times. One of the 23 relatives got 7 correct and another got 5 right. What is the probability that random guessing will get 5 or more correct? (Do not use a normal approximation.) What is the probability that 2 or more people out of 23 random guessers will get 5 or more correct? When he repeated the experiment for those who got more than 5 correct, one got 2 right and the other none; how do you explain this curious reversal?

11.57 The *Billboard Book of Top Forty Hits* lists the peak rating of every song that has made the weekly Top 40. A number 1 song was, during some week, rated the most popular song in the United States. A number 15 song was never rated higher than 15. A 1990 compilation of Top 40 songs calculated that the peak rating averaged 17.83 with a standard deviation of 11.84.[24] The Beatles had 49 Top 40 songs with an average peak rating of 9.04. Test the null hypothesis (at the 1 percent level) that 49 songs selected at random from a distribution with a mean rating of 17.83 and a standard deviation of 11.84 would have an average rating of 9.04 or less. Do you have to assume a normal distribution?

11.58 *(continuation)* Make similar calculations for The Supremes (33 songs with an average rating of 11.67) and The Rolling Stones (39 songs with an average rating of 12.00).

11.59 *(continuation)* What weaknesses do you perceive in these statistical tests of a group's popularity?

11.60 In the 1977 case of *Ballew* v. *Georgia,* the Supreme Court opinion, written by Justice Blackmun, stated

> Statistical studies suggest that the risk of convicting an innocent person [Type I error] rises as the size of the jury diminishes. Because the risk of not convicting a guilty person [Type II error] increases with the size of the panel, an optimal jury size can be selected as a function of the interaction between the two risks.

What is the null hypothesis?

• **11.61** *(continuation)* Why do you suppose that an increase in jury size reduces the probability of Type I error but increases the probability of Type II error?

11.62 The results of hypothesis tests are often misinterpreted in the popular press. For instance, explain how this 1985 *Business Week* summary of a Data Resources Incorporated (DRI) study of insider trading in the stock market is misleading.

> DRI also compared the action of a sample of takeover stocks in the month before the announcement with the action of those same stocks in other, less significant one-month intervals. The conclusion: There was only 1 chance in 20 that the strength of the takeover stocks was a fluke. In short, the odds are overwhelming that inside information is what made these stocks move.[25]

11.63 The U.S. Color Additive Amendments of 1960 require that a color additive be "deemed unsafe" unless the Commissioner of Food and Drugs has issued a regulation stating that the additive has been found suitable and safe for general use or for a specified particular use. Why do you suppose that, implicitly, the null hypothesis in tests of color additives is that the additive is unsafe? That is, why is a color additive presumed unsafe until proven safe, rather than assumed safe until proven unsafe?

11.64 Explain the error in this explanation of how courts should interpret statistical significance levels in legal evidence: "The findings of statistical significance at the $P < .05$, $P < .04$, and $P < .02$ levels indicate that the court can be 95%, 96%, and 98% certain, respectively, that the null hypothesis . . . should be rejected."[26]

11.65 In 1989 the *New York Times* reported that, in 12 of the previous 14 years, if the Dow Jones Industrial Average increased between the end of November and the time of the Super Bowl, the football team whose city comes second alphabetically won the Super Bowl.[27] Thus the Dow increased from 2,115 to 2,235 between November 30, 1989, and the day of the January 1990 Super Bowl and San Francisco beat Cincinnati. How would you explain the success of this theory? Why did they choose the end of November for their starting date?

11.66 An ESP study involved 20 trials with 120 college students, 2,400 trials in all. In each trial, the researcher shuffled four cards (marked with a circle, square, triangle, and star), selected one card, and studied it. The student participant was asked to determine the shape that had been selected by concentrating on the researcher's thoughts.[28] In all, 644 of the student responses were correct and 1,756 were incorrect. Use a normal approximation to the binomial to calculate the Z-value and P-value under the null hypothesis that there is no ESP. Does this evidence reject the null hypothesis at the 1 percent level using a one-tailed test? Does the difference between the students' performance and random guessing seem substantial to you?

11.67 *(continuation)* Each student did 20 trials. Use a binomial distribution to determine the number of correct answers that a single subject must make to reject the null hypothesis that there is no ESP at the 1 percent level using a one-tailed test.

11.68 *(continuation)* The researcher wrote that "individual scores, however, were often surprisingly high. Ten students, for example, correctly identified [exactly] eight of the twenty cards." Two other students correctly identified nine cards; none identified more than nine. Use the binomial distribution to determine the probability of correctly identifying at least eight of twenty cards by guessing alone. In tests of 120 guessers, what is the expected value of the number of volunteers who will correctly identify eight or more cards? Do you agree that the individual scores in the reported tests were surprisingly high?

11.69 *(continuation)* Thirty-two subjects who correctly identified 7 or more cards were retested and, in the aggregate, correctly identified 183 of 640 cards. Use a normal approximation with a continuity correction to calculate the Z-value and P-value for these data. Is the aggregate reported ESP ability statistically significant using a one-tailed test at the 1 percent level?

11.70 A reader of the ESP experiment described in Exercise 11.66 wrote: "In conducting 2,400 trials, your 'random shuffle' may subconsciously have been intended mainly to get the previously selected card away from the top of the deck. If you announce the outcome of the previous trial and the subjects believe in the fallacious law of averages, this nonrandom shuffle will give them an advantage." Explain.

11.71 A psychologist has written that "extraordinary claims require extraordinary proof."[29] Does this argument imply that an ESP test with no ESP as the null hypothesis should be designed to allow a high or low probability of Type I or Type II error?

11.72 Home teams won 1,157 of 2,106 games during the 1989 major league baseball season. Assume that these are a random sample of all professional baseball games and use these data to test at the 5 percent level the null hypothesis that the winner of a randomly selected game is equally likely to be the home team or the visitor. (Test this hypothesis by calculating the Z-value and the P-value.) Does the observed frequency with which the home team wins seem substantially different from .5?

11.73 In 1984, *Sports Illustrated* reported that a good way to win a championship is to beat Houston.[30] As evidence, they cited the following:

1. In the 1981 National Basketball Association playoffs, the Houston Rockets lost to the Boston Celtics, who went on to win the championship.
2. In the 1980 and 1981 National League Baseball playoffs, the Houston Astros lost to the Philadel-

phia Phillies and the Los Angeles Dodgers, who went on to win the World Series.
3. In the 1978, 1979, and 1980 National Football League playoffs, the Houston Oilers lost to the team that won the Super Bowl.
4. The University of Houston basketball team qualified for the NCAA Final Four five times and, each time, lost to the eventual champion.

Are you persuaded by this statistical evidence? Explain your reasoning.

11.74 A magazine advertisement once claimed that a famous psychic could identify the sex of an unborn baby from a photo of the mother. The price was $10 with a money-back guarantee if the psychic's prediction was wrong. Do you suppose that the psychic made money on this offer?

11.75 A law professor has written that "Statisticians typically assess the statistical significance of a result before examining its practical significance. A lawyer might find it easier to distinguish between the two by identifying statistical significance with the credibility of the evidence and practical significance with the relevance and materiality of the evidence."[31] Illustrate this distinction with a specific hypothetical legal example of evidence that is of statistical but not practical significance.

11.76 *(continuation)* For the hypothetical legal case chosen in Exercise 11.75, give an example of evidence that the opposing side might present that is of practical but not statistical significance. If you were a juror hearing these two arguments, would you be more swayed by evidence that is statistically or practically significant?

11.77 A person once ran an investment advisory service based on his readings of comic strips. How would you explain any successes he might have had in predicting stock prices?

11.78 Explain the following observation:

Suppose Juliet had asked Romeo if he loved her and he had responded, "I'm 99% sure." Would that have been enough? No way.... Yet most scientists would jump with glee if they thought they could prove anything with Romeo's 99% level of certainty. In fact, the word *proof* practically never appears in scientific literature.[32]

11.79 Three psychiatrists wanted to identify observed characteristics that distinguish schizophrenic persons from the nonschizophrenic.[33] They considered 77 different characteristics, and 2 of their 77 tests turned up statistically significant differences at the 5 percent level. Why was the publication of their results, emphasizing the statistical significance of these two characteristics, potentially misleading?

11.80 In criminal cases, the usual legal standard is "proof beyond a reasonable doubt." In civil cases where monetary damages are at stake, the general legal standard is a "preponderance of evidence." When judges in the Eastern District of New York were asked to assign probabilities that correspond to proof beyond a reasonable doubt, their answers ranged from 76 percent to 95 percent; for a preponderance of evidence, seven judges answered "+50 percent" with the others responding 50.1 percent, 51 percent, and 51 percent.[34] Use the concepts of Type I and Type II errors to explain these different legal standards.

11.81 A researcher wanted to see how sleep patterns are affected by the common practice of a husband and wife sleeping together. After a 12-month search, this researcher found 10 husbands and wives who normally sleep together and were willing to sleep apart for one night in a university laboratory, with their sleep monitored by electronic equipment. For each of these 10 pairs, the researcher compared the breathing patterns, heart rate, body temperature, and sleep state of the husband and wife. Explain several reasons why the results of this study are unconvincing.

11.82 How would you explain the success of a Minneapolis stock broker who, over a five-month period in 1976, selected stocks by spreading the *Wall Street Journal* on the floor and buying the stock touched by the first nail on the right paw of his golden retriever?

11.83 Read the following observation, and then explain why people usually find a pattern and why this is sad.

> There have always been a considerable number of pathetic people who busy themselves examining the last thousand numbers which have appeared on a roulette wheel, in search of some repeating pattern. Sadly enough, they have usually found it.[35]

11.84 Consider a patient with advanced breast cancer who is considering a combination of bone marrow transplants and high doses of anticancer drugs. Exercise 10.68 discusses some research indicating that $m = 150$ out of $n = 259$ patients who received this treatment were free of cancer after 12 months, as compared to 30 percent of those who received conventional treatment. Create some hypothetical values for m and n that might persuade a patient to undergo this combination treatment, even though the results are not statistically significant. Create alternative values for m and n that a patient might consider unpersuasive, even though statistically significant.

11.85 A mail order firm that has historically filled orders in an average of 78 hours with a standard deviation of 30 hours wants to test a new order procedure on a random sample of n orders. Give an example of results that might be considered impressive even though not statistically significant. Give an example of results that seem inconsequential, though statistically significant.

11.86 There were three Friday-the-13ths during 1987. That year, the chief economist at a Philadelphia bank reported that in the past 40 years there had been 6 other years with three Friday-the-13ths and that a recession started in 3 of those years.[36] Explain why this evidence does not convince you that recessions are more likely in years with 3 Friday-the-13ths.

11.87 The chapter tells of 20 coins flips and a theory (that a flipped coin tends to give a tail followed by two heads) that successfully explains 15 of those flips. Flip a coin 20 times yourself, recording each outcome, and then find a simple pattern that explains at least 15 of these 20 flips. Do you think that this theory will accurately predict at least 15 of the next 20 flips? Flip your coin 20 more times and see if it does.

11.88 In a typical two-tailed hypothesis test at the 5 percent level regarding a population mean, explain the effect of an increase in the sample size on the **a.** standard deviation of the sample mean. **b.** probability of a Type I error. **c.** probability of a Type II error. **d.** width of the confidence interval for the mean.

11.89 Identify the implicit, sometimes erroneous, assumption in this explanation of a standard deviation (S.D.) by a Federal Trade Commission hearing examiner: "The S.D. is a statistical criterion based upon the fact that in plotting a normal distributional curve . . . , one observes a curve which is in the shape of a bell. The S.D. is one-sixth of the base line across the curve so plotted."[37]

11.90 The United States has won many Nobel prizes. Use a statistical argument to explain why, even if the average scientist in the United States is no better than the average scientist in England, we would expect the best scientist in the United States to be much better than the best scientist in England.

11.91 In the first half of the baseball season, John batted .300 and Joe batted .250. During the second half of the season, John batted .400 and Joe batted .350. Yet for the season as a whole, both had 200 times at bat and Joe had the higher batting average .330 to .320. Is this possible?

11.92 There was a players' strike in the middle of the 1981 baseball season. Each of the four divisions determined its season winner by having a playoff between the winner of the first part of the season, before the strike, and the winner of the second part of the season. Is it possible for a team to have the best winning percentage for the season as a whole, yet not win either half of the season and consequently not have a chance of qualifying for the World Series?

11.93 An advertisement for a computer program claimed that this program had picked 14 stocks, 13 of which went up during a four-month period:

> Each disk is valid for only four months. You pay only $199 for the program . . . and purchasing the disk entitles you to renew the disk every four months thereafter for the same $199 per disk. . . .
>
> Our guarantee of satisfaction is very compelling. If, after the four months, you are not satisfied with the program or have not seen it pay for itself many times over, please return it and get a full refund of your $199 investment. You can't lose.[38]

Is it true that you can't lose? Why do they want you to renew the disk every four months? How could an unscrupulous company write a very simple program for predicting the direction of the stock market, guaranteeing that half its customers will be satisfied?

• 11.94 A researcher interested in global warming has five years of data on the average temperature on February 8 in New Haven, Connecticut. Recognizing the limitations of a statistical test with so few data, he divides each day into 30-second intervals. Now, instead of 5 observations, he has $5(2)(60)(24) = 14,440$ observations. A colleague points out that dividing the day into nanoseconds would give 432 trillion observations. What's the catch? What assumption underlying the calculation of a Z-value is violated by this quick-and-easy data multiplication?

• 11.95 Federal courts have distinguished between statistical significance and practical importance. In *United States* v. *Goff* (1975), two defendants appealed their mail-fraud conviction, arguing that the 23-person grand jury was selected from lists of registered voters of the Eastern District of Louisiana, of whom 21.06 percent were black, rather than from the voting-age population, of whom 26.33 percent were black. The district and appellate courts ruled that this difference "was not substantial." If you were a judge, how would you decide whether the difference was substantial?

11.96 The U.S. Department of Health, Education, and Welfare requires that state Medicaid payments to nursing homes cover the costs of "an efficient and economically operated facility." Alabama set their payments equal to the 60th percentile of nursing home rates in Alabama; that is, 60 percent of Alabama nursing homes charge less than this amount, 40 percent charge more. In an unsuccessful 1979 lawsuit, the Alabama nursing home association argued that "the lowest ceiling which would meet the requirement of reasonable cost-related reimbursement would be the mean cost plus one-half standard deviation."[39] If costs are normally distributed, how does this rule compare to the 60th percentile rule?

11.97 *(continuation)* Do you think that nursing home costs are normally distributed? Explain your reasoning.

11.98 Explain why you agree or disagree with the following argument:

> It's been estimated that, because of the exponential growth of the world's population, between 10 and 20 percent of all the human beings who have ever lived are alive now. If this is so, does this mean that there isn't enough statistical evidence to conclusively reject the hypothesis of immortality?[40]

11.99 At the start of the semester, Professor Smith flips a coin 10 times and asks his statistics students to try to read his mind and write down what they believe to have been the outcome of each flip. On average, one student will display either positive ESP by correctly identifying 9 or 10 of the flips or negative ESP by getting 9 or 10 wrong. Does such a performance reject at the 5 percent level the null hypothesis that this student has a .5 probability of identifying the outcome of a Smith coin toss? (Use a two-tailed test and the binomial distribution.) About how many students do you think Professor Smith has, on average, in his statistics class—10 to 20, 20 to 40, 40 to 60, or 60 to 100?

11.100 In 1990 a Stanford professor of physics and philosophy said that for many potential disasters, such as global warming, "we're going to have to make decisions on a policy level before we have certainty. . . . You have to be able to understand what the confidence level is, and that's hard. It's easy to convey certainty to somebody."[41] Suppose that we think of the average temperature on the surface of the earth on November 11, 2020, as a random variable X with a mean μ. Carefully explain the difference between a certainty, a probability, and a confidence level for the value of X.

12

A Few More Tests

A friend of mine once remarked to me that if some people asserted that the earth rotated from east to west and others that it rotated from west to east, there would always be a few well-meaning citizens to suggest that perhaps there was something to be said for both sides, and that maybe it did a little of one and a little of the other; or that the truth probably lay between the extremes and perhaps it did not rotate at all.

Maurice G. Kendall
"On the Reconciliation of Theories of Probability"

TOPICS

Often, we are interested in a comparison of two populations: whether cigarette smokers have a shorter life expectancy than nonsmokers; whether children who watch television frequently are more violent than children who seldom watch television; whether people who take large doses of vitamin C have fewer colds than other people. Educators compare the performance of students who have been taught in different ways; engineers compare the reliability of different missile designs; and economists compare the rates of return on different assets. We will begin this chapter by extending the tests that you've already learned to encompass these sorts of comparative situations. Then we'll look at a new kind of test, which can be used for comparisons and much more.

12.1   THE DIFFERENCE BETWEEN TWO MEANS

In the math SAT example, we had a benchmark mean score of 500 against which to measure the performance of students who take the cram course. Sometimes, however, there is no such benchmark. We may want to compare two groups and we may not know the population mean for either group. Students may take two different crash courses, and we may want to test to see if there is a significant difference in their performance. We may want to test to see if there is a significant difference in the life expectancies of two groups who have different lifestyles. Or we may want to test to see if there is a significant difference in the number of miles per gallon for two different types of cars. There are many interesting questions of this sort, and we have to modify our procedures slightly to handle them.

As an illustration, consider the question of whether there is a significant difference in the stock market forecasts of economists and stockbrokers. The objective is to forecast the Dow Jones Industrial Average for one year from now. The economists' forecasts have an unknown mean μ_1 and standard deviation σ_1, while the brokers' forecasts have an unknown mean μ_2 and standard deviation σ_2. A random sample of 36 economists and 25 stockbrokers yields the following information:

Forecasters	Number	Average Forecast	Standard Deviation
Economists	36	2408	150
Stockbrokers	25	2324	75

These economists are more optimistic than the stockbrokers who were interviewed. Is this difference statistically significant, indicating that economists and stockbrokers as a whole have different views about the market, or can the reported difference be attributed to sampling error?

A statistical answer can be provided by a test of these competing hypotheses:

$$H_0: \mu_1 = \mu_2$$

$$H_2: \mu_1 \neq \mu_2$$

The null hypothesis is that the mean forecast of the stockbroker population and the mean forecast of the economist population are equal. Any difference that is observed in a sampling of these two groups is due to sampling error. The alternative hypothesis is that the two populations have different means. It is a two-sided hypothesis because we have no a priori reason for believing that economists as a whole are more optimistic than stockbrokers, or vice versa.

These hypotheses pose the right question, but you are not quite prepared to answer it. You have learned to test the null hypothesis that a parameter is equal to a specified number. Now, we have a null hypothesis that two parameters are equal to each other. Fortunately, a simple trick transforms the new hypothesis into the familiar format:

$$H_0: \mu_1 - \mu_2 = 0$$

$$H_1: \mu_1 - \mu_2 \neq 0$$

The question of whether or not μ_1 is equal to μ_2 can be rephrased as a question of whether or not the parameter $\mu_1 - \mu_2$ is equal to zero.

The difference in the population means can be tested by looking at the difference in the sample means. If the samples are reasonably sized, then the central limit theorem tells us that each sample mean is normally distributed:

$$\overline{X}_1 \sim N[\mu_1, \sigma_1/\sqrt{n_1}]$$

$$\overline{X}_2 \sim N[\mu_2, \sigma_2/\sqrt{n_2}]$$

where n_1 and n_2 are the respective sample sizes. If the samples are independent, then the difference in the sample means is also normally distributed, with an expected value equal to the difference in the separate expected values and a variance equal to the *sum* of the separate variances:

$$\overline{X}_1 - \overline{X}_2 \sim N\left[\mu_1 - \mu_2, \sqrt{\frac{\sigma_1^2}{n_1} + \frac{\sigma_2^2}{n_2}}\right] \tag{12.1}$$

The individual variances are summed because squaring converts the negative sign into a positive sign.

Now let's apply this distribution. Under the null hypothesis, the difference in the population means is zero. We don't know the population standard deviations, but the samples are large enough so that the sample standard deviations, 150 and 75, should be satisfactory estimates. Thus, if the null hypothesis is true, an observed difference in the sample means comes from this distribution:

$$\overline{X}_1 - \overline{X}_2 \sim N[0, \sqrt{(150^2/36) + (75^2/25)}]$$

$$\sim N[0, 29.15]$$

The difference in the sample means is normally distributed with an expected value of zero (according to the null hypothesis) and a standard deviation of 29.15. If the null hypothesis is true, then our observed difference should almost certainly (.95 probability) be within two standard deviations of zero: between -58.3 and $+58.3$. The actual difference is $2408 - 2324 = 84$, which is far outside this range. Because this 84-point difference is too large to be plausibly explained by sampling error, we conclude that the null hypothesis is incorrect.

A second way of looking at these data is to calculate the value of the Z-statistic under the null hypothesis $\mu_1 - \mu_2 = 0$:

$$Z = \frac{(\overline{X}_1 - \overline{X}_2) - (\mu_1 - \mu_2)}{\sqrt{(\sigma_1^2/n_1) + (\sigma_2^2/n_2)}} \tag{12.2}$$

Here

$$Z = \frac{(84 - 0)}{29.15}$$

$$= 2.88$$

The observed 84-point difference in the sample means is 2.88 standard deviations away from zero. This 2.88 Z-value implies a P-value of .002. There is only a .002 chance that a normally distributed variable will be this far from its expected value. The improbability of this occurrence leads us to conclude that zero is not, in fact, the correct expected value.

A third way to examine the data is to estimate the true difference in the population means. A 95 percent confidence interval is

$$.95 = P[(\overline{X}_1 - \overline{X}_2) - 2\sigma_{\overline{X}_1 - \overline{X}_2} < (\mu_1 - \mu_2) < (\overline{X}_1 - \overline{X}_2) + 2\sigma_{\overline{X}_1 - \overline{X}_2}] \tag{12.3}$$

Here,

$$25.7 < (\mu_1 - \mu_2) < 142.3$$

Since zero is outside this confidence interval, we rule out the hypothesis that the stock market forecasts of stockbrokers and economists have the same mean.

What if one of the samples is small, say fewer than 25 observations, or perhaps 30 to be more conservative? If we are willing to assume that both are random samples from normal distributions with the *same* standard deviation, then we can use the t-statistic

$$t = \frac{(\overline{X}_1 - \overline{X}_2) - (\mu_1 - \mu_2)}{\sqrt{(s_P^2/n_1) + (s_P^2/n_2)}} \tag{12.4}$$

Experimental Cancer Tests

A half million people die of cancer in the United States each year, as the search goes on for an anticancer drug. In one laboratory test, cancerous tumors were implanted in nine mice and three were given a potential anti-cancer drug.[1] The tumors were later removed and weighed to see if the drug had slowed or stopped their growth. Here are the tumor weights (in grams):

	Control	Treatment
	1.29	.96
	1.60	1.59
	2.27	1.14
	1.31	
	1.88	
	2.21	
Mean	1.76	1.23
Standard deviation	.43	.32

The tumors did weigh less, on average, in the treatment group; but the samples are small and the standard deviations are relatively large. If we are willing to assume normality and a common standard deviation, then we can make a t-test of the null hypothesis that there is no difference in the population means. The pooled sample variance is

$$s_P^2 = \frac{5(.39^2) + 2(.26^2)}{6 + 3 - 2}$$
$$= .162$$

and

$$t = \frac{(1.76 - 1.23) - 0}{\sqrt{(0.162/6) + (0.162/3)}}$$
$$= 1.86$$

less than the 2.37 t-value required for a two-tailed test at the 5 percent level. Thus the sample is too small and the standard deviations are too large to reject the hypothesis that the observed difference in mean tumor weights is due just to sampling error.

with $n_1 + n_2 - 2$ degrees of freedom, where s_P^2 is an estimate of the common or pooled variance

$$s_P^2 = \frac{(n_1 - 1)s_1^2 + (n_2 - 1)s_2^2}{n_1 + n_2 - 2}$$

In our forecasting example, let's say that we're nervous about having sampled only 25 stockbrokers, but are willing to assume that both samples are from normal distributions with a common variance. The pooled estimate is

$$s_P^2 = \frac{35(150^2) + 24(75^2)}{36 + 25 - 2}$$

$$= 15635.9$$

Thus the t-value under the null hypothesis that the population means are equal is

$$t = \frac{(2408 - 2324) - 0}{\sqrt{(15635.9/36) + (15635.9/25)}}$$

$$= 2.58$$

substantially larger than the 2.00 cutoff for a two-tailed 5 percent test with $36 + 25 - 2 = 59$ degrees of freedom.

Assistance for Handicapped Children

We can also test the difference between two binomial probabilities. As an illustration, consider the implementation of federal laws concerning special education for handicapped children. In California, a student can qualify for government aid or for special schooling if there is a statistically significant discrepancy between the student's scores on an intelligence test and on an achievement test. The idea is that special assistance may be needed when performance is far below capability.[2]

For a stylized example, consider an intelligence test with n_1 questions and a reading test with n_2 questions. We can think of the test questions as random samples from the populations of all the questions that might be asked on these sorts of tests. If this student can correctly answer a fraction π of all such questions, then the probability of correctly answering a randomly selected question is also π. Let π_1 be the probability of correctly answering an intelligence question and π_2 be the probability of correctly answering a reading question. We can see whether or not there is a statistically significant difference in the student's intelligence and reading achievement by testing the hypotheses

$$H_0\colon \pi_1 - \pi_2 = 0$$

$$H_1\colon \pi_1 - \pi_2 > 0$$

The null hypothesis is that the student's intelligence and achievement are equal. The alternative, one-sided hypothesis is that the student has not learned as much as he or she could.

It is natural to choose between these two hypotheses based on the student's scores on the two tests. Let

$x_1 = $ number of correct answers on intelligence test

$x_2 = $ number of correct answers on achievement test

$p_1 = x_1/n_1$, fraction correct on intelligence test

$p_2 = x_1/n_2$, fraction correct on achievement test

With a reasonable number of questions, the student's score will be normally distributed,* measured either in terms of absolute numbers

$$x_1 \sim N[\pi_1 n_1, \ \sqrt{\pi_1(1 - \pi_1)n_1}]$$

$$x_2 \sim N[\pi_2 n_2, \ \sqrt{\pi_2(1 - \pi_2)n_2}]$$

or, equivalently, in fractions

$$p_1 \sim N[\pi_1, \ \sqrt{\pi_1(1 - \pi_1)/n_1}]$$

$$p_2 \sim N[\pi_2, \ \sqrt{\pi_2(1 - \pi_2)/n_2}]$$

The difference in the two scores, whether absolute or fractional, will also come from a normal distribution, with an expected value that is equal to the difference in the expected values and a variance equal to the sum of the variances:

$$x_1 - x_2 \sim N[\pi_1 n_1 - \pi_2 n_2, \ \sqrt{\pi_1(1 - \pi_1)n_1 + \pi_2(1 - \pi_2)n_2}] \qquad (12.5)$$

$$p_1 - p_2 \sim N[\pi_1 - \pi_2, \ \sqrt{\pi_1(1 - \pi_1)/n_1 + \pi_2(1 - \pi_2)/n_2}] \qquad (12.6)$$

These formulas are very general and can be applied to any problem of this type.

The Z-statistic is, as usual, the difference between the sample value and its expected value under the null hypothesis, divided by the appropriate standard deviation:

$$Z = \frac{(x_1 - x_2) - (\pi_1 n_1 - \pi_2 n_2)}{\sqrt{\pi_1(1 - \pi_1)n_1 + \pi_2(1 - \pi_2)n_2}} \qquad (12.7)$$

or, equivalently,

$$Z = \frac{(p_1 - p_2) - (\pi_1 - \pi_2)}{\sqrt{\pi_1(1 - \pi_1)/n_1 + \pi_2(1 - \pi_2)/n_2}} \qquad (12.8)$$

A 95 percent confidence interval is

$$.95 = P[(p_1 - p_2) - 2\sigma_{p_1 - p_2} < (\pi_1 - \pi_2) < (p_1 - p_2) + 2\sigma_{p_1 - p_2}] \qquad (12.9)$$

---

*Do not confuse these distributions with the distribution of scores across students. Each student's score has a probability distribution because there is sampling error in measuring intelligence or achievement. If you take several IQ tests, your scores will vary slightly from one test to the next. There is also a distribution of scores across people, in that abilities vary from person to person. Because there are many inherited and environmental factors that influence ability, this variation across individuals is approximately normally distributed, too. The standard deviation of an individual's score is much smaller than the standard deviation of scores across individuals.

The absolute and fractional scores are just two different ways of looking at the same thing. In some situations, such as political polls, it is most natural to look at fractions or percentages. Here, it is more natural to work with the absolute scores and we will do so. Let's say that there are 50 questions on each test. To obtain the all-important standard deviations, we also need to put in some values for π_1 and π_2. We could estimate these from the student's performance, but to be conservative (and to derive a test that can be readily applied to every student), let's use .5 for these probabilities. Under the null hypothesis that $\pi_1 = \pi_2$, Equation 12.5 implies that

$$x_1 - x_2 \sim N[(\pi_1)50 - (\pi_2)50, \sqrt{(.5)(.5)50 + (.5)(.5)50}]$$

$$\sim N[0,5]$$

If a student's intelligence and achievement are comparable, then the difference in scores on two 50-question tests should follow a normal distribution with an expected value of 0 and a standard deviation of 5.

How large a difference in the two scores can be explained as sampling error, the inevitable variation in scores from one test to the next? We have a one-tailed test here, because we are only interested in the case where the achievement score is lower than the intelligence score. Thus, we use 1.65 standard deviations, which work out to be 1.65(5) = 8.25 points. Because questions cannot be partially right, we can continue being conservative by using 9 as our decision rule.

> If a student gets 9 more questions correct on a 50-question intelligence test than on a 50-question achievement test, then this disparity is statistically sufficient to reject the hypothesis that the student's achievements and intelligence are equal.

Matched-Pair Samples

Many factors influence the outcome of any sampling experiment. For instance, a company may want to test consumer reaction to two cola formulas. One sampling design is to select two random samples of size 10. The people in the first sample test and rate one cola, while the second sample tests and rates the other cola. The mean ratings can then be compared in the manner you have learned so far in this chapter.

Perhaps the ratings have the results shown in Table 12.1. The first cola seems to be the more popular of the two, but a statistical comparison must take into account sampling error, which arises because of the inevitable variation in people's tastes. Some people like cola and some don't and, in addition, each cola has its own partisans. The second cola may have received the lower ratings simply because, by the luck of the draw, its sample consisted primarily of cola haters. To take this possible sampling error into account, our statistical test looks at the standard deviation of the ratings and the size of the samples, to gauge the chance of drawing a

Job Discrimination: Statistical Significance and the 80 Percent Rule

Difference-in-means tests have frequently been accepted by U.S. courts as prima facie evidence of employment discrimination that violates Title VII of the Civil Rights Act of 1964. In *Griggs* v. *Duke Power Company* (1971) the Supreme Court ruled that it is illegal to use a job qualification test that has a disparate impact on white and black applicants unless the employer can show a business necessity for the test (for example, an eye examination for pilots). In *Hazelwood School District* v. *United States* (1977) the Supreme Court stated that "gross statistical disparities" can be sufficient to prove discrimination.

However, courts have also recognized that, as explained in Chapter 11, a statistically significant difference may not be substantial. For instance, in *Moore* v. *Southwestern Bell* (1979) a promotion test for clerks was passed by 453 of 469 whites (96.6 percent) and by 248 of 277 blacks (89.5 percent). The Z-value is 3.49, which has a P-value of .00003:

$$Z = \frac{.966 - .895}{\sqrt{\dfrac{.966(.034)}{469} + \dfrac{.895(.105)}{277}}} = 3.49$$

The district court ruled that although the difference between the 96.6 percent and 89.5 percent pass rates was statistically significant, it was not substantial enough to prove discrimination.

The U.S. Equal Employment Opportunity Commission (EEOC) has argued and the Supreme Court endorsed (in *Connecticut* v. *Teal*, 1982) the use of an "80 percent rule" in addition to tests of statistical significance in evaluating evidence of employment discrimination against members of a specified ethnic, racial, or sexual group. The 80 percent rule considers whether the selection rate for this group is less than 80 percent of the selection rate for the most favored group.

For instance, in the case of *Moore* v. *Southwestern Bell* cited previously, the ratio of the selection rate for blacks to the selection rate for whites is .895/.966 = .927. Although the difference is statistically significant, it is not, by the 80 percent rule, substantial. The opposite situation occurred in *Chicano Police Officers Association* v. *Stover* (1975), where 3 of 26 Spanish surnamed officers passed a promotion test, a selection rate of 3/26 = .115, while 14 of 64 other officers passed, a selection rate of 14/64 = .219. The ratio .115/.219 = .53 is less than .80, indicating a substantial difference in selection rates. However, this difference is not statistically significant at the 5 percent level, since the Z-value is 1.27 (and the P-value is .102):

$$Z = \frac{.219 - .115}{\sqrt{\dfrac{.219(.781)}{64} + \dfrac{.115(.885)}{26}}} = 1.27$$

Courts have found statistical evidence of discrimination to be most persuasive when the observed differences are both statistically significant and substantial.

Table 12.1 *Cola Ratings for Two Independent Samples*

	First Sample's Ratings of First Cola X_1	*Second Sample's Ratings of Second Cola* X_2
	5.9	6.0
	8.0	7.1
	6.4	9.2
	6.9	6.0
	8.3	5.9
	6.5	5.2
	6.7	7.9
	6.7	6.4
	6.5	5.6
	6.1	6.7
Mean	6.80	6.60
Standard deviation	0.77	1.20

sample that is disproportionately full of cola haters. If the standard deviation is small and/or the sample is large, there is little chance that an imbalance of cola haters will sway the results.

In accordance with Equation 12.1 the standard deviation of the difference in the sample means is

$$\sqrt{(\sigma_1^2/n_1) + (\sigma_2^2/n_2)}$$

Using the data shown in Table 12.1, the estimated value of this standard deviation is

$$\sqrt{(0.77^2/10) + (1.20^2/10)} = 0.45$$

which is sufficiently large so that the observed 0.2 difference in the mean ratings is not statistically significant—it's less than half a standard deviation away from zero. The samples are so small and the sample standard deviations are so large that we cannot rule out sampling error as the explanation; the second cola may have been victimized by bad luck.

Another way to control this possible source of sampling error, that some people like cola while others don't, is to do the Pepsi Challenge—have the same people, a single sample, rate both colas. Perhaps a sample of 10 tasters yields the results shown in Table 12.2. Each observed difference in the ratings can be treated as a single observation x in a random sample. If the underlying individual ratings are normally distributed or if the sample is sufficiently large, then the sample mean

Table 12.2   *A Single Sample's Ratings of Both Colas*

First Cola	Second Cola	Difference x
7.4	5.4	2.0
7.6	7.2	0.4
6.0	6.7	−0.7
5.9	4.9	1.0
5.9	4.7	1.2
8.4	7.7	0.7
8.2	7.7	0.5
5.1	5.6	−0.5
7.8	7.4	0.4
6.0	4.4	1.6
	Mean	0.66
	Standard deviation	0.85

comes from a normal distribution and a standard test can be made of the null hypothesis that its expected value is zero, that is, that there is no difference in the average population ratings of the two colas.

Here, the estimated standard deviation of the sample mean is $0.85/\sqrt{10} = 0.27$, which implies that the observed sample mean of 0.66 is statistically significantly different from zero.*

Notice that the 0.27 standard deviation for the single comparison sample is much smaller than the 0.45 standard deviation for two independent samples. The reason is that the comparison sample controls for one source of sampling error—that one cola may draw mostly cola haters while the other gets cola lovers. (There are, of course, still reasons for possible sampling error. In particular, each cola has its own partisans, and a sample may draw disproportionately from either group.)

A comparison sample can be used in all sorts of taste tests and in many other situations, too; for instance, two machines can be compared by giving identical tasks to each one and two forecasters can be matched in head-to-head competition. Sometimes, however, a comparison test is unfair or impractical. In taste tests, the first item may leave a lingering taste that distorts the rating of the second item. In comparing the opinions of Republicans and Democrats or any other two groups, two samples are required by the very nature of the poll. In testing bombs dropped from different altitudes, the same bomb cannot be tested twice. In medical experiments, two drugs cannot be compared by giving both to the same patient. Instead, the medical researcher can try to construct two samples containing **matched pairs.**

---

*Because the sample is so small, the *t* distribution should be used. With $n - 1 = 9$ degrees of freedom, the critical *t*-value for a 5 percent test is 2.262; the observed mean must be 2.262 standard deviations away from the null hypothesis to reject that hypothesis. Here, it is 2.44 standard deviations away from zero.

If truly identical twins could be found, then it would be as if a single person had made a comparison and, hence, the two matched samples could be used as a single sample of differences.

To illustrate the procedure, consider researchers who are interested in seeing if a certain cholesterol-reducing drug reduces the risk of a heart attack. Other factors figure in, too—including age, gender, smoking habits, and blood pressure—and so if two independent samples are taken, the observations will have a substantial standard deviation. Statistically conclusive results will consequently require a large sample. There is a chance, for example, that the drug might be given mostly to young women, while the control group consists mostly of old men. In this way, a lower incidence of heart problems might be attributed to the drug when it is, in fact, at least partly due to age and gender differences between the two samples. A large sample is needed to reduce the chance of this sort of sampling error.

A single-sample comparison would eliminate some of the sources of sampling error, but we can hardly give the drug to a person and simultaneously use the same person as the control group. Instead, matched pairs can be assembled and this is just what researchers at the National Heart, Lung, and Blood Institute did. They took a sample of 3806 middle-aged males who had high cholesterol levels but no history of heart disease. These were divided into two samples, with each person in the treatment group matched with someone in the control group according to age, cholesterol level, smoking habits, blood pressure, and other factors that were thought to affect one's chances of a heart attack. These matched pairs served as twins, as if a single person were simultaneously in the treatment and control groups.

The patients were selected by doctors at 12 medical centers throughout the United States. Those in the treatment group were given a cholesterol-reducing drug over a ten-year period, while those in the control group were given a placebo. Neither the patients nor their doctors knew if they were using the drug or the placebo. The results over this ten-year period were that those in the treatment group had 19 percent fewer heart attacks, 20 percent fewer chest pain attacks, and 21 percent fewer coronary-bypass operations.[3]

As in this medical study, there are often several advantages of matched-pair sampling. Most important, it allows for a smaller sample, because controlling for some factors means that the researcher doesn't have to rely solely on sample size to reduce sampling error. In addition, as a practical matter, it defuses the criticism that the samples may have slanted the results. Critics cannot make the easy claim that one cola may have been tasted by cola haters or that one medication may have been given only to people with low blood pressure, if, in fact, the same people tasted both colas and similar people took both medications.

On the other hand, matching pairs is a time-consuming process if done properly. It requires a considerable knowledge of the important factors that need to be controlled and a careful attention to the matching process. Even though it allows for testing a smaller sample, a large initial sample may be needed to find close matches. The matching process itself is subjective and prone to error. Some factors may have been held constant at unrepresentative levels; for example, a sloppy researcher might test a cholesterol-reducing drug on young women—who are not the target patients. Other factors might be inadvertently neglected, so that the matched pairs aren't really that similar after all; for example, the sloppy researcher might not take smoking habits or blood pressure into account. Sometimes matching simply isn't practical because the researchers don't have the necessary information. For instance, if the "lingering taste" issue argued against having people taste both colas, we would hardly know where to begin finding "twins." However, when it is feasible and done carefully, matched-pair samples offer an attractive sample design.

EXERCISES

12.1 A study of the amount of time network newscasts devoted to a reported increase or decrease in the unemployment rate used these data for 1973 through 1984:[4]

	Increase in Unemployment	Decrease in Unemployment
Number of observations	171	170
Average news time (seconds)	161.8	123.6
Standard deviation	110.8	103.9

Is the observed difference in average news time statistically significant at the 1 percent level?

12.2 In 1987, a small data processing firm tried to persuade a large Southern California grocery chain that it could reduce its costs of handling workers' compensation bills. Both this new firm and the firm currently handling the claims based their charges on complex computerized formulas, depending on the number of lines in the doctors' bills and other factors. The only way to tell which was really less expensive was to compare their charges for actual bills. As a test, the firm currently handling the claims processed a random sample of 3,800 bills and charged an average of $9.26 per bill, with a standard deviation of $9.70; the new firm processed a random sample of 3,500 bills and charged an average of $7.35 per bill, with a standard deviation of $5.26. A consultant advised the grocery that it should stick with its current firm, because the observed difference could easily be explained by sampling error. Do you agree?

12.3 In a 1983 survey,[5] 66 of 110 company presidents and 172 of 302 business school professors agreed with the statement that "The ideal MBA program should focus strongly on quantitative analysis." Is this observed difference between the presidents and professors statistically significant?

12.4 A study found that derelicts less often are balding than are college professors:[6]

	Number Studied	Median Age	Percent Balding
Derelicts	141	47.5	36
Professors	49	47.5	71

Assuming these to be scientific random samples, test the null hypothesis that balding is equally likely for derelicts and professors.

12.5 A *Los Angeles Times* baseball story ("The Honeymoon Might be Over for Steve Garvey," September 18, 1983) quoted the manager of the San Diego Padres as saying, "I'm not taking anything away from Steve [Garvey], but we've played better won-lost percentage ball without him." At the time, the Padres had won 50 and lost 52 with Garvey and won 24 and lost 22 without him. Is this difference statistically significant using a binomial test?

12.6 The Federal Tobacco Commission measured the nicotine content (in mg.) of 10 cigarettes of two different brands, Life and Death:

Life		Death	
.45	.60	.35	.40
.35	.35	.55	.60
.30	.55	.50	.40
.65	.25	.30	.75
.40	.40	.45	.30

Assuming normality and a common standard deviation, calculate the pooled sample variance and make a *t*-test at the 1 percent level of the null hypothesis that there is no difference in the population means.

12.7 In 1973, the U.S. Public Health Service studied the effects of vitamin C on the incidence of colds at a Navajo boarding school in Arizona.[7] The study covered a 14-week period, from February to May; 321 children were given daily doses of from 1 to 2 grams of vitamin C, while 320 were given placebos, which were identical in taste and appearance to the vitamin C tablets. They found that 143 of those who took vitamin C and 92 of those on placebos had no sick days during this 14-week period. Is this difference statistically significant at the 1 percent level?

12.8 In 1973, two Columbia University social scientists studied the effects of prejudicial news coverage on jury verdicts. They selected people from actual jury pools and used real lawyers to reenact a murder trial. Before the mock trial, some jurors read prejudicial news stories concerning the case, while others read objective reports. In one case, 47 of 60 jurors who read the prejudicial stories voted guilty, while 33 of 60 jurors who read objective reports voted guilty. Is this difference statistically significant at the 5 percent level?

12.9 A study was made of 10,590 men who had vasectomies, each being matched to a man of approximately the same age, race, and marital status who had not been vasectomized.[8] Over the 5 to 14 years covered by the study, 212 of the vasectomized men and 326 of the nonvasectomized men died. If these were assumed to be independent random samples, would

this observed difference be statistically significant at the 5 percent level? If we took into account the fact that these were matched pairs, would the calculated Z-value be higher or lower? The authors caution that "Exposure to vasectomy cannot be considered random; the men chose to be vasectomized." Explain why this is a good reason for caution.

12.10 Seven psychologists tried two different approaches to reducing the cigarette consumption of heavy smokers who professed a desire to quit smoking.[9] One group was given supportive counseling (verbal encouragement) and the others were given electrical shocks while they smoked. Here are the posttreatment cigarette consumption data, expressed as a percentage of pretreatment consumption:

Psychologist	Supportive	Punitive
1	125.0	59.5
2	59.0	14.0
3	43.0	72.0
4	0.0	27.5
5	50.0	80.5
6	18.5	17.5
7	66.0	83.0

Which approach reduced cigarette consumption the most, on average? Use a matched-pair test to see whether the observed difference is statistically significant at the 1 percent level.

12.11 If the cola ratings in Table 12.2 had represented two independent random samples, would they have rejected decisively the null hypothesis that the two colas are rated equally by the population as a whole? Why does the test described in the text not come to an identical conclusion?

12.2 CHI-SQUARE TESTS

So far, our hypothesis tests have been concerned with averages. We looked at the average performance of a sample of students to gauge the average performance of all students. We looked at the average weight of a sample of cracker boxes to gauge the average weight of the population. And so on. But often, we are interested in the fine detail that an average misses—we don't want to suffer the fate of the statistician who drowned crossing a river that had an average depth of 3 feet.

For instance, we may suspect that a certain gambling casino is using loaded dice to affect the frequencies with which various numbers appear. It's not much consolation to the bettor if a pair of loaded dice have an average sum of 7; instead, bettors are interested in how frequently specific numbers appear. The long-run average sum will be 7 if we roll two fair dice, dice that always sum to 7, or dice that sum to 7 half the time and sum to 2 to 12 the other half. These three sets of dice have the same expected value, but very different implications for dice games.

No casino would dare use dice that were so obviously rigged, but a crooked casino might use dice that were loaded to have more subtle effects. Let's say that we carefully record 720 rolls of a suspicious casino's dice. Table 12.3 shows the theoretical expected frequencies for the various possible numbers, along with the actual observed frequencies. These expected frequencies show how often, on average, we can expect each number to appear if we repeat this 720-roll experiment many times. For example, because the probability of a 2 with fair dice is 1/36, the number of 2s we should average in every 720 rolls is $(1/36)720 = 20$. In any particular 720 rolls, the number of 2s will most likely be somewhat higher or lower than 20. The statistical question is how many more or fewer 2s are needed to fuel our suspicion that the dice are rigged. In particular, does the fact that the number of 2s in these 720 rolls was 26 show that these dice are crooked, or is that the sort of chance deviation that we should anticipate?

Our earlier test procedures suggest calculating the standard deviation to gauge the difference between observed and expected values. If the probability of a 2 is $\pi = 1/36$, then the number of 2s obtained in 720 rolls has an expected value and standard deviation,

$$\mu = \pi n = (1/36)(720) = 20$$

$$\sigma = \sqrt{\pi(1 - \pi)n} = (1/36)(35/36)(720) = 4.41$$

Table 12.3   *Actual and Expected Outcomes for 702 Rolls of a Pair of Dice*

Number	Expected Frequency E_i	Observed Frequency O_i	$O_i - E_i$	$(O_i - E_i)^2$	$\dfrac{(O_i - E_i)^2}{E_i}$
2	20	26	+6	36	1.80
3	40	38	−2	4	0.10
4	60	57	−3	9	0.15
5	80	74	−6	36	0.45
6	100	113	+13	169	1.69
7	120	106	−14	196	1.63
8	100	101	+1	1	0.01
9	80	77	−3	9	0.11
10	60	65	+5	25	0.42
11	40	41	+1	1	0.03
12	20	22	+2	4	0.20
Total	720	720	0		6.59

Because there is about a .95 probability of being within two standard deviations of the expected value, there is a .95 probability that the number of 2s obtained in 720 rolls of fair dice will be between 11 and 29 (rounding off to whole numbers). Thus, the occurrence of twenty-six 2s is not sufficient to reject the null hypothesis that the probability of a 2 is 1/36.

This analysis is fine as far as it goes, but we want to go farther. We want to look at all 11 possible numbers, 2 through 12, and yet there are several reasons for not examining them one-by-one. First, the outcomes are interdependent because they must add up to 720. If we have six more 2s than expected, then we must have six fewer of some other numbers. We are, in a sense, repeating ourselves in that every number we test tells us a little more about the outcomes that we haven't yet tested. At the end, after we've tested all but one number, we'll know how many times that last number appeared even before we look.

Second, if we make several tests, each with a .05 probability of Type I error, then the overall probability of committing a Type I error will be considerably greater than 5 percent. The more tests we make, the more likely it is that we will draw at least one wrong conclusion. We would like to adjust our tests somehow, while also taking into account their interdependence, so as to put the overall probability of Type I error at 5 percent.

Third, there is a strong temptation to look at the data and pick out anomalies to be tested. The casual statistician might look at Table 12.3, notice that there were fourteen fewer 7s than expected, thirteen extra 6s, and six extra 2s, and then see if one of these disparities is statistically significant. But this is plain and simple data mining; in choosing to write down a test of the number of 7s, 6s, or 2s, this statistician mentally tested each of the other nine numbers and concluded that their differences were less statistically significant. As with the statistician who meticulously tests each number, the probability of Type I error is much larger than 5 percent.

The Chi-Square Statistic

In 1900, Karl Pearson developed a test that avoids these problems. It is a single test that simultaneously compares the observed and expected values for each of the possible outcomes. It can be motivated as follows. In testing a single outcome, such as the number 2, we look at the Z-statistic

$$Z = \frac{(X - \pi n)}{\sqrt{\pi(1 - \pi)n}}$$

which tells us how many standard deviations the number of observations X is away from its expected value πn. The square of this statistic can be rearranged to give

$$Z^2 = \frac{(X - \pi n)^2}{\pi(1 - \pi)n}$$

(continued)

$$= (X - \pi n)^2 \left\{ \frac{1}{\pi n} + \frac{1}{(1 - \pi)n} \right\}$$

$$= \frac{(X - \pi n)^2}{\pi n} + \frac{\{(n - X) - (1 - \pi)n\}^2}{(1 - \pi)n}$$

$$= \frac{(O_1 - E_1)^2}{E_1} + \frac{(O_2 - E_2)^2}{E_2}$$

The O_i are the observed values of the two possible outcomes: 2s or non-2s. The E_i are the corresponding expected values. What the algebra shows is that a squared Z-statistic can be rearranged to give another, quite logical way of looking at the discrepancies between the observed and expected values: take the observed number of 2s, subtract the expected number, and square this difference. Then scale this squared deviation by dividing by the expected number of 2s. Do the same thing for non-2s and then add the two scaled squared-deviations together. This is a logical way of measuring the extent to which the observed data differ from their expected values. Because it is a Z-statistic squared, we can calculate just how likely various values are.

Karl Pearson carried this logic one step further by suggesting that statisticians can make the same scaled squared-deviation calculation when there are more than two possible outcomes. Table 12.3 shows the arithmetic.

1. *For each possible outcome, 2 through 12, calculate the deviation $O_i - E_i$* between the observed and expected occurrences. It would do no good to calculate the average such disparity, because the deviations sum to zero.

2. *Square each deviation, instead.* Now, positive and negative deviations count equally and they don't cancel each other out. In addition, large disparities are more important than small disparities. This weighting makes sense, because the improbability of large deviations makes their presence convincing evidence that more than chance is involved.

3. *After squaring the deviations, divide by the expected number of occurrences,* E_i. This scaling is needed to put the squared deviations into perspective and to make different studies comparable. The fact that there were thirteen extra 6s in this study is actually less surprising than the fact that there were six extra 2s, because 6s are five times as likely as 2s. There were only 13 percent more 6s than expected, but 30 percent more 2s than expected. Dividing by E_i scales each squared deviation relative to the expected number of occurrences. In addition, scaling is absolutely essential to calibrate studies using different units, such as millions versus billions.

4. *Add up the scaled squared-deviations.* This is like calculating the average scaled squared-deviation. We just neglect to divide by 11, the number of possible outcomes.

Pearson called the end result a **chi-square statistic**

$$\chi^2 = \frac{(O_1 - E_1)^2}{E_1} + \frac{(O_2 - E_2)^2}{E_2} + \cdots + \frac{(O_m - E_m)^2}{E_m}$$ (12.10)

where O_i and E_i are the observed and expected occurrences for m (exhaustive and mutually exclusive) possible outcomes.

Table 12.3 shows that the chi-square statistic for these dice data is 6.59. Now, the question is whether this 6.59 is sufficiently large to persuade us that it is unreasonable to attribute the deviations to chance alone. Are deviations this large sufficiently improbable to make rigged dice the more plausible explanation? To answer this question, we need to know the probability distribution for the chi-square statistic.

The Chi-Square Distribution

It takes a high-powered computer to determine the exact probability distribution for the chi-square statistic. Pearson didn't have such a computer, so he figured out an approximation. I've shown you how the square of a single standardized binomial leads to a chi-square statistic with $m = 2$. Because the standardized binomial can be approximated by the standardized normal distribution if the sample is large, it follows that the general chi-square statistic, Equation 12.4, can be approximated by the sum of $m - 1$ squared, standardized normal distributions. This is what Pearson did.*

> The **chi-square distribution,** with $m - 1$ degrees of freedom, is the distribution of the sum of $m - 1$ independent, squared standardized normal variables $Z_1^2 + Z_2^2 + \cdots + Z_{m-1}^2$.

When there are m possible outcomes, there are only $m - 1$ degrees of freedom because we know the number of occurrences of the last outcome if we know the occurrences of the other $m - 1$ outcomes.

The chi-square distribution can be calculated readily and is shown in Table 5 at the end of the book. Figure 12.1 shows three chi-square distributions, for $m - 1 = 4, 6$, and 10 degrees of freedom. Notice that negative values are not possible, because χ^2 is a sum of squares. Notice also that the chi-square distribution is positively skewed. As $m - 1$ increases, though, the chi-square distribution becomes increasingly symmetrical and approaches a normal distribution.

Pearson used the (continuous) chi-square distribution to approximate the (discrete) distribution of the chi-square statistic. The generally accepted rule of thumb

---

*The chi-square distribution was first discovered in 1876 by F.R. Helmert (1843–1917), a German physicist, and then rediscovered and popularized by Pearson.

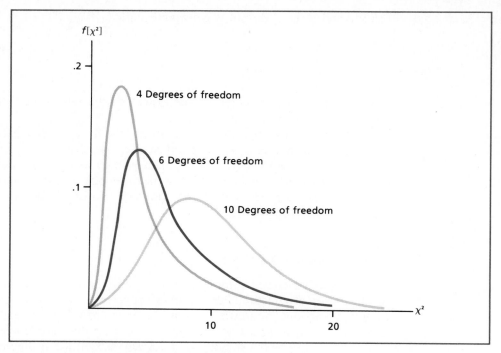

Figure 12.1   *Three Chi-Square Distributions*

is that this approximation is satisfactory as long as there are at least 10 (some say 5) expected occurrences of every possible outcome; that is, none of the E_i are less than 10. (If any E_i are less than 10, they can be boosted by combining categories.)

In our dice example, the smallest value of E_i is 20, so the chi-square distribution should be satisfactory. Because there are $m = 11$ possible outcomes, there are $m - 1 = 10$ degrees of freedom. Figure 12.2 shows the chi-square distribution for 10 degrees of freedom. If the value of the chi-square statistic were zero, the observed and expected frequencies would all coincide, providing no evidence whatsoever against the null hypothesis that the dice are fair. Doubt is cast on this hypothesis when the value of the chi-square statistic is large; this indicates that the deviations are too large to be attributed to chance alone.

Figure 12.2 shows that, with 10 degrees of freedom, there is only a 5 percent chance that the chi-square statistic will exceed 18.31. Therefore, in order to set the probability of Type I error at 5 percent, we require a chi-square value of 18.31 before we reject the null hypothesis that the dice are fair. Table 12.3 shows that the actual value of the chi-square statistic turns out to be 6.59, which is far below the 18.31 needed to reject the null hypothesis. With 10 degrees of freedom a chi-square statistic of 6.59 is not at all surprising; indeed, there is about a .75 probability that the chi-square statistic will be that high or higher. Thus, these deviations between

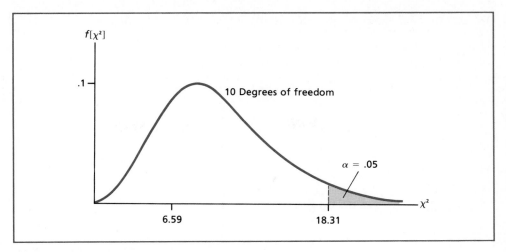

Figure 12.2 *Testing for Fair Dice*

the observed and expected frequencies can be explained by chance and do not provide statistical evidence that the dice are rigged.

Can the Famous Postpone Death?

In 1972, David Phillips published a provocative statistical investigation.[10] He began his paper by observing that

> In the movies and in certain kinds of romantic literature, we sometimes come across a deathbed scene in which a dying person holds onto life until some special event has occurred. For example, a mother might stave off death until her long-absent son returns from the wars. Do such feats of will occur in real life as well as in fiction?

Phillips went out and collected readily available data—the birth and death dates of famous people—his idea being that people may postpone death until after the celebration of their birthday. Famous people were used because their birth and death dates were easily obtained. For his first sample, he chose 348 deceased people from the book *Four Hundred Notable Americans.* Then, he compared their death and birth dates to see how many months separated the two. Table 12.4 shows his data.

The null hypothesis that birth and death dates are unrelated implies that a person's death is equally likely to occur one month before a birthday, one month after a birthday, or in any other month. (This is not quite right, because some months have slightly more days than others, but it is close enough.) If 348 deaths were equally divided over twelve months (measured relative to birthdays), there would be $348/12 = 29$ deaths in each month. Table 12.4 shows that, in fact, there were

Table 12.4   *Phillips's Data on Deaths Before, During, and After the Birth Month*

Number of Months Before (−) or After (+) Birth Month	Expected Number of Deaths E_i	Actual Number of Deaths O_i	$O_i - E_i$	$(O_i - E_i)^2$	$\dfrac{(O_i - E_i)^2}{E_i}$
−6	29	24	−5	25	0.86
−5	29	31	+2	4	0.14
−4	29	20	−9	81	2.79
−3	29	23	−6	36	1.24
−2	29	34	+5	25	0.86
−1	29	16	−13	169	5.83
0	29	26	−3	9	0.31
+1	29	36	+7	49	1.69
+2	29	37	+8	64	2.21
+3	29	41	+12	144	4.97
+4	29	26	−3	9	0.31
+5	29	34	+5	25	0.86
Total	348	348	0		22.07

only 16 deaths in the month immediately preceding a birthday, but 36 deaths in the month after. There are also fewer-than-expected deaths during the birth month itself and three to four months before a birthday, and more-than-expected deaths two and three months after a birthday. Is this "death dip" before a birthday large enough to be statistically significant, or is it explainable by chance fluctuations?

Table 12.4 shows that the chi-square statistic works out to be 22.1. With 29 expected occurrences in each month, the chi-square distribution is usable. The twelve monthly divisions give $m = 12$, so that there are $m - 1 = 11$ degrees of freedom. Table 5 at the end of this book shows that, with 11 degrees of freedom, there is a .05 probability that a chi-square statistic will exceed 19.7 and a .01 probability that it will exceed 24.7. There is about a 2.5 percent chance that a chi-square statistic will exceed 22.1. Thus, the observed relation is significant at the 5 percent level, but it is not "highly significant" in that it is not significant at the 1 percent level.

Phillips' own description of his results is "an unexpected connection," and most people are surprised by his evidence. Phillips' explanation is that some people can successfully postpone death until after the celebration of a birthday. An alternative explanation is that the excessive food, drink, and excitement on a birthday are bad for one's health. Another possibility is that it is just a fluke in the data, one of those one-out-of-twenty occasions when the null hypothesis is incorrectly rejected. We should be wary of "unexpected connections," because statistical relationships without a convincing explanation are often just due to chance.

This latter possibility is supported by the fact that Phillips gathered some other data that were not so kind to his theory. He took three more samples that were drawn from people listed in *Who Was Who in America*, who also had their surnames listed as among "The Foremost Families of the U.S.A." There is a slight death dip in each of these other three samples, but the deviations are not statistically significant. If we put the four samples together, the chi-square statistic falls to 17.03, which is not statistically significant either. Phillips naturally prefers to emphasize his first sample, because it gives a surprising, publishable result. His stated rationale is that these people are more famous and that famous people have more reason to postpone death until after a birthday, because they receive larger gifts and more public attention. However, choosing the sample that gives the most favorable result is a statistically treacherous practice. Those who are intrigued by Phillips' theory should go out and gather some fresh data.

EXERCISES

12.12 Three-fourths of an animal population have a certain trait A, while the remaining one-fourth have trait a. Similarly, three-fourths of this animal population have trait B, while one-fourth have trait b. If these traits are inherited independently (in accordance with Mendel's law), then the probabilities of various combinations of these traits are $P[A \text{ and } B] = 9/16$, $P[A \text{ and } b] = 3/16$, $P[a \text{ and } B] = 3/16$, and $P[a \text{ and } b] = 1/16$. A breeding experiment yielded the following data:

	Number of Observations
A and B	216
A and b	83
a and B	68
a and b	33

Use a chi-square test at the 5 percent level to see if these data are consistent with Mendel's law.

12.13 Karl Pearson reported dealing 3400 thirteen-card hands from an ordinary deck of playing cards.[11] He kept track of whether each hand had few trumps (0–2), a moderate number of trumps (3–4), or a large number of trumps (5–13). His results were

Number of Trumps	Expected Number of Hands	Actual Number of Hands
0–2	1015.67	1021
3–4	1784.79	1788
5–13	599.54	591
Total	3400	3400

Are these results consistent at the 1 percent level with the null hypothesis that the cards are fairly dealt? Show how you would check if his results are too good to be true.

12.14 An economist claims that each day the stock market has a .5 probability of going up and a .5 probability of going down, regardless of what has happened on previous days. If this economist is right, what is the probability that, in any given week, the market will go up exactly 5 days? 4 days? 3 days? 2 days? 1 day? 0 days?

A study of 500 five-day weeks yielded the following data:

Number of Days Market Went Up During Week	Number of Occurrences
5	32
4	76
3	136
2	142
1	88
0	26
Total	500

Use a chi-square rest to see if these data are consistent at the 1 percent level with the economist's claim.

12.15 A study of the effects of the alphabetical listing of candidates in Irish elections obtained the data in Table 12.5.[12]

Test the null hypothesis (at the 1 percent level) that the candidates are random draws from all registered voters, without regard for surname. Make a sim-

Table 12.5   *Data for Exercise 12.15*

First Letter of Surname	Fraction of Irish Voters	Number of Candidates	Number Elected
A–C	.203	91	47
D–G	.179	63	32
H–L	.172	59	23
M–O	.253	75	22
P–Z	.194	47	20

ilar test for the null hypothesis that those elected are randomly drawn from all candidates without regard for surname.

12.16 A study of the relationship between auto weight and accident frequency used the following data:[13]

Auto Weight (pounds)	Registration Frequency	Accident Frequency
<3,000	21.04%	162
3,000–4,000	46.13%	318
4,000–5,000	31.13%	689
>5,000	1.70%	35
Total	100.00%	1204

Test the hypothesis that auto weight and accident frequency are statistically independent at the 5 percent level.

12.17 W.F.R. Weldon carefully recorded the number of dice showing a 5 or 6 when 12 dice were rolled 26,306 times:[14]

Number of Dice Showing a 5 or 6	Expected Frequency	Observed Frequency
0	203	185
1	1,217	1,149
2	3,345	3,265
3	5,576	5,475
4	6,273	6,114
5	5,018	5,194
6	2,927	3,067
7	1,254	1,331
8	392	403
9	87	105
10 or more	14	18
Total	26,306	26,306

Make a chi-square test of the null hypothesis that these dice were fair.

12.18 Use a computer's random number generator to simulate 60 rolls of a single six-sided die, and make a chi-square test of the null hypothesis that each of the six possible numbers is equally likely to be rolled. If this hypothesis were rejected at the 5 percent level, what course of action would you recommend?

12.19 Chapter 10 included a discussion of the American Stove Company's process control procedure for the height of a metal piece they manufactured. This procedure assumed that when the production process was operating properly, the heights would come from a normal distribution with a mean of 0.8312 inch and a standard deviation of 0.006169 inch. Measurements of a random sample of 145 pieces yielded the following sample distribution:[15]

Height (inches)	Number of Pieces
<0.8245	14
0.8245–0.8274	14
0.8275–0.8304	21
0.8305–0.8334	55
0.8335–0.8364	23
0.8365–0.8394	7
>0.8394	11

Make a chi-square test at the 1 percent level of the null hypothesis that these data came from the assumed population.

• **12.20** Exercise 6.16 shows Bowditch's data on the heights of 1,253 twelve-year-old boys. Use a chi-square test to see if these data are consistent with the null hypothesis that the population of heights is normally distributed.

• **12.21** Exercise 7.11 shows the distribution of 10,720 GRE Advanced Economics Test scores. Make a chi-square test at the 5 percent level of the null hypothesis that these data are normally distributed with a mean of 606 and a standard deviation of 108.

12.3 CONTINGENCY TABLES

The nice thing about Pearson's chi-square test is that it can be applied to a wide variety of situations. All we need are a set of mutually exclusive possible outcomes and a null hypothesis giving the probabilities that these outcomes will occur. We can then calculate how often each outcome would be expected to occur if the null hypothesis were true and compare these expected frequencies to the actual frequencies with which the outcomes occur. Pearson's chi-square test gauges whether the deviations between the observed and expected frequencies can reasonably be attributed to chance, or whether they are so large that the null hypothesis should be rejected.

The framework is so general that it is often given the informative label, **goodness-of-fit tests.** Pearson's chi-square test is a way of statistically gauging how closely the data agree with the detailed implications of a null hypothesis. One application is, as we have seen, where there is a single list of possible outcomes, such as dice rolls, heights, or death months relative to birth months. Another application, which I will now explain, is where the data are labeled in more than one way.

For example, the number 7 can be either a winning roll or a losing roll in craps, depending on the outcomes of previous rolls. In looking for a crooked dice game, we might ask not only how often 7 comes up but, more specifically, how often 7 comes up when it is a winning roll and when it is a losing roll. We are then labeling the outcomes in two ways: by the numbers rolled and by when they are rolled. A natural null hypothesis is that these two facets of the data are unrelated and that the odds of rolling a 7 do not depend on whether 7 is a winning or losing roll. If the data are not consistent with this null hypothesis, then they provide statistical evidence that the game is crooked, in that the casino may switch dice as needed to increase its edge. This line of thought leads to a very useful application of chi-square tests.

If we want to test whether or not two factors are independent, we can label the data by these two factors, let the null hypothesis be that they are independent, and then use a chi-square test to see how closely the data agree with this independence hypothesis.

Income and Political Opinions

Consider a hypothetical poll that rates a Republican president's performance:

President's Rating	Number of People
Excellent	200
Good	250
Fair	300
Poor	250
Total	1,000

This poll shows a pretty evenly balanced evaluation of the President's performance, but we may suspect that the President's support is not uniform throughout society. Beneath the surface, there may be significant differences in the evaluations made by different segments of society: rich versus poor, young versus old, or along other lines. Perhaps we believe that the high-income and low-income people have very different views of the President's performance. If so, then the poll results can be labeled in a second direction, according to the income of the respondent. In Table 12.6, the sample is labeled in two ways, by income class and by performance evaluation. Statisticians call such a table a **two-way contingency table.**

It is easy to think of reasons why a person's income and political opinions may be related. People who are down and out tend to blame the government for their misfortune. Those who are well off are more likely to be satisfied with the present administration and reluctant to make changes. In addition, presidents often pursue policies that have uneven effects on the different income classes. Franklin Roosevelt was perceived to favor the poor and the unemployed, and these low-income groups were big Roosevelt supporters. Ronald Reagan, on the other hand, was believed to favor the well-to-do, and they responded by favoring him. Other presidents have been perceived as being more even-handed.

A brief study of Table 12.6 reveals that low-income people tend to judge this President more harshly, while high-income people are more favorably impressed. The statistical question is whether this disparity can reasonably be explained by the

Table 12.6 *Poll Responses Classified by Rating and Income*

Rating	Respondent's Income Class			
	Low-Income	Middle-Income	High-Income	Total
Excellent	32	102	66	200
Good	55	118	77	250
Fair	114	148	38	300
Poor	99	132	19	250
Total	300	500	200	1,000

luck of the draw in selecting this particular sample. If not, then we conclude that the poll shows a population relationship between income-class and evaluation of the President. This statistical question can be answered by Pearson's chi-square test. The competing hypotheses are

$$H_0: \text{income and opinion are unrelated}$$

$$H_1: \text{income and opinion are related}$$

The null hypothesis implies certain expected values for the entries in Table 12.5. Once these are calculated, they can be compared to the actual entries and a chi-square statistic computed.

First take the totals from 12.6:

	Low-Income	Middle-Income	High-Income	Total
Excellent				200
Good				250
Fair				300
Poor				250
Total	300	500	200	1,000

If income-class and rating were independent, how would we expect these totals to be allocated across the table? Well, 20 percent of all 1,000 respondents gave the President an "excellent" rating. If income class and rating were unrelated, we would expect 20 percent of the 300 low-income people, 20 percent of the 500 middle-income people, and 20 percent of the 200 high-income people to give the President an excellent rating:

	Low-Income	Middle-Income	High-Income	Total
Excellent	.2(300) = 60	.2(500) = 100	.2(200) = 40	200
Good				250
Fair				300
Poor				250
Total	300	500	200	1,000

Now, let's fill in the row of "good" ratings. Twenty-five percent of all 1,000 respondents rated the President's performance "good." If income class and rating were unrelated, then 25 percent of the low-, middle-, and high-income groups could be expected to give a "good" rating:

	Low-Income	Middle-Income	High-Income	Total
Excellent	60	100	40	200
Good	.25(300) = 75	.25(500) = 125	.25(200) = 50	250
Fair				300
Poor				250
Total	300	500	200	1,000

Continuing with this reasoning, we can complete the entire Table 12.7.*

Table 12.7   *The Expected Frequencies if Ratings and Income Class Are Unrelated*

	Low-Income	Middle-Income	High-Income	Total
Excellent	60	100	40	200
Good	75	125	50	250
Fair	90	150	60	300
Poor	75	125	50	250
Total	300	500	200	1,000

Now we can make an informed scrutiny of the disparities between the actual responses and the responses that we would expect if income and rating were independent. A comparison of Tables 12.6 and 12.7 reveals that there are some large deviations. Instead of 60 low-income people giving the President an excellent rating, only 32 did. In sharp contrast, 66 high-income people rated the President's performance excellent, instead of the expected 40. Are these and the other disparities statistically significant, or can they be explained as chance variation in the random selection of a sample? We can answer this question by calculating Pearson's chi-square statistic, Equation 12.10, using the data in Table 12.6 as the observed frequencies O_i and the data in Table 12.7 as the expected frequencies E_i:

$$\chi^2 = \frac{(32 - 60)^2}{60} + \frac{(102 - 100)^2}{100} + \frac{(66 - 40)^2}{40}$$
$$+ \frac{(55 - 75)^2}{75} + \cdots + \frac{(19 - 50)^2}{50} = 92.10$$

To see if this chi-square statistic is statistically significant, we need to know which degrees of freedom to use in Table 5 at the end of this book. We've seen that when the data are labeled in one direction, with m possible outcomes, there are $m - 1$ degrees of freedom because there is one constraint on the data. Here, the data are labeled in two directions, with three possibilities one way and four possibilities the other way. Our chi-square statistic is the sum of $(3)(4) = 12$ squared deviations. But there are also constraints in both directions. Only two of the three entries in any row of Tables 12.6 or 12.7 can be independently specified; given those two, the third entry is known immediately. And only three of the four entries in any column

---

*We filled the table row-by-row. What if we instead work column by column? We get exactly the same answers. For example, 30 percent of the 1,000 respondents are low-income. If income and rating were independent, then we would expect 30 percent of the 200 excellent ratings to come from low-income people: .3(200) = 60, the same answer as when we reasoned row by row.

of Tables 12.6 or 12.7 are independent. Overall, there are only $(3 - 1)(4 - 1) = 6$ independent entries in these tables and, thus, only 6 degrees of freedom.*

> The degrees of freedom for a contingency table is
> (number of rows − 1)(number of columns − 1)

Table 5 at the end of this book tells us that, with a 5 percent significance level, the cutoff point for a chi-square statistic with 6 degrees of freedom is 12.59. Our computed chi-square value 92.10 has a *P*-value far below .005, the smallest probability given in Table 5. There is virtually no chance that the luck of the draw would give a sample with such large deviations between the observed and expected frequencies, so we reject the null hypothesis that rating and income class are independent and conclude that, among the population as a whole, different income classes rate this President differently.

Do John McEnroe's Grunts Help?

An article entitled "Should John McEnroe Grunt?" began

> A third of a million people munched something like 11 tonnes of fresh strawberries and drank 17,000 bottles of champagne as John McEnroe sweated and grunted his way to the men's singles title at Wimbledon . . . His winning way with officials and loud service grunt make him one of the most audible players in the modern game. But does the McEnroe grunt affect his game?[16]

Dr. Dennis Lendrum, an animal-behavior analyst at Nottingham University in England, kept careful records of McEnroe's grunts during the 1983 Wimbledon matches and drew some interesting conclusions.

If McEnroe's grunts were a sign of extra effort, then we might expect more grunting on the first serve than on the second, when the primary objective is to hit the ball accurately. Lendrum's data were

Serves	Grunts	Silent	Total
First	171	69	240
Second	62	26	88
Total	233	95	328

*Another explanation is that the sum of 12 squared deviations has 11 degrees of freedom, because it represents the sum of 11 independent squared normal variables. But this chi-square statistic requires the estimation of five parameters: the fractions of the population that are low-income and middle-income, and the fractions that rate the President as excellent, good, and fair. (The other two fractions, high-income and poor rating, can be computed from these five.) Because we lose 1 degree of freedom for every parameter that must be estimated from the data, the degrees of freedom fall by 5, from 11 to 6.

Statistical Evidence on the Death Penalty

An earlier example in this chapter discussed the use of statistical evidence in enforcing Title VII of the Civil Rights Act of 1964. Courts have been more reluctant to accept statistical evidence in some cases involving important constitutional issues.

In the 1987 case of *McCleskey* v. *Kemp,* the U.S. Supreme Court upheld the constitutionality of Georgia's death penalty, rejecting statistical evidence that blacks who murder whites are disproportionately likely to be sentenced to death. Table 1 gives some of the data considered by the Court regarding 2,475 Georgia murder convictions between 1973 and 1980.

Overall, 5.2 percent of those convicted of murder were sentenced to death. For a chi-square test, the expected value for the category white defendant/black victim receives death sentence works out to be only 3.31 and so should be combined with another category. It seems reasonable to combine white defen-

dant/black victim with white defendant/white victim, and the contingency table shown as Table 2 results. The expected values under the null hypothesis that death sentences are independent of race are given in Table 3. Observed death sentences occurred 56.37 fewer times than expected for black defendants with black victims, 38.21 more times than expected for black defendants with white victims, and 18.16 more times than expected for white defendants. The value of the chi-square statistic is 183.93, which, with $(2 - 1)(3 - 1) = 2$ degrees of freedom, is highly statistically significant.

If we group the data solely by the race of the defendant, the differences between the observed values,

Death Sentence	Black Defendant	White Defendant	Total
Yes	68	60	128
No	1,598	749	2,347
Total	1,666	809	2,475

Table 1

Death Sentence	Black Defendant		White Defendant	
	Black Victim	White Victim	Black Victim	White Victim
Yes	18	50	2	58
No	1,420	178	62	687
Rate	1.2%	21.9%	3.1%	7.8%

Table 2

Death Sentence	Black Defendant		White Defendant	Total
	Black Victim	White Victim		
Yes	18	50	60	128
No	1,420	178	749	2,347
Total	1,438	228	809	2,475

and the expected values,

Death Sentence	Black Defendant	White Defendant	Total
Yes	86.16	41.84	128
No	1,579.84	767.16	2,347
Total	1,666	809	2,475

seem less pronounced, though the value of the chi-square statistic is 12.35, which is still highly statistically significant.

The Supreme Court ruled, by a narrow 5–4 margin, that there was no direct evidence that the defendant in this particular case, a black man convicted of killing a white Atlanta policeman, had been treated unfairly. Although the Court acknowledged that the data suggested imperfections in Georgia's system of capital punishment, the majority ruled that this evidence did "not demonstrate a constitutionally significant risk of racial bias."

Table 3

Death Sentence	Black Defendant		White Defendant	Total
	Black Victim	White Victim		
Yes	74.37	11.79	41.84	128
No	1,363.63	216.21	767.16	2,347
Total	1,438	228	809	2,475

The proportions of the first and second serves accompanied by grunts were remarkably similar

$$\text{grunts on first serve: } 171/240 = 0.71$$
$$\text{grunts on second serve: } 62/88 = 0.70$$

For a chi-square test, we can compare the observations to the expected values, if the grunting is unrelated to whether it is a first or second serve:

Serves	Grunts	Silent	Total
First	170.5	69.5	240
Second	62.5	25.5	88
Total	233	95	328

Figure 12.3 shows the application of the MINITAB and MYSTAT computer programs to these data. The value of the chi-square statistic is virtually zero, 0.02. Thus, there is no evidence that McEnroe grunts significantly more on his first serves. His grunting may be psychological, rather than a reflection of extra effort.

Dr. Lendrum inquired whether the grunt serves were more effective than the silent ones:*

Serves	Grunts	Silent	Total
Aces	61	35	96
Errors	32	8	40
Other	144	53	197
Total	237	96	333

A calculation of the proportions,

Serves	Grunts	Silent
Aces	25.7%	36.5%
Errors	13.5%	8.3%
Other	60.8%	55.2%
Total	100.0%	100.0%

shows that grunt serves were less often aces and more often errors. For a formal chi-square test, the expected values if grunts didn't matter are

Serves	Grunts	Silent	Total
Aces	68.3	27.7	96
Errors	28.5	11.5	40
Other	140.2	56.8	197
Total	237	96	333

*There are slightly more serves here, because of the inclusion of 5 third serves.

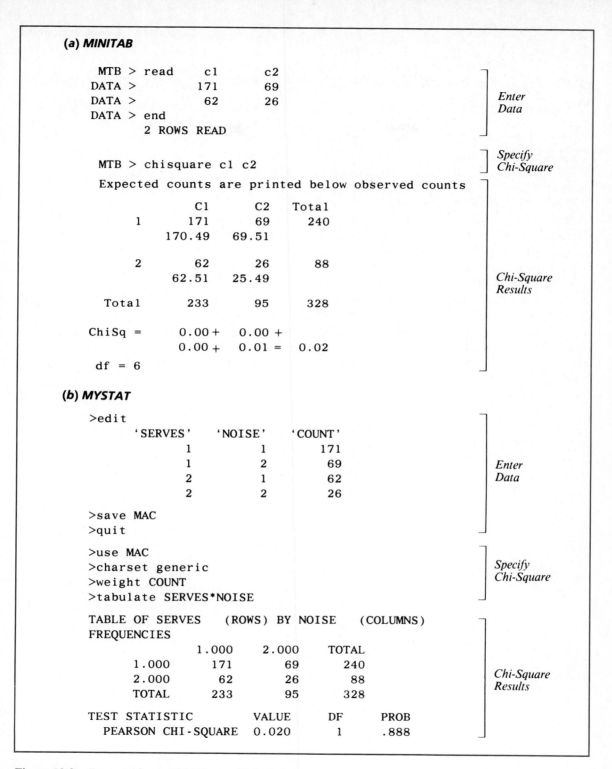

(a) MINITAB

```
    MTB > read      c1        c2
    DATA >          171       69
    DATA >          62        26
    DATA > end
          2 ROWS READ

    MTB > chisquare c1 c2

    Expected counts are printed below observed counts

                 C1       C2      Total
         1       171      69      240
                 170.49   69.51

         2       62       26      88
                 62.51    25.49

    Total        233      95      328

    ChiSq =       0.00 +   0.00 +
                  0.00 +   0.01 =   0.02
    df = 6
```

Enter Data

Specify Chi-Square

Chi-Square Results

(b) MYSTAT

```
    >edit
           'SERVES'      'NOISE'      'COUNT'
               1            1           171
               1            2           69
               2            1           62
               2            2           26
    >save MAC
    >quit

    >use MAC
    >charset generic
    >weight COUNT
    >tabulate SERVES*NOISE

    TABLE OF SERVES    (ROWS) BY NOISE    (COLUMNS)
    FREQUENCIES
                   1.000     2.000      TOTAL
         1.000      171        69        240
         2.000      62         26        88
         TOTAL      233        95        328

    TEST STATISTIC          VALUE    DF      PROB
       PEARSON CHI-SQUARE   0.020     1      .888
```

Enter Data

Specify Chi-Square

Chi-Square Results

Figure 12.3 *Excerpts from a Chi-Square Calculation*

Does Smoking Cause Sex?

Contingency tables are a useful way of organizing data and looking for dependencies, but the discovery of a statistical dependency does not tell us its cause. Two variables may be related because the first influences the second, because the second influences the first, because some other factor influences both, or because of sampling error. Only additional data or common sense can sort out the reason for a statistical dependency. Additional data may reveal that it was an accidental relation, a mere sampling error. Or, in an experimental context, if we suspect that some third influence is responsible, we may be able to hold that factor constant and see what happens. In many cases, common sense suffices.

A good example is provided by the following deliberate misinterpretation, which was intended to show the need for common sense (and to provoke a chuckle, too).[17] A survey asked 508 girls aged 14–17 whether or not they smoked and whether or not they used oral contraceptives, thus indicating sexual activity. The results of this survey can be arranged in a contingency table as follows:

Smoker	Sexually Active	Not Sexually Active	Total
Yes	24	122	146
No	11	351	362
Total	35	473	508

Sixteen percent of the smokers are sexually active, while only 3 percent of the nonsmokers are. There is a very high "risk of sex for smokers." For a chi-square test, we compare these observed data with the expected values if sex were independent of smoking:

Smoker	Sexually Active	Not Sexually Active	Total
Yes	10.1	135.9	146
No	24.9	337.2	362
Total	35	473	508

The value of the chi-square statistic is

$$\frac{(24 - 10.1)^2}{10.1} + \frac{(11 - 24.9)^2}{24.9}$$
$$+ \frac{(122 - 135.9)^2}{135.9} + \frac{(351 - 337.2)^2}{337.2}$$
$$= 28.88$$

which is highly significant. Thus, the author concludes that smoking not only causes cancer and cardiovascular disease, but sexual activity as well.

The data, of course, show only association and not causation. It may be that smoking encourages sexual activity, that sexual activity encourages smoking, or that girls who smoke also happen to be sexually active, without either activity causing the other. Common sense is needed to sort through these competing explanations of the data. I'll put my money on the third explanation.

The value of the chi-square statistic works out to be 4.56, which (with $(2 - 1) \times (3 - 1) = 2$ degrees of freedom) is statistically significant at the 5 percent level. Dr. Lendrum concludes that McEnroe should "shut up when serving."

The Unshuffled Draft Lottery

For another example, consider the national lottery that was held in 1969 to determine the order in which men born in 1950 would be drafted to fight in the Viet Nam War. The Selective Service System prepared 366 capsules, each containing a birthdate, January 1, January 2, and so on. Each month's capsules were carried to the drawing in a separate box. At the appropriate time, the boxes were emptied into a large drum, one after another—first January, then February, and on through the months with December poured in last. The drum was rotated a few times and then the capsules were drawn out, one by one. Those capsules drawn first would be drafted first, those drawn last would most likely not be drafted at all. The results are summarized in Table 12.8.

A casual inspection suggests that the late months, August through December, got more than their share of the low, "first-drafted" numbers. The early months, January through May, got more than their share of the late, "not-drafted-at-all" numbers. A plausible explanation is that a few rotations of the drum did not do a very good job of mixing up the capsules. The early months stayed at the bottom of the heap, while the late months stayed on top, ready to be picked out first. Is this suspicion an unwarranted criticism from a desperate statistician with a December birthday? After all, someone has to go first and, by the luck of the draw, the first

Table 12.8 *The 1969 Draft Lottery*

	Priority Numbers			
Month	*1–122* (First-Drafted)	*123–244* (Middle)	*245–366* (Last-Drafted)	*Total*
January	9	12	10	31
February	7	12	10	29
March	5	10	16	31
April	8	8	14	30
May	9	7	15	31
June	11	7	12	30
July	12	7	12	31
August	13	7	11	31
September	10	15	5	30
October	9	15	7	31
November	12	12	6	30
December	17	10	4	31
Total	122	122	122	366

few choices may well turn out to be November and December days. Are the deviations seen in Table 12.8 so large that they cannot be explained by chance alone?

This is the sort of question a chi-square test is designed to answer. The null hypothesis is that the capsules were well-shuffled, so that each date was equally likely to be chosen. If this null hypothesis were true, we would expect each month to have its days equally divided among the three categories 1–122, 123–244, and 245–366. For November and other months with 30 days, the expected number of days in each of these three categories is 10. Months with 31 days would average 31/3 = 10.33 days per category. February, with 29 days, would average 29/3 = 9.67 days per category in the long run, over many such drawings. To gauge how severely the 36 actual frequencies shown in Table 12.8 depart from these expected frequencies, we calculate the chi-square statistic

$$\chi^2 = \frac{(9 - 10.33)^2}{10.33} + \frac{(12 - 10.33)^2}{10.33} + \cdots + \frac{(4 - 10.33)^2}{10.33}$$

$$= 37.2$$

Because Table 12.8 has 12 rows and 3 columns, there are $(12 - 1)(3 - 1) = 22$ degrees of freedom. A chi-square statistic with 22 degrees of freedom has a 5 percent chance of exceeding 33.9 and a 1 percent chance of exceeding 40.3. There is only about a 2 percent chance of a chi-square value being as large as the observed 37.2. Thus, at the 5 percent level at least, these data reject the hypothesis that the capsules were well shuffled in the 1969 draft lottery. In response to such criticism, the Selective Service was more careful the following year, when they conducted a draft lottery for men born in 1951. In the 1970 lottery, two drums were used—one with dates and another with numbers—and each drum was loaded in a random order.[18]

EXERCISES

12.22 A study investigated whether a mouse raised by a foster mother would be more likely to fight with other mice than would a mouse raised by its natural mother:[19]

Mother	Fighters	Nonfighters
Natural	27	140
Foster	47	93

Is there a statistically significant relationship? Use a chi-square test with a 1 percent significance level.

12.23 Yale University admitted its first women undergraduates in 1969. For the first few years, male and female students were admitted separately, in an effort to preserve Yale's tradition of "graduating 1,000 fine young men each year." However, a separate admission policy can cause disparities between the caliber of the male and female students. A University Committee on Coeducation compared the academic performance of Yale male and female undergraduates, using data such as these:

Course Grades, Spring 1972	Males	Females
Honors	6,591	1,593
High Pass	7,573	1,655
Pass	3,108	539
Other	590	117

Are grades and gender unrelated?

12.24 Four toothpastes were tested by randomly dividing 400 children into four groups of 100. Each group used a different brand of toothpaste for one year, and then the following results were observed:

Brand of Toothpaste	Number of New Cavities		
	0–1	2–3	4 or more
A	50	28	22
B	41	30	29
C	39	23	38
D	36	27	37

Use a chi-square test at the 5 percent level to see if these data are consistent with the null hypothesis that there are no significant differences in the effectiveness of these toothpastes in preventing cavities.

12.25 After attempting to imitate a popular television character, a young man concluded that whether one is right-handed or left-handed affects how far apart the two middle fingers on each hand can be spread. If a wider "V" is made with the right hand, then the person is probably left-handed. If a wider V is made with the left hand or there is no difference, then the person is probably right-handed. After an article explaining this theory appeared in the February 20, 1974 issue of *Current Science,* 3,225 readers reported the following results:

Readers	Test Correct	Test Wrong
Right-Handed	2,097	783
Left-Handed	289	56

Find the probabilities that a person picked from this group will be: **a.** right-handed **b.** left-handed **c.** right-handed if the test predicts right-handed **d.** left-handed if the test predicts left-handed **e.** correctly identified by this test. Use a chi-square test at the 1 percent level to see if there is a statistically significant relationship between handedness and the test predictions.

12.26 A Gallup poll asked 260 males and 263 females whether male or female students are more assertive in class:[20]

Assertiveness	Males	Females
"Men Are"	80	61
"Women Are"	104	112
"No Difference"	76	90
Total	260	263

Make a chi-square test of the null hypothesis that the answer given does not depend on whether the student is a male or female.

12.27 *(continuation)* This same Gallup poll divided the answers to this question by class, which are shown in Table 12.9. Make a chi-square test at the 1 percent level of the null hypothesis that the answers do not depend on a student's class.

12.28 Roll two dice 50 times, each time rolling first one die with the left hand and then the other die with the right hand. For each turn, record whether the left-hand die is **a.** greater than 3, or **b.** less than or equal to 3, and record whether the sum is **c.** greater than 7, or **d.** less than or equal to 7. Make a chi-square test at the 5 percent level of the null hypothesis that the sum is independent of the outcome of the first roll.

12.29 Based on their answers to questions regarding such activities as school truancy, theft, and gang fights, 1,137 boys were separated into two categories: "most delinquent" and "least delinquent." The researcher then investigated whether their delinquent behavior was related to their birth order.[21] The data are shown in Table 12.10. Which boys seem to

Table 12.9 *Data for Exercise 12.27*

Assertiveness	Freshman	Sophomore	Junior	Senior
"Men Are"	26	29	44	48
"Women Are"	72	60	38	39
"No Difference"	37	44	43	43
Total	135	133	125	130

Table 12.10   *Data for Exercise 12.29*

Delinquent	Oldest	In-Between	Youngest	Only Child
Most	127	123	93	17
Least	345	209	158	65

be unusually delinquent? Make a chi-square test of the null hypothesis that birth order and delinquency are independent.

12.30 A company's annual report to its shareholders typically begins with a brief letter from the company's president or chairman, usually optimistic and often regarded as mere fluff. To test this fluff theory, three researchers randomly selected 20 companies whose price rose substantially in one of the years 1980–1983 and 20 companies whose price fell sharply. In each case, they looked at the president's letter in the preceding year to see if it contained signals of the good or bad times ahead.[22] In one of their tests, they looked at whether the letters mentioned imminent gains or losses for the company, as shown

in Table 12.11. Make a chi-square test at the 1 percent level of the null hypothesis that the mention of gains, losses, or neither is unrelated to the performance of the stock in the subsequent year. (These researchers say that "Because of empty cell, chi-square is not appropriate." Why are they wrong?)

12.31 A sample of professional economists was asked whether they agreed or disagreed with each of 27 statements, to see if there was consensus or dissension within the profession.[23] Table 12.12 gives 96 their responses, by country, to the statement "Inflation is primarily a monetary phenomenon." Make a chi-square test at the 1 percent level of the null hypothesis that the responses are independent of the economist's nationality.

Table 12.11   *Data for Exercise 12.30*

Stock Performance	Imminent Gains	No Mention	Imminent Losses
Up	13	7	0
Down	4	8	8

Table 12.12   *Data for Exercise 12.31*

Country	Generally Agree	Agree with Provisions	Generally Disagree	Total
United States	55	61	87	203
Austria	12	25	51	88
France	17	30	110	157
Germany	67	84	117	268
Switzerland	62	70	65	197
Total	213	270	430	913

12.4 SUMMARY

We often want to test if two population means μ_1 and μ_2 are equal. We can do so by testing the null hypothesis that the difference between the means is equal to zero,

$$H_0: \mu_1 - \mu_2 = 0$$

$$H_1: \mu_1 - \mu_2 \neq 0$$

The test is based on the statistical fact that the means $\overline{X}_1$ and $\overline{X}_2$ of two independent random samples of size n_1 and n_2 tend to be normally distributed, as does their difference,

$$\overline{X}_1 - \overline{X}_2 \sim N[\mu_1 - \mu_2, \sqrt{(\sigma_1^2/n_1) + (\sigma_2^2/n_2)}]$$

Normality is assured if X_1 and X_2 are normally distributed and approximated in accordance with the central limit theorem, if the samples are reasonably sized.

The test itself can be conducted by setting a cutoff point for $\overline{X}_1 - \overline{X}_2$, by calculating a Z-value and corresponding P-value for the observed $\overline{X}_1 - \overline{X}_2$, or by constructing a confidence interval.

A similar logic applies to the difference in two success probabilities π_1 and π_2, because the observed success proportions p_1 and p_2 can be interpreted as sample means. The only added wrinkle is that the expected value and standard deviation of the p_i are π_i and $\sqrt{\pi_i(1 - \pi_i)/n_i}$. Thus

$$p_1 - p_2 \sim N[\pi_1 - \pi_2, \sqrt{\pi_1(1 - \pi_1)/n_1 + \pi_2(1 - \pi_2)/n_2}]$$

A paired-sample test looks at differences for individual observations within a single sample, or for "twins"—matched pairs in the sample. Individuals might compare two colas, or two medicines might be compared by administering them to similar individuals. The observed differences are then the random variable X, and we can use the sample mean to test in the usual way whether or not the expected value of X is zero.

Chi-square tests provide a handy procedure for testing a wide variety of hypotheses. The basic idea is to identify m mutually exclusive categories, compute the expected number of observations E_i in each category under the null hypothesis, and then make a comparison with the actual number of observations O_i by computing the value of the chi-square statistic,

$$\chi^2 = \Sigma(O_i - E_i)^2/E_i$$

If all of the $E_i \geq 10$, then the distribution of this statistic can be approximated by the chi-square distribution with $m - 1$ degrees of freedom shown in Table 5 at the end of this book. These are called *goodness-of-fit* tests because they gauge how closely the sample distribution agrees with the population distribution under the null hypothesis.

A special application of chi-square tests is two-way contingency tables. The data are labeled by two factors—for instance, opinion poll responses that are classified by both the rating given a president's performance and the income of a respondent. A chi-square statistic can be used to test the null hypothesis that these two factors are independent, by comparing the observed data with the expected values if the factors are independent.

REVIEW EXERCISES

12.32 A study of changes over a 100-year period in the lead content of people's hair, measured in micrograms (millionths of a gram) of lead per gram of hair, used the data in Table 12.13.[24] Are there statistically significant differences in average lead content between **a.** children 1871–1923 and adults 1871–1923? **b.** children 1924–1971 and adults 1924–1971? **c.** children 1871–1923 and children 1924–1971? **d.** adults 1871–1923 and adults 1924–1971?

12.33 There has been considerable concern that activity in the options market has been affecting the stock market ("the tail wagging the dog"). One study compared the percentage price change in the stock market during the last hour of trading on days when S&P 500 Index options and futures both expired with days when neither expired.[25] Is the observed difference statistically significant at the 1% level?

	Expiration Day	No Expiration
Mean return	−.352	.061
Standard deviation	.641	.211
Number of observations	10	97

12.34 Jack and Jill each take a 200-question IQ test. The number of correct answers is the test's estimate of a person's IQ. As it turns out, Jack's IQ score is 110 and Jill's is 132. Is this difference statistically significant at the 5 percent level? Why, in answering this question, is it of no help to know that IQ scores on this test are normally distributed with a mean of 100 and a standard deviation of 16?

12.35 To see whether persons with glaucoma have abnormally thick corneas, the corneas were measured in eight persons with glaucoma in one eye but not in the other.[26] Here are the thicknesses, in microns:

Glaucomatous Eye	Other Eye
488	484
478	478
480	492
426	444
440	436
410	398
458	464
460	476

Table 12.13 *Data for Exercise 12.32*

Subjects	Sample Size	Sample Mean	Standard Deviation
Children, 1871–1923	36	164.2	124.0
Adults, 1871–1923	20	93.4	72.9
Children, 1924–1971	119	16.2	10.6
Adults, 1924–1971	28	6.6	6.2

In what sense are these a matched-pair sample? Are the observed differences in corneal thickness statistically significant at the 1 percent level?

• **12.36** A school psychologist administers tests to see if there is a statistically significant difference (at the 5 percent level) in a student's intelligence and achievement scores. What is the probability of incorrectly concluding that there is a statistically significant difference if **a.** just one pair of tests (intelligence and reading) is administered? **b.** five pairs of tests (intelligence and reading, intelligence and writing, intelligence and math, and so on) are administered? (Assume independence.) If five pairs of independent tests are administered, how can we reduce the overall probability of making a Type I error to 5 percent? Do you think that independence is a reasonable assumption here?

12.37 Three researchers suspected that the Federal Trade Commission (FTC) is more likely to dismiss a case brought against a firm if the member of Congress from that district is on a committee overseeing the FTC.[27] Among the data they examined, they found that, between 1961 and 1969, in districts where the representative was on a related House subcommittee, 84 of 1104 cases were dismissed, while in other districts 81 of 1371 cases were dismissed. Is this difference statistically significant at the 1 percent level?

12.38 A consumer advisory group measured the lifetimes in hours of 15 light bulbs of two different brands, Longlast and Lastlong.

Longlast			Lastlong		
781	825	893	745	920	821
648	924	680	886	981	1071
728	565	724	940	785	804
1002	710	812	615	833	767
850	903	709	830	678	732

Assuming normality and a common standard deviation, calculate the pooled sample variance and make a t-test at the 1 percent level of the null hypothesis that there is no difference in the mean population lifetimes.

12.39 Here are data on the hair and eye color of 6,800 Germans:[28]

	Eye Color		
Hair Color	Blue	Brown	Grey or Green
Fair	1768	115	946
Brown	807	438	1387
Black	189	288	746
Red	47	16	53

Do these data show a statistically significant relationship between hair color and eye color?

12.40 A 1954 study of the effects of the original Salk vaccine on the incidence of paralytic polio yielded the following data (from Table 8.1):

Children Inoculated	Number Inoculated	Number of Polio Cases
With Salk vaccine	200,745	28
With placebo	201,229	71

Is there a statistically significant difference in polio incidence between the two groups of children?

12.41 A 1986 *New York Times*/CBS News poll of 168 Americans of Hispanic descent and a similar number of American blacks found that 43 percent of the Hispanics and 27 percent of the blacks labeled themselves politically "conservative."[29] Is this difference statistically significant at the 1 percent level?

12.42 To gauge the effects of writing deficiencies on college performance, Pomona College examined the freshman records of those 45 students in the 1983 freshman class of 409 with the lowest verbal SAT scores. Of the 66 students who received low grade notices in the fall term, 23 had low verbal SATs. Of the 14 students placed on academic probation, 8 had low verbal SATs. By each measure, are the differences between the low-verbal-score students and the remaining freshmen too large to be explained by sampling error? From what population are these students a sample?

12.43 A psychic claims an ability to "destabilize" coins so that heads and tails do not come up equally often. A "destabilized" coin is flipped 200 times and comes up heads 93 times. Construct a 95 percent confidence interval for the probability π of obtaining a head with this coin. Now test these hypotheses (at the 5 percent level)

$$H_0: \pi = .5$$
$$H_1: \pi \neq .5$$

First, use a normal approximation to the binomial and, second, use a chi-square test. Do both tests give the same conclusion?

12.44 David Phillips collected data on the relationship between death and birth dates, shown in Table 12.14. For each of these three samples, test the hypothesis that the death month is unrelated to the birth month.

• **12.45** Make a chi-square test of the null hypothesis that the batting averages shown in Table 1 on page 252 come from a normal distribution.

12.46 Two surveys, conducted 20 years apart, asked five questions to gauge people's trust in government (e.g., "Are quite a few of the people running the government a little crooked?").[30] Those surveyed were labeled as follows:

Rating	1958	1978
Cynical	200	1198
Middle	456	599
Trusting	1057	438
Total	1713	2235

Is there a statistically significant difference (at the 1 percent level) in the 1958 and 1978 responses?

12.47 To study social mobility in Great Britain, a 1954 sample of 3,497 male workers determined the occupations of the workers and of their fathers. These occupations were then divided into upper, middle, and lower class:[31]

Father's Status	Worker's Status		
	u	m	l
u	588	395	159
m	349	714	447
l	114	320	441

If there was no social mobility whatsoever, what would the data look like? If there was completely free social mobility, what would you expect the data to look like? Make a chi-square test of the null hypothesis that worker and father status are independent.

12.48 U.S. soldiers in World War II were taller, heavier, and stronger than those in World War I. To see if the same was true of U.S. women in those periods, a statistics class at Mount Holyoke College collected the data shown in Table 12.15.[32] For each of these three physical characteristics, calculate a 95 percent confidence interval for the difference in the pop-

Table 12.14   *Data for Exercise 12.44*

Number of Months Before (−) or After (+) Birth Month	Observed Number of Deaths		
	Sample 2	Sample 3	Sample 4
−6	17	10	39
−5	23	14	32
−4	26	12	29
−3	27	11	35
−2	28	8	31
−1	28	12	30
0	42	15	36
+1	32	15	35
+2	31	15	38
+3	34	13	26
+4	36	20	31
+5	30	13	29
Total	354	158	391

Table 12.15   *Data for Exercise 12.48*

Measurements	1918 (250 freshmen)		1943 (308 freshmen)	
	Mean	*Std. Dev.*	*Mean*	*Std. Dev.*
Height (cm)	161.7	6.3	164.8	5.9
Weight (lb)	118.4	16.6	128.0	17.2
Grip (kg)	30.9	4.6	32.3	5.4

ulation means. Do these data show statistically significant differences in the heights, weights, and grips of 1918 and 1943 students?

12.49 As part of the investigation of the inheritance of various traits, the athletic skills of brothers were evaluated. The data are shown in Table 12.16.[33] Make a chi-square test at the 1 percent level of the null hypothesis that the athletic skills of brothers are independent.

12.50 In theory, companies with high profit rates should reinvest most of their earnings (and borrow to invest even more) while those with low profit rates should refrain from expansion and pay most of their earnings out as dividends. To see whether firms behave in this way, a Harvard Business School professor looked at 1,448 U.S. industrial companies during the ten-year period 1966–1975.[34] Use his data, given in Table 12.17, to test at the 1 percent level the null hypothesis that profit rate and reinvestment policy are independent.

12.51 A 20-year study in Ypsilanti, Michigan, followed the lives of 66 poor children who attended

Table 12.16   *Data for Exercise 12.49*

Second Brother	First Brother			
	Athletic	*Betwixt*	*Nonathletic*	*Total*
Athletic	906	20	140	1066
Betwixt	20	76	9	105
Nonathletic	140	9	370	519
Total	1066	105	519	1690

Table 12.17   *Data for Exercise 12.50*

Earnings Reinvested	Average Profit Rate				
	<8%	*8%–12%*	*12%–18%*	*>18%*	*Total*
<40%	107	36	27	13	183
40%–60%	45	92	75	25	237
60%–80%	48	119	133	37	337
80%–120%	68	144	154	31	397
>120%	158	83	46	7	294
Total	426	474	435	113	1,448

preschool in the early 1960s and 57 who did not. Four of the findings follow. In each case, test whether the reported difference is statistically significant at the 5 percent level. (Use the sample frequencies when calculating the standard deviation.) **a.** 44 of the preschool group and 28 of the other group completed high school. **b.** 12 of the preschool group and 18 of the other group were on welfare at age 19. **c.** 20 of the preschool group and 29 of the other group had been arrested at some time. **d.** 33 of the preschool group and 18 of the other group were employed.

12.52 Chapter 10 discusses a test of a vaccine for the common cold, using 548 college students who believed themselves especially susceptible to colds. Half were given the experimental vaccine and half, the control group, were given plain water. Here are the test results:

	Vaccine	Control Group
Number of persons	272	276
Colds per person		
Mean	1.6	2.1
Standard deviation	0.8	1.0

Calculate the Z-value and P-value for a difference-of-means test. Is the observed difference statistically significant at the 1 percent level? The researchers concluded that "the effect is too small to be of practical importance."[35] What is the basis for this conclusion?

12.53 Exercise 8.75 relates how market research firms gauge the effectiveness of commercials by allowing volunteers to choose a free product from a list that includes the advertised product. In one test, the advertised product was chosen by 83 of 400 volunteers who watched the commercial and by 57 of 400 who did not. Is this difference statistically significant at the 1 percent level? (Use the sample frequencies to calculate the standard deviation.)

12.54 A compilation of over 12,000 times at bat by major league baseball players in 1951 and 1952 separated the data into four categories, depending on whether the pitcher and batter were right- or left-handed.[36] Calculate the batting averages for each of these four groups. Then combine the data into two groups, depending on whether the pitcher and batter are same-handed or opposite-handed, and, assuming these data to be Bernoulli trials, test at the 1 percent

level the null hypothesis that the probability of getting a hit does not depend on whether the pitcher and batter are same-handed or opposite-handed.

Batter	Pitcher	Times at Bat	Hits
Right	right	5,197	1,201
Left	left	1,164	270
Left	right	4,002	1,055
Right	left	2,245	590

12.55 In 1983, 22,000 doctors volunteered to participate in an experiment testing the effects of aspirin on the incidence of heart attacks.[37] Half of these doctors were given a single aspirin tablet every other day; the other half took a placebo every other day. Neither the volunteers nor the scientists conducting the experiment knew which volunteers were taking aspirin. After five years of what was intended to be a seven-year test, the experiments were stopped by an outside committee monitoring the results. During this period, fatal heart attacks were suffered by 18 of those taking placebos and by 5 of those taking aspirin, and nonfatal heart attacks were experienced by 171 of those taking placebos and by 99 of those taking aspirin. Using the sample frequencies to estimate the standard deviations, calculate the Z-value and the P-value and determine if the difference in fatal heart attacks is statistically significant by a two-tailed test at the 1 percent level. What about the differences in nonfatal heart attacks? Why do you suppose that the experiment was stopped two years early?

12.56 In 1990 New Zealand researchers reported that 10 of 135 heart-attack victims who were given streptokinase to dissolve clots in the arteries died within 30 days, while 5 of 135 heart-attack victims who were given TPA died within 30 days.[38] Estimate the standard deviations with these sample frequencies and use a normal approximation to calculate the Z-value and the P-value and determine if this difference is statistically significant by a two-tailed test at the 1 percent level.

12.57 In another study reported in 1990, half of 20,800 patients in thirteen countries who had been hospitalized for heart attacks were given streptokinase and half were given TPA.[39] By the end of the average two-week hospitalization, 8.5 percent of the streptokinase group had died, compared to 8.9 percent of the TPA group. Estimate the standard deviations with these sample frequencies, and use a normal approximation to calculate the Z-value and the P-value and determine if this difference is statistically significant at the 1 percent level.

12.58 A 1984 Associated Press news story on the University of Texas basketball team said, "The Longhorns went 6–22 [6 wins, 22 losses] in [Coach] Weltlich's first season. Last season, they improved to 7–21." Assuming Bernoulli trials and using a normal approximation (with $\pi = .5$ for estimating the standard deviation), is this improvement statistically significant at the 5 percent level? Why might the Bernoulli trials assumption not be appropriate?

12.59 Exercise 9.87 shows Michael Jordan's regular-season shooting accuracy during his first five years in the National Basketball Association. Shown in the following table are comparable data for playoff games during this same period. In each of these three categories, use a difference-in-means Z-value and P-value to test the null hypothesis that his probability of making a shot is the same during the regular season as in the playoffs. Assume that these shots are Bernoulli trials, use a normal approximation, and use the observed frequencies when calculating the standard deviation. Are any of these differences statistically significant using a two-tailed test at the 5 percent level?

Category	Attempted	Made
3-point shots	52	15
2-point shots	907	454
Free-throws	464	386

12.60 The 794 students in Pomona College's freshman classes of 1987–1988 and 1988–1989 were separated into two groups: 552 who had attended public high schools and 242 who had attended private schools.[40] Of those from public schools, 48 were Pomona Scholars their freshman year (an A− grade-point average); of those from private schools, 13 were Pomona Scholars. Estimate the standard deviations with the sample frequencies and apply a difference-in-means test at the 5 percent level to the null hypothesis that Pomona freshmen from public and private schools are equally likely to be Pomona Scholars.

The author concluded that although "two classes of Pomona freshmen proved the null hypothesis to be true, . . . the sample size is minuscule in comparison to the population of college and university students."

Did she prove the null hypothesis to be true? Are her results invalidated by the small sample size?

12.61 Gallup polls periodically ask, "Do you approve of the way . . . is handling his job as President?" Richard Nixon's approval rating jumped from 27 percent to 31 percent after Gerald Ford was confirmed as vice president, replacing Spiro Agnew. If 1,000 people were included in each poll, was this change statistically significant at the 5 percent level? (Use a normal approximation, a value of $\pi = .5$ when calculating the standard deviation, and a two-tailed test.)

12.62 A student randomly selected 128 five-day periods and, for each five-day period, calculated the number of days that the Dow Jones Industrial Average of stock prices went up.[41] Use her data to test the null hypothesis that daily changes in the Dow Jones Industrial Average are Bernoulli trials with $\pi = .5$

Number of Days Market Up	Number of Occurrences
5	3
4	27
3	48
2	31
1	17
0	2
Total	128

12.63 The Federal Tobacco Commission has estimated the average nicotine content of two brands of cigarettes, Life and Death:

Brand	Number Tested	Average Nicotine	Standard Deviation
Life	16	0.3 mg	0.3
Death	16	0.5 mg	0.4

Calculate the pooled variance and use a t-test to determine if this observed difference is statistically significant at the 5 percent level. Be sure to state your null hypothesis.

12.64 Believing that 3 is a special number, a student set up mock ESP experiments in which 40 students were asked to predict which one of four cards, numbered 1 to 4, would be drawn by the researcher.[42]

The results were

Number Predicted	Number of Students
1	2
2	13
3	21
4	4
Total	40

Test at the 1 percent level the null hypothesis that each of these four numbers is equally likely to be selected by students.

12.65 Two hundred college students were asked to pick a number from 1 to 10.[43] Use the data in the following table to test at the 1 percent level the null hypothesis that each of these 10 numbers is equally likely to be chosen. How do you explain the differences between the observed number of selections and the expected number under the null hypothesis?

Number	Responses
1	4
2	8
3	31
4	37
5	13
6	16
7	57
8	24
9	6
10	4

12.66 A 1989 poll asked 418 Pomona students whether existing campus fraternity rooms should be

used exclusively by fraternities
open to other groups, with fraternity control
open to all organizations, fraternities removed

Here are the results, combining the first two categories, and grouping the data according to whether the student attended an earlier public forum on the issue:[44]

Fraternities	Attended Forum	Did Not Attend Forum
Exclusive or control	77	245
Removed	39	57
Total	116	302

a. Which group was more in favor of removing the fraternities? **b.** Use a two-tailed difference-in-means test to see if the observed difference between these two groups is statistically significant at the 5 percent level. **c.** Why do we test for statistical significance rather than simply compare the observed fractions? **d.** What is the null hypothesis? **e.** If the null hypothesis is rejected, does this show that those who attended the forum were persuaded by the debate? Give another plausible interpretation.

12.67 Exercise 11.66 describes an ESP study involving 120 college students, each of whom participated in 20 trials attempting to identify a selected card (of four possibilities marked with a circle, square, triangle, and star). Two students were correct 9 times and ten were correct 8 times. Here is the complete distribution:

Number Correct	Number of Subjects
9	2
8	10
7	21
6	20
5	25
4	28
3	11
2	3

Make a chi-square test (at the 1 percent significance level) of the null hypothesis that these data are from a binomial probability distribution with $\pi = .25$.

12.68 Each of twelve college intramural basketball players was asked to take 30 shots 15 feet from the basket—10 shots from the free-throw line, 10 shots from a 45° angle to the basket, and 10 shots from the baseline.[45] Use the following results to make a chi-square test, at the 5 percent level, of the null hypothesis that hits and misses are independent of location:

Shot	Hit	Miss
Free-throw line	84	36
45° angle	62	58
Baseline	53	67

12.69 A 1989 survey of 60 Claremont McKenna College students grouped their freshman class schedules into stereotypical masculine and feminine courses to see whether there was a relationship between gender and courses selected:[46]

masculine: accounting, computing, economics, government, international relations, math

feminine: acting, history, language, literature, philosophy, psychology

Required courses were omitted from her data:

Students	Courses Masculine	Feminine	Total
Males	103	97	200
Females	78	122	200
Total	181	219	400

Is there a statistically significant relationship at the 5 percent level? What is your null hypothesis?

12.70 *(continuation)* Identification of these students' eventual majors yielded the data in Table 12.18. Is there a statistically significant relationship at the 5 percent level?

12.71 A student surveyed 104 out of 1,579 students at Pomona College to see if religious beliefs varied by class level, suspecting that students who attend college become less religious between their freshman and senior years. The data are given in Table 12.19. When he did not find a statistically significant relationship, he "used the sample mean to estimate the population mean," which also appears in Table 12.19. When he recalculated the chi-square statistic, he obtained a value of 102.29, which is statistically significant. What is the flaw in his procedure?

Table 12.18 *Data for Exercise 12.70*

Students	Major Masculine	Feminine	Both	Total
Males	21	5	4	30
Females	12	6	12	30
Total	33	11	16	60

Table 12.19 *Data for Exercise 12.71*

Sample data:

Class	Believing	Agnostic	Unbelieving	Total
Freshmen	11	7	12	30
Sophomore	8	9	8	25
Junior	17	5	6	28
Senior	10	6	5	21
Total	46	27	31	104

Estimated population:

Class	Believing	Agnostic	Unbelieving	Total
Freshmen	167	106	182	455
Sophomore	121.5	137	121.5	380
Junior	258	76	91	425
Senior	152	91	76	319
Total	698.5	410	470.5	1,579

12.72 One hundred randomly selected students were asked to rate the qualifications of two job candidates based on a résumé.[47] Half were given the résumé for a mythical John Benson and half were given another résumé, identical to the first in all respects except for the name Mary Benson. Is the difference in the average rating statistically significant using a two-tailed test at the 5 percent level?

	John Benson	Mary Benson
Number of ratings	50	50
Average rating	8.23	8.02
Standard deviation	1.15	2.18

12.73 A student asked a random sample of 54 of her classmates whether they were nearsighted and whether their grade-point average (GPA) was above or below B+:

GPA	Nearsighted	Not Nearsighted
< B+	14	15
≥ B+	17	8

Do these data reject (at the 1 percent level) the null hypothesis that GPA and nearsightedness are independent?

12.74 *(continuation)* This student then multiplied all of her data by 3:

GPA	Nearsighted	Not Nearsighted
< B+	42	45
≥ B+	51	24

"In doing this, I made the assumption that I had originally picked a perfectly random sample of students, and that if I were to have polled three times as many people, my data would have been greater in magnitude but still distributed on the same normal distribution." Does this procedure make sense to you? How do you think it affected her chi-square value?

12.75 A student watched randomly selected daytime soap operas and nighttime news programs to see if there was any relationship between the type of show and whether the women in commercials were portrayed in a family setting or in either an independent or a workplace setting.[48] What is the null hypothesis? Is it rejected at the 5 percent level by these data?

	Portrayal		
Show	*Family*	*Independent*	*Total*
Daytime	41	47	88
Evening	36	36	72
Total	77	83	160

12.76 A study of 2,000 major league baseball games during the 1959 season tabulated the number of runs scored by inning.[49] The data in Table 12.20 are for the first inning of all 2,000 games and for the ninth inning of 1,576 games in which both teams batted. (The game ends if the home team, which bats second, is ahead after the visiting team has batted in the ninth inning.) Use these data to make a chi-square test (at the 1 percent level) of the null hypothesis that run scoring does not depend on whether it is the first or last inning of a baseball game.

12.77 *(continuation)* The number of runs scored in the first inning averaged 0.54 with a standard deviation of 1.03. The number of runs scored in the ninth inning averaged 0.40 with a standard deviation of 0.87. Use a normal approximation to calculate the Z-value and test the null hypothesis that this observed difference in means is statistically significant by a two-tailed test at the 1 percent level.

Table 12.20 *Data for Exercise 12.76*

	Runs Scored					
Inning	*0*	*1*	*2*	*3*	*4*	*>4*
First	1,400	318	162	62	36	22
Ninth	1,213	203	95	43	16	6

12.78 In his Ph.D. dissertation, Kenneth Bayley reported these data on Alaskan defendants released on bail:[50]

Race	Released	Not Released
Black	61	208
Native	137	240
Other	377	958

Calculate the expected values under the null hypothesis that being released on bail is unrelated to the defendant's race, and make a chi-square test of this null hypothesis at the 5 percent level.

12.79 A Canadian government agency fed 100 rats a 5 percent saccharin diet (equivalent to 800 cans of diet soda a day) and fed 100 rats a saccharin-free diet. Three rats in the first group developed bladder cancer, none in the second group. Is this difference statistically significant at the 5 percent level? Is it substantial?

12.80 The deathday-birthday connection was investigated using 400 randomly selected names from *Biography Almanac,* which lists more than 20,000 internationally famous newsmakers.[51] On every third page, the person with the longest biographical sketch was selected. The relationship between birth and death months is shown in the following table. Make a chi-square test at the 5 percent level of the null hypothesis that each month is equally likely to be the month of death.

Number of Months Before (−) or After (+) Birth Month	Number of Deaths
−6	30
−5	40
−4	29
−3	25
−2	36
−1	30
0	35
+1	32
+2	40
+3	42
+4	28
+5	33

12.81 A random sample of 239 Pomona College students obtained these data on birth order and grade-point average (GPA):[52]

Birth Order	GPA > B+	GPA ≤ B+
Only child	12	10
Oldest child	45	78
Not oldest	31	63

(The data on oldest and not oldest are for families with more than one child.) Make a chi-square test at the 1 percent level of the null hypothesis that birth order and GPA are independent.

12.82 A security analyst looked at the prices of 861 stocks that were listed on the New York Stock Exchange at the end of both 1949 and 1957 and found[53]

10 rose in 8 of 8 years
55 rose in 7 of 8 years
177 rose in 6 of 8 years
228 rose in 5 of 8 years
225 rose in 4 of 8 years
115 rose in 3 of 8 years
43 rose in 2 of 8 years
8 rose in 1 of 8 years
0 rose in 0 of 8 years

Overall, aggregating all stocks and years, there were 4,006 cases in which a stock's price went up and 2,882 cases in which a stock's price went down. Apply a goodness-of-fit test to the null hypothesis that these data come from a binomial distribution with success probability $4{,}006/6{,}888 = .5819$. What are the assumptions embedded in the binomial distribution? Interpret these assumptions in the context of the stock market.

12.83 To illustrate its argument that Germans have become less obsessed with work, a 1984 newspaper article compared polls of "about 1,000" housewives in 1969 and 1977.[54] In the first poll 61 percent said they did the dishes immediately after dinner because they couldn't stand "having something dirty lying around"; in 1977 only 51 percent said this. Is this difference statistically significant at the 5 percent level? Use a two-tailed test and a conservative value of $\pi = .5$ when estimating the standard deviation.

12.84 Law and order was a major campaign issue for 1968 presidential candidate George Wallace. A preelection survey compared cars bearing George Wallace and Hubert Humphrey bumper stickers in Nashville-Davidson County, Tennessee, to see how many complied with an ordinance requiring a $15.00 windshield Metro sticker.[55] To control for the fact that Wallace supporters may have been less able to afford the $15.00 sticker, the survey also looked at cars parked adjacent to Wallace supporters—the idea being that people in similar financial condition tend to live, work, and shop in similar places:

	With Stickers	Without	Total
Humphrey supporters	154	24	178
Wallace supporters	270	91	361
Adjacent to Wallace	303	58	361

Use these data for two-tailed difference-in-means tests at the 5 percent level of the null hypotheses that among the entire population of this county, **a.** equal proportions of Humphrey and Wallace supporters obey this sticker law. **b.** equal proportions of Wallace and adjacent cars obey the law.

12.85 A total of 352 people were asked if they had nightmares "seldom" or "often" (at least once a month):[56]

Subjects	Seldom	Often	Total
Men	105	55	160
Women	132	60	192

Estimate the proportion of all men and women who often have nightmares and then test (at the 1 percent level) the null hypothesis that men and women are equally likely to have frequent nightmares. Are there any reasons why you might suspect that this survey is not representative of the population as a whole?

• **12.86** The bowlers on two teams have the following expected scores and standard deviations:

Team	Expected Score	Standard Deviation
Boomers		
Bob	200	30
Ralph	180	30
Mario	190	30
Bouncers		
Arv	170	20
Mike	190	30
Walt	180	30

For each game, the teams add their three bowlers' scores together to determine which team wins. Assuming that the scores are independent, what is the probability of the Boomers winning a game between these two teams? If they play three games, what is the probability that the Boomers will win all three games? Two games? One game? No games? State any necessary assumptions.

• **12.87** *(continuation)* The Boomers and the Bouncers agree to a showdown to determine which is the better team. Each team member will bowl n games, and the team with the highest average over these n games is the winner. What is the probability that the Bouncers will win if $n = 1$? If $n = 3$? If $n = 5$? If $n = 25$? How high does n have to be for 99 percent certainty that the better team, the Boomers, will win?

12.88 At Small College, statistics is taught in three sections of 50 students each. The students are randomly assigned to a section. Each section has a different instructor, but all sections use the same textbook and give the same tests. The following final grades were received in this course. Do you feel that these data support or reject the hypothesis that student performance is independent of the instructor?

	Grade	
Section	A or B	C, D, or F
1	30	20
2	16	34
3	38	12
Total	84	66

12.89 Based on accepted predictive criteria (such as a history of hypertension), the Veteran's Administration Hospital identified 177 patients with a high risk of narrowed coronary arteries.[57] Of these, random procedures were used to select 94 that were given medication and 83 that were given coronary bypass operations. Here are the number that survived after 5 and 11 years:

Survival	Medication	Surgery
5 years	59	71
11 years	34	41

Use a normal approximation and a two-tailed test to determine whether each of these observed differences between medication and surgery is statistically significant at the 5 percent level. Use a two-tailed test and a conservative $\pi = .5$ for calculating the standard deviation.

• **12.90** Exercise 6.43 gives the observed heights of 1,052 mothers. Calculate the expected number of women in each category, assuming heights are normally distributed with a mean of 62.484 inches and a standard deviation of 2.3904 inches. Combine the categories in the tails so that the expected number of mothers in each category is at least 10.0. Now use the differences between the observed and expected number of women in each category to make a chi-square test at the 5 percent level of the null hypothesis that these heights came from a normal distribution with the specified mean and standard deviation.

12.91 A British chemical company wanted to see if a chlorinating agent would increase the abrasion resistance of rubber.[58] The researchers sampled ten matched pieces of rubber by cutting ten pieces of rubber in half and flipping a coin to determine which half would be treated and which would not. Here are the observed differences in abrasion resistance, treated minus untreated:

$$2.6 \quad 3.1 \quad -0.2 \quad 1.7 \quad 0.6$$
$$1.2 \quad 2.2 \quad 1.1 \quad -0.2 \quad 0.6$$

Are these differences statistically significant at the 5 percent level? What assumptions are necessary for your test?

12.92 A study of the effects on cholesterol levels of Lovastatin alone or in combination with colestipol obtained these data:[59]

	Control Group	Lovastatin	Lovastatin + Colestipol
Mean	258	191	165
Standard deviation	23.8	21.2	18.5

There were seven patients in each group. Assuming that these data are random samples from a normal distribution, test the following null hypotheses (with a two-tailed test at the 5 percent level): **a.** Lovastatin alone has no effect. **b.** The addition of colestipol makes no difference. The seven patients in each group were, in fact, the same seven people tested at different times. If this were taken into account by using a matched-pair test, do you think that the t-values would increase or decrease?

12.93 In 1986 the Boston Red Sox won the American League East championship with 108 wins and 54 losses; the Cleveland Indians finished a distant fifth with 84 wins and 78 losses. Assuming (unrealistically) that baseball games are Bernoulli trials, does this evidence reject (at the 5 percent level) the null hypothesis that Boston and Cleveland had the same probability of winning a baseball game?

12.94 Two polling organizations report these voter preferences on the eve of an election:

Poll	Angela	Karen	Sample Size
A	55%	45%	600
B	52%	48%	1,200

Is the difference in these poll results statistically significant at the 5 percent level, or can it be explained reasonably well by sampling error? If you have no reason to believe that one poll is more reliable than the other, what would you use for a prediction of each candidate's percentage of the votes?

• **12.95** The heights of adult U.S. males are approximately normally distributed with a mean of

68.8 inches and a standard deviation of 2.65 inches. The heights of Norwegian males are approximately normally distributed with a mean of 69.9 inches and a standard deviation of 2.70 inches. If an adult U.S. male and an adult Norwegian male are selected at random, what is the probability that the Norwegian is the taller of the two?

• **12.96** *(continuation)* The heights of adult U.S. females are approximately normally distributed with a mean of 63.6 inches and a standard deviation of 2.50 inches. What is the probability that a randomly selected adult U.S. male is taller than a randomly selected adult U.S. female?

12.97 In the court case *Cormier* v. *P.P.G. Industries* (1981) data were presented showing that 86 of 2,499 white applicants were hired for utility crew jobs, while 24 of 1,050 black applicants were hired. Is the observed difference in hiring rates statistically significant at the 1 percent level? Substantial by the 80 percent rule discussed in the text?

12.98 In *Connecticut* v. *Teal* (1982) it was determined that 206 of 259 white employees passed a test required for promotion, compared to 26 of 48 black employees. Is the observed difference in passing rates

statistically significant at the 1 percent level? Substantial by the 80 percent rule?

12.99 Exercise 8.81 describes a study of the effect on viewer recall of the zapping of commercials on television shows that have been recorded on VCRs. The viewers' ability to make unaided brand recall responses was tested by asking them to list brand names that had been advertised during the program. The zappers were able to make correct unaided brand recall responses in 4 out of 139 possible cases, while nonzappers were correct in 40 of 122 cases. Is this difference statistically significant at the 1 percent level? Use a conservative value of $\pi = .5$ when calculating the standard deviation.

12.100 *(continuation)* The viewers were also given a recognition test in which they were shown the product labels for the advertised product and for four competing products and asked to identify the one that had been advertised during the program. The zappers were correct in 47 out of 139 cases; the nonzappers were correct in 92 of 122 cases. Is this difference statistically significant at the 1 percent level? Compute the Z-value and use a conservative value of $\pi = .5$ when calculating the standard deviation.

13

The Idea of Regression Analysis

Econometricians, like artists, tend to fall in love with their models.

Edward Leamer

Scientists use models to explain various phenomena. For example, Newton said that the force applied by a body in motion depends on its mass and acceleration. Agronomists say that crop yields depend on weather, soil conditions, and nutrients. Economists say that the demand for a product depends on its price. In each case, there is a simple model that purports to explain a complex reality. Statistics can be used to quantify such theories and to give them practical application. For a detailed example, let's look at one of the most famous economic theories.

13.1   KEYNES' CONSUMPTION FUNCTION

John Maynard Keynes, a brilliant British economist, wanted to explain fluctuations in consumer spending. He believed that consumer spending was one of the keys to understanding economic booms and busts, and so he naturally wondered what it was that caused consumer spending to go up or down. Keynes hypothesized that household income was the primary determinant of household spending. When income goes up, people spend more; when their income drops, they spend less. He wrote that this is a " . . . fundamental psychological law, upon which we are entitled to depend both *a priori* from our knowledge of human nature and from the detailed facts of experience."[1]

A Model

A simple algebraic representation of Keynes' theory is

$$Y = \alpha + \beta X \qquad (13.1)$$

where Y is consumer spending and X is income. In regression models, α and β are not the probabilities of Type I and Type II errors. Instead, α and β are two unknown parameters that describe the relationship between income and consumption. Income is the **explanatory variable,** because changes in income *explain* changes in spending. Spending is the **dependent variable,** because spending *depends* on income.

> Models usually embody an assumed **causal relationship;** the researcher believes that changes in the explanatory variable *cause* changes in the dependent variable.

The sign of the parameter β tells us whether there is a positive or negative relationship between the explanatory and dependent variables. Figure 13.1 shows three possibilities: a positive relationship, a negative relationship, and no relationship at all. In the consumption model, the parameter β, which Keynes called "the marginal propensity to consume," tells us the effect of income changes on spending. Keynes argued that β is surely positive, but undoubtedly less than 1.0. If β is .8, then $1 of additional income increases spending by 80 cents and, similarly, a $1 decline in income reduces spending by 80 cents.

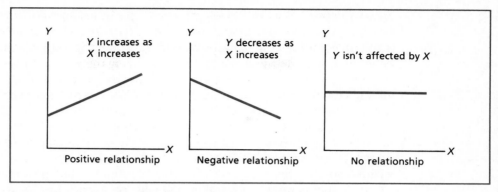

Figure 13.1 *Three Possible Relationships Between X and Y*

Equation 13.1 is conveniently linear, but it is unlikely that the real world is this simple. We should think of Equation 13.1 as a linear approximation to a possibly nonlinear relationship. Figure 13.2 compares a hypothetical "true" relationship with a linear approximation that has a slope β and intercept α. This linear approximation is accurate enough as long as we confine our attention to small changes in income. As income goes up or down from its current level, consumption goes up or down by approximately β times the change in income. However, the linear

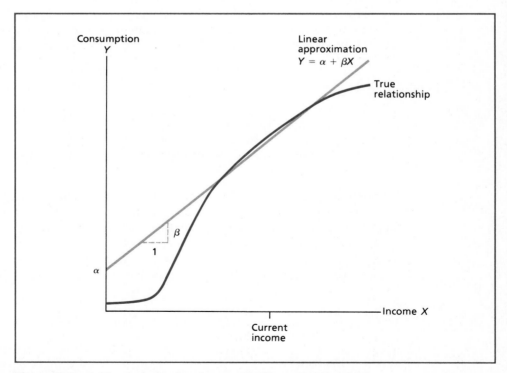

Figure 13.2 *A Linear Approximation to a Nonlinear Relationship*

approximation goes awry for very large changes in income. For instance, the intercept α gives a very inaccurate prediction of spending when income goes down to zero. Unless we're confident that the true relationship really is linear all the way down to $X = 0$, we shouldn't think of α as being the level of consumption at $X = 0$. Instead, we should interpret α as being the intercept of a linear approximation, an intercept that places the linear approximation in a position to predict consumption accurately for values of income near the current level.

Economists are rightfully wary of predicting what consumption would be if a nation's income were zero. Presumably, such an economic disaster would have drastic effects on people's behavior that are hardly imaginable to our stable and well-fed society. Similarly, we should be cautious about predicting what grade-school dropouts would earn if they had gone on to earn Ph.D.s, what the productivity of labor will be 200 years from now, or what my batting average would be if I played professional baseball. Linear approximations are very useful and usually quite satisfactory as long as we don't try to extrapolate them to situations for which we have little knowledge or evidence.

Some Data

The model, Equation 13.1, is Keynes' explanation of consumption. If we had values for the parameters α and β, then the model could be used to predict consumer spending for various values of income. It is natural to test and implement this model by collecting some data and seeing what, if any, relationship there is between consumption and income. Here are some data on real per capita household spending and disposable income (net of taxes) for the years 1977 through 1981:*

Year	Disposable Income X	Household Spending Y
1977	$3,319	$3,042
1978	$3,421	$3,124
1979	$3,404	$3,108
1980	$3,276	$2,994
1981	$3,271	$2,971

Figure 13.3 plots these data in what is called a **scatter diagram.** If, as is illustrated in Figure 13.4, there is too much scatter in the data, then there isn't a convincing empirical relationship between X and Y. In Figure 13.3, there seems to be a fairly close positive relationship between income and spending. If we fit a line to these data, we can obtain values for the parameters α and β and thereby quantify Keynes' consumption function. I've used a transparent ruler and some squinting to draw

*The economic term *real* means that an adjustment has been made for inflation. These data are all in terms of 1967 dollars—what a dollar would have bought in 1967. In 1981, commodities cost, on the average, almost three times as much as they did in 1967. Thus 9,000 1981-dollars are about the same as 3,000 1967-dollars.

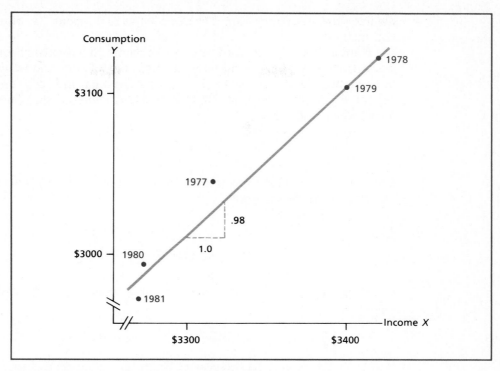

Figure 13.3 *Some Data on Household Spending and Income*

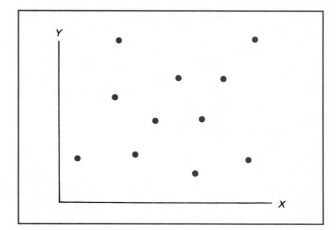

Figure 13.4 *If There Is Too Much Scatter in a Scatter Diagram, Then There Is No Apparent Relationship between X and Y*

the line shown in Figure 13.3. The slope seems to be about .98, while the intercept appears to be −227.*

Economic forecasters used models like Figure 13.3 to predict consumer spending in 1982. Let's imagine that it is the beginning of 1982 and we want to make a forecast too. We specify our model, gather some data, and come up with an intercept of −227 and a slope of .98. Our predicted value for spending $\hat{Y}$ is given by the fitted equation

$$\hat{Y} = -227 + .98X \qquad (13.2)$$

If we believe household income will be $3,300 in 1982, then our model predicts that household spending will be

$$\hat{Y} = -227 + .98(3,300)$$
$$= 3,007$$

If per capita income is $100 higher in 1982, spending will be .98($100) = $98 higher, or

$$\hat{Y} = -227 + .98(3,400)$$
$$= \$3,105$$

These sorts of calculations can describe a variety of possible scenarios, ranging from pessimistic to optimistic forecasts, or they can be used to determine the effects of government policy decisions. If, for instance, Congress is debating a tax cut, an economic model like this can predict the consequences for consumer spending. According to our model, a tax cut that raises per capita disposable income by $100 will raise consumer spending by $98.

The model can also answer hypothetical historical questions. In 1980, per capita income was $3,276 and spending was $2,994. If Congress had enacted a tax increase in 1980 that reduced per capita disposable income by $100, what effect would that tax hike have had on consumer spending? Our model predicts that spending would have been reduced by $98 per person.

Economists do, in fact, build models to answer just these sorts of questions. In practice, of course, empirical economic models include many equations, which predict not only consumer spending, but also prices, interest rates, unemployment, and income itself.

---

*To show the variation in these data, Figure 13.3 focuses on a small range of income and spending. Using the "guesstimated" slope .98 and the fact that the fitted line seems to run through the point $X = 3,300$, $Y = 3,007$, we can estimate the intercept: $3,007 = a + .98(3,300)$ implies $a = -227$.

EXERCISES

13.1 Here is a plot of the midterm and final exam scores for Alice, Bill, Carrie, David, and Elaine.

a. Which student earned the highest score on the midterm? **b.** Which student earned the highest score on the final? **c.** Which student's score increased the most between the midterm and the final? **d.** Which student's score fell the most between the midterm and the final? **e.** Was the average score higher on the midterm or on the final? **f.** Was there a greater dispersion of scores on the midterm or on the final? **g.** Select one of the following statements:

"Midterm and final scores are positively related";
"Midterm and final scores are negatively related"; or
"Midterm and final scores are unrelated."

13.2 Professor Smith claims that a student's score on the final exam can be predicted from his or her score on the midterm. Here are some midterm and final exam scores:

Students	Midterm X	Final Y
Fred	92	87
Gail	65	71
Hannah	75	75
Irene	83	84
Joshua	95	93

a. What are the average midterm and final scores? **b.** Plot these data with midterm scores on the horizontal axis and final scores on the vertical axis. **c.** Does there seem to be any evidence of a relationship between final and midterm scores? **d.** Using only your judgment, draw a line that appears to fit these data, and then calculate the (approximate) intercept and slope.

13.3 Explain why you think that each of the following pairs of data are positively related, negatively related, or essentially unrelated. If you think that there is a relationship, which is the dependent variable? **a.** the height of a father and his eldest son **b.** the age of a mother and her oldest child **c.** the income and age of male lawyers aged 21 to 54 **d.** the height and weight of 21-year-old males **e.** a woman's age and the cost of her life insurance.

13.4 Explain why you believe that each of the following pairs of data are positively related, negatively related, or essentially unrelated. If there is some relationship, which is the dependent variable? **a.** a child's Little League batting average and the number of hours spent practicing **b.** the temperature in Houston and attendance at the city's Fourth of July parade **c.** the number of automobiles on a freeway and their average speed **d.** the weight of a car and its miles per gallon **e.** the mileage on a five-year-old Chevrolet and its current market price.

13.5 Explain why you believe that each of the following pairs are positively related, negatively related, or not related at all. If there is a relationship, which is the dependent variable? **a.** advertising expenditures and sales **b.** the winter price of corn and the number of acres of corn planted by Iowa farmers the following spring **c.** the price of steak and the amount purchased by your college dining hall **d.** the price of gasoline and sales of large cars **e.** the number of wins by the New York Yankees and the price of tea in China.

13.2   LEAST SQUARES ESTIMATION

Fitting a line by hand to scattered data points is a hopelessly crude method for estimating values for the intercept and slope. It's hard to know where to draw the line and it is hard to read the intercept and slope from a smudged piece of graph paper. It would be much better to have some standard formula, one analogous to estimating a population mean by calculating the sample mean. But here we are trying to estimate two parameters, α and β simultaneously, and a natural formula is not apparent.

Least Squares Estimates of a Population Mean

In Chapter 9, you were introduced to Gauss' justification for using the sample mean to estimate a population mean. Gauss extended this justification to obtain a method for estimating two parameters, like α and β. Gauss' initial concern was with using physical measurements Y to estimate a distance μ. The individual measurements can be distinguished by using an i subscript:

Observation i	Measurement Y_i
1	115.1
2	114.4
3	112.5
4	116.2
5	114.8

The difference between each measurement Y_i and the true distance μ is the (unobserved) measurement error ϵ_i. That is,

measured value = true value + measurement error

In symbols,

$$Y_i = \mu + \epsilon_i \qquad (13.3)$$

We can interpret Gauss' equation as a model similar to our model of household spending. Equation 13.3 says that measurements depend on the true value and on the measurement error. We can view μ as an unknown parameter, like α and β, that we want to estimate. Our estimate $\hat{\mu}$ of this parameter is a prediction $\hat{Y} = \hat{\mu}$ of the measurements that people will obtain. The error term ϵ_i allows for the fact that the measurements inevitably diverge from the true value in ways that can't be predicted ahead of time.

Gauss argued that we should use the estimate that minimizes the sum of the squared prediction errors for the data we have:*

---

*Notice that Gauss did not try to minimize the difference between the true value and the estimate, $(\mu - \hat{\mu})^2$, because the true value μ is unknown.

$$(Y_1 - \hat{Y})^2 + (Y_2 - \hat{Y})^2 + \cdots + (Y_n - \hat{Y})^2$$
$$= (Y_1 - \hat{\mu})^2 + (Y_2 - \hat{\mu})^2 + \cdots + (Y_n - \hat{\mu})^2$$
$$= (115.1 - \hat{\mu})^2 + (114.4 - \hat{\mu})^2 + \cdots + (114.8 - \hat{\mu})^2$$

Gauss showed that the sample mean, or the average of our measurements, is in fact the estimator $\hat{\mu}$ that minimizes the sum of these squared prediction errors. The mean of these particular five measurements is

$$\overline{Y} = \frac{115.1 + 114.4 + 112.5 + 116.2 + 114.8}{5}$$

$$= 114.6$$

The sample mean, $\overline{Y} = 114.6$, is an estimate of the true distance μ. Inevitably, measurements will sometimes be above and other times be below the true distance, but there is no way to predict the size of these measurement errors. For the five observations we have, the sample mean $\overline{Y} = 114.6$ is the measurement predictor that has the smallest sum of squared prediction errors. Thus, the sample mean is called the **least squares estimator.**

The Least Squares Logic

A graph may clarify and reinforce this logic. Figure 13.5 shows the five measurements, together with the sample mean. The horizontal axis lists the observations and the vertical axis gives their distance measurements; our prediction line is horizontal with an intercept equal to the sample mean, $\overline{Y} = 114.6$. (There is no slope, because we assume that the order in which the observations are numbered is unimportant.) Our prediction errors e_i are the vertical distances between the actual measurements Y_i and our prediction $\hat{Y}_i = \overline{Y}$:

$$e_i = Y_i - \hat{Y} \tag{13.4}$$

Gauss' least squares criterion is to minimize the sum of the squared prediction errors, that is, the sum of the vertical discrepancies between the data and our fitted line in Figure 13.5.

Squaring the prediction errors means that positive and negative errors count equally. We don't care if our prediction is too high or too low, only how far off we are. Squaring the errors also implies that large errors are more worrisome than small errors. We prefer an estimate that has 99 one-inch errors to an estimate that has 1 ten-inch error. This logic has considerable appeal. In addition, as Gauss discovered, least-squares mathematics are very tractable and least squares estimators (like the sample mean) are quite sensible.

One final note is that these are not true forecasts in the sense that we make them before we see the data. We specify a model beforehand, but we cannot esti-

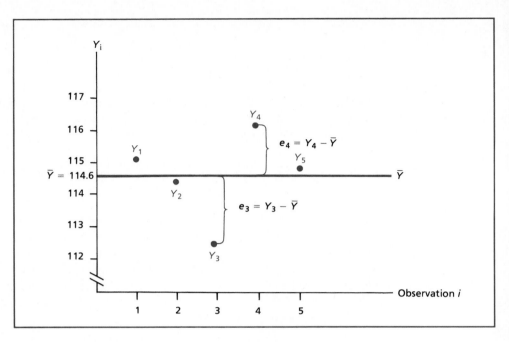

Figure 13.5   *Data, an Estimate, and Prediction Errors*

mate the model's parameters without looking at the data. The least squares logic assumes that a specific model is to be used for predictions. For the data we have, we determine the parameter estimates that give the smallest sum of squared prediction errors. The real test of the model will, of course, come when we use it to predict data that we have not yet seen.

Least Squares Estimates of a Linear Equation

This logic can be extended to a variety of situations. One example is a linear equation, such as Keynes' consumption function,

$$Y = \alpha + \beta X \tag{13.5}$$

We want to predict consumer spending Y for various levels of income X, but we recognize that other factors besides income influence household spending decisions. We put these other influences into a catch-all error term ϵ:

$$\boxed{Y = \alpha + \beta X + \epsilon} \tag{13.6}$$

It is mathematically convenient to treat the error term ϵ as if it were the chance result of rolling dice. However, consumer spending is determined not by dice rolls, but by households purposefully acting on the basis of all sorts of real and fanciful information. The error term ϵ includes the effects of these neglected influences. We invariably omit some factors—such as credit availability, inflation expectations, and consumer confidence—because we don't think that they are all that important, because we don't have reliable data on them, or because we simply haven't thought of them. Some influences are so poorly understood—what Keynes called "animal spirits"—that we can consider them essentially random variation. In addition, measurement errors are made in tabulating household spending, and our data are based on *some* rather than *all* households and are therefore subject to sampling error.

> We do not observe all of the factors that influence household spending; nor do we observe all households or measure spending perfectly. Because of these errors and omissions, our predictions won't be perfect and we have to introduce an **error term** ϵ to allow for this imperfection.

Figure 13.6 shows the least squares logic. The data points are the observed levels of income X_i and spending Y_i, while the fitted line gives our spending predictions $\hat{Y}_i$. The difference is our prediction error,

$$e_i = Y_i - \hat{Y}_i$$

Figure 13.6 *Fitting a Line to Spending and Income Data*

Okun's Law

In the early 1960s, President John F. Kennedy's Council of Economic Advisers tried to convince a reluctant Congress of the benefits of a tax cut to "get the country moving again." The Council argued that the prevailing unemployment rate of 6 to 7 percent was unnecessarily high, and that, in addition to the individual hardship suffered by unemployed workers and the owners of idle factories, the nation lost a considerable amount of potential output. Arthur Okun tried to quantify this output loss by estimating an empirical relationship between changes in the unemployment rate (Y) and changes in the gross national product (X),

$$Y = \alpha + \beta X + \epsilon$$

(The equation is estimated with unemployment as the dependent variable because business decisions to hire and fire workers depend on sales.)

A scatter diagram of annual data for the period studied by Okun is shown in Figure 1. As anticipated, there appears to be a *negative* relationship—increases in output are associated with declines in the unemployment rate. The least squares fitted line is

$$\hat{Y} = 1.18 - 0.33X$$

indicating that a 1 percent increase in output is associated with about a 0.33 percent decline in the unemployment rate. Thus, each percentage point of excessive unemployment signals that the nation's output is about 3 percent below its potential. This three-for-one rule became known as *Okun's Law* and, as time passed, proved to be a remarkably accurate rule of thumb.[2]

Figure 1   *Okun's Law, 1947–1960*

Gauss' least squares procedure says that we should choose the prediction line that minimizes the sum of these squared prediction errors; that is, we should estimate the intercept α and the slope β by minimizing the sum of the squared errors in predicting spending. This is a logical procedure, but how do we do it?

The Least Squares Estimators of α and β

We want to determine the estimates a and b of the parameters α and β that minimize the sum of squared prediction errors for the data we have. The formulas can be derived using calculus, as shown in the appendix to this chapter. Here, we will simply discuss and apply these equations.

The formula for the least squares slope estimator

$$b = \frac{\Sigma(X_i - \overline{X})(Y_i - \overline{Y})}{\Sigma(X_i - \overline{X})^2} \qquad (13.7)$$

is a complex reckoning of the interrelationship between X and Y.* Once b has been calculated, it is easy to compute the intercept estimator, because one property of least squares estimates is that the fitted line goes through the average values of the data: $\overline{Y} = a + b\overline{X}$. Thus,

$$a = \overline{Y} - b\overline{X} \qquad (13.8)$$

Table 13.1 illustrates a detailed application of these least squares formulas to our consumption data. The first thing to do is to write down the paired values of X and Y. Second, compute their average values $\overline{X}$ and $\overline{Y}$. Third, calculate the deviations about these averages, $X - \overline{X}$ and $Y - \overline{Y}$. Fourth, find the products $(X - \overline{X})(X - \overline{X})$ and $(X - \overline{X})(Y - \overline{Y})$. Fifth, add these two products up and take their ratio to determine the least squares estimate of β, in accordance with Equation 13.7. Finally, use this estimate, along with the average values of X and Y, to determine the least squares estimate of α, in accordance with Equation 13.8. In practice, we no longer do these calculations by hand because the least squares formulas are built into convenient programs for computers ranging from mainframes to pocket calculators.

As shown in Table 13.1, the least squares estimates turn out to be

$$a = -114.7$$

$$b = 0.947$$

*An alternative formula for the slope, which is less logical but often easier to compute, is

$$b = \frac{\Sigma X_i Y_i - n\overline{X}\,\overline{Y}}{\Sigma X_1^2 - n(\overline{X})^2}$$

The Relationship Between Height and Weight

Table 1 gives the average weights of 18- to 24-year-old U.S. males, by height (measured as inches above 5 feet).

As shown in Figure 1, a least squares regression confirms that, on average, weight rises by about 4½ pounds for every additional inch of height.

$$\hat{Y} = 120.7 + 4.55X$$

(Incidentally, the American Heart Association says that the ideal male weight is $110 + 5.0X$, putting the average 18- to 24-year-old male some 5 to 10 pounds overweight; the ideal female weight is $100 + 5.0X$)

Guindon/The Studio

Harry has a height problem. According to the weight chart he should be 7 feet tall.

Table 1 *Weight by Height Data*

X Height (inches)	Y Weight (pounds)
2	130
3	135
4	139
5	143
6	148
7	152
8	157
9	162
10	166
11	171
12	175
13	180
14	185

Figure 1

Table 13.1 *A Least Squares Computation of Income and Consumption Data*

Year	X	Y	$X - \overline{X}$	$Y - \overline{Y}$	$(X - \overline{X})^2$	$(X - \overline{X})(Y - \overline{Y})$
1977	3319	3042	−19.2	−5.8	368.64	111.36
1978	3421	3124	82.8	76.2	6855.84	6309.36
1979	3404	3108	65.8	60.2	4329.64	3961.16
1980	3276	2994	−62.2	−53.8	3868.84	3346.36
1981	3271	2971	−67.2	−76.8	4515.84	5160.96
	$\overline{X} = 3338.2$	$\overline{Y} = 3047.8$	0.0	0.0	19,938.8	18,889.2

$$b = \frac{(X_1 - \overline{X})(Y_1 - \overline{Y}) + \cdots + (X_n - \overline{X})(Y_n - \overline{Y})}{(X_1 - \overline{X})^2 + \cdots + (X_n - \overline{X})^2}$$

$$= \frac{18,889.2}{19,938.8}$$

$$= 0.9474$$

$$a = \overline{Y} - b\overline{X}$$

$$= 3047.8 - (.9474)(3338.2)$$

$$= -114.7$$

Thus, $\hat{Y} = -115 + .947X$

so that the least squares fitted line is

$$\hat{Y} = -114.7 + 0.947X$$

This is certainly close to the line I drew with a clear ruler, but the differences are not negligible. A precise calculation reveals that the sum of squared errors is 440 with my hand-drawn line and only 362 with the least squares line.*

EXERCISES

13.6 There are two competing academic theories regarding the relationship between student course evaluations and the number of years that a professor has been teaching. One theory holds that the most experienced professors will get the highest ratings. The second theory says that the young and energetic, although inexperienced, professors will score highest.

What would a scatter diagram of evaluations (Y) and experience (X) look like if the first theory is correct? (Make a rough sketch.) What would the scatter diagram look like if the second theory is right? In either case, why might we anticipate that the data will not lie exactly on a line $Y = \alpha + \beta X$?

A college dean believes that there is no relation-

*There is another practical reason why calculated estimates are better than hand-drawn ones. Researchers often estimate equations with more than one explanatory variable, but it is difficult or impossible to draw graphs with more than two dimensions.

ship at all between course evaluations and teaching experience. If this dean is right, what would a scatter diagram of evaluations and years look like?

13.7 In Exercise 13.2, you plotted a scatter diagram of midterm (X) and final exam (Y) scores, and used your judgment to fit a line to these data. Now calculate the least squares estimates of the parameters α and β for the equation

$$Y = \alpha + \beta X + \epsilon$$

Are the two fitted lines similar?

13.8 A certain armchair economist argues that there is a negative relationship between world coffee prices and sugar prices: "When the price of coffee is low, people will buy more coffee and more sugar to put in their coffee, driving up the price of sugar." Here are some data to test this theory:

Year	Coffee Price ($/lb)	Sugar Price ($/lb)
1975	0.68	0.245
1976	1.21	0.126
1977	1.92	0.092
1978	1.81	0.086
1979	1.55	0.101
1980	1.87	0.223
1981	1.56	0.212

Plot these data on a scatter diagram with coffee price (X) on the horizontal axis and sugar price (Y) on the vertical axis. Do these data appear to confirm the theory? Calculate the least squares estimates of the parameters of the equation $Y = \alpha + \beta X + \epsilon$. Do these estimates support the theory?

13.9 Dorothy Brady estimated Keynes' consumption function with the following cross-section budget data on family income and spending during 1935–1936:[3]

Income X	Spending Y
$292	$493
$730	$802
$1,176	$1,196
$1,636	$1,598
$2,292	$2,124
$3,243	$2,814

Income X	Spending Y
$4,207	$3,467
$6,598	$4,950
$22,259	$12,109

Calculate the least squares estimates of the parameters α and β in the equation $Y = \alpha + \beta X + \epsilon$.

13.10 Plot the following data and calculate the least squares estimates of the parameters α and β in the equation $Y = \alpha + \beta X + \epsilon$:

X	1	3	5	7	9
Y	5	5	5	5	5

13.11 Plot the following data and calculate the least squares estimate of the parameter β in the equation $Y = \alpha + \beta X + \epsilon$:

X	5	5	5	5	5
Y	1	3	5	7	9

13.12 Estimate the parameters α and β in the linear equation

$$Y = \alpha + \beta X + \epsilon$$

using the following data

X	Y	X	Y
0	5	-3	-4
3	4	0	-5
-4	-3	3	-4
5	0	4	3
-3	4	-5	0
4	-3	-4	3

Now plot these data and draw in your fitted line. Do you think that the fitted line accurately describes the relationship between X and Y?

• ▩**13.13** The text estimates Okun's Law using annual data for 1947–1960, the years originally scrutinized by Okun. Reestimate this equation using data for 1950–1989 from Table 8 at the end of the book.

•• **13.14** We are trying to estimate the average age μ of people who are attending a certain movie by asking a random sample of four people for their ages: X_1, X_2, X_3, and X_4. Show that the sample mean is the estimator $\hat{\mu}$ that minimizes the sum of squared errors

$$(X_1 - \hat{\mu})^2 + (X_2 - \hat{\mu})^2 + (X_3 - \hat{\mu})^2 + (X_4 - \hat{\mu})^2$$

13.3   PROBABILITY DISTRIBUTIONS FOR THE ESTIMATES

It is tempting to think that a given set of historical data cannot reflect random variation. This is another of those temptations that should be resisted. An analogous error is to ignore the effect of random variation on a sample mean. The researcher sees one particular sample mean, but must remember that this is just one of many samples that might have been selected. The same is true of the dependent variable Y in a regression model. The researcher sees a particular historical value of Y that occurred, but must remember that it is but one of many that might have occurred.

In regression models, the randomness that underlies historical data comes from the error term ϵ, which reflects inaccurate measurements, omitted influences, and sampling error. Because the exact effects of these inaccuracies, omissions, and luck of the draw are uncertain, we can conceive of a probability distribution for ϵ. The simplest assumptions are:

1. The expected value of ϵ is zero.
2. The standard deviation of ϵ is σ, which does not vary from observation to observation.
3. The values of ϵ are independent of X and each other.

These are just the Bernoulli-trial assumptions (Chapter 5) that are so useful throughout probability and statistics. An irreverent caricature goes as follows. The model $\alpha + \beta X$ determines an appropriate value of Y. However, to keep things interesting, a bemused god rolls a pair of dice to push the observed Y up or down a bit. If the dice come up 7, then ϵ is zero. If an 8 is rolled, then $\epsilon = +1$; if a 6, then $\epsilon = -1$. If 9 is rolled, then $\epsilon = +2$; if 5, then $\epsilon = -2$, and so on. Before the dice are rolled, the value of ϵ is uncertain. It has a mean of zero and a constant standard deviation, and is independent both of X and of the values that come up on other rolls.

While the Bernoulli-trial assumptions are certainly familiar and convenient, we need to consider their applicability. The first assumption that was made simply reminds us that we are using the intercept to locate the model. The value of the intercept α is chosen so that the expected value of ϵ is zero. This assumption is harmless.

The second assumption is also very natural. We know little or nothing about what the value of ϵ will turn out to be for each observation. The standard deviation of ϵ is a measure of the degree of our uncertainty, and we assume that there is no reason to be any more or less uncertain about ϵ from one observation to the next. While this is an appealing assumption, we will later consider cases where it is violated.

The third assumption is the strongest of the group and the one most likely to be inappropriate. To simplify our computations, we assume that the error terms for different observations are independent of each other. However, these error terms involve omitted variables whose values may not be independent. For instance, the rate of inflation may be one of the omitted variables, and inflation is not determined by independent dice rolls. Whether the rate of inflation is high or low this year may depend on whether the rate of inflation was high or low last year. An optional section of Chapter 14 shows how such correlations between errors can be detected and how the least squares procedure can be modified to account for this additional information. Advanced courses provided further details.

The third assumption states that the values of the error term ϵ are not only independent of each other but are also independent of the values of X. This, too, may not be the case. Perhaps, in our consumption function example, one of the omitted variables in ϵ is the rate of inflation and is correlated with the explanatory variable, income. If so, the third assumption is inappropriate. Because this possibility has very important, but complicated, implications, we'll postpone a fuller discussion until the next chapter.

The Distribution of the Parameter Estimates

Although we don't know the values of the error term that occur, we can be certain that their presence affects our estimates.

> The **parameter estimates** depend on the observed values of Y, which, in turn, depend on the signs and magnitudes of the error term ϵ. Thus, the parameter estimates depend on what the values of ϵ turn out to be.

For any given value of X, the expected value of ϵ is zero and therefore the expected value of Y is $\alpha + \beta X$. The actual value of Y will diverge from this expected value to the extent that ϵ is not zero. Because the error term is the sum of many separate influences that involve measurement errors, omitted variables, and sampling error, the central limit theorem suggests that the probability distribution for ϵ may be approximately normal, as is shown in Figure 13.7. For any value of X, the expected value of ϵ is zero and the expected value of Y is $\alpha + \beta X$. If ϵ is positive, then Y will be greater than $\alpha + \beta X$. If ϵ is negative, then Y will be less than $\alpha + \beta X$.

Figure 13.8 shows one possible set of data. For the first observation X_1, the error term ϵ_1 turns out to be positive and, hence the observed value of Y_1 is above $\alpha + \beta X_1$. For the second observation X_2, the error happens to be negative, so that Y_2 is below $\alpha + \beta X_2$. Of course, we have no way of knowing this in practice, because the ϵs are not observed. All that we see are the paired observations (X_1, Y_1) and (X_2, Y_2). The line $\hat{Y} = a + bX$ that we fit to these particular data underestimates the slope and overestimates the intercept.

This is just one possible outcome, which results from a positive value for ϵ_1 and a negative value for ϵ_2. If, instead, it happens that ϵ_1 is negative while ϵ_2 is positive, then the resulting data will lead us to overestimate the slope and underestimate the intercept. If the ϵs turn out to be both positive or both negative, then our estimates of α and β may be either both too high or both too low. And the sizes of our misestimates depend on the magnitudes of the ϵs.

Just as there is a probability distribution for ϵ, so are there probability distri-

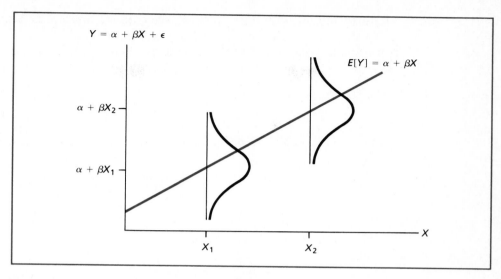

Figure 13.7   *The Theoretical Determination of Y*

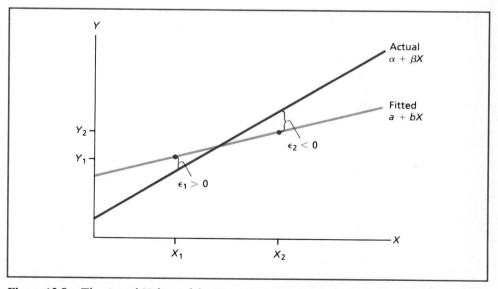

Figure 13.8   *The Actual Values of the Error Term Determine the Observed Values of Y and, Hence, the Estimates of α and β*

butions for our estimates of α and β. The estimation formulas, Equations 13.7 and 13.8 show that a and b are linear combinations of the observed values for Y, which, in turn, depend on ϵ. Thus it can be shown that, if $\epsilon \frown N[0, \sigma]$, then

$$b \frown N[\beta, \sigma/(\sqrt{n}\, s_X)] \tag{13.9}$$

$$a \frown N[\alpha, \sigma\sqrt{(\overline{X^2})}/(\sqrt{n}\, s_X)] \tag{13.10}$$

where σ is the standard deviation of ϵ and n is the number of observations. The term

$$s_X = \sqrt{\frac{\Sigma(X_i - \overline{X})^2}{n}}$$

is the standard deviation of the observations of X, and

$$\overline{X^2} = \frac{\Sigma X_i^2}{n}$$

is the average value of X^2.

Notice first that a and b are each unbiased estimators of the respective parameters α and β. A researcher's estimate of β will sometimes be too high and sometimes be too low. If the model's assumptions are satisfied, however, then it is very unlikely that researchers will continually underestimate or overestimate the parameters. On the average, over many studies using fresh data, the underestimates should be balanced out by overestimates.

However, being right on average is not all there is to it. Again, we would like something better than the lawyer who allows innocent people to be convicted but wins acquittals for the guilty. It is of little comfort to know that the statistician's estimate is just as likely to miss by a mile to one side as to miss by a mile to the other side. We would prefer the statistician who seldom, if ever, misses by a mile to either side.

To gauge the dispersion in the estimates, we can look at their standard deviations. Equation 13.9 shows that

1. the standard deviation of b is smaller the lower is the standard deviation of ϵ. If the observed ϵ are close to zero, then the fitted line is almost certain to be very close to the true relationship. If, however, the ϵ are far from zero, then wild estimates of α and β may result.

2. the standard deviation of b is smaller the more data we have. With lots of data, the observed errors are almost certain to balance each other out and yield a fitted line that is close to the true line. If, however, there are only a few observations, then almost anything can (and most likely will) happen.

3. the standard deviation of b is smaller the larger is the standard deviation of the X data. Figure 13.9 illustrates this logic. If there is little dispersion in the values

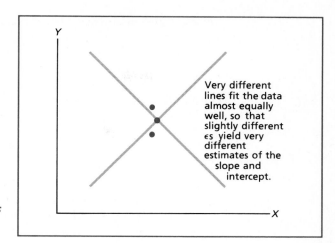

Figure 13.9 *There Is Considerable Uncertainty About α and β When There Is Little Dispersion in the Observed Values of X*

of X, then almost any slope and intercept are consistent with the data, and slight variations in the ϵ can then have a considerable impact on our estimates. We can hardly hope to gauge the effects of income on spending if we have never seen income change.

Equation 13.10 shows that the standard deviation of the intercept estimate depends on these very same factors, plus one additional consideration—the average squared value of X. The average squared value of X tells us whether the data are close to the Y-axis or far from it. If the data are far from the Y-axis, then small errors in estimating the slope will cause very large errors in estimating the Y-intercept. Thus, the larger the average squared value of X is, the bigger the standard deviation of a.

Estimating the Standard Deviation of the Error Term

In practice, we don't know the value of the standard deviation of ϵ, because we don't know precisely what comprises the error term. But there is a natural way of estimating it. The (unobserved) error term is

$$\epsilon = Y - (\alpha + \beta X)$$

while our (observed) prediction error is

$$e = Y - \hat{Y}$$
$$= Y - (a + bX)$$

If we use a and b to estimate α and β, then we can use the standard deviation of e to estimate the standard deviation of ϵ.

We noted earlier that the least squares line goes through the average values of the data $\overline{Y} = a + b\overline{X}$. Therefore, the average value of e is equal to zero:

$$\overline{e} = \overline{Y} - (a + b\overline{X}) = 0$$

The standard deviation of e is called the **standard error of estimate (SEE)**:

$$SEE = \sqrt{\frac{e_1^2 + e_2^2 + \cdots + e_n^2}{n - 2}}$$

where $e_i = Y_i - \hat{Y}_i$ is the prediction error for the ith observation. Notice that the sum of the squared prediction errors is divided by $n - 2$ rather than by n or $n - 1$. This is because the estimation of two parameters (α and β) uses up two degrees of freedom.* With the division by $n - 2$, the SEE^2 is an unbiased estimator of σ^2.

In our consumption function example, Table 13.1 shows that $\Sigma(X - \overline{X})^2 = 19{,}938.8$ so that $s_x = \sqrt{19{,}938.8/5} = 63.15$. The prediction errors shown in Table 13.2 imply $SEE = \sqrt{361.93/3} = 10.98$. Thus, using Equation 13.9, the estimated standard deviation of b is

$$s_b = \frac{SEE}{\sqrt{n}\ s_X}$$

$$= \frac{10.98}{\sqrt{5}\ (63.15)}$$

$$= 0.078$$

Similarly, using Equation 13.10, the standard deviation of a works out to be

$$s_a = 260$$

Table 13.2   *The Consumption Model's Prediction Errors*

Income X	Actual Spending Y	Predicted Spending $\hat{Y}$	Prediction Error $e = Y - \hat{Y}$	Squared Prediction Error
3,319	3,042	3,029.61	+12.39	153.49
3,421	3,124	3,126.24	−2.24	5.02
3,404	3,108	3,110.14	−2.14	4.56
3,276	2,994	2,988.87	+5.13	26.28
3,271	2,971	2,984.14	−13.14	172.58
Total			0.00	361.93

*One interesting way to see this is to consider the case where there are only two observations. The least squares line goes through both points and there are no prediction errors. The division by $n - 2$ correctly warns us that the standard deviation of ϵ cannot be estimated from just two observations.

We calculate a single pair of estimates of α and β, using the observed data. These estimates depend on the particular values of ϵ that happened to occur and thus have standard deviations because there are many different values of ϵ and, correspondingly, many different estimates that might have occurred.

Confidence Intervals for the Parameters

Once we know the distributions of a and b, it is a straightforward matter to construct confidence intervals. We will use b as an example. Our estimate b comes from a normal distribution with expected value β and standard deviation σ_b, whose formula is given in Equation 13.9. If b is normally distributed, then (as shown in Figure 13.10) there is a .95 probability that this estimate will be within (approximately) two standard deviations of the true parameter value β. We can write this statistical relationship as

$$.95 = P[\beta - 1.96\sigma_b < b < \beta + 1.96\sigma_b]$$

Rewriting this as a confidence interval, we have

$$.95 = P[b - 1.96\sigma_b < \beta < b + 1.96\sigma_b] \tag{13.11}$$

If there is a .95 probability that the estimate b will be within 1.96 standard deviations of β, then there is correspondingly a .95 probability that $b \pm 1.96$ standard deviations will encompass β. In the consumption function example, the 0.947 estimate of b and 0.078 standard deviation imply a 95 percent confidence interval 0.947 ± 0.153.

There is an analogous relationship between a and α, in that there is a .95 probability that a will be within 1.96 standard deviations of α. For $\alpha = -115$ with a

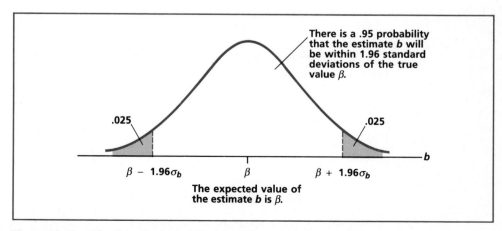

Figure 13.10 *The Distribution of the Least Squares Estimate of β*

standard deviation of 260, the 95 percent confidence interval stretches from -625 to 395.

Because the standard deviations of a and b must be estimated, the Student's t distribution with $n - 2$ degrees of freedom should be used in place of the normal distribution. However, with any sort of reasonably sized sample, the exact cutoff will be close to 2 and this is why researchers routinely use a two-standard-deviations rule of thumb.*

Hypothesis Tests for the Parameters

It is one short step from confidence intervals to hypothesis tests for the parameters. The most common null hypothesis is that β equals zero:

$$H_0: \beta = 0$$

$$H_1: \beta \neq 0$$

If β does equal zero, then the explanatory variable X has no influence on Y. Thus, $\beta = 0$ is usually a *straw hypothesis* that the researcher hopes to disprove. The researcher begins with the theory that X influences Y and wants to confirm this theory statistically by rejecting the possibility that $\beta = 0$.

A null hypothesis can be tested by computing the value of the t-statistic,

$$t = \frac{\text{estimated value of } \beta - \text{value of } \beta \text{ under } H_0}{\text{standard deviation of estimate of } \beta}$$

This t-value tells us how many standard deviations the estimated value of β is from its value under the null hypothesis. For the null hypothesis $\beta = 0$, the computed value of the t-statistic reduces to simply

$$t = \frac{b}{s_b}$$

The null hypothesis is rejected at a specified significance level if the t-value is larger than the appropriate cutoff, using a t-distribution with $n - 2$ degrees of freedom. For a test at the 5 percent level with a reasonable number of observations, the cutoff is approximately two. Thus the null hypothesis $\beta = 0$ is usually rejected if the ratio of the absolute value of the estimate of b to its standard deviation is larger than two.

---

*To focus our attention on the logic rather than the numbers, I used very few data ($n = 5$) in the consumption example. With $n - 2 = 3$ degrees of freedom, the appropriate t-value is 3.182 rather than 2.0. In practice, we would almost always use more data.

The quick-and-easy test of the commonplace hypothesis $\beta = 0$ is to see whether or not the absolute value of the ratio of the estimate to its standard deviation is greater than two.

Using the consumption function again as an example, the results might be reported as follows:

Keynes' consumption function was estimated by ordinary least squares, using annual data for the years 1977–1981. The results are

$$\hat{Y} = -115 + .947X$$
$$\quad\quad (260) \quad (.078)$$

The standard deviations of the parameter estimates are in parentheses under the estimates.

Most readers will automatically take the ratio of the estimated slope to its standard deviation to see whether or not this ratio is greater than 2. If so, then we can conclude that income has (at the 5 percent level) a statistically significant effect on consumption; in other words, zero is outside a 95 percent confidence interval for this estimate and, so, we reject the null hypothesis that $\beta = 0$.

Here, the value of the t-statistic for b is $0.947/0.078 = 12.2$, giving a decisive rejection of $\beta = 0$. For α, on the other hand, the t-value is only $-115/260 = -0.442$. We cannot reject the null hypothesis that the intercept is zero.

This is such a standard operating procedure that most researchers report their t-values along with their parameter estimates, to save readers the trouble of doing the division themselves. Thus, another standard reporting format is

Using annual data for 1977–1981, the least-squares estimates of Keynes' consumption function turned out to be

$$\hat{Y} = -115 + 0.947X$$
$$\quad\quad [0.44] \quad [12.2]$$

[   ]: t values

(The absolute values of the t-ratios are typically reported because the signs are unimportant.) Some researchers only report their t-values, for example,

To test Keynes' theory that income influences consumption, a regression was run using annual data for 1977–1981. The t-value turned out to be 12.2, which decisively confirms Keynes' theory.

Of course, $\beta = 0$ is not the only possible null hypothesis. Any hypothesis can be tested by finding the appropriate cutoff points, the P-value, or a confidence interval. For instance, Keynes argued that the slope β is almost certainly less than 1.0. This hypothesis can be tested by seeing whether or not 1.0 is inside a confidence interval for β. In this case it is, and so these data do not reject the possibility that β is actually greater than 1. Notice that these data do not reject Keynes' hypothesis

The Cost of Another Dozen Stockings

In the 1930s, Joel Dean pioneered the use of regression models to analyze business production costs.[4] One of his classic studies involved a mill for knitting silk stockings. Using silk prepared by another plant, this mill knitted stockings, which were then sent on to another plant for dyeing and finishing. Professor Dean used monthly data for the period 1935–1939 to see how this knitting mill's production costs (Y) depended on its output (X). The scatter diagram of his data shown in Figure 1 reveals a positive, linear relationship. Using least squares, Dean estimated the following equation,

$$\hat{Y} = 2935.59 + 1.998X$$
$$(0.017)$$

( ): standard deviation

where costs Y are recorded in dollars and output X is measured in dozens of pairs of stock-ings. The t-value of $1.998/0.017 = 117.5$ decisively rejects the null hypothesis that the mill's costs do not depend on its output. It cost this mill about \$2 to knit an extra dozen pair of stockings. A 95 percent confidence interval for this marginal cost is 1.998 ± 0.034, or \$1.964 to \$2.032.

Figure 1   *Production Costs and Output at a Hosiery Mill*

that β is less than 1. Indeed, the preferred estimate, $b = 0.947$, is less than 1, but (because the sample is so small) the confidence interval is sufficiently wide, stretching from 0.791 to 1.103, that we cannot rule out $\beta > 1$.

Confidence Intervals for $\alpha + \beta X$

Models are usually intended for predictions, either in actual or hypothetical situations. We may want to predict consumption spending next year or how consumption would be affected by a purely hypothetical change in income. Our estimates of α and β are the basis for such predictions and, because these estimates are of uncertain accuracy, so are our predictions. The exact estimates a and b that we obtain depend on what the error term ϵ happens to be in the sample data. Different values for ϵ give different data, parameter estimates, and predictions.

Suppose that we want to predict what Y will be for a value X_0 for the explanatory variable. Because a and b are unbiased estimators of α and β, our prediction $\hat{Y}_0 = a + bX_0$ is an unbiased estimator of $\alpha + \beta X_0$, with a standard deviation that can be computed, most easily by computer. If the error ϵ is normally distributed, then so are a, b, and $\hat{Y}_0$:

$$\hat{Y}_0 \backsim N\left[\alpha + \beta X_0, \left(\frac{\sigma}{\sqrt{n}}\right)\sqrt{1 + \frac{(X_0 - \overline{X})^2}{s_X^2}}\right] \qquad (13.12)$$

The exact formula for the standard deviation of $\hat{Y}_0$,

$$\sigma_{\hat{Y}_0} = \left(\frac{\sigma}{\sqrt{n}}\right)\sqrt{1 + \frac{(X_0 - \overline{X})^2}{s_X^2}} \qquad (13.13)$$

is far from obvious, but its implications are quite logical. Our predictions are less certain (and more likely to be inaccurate) if:

1. the standard deviation σ of ϵ is large, because then there is a greater chance that the data will yield misleading estimates;
2. the number of observations n is small, because few data provide little information to use in our estimates;
3. the standard deviation s_X of X is small, because it is hard to gauge the effects of changes in X on Y, if we hardly observe any change in X; and
4. the value of X_0 for which we are making a prediction is far from the average value of X, because our uncertainty is greatest on the fringes of our data.

The first three points are easily understood—these are all reasons why the estimates of α and β may be inaccurate. The fourth point is new, but it makes good sense, too. Figure 13.11 illustrates this logic. Unless the estimates are perfect, there will always be a substantial prediction error as we move far away from the mean of our data. A large prediction error near the mean, on the other hand, requires an unlikely string of bad luck in the ϵs. Thus, the researcher can be much more confident about a model's implications and predictions near the center of the data. Caution is called for when you are working on the fringes of the data.

Equation 13.12 can be used to construct a confidence interval for $\alpha + \beta X_0$. Because $\hat{Y}_0$ has about a 95 percent chance of being within (approximately) two standard deviations of $\alpha + \beta X_0$, it follows that there is a .95 probability that $\hat{Y}_0 \pm$ two standard deviations will encompass $\alpha + \beta X_0$:

$$\alpha + \beta X_0 = \hat{Y}_0 \pm 1.96\sigma_{\hat{Y}_0}$$

where the standard deviation of $\hat{Y}_0$ is given by Equation 13.13 and σ can be estimated by the standard error of estimate. For a more precise confidence interval, we can replace 1.96 with the appropriate value from a t distribution with $n - 2$ degrees of freedom.

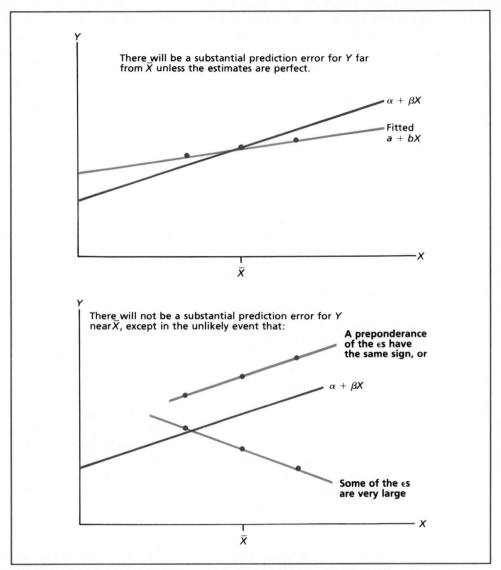

Figure 13.11   *The Prediction Errors Are Unlikely to Be Large Near the Mean of the Observations*

A Prediction Interval for Y

You've now seen how to construct a confidence interval for $\alpha + \beta X_0$. Typically, however, it is $Y_0 = \alpha + \beta X_0 + \epsilon_0$ that we are interested in predicting. One of the reasons for inaccurate predictions is, as I've just explained, the luck of the draw in our observed data. The specific values of ϵ that happen to occur cause imperfect estimates of α and β and, hence, imperfect predictions of $\alpha + \beta X_0$. In addition, our predictions may be off because of what ϵ turns out to be when the prediction is made. Unless ϵ happens to be zero during the forecast, we won't make perfect forecasts of Y, even if our estimates of α and β are perfect.

We can take this added source of prediction error into account by using the fact that our prediction error $Y_0 - \hat{Y}_0$ has an expected value of zero and a variance that is equal to the variance of $\hat{Y}_0$ plus the variance of ϵ_0:*

$$\text{variance of } Y_0 - \hat{Y}_0 = (\text{variance of } \hat{Y}_0) + (\text{variance of } \epsilon_0) \qquad (13.14)$$

The variance of $\hat{Y}_0$ is given by Equation 13.13, while the SEE provides an estimate of the variance of ϵ_0. If there is a .95 probability that $Y_0 - \hat{Y}_0$ will be within approximately two standard deviations of zero, then there is a .95 probability that $\hat{Y}_0$ will be within two standard deviations of Y_0. Thus, a 95 percent confidence interval for Y_0 is

$$\boxed{\hat{Y}_0 \pm 1.96 \text{ standard deviations of } Y_0 - \hat{Y}_0}$$

Again, for a more exact interval, the 1.96 value (or two-standard-deviations rule of thumb) can be replaced by the value from a t distribution with $n - 2$ degrees of freedom.

This confidence interval is commonly called a **prediction interval.** Table 13.3 shows 95 percent prediction intervals for selected values of X_0, while Figure 13.12 on page 540 shows a more complete picture. Notice that the confidence interval is narrowest at $\overline{X}$, and then widens as we move away from the mean of our data.

Table 13.3 *Ninety-Five Percent Prediction Intervals for Consumer Spending*

Value of Income X_0	Predicted Spending Y_0	Standard Deviation of Prediction	95% Prediction Interval
3,038.2	2,763.6	26.3	2,711.1 to 2,816.1
3,138.2	2,858.3	19.7	2,819.0 to 2,897.7
3,238.2	2,953.1	14.3	2,924.4 to 2,981.7
3,338.2	3,047.8	12.0	3,023.7 to 3,071.9
3,438.2	3,142.5	14.3	3,113.9 to 3,171.2
3,538.2	3,237.3	19.7	3,197.9 to 3,276.6
3,638.2	3,332.0	26.3	3,279.5 to 3,384.5

*The prediction error can be rearranged as

$$Y_0 - \hat{Y}_0 = (Y_0 - E[Y_0]) + (E[Y_0] - \hat{Y}_0)$$

with a variance equal to the sum of the variances of the two parenthetical terms. The first term is ϵ_0 and so, has its variance. The variance of the second term is equal to the variance of $\hat{Y}_0$.

Estimating Western Union's Cost of Equity

In some industries, such as the provision of water or electricity to households, there is a natural monopoly in that it would be very inefficient to have homes connected to multiple suppliers of water or electricity. However, a single supplier of a vital product could exploit its monopolistic position by charging exorbitant prices. Recognizing this conflict, governments either run the public utility themselves or permit noncompetitive monopolies to exist but require that their prices be determined by regulatory agencies—thus the generic label *regulated utility*.

Regulators typically look at the firm's costs and set prices so that the rate of return on the firm's assets is approximately equal to that of comparable businesses. As explained in Chapter 4's example "Risk Has Its Rewards," relatively risky investments are generally priced to have relatively high average returns. Thus measures of a regulated firm's risk often are used to identify comparable businesses.

For example, in a hearing to determine Western Union's telegraph rates,[5] the Federal Communications Commission wanted to estimate Western Union's cost of equity—the rate of return that shareholders require on their stock. The price-earnings ratio is often used for this purpose, but Western Union argued that its earnings had been temporarily depressed, making its price-earnings ratio meaningless. Instead, they proposed an examination of the cost of equity in similar industries, adjusting for differences in risk as measured by the variability of profits. (Profit variability was calculated by dividing the standard deviation of profits about a trend line by average profits.) The data are shown in Table 1.

Table 1

Related Industries	Profit Variability	Cost of Equity
AT&T	3.6%	9.9%
Electric utilities	7.1	10.4
Independent telephones	10.6	10.4
Gas pipelines	11.9	11.4
Gas distribution	12.9	12.2
Water utilities	17.1	10.0
Truckers	44.0	16.5
Airlines	76.0	18.9

Western Union's profit variability was 27.0 percent. To estimate the relationship between profit variability (X) and cost of equity (Y), an expert witness used these data to estimate the regression equation

$$\hat{Y} = 9.459 + 0.1312\,X, \qquad R^2 = .922$$
$$[18.6] \quad [8.43]$$

The scatter diagram is shown in Figure 1. The estimated effect of profit variability on the cost of equity is, as expected, positive and highly statistically significant.

Western Union's cost of equity was estimated by calculating the predicted value of Y for $X = 27.0$:

$$\hat{Y} = 9.459 + 0.1312(27.0) = 13.001$$

The standard deviation for this prediction works out to be 1.083. With only eight observations (and $n - 2 = 6$ degrees of freedom), the appropriate t-value for a 95 percent prediction interval is 2.447, substantially larger than a two-standard-deviations rule of thumb. A 95 percent prediction interval for Y for $X = 27.0$ is

$$13.001 \pm 2.447(1.083)$$
that is, $\qquad$ 100.350 to 15.651

Thus least squares regression provided not only an estimate of Western Union's cost of equity but also a means of quantifying uncertainty about this estimate.

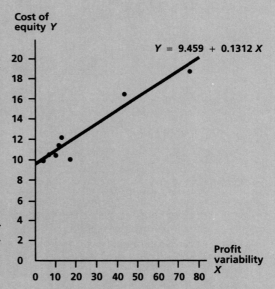

Figure 1 *Cost of Equity and Profit Variability for Regulated Utilities*

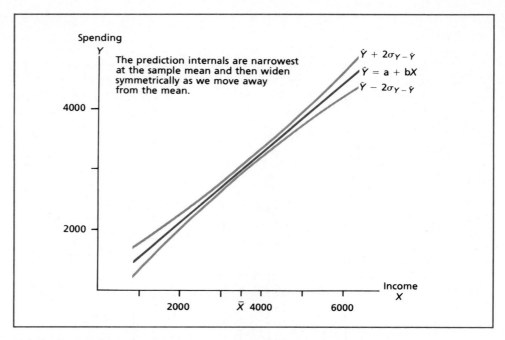

Figure 13.12   *Ninety-five Percent Prediction Intervals for Consumer Spending*

EXERCISES

13.15 A marketing consultant has been hired to evaluate the effect of advertising expenditures on the sales of a certain beer. Differing advertising expenditures in eight similar cities gave these results:

Advertising ($1,000)	Sales ($1,000)
20	370
25	350
25	330
30	350
30	400
35	380
35	410
40	400

Use these data to determine the least squares estimates of the slope and intercept of the equation $Y = \alpha + \beta X + \epsilon$, where X is advertising expenditures and Y is sales.   **a.** What is the estimated effect of advertising on sales?   **b.** If this estimate is correct, does an extra $1 of advertising pay for itself? (Assume that, not counting advertising costs, the firm makes a 20-cent profit on every $1 of sales.)   **c.** Is the estimated effect of advertising on sales statistically significant at the 5 percent level?   **d.** What does "statistically significant" mean?   **e.** Calculate a 95 percent confidence interval for the effect of advertising on sales.   **f.** If you were this consultant, what would your recommendation be?

13.16 An ice cream store wants to investigate the relationship between its sales and the average temperature. The following data are collected:

Month	Temperature (°F)	Sales (gal)
Jan	28.4	25
Feb	30.3	27
Mar	39.2	82

errors with the sum of the squared deviations of Y about its mean,

$$\frac{\Sigma(Y_i - \hat{Y}_i)^2}{\Sigma(Y_i - \overline{Y})^2}$$

This specific ratio can be justified as follows. If we thought that X had no influence on Y, then we would set β equal to zero, leaving our model as

$$Y = \alpha$$

Because $\overline{Y} = a + b\overline{X}$, the least squares estimate of α when $b = 0$ is the mean of Y,

$$a = \overline{Y}$$

and the sum of the squared prediction errors is

$$\Sigma(Y_i - \overline{Y})^2$$

Thus, the ratio

$$\frac{\Sigma(Y_i - \hat{Y}_i)^2}{\Sigma(Y_i - \overline{Y})^2}$$

compares the sum of the squared prediction errors for the model with what the sum of squared prediction errors would be if b were set equal to zero. This ratio is a measure of how much prediction accuracy is improved by allowing for the possibility that Y depends on X (that β is not zero).

Because the numerator and denominator are both sums of squared errors, their ratio cannot be negative. Also, because least squares chooses the value of b that minimizes the prediction errors, this ratio cannot exceed 1. Thus

$$0 \leq \frac{\Sigma(Y_i - \hat{Y}_i)^2}{\Sigma(Y_i - \overline{Y})^2} \leq 1$$

The relevant question is whether this ratio is near 1 or 0. If the ratio is close to 1, then not much is gained by allowing for the possibility that X influences Y. If the ratio is close to 0, then there is a relatively large improvement in predictive accuracy by allowing for the effect of X on Y.

In practice, statisticians subtract this ratio from 1, giving the **coefficient of determination**:

$$R^2 = 1 - \frac{\Sigma(Y_i - \hat{Y}_i)^2}{\Sigma(Y_i - \overline{Y}_i)^2} \qquad (13.15)$$

This is commonly called **R-squared.** The square root of the coefficient of determination is called the **correlation coefficient, R.** Because a square root can be either positive or negative, the convention is to let R be positive when b is positive (indicating a positive relationship between Y and X) and to let R be negative when b is negative (indicating a negative relationship).

Yet another measure is to adjust R-squared for the degrees of freedom:

$$R_a^2 = 1 - \frac{\Sigma(Y_i - \hat{Y}_i)^2/(n-2)}{\Sigma(Y_i - \overline{Y})^2/(n-1)}$$

The model's errors are divided by $n-2$ because two parameters (α and β) are estimated from the n observations, while the denominator uses $n-1$ because only one parameter (the mean of Y) is estimated.

Another interpretation of R-squared is as follows. As shown in Figure 13.13, the variation of Y about its mean $\overline{Y}$ can be separated into two parts, the deviation of the prediction from the mean of Y and the prediction error $Y - \hat{Y}$:

$$Y - \overline{Y} = (\hat{Y} - \overline{Y}) + (Y - \hat{Y})$$

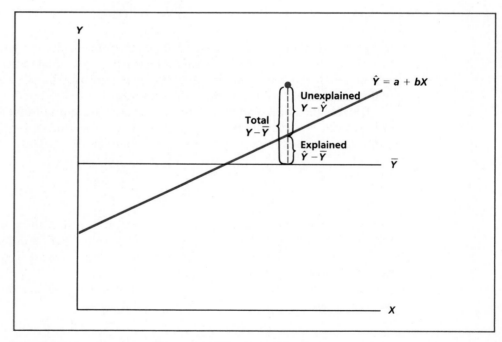

Figure 13.13   *The Separation of $Y - \overline{Y}$ into Explained and Unexplained*

Can Anyone Suggest a Good Hedge Against Inflation?

Inflation erodes the real purchasing power of money, bonds, and other assets with fixed-dollar values. Consider, for instance, $100 deposited in a bank at a fixed 5 percent interest rate, which pays $105 after a year. If prices increase by 5 percent during the year, then this $105 at the end of the year will buy no more than $100 would now. If prices increase by 10 percent, $105 at the end of a year buys 5 percent less than $100 does now.

Investors try, of course, to find investments that keep ahead of inflation. Reluctant to settle for a 5 percent return in the face of 10 percent inflation, they shop around for bonds that pay higher yields. If this fails, they buy land and other physical assets that may appreciate during inflation or they buy stock in firms that own assets that may appreciate. This last possibility is investigated in Figure 1, which is a scatter diagram of annual inflation rates and stock returns (dividends plus capital gains) over the period 1926 through 1980.[7] Stock returns have averaged more than inflation (the numbers work out to be 11.7 percent versus 3.0 percent), but they have also been much more variable and appear to bear little or no relation to the rate of inflation. Whether it is a year of deflation or double-digit inflation doesn't seem to affect the stock market's chances of collapse or boom. A least squares regression of the percent return (Y) on the rate of inflation (X) confirms that there is little relation between the two:

$$\hat{Y} = 11.72 - .009X, \qquad R^2 = .000004$$
$$[3.36] \quad [.02]$$

[ ]: t-values.

The estimated coefficient of X, the t-value, and R-squared are all virtually zero. There is no evidence whatsoever here that stocks provide a hedge against inflation.

Figure 1 *Stock Returns and Inflation, 1926–1980*

If we square this and sum it over the different values of Y, the cross-product term $(\hat{Y} - \overline{Y})(Y - \hat{Y})$ drops out and we get

$$\Sigma(Y - \overline{Y})^2 = \Sigma(\hat{Y} - \overline{Y})^2 + \Sigma(Y - \hat{Y})^2$$

or
$$\frac{\text{total sum}}{\text{of squares}} = \frac{\text{explained}}{\text{sum of squares}} + \frac{\text{unexplained}}{\text{sum of squares}}$$

The variation in the predictions is the "explained" sum of squares, because this is the variation in Y that is attributed to the model. The prediction errors are the unexplained variation in Y. In ratio form,

$$1 = \frac{\text{explained}}{\text{total}} + \frac{\text{unexplained}}{\text{total}}$$

Because R-squared is 1 minus the ratio of the sum of squared prediction errors to the sum of squared deviations about the mean,

$$R^2 = 1 - \frac{\text{unexplained}}{\text{total}} = \frac{\text{explained}}{\text{total}}$$

Thus, R-squared can be thought of as the fraction of the variation in the dependent variable that is explained by the regression model.

R-squared gauges what fraction of the variation in Y is explained by allowing for a nonzero value of β. Specifically, it tells us how successful the model is in predicting Y, relative to the benchmark of setting $b = 0$ and using the sample mean to predict Y. Figure 13.14 shows three examples. If there is little or no improvement

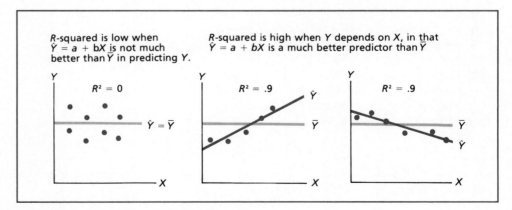

Figure 13.14   *Three Correlation Coefficients*

Mutual Fund Consistency

Mutual funds pool investor savings in order to acquire a large diversified portfolio of assets. The investors' return is generally not guaranteed since mutual funds buy securities on behalf of their shareholders and pass along the investment profits, less management fees and other expenses. Investors flock to funds with impressive records, hoping that a fund's past success ensures future success. Unfortunately, there is little or no consistency to mutual fund performance. Funds with good records are no more likely to do well in the future than are funds that have had disappointing records.

For example, data were obtained for 64 mutual funds that, according to the Wiesenberger Service, seek to maximize capital gains.[8] Each fund's annual percentage return, dividends plus capital gains, was calculated over the five-year period January 1, 1977, to January 1, 1982. This is the performance record that investors might have used to choose a mutual fund in 1982. Each fund's annual percentage return was then calculated over the subsequent five-year period, from January 1, 1982 to January 1, 1987. This is the five-year return for an investor who bought in 1982 on the basis of each fund's performance during the preceding five years.

To test how well past performance predicts future performance, the regression equation $Y = \alpha + \beta X + \epsilon$ was estimated, where Y = 1982–1987 return and X = 1977–1982 return. An estimate of β that is positive (perhaps close to 1) and statistically significant, and has a high value for R^2, would provide

evidence that past performance is a reliable predictor of future performance. The results of a least squares regression, however, are

$$\hat{Y} = \begin{matrix} 15.267 \\ (3.118) \\ [4.9] \end{matrix} - \begin{matrix} 0.0387\,X, \\ (0.1513) \\ [0.3] \end{matrix} \qquad R^2 = .001$$

The estimate of β has the wrong sign and is far from statistically significant. The value of R^2 is virtually 0, confirming the visual impression conveyed by the scatter diagram in Figure 1: There is no correlation between fund performance in one five-year period and in the next five-year period.

Figure 1 *Mutual Fund Performance, 1982–1987 Versus 1977–1982*

in our predictions when we take into account the effect of X on Y, then the ratio of the prediction errors will be almost 1.0 and, thus, R-squared will be close to 0. If taking the effect of X on Y into account improves our predictions dramatically, then the ratio of prediction errors will fall almost to 0, and R-squared will be close to 1.

Our consumption model is quite successful by this standard. Table 13.2 shows the predicted and actual values of consumption. The model's sum of squared prediction errors is 361.93. If X is omitted and the mean of Y is used for predictions, then the sum of squared errors leaps to 19,048.83. The ratio is .019 and the value of R-squared is .981.

EXERCISES

13.23 A classic study of the IQ scores of 34 identical twins who were raised apart obtained these intelligence scores.[9]

1st Born	2nd Born	1st Born	2nd Born
22	12	27	28
36	34	22	23
13	10	15	9
30	25	24	33
32	28	6	10
26	17	23	21
20	24	38	27
30	26	33	26
29	35	16	28
26	20	27	25
28	22	4	2
21	27	19	9
13	4	41	41
32	33	40	38
30	34	12	9
27	24	13	22
32	18	29	30

Calculate the R-squared for a regression of the intelligence score of the second born on the intelligence score of the first born. (This R-squared has been used to gauge the influence of genes on intelligence, while $1 - R$-squared measures the influence of one's environment.)

13.24 In the Heightlands, men always marry women who are exactly six inches shorter than themselves. Write down an equation predicting the height of a woman Y from the height of her husband X. What is the correlation coefficient for this equation?

13.25 Table 1.3 gives data on the annual returns for 1984, 1985, and 1986 for 24 randomly selected mutual funds. Letting Y be the return in 1986 and X the return in 1985, use least squares to see if there is a statistically significant relationship at the 5 percent level between the fund returns in these two years. Now let Y be the return in 1985 and X the return in 1984 and use least squares to see if there is a statistically significant relationship between the fund returns in these two years. (Overall, the market rose by more than 30 percent in both 1985 and 1986, but fell by 5 percent in 1984.)

13.26 Use the 1950–1989 data on the rate of inflation (X) and the rate of return on Treasury bills (Y) in Table 8 at the end of this book to estimate the equation $Y = \alpha + \beta X + \epsilon$. Compare this estimated equation with that for stock returns reported in the text. Imagine that you are an investment counselor and write a paragraph summarizing your conclusions.

13.27 A businessperson is trying to estimate the relationship between price (X) and sales (Y) of a certain line of clothing. Tests in similar cities throughout the country have yielded these data:

Price (X)	Sales (Y)
$20	10,300
$25	9,100
$30	8,200
$35	6,500
$40	5,100
$50	2,300

Use these data to estimate the equation $Y = \alpha + \beta X + \epsilon$. What are the estimates of α and β? What is the

Standard Error of Estimate (SEE)? What is the R-squared?

13.28 *(continuation)* If sales are measured in thousands (10.3, 9.1, etc.), do we get the same estimates of α and β, SEE, and R-squared? Are our conclusions affected? If the price is measured in cents instead of dollars (2,000, 2,500, etc.), do we get the same estimates of α and β, SEE, and R-squared? Are our conclusions affected? (Use the original units for sales: 10,300, 9,100, and so on.)

13.29 *(continuation)* If the price is measured in the amount above \$20 (\$0, \$5, and so on), do we get the same estimates of α and β, SEE, and R-squared? Are our conclusions affected? Summarize the moral of Exercises 13.28 and 13.29.

13.30 In 1983, an investor interested in the relationship between the earnings yield and the risk-free rate of return reported his estimation of the following equation, using annual data for the preceding 16 years:[10]

$$(E/P) = 1.22 + .85(I + 3)$$

where (E/P) is the earnings-price ratio for the Dow Jones Industrials, I is the inflation rate, and the real interest rate is assumed to be a constant 3 percent, implying a nominal interest rate of $I + 3$. Interpret his results.

The journal's editor reestimated the equation, obtaining these results,

$$(E/P) = 3.77 + .85I$$

Are these two estimated equations consistent or contradictory? Which do you think has the higher R-squared?

13.5   SUMMARY

In the regression model $Y = \alpha + \beta X + \epsilon$, changes in Y (the *dependent* variable) are explained by changes in X (the *explanatory* variable). The error term ϵ reflects measurement errors, omitted influences on Y, and sampling error. Because these factors cannot be predicted in advance, ϵ is a random variable and Y is too. This random variation implies that the parameters α and β cannot be known with certainty, although they can be estimated.

The least squares estimation procedure determines the estimates a and b that minimize the sum of squared prediction errors $\Sigma(Y_i - \hat{Y}_i)^2$ for the data at hand. The formulas are

$$b = \frac{\Sigma(X_i - \overline{X})(Y_i - \overline{Y})}{\Sigma(X_i - \overline{X})^2}$$

$$a = \overline{Y} - b\overline{X}$$

Because the error term is unknown, it can be described by a probability distribution. The usual assumptions are that this probability distribution is normal with a zero expected value and constant variance, and that the values of ϵ are independent of each other and of X. With these assumptions, the least squares estimates are normally distributed, too,

$$b \sim N[\beta, \sigma/(\sqrt{n}s_X)]$$

$$a \sim N[\alpha, \sigma\sqrt{\overline{X}^2}/(\sqrt{n}s_X)]$$

where σ is the standard deviation of ϵ; n is the number of observations;

$$s_X = \sqrt{\frac{\Sigma(X_i - \overline{X})^2}{n}}$$

is the standard deviation of the observations of X; and

$$\overline{X}^2 = \frac{\Sigma X_i^2}{n}$$

is the average value of X^2. The standard deviation σ of the error term is usually unknown, but it can be estimated by the standard error of estimate (SEE),

$$SEE = \sqrt{\frac{\Sigma(Y_i - \hat{Y}_i)^2}{n - 2}}$$

The probability distributions for the estimates provide the information needed to construct confidence intervals and conduct hypothesis tests. For instance, there is a .95 probability that the estimate b will be within approximately two standard deviations of its expected value β. To test the null hypothesis that $\beta = 0$ (that X has no influence on Y), we can calculate b's t-value—the absolute value of the estimate divided by its standard deviation. If the t-value is greater than (approximately) 2.0, then the estimate is more than two standard deviations away from zero and, hence, zero is outside of a 95 percent confidence interval and can be rejected as a null hypothesis at the 5 percent level. (The exact cutoff is given by a t distribution with $n - 2$ degrees of freedom.)

A prediction of Y for some particular value X_0 is provided by $\hat{Y}_0 = a + bX_0$. This prediction has a 95 percent confidence interval (or *prediction interval*),

$$Y_0 = \hat{Y}_0 \pm 2 \text{ (standard deviations of } Y_0 - \hat{Y}_0)$$

where the standard deviation is a complicated formula spelled out in Equations 13.13 and 13.14.

The predictive accuracy of the model can be gauged by the

$$\text{coefficient of determination} = R^2 = 1 - \frac{\Sigma(Y_i - \hat{Y}_i)^2}{\Sigma(Y_i - \overline{Y}_i)^2}$$

R-squared compares the sum of squared prediction errors for the model and for the assumption that β is zero, that X does not influence Y. If allowance for the influence of X on Y substantially reduces the prediction errors, then R-squared will be close to 1.0. If not, then R-squared will be close to zero.

APPENDIX: THE LEAST SQUARES SOLUTION

We want to minimize the sum of squared prediction errors,

$$S = e_1^2 + \cdots + e_n^2 = (Y_1 - a - bX_1)^2 + \cdots + (Y_n - a - bX_n)^2$$

This minimum is found by setting the partial derivatives with respect to a and b both equal to zero. The derivative with respect to a yields

$$0 = \frac{\delta S}{\delta a} = -2(Y_1 - a - bX_1) - \cdots - 2(Y_n - a - bX_n)$$

This implies

$$Y_1 + \cdots + Y_n = na - b(X_1 + \cdots + X_n)$$

so that

$$a = \overline{Y} - b\overline{X}$$

where $\overline{Y}$ and $\overline{X}$ are the average values of Y and X.
The derivative with respect to b yields

$$0 = \frac{\delta S}{\delta b} = -2X_1(Y_1 - a - bX_1) - \cdots - 2X_n(Y_n - a - bX_n)$$

This implies, using $a = \overline{Y} - b\overline{X}$,

$$0 = X_1\{(Y_1 - \overline{Y}) - b(X_1 - \overline{X})\} + \cdots + X_n\{(Y_n - \overline{Y}) - b(X_n - \overline{X})\}$$

so that

$$b = \frac{X_1(Y_1 - \overline{Y}) + \cdots + X_n(Y_n - \overline{Y})}{X_1(X_1 - \overline{X}) + \cdots + X_n(X_n - \overline{X})}$$

Finally, taking into account the computational fact that

$$X_1(Y_1 - \overline{Y}) + \cdots + X_n(Y_n - \overline{Y})$$
$$= (X_1 - \overline{X})(Y_1 - \overline{Y}) + \cdots + (X_n - \overline{X})(Y_n - \overline{Y})$$

for either Y or X, we have

$$b = \frac{(X_1 - \overline{X})(Y_1 - \overline{Y}) + \cdots + (X_n - \overline{X})(Y_n - \overline{Y})}{(X_1 - \overline{X})(X_1 - \overline{X}) + \cdots + (X_n - \overline{X})(X_n - \overline{X})}$$

REVIEW EXERCISES

13.31 Table 13.4 gives some data on the normal precipitation P (in inches) and average daily temperature T (in °F) in Chicago, Los Angeles, and New York during the twelve months of the year. The equation $P = \alpha + \beta T + \epsilon$ can be used to see whether there is a relationship between temperature and precipitation over the course of a year. Estimate this equation using data for **a.** Chicago **b.** Los Angeles **c.** New York. What conclusions do you draw from these results?

13.32 (*continuation*) Now use these data to see whether there is any relationship between these cities' monthly temperatures, by estimating the equation $Y = \alpha + \beta X + \epsilon$, where **a.** X = Chicago temperature, Y = L.A. temperature **b.** X = Chicago temperature, Y = N.Y. temperature **c.** X = L.A. temperature, Y = N.Y. temperature. What conclusions do you draw from these results?

13.33 (*continuation*) Now examine the relationship between these cities' monthly precipitation by estimating $Y = \alpha + \beta X + \epsilon$, where **a.** X = Chicago precipitation, Y = L.A. precipitation **b.** X = Chicago precipitation, Y = N.Y. precipitation. **c.** X = L.A. precipitation, Y = N.Y. precipitation. What conclusions do you draw from these results?

13.34 The U.S. National Center for Health Statistics reported the following average heights and weights for 18- to 24-year-old U.S. males:

Height (Inches) X	Weight (Pounds) Y	Height (Inches) X	Weight (Pounds) Y
62	130	69	162
63	135	70	166
64	139	71	171
65	143	72	175
66	148	73	180
67	152	74	185
68	157		

Use these data to obtain the least square estimates of the equation $Y = \alpha + \beta X + \epsilon$. Give the SEE and R-squared, too. On average, an extra five inches of height is associated with how many additional pounds?

13.35 (*continuation*) Now use these data to find the least-squares estimate of the equation $Y = \alpha + \epsilon$. What are the SEE and R-squared for this equation? Is this equation a better or worse predictor than the equation estimated in the preceding exercise? Is the

Table 13.4   *Data for Exercise 13.31*

Month	Chicago T	Chicago P	Los Angeles T	Los Angeles P	New York T	New York P
Jan	22.9	1.70	54.5	2.52	32.2	2.71
Feb	26.1	1.30	55.6	2.32	33.4	2.92
Mar	35.7	2.52	56.5	1.71	41.1	3.73
Apr	48.8	3.38	58.8	1.10	52.1	3.30
May	58.4	3.41	61.9	0.08	62.3	3.47
June	68.1	4.15	64.5	0.03	71.6	2.96
July	71.9	3.46	68.5	0.01	76.6	3.68
Aug	71.1	2.73	69.5	0.02	74.9	4.01
Sept	63.7	3.01	68.7	0.07	68.4	3.27
Oct	53.8	2.32	65.2	0.22	58.7	2.85
Nov	39.2	2.10	60.5	1.76	47.4	3.76
Dec	27.1	1.64	56.9	1.75	35.5	3.53

relationship between height and weight statistically significant at the 5 percent level?

13.36 The U.S. National Center for Health Statistics reported the following average heights and weights for 18- to 24-year-old U.S. females:

Height (Inches) X	Weight (Pounds) Y	Height (Inches) X	Weight (Pounds) Y
57	111	63	131
58	114	64	134
59	118	65	137
60	121	66	141
61	124	67	144
62	128	68	147

Use these data to obtain the least squares estimates of the equation $Y = \alpha + \beta X + \epsilon$. Give the SEE and R-squared, too. Briefly compare the empirical relationships between height and weight for males and females found in Exercises 13.34 and 13.36.

13.37 Explain why you either agree or disagree with this reasoning: "A person with zero height will have zero weight. Therefore, a regression of weight on height must go through the origin."

13.38 Use the 1980 data, in Table 13.5, from the U.S. Department of Transportation to see whether there is a statistically significant relationship at the 1 percent level between highway fatalities per 100 million miles of travel (Y) and the percentage of all vehicles clocked at over 55 miles per hour (X). (Florida data are not available.)

Table 13.5   *Data for Exercise 13.38*

State	% Over 55 mph	Fatalities	State	% Over 55 mph	Fatalities
New Mexico	67.4	5.42	Iowa	48.6	3.30
Texas	64.7	4.01	West Virginia	48.4	5.03
Arizona	63.2	5.03	Massachusetts	48.3	2.49
Utah	61.9	3.09	Michigan	47.9	2.86
California	61.7	3.52	Oklahoma	47.9	3.55
North Dakota	58.3	2.86	Maine	46.0	3.50
Kansas	58.1	3.44	Pennsylvania	46.0	2.92
Delaware	57.9	3.61	New Jersey	45.9	2.23
Montana	57.8	4.91	Arkansas	45.9	3.62
Wyoming	57.5	4.89	South Carolina	45.5	3.78
Indiana	56.6	3.17	New York	45.4	3.35
Nevada	56.0	5.67	Illinois	45.0	3.08
Minnesota	55.2	3.03	North Carolina	44.7	3.63
Ohio	55.2	2.82	Kentucky	43.1	3.25
Vermont	54.9	3.73	Louisiana	41.3	4.86
Alabama	53.7	3.28	Virginia	39.9	2.71
Nebraska	53.1	3.52	Idaho	39.0	4.77
Mississippi	52.6	4.22	Hawaii	38.7	3.29
Rhode Island	52.4	2.38	Oregon	38.1	3.38
Wisconsin	50.2	3.11	Georgia	37.6	3.48
Missouri	50.1	3.45	Colorado	35.9	3.16
South Dakota	50.0	3.68	New Hampshire	35.1	3.02
Tennessee	49.7	3.29	Alaska	30.3	3.30
Washington	48.9	3.34	Maryland	27.0	2.71
Connecticut	48.8	2.87			

13.39 Some child development researchers tabulated the amount of crying by 38 babies, aged four to seven days, and compared these cry counts to each baby's measured IQ at three years of age:[11]

Cry Count	IQ Score	Cry Count	IQ Score
9	103	20	132
9	119	20	90
10	87	19	120
10	109	19	103
12	94	18	112
12	97	18	109
12	103	18	109
12	119	17	141
12	120	17	94
13	104	16	136
13	162	33	159
14	106	31	135
15	112	30	155
15	114	27	108
15	133	23	113
16	100	23	103
16	106	22	157
16	118	22	135
16	124	21	114

Plot these points on a scatter diagram and use least squares to estimate the regression line $Y = \alpha + \beta X + \epsilon$, where Y is IQ and X is the cry count. Is the relationship statistically significant at the 5 percent level? If there is a positive, statistically significant relationship, does this mean that you should make your baby cry?

13.40 In 1987, Mark Hulbert graded the annual performances of the ten largest stock market newsletters:[12] The data are given in Table 13.6. Convert each of these letter grades to a numerical score (A = 4, B = 3, C = 2, D = 1 and F = 0) and compare the scores with those of a year earlier (1986 with 1985, 1985 with 1984, and so on) by estimating the equation $Y = \alpha + \beta X + \epsilon$, where Y = letter's score and X = letter's score a year earlier. You will estimate one equation with 40 observations. Is the relationship statistically significant at the 5 percent level? What is the predicted 1987 score (and a 95 percent prediction interval) for a fund with a score of 2 in 1986? Summarize your results in two sentences.

13.41 In Table 13.7, there are some 1960 data for 43 states and the District of Columbia on annual cigarette consumption (estimated from state cigarette taxes) and deaths from lung cancer (the remaining states did not levy cigarette taxes).[13] Letting cigarette consumption be the independent variable and lung cancer deaths the dependent variable, estimate a least squares regression equation to see if the relationship is statistically significant at the 1 percent level. What is the equation's predicted lung cancer death rate for a per capita cigarette consumption of 2,000? Of 3,000? Does this difference in predicted death rates seem substantial?

13.42 In the standard regression model, what happens to the standard deviation of $Y_0 - \hat{Y}_0$ as the number of observations n becomes very large? What is the explanation?

Table 13.6   *Data for Exercise 13.40*

Newsletter	1982	1983	1984	1985	1986
1	A	A	D	A	B
2	A	A	D	A	B
3	C	A	B	B	D
4	C	C	C	B	B
5	B	F	B	A	C
6	D	B	C	B	B
7	A	B	D	D	C
8	D	D	C	C	D
9	B	F	F	D	D
10	F	F	B	F	F

Table 13.7 *Data for Exercise 13.41*

	Cigarettes per Capita	Deaths per 100,000		Cigarettes per Capita	Deaths per 100,000
Alabama	1,820	17.05	Missouri	2,756	20.98
Alaska	3,034	25.88	Montana	2,375	19.50
Arizona	2,582	19.80	Nebraska	2,332	16.70
Arkansas	1,824	15.98	Nevada	4,240	23.03
California	2,860	22.07	New Jersey	2,864	25.95
Connecticut	3,110	22.83	New Mexico	2,116	14.59
Delaware	3,360	24.55	New York	2,914	25.02
District of Columbia	4,046	27.27	North Dakota	1,996	12.12
Florida	2,827	23.57	Ohio	2,638	21.89
Idaho	2,010	13.58	Oklahoma	2,344	19.45
Illinois	2,791	22.80	Pennsylvania	2,378	22.11
Indiana	2,618	20.30	Rhode Island	2,918	23.68
Iowa	2,212	16.59	South Carolina	1,806	17.45
Kansas	2,184	16.84	South Dakota	2,094	14.11
Kentucky	2,344	17.71	Tennessee	2,008	17.60
Louisiana	2,158	25.45	Texas	2,257	20.74
Maine	2,892	20.94	Utah	1,400	12.01
Maryland	2,591	26.48	Vermont	2,589	21.22
Massachusetts	2,692	22.04	Washington	2,117	20.34
Michigan	2,496	22.72	West Virginia	2,125	20.55
Minnesota	2,206	14.20	Wisconsin	2,286	15.53
Mississippi	1,608	15.60	Wyoming	2,804	15.92

• **13.43** A researcher has a model $Y = \alpha + \beta X + \epsilon$, where the error term ϵ has a zero expected value and is independent of X. Show that the least squares estimates of α and β are unbiased, that $E[a] = \alpha$ and $E[b] = \beta$.

• **13.44** The least squares estimates of α and β in the equation $Y = \alpha + \beta X + \epsilon$ are not uncorrelated. If we knew that our estimate of β was too high, we would also want to revise our estimate of α. Would we want to revise it upward or downward?

•• **13.45** (*continuation*) Find the covariance $E[(a - \alpha)(b - \beta)]$. Does its sign agree with your intuition?

•• **13.46** Researcher A believes that egg production E depends on the number of chickens C:

$$E = \alpha + \beta C + \epsilon$$

Researcher B argues that the number of chickens C depends on the number of eggs E:

$$C = -(\alpha/\beta) - (1/\beta)E - (\epsilon/\beta)$$

If they each run a regression using the same data, will they agree on their estimate of β? Will they get the same correlation coefficient?

• **13.47** Two statisticians, who are working independently with different data, each used 10 observations to calculate least squares estimates of the equation $Y = \alpha + \beta X + \epsilon$. The values of the sample means $\overline{X}$ and $\overline{Y}$ are the same in each data set. We'll call the first statistician's estimate of the slope b_1 and the second statistician's estimate b_2. If they were to get together and pool their data, they would obtain a least squares estimate b. **a.** How is b related to b_1 and b_2? **b.** Is b the same as the average of the two separate estimates, $(b_1 + b_2)/2$?

•• 13.48 (*continuation*) Find the expected value and variance of the pooled estimate b and of the average estimate $(b_1 + b_2)/2$. Which would you use?

• 13.49 An economist regressed Gross National Product (Y) on employment (X) and got an R-squared of .99. Another economist used the same data, but regressed the logarithm of Y on the logarithm of X and got an R-squared of only .95. Should we conclude that the first economist has the right formulation or, at least, fits the data better? If not, how would you compare the fit of these two models?

•• 13.50 There is a linear relationship between Y and X

$$Y = \alpha + \beta X + \epsilon$$

The errors ϵ are independent, with mean 0 and a constant variance σ^2. Three estimators of β have been proposed:

1. least squares
2. the average change in Y divided by the average change in X

$$\frac{\{(Y_2 - Y_1) + (Y_3 - Y_2) + \cdots + (Y_n - Y_{n-1})\}/n}{\{(X_2 - X_1) + (X_3 - X_2) + \cdots + (X_n - X_{n-1})\}/n}$$

3. the slope of a line between the first data point and the last data point

$$\frac{Y_n - Y_1}{X_n - X_1}$$

Find the expected values and variances of these three estimators. **a.** Which are unbiased? **b.** Which has the smallest variance?

13.51 For each of the following pairs, explain why you believe that they are positively related, negatively related, or essentially unrelated; if related, identify the dependent variable. **a.** Among students in a chemistry class, hours spent studying and course grade **b.** Among boys under 18, height and time for 100-meter dash **c.** Among men over 30, age and number of hairs on head **d.** Among women over 30, age and the number of grandchildren.

13.52 For each of the following pairs, explain why you believe that they are positively related, negatively related, or essentially unrelated. If they are related, identify the dependent variable. **a.** The price of corn and the price of Iowa farmland **b.** The

cost of constructing a house and its market price **c.** Stock prices and stock dividends **d.** The probability of a bond defaulting and its market interest rate.

13.53 For each of the following pairs, explain why you believe that they are positively related, negatively related, or essentially unrelated; if related, identify the dependent variable. **a.** The rate of increase in a nation's money supply and its rate of inflation **b.** A nation's rate of inflation and interest rates **c.** The productivity of a nation's workers and their wages **d.** A nation's birth rate and the average age of its citizens.

13.54 A regression of Y = cricket chirps per minute for one cricket species on X = temperature in degrees Fahrenheit gave this estimated equation:[14]

$$\hat{Y} = -137.2 + 3.777\,X$$

Which is the dependent and which the independent variable? Does it seem more plausible to you that X influences Y or that Y influences X? The temperature in degrees Celsius Z is related to the temperature in degrees Fahrenheit X by $Z = (5/9)(F - 32)$. What estimated coefficients do ou predict would be obtained if we estimated the regression equation $Y = \alpha + \beta Z + \epsilon$?

13.55 (*continuation*) What is the predicted value of Y when $X = 0$? When $X = 300$? Do these predictions seem plausible? Can we judge the reasonableness of the model by the plausibility of these predictions?

13.56 Following are some data on cricket chirps (Y) and temperature (X) for the striped ground cricket.[15] Draw a scatter diagram; does the relationship seem linear? Estimate the equation $Y = \alpha + \beta X + \epsilon$ by ordinary least squares and draw this fitted line in your scatter diagram. Compare your results with the equation reported in Exercise 13.54.

Cricket Chirps per Second	Temperature (°Fahrenheit)
20.0	88.6
16.0	71.6
19.8	93.3
18.4	84.3
17.1	80.6
15.5	75.2
14.7	69.7

Cricket Chirps per Second	Temperature (°Fahrenheit)
17.1	82.0
15.4	69.4
16.2	83.3
15.0	79.6
17.2	82.6
16.0	80.6
17.0	83.5
14.4	76.3

13.57 *(continuation)* If we were to use the frequency of cricket chirping to predict the temperature, we might estimate the equation $X = \gamma + \delta X + \epsilon$. Use the data in Exercise 13.56 to estimate this equation. Is your estimate of δ equal to one divided by your estimate of β? Explain why or why not.

13.58 A study of the effect of driving speed on traffic fatalities proposed estimating the following model:

$$Y = \alpha + \beta X = \epsilon$$

where
Y = deaths per million miles driven
X = average speed, miles per hour

The researcher argued that "supporters of lower speed limits would expect β to be close to one, reflecting a strong, direct effect of X on Y." What is wrong with this argument?

She collected the following data:

Year	X	Y
1980	57.0	1.6
1984	59.2	1.4
1985	59.5	1.3

Do you think that her ordinary least squares estimate of β will be positive or negative? Why would you be cautious in interpreting her results?

13.59 A zero-coupon bond (a zero) pays a specified amount at maturity and nothing before the bond's maturity. Financial theory teaches that the sensitivity of a zero's price to a general rise in interest rates is proportional to its maturity. Use the following data[16] on the standard deviation of the prices of U.S. Treasury zeros during the years 1984–1987 to estimate the equation $Y = \alpha + \beta X = \epsilon$, where Y = standard deviation and X = maturity. Does the estimated coefficient of X have the right sign? Does it seem substantial? Is its value statistically significant at the 5 percent level?

Maturity (Years)	Standard Deviation (%)
5	10.8
10	20.7
15	28.8
20	38.1

13.60 *(continuation)* Give a 95 percent confidence interval for β. Does your estimate of β reject at the 5 percent level the null hypothesis that $\beta = 1$?

13.61 A nation's monetary base consists of the reserves held by banks and the currency outside of banks; M1 consists of currency outside banks and checkable deposits. The monetary base and M1 are linked in the United States (and in many other countries) because banks are required to hold reserves equal to a specified fraction of their checkable deposits (approximately 12 percent in the United States) and hold very little additional reserves. This linkage is not perfect because the ratio of M1 to the monetary base is affected by changes in bank reserves relative to deposits and by the public's holding of currency relative to deposits. To see whether there is a close relationship between changes in the U.S. monetary base and M1, use the data shown below to estimate the regression equation $Y = \alpha + \beta X + \epsilon$, where Y = annual percentage change in M1 and X = annual percentage change in the monetary base. Is the estimated relationship positive? Substantial? Statistically significant?

Year	Annual Percentage Changes		
	Monetary Base	M1	M2
1980	8.2%	6.8%	8.9%
1981	5.6	6.5	10.0
1982	7.4	8.5	8.8
1983	8.8	9.6	11.8
1984	7.2	5.7	8.3
1985	8.7	12.4	8.5
1986	10.0	17.0	9.5
1987	6.9	3.6	3.5
1988	6.8	4.7	5.6
1989	3.5	0.9	4.9

13.62 *(continuation)* Use the data in Exercise 13.61 to make a scatter diagram with the percentage change in the monetary base on the horizontal axis and the percentage change in M1 on the vertical axis. Does the relationship look close or loose? Move five of the points to give ten hypothetical observations that appear to you to have an R^2 of about .9. Read the values of these ten observations from the axes and make a table showing these ten observations. Now use these data to estimate the equation $Y = \alpha + \beta X + \epsilon$ and see if the value of R^2 is approximately .9.

13.63 *(continuation)* Redo Exercise 13.62, this time moving three of the points in the original scatter diagram to obtain ten observations that appear to have an R^2 of about .0.

13.64 *(continuation)* The monetary aggregate M2 is a broad measure of money that includes not only currency and checkable deposits but also such items as savings accounts, small time deposits, and money market funds. Use the data in Exercise 13.61 to estimate the regression equation $Y = \alpha + \beta X + \epsilon$, where Y = annual percentage change in M2 and X = annual percentage change in the monetary base. Is the estimated relationship positive? Substantial? Statistically significant?

13.65 *(continuation)* As judged by R^2, is there a closer relationship between the monetary base and M1, the monetary base and M2, or M1 and M2?

13.66 In theory, a stock's price-earnings (P/E) ratio should be high if the company's earnings are expected to increase rapidly in the future. Two finance professors calculated the average five-year earnings growth rates forecast by professional security analysts for 900 companies in 1983.[17] They then divided these companies into five groups, ranging from those companies with the highest predicted growth rates to those with the lowest, and calculated the average P/E ratio for the stocks in each group. They repeated these calculations in 1984 and 1985. The following table averages their results. Use these data to estimate the regression equation $Y = \alpha + \beta X + \epsilon$, where Y = P/E and X = predicted growth rate. Does the estimate of β have the anticipated sign? Is the estimate of β statistically significantly different from zero at the 5 percent level? Does the estimate of β seem substantial to you?

Stock Group	Average Predicted Growth Rate	Average P/E Ratio
1	27.3%	15.7%
2	21.0	12.4
3	18.1	11.4
4	15.5	10.1
5	9.3	7.7

13.67 The December 1987 issue of *Road and Track* magazine has data for 108 cars on their price (X) and the time (Y) in seconds for each car to accelerate from 0 to 60 miles per hour. The average value of X was $23,429, and the average value of Y was 8.9 seconds. Interpret and evaluate this least squares regression:[18]

$$\hat{Y} = 10.31 - 0001\, X, \qquad R^2 = .32$$
$$[38.6] \qquad [7.0]$$

[ ]: *t*-values

13.68 A random sample of college students were asked to rate their courses on a scale of 1 to 10 (10 being best) and to state their grade in the course on a scale of 1 to 12 (A = 12, A− = 11, and so on).[19] The average rating was 5.53 and the average grade 9.69. A least squares regression of rating (Y) on grade (X) yielded

$$\hat{Y} = -2.14 + 0.79\, X$$
$$(2.91) \qquad (.296)$$

( ): standard deviations

What does a least squares regression tell us that a comparison of the average values, 5.53 versus 9.69, does not? How do you interpret the estimate −2.14? How do you interpret the estimate 0.79? Is the estimated relation statistically significant at the 1 percent level? Do you think that grades influence ratings or that ratings influence grades? What difference does it make to a regression equation?

13.69 Some people believe that the psychological stress that accompanies major events (such as a new job, divorce, or death of a relative) can affect one's health. A number of patients hospitalized for 17 various chronic ailments ranging from dandruff to cancer were asked if they had experienced any of 42 specified major events during the preceding two years.[20] Each patient was given a numerical score based on the frequency and severity of these events, and the average event score X was calculated for patients with each identified ailment. Each of the 17 ailments was given a medical score Y based on its severity—17 for dandruff, 468 for arthritis, 1,020 for cancer. Use the data in the following table to estimate the regression equation $Y = \alpha + \beta X + \epsilon$. Is the relationship statistically significant? What is the value of R^2?

X	Y	X	Y
26	21	405	520
130	173	599	621
317	174	357	636
231	204	688	688
325	312	443	733
816	393	609	776
563	454	772	824
312	468	777	1020
603	500		

13.70 Some economists believe that a one-percentage-point increase in the rate of inflation X tends to increase interest rates Y by one percentage point. Table 8 at the end of this book gives forty years of annual data on the nominal interest rate on one-year Treasury bills and the rate of inflation. Use these data to estimate the equation $Y = \alpha + \beta X + \epsilon$. Is there a statistically significant (at the 5 percent level) effect of inflation on interest rates? Does the estimated magnitude of β seem plausible to you? Judging from the value of R^2, would you say that there is a tight or loose relationship between inflation and interest rates?

13.71 *(continuation)* Use the regression equation estimated in the preceding exercise to find a 95 percent confidence interval for β. Do these data reject at the 5 percent level the null hypothesis that $\beta = 1$?

13.72 Soccer enthusiasts have long debated how to decide tournament games in which the score is a tie but a winner must be determined. If the game is allowed to continue until one of the teams scores a deciding goal, this team may be too tired to play well in the next round; yet other criteria (such as a succession of penalty kicks) seem arbitrary. Some have argued that a team's dominance can be measured by the number of corner kicks it is awarded during the game. To test this theory, a student estimated this equation using data from the 17 games played by the Pomona-Pitzer men's soccer team during the 1987 season:[21]

$$\hat{Y} = 0.376 + 0.200X, \qquad R^2 = .147$$
$$(0.594) \quad (0.125)$$

where
- Y = goals by Pomona minus goals by opponent (there were no ties.); average value 0.59
- X = corner kicks by Pomona minus corner kicks by opponent; average value = 1.06
- ( ) = standard deviation

Write a paragraph summarizing his results.

13.73 A study of velocity and distance of certain extra-galactic nebulae used the data in Table 13.8.

Table 13.8 *Data for Exercise 13.73*

Constellation in Which Nebula Is Situated	Mean Velocity Y (Kilometers/Second)	Distance X (Millions of Parsecs)
Isolated Nebula II	630	1.20
Virgo	890	1.82
Isolated Nebula I	2,350	3.31
Pegasus	3,810	7.24
Pisces	4,630	6.92
Cancer	4,820	9.12
Perseus	5,230	10.97
Coma	7,500	14.45
Ursa Major	11,800	22.91
Leo	19,600	36.31

Plot these data and then determine the least squares estimates of the slope and intercept for the equation $Y = \alpha + \beta X + \epsilon$. Interpret your estimates.

13.74 To investigate the relationship between federal budget deficits and interest rates, the following equation was estimated by least squares, using quarterly data for the period 1970–1983:[22]

$$\hat{Y} = 0.103 - 0.079\,X, \quad SEE = .523, \quad R^2 = .00$$
$$\quad\;\;(0.092) \quad (0.079)$$

where Y = interest rate on Aaa bonds; average value about 10 percent

   X = federal deficit as a percent of gross national product; average value about 3 percent

   ( ) = standard deviation

a. Give a 95 percent confidence interval for the effect of X on Y.   **b.** Is the estimated effect of X on Y statistically significant at the 5 percent level?   **c.** Is this estimated effect substantial?   **d.** It is widely believed that an increase in the federal deficit increases interest rates. Do these results disprove that belief?   **e.** What does SEE measure?   **f.** What does R^2 measure?

13.75 A researcher compared the 1960 regular season and World Series batting averages of those players on the New York Yankees and the Pittsburgh Pirates with at least 200 times at bat during the regular season.[23] The data are given in Table 13.9. For each team, see whether a player's regular season batting

average X is a useful predictor of the player's World Series batting average Y by estimating the regression equation $Y = \alpha + \beta X + \epsilon$. Are the results for each team similar or dissimilar?

13.76 *(continuation)* Redo the two regressions in the preceding exercise, this time omitting the three players on each team who had the fewest times at bat during the World Series: Lopez, Howard, Boyer, Nelson, Smith, and Skinner. Do your results change substantially?

13.77 Redo Exercise 3.75, using data for the most recent World Series. These data will be in a book such as *The Baseball Encyclopedia.*

13.78 Table 1.9 in Chapter 1 gives the annual percentage returns for four assets from 1950 through 1988. Do you think that the returns from long-term Treasury bonds and high-grade corporate bonds are positively or negatively correlated? Do you think that the correlation coefficient is closer to 0 or 1? See if your prediction is correct by estimating a regression equation, letting X be the return on long-term Treasury bonds and Y the return on high-grade corporate bonds.

13.79 *(continuation)* Do you think that the returns from long-term Treasury bonds and corporate stock are positively or negatively correlated? Do you think that the correlation coefficient is closer to 0 or 1? Check your prediction by estimating a regression equation, letting X be the return on long-term Trea-

Table 13.9   *Data for Exercise 13.75*

	Yankees			Pirates	
	Regular Season	*World Series*		*Regular Season*	*World Series*
B. Skowron	.309	.375	D. Groat	.325	.214
H. Lopez	.284	.429	R. Clemente	.314	.310
R. Maris	.283	.267	R. Nelson	.300	.333
Y. Berra	.276	.318	H. Smith	.295	.375
M. Mantle	.275	.400	S. Burgess	.294	.333
T. Kubek	.273	.333	D. Hoak	.282	.217
G. McDougald	.258	.278	B. Mazeroski	.273	.320
B. Richardson	.252	.367	B. Skinner	.273	.200
B. Cerv	.250	.357	G. Cimoli	.267	.250
E. Howard	.245	.462	B. Virdon	.264	.241
C. Boyer	.242	.250	D. Stuart	.260	.150

sury bonds and Y the return on NYSE corporate stock.

13.80 *(continuation)* Use R^2 to gauge whether the returns on corporate stock are more closely correlated with the return on short-term Treasury bills or long-term Treasury bonds.

13.81 The theory of purchasing power parity implies that the percentage change in a nation's exchange rate relative to another nation's currency (for example, yen/dollar) is equal to the percentage change in that country's prices relative to prices in the second country. Thus if prices rise by 3 percent in Japan and by 5 percent in the United States, the yen/dollar exchange rate should change by .03 − .05 = −.02, representing a 2 percent increase in the value of yen relative to the dollar. Test this theory by using the following data (for the period 1974–1988) to estimate the regression equation $Y = \alpha + \beta X = \epsilon$, where $Y =$ change in exchange rate relative to the U.S. dollar and $X =$ percentage change in consumer prices relative to U.S. consumer prices. Write a paragraph summarizing your results.

Country	Change in Prices Relative to U.S. Prices	Change in Exchange Rate Relative to Dollar
Australia	51.2%	84.8%
Austria	−22.1	−34.0
Belgium	−6.5	−5.6
Canada	13.6	25.9
Denmark	23.6	10.5
Finland	38.8	11.2
France	31.4	23.9
Ireland	90.0	53.9
Italy	136.8	100.3
Japan	−25.47	−56.1
Netherlands	−24.3	−26.5
New Zealand	134.2	116.0
Norway	33.2	18.0
South Korea	84.6	80.5
Spain	138.1	102.3
Sweden	33.0	38.3
Switzerland	−35.9	−50.9
United Kingdom	61.8	31.6
West Germany	−34.1	−32.1

13.82 *(continuation)* Give a 99 percent confidence interval for β (i.e., for the effect on Y of a 1-point increase in X). Give a 99 percent confidence interval for 5β (i.e., for the effect on Y of a 5-point increase in X). Compare these two confidence intervals.

13.83 *(continuation)* Draw a scatter diagram, using the data in Exercise 13.81, and label each point with the country's name. Draw the regression line on this scatter diagram. Does the relationship look approximately linear or obviously nonlinear?

13.84 *(continuation)* Looking at the scatter diagram drawn in Exercise 13.83, do you think that the fitted line will change substantially if Japan and South Korea are omitted from the regression? Reestimate the equation, using the data for the other 17 countries, and see if you are right.

13.85 *(continuation)* Looking at the scatter diagram drawn in Exercise 13.83, do you think that the fitted line will change substantially if New Zealand, Spain, and Italy are omitted from the regression? Reestimate the equation, using the data for the other 16 countries, and see if you are right.

13.86 How do you suppose that least squares coefficient estimates, standard deviations, t-values, and R^2 would be affected by the inclusion of one additional observation, with the values of X and Y each equal to the respective average values of X and Y?

13.87 *(continuation)* Use the data in Exercise 13.81 to test your theory, estimating the equation once with the original 19 observations and once with a 20th observation that is equal to the average values of X and Y.

13.88 How do you suppose that least squares coefficient estimates, standard deviations, t-values, and R^2 would be affected by a doubling of the number of observations, with the new data exactly replicating the original data? Use the data in Exercise 13.81 to test your theory.

• **13.89** The seismic energy E (in ergs) released by an earthquake of magnitude M on the Richter scale is given by this estimated empirical equation[24]

$$\log[E] = 11.4 + 1.5M$$

In the spring of 1990 a 5.5 earthquake rocked Claremont, California. How much energy was released by this earthquake in comparison with 8×10^{20} ergs released by the atomic bomb dropped on Hiroshima? How much energy is released by a magnitude 8.3 earthquake (such as hit San Francisco in 1906), in comparison with a magnitude 5.5 earthquake?

13.90 The text discusses the use of regression analysis to estimate Western Union's cost of equity. A rebuttal witness argued that truckers and airlines should be excluded from this analysis because, unlike the other groups, they do not have a territorial monopoly. Reestimate the regression equation using data for the six groups other than truckers and airlines. Does their exclusion make much difference to the results?

13.91 Estimate the parameters in the linear equation $Y = \alpha + \beta X = \epsilon$, using the following data:

X	1.0	2.0	3.0	4.0	5.0	30.0
Y	0.8	1.9	3.0	4.1	5.2	3.0

Now plot these data and draw in your fitted line. Explain why you either do or do not think that the fitted line gives an accurate description of the relationship between X and Y.

13.92 A researcher recently reported that his investigation of the relationship between advertising expenditures and sales obtained "a t-value much larger than 2." Why would he be interested in knowing his t-value?

13.93 James Tobin argued that the stock of very profitable firms tends to have a high market value relative to the firm's book value. Use the data in Exercise 1.65 in Chapter 1 to test this theory by estimating the regression equation

$$Y = \alpha + \beta X = \epsilon$$

where Y is the value of Tobin's q and X is the firm's return on equity. Are these data consistent or inconsistent with Tobin's theory? What criteria did you use to answer that question?

13.94 An economics professor gives her students a final exam consisting of 100 multiple-choice questions. In her grade book, she writes down the number of questions Y that each student got right and the number X that each student got wrong. If she uses these data to estimate the equation $Y = \alpha + \beta X + \epsilon$, what will be her estimates of α and β, and the value of R^2?

13.95 Table 1.11 in Chapter 1 shows the percent discount from net asset value for 15 closed-end stock investment companies. Just by looking at these data, does there seem to be a positive or negative relationship between the size of the fund (as measured by its net asset value) and the size of its percent discount? Make a scatter diagram with net asset value on the horizontal axis and percent discount on the vertical axis and answer this same question. Now fit a line to these data by estimating the equation $Y = \alpha + \beta X = \epsilon$ where Y is percent discount and X is the fund's net asset value. Is the estimate of β positive or negative? Is it statistically significantly different from zero at the 5 percent level? Does the value of R^2 seem large or small to you? Sketch the estimated equation on your scatter diagram.

13.96 Following is a table of values for X and Y with seven observations missing. Copy this table and make a scatter diagram, filling in the seven missing values in the table as you plot the points, trying to obtain eight observations that will given an R^2 of approximately .5. When you have your eight hypothetical observations, calculate R^2 and see how close it is to .5.

X	1	2	2	4	5	7	7	8
Y	3							

13.97 *(continuation)* Repeat Exercise 13.96, this time trying to obtain an R^2 of .9. Do this exercise once again, this time aiming for an R^2 of 1.0.

13.98 *(continuation)* Repeat Exercise 13.96, trying for an R^2 of .0.

13.99 *(continuation)* What do you think the value of R^2 is if the first seven values of Y are 3 and the eighth value is 3.01? Calculate its value and see if you are right.

13.100 *(continuation)* In Exercise 13.98 there are two observations with $X = 2$; did you use the same Y-value for each of these two observations? There are also two observations with $X = 7$; did you use the same Y-value for each of these two observations? Explain why the values of Y either must be the same or need not be the same when the values of X are the same.

14

The Practice of Regression Analysis

"The cause of lightning," Alice said very decidedly, for she felt quite sure about this, "is the thunder—no, no!" she hastily corrected herself, "I meant it the other way."

"It's too late to correct it," said the Red Queen, "When you've once said a thing, that fixes it, and you must take the consequences."

Lewis Carroll
Alice in Wonderland

TOPICS

Regression analysis is one of the most powerful and useful statistical tools. This chapter will examine several detailed applications, to show the variety of questions that regression analysis can answer. You will see just how simple and yet powerful a tool it is; you will also be warned of several pitfalls that endanger an incautious user of regression analysis.

14.1 IS A FINAL EXAM NECESSARY?

Professors generally see a consistency in student scores. Those students who do well on the midterm usually do well on the final examination, while those who do poorly on the midterm have trouble with the final exam, too. If this relationship were perfect, professors could dispense with final exams, pleasing everyone (except those optimistic students who think that their low midterm scores were an accident).

Table 14.1 shows some data on midterm (X) and final (Y) scores for one randomly selected statistics class. Figure 14.1 shows how I entered these data into two

Table 14.1 *Midterm and Final Exam Scores*

Student	Midterm	Final
AB	76	81
CC	75	84
GC	66	81
CD	61	62
HG	72	60
JG	56	69
BH	72	82
CH	85	82
FH	63	85
KK	60	53
CL	80	81
JL	25	49
SM	66	84
BM	76	56
DO	89	83
RP	59	70
SR	78	65
AS	68	91
OS	40	48
MS	82	86
CT	77	88
JV	76	83
MW	90	94
VW	76	79
LW	56	67

```
   MTB > read    c1    c2
DATA >         76    81
DATA >         75    84
DATA >         66    81
DATA >         61    62
DATA >         72    60
DATA >         56    69
DATA >         72    82
DATA >         85    82
DATA >         63    85
DATA >         60    53
DATA >         80    81
DATA >         25    49
DATA >         66    84
DATA >         76    56
DATA >         89    83
DATA >         59    70
DATA >         78    65
DATA >         68    91
DATA >         40    48
DATA >         82    86
DATA >         77    88
DATA >         76    83
DATA >         90    94
DATA >         76    79
DATA >         56    67
DATA > end
      25 ROWS READ
```

Enter Data

```
MTB >name c1 'FINAL' c2 'MIDTERM'
MTB >regress c1 1 c2
MTB >predict 70;
MTB >predict 90.
```

Specify Regression

```
The regression equation is
FINAL = 31.308 + 0.627 MIDTERM
```

Predictor	Coef	Stdev	t-ratio	P
Constant	31.308	9.8612	3.18	0.001
MIDTERM	0.6270	0.1403	4.48	0.000

```
s=10.090     R-sq=46.6%
```

Fit	Stdev.Fit	95% P.I.	
75.20	10.29	(53.90,	96.49)
87.74	10.70	(65.59,	109.88)

Regression Results

Figure 14.1 *Excerpts from a Least Squares Regression of Final Exam Scores on Midterm Scores*

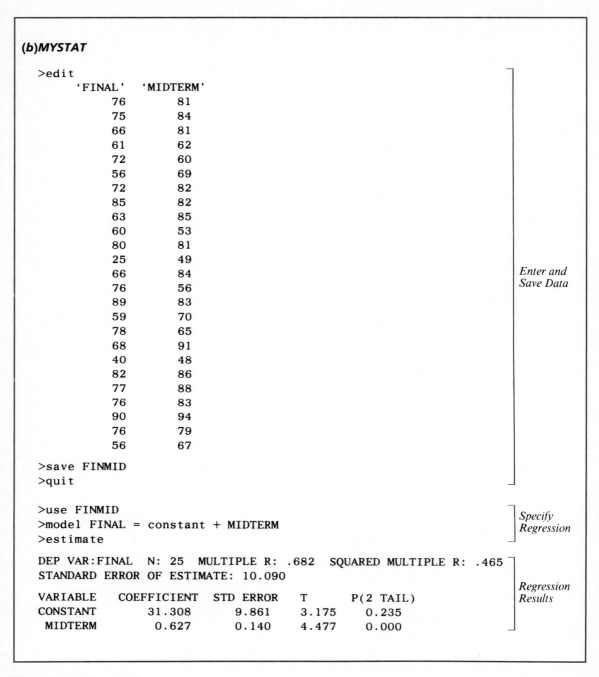

(b)MYSTAT

```
>edit
     'FINAL'   'MIDTERM'
          76        81
          75        84
          66        81
          61        62
          72        60
          56        69
          72        82
          85        82
          63        85
          60        53
          80        81
          25        49
          66        84
          76        56
          89        83
          59        70
          78        65
          68        91
          40        48
          82        86
          77        88
          76        83
          90        94
          76        79
          56        67

>save FINMID
>quit

>use FINMID
>model FINAL = constant + MIDTERM
>estimate

DEP VAR:FINAL  N: 25  MULTIPLE R: .682  SQUARED MULTIPLE R: .465
STANDARD ERROR OF ESTIMATE: 10.090

VARIABLE    COEFFICIENT   STD ERROR    T       P(2 TAIL)
CONSTANT       31.308       9.861    3.175     0.235
 MIDTERM        0.627       0.140    4.477     0.000
```

Enter and Save Data

Specify Regression

Regression Results

Figure 14.1   *Continued*

different computer programs—MINITAB and MYSTAT. After the data are entered, the computer makes the necessary (but tedious) calculations very quickly and prints out the regression results shown in Figure 14.1.

I have plotted the estimated equation

$$\hat{Y} = 31.3 + 0.627\, X$$

together with the data in Figure 14.2. There is a definite positive relationship, but the slope is not 1.0 and the fit is not perfect. The biggest deviation from the fitted line is the fourteenth student, who got a 76 on the midterm and was predicted to get a 79 on the final, but only earned a score of 56—a 23-point shortfall! On the other side is the eighteenth student, who got a 68 on the midterm and was predicted to get a 74 on the final, but actually got the second-highest grade on the final, 91, a full 17 points better than expected. The predicted final exam score was off by more than 5 points for 14 of these 25 students. Apparently, final exams do contain useful information, because student scores are not predicted very accurately from midterm performance.

The computer output shows the standard error of estimate (a fairly large 10.090), the R-squared (0.466, not nearly 1.0), and the standard deviations and t-values for the estimated coefficients. Even though final exam scores cannot be predicted perfectly from midterm scores, there is a strong statistical relationship between the two. The 4.477 t-value convincingly rejects the null hypothesis that midterm scores are of no value in predicting final exam scores.

In the MINITAB computer program, I used the estimated equation to predict the values of Y for $X = 70$ and $X = 90$. You can see from the wide prediction intervals given in response that midterm scores are far from a precise predictor. For a student who scores 70 on the midterm, the regression equation predicts a score of 75.2 on the final, with a 95 percent prediction interval of ± 20 points. We can

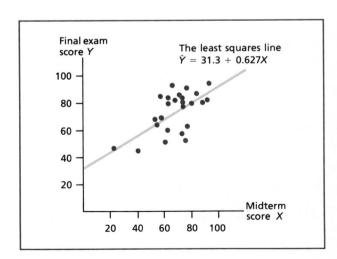

Figure 14.2 *A Scatter Diagram of the Relationship Between Midterm and Final Exam Scores*

only say that there is .95 chance that the interval 53.9 to 96.5 will encompass the final score. The prediction interval for $X = 90$ reminds us that we have to exercise some common sense when we get away from the middle of the data. Mathematically, the prediction interval is 65.6 to 109.9, but no student is going to score more than 100 points on my final, because only 100 points are possible.

Overall, the bottom line on this statistical exercise is that there is, as expected, a statistically significant relationship between midterm scores and final scores. A student who does well on the midterm can also be expected to do well on the final exam, but this relationship is far from perfect. Many students do far better or worse than might be predicted.

EXERCISES

14.1 A certain college wants to see if there is any relationship between math SAT scores and grades in a required freshman math course. A random sample of 10 students gave these data:

Math SAT X	Math Course Y
480	82
540	73
660	79
720	95
570	85
420	54
590	71
630	92
520	61
550	82

Plot these data and then use them to estimate the equation

$$Y = \alpha + \beta X + \epsilon$$

Sketch this least squares fitted line on your scatter diagram. **a.** What are the least squares estimates of α and β? **b.** What are the standard error of estimate and R-squared? **c.** Does there appear to be a positive relationship, a negative relationship, or no relationship at all between math SAT scores and course scores? **d.** Are math SAT scores a good predictor of math course scores? **e.** Could the college eliminate this math course and just give math grades on the basis of SAT scores?

14.2 A dean at SML college estimated the relationship between math SAT scores (X) and freshman math course scores (Y):

$$Y = 25.0 + 0.10\,X, \qquad R^2 = .21$$
$$(8.1) \quad (0.06)$$

( ): standard deviations

The average SAT score for these students was 580. **a.** What is the students' average score in this course? **b.** According to these least squares estimates, what is the predicted math course score for a student with a math SAT score of 500? Of 600? Of 700? **c.** If there is a 100-point difference in the math SAT scores of two students at this college, what is the predicted difference in their math course scores? **d.** Is this difference substantial? **e.** Is it statistically significant at the 5 percent level?

14.3 *(continuation)* A dean at BIG college also estimated the relationship between math SAT scores (X) and freshman math course scores (Y):

$$Y = 70.0 + 0.02\,X, \qquad R^2 = .74$$
$$(100.0)\ (0.005)$$

( ): standard deviations

The average math SAT score was 500 for these students. Answer the same five questions asked in the preceding exercise.

14.4 A certain college administrator is interested in seeing whether math or verbal SAT scores are the better predictor of performance in a required freshman history course. Here are data for a random sample of 20 students:

Math SAT X_1	Verbal SAT X_2	History Course Y
520	570	71
650	740	88
540	700	92
490	520	74
410	420	59
530	650	83
690	790	97
740	570	73
500	370	47
570	580	77
660	710	94
600	560	81
570	430	51
550	510	82
790	760	91
640	640	85
530	480	68
650	610	79
490	520	72
720	620	87

Use these data to obtain the least squares estimates of each of these equations

$$Y = \alpha_1 + \beta_1 X_1 + \epsilon_1 \quad \text{and} \quad Y = \alpha_2 + \beta_2 X_2 + \epsilon_2$$

a. Is there a statistically significant relationship (at the 5 percent level) between math SAT scores and history course scores? Is it a positive or negative relationship? **b.** Is there a statistically significant relationship (at the 5 percent level) between verbal SAT scores and history course scores? Is it a positive or negative relationship? **c.** Is the math or verbal SAT score a better predictor of success in this history course?

14.5 A random sample of ten male soccer players yielded these data on heights and weights:

Height (Inches) X	Weight (Pounds) Y	Height (Inches) X	Weight (Pounds) Y
65	142	70	158
67	154	71	179
67	143	72	157
68	155	72	181
69	161	75	190

Use these data to estimate the equation

$$Y = \alpha + \beta X + \epsilon$$

where ϵ is assumed to be normally distributed with an expected value of zero and a variance σ^2. **a.** What are the least squares estimates of α and β? **b.** What are the standard error of estimate and R-squared? **c.** Test (at the 5 percent level) the null hypothesis that $\beta = 0$. **d.** Give a 95 percent confidence interval for β.

● 14.6 *(continuation)* Use your data and estimates for the preceding exercise to answer these questions: **a.** What is the predicted weight for a 72-inch tall soccer player? **b.** What is a 95 percent confidence interval for this weight? **c.** What is the predicted weight for an 80-inch tall soccer player? **d.** What is a 95 percent confidence interval for this weight?

14.2 THE CAPITAL ASSET PRICING MODEL

One of the primary lessons of portfolio theory is that risk-averse investors can reduce their risk by choosing diversified investment portfolios. A portfolio of several stocks can be safer than any individual stock in the portfolio. Think of each investment as a coin flip—heads you win 100 percent, tails you lose 50 percent. If you put all of your eggs in a single basket by investing in a single coin flip, then there is a .5 probability that you will lose 50 percent of your wealth. If instead you *diversify* by investing half of your wealth in one coin flip and half in a second flip, then you won't lose half of your wealth unless you lose both coin flips. Diversification reduces the probability of a 50 percent loss from .50 to .25. With 100 coin flips, the probability of losing half of your wealth drops to $(.5)^{100}$.

For a single flip, you will be either a big winner or a poor loser. But with many flips, you are almost certain to have about half heads and half tails. Bad luck with some investments will undoubtedly be offset by good luck with others. It is very unlikely that you will have bad luck with almost all of your investments.

This coin flip analogy assumes that the outcomes are independent, in that whether you have good or bad luck on one investment does not depend on how your other investments turn out. In practice though, when people buy stocks, bonds, and other assets, the investment outcomes are not independent, because most assets are affected to some extent by the overall state of the economy and by interest rates. Most stocks and bonds tend to go up and down together.

An extreme analogy is two bets on the same flip of a coin. If you bet "heads" with Jack and "heads" with Jill, too, there are no gains from diversification, because there is no chance that good luck on the second investment will offset bad luck on the first. The opposite, and equally extreme, situation is a perfect inverse relationship between the results of two investments: If you have bad luck on the first, then you are sure to have good luck on the second. For instance, you could bet "heads" with Jack and "tails" with Jill on a single flip. No matter which way the coin lands, you are sure to have a 100 percent gain on one investment and a 50 percent loss on the other—a guaranteed 25 percent return overall. This, in fact, is exactly the hedged bet that risk-averse bookies covet.

Most investments are not perfectly correlated, either positively or negatively. But the moral remains, that the advantages of diversification depend on the correlation between asset returns. There is little diversification to be gained from putting half of your wealth in one international oil company and half of your wealth in another. Investing in one oil company and one computer firm reduces risk much more effectively.

Overall, if the stock market is dominated by institutions that hold enormous portfolios, which stocks will be considered risky and which will be considered safe? The lesson of portfolio theory is that the riskiness of an asset depends on how correlated its return is with other assets in the portfolio. An asset that always does poorly when other assets are doing poorly is a risky asset, because it has no diversification potential. An asset whose return is independent of others or, even better, does well when others are doing poorly, reduces portfolio risk. Thus, the proper gauge of risk for portfolio managers is some measure of how an asset's return is related to the returns on other assets.

The *Capital Asset Pricing Model* (CAPM) formalizes these ideas. Suppose there are n assets, each with some (uncertain) percentage return r_i ($i = 1, 2, \ldots, n$). The overall "market" return on all n assets is r_M. Let's assume that there is a linear relationship between r_i and r_M:

$$r_i = \alpha_i + \beta_i r_M + \epsilon_i \qquad (14.1)$$

The parameter β_i measures the extent to which the return on this particular asset moves with the returns on all assets. The parameter α_i is the intercept of this relationship. The error term ϵ_i is a random variable, reflecting the fact that, no matter what happens to the market, the return on this particular asset may be pushed up or down by various unexpected developments, such as technological breakthroughs, the loss of a key executive, favorable tax changes, or legal setbacks. The i subscripts indicate that the parameters α_i and β_i and the error term ϵ_i vary from asset to asset.

Now consider a portfolio with n assets: $i = 1, 2, \ldots, n$. If a fraction λ_i of wealth is invested in asset i, then the overall percentage return on the portfolio will be

$$r = \lambda_1 r_1 + \lambda_2 r_2 + \cdots + \lambda_n r_n$$
$$= (\lambda_1 \alpha_1 + \cdots + \lambda_n \alpha_n) + (\lambda_1 \beta_1 + \cdots + \lambda_n \beta_n)r_M + (\lambda_1 \epsilon_1 + \cdots + \lambda_n \epsilon_n)$$
$$= \alpha + \beta r_M + \epsilon$$

The portfolio intercept α and slope β are weighted averages of the intercepts and slopes of the assets in the portfolio. The portfolio error term, too, is a weighted average of the underlying error terms. However, to the extent that these individual errors are independent, their weighted sum, ϵ, will be close to zero!

Intuitively, we can think of each ϵ_i as the flip of a coin—heads is a pleasant surprise ($\epsilon_i > 0$) and tails an unpleasant one ($\epsilon_i < 0$). With a large number of flips, the average outcome should be close to zero—no surprise at all.*

Thus, the return on a diversified portfolio depends on the portfolio's α and β, and on how well the market does. The critical uncertainty for a diversified portfolio is what r_M will be. A portfolio manager can't do anything about r_M, but can do something about the portfolio's β, which is what determines how sensitive the portfolio is to market fluctuations. A portfolio with a β of 1.0 is about as risky as the overall market. If the market return turns out to be 10 percent higher than expected, then this portfolio should do about 10 percent better than expected; if the market drops 10 percent, then this portfolio will probably drop about 10 percent, too. A portfolio with a β greater than 1.0, in contrast, is aggressive in that it has above-average risk. If the market return turns out to be either 10 percent better or 10 percent worse than expected, then a portfolio with a β of 2.0 should do 20 percent better or worse than expected. Finally, a portfolio with a β less than 1.0 is a conservative portfolio. Market fluctuations, up or down, of 10 percent only cause

---

*More formally, consider an equal amount invested in each asset, $\lambda_i = 1/n$, and let each error term have the same standard deviation σ. If the error terms are independent, then the variance of the overall ϵ is

$$\left(\frac{1}{n}\right)^2 \sigma^2 + \cdots + \left(\frac{1}{n}\right)^2 \sigma^2 = \left(\frac{1}{n}\right)\sigma^2$$

which approaches zero as n increases.

about 5 percent fluctuations in a portfolio with a β of .5. *Thus for portfolio managers, β is the appropriate measure of risk.* High-β stocks are risky because they give risky, high-β portfolios. Low-β stocks are safer, because they give low-β portfolios. It is not surprising, then, that portfolio managers routinely calculate the βs for their portfolios and also look at the βs of individual stocks that they consider putting into their portfolios.

Figure 14.3 shows a scatter diagram of the annual percentage returns (dividends plus capital gains) for Merrill Lynch stock and for the market as a whole over the period 1972 through 1980. A least squares regression yielded the following results:

$$\hat{r}_{ML} = -2.26 + 1.82\, r_M, \qquad \text{SEE} = 34.2, \qquad R^2 = .61$$
$$\quad\;\; [3.09] \quad\; [3.28]$$

[   ]: t-values

Merrill Lynch is the nation's largest brokerage firm and its success is heavily dependent on the course of the stock market. When the market rises, lots of eager inves-

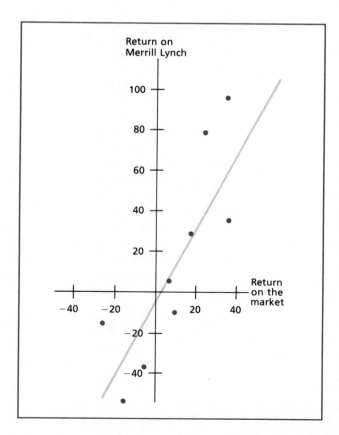

Figure 14.3 *The Returns on Merrill Lynch Stock and on the Market as a Whole*

tors buy stocks and pay commissions to Merrill Lynch. When stock prices slump, investors leave the market and Merrill Lynch suffers too. The 1.82 beta coefficient indicates that Merrill Lynch stock tends to go up or down almost twice as fast as the market as a whole. Notice though, that while this relationship is statistically significant, it is far from perfect. The scatter diagram, Standard Error of Estimate (SEE), and R-squared all tell us that, even if we knew how well the market was going to do, we still wouldn't know how well Merrill Lynch would do; other things matter, too.

Table 14.2 shows the data behind this regression and some similar data for three other stocks. IBM and General Mills are more typical companies that tend, on the average, to go up or down about as fast as the market as a whole: they have βs near 1. Least square regressions yielded these estimates:

$$\hat{r}_{IBM} = \begin{array}{cc} -2.84 & + 0.87\ r_M, \\ [0.59] & [4.08] \end{array} \qquad SEE = 1.31, \qquad R^2 = .70$$

$$\hat{r}_{GMills} = \begin{array}{cc} -0.62 & + 0.96\ r_M, \\ [0.08] & [2.88] \end{array} \qquad SEE = 20.7, \qquad R^2 = .54$$

[ ]: t-values

American Telephone (AT&T), on the other hand, was a heavily regulated utility over this period and the profits of utility companies depend more on regulatory commissions than the overall economy. Thus, the return on AT&T's stock is not as volatile as most companies (a lower variance) and is insulated from overall market swings (lower beta). For this particular period, the estimated beta coefficient

Table 14.2 *Some Annual Stock Returns, Percent Dividends Plus Capital Gains*

Year	Market	Merrill Lynch	IBM	General Mills	AT&T	Portfolio
1972	18.98	−8.0	21.1	57.4	23.9	23.6
1973	−14.66	−55.0	−21.9	−13.0	0.5	−22.5
1974	−26.47	−12.8	−29.7	−24.4	−4.5	−17.9
1975	37.20	37.2	37.4	50.4	21.6	36.7
1976	23.84	79.7	28.1	17.3	32.3	39.4
1977	−7.18	−36.7	1.3	−9.3	1.9	−10.7
1978	6.56	7.5	13.6	0.7	7.6	7.4
1979	18.44	31.2	−9.1	−11.8	−5.6	4.7
1980	32.42	98.3	10.8	13.1	1.4	30.9

The market data, giving the return on Standard and Poor's 500 stocks, is from Roger G. Ibbotson and Rex A. Sinquefield, *Stocks, Bonds, Bills, and Inflation: Historical Returns (1926–1978)* (Charlottesville, Virginia: Financial Analysts Research Foundation, 1989). The portfolio return is for an equal dollar investment in Merrill Lynch, IBM, General Mills, and AT&T.

turns out to be only 0.35:

$$\hat{r}_{AT\&T} = 5.34 + 0.35\ r_M, \qquad SEE = 12.2, \qquad R^2 = .31$$
$$[1.19]\quad[1.77]$$

[   ]: *t*-values

Now, let's look at a portfolio consisting of an equal investment in each of these four stocks. A well-diversified portfolio would have at least 20 or 30 carefully selected stocks, but this simple four-stock portfolio can hint at the effects of diversification. The portfolio return r_{port} is shown in Table 14.2 and a scatter diagram is shown in Figure 14.4. A least squares regression of this return on the market return yielded these estimates:

$$\hat{r}_{port} = 0.19 + 1.01\ r_M \qquad SEE = 9.3, \qquad R^2 = .87$$
$$[0.06]\quad[6.73]$$

[   ]: *t*-values

The portfolio β is the average of the βs for the securities in the portfolio; here it happens to be almost exactly 1.0. More importantly, the portfolio has a lower Standard Error of Estimate (SEE) and higher *R*-squared than any of the four stocks in the portfolio. For any given overall performance of the stock market, how well the portfolio does is more certain than the behavior of any of the individual stocks in the portfolio. This is why βs are so important for portfolio managers.

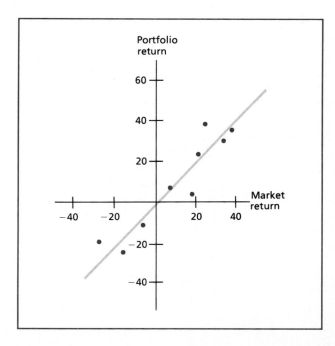

Figure 14.4 *A Portfolio of Merrill Lynch, IBM, General Mills, and AT&T*

Table 14.3 *Mutual Funds Have βs That Are Generally Consistent with Their Stated Objectives*

Investment Objective	Number of Funds	Average β
Maximum capital gain	117	1.23
Growth	85	1.04
Long-term growth and income	13	1.01
Growth and current income	31	0.99
Growth, income, and stability	14	0.90
Income and growth	12	0.85
Income, growth, and stability	12	0.86
Income, stability, and growth	8	0.71
Income	20	0.71
Stability, income, and growth	9	0.74

Table 14.3 shows that there is a correspondence between the stated objectives of mutual fund managers and the portfolio βs they choose. A researcher examined the investment objectives given by 321 mutual funds and divided these goals into ten categories ranging from most aggressive ("maximum capital gain") to most defensive ("stability, income, and growth"). This researcher then examined the portfolio βs to see if funds with more aggressive intentions do, in fact, have high-β portfolios. Table 14.3 shows that this is indeed the case.[1]

Risk-averse investors will not choose high-β securities unless the expected return is higher than for low-β securities. Therefore, if investors are risk-averse, high-β securities must be priced to offer above-average expected returns to compensate for their greater risk. Various studies have found that high-β securities do, on the average, earn higher returns than low-β securities. Table 14.4 shows one such study that compares mutual fund performance over the same period as Table 14.3. Aggressive high-β funds took greater risks, but did better on average than more conservative funds.[2]

Table 14.4 *Mutual Funds with High-β Portfolios Have Done Better, on Average, Than Those with Low-β Portfolios*

β Range	Number of Funds	Assets ($ Millions)	Average β	Average Monthly Return
0.0–0.4	3	27.3	0.23	0.43%
0.4–0.8	35	94.3	0.68	0.63
0.8–1.0	44	137.4	0.91	0.79
1.0–1.2	30	73.7	1.07	0.86
1.2+	13	90.8	1.33	1.05
Average		102.6	0.91	0.78

EXERCISES

14.7 Here are some data on the annual percentage returns for the market as a whole and for three selected stocks:

Year	Market	U.S. Home	Kloof	Consolidated Natural Gas
1970	4.01	9.6	38.5	23.0
1971	14.31	51.4	6.3	7.1
1972	18.98	−17.2	53.6	7.3
1973	−14.66	−77.3	107.9	−13.4
1974	−26.47	−48.8	1.1	−3.7
1975	37.20	128.0	−38.1	25.1
1976	23.84	56.5	−45.1	59.3
1977	−7.18	−8.8	58.2	29.0
1978	6.56	22.7	16.2	−8.3
1979	18.44	113.1	304.4	15.0

U.S. Home is the nation's largest homebuilder and its profits are very sensitive to the state of the economy. A weak economy and high interest rates are bad for the stock market and very bad for U.S. Home. What kind of results do you think we would obtain if we estimated the equation

$$r = \alpha + \beta r_M + \epsilon$$

where r is the return on U.S. Home stock and r_M is the overall market return? Use the data to estimate this equation and see if you are right.

14.8 (*continuation*) Kloof is a South African gold mining stock and Consolidated Natural Gas is a regulated utility company. Use the data in Exercise 14.7 to estimate the β coefficients for each stock and interpret your results.

14.9 (*continuation*) If a portfolio were constructed by investing an equal number of dollars in

each of these three stocks, what would be this portfolio's β? Use the data to calculate the annual returns on such a portfolio over the ten years 1970–1979. Estimate the equation

$$r = \alpha + \beta r_M + \epsilon$$

where r is the portfolio return and r_M is the market return. Interpret your results.

• **14.10** (*continuation*) How could a portfolio that has a β of zero be constructed out of these three stocks?

14.11 One researcher calculated the annual returns from 1954 through 1978 on an investment in 15 representative modern prints and found that these returns were negatively correlated with annual stock market returns.[3] What does this tell us about the β coefficient for modern prints? Why might someone invest in modern prints even if the expected return were very modest?

14.12 Explain why you either agree or disagree with this statement

> A stock with a low β is safer than a high-β stock, because the lower β is, the less variation there is in a stock's return.

• **14.13** Many investors estimate the β and R-squared of their portfolios. It has been argued that

> If the market was expected to rise, an investor might increase the R-square of the portfolio. If the market was expected to decline, a portfolio with a low R-square would be appropriate.[4]

Explain why you agree or disagree with this advice.

14.3   THE POLITICAL BUSINESS CYCLE

Economic events often seem to have a decisive influence on political elections, in that people take into account their economic well-being when casting their votes. If they are prospering, they are apt to think that the incumbent party is doing a pretty good job and should be allowed to continue with more of the same; there's no reason to take a chance on something new. As folk wisdom says, "If it ain't broke, don't fix it." If, on the other hand, people are out of work or in fear of soon losing their jobs, they are more apt to vote for a change in policy: "Throw the bums out and give someone else a chance."

Now, of course, the state of the economy isn't the only item on the minds of the electorate. International issues of war and peace are obviously important, and so are the candidates' individual personalities—particularly in local elections. But economists and political scientists tell us that national elections are often a referendum on the incumbent party's economic policies. In particular, it seems that the incumbent party is more likely to win congressional and presidential elections if voters perceive that the economy is improving.[5]

For a simple investigation of this idea, Table 14-5 shows some relevant data for 23 presidential elections since the turn of the century. Here we will just look at the 16 elections since 1928; an analysis of all 23 elections is left as an exercise.*

The direction in which the economy is moving is gauged by the change in the unemployment rate. The incumbent party's vote is calculated as a percentage of the total vote cast for the two major-party candidates.

What can we conclude from these data? First, let's divide the 16 elections from 1928–1988 into years in which the unemployment rate went up and years in which it went down:

Unemployment Up		Unemployment Down	
Year	% Vote	Year	% Vote
1928	58.8	1936	62.5
1932	40.9	1940	55.0
1980	44.7	1944	53.8
		1948	52.4
		1952	44.6
		1956	57.8
		1964	61.3
		1968	49.6
		1972	61.8
		1976	48.9
		1984	59.2
		1988	53.9
Average	48.1		55.1
Standard Deviation	9.4		5.7

Notice the imbalance in these data. The unemployment rate has gone up in an election year only once since 1932 (during Jimmy Carter's ill-fated 1980 reelection bid). Overall, considering both election and nonelection years, the unemployment rate goes up about as often as it goes down. The fact that the unemployment rate nowadays almost always seems to drop in election years suggests that the politi-

*In the 1912 and 1924 elections, there were three major candidates and it is not obvious how the votes should be counted for and against the incumbent party in a way that would be comparable to two-party elections.

Table 14.5   *Some Data on the Economy and Election Results*

Election Year	Unemployment Change* X	Incumbent % of Vote† Y
1900	−1.5	53.2
1904	+1.5	60.0
1908	+5.2	54.5
1912‡	−2.1	54.7
1916	−3.4	51.7
1920	+3.8	36.1
1924‡	+2.6	54.3
1928	+0.9	58.8
1932	+7.7	40.9
1936	−3.2	62.5
1940	−2.6	55.0
1944	−0.7	53.8
1948	−0.1	52.4
1952	−0.3	44.6
1956	−0.3	57.8
1960	0.0	49.9
1964	−0.5	61.3
1968	−0.2	49.6
1972	−0.3	61.8
1976	−0.8	48.9
1980	+1.3	44.7
1984	−2.1	59.2
1988	−0.7	53.9

*unemployment change = unemployment rate in election year minus unemployment rate in preceding year

†% vote = fraction of the total presidential vote for the Democratic and Republican candidates received by the party in office at the time of the election.

‡In 1912, there were three major candidates: Roosevelt, Taft, and Wilson. The Roosevelt and Taft votes were taken to be votes for the incumbent party. In 1924, the major candidates were Davis, LaFollette, and Coolidge and the Coolidge votes were counted as votes for the incumbent.

Source: Ray C. Fair, "The Effect of Economic Events on Votes for President," *The Review of Economics and Statistics* 60 (May 1978), pp. 159–173 and various issues of the *New York Times*.

cians, at least, believe that a stronger economy wins votes. They apparently choose policies that are designed to boost the economy during the election years—causing political scientist Edward Tufte to joke that we could eliminate recessions by holding presidential elections every year.

Does it work? Well, the incumbent won 9 of the 12 elections when the unemployment rate went down and only 1 of 3 when the unemployment rate went up. This difference is suggestive, but not conclusive.* Similarly, the average vote for the incumbent was only 48.1 percent when the unemployment rate was increasing, but 55.1 percent when the unemployment rate was falling. This also is suggestive, but the standard deviations are too large to rule out chance variation. (The actual test is left as an exercise.)

An inadequacy of these first two comparisons is that we paid no attention to the size of the change in the unemployment rate, only its sign. It is likely that the effect (positive or negative) on the election outcome is stronger when the change in the unemployment rate is larger. Regression analysis offers a way to take this kind of quantitative consideration into account. Let's use the linear model

$$Y = \alpha + \beta X + \epsilon$$

where Y is the incumbent vote, X is the change in the unemployment rate, and ϵ is the error term that reflects all of the other factors that affect election outcomes.†

A least squares regression yields

$$\hat{Y} = 53.24 - 1.75\ X, \qquad R^2 = .40$$
$$[40.0] \quad [3.04]$$

[ ]: t-values

According to these estimates, the change in the unemployment rate has a substantial (and statistically significant) effect on election outcomes. If the unemployment rate is steady ($X = 0$), then the model predicts that the incumbent will get 53.24 percent of the vote. A 2 percent increase in the unemployment rate during the election year reduces the predicted vote by 3.5 percent, down to 49.73 percent—the difference between victory and defeat!

There are some reasons for caution in interpreting these estimates, however.

*If the election outcomes were Bernoulli trials with probabilities π_u when the unemployment rate goes up and π_d when it goes down, then a statistical test of the null hypothesis that $\pi_u = \pi_d = \pi$ could be based on the statistical fact that the difference in the sample proportions p_u and p_d would be distributed normally with a mean zero and variance $\pi(1 - \pi)(1/n_u + 1/n_d)$. Using 10/15 for an estimate of π, the estimated variance is $(10/15)(5/15)(1/3 + 1/12) = .093$ and the estimated standard deviation is .304. Thus, the observed .42 difference between the sample proportions 9/12 and 1/3 is not statistically significant.

†This error term can be interpreted as a sample from the probability distribution describing the many possible extraneous influences on Y. In any particular election year with any given change in the unemployment rate, many things (war and peace, for instance) can happen that influence the outcome. The observed data reflect what actually happened to occur that year.

First, even though the slope estimate is statistically significant, the prediction intervals are pretty wide. The 53.24 percent prediction works out to be ±11.0 percent, while the 49.73 percent prediction is ±11.2 percent. So, even while the data say that the unemployment rate makes a difference, they also say that a lot of other things matter, too. The incumbent has the edge and a healthy economy helps, but a healthy economy is no guarantee of victory and a sick economy does not ensure defeat.

The second reason for caution can be seen from the scatter diagram shown in Figure 14.5. Cover up the 1932 result (when, in the depths of the Great Depression, Franklin Roosevelt clobbered the much maligned Herbert Hoover) and try to imagine where you would draw a line that best fit the remaining points. The answer is far from obvious. An outlier like the 1932 election can have a decisive influence on a regression line because, in least squares estimation, many small prediction errors are more tolerable than one big error. The fitted line will avoid large prediction errors by going very close to the outlier. Figure 14.6 shows an extreme example in which there is a perfect negative relationship between X and Y, except for the outlier—which causes the fitted line to have a positive slope.

In these sorts of situations, the researcher must decide whether it is reasonable to report results that are so dependent on a few observations. If, as was shown in Figure 14.6, all of the data except for this one outlier suggest a completely different slope, then there is reason for suspicion. Perhaps the outlier is wrong—the unfortunate consequence of a typographical error somewhere along the line. Or perhaps something extraordinary happened that year that caused a very misleading value for the dependent variable. If either of these things is true, then the outlier can be corrected or discarded.

These sorts of obvious errors and extraordinary circumstances do happen. But another possibility is that most of the data show so little variation in the explanatory variable that it is difficult to fit any sort of line. Perhaps the unemployment rate is usually up or down a mere 0.1 percent, and it is hard to tell whether or not

Figure 14.5   *The State of the Economy and Election Outcomes*

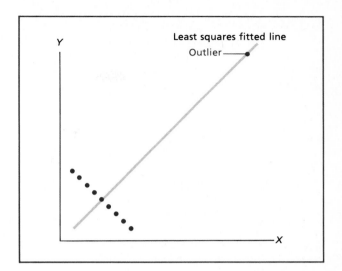

Figure 14.6 *Outliers Can Strongly Influence Least Squares Estimates*

this is having any effect on the voters. Then, in one particular year, there is a large change in the unemployment rate: it jumps 7.7 percent and the incumbent party is soundly defeated. Doesn't this sort of outlier answer the very question that we are posing—whether a substantial change in the unemployment rate has a substantial effect on the election outcome? In this situation, we wouldn't want to discard the outlier. Indeed, it's the most valuable piece of data we have!

Here, I reestimated the regression equation excluding 1932 and obtained these results:

$$\hat{Y} = 52.84 - 2.24\ X, \qquad R^2 = .21$$
$$[32.7] \quad [1.83]$$

[ ]: *t*-values

If we exclude the three outliers 1932, 1936, and 1940, then we get

$$\hat{Y} = 52.89 - 2.41\ X, \qquad R^2 = .11$$
$$[30.8] \quad [1.17]$$

In this particular example, leaving out an outlier (or three) doesn't make a lot of difference to the estimated slope, but it makes a great deal of difference as to whether or not the relationship is statistically significant. Without the outliers, *R*-squared and the *t*-value for the slope drop dramatically. Since 1940, politicians have apparently managed to manipulate the economy so that the unemployment rate almost always falls slightly during an election year. It is up to the reader to decide whether 1932, 1936, and 1940 are misleading outliers or the most convincing evidence of the effects of the economy on voters.

EXERCISES

14.14 Professor McTower has made a study of the relationship between the unemployment rate and the outcomes of presidential elections. Her results are

$$V = 51.0 - 2.00\ U$$
$$[3.5]\quad [4.0]$$

where V = percent of vote received by the incumbent party; average value = 53 percent

U = change in the unemployment rate in the year of the election, average value = -1.0

[ ] = t-values

a. What is her estimate of the effect of unemployment on presidential elections? Is her estimate substantial?   **b.** What would be a 95 percent confidence interval for her estimate? Is her estimate statistically significant at the 5 percent level, in that we can reject the hypothesis that there is no effect of unemployment changes on presidential elections?   **c.** If we were to use her equation to predict the outcome of the next presidential election, for which value of U would we be most confident about our prediction? Why? **d.** According to her equation, what fraction of the vote will the incumbent party receive if the unemployment rate falls by 15 percent ($U = -15$) during the election year? Is there any special statistical reason for caution about such a forecast?

14.15 The text used presidential election data for the years 1928 through 1988 to test the idea that the state of the economy influences voters. How would the results be affected if we threw out, as outliers, the five elections in which the absolute value of the change in the unemployment rate was greater than 1.0? Does this discarding make sense?

14.16 *(continuation)* The data for 1900 through 1924 shown in Table 14.5 were not used because of some uncertainty about how to calculate the incumbency vote in a three-person race. How are the results affected if we use all 23 elections, 1900–1988? What if we use the elections from 1900 to 1988, excluding 1912 and 1924? Are the estimates of β substantial and/or statistically significant at the 5 percent level?

• **14.17** *(continuation)* Let μ_u and μ_d be the expected percentage votes for the incumbent party when the unemployment rate goes up and down, respectively. If the data for 1928–1988 in Table 14.5 are considered a random sample, construct a statistical test of the null hypothesis that $\mu_u = \mu_d$.

14.18 Here are some data on the unemployment rate and per capita spending (in 1980-dollars) at motion picture theaters:

Year	% Unemployment Rate (X)	Motion Picture Spending (Y)
1975	8.5	$10.51
1976	7.7	$11.59
1977	7.1	$14.70
1978	6.1	$15.99
1979	5.8	$14.89
1980	7.1	$12.76
1981	7.6	$10.87

Do you think that there may be a relationship between these variables? Is it a positive or negative relationship? Explain your reasoning and then check your logic by estimating the equation $Y = \alpha + \beta X + \epsilon$.

14.19 Corporate dividends (Y) are paid out of the firm's earnings (X). To examine the relationship between the two, use the aggregate S&P 500 dividend and earnings data for 1950–1989 shown in Table 8 at the back of the book to estimate the regression equation $Y = \alpha + \beta X + \epsilon$. Report your results and write a paragraph summarizing your conclusions.

14.4   SOME REGRESSION PITFALLS

Although regression analysis is very simple and powerful, it can also be misused—yielding inaccurate or misleading conclusions. Let's look at several pitfalls that should be avoided.

Careless Extrapolation

Researchers are often interested in situations for which they have little or no data. One possible approach is to analyze a situation that has plentiful data and then try to extrapolate the results to cover the situation of interest. For example, some years ago, a group of scientists got together to project population and income hundreds of years from now. They used data for the present and then extrapolated their calculations into the future.

Sometimes the extrapolation is almost an afterthought. A researcher may fit a regression model in the usual way and then, to explore some of the implications of the results, he or she may calculate the predicted values of the dependent variable for various assumed values of the explanatory variable. These calculations may wander off into virgin territory, where there are no data.

Extrapolators should proceed cautiously. It is not enough to plug in a value of X and mechanically compute the predicted value of Y. The calculation of a confidence interval for the prediction is a bit of extra work, but it may well be eye-opening. The prediction interval widens as we get away from the center of the data, and provides a fair warning that predictions on the fringes of the data are less trustworthy. If we find that the prediction is $Y = 1.235 \pm 232.6$, then we may realize that our data don't really tell us very much about Y for this particular value of X.

Even the widening confidence intervals may understate the unreliability of our extrapolations, because our model is probably not appropriate for any and all values of X. Linearity, for example, is a very convenient assumption and a satisfactory approximation as long as we confine our attention to modest changes in the explanatory variable. However, if we are extrapolating our model to large changes in X, then we should stop and think about the range over which linearity is a reasonable assumption. A linear model assumes that the effect on Y of a change in X is constant, no matter what the value of X. But for most phenomena, there comes a point where the effect of X on Y either becomes stronger or diminishes, and violates the linearity assumption.

For instance, the application of fertilizer increases crop yields, but there comes a point when the effects get smaller and smaller and then turn negative. Too much fertilizer damages crops. Similarly, plants cannot grow without water or light, and the application of either stimulates plant growth. But again, there comes a point where additional water or light doesn't help much and another point, farther along, where more water or light actually will stunt growth or even kill plants.

Keynes' consumption function is a useful approximation to the relationship between household income and spending. People tend to spend more when their income goes up and to spend less when their income goes down. However, their behavior would change drastically if the economy collapsed and their incomes fell to zero. Who would be foolish enough to believe that the intercept of a linear consumption function (perhaps zero, perhaps a negative number) accurately predicts consumer behavior in such an economic disaster? And yet, the authors of some economics textbooks say just that foolish thing.

There are other examples that are so ludicrous that everyone recognizes the error of incautious extrapolation. In 1940, there were an average of 3.2 people in each car on the highway in the United States. By 1960 this average had dropped to 1.4. At this rate, by the year 2000, three out of four cars on the road will be empty! The relationship between height and weight was estimated in Chapter 13; a hasty extrapolation to someone 33 inches tall predicts a weight of zero. Another researcher, with tongue firmly in cheek, extrapolated the observation that automobile deaths declined after the maximum speed limit in the United States was reduced to 55 miles per hour:

> . . . To which Prof. Thirdclass of the U. of Pillsbury, stated that to reach zero death rate on the highways, which was certainly a legitimate goal, we need only set a speed limit of zero mph. His data showed that death rates increased linearly with highway speed limits, and the line passing through the data points, if extended backwards, passed through zero at zero mph. If fact, if he extrapolated even further to negative auto speeds, he got negative auto deaths, and could only conclude, from his data, that if automobiles went backwards rather than forwards, lives would be created, not lost.[6]

Other tongue-in-cheek extrapolations have found that:

1. If the number of microscope specimen slides acquired by a certain St. Louis hospital continues to increase at the current rate, then St. Louis will be buried under three feet of glass by the year 2224.
2. If *National Geographic* magazines continue to accumulate in basements and garages at the present rate, the North American continent will sink beneath the seas.

More than one hundred years ago, Mark Twain came up with this one:

> In the space of one hundred and seventy-six years the Lower Mississippi has shortened itself two hundred and forty-two miles. This is an average of a trifle over one mile and a third per year. Therefore, any calm person, who is not blind or idiotic, can see that in the Old Oolitic Silurian Period, just a million years ago next November, The Lower Mississippi River was upward of one million three hundred thousand miles long, and stuck out over the Gulf of Mexico like a fishing rod. And by the same token any person can see that seven hundred and forty-two years from now the Lower Mississippi will be only a mile and three-quarters long, and Cairo and New Orleans will have joined their streets together. . . . There is something fascinating about science. One gets such wholesale returns of conjecture out of such a trifling investment of fact.[7]

Regression Toward the Mean

Regression toward the mean is a phenomenon that was first noticed by Sir Francis Galton in a study of the relationship between the heights of fathers and their sons. And, indeed, Galton is the source of the label *regression* for the least squares estimation technique. Figure 14.7 shows a similar line estimated by Karl Pearson and Alice Lee, as part of a seminal study of the inheritance of a wide variety of physical traits.[8] Pearson and Lee grouped the heights of 1078 fathers into 17 intervals, rang-

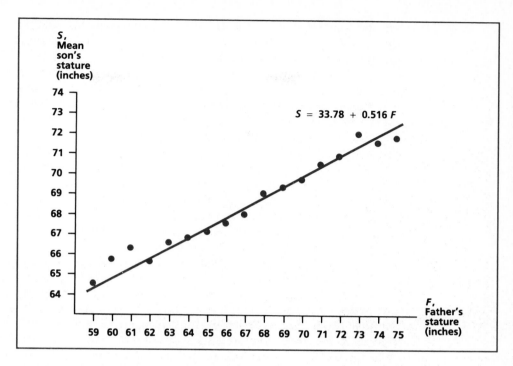

Figure 14.7 *A Regression Line of the Heights of Fathers and Their Sons*

ing from 59 to 75 inches. For each interval, they calculated the average height of these fathers' sons and fit a least squares line to these 17 points, as shown in Figure 14.7. Notice that the sons of 59-inch fathers average 64-1/2 inches while the sons of 75-inch fathers average less than 72 inches.

Galton, too, found that unusually tall fathers tend to have sons who are shorter than themselves, while very short fathers usually have somewhat taller sons. He concluded that the world is "regressing towards mediocrity." However, these data do not mean that soon everyone will be the same height; instead, they reflect the statistical consequences of two chance factors.

The first factor is genetic; a person's height depends on the heights of both the mother and the father, and very tall men generally marry women who are not exceptionally tall—not because they prefer short women, but because height is not the only factor that enters into a marriage. A man who is two standard deviations taller than the average male may well marry a woman who is two standard deviations taller than the average female. But, because attributes other than height matter, she is very likely to be a woman who is less than two standard deviations taller than the average female, just because there are so many more of them. For every woman whose height is more than two standard deviations above the average, there are 40 women whose height is less than two standard deviations above the average. Thus, most very tall men marry women who are not as exceptionally tall as they are and, thus, have sons who are not as exceptionally tall either. In the same way, most very small men marry women who are not exceptionally short and consequently have somewhat taller sons.

The second chance factor is that diet, exercise, and other environmental factors influence heights, so that observed height is not a perfect reflection of one's genes and, hence, not a perfect predictor of the genetically expected height of one's children. Someone who is very tall is much more likely to be the unusually tall result of more average genes than the unusually short result of extremely tall genes, just because there are so few extremely tall genes. Thus, the observed height of a very tall person is usually an overstatement of his or her genetic height, which is what determines the expected height of the child. In the same way, a short person is much more likely to have been the unusually short child of more average parents than the unusually tall child of exceptionally short parents. A short parent's genetic height is probably somewhat taller and, so, the children will be taller too.

These explanations do not mean that we will soon all be the same height. Indeed, we could just as well turn the argument upside down by noting that most very tall people had somewhat shorter parents, while most very short people had somewhat taller parents—does this mean that heights are diverging? No, heights are neither converging nor diverging. There will always be unusually tall and unusually short people. What we must recognize is that heights are influenced by chance, so that those who are unusually tall most likely had chance influences that pulled them above their genetically expected height. The *regression-toward-the mean fallacy* is to conclude that some important trend is at work, when it is really just the complex implications of chance.

The midterm/final example that was analyzed earlier is another good illustration of the regression-toward-the-mean phenomenon at work. Our estimated equation is

$$\hat{Y} = 31.3 + .627\,X$$

where Y is the score on the final exam and X is the score on the midterm. Here are the predicted final exam scores for three different midterm grades:

Midterm Score	Predicted Final Exam Score
59	68.2
69	74.6
79	80.8

A student who gets the average score on the midterm (69) is predicted to receive an average score on the final (about 75). A student who scores 10 points below the class average on the midterm is predicted to score only 6 points below the average on the final. A student who is 10 points above the class average on the midterm is predicted to be only 6 points above the average score on the final.

Does this mean that students converge to a depressing mediocrity, with the weak getting stronger while the strong get weaker? Or, turning the argument on its head, does the fact that the highest scorers on the final did not do as well on their midterms show that scores are diverging from the mean? The answer to both questions is no! The most plausible explanation is that high scores involve a bit of good luck. Those students with the highest scores on any particular test are mostly above-average students who did exceptionally well because the questions that were asked happened to be ones that they were well prepared to answer. It is more probable that they are good students who did unusually well than that they are great students who had an off day. Thus, the students with the highest scores on the midterm are more likely to have somewhat lower scores on the final than to have even higher scores. Similarly, those students with the highest scores on the final probably had somewhat lower scores on the midterm. The highest scorers on any given test most likely did not do as well on their last exam and will not do as well on their next exam, but this does not mean that scores are either diverging or converging.

Regression toward the mean involves a subtle sort of data mining. If, beforehand, we select students at random from some population (perhaps those who have had special counseling), their average score will be an unbiased estimate of the population mean. But if, after the test, we pick out those students who happened to have done exceptionally well, these are not unbiased estimates of some population's ability because they are not a random sample; they were selected because they had the highest scores in the sample. In any population distribution, the highest scores are an overestimate of that population mean. If we want to estimate some expected score or see whether the dispersion in the scores is increasing or decreasing, then we need to use a random sample that is not chosen based on the scores themselves.

We also see this regression toward the mean in IQ scores:

> . . . 4 year olds with IQs of 120 typically have adult scores around 110. Similarly, 4 year olds with scores of 70 have an average adult score of 85. . . . This does not mean that there will be fewer adults than children with very high or very low IQs. While those who start out high or low will usually regress toward the mean, their places will be taken by others who started closer to the mean.[9]

Similarly, those who get relatively low scores on the Preliminary Scholastic Aptitude Test (PSAT) taken by high school juniors get, on average, somewhat higher scores when they take the Scholastic Aptitude Test (SAT) their senior year, while the high scorers on the PSAT, on average, get somewhat lower scores on the SAT.

Another example involves Air Force flight instructors who observed that a good landing that was praised was usually followed by a poorer landing, and that a poor landing that was criticized harshly was usually followed by a somewhat better landing. Falling for the regression toward the mean fallacy, the flight instructors concluded, contrary to well-accepted learning research, that praise is detrimental and severe criticism is helpful.[10]

An economic example occurs in the estimation of household consumption functions. Researchers find that people with relatively high incomes one year tend to have relatively less income the next year, while those with low incomes improve their relative position. This does not mean that incomes are becoming more and more alike. A person who is observed to have had a relatively high income is more likely to have had an unusually good year than an atypically bad year. A normal year would see relatively less income. Of course, their places will be taken by other people who have unusually good years and atypically high incomes. At the other end, people with relatively low incomes are more likely to have had a bad year than a good year. A normal year would be one with a somewhat higher income.

Another economic example is provided by a book published some time ago with the provocative title *The Triumph of Mediocrity in Business*. The author discovered that businesses with exceptional profits in any given year tended to have smaller profits the following year, while most firms with very low profits did somewhat better the next year. From this evidence, he concluded that strong companies were getting weaker, and the weak stronger, so that soon they would all be mediocre. The author's fallacy is now obvious. A famous statistician explained it as follows:

> . . . while [businesses] at the margins . . . often go towards the center, those in the center of the group also go towards the margins. Some go up and some down; the average of the originally center group may, therefore, display little change, since positive and negative deviations cancel in averaging; while for an extreme group, the only possible motion is toward the center.[11]

Yet a best-selling investments textbook seemingly continues to make the same error.[12] While discussing a model of stock prices, the author asserts that, "Multistage dividend discount models assume that, ultimately, economic forces will force convergence of the profitability and growth rates of different firms." To support this assumption, the author looks at the 20 percent of firms with the highest profit rates in 1966 and the 20 percent with the lowest profit rates. Fourteen years later, in 1980, the profit rates of both groups are more nearly average: "convergence toward an overall mean is apparent. . . . the phenomenon is undoubtedly real." The phenomenon is most likely regression toward the mean, and the explanation is statistical, not necessarily economic.

Correlation Is Not Causation

Usually the basis for a regression equation is the belief that changes in *X cause* changes in *Y*. For example, our estimation of a consumption function was based on Keynes' argument that income affects spending. Similarly, our investigation of the political business cycle began with the idea that the state of the economy influences the outcome of presidential elections. A strong economy wins votes for the incumbent party, while a weak economy is politically costly.

It is sound practice to have a logically plausible model that motivates the regression equation. However, this sound practice is not always followed, in that often researchers accidentally stumble upon unexpected correlations. The art of regression analysis is then to interpret these empirical results. This is what Ed Lea-

Do Champions Choke?

As of 1990, there had been 24 Super Bowls and only 4 champions were able to repeat their feat the following year. In baseball, of the last 26 World Series champions (1964–1990), only 4 have repeated. No professional basketball team repeated as NBA champion between 1969 (the Boston Celtics) and 1988 (the Los Angeles Lakers). This doesn't mean that champions choke or that the teams in each of these sports are becoming increasingly equal in ability. Instead, these can be explained as regression toward the mean. There is a considerable amount of luck involved in winning a championship. Injuries, bad bounces, and questionable officiating all play a role. The team that ends up as champion is more likely to have had its share of good fortune. The next year, some other team will probably get the breaks.

There is plenty of other more detailed evidence of regression toward the mean in sports.[13] Of those major league baseball teams that finish a season with winning records, two-thirds win fewer games the next season. Of those teams with losing records, two-thirds do better the next season. Of those teams that win more than 100 out of 162 games in a season, 90 percent will find their record deteriorating the next season. Looking at individuals, there were 28 major league baseball players who batted above .300 in 1979, and 23 of these 28 had lower batting averages in 1980 than in 1979. This is to be expected because someone who bats .320 in any given year is more likely to be a .300 hitter having a good year than a .340 hitter having a bad year, just because there are so few (if any) .340 hitters around. The regression-toward-the-mean fallacy is to conclude that good hitters' skills deteriorate. The correct conclusion is that those with the highest batting averages in any particular year aren't really as skillful as their high batting averages suggest, in that they undoubtedly had more than their share of good luck that year.

mer calls "Sherlock Holmes inference" and, as I pointed out in earlier chapters, it can lead to both useful and useless models.

There are three major reasons why empirical correlations can be misleading:

simple chance
reverse causation
omitted factors

The first explanation, simple chance, involves the observation that even unrelated things may, by historical accident, be statistically correlated. In the model $Y = \alpha + \beta X + \epsilon$, we say that X has a statistically significant effect on Y at the 5 percent level if there is less than a 5 percent chance that our estimate of β would be so far from zero, if β really were zero. What we must realize is that even if β really is zero, there is a 5 percent chance that the data will happen to indicate a statistically significant relationship. If we spend all of our lives looking at unrelated variables, we will still find statistical significance on about one out of every twenty tries. Or, turning it around, if we run twenty regressions and report the best one, we may have nothing more than chance correlation.

The second reason why correlations can be misleading is that perhaps it is Y that determines X, rather than the other way around. Let's say, for instance, that it is a Sunday afternoon and, with nothing better to do, we are rummaging through some data looking for impressive correlations. We stumble upon a regression of the change in the unemployment rate (Y) on the vote for the incumbent party (X)— the reverse of the regression run earlier.

Using the data shown in Table 14.5, the results turn out to be

$$\hat{Y} = 11.7 - 0.22\,X, \qquad R^2 = .37$$
$$[2.67]\quad[2.67]$$

[ ]: t-values

This statistically significant relationship tempts us to find some logical explanation. Why would the reelection of an incumbent cause the unemployment rate to fall, *preceding* the reelection? Perhaps, the jubilant incumbent who foresees reelection celebrates with expansionary policies, while the bitter incumbent who foresees defeat punishes the electorate with a recession. Or perhaps businesspeople love stability and when they believe that the incumbent is going to be reelected, they happily increase employment and production. However, when businesspeople see a new administration coming, they grow fearful and cut back. Interesting theories! But none so plausible as the possibility that it is not election prospects that determine unemployment, but rather unemployment that influences election outcomes.

There are plenty of other examples, some less obvious and some more so. For clarity, we will consider some obvious ones. There are empirical correlations between people's ages and their insurance rates, between the nylon content of ropes and their breaking strength, and between fertilizer applications and crop yields. In each case, we have to use our powers of reasoning to decide what is affecting what. Does increasing age cause insurance companies to charge more? Or do high insurance rates cause people to age faster? Does nylon increase breaking strength or does breaking strength create nylon? Does additional fertilizer increase yields or does productive land persuade farmers to reward the soil with a fertilizer treat?

The third reason for misleading correlations is that there may be some omitted factor that influences both X and Y. In the regression model

$$Y = \alpha + \beta X + \epsilon$$

we assume that the omitted factors collected in the error term ϵ are uncorrelated with the explanatory variable X. If they are not, then we may misestimate the effect of X on Y, or even identify an effect when there really is none. In this unfortunate situation, Y may appear to depend on X, when in fact, Y depends on ϵ, which happens to be correlated with X.

Think of Keynes' consumption function again, with Y being consumer spending and X being income. Perhaps consumer spending depends not only on income but also on prices: high income encourages spending, while high prices discourage spending. By using income as the explanatory variable, we relegate prices to the error term. Now, what if prices tend to be high when income is high, and to be low when income is low? If so, then we will be fooled into underestimating the effect of income on spending, because whenever income goes up, the positive effects of income on spending will be partly offset by the negative effects of the higher prices. Even worse, the negative effects of higher prices may be so strong that spending actually falls whenever income goes up. We would erroneously conclude that income discourages spending, when in fact, it is higher prices that discourage spending.

Our mistakes can also go in the other direction. Perhaps spending doesn't really depend on income; instead, spending depends solely on what happens in the stock market. When the market goes up, buoyed customers spend happily, and when the market crashes, depressed consumers cut back. Now it may so happen that in the particular data we examine, stock prices increased in years when income was high and fell in years when income was low. If so, then when we see income and spending going up and down together, we will erroneously conclude that spending depends on income, when spending really depends on the stock market. Or, to make it sillier, perhaps the National Football Conference happens (by pure chance) to have won the Super Bowl in the same years that the stock market went up and to have lost in those years when the market went down. If so, then a regression of consumer spending on Super Bowl scores will lead us to conclude erroneously that consumer spending depends on which conference wins the Super Bowl.

A Few Examples

A blurry causation does not necessarily render a regression equation worthless. The estimated equation may still be useful in describing the nature of the data. The error is in misperceiving causation. For example, Exercise 13.31 in the previous chapter gave some monthly data on precipitation ($\hat{P}$) and temperature (T) in Chicago, Los Angeles, and New York. A least-squares regression using the Chicago data yields

$$\text{Chicago: } \hat{P} = 0.71 + 0.04\ T, \quad R^2 = .68$$
$$[1.62]\quad [4.67]$$

[   ]: t-values

There is a statistically significant positive relationship between temperature and precipitation. A 10-degree increase in temperature is associated with about a half-inch of additional precipitation.

Fine enough. But if we use the Los Angeles data, we get

$$\text{Los Angeles: } \hat{P} = 11.2 - 0.17\ T, \quad R^2 = .82$$
$$[7.43]\quad [6.81]$$

indicating that a 10-degree increase in temperature is associated with an almost 2-inch *decrease* in precipitation! The New York data gives yet different results:

$$\text{New York: } \hat{P} = 2.95 + 0.01\ T, \qquad R^2 = .10$$
$$\qquad\qquad\quad [6.79] \quad [1.05]$$

Here, precipitation is apparently unrelated to temperature.

These three conflicting results are only contradictory if we attempt to find a causal relationship between temperature and rainfall. Temperature and precipitation are complex climatological phenomena that depend critically on a city's geographic location. Chicago is so located that it happens to have a rainy season during the warm summer months. This doesn't mean that the sun causes rain. Los Angeles, in contrast, has its rainy season during the cool winter months. And New York gets almost the same amount of precipitation every month—winter, spring, summer, and fall. That is all these three regressions are telling us: that it rains a lot in Chicago during the summer; that Los Angeles gets most of its rain in the winter; and that New York rainfall is pretty evenly distributed throughout the year. These conclusions are informative and correct. We just have to avoid the incorrect conclusion that rain is caused by sunshine, or vice versa.

Because regression results are often just descriptive, cautious researchers use language such as this: "The results show that changes in Y are associated with changes in X." The readers are left to draw their own conclusions and it is hoped that they will remember that a sound theory is needed for a causal interpretation of the results.

Our example of final and midterm scores is similar. We don't really believe that it is the score on one test that determines the score on another; it is the person taking the test who determines the scores. However, there is some noticeable consistency in performance in that a bright, conscientious student does well on most tests, while a goof-off does poorly. Their scores vary from test to test, because there is some randomness in the specific questions asked, in how the professor grades the questions, and in each student's fitness on test day. But, logically, there is some consistency, too, and that consistency is what we tried to identify with our regression estimates—the extent to which final exam scores can be predicted from midterm scores. Our results were interesting and informative: there is some consistency in scores; there is some regression toward the mean; and there is substantial variation in scores from one test to the next. As long as we are clear about causation, we are okay.*

Unfortunately, there are lots of examples of incorrect conclusions that can be drawn by seeing causation when there is only correlation. Spurious correlations are

---

*The same is true of our estimated relationships between the market return and the returns on individual stocks. We don't believe that there is a living, breathing "market return" that decides how well each individual stock will do. The market return is nothing more than the average return on all stocks. However, what we do believe is that stocks exhibit varying degrees of sensitivity to the economic events that buffet the economy. We try to gauge these relative sensitivities by regressing individual returns on the calculated market return.

The Radiophobic Farmer

Darrell Huff tells the story of a farmer who claimed that his fruit trees were being damaged by radio waves from a nearby station. He put a wire fence around some of his trees to "shield" them from these radio waves and, sure enough, they quickly recovered, while the unshielded trees still suffered. About the same time, many citrus trees throughout the country were threatened by the spread of "little leaf" disease. Then some farmers in Texas found that a solution of iron sulphate cured this disease. However, it didn't always work in Texas and it almost never worked in Florida or California. Huff explains that these mysteries were solved when it was discovered that the real problem was a zinc deficiency in the soil:

The radiophobic farmer's fence wire was galvanized; enough zinc washed off the wire to give the trees the tiny trace of zinc they needed. The iron sulphate did nothing for the other trees, but the zinc-coated buckets in which it was carried saved them. In regions where buckets other than galvanized were used, the trees went right on ailing.[14]

undoubtedly the most serious hazard in building and estimating useful regression models. The only real safeguard is to use common sense in specifying models and to reexamine suspicious results with fresh data. For instance, a positive correlation has been reported between storks' nests and births in northwestern Europe. Few would conclude that storks bring babies. A more logical explanation is that storks like to build their nests on buildings. Where there are more people, there are usually more buildings and, hence, more places for storks to nest.

Another study found that a certain state had both an unusually high milk consumption and an abnormally high cancer rate. Does this show that milk causes cancer? It turns out that many elderly people live in this state and that cancer strikes the elderly more than the young. Because of this state's large elderly population, it had an unusually high cancer rate. Anything else that was unusual about this state might seem to be an explanation for the high cancer rate, but it was really just the unusual number of elderly citizens.

Similarly, Arizona has the highest death rate in the nation from bronchitis, emphysema, asthma, and other lung diseases, but this doesn't mean that Arizona's climate is hazardous to one's lungs. On the contrary, doctors recommend that patients with lung problems move to Arizona for its beneficial climate. Many do move to Arizona and benefit from the dry, clean air. Although some die of lung disease in Arizona, it is not because of the Arizona climate.

Data Grubbing

If we have no logically sound model to guide us, then we may well fall into one of these correlation-is-not-causation pitfalls. The usual route to a pitfall is an overly zealous researcher who gives the computer a good workout by running hundreds or even thousands of regressions, and then reports those results that are most statistically significant. Because this pure and simple data grubbing is sure to turn up something that is statistically significant, we do not know whether the reported result validates an interesting model or merely demonstrates the researcher's endurance. Only common sense and fresh data can tell the difference.

To illustrate the dangers of data grubbing, one economist ran several regressions and eventually found a striking correlation between stock prices and the number of strikeouts by the Washington Senators baseball team.[15] British economists found a high correlation between the annual rate of inflation and the number of dysentery cases in Scotland the previous year, and a "spectacular" correlation between Britain's price level and cumulative rainfall.[16]

Many business and economic variables grow over time with the overall economy. If we pick two such variables at random, they will usually appear correlated, simply because they have both been growing. One way of seeing this logic is to think of time itself as the omitted factor. Let's say that X and Y are two unrelated variables, each of which grows over time, plus or minus some random disturbance:

$$Y = \alpha_1 + \beta_1 t + \epsilon_1 \tag{14.2}$$

$$X = \alpha_2 + \beta_2 t + \epsilon_2 \tag{14.3}$$

The new variable t is a measure of time; for instance,

$$t = 1 \text{ in } 1971$$

$$t = 2 \text{ in } 1972$$

and so on. Each year, the passage of time increases Y by β_1 and X by β_2. In addition the random disturbances ϵ_1 and ϵ_2 cause Y and X to diverge somewhat from their long-run growth paths. If ϵ_1 and ϵ_2 are independent, then we would say that X and Y are two unrelated variables that follow their separate courses.

Figure 14.8 shows an example of this. I set $\alpha_1 = 10$, $\beta_1 = 2$, $\alpha_2 = 5$, and $\beta_2 = 1$. Twenty observations for Y and X (for $t = 1$ to 20) were generated by flipping a coin 40 times, 20 for ϵ_1 and 20 for ϵ_2:

$$\epsilon_1 = \begin{array}{l} +2 \text{ if heads} \\ -2 \text{ if tails} \end{array} \qquad \epsilon_2 = \begin{array}{l} +1 \text{ if heads} \\ -1 \text{ if tails} \end{array}$$

From the way the data are constructed, Y and X are clearly unrelated. We would not say that X determines Y, or that Y determines X. Instead, each wanders along its own path; these paths are shown in Figure 14.8.

But what happens if we estimate the regression equation $Y = \alpha + \beta X + \epsilon$? It turns out that we get

$$\hat{Y} = -1.75 + 2.07\,X, \qquad R^2 = .95$$
$$\quad\;\; [0.98] \quad\; [19.3]$$

[ ]: t-values

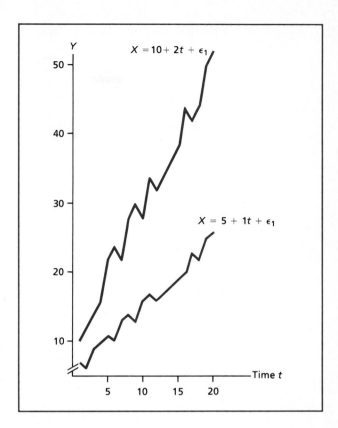

Figure 14.8 *Two Unrelated Variables that Grow over Time*

Although Y and X are generated independently, there turns out to be a highly statistically significant relationship between them! The reason is that both grow over time; that is, they both depend on the same third factor, the passage of time.

A second way of seeing this important point is to remember what R-squared measures—the mean squared forecasting error of the estimated model $Y = \alpha + \beta X + \epsilon$ relative to the mean squared forecasting error when the average value of Y is used to forecast Y.* If Y has been growing over time, then the average value of Y will be such a poor forecaster of Y that any variable X that also has been growing over time will seem to be a wonderful predictor by comparison.

Here is a simple example that some data grubbing turned up. Below are annual data on the number of golfers (X) in the United States and the total number of

*Recall that the formula for R-squared is

$$R^2 = 1 - \frac{\Sigma(Y_i - \hat{Y}_i)^2}{\Sigma(Y_i - \overline{Y}_i)^2}$$

missed days at work due to reported injury or illness (both in 1000s):

Year	X	Y
1960	4,400	370,000
1965	7,750	400,000
1970	9,700	417,000
1975	12,036	433,000
1980	13,000	485,000
1985	14,700	500,000

A least squares regression gives

$$\hat{Y} = 304{,}525 + 12.6\ X, \qquad R^2 = .91$$
$$\quad\ [14.2] \qquad\ [6.4]$$

[ ]: t-values

The relationship is statistically significant and very substantial—every extra golfer leads to 12.6 missed days of work. A simple explanation is, of course, that the number of golfers and the number of missed work days both have been growing along with the population.

Edward Ames and Stanley Reiter did a very interesting generic data mining experiment in which they randomly selected 100 economic variables.[17] They then ran regressions, trying to predict each of these variables by selecting an explanatory variable at random. This was intended to mimic the behavior of a completely uninformed data miner. They found that it usually took only from two to six tries to find an explanatory variable that gave an R-squared of at least .5. This is *not* a recommended research strategy! It is intended to show that we should not be overly impressed by high R-squareds and t-values, because these are so easy to obtain. Rather, we must exercise common sense and, if there is doubt, gather some fresh data.

EXERCISES

14.20 An agronomist has collected these data on the effects of fertilizer on crop yields:

Fertilizer X (Pounds/Acre)	Yield Y (Bushels/Acre)
0	5
50	12
100	18

Determine the least squares estimates of the parameters α and β in the equation $Y = \alpha + \beta X + \epsilon$.

Encouraged by these results, the agronomist collects some more data:

Fertilizer X (Pounds/Acre)	Yield Y (Bushels/Acre)
100	19
150	15
200	9

Use just these three new observations to estimate the parameters α and β in the equation $Y = \alpha + \beta X +$

ϵ. Now use all six observations to estimate α and β. Compare your three sets of estimates. What should the agronomist conclude?

14.21 Here are annual U.S. data on beer production (X) and the number of married people (Y):

Year	X	Y
1960	95	84.4
1965	108	89.2
1970	135	95.0
1975	158	99.7
1980	193	104.6
1985	200	107.5

Estimate the equation $Y = \alpha + \beta X + \epsilon$ to see if there is a statistically significant relationship at the 5 percent level between X and Y. If so, do you think that beer drinking leads to marriage, marriage leads to drinking, or what?

14.22 A business is interested in the effects of room temperature on worker productivity. It would like to reduce its heating bills, but not if lower temperatures substantially impair productivity. A consulting firm conducts an experiment, setting the temperature at 50°F in one of the business's office buildings, 60°F in a second building, and 70°F in a third. The observed average typing speeds (words per minute) turn out to be

Temperature X	Typing Speed Y
50	63
60	74
70	79

Calculate the least squares estimates of the parameters α and β in the equation $Y = \alpha + \beta X + \epsilon$. Is the estimate of β substantial? What are the predicted typing speeds for $X = 30$ and for $X = 90$? Would you use these predictions in making recommendations?

14.23 A study was made of the average sentence lengths of British public speakers over the period 1598 to 1940.[18]

Speaker	Year X	Words/ Sentence Y
Francis Bacon	1598	72.2
Oliver Cromwell	1654	48.6
John Tillotson	1694	57.2
William Pitt	1777	30.0
Benjamin Disraeli	1846	42.8
David Lloyd George	1909	22.6
Winston Churchill	1940	24.2

Find the least squares estimates of the equation

$$Y = \alpha + \beta X + \epsilon$$

What is the predicted value of Y for **a.** $X = 1984$? **b.** $X = 2100$? **c.** $X = 2200$?

14.24 A history teacher finds that several students do poorly on his midterm. He aims his lectures "a notch lower" and is pleased to find that these weaker students do better on the final exam; this success is tainted only slightly by the fact that his best students on the midterm do not do as well on the final exam. Does his experience show that weak students can handle basic material, but that strong students have trouble with elementary concepts?

14.25 In professional baseball, a National League and American League "Rookie of the Year" is selected each year. Most Rookies of the Year do not do as well in their second year of professional baseball as they did in their first year. Sportswriters call this the "sophomore slump" and attribute it to the pressure put on a star and to the distractions that fame brings. Is there any simpler, purely statistical explanation? Explain your reasoning carefully and clearly.

14.26 Here are some batting averages for the 1983 and 1984 seasons:

Player	1983	1984
Baker	.320	.295
Fred	.240	.220
Jones	.260	.250
Marv	.220	.240
Moose	.250	.260
Slim	.295	.320

Do Baker and Marv show that batting averages are converging to the mean? Or do the batting averages of Fred and Slim show that batting averages are diverging from the mean?

14.27 At Harvey Mudd College, students are given an achievement test when they enter as freshmen and then they are given the same test again as graduating seniors. The college has found that those students who do exceptionally well as freshmen do not do as well when they are seniors. Is a college education less useful to the brightest students? Do the very intelligent unlearn at college, perhaps by exposure to the less intelligent? Or what?

14.28 A study of U.S. males ages 21–34 found a positive correlation between income and weight. Does extra weight increase one's earning power? Do high-income people overeat? Or what?

14.29 A 1978 Gallup poll found that 30.3 percent of the people with a college education smoke cigarettes, while 43.5 percent of those with only a high school education smoke. Do these data show that smoking makes it harder to finish college?

14.30 People who have a college education earn much more, on average, than those who don't. Here are some census data on the median incomes of U.S. males, 18 years and older:

Years of Schooling	Approximate Years X	Median Income Y
less than 8	6	$7,035
8	8	$8,960
9–11	10	$9,924
12	12	$14,583
13–15	14	$15,674
16	16	$22,173
more than 16	18	$26,927

I have put in approximate years-of-education numbers so that you can use least squares to estimate the parameters of the regression equation $Y = \alpha + \beta X + \epsilon$. Is there a statistically significant relationship between education and income? According to your estimates, a person with an extra year of education earns, on average, how much extra income? Can we infer the value of another year of education from this estimate? Explain why you either would or would not use this estimate to advise a friend who is thinking of quitting school. According to your estimated equation, what is the predicted income of a professional student, say someone who went to school for 30 years?

14.5   SOME REGRESSION DIAGNOSTICS (OPTIONAL)

As explained earlier, a statistical analysis of the least squares estimates of the regression equation

$$Y = \alpha + \beta X + \epsilon$$

assumes that the error term ϵ has a constant standard deviation and that the values of ϵ are independent of each other. Violations of these assumptions can sometimes be detected by an examination of the prediction errors, or residuals,

$$e = Y - \hat{Y}$$
$$= Y - (a + bX)$$

Nobel Laureate Paul Samuelson has forcefully advised scientific forecasters to "always study your residuals." We have already seen examples of the value of this advice. A look at a scatter diagram and the prediction errors can reveal outliers in the data—perhaps an atypical observation or a typographical error. Here, we will look at how the residuals can be used to check the assumptions that ϵ has a constant standard deviation and that the values are independent of each other. These procedures work equally well for simple regression models with one explanatory variable X and for multiple regression models with more than one explanatory variable.

Checking the Standard Deviation

Suppose that we are investigating the variation among cities in the annual sales of weekly newsmagazines. Believing that the total sales Y in a city depend primarily on the city's aggregate income X, we collect the following data for a sample of ten cities:

City	Aggregate Income X ($ Millions)	Aggregate Sales Y (Thousands)
1	$620	964
2	340	462
3	7,400	10,440
4	3,000	4,200
5	1,050	1,310
6	2,100	2,610
7	4,300	5,220
8	660	948
9	2,300	2,830
10	1,200	1,560

Least squares estimation of the regression equation $Y = \alpha + \beta X + \epsilon$ yields

$$\hat{Y} = -\underset{[0.8]}{109.1} + \underset{[31.0]}{1.377} X, \qquad R^2 = .992$$

[ ] t-values

The results seem very satisfactory. The relationship between income and sales is, as expected, positive and, as hoped, statistically significant. The value of R^2 is nearly 1.

Following Samuelson's advice, the prediction errors are plotted in Figure 14.9 with income on the horizontal axis. It appears from this scatter diagram that the residuals are larger for high-income than for low-income cities. On reflection, this relationship is plausible. It is reasonable to anticipate that the errors in predicting aggregate magazine sales in a large city will typically be larger than the errors in predicting aggregate magazine sales in a small city. If so, the standard deviation is larger too, violating our assumption that the standard deviation of the error term is constant.

If the error term ϵ has a constant standard deviation, it is said to be homoskedastic; otherwise, it is heteroskedastic and the model is said to have a heteroskedasticity problem. Figure 14.9 strongly suggests a heteroskedastic error term. A variety of formal statistical procedures for testing for heteroskedasticity are

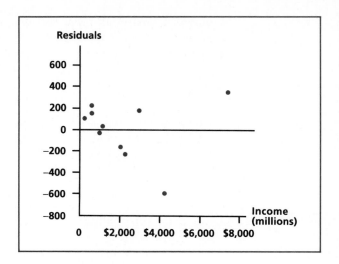

Figure 14.9   *Sales Residuals*

described in more advanced courses. Here, we will discuss two kinds of solutions to a heteroskedasticity problem.

The error term encompasses the effects of omitted variables; the magnitude and standard deviation of these omitted variables may not be constant. Particularly with time series data, the size and standard deviation of the omitted variables may increase over time, causing a heteroskedasticity problem. The solution is to include these omitted variables in the equation, using the multiple regression procedures described in Chapter 15.

A second solution is to respecify the model, rescaling the variables so that error term has a constant standard deviation. In our magazine sales example, the hetero-skedasticity problem apparently is caused by the fact that some cities are much larger than others. To give the data a comparable scale, we can respecify our model in terms of per capita income and per capita magazine sales, using the following population data:

City	Population P (Thousands)	Per Capita Income X/P (Thousands)	Per Capita Sales Y/P
1	40	$15.5	24.1
2	20	17.0	23.1
3	400	18.5	26.1
4	150	20.0	28.0
5	50	21.0	26.2
6	100	21.0	26.1
7	200	21.5	26.1
8	30	22.0	31.6
9	100	23.0	28.3
10	50	24.0	31.2

Least squares estimation of the regression equation ($Y/P = \alpha + \beta(X/P) + \epsilon$) yields

$$(Y/P) = \underset{[2.3]}{10.207} + \underset{[3.8]}{0.829} (X/P), \qquad R^2 = .641$$

[ ]: t-values

The plot of the residuals in Figure 14.10 shows no indication of heteroskedasticity. Evidently, the heteroskedasticity problem was caused by the use of aggregate data for cities of very different sizes and was solved by a rescaling to per capita data.

Notice that this rescaling caused a substantial change in the estimated slope of the relationship. Our initial regression states that a $1 million increase in aggregate income will increase aggregate sales by 1.377 thousand—implying an additional 1.377 magazines per $1,000 of income. The rescaled regression gives an estimate that is 40 percent lower: an additional 0.829 magazine per $1,000 of income. If our rescaled estimates are correct, then our initial estimates are very misleading.

On the other hand, some may be concerned by the observation that the rescaling caused a substantial decline in the t-value and R^2, suggesting that the rescaled model was not an improvement after all. However, we can't compare these statistical measures directly since the explanatory variables are different. One R^2, for example, gauges the accuracy in predicting aggregate sales; the other R^2 measures the accuracy in predicting per capita sales. If, as here, aggregate sales vary more than per capita sales, the per capita model could be more accurate and yet have the lower R^2!

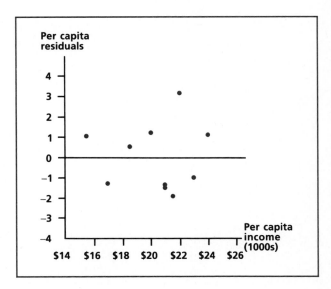

Figure 14.10 *Per Capita Sales Residuals*

To compare the predictive accuracy of these two models, we need to focus on a single variable—either aggregate sales or per capita sales. For the aggregate model the predictions are given by the regression equation; for the per capita model we can multiply predicted per capita sales by the city's population:

aggregate model: $\hat{Y} = -109.1 + 1.377\,X$

per capita model: $\hat{Y} = P\widehat{(Y/P)}$ where $\widehat{(Y/P)} = 10.207 + 0.829\,(X/P)$

Table 14.6 shows these predictions. The per capita model generally has smaller prediction errors, often substantially so. The standard error of estimate (the square root of the sum of the squared prediction errors, divided by $n - 2$) is 290.4 for the aggregate model and 188.3 for the per capita model. Despite its lower t-value and R^2, the per capita model actually fits the data better than the aggregate model.

This example provides a good illustration not only of one possible technique for solving a heteroskedasticity problem but also of the dangers of overrelying on t-values and R^2 to judge a model's success. A successful model should have a plausible estimate of β, reasonably accurate predictions, and residuals that are consistent with the model's assumptions.

Independent Error Terms

Sometimes, an examination of the residuals provides evidence that the values of the residuals are not independent of each other. Figure 14.11 shows one example. Instead of a randomness, there is a definite pattern to these residuals: consistently negative for low values of X, then consistently positive, and then negative again. The residuals are not independent because there are clumpings of positive and negative values.

In this case, as shown in Figure 14.12, the explanation is that there is a pronounced nonlinear relationship between X and Y, and the inappropriate use of a

Table 14.6 *Aggregate Predictions of the Aggregate and per Capita Models*

Actual Sales	Aggregate Model Predicted $(\hat{Y})$	Error	Per Capita Model Predicted $P\widehat{(Y/P)}$	Error
964	744.8	219.2	922.3	41.7
462	359.2	102.8	486.0	−24.0
10,440	10,082.2	357.8	10,218.3	221.7
4,200	4,022.5	177.5	4,018.4	181.6
1,310	1,337.0	−27.0	1,380.9	−70.9
2,610	2,783.1	−173.1	2,761.9	−151.9
5,220	5,812.9	−592.9	5,606.6	−386.6
948	799.9	148.1	853.4	94.6
2,830	3,058.5	−228.5	2,927.7	−97.7
1,560	1543.6	16.4	1,505.3	54.7

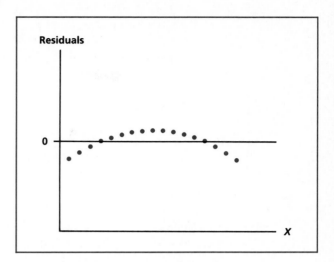

Figure 14.11 *A Suspicious Residual Graph*

linear approximation gives the residuals a telltale pattern. The solution in this case is to fit a nonlinear equation to these nonlinear data. Perhaps the replacement of X and Y with the logarithms of X and Y is appropriate:

$$\ln[Y] = \alpha + \beta\ln[X] + \epsilon$$

Or perhaps we can use the multiple regression procedures discussed in Chapter 15 to fit a quadratic equation:

$$Y = \alpha + \beta_1 X_1 + \beta_2 X^2 + \epsilon$$

In either case, one test of the appropriateness of the nonlinear specification is whether the residuals reveal a nonrandom pattern.

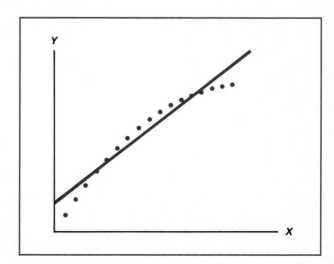

Figure 14.12 *The Solution to the Suspicious Graph*

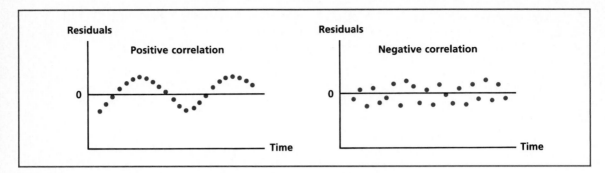

Figure 14.13   *Serially Correlated Residuals*

Figure 14.13 shows two examples of nonindependent residuals in regression equations that use time series data. In these cases, the residuals have been plotted, not against the explanatory variable, but against time, in order to show the sequence in which the residuals appear. If there is a nonrandom pattern in the sequence of residuals, as in Figure 14.13, the residuals are said to show serial correlation.

In the first case, the residuals are positively correlated in that positive values tend to be followed by positive values and negative values by negative values. Negatively correlated residuals, in contrast, zigzag back and forth, alternating between positive and negative values. A number of formal tests for serial correlation exist and are discussed in more advanced texts.

The usual cause of serially correlated residuals is the omission of an important explanatory variable that is itself serially correlated—usually positively so. For example, in an economic model, an important omitted variable might be gross national product or the unemployment rate. As the economy goes through business cycles, so do these variables, and their presence in the error term of a regression equation can give the positively serially correlated residuals shown in Figure 14.13.

The best cure for serial correlation is to identify the omitted variable, include it in the equation, and then apply the multiple regression procedures described in Chapter 15. Alternatively, sometimes the variables can be respecified to eliminate serial correlation. Researchers often use a t subscript to label the time period (daily, monthly, annual) of each observation,

$$Y_t = \alpha + \beta X_t + \epsilon_t \tag{14.4}$$

and assume that the error term ϵ_t has a first-order autoregressive structure,

$$\epsilon_t = \rho \epsilon_{t-1} + u_t \tag{14.5}$$

where ϵ_t = the error term in period t
        ϵ_{t-1} = the error term in the previous period
        ρ = the autocorrelation parameter, $-1 < \rho < 1$
        u_t = an independent error term with a constant standard deviation

If $\rho = 0$, the ϵ_t are independent; $\rho > 0$ shows positive serial correlations, and $\rho < 0$ is negative serial correlation. It can be shown that the least squares estimation of Equation 14.4 has the following consequences when there is serial correlation:

1. The estimates a and b of the parameters α and β are still unbiased, but their standard deviations are unnecessarily large.
2. The standard error of estimate SEE underestimates the standard deviation of the error term, causing us to underestimate the standard deviations of a and b and to overestimate their t-values.

Thus serial correlation causes the estimates to be less precise, while leading us to believe that the estimates are more precise than they really are.

The serially correlated error term ϵ_t can be eliminated by manipulating the model algebraically. Compare period t with the previous period:

period t: $Y_t = \alpha + \beta X_t + \epsilon_t$ implies $Y_t = \alpha + \beta X_t + \rho \epsilon_{t-1} + u_t$

period $t - 1$: $Y_{t-1} = \alpha + \beta X_{t-1} + \epsilon_{t-1}$ implies $\rho Y_{t-1} = \rho \alpha + \rho \beta X_{t-1} + \rho \epsilon_{t-1}$

Now take the difference between these two equations:

$$(Y_t - \rho Y_{t-1}) = (\alpha - \rho \alpha) + \beta(X_t - \rho X_{t-1}) + u_t \qquad (14.6)$$

Thus the model has been transformed so that the serially correlated error term ϵ_t is replaced with a well-behaved error term u_t.

In the special case $\rho = 1$, the transformed model simplifies to

$$(Y_t - Y_{t-1}) = \beta(X_t - X_{t-1}) + u_t$$

Thus a quick-and-dirty procedure that often works reasonably well is to omit the constant term and use the change in Y and the change in X as variables, rather than their levels.

Alternatively, we can estimate the value of ρ from the following three-step procedure:

1. Estimate Equation 14.4: $Y_t = \alpha + \beta X_t + \epsilon_t$.
2. Calculate the residuals $e_t = Y_t - (a + bX_t)$ and use these in place of the unobserved error term ϵ_t to estimate Equation 14.5: $\epsilon_t = \rho \epsilon_{t-1} + u_t$. Use the usual t-value to determine if the estimate of ρ is statistically significantly different from 0.
3. Use the estimate of ρ to transform the data and estimate Equation 14.6:

$$(Y_t - \rho Y_{t-1}) = (\alpha - \rho \alpha) + \beta(X_t - \rho X_{t-1}) + u_t$$

Most researchers prefer the following slightly different, but more powerful, method:

1. Estimate Equation 14.4: $Y_t = \alpha + \beta X_t + \epsilon_t$.
2. Calculate the residuals $e_t = Y_t - (a + bX_t)$ and use the Durbin-Watson test

described in the appendix to this chapter to determine if the estimate of ρ is statistically significantly different from 0.

3. If the null hypothesis $\rho = 0$ is rejected, then apply a generalized least squares procedure to Equation 14.6 to find the estimates of α, β, and ρ simultaneously that minimize the sum of squared prediction errors of Y_t.

In either case, the three steps are not carried out by hand, but by a computer program that incorporates these procedures.

EXERCISES

14.31 Many economists believe that there is an inverse relationship between the unemployment rate (X) and the rate of inflation (Y). This is why the government sometimes "fights inflation with unemployment." Estimate the equation $Y = \alpha + \beta X + \epsilon$ using the following data:

Year	% Unemployment	% Inflation
1960	5.5	1.6
1961	6.7	1.0
1962	5.5	1.1
1963	5.7	1.2
1964	5.2	1.3
1965	4.5	1.7
1966	3.8	2.9
1967	3.8	2.9
1968	3.6	4.2
1969	3.5	5.4

According to your estimates, how large a change in the inflation rate (give a 95 percent confidence interval) is associated with a one percentage point increase in the unemployment rate? Plot a scatter diagram of the prediction errors $Y - (a + bX)$ versus X. Do these errors seem to be scattered randomly about the line $e = 0$, or is there a discernible pattern? If there is a pattern, how do you explain it?

14.32 The boxed example in Chapter 13 entitled "Can Anyone Suggest a Good Hedge Against Inflation" (p. 000) reports the results of using data for 1926 to 1980 to estimate the relationship between stock returns (Y) and the rate of inflation (X). Estimate the equation $Y = \alpha + \beta X + \epsilon$, using the 1950–1989 data in Table 8 at the back of the book. Compare your results with those reported in the text for 1926–1980.

Make a scatter diagram with the regression residuals $Y - (a + bX)$ on the vertical axis and X on the horizontal axis. Does this graph seem consistent with the least squares assumption that the standard deviation of ϵ is a constant, independent of the value of X?

14.33 *(continuation)* Make a scatter diagram with the regression residuals $Y - (a + bX)$ on the vertical axis and the year of the observation on the horizontal axis. Does this graph seem consistent with the least squares assumption that the standard deviation of ϵ is a constant? Does there appear to be serial correlation in the residuals? If the computer program that you are using reports the value of Durbin-Watson's d-statistic, test at the 5 percent level the null hypothesis that $\rho \leq 0$. Estimate the equation $(Y_t - Y_{t-1}) = \beta(X_t - X_{t-1}) + u$ and compare your estimate of β and the associated t-value with the values obtained in Exercise 14.32. Explain why the values of R^2 either are or are not comparable.

14.34 Estimate the consumption function $Y = \alpha + \beta X + \epsilon$, where Y = real per capita consumption and X = real per capita disposable income, using the 1950–1989 data in Table 8 at the back of the book to calculate the values of Y and X. Make a scatter diagram of Y and X. Does the relationship look linear to you? Take the logarithms of Y and X and estimate the equation $\ln[Y] = \gamma + \delta\ln[X] + \nu$. Compare the accuracy of these two models in predicting Y.

14.35 *(continuation)* Calculate the prediction errors $e = Y - (a + bX)$ for each of these 40 years and make a scatter diagram with e on the vertical axis and the year on the horizontal axis. Does the standard deviation of the residuals appear to be constant, to increase over time, or to decrease over time? Does there appear to be serial correlation in the residuals? If your computer program gives the value of Durbin-Watson's d-statistic, test at the 5 percent level the null hypothesis that $\rho \leq 0$. Estimate the equation $(Y_t - Y_{t-1}) = \beta(X_t - X_{t-1}) + u$, and compare your results with those obtained in Exercise 14.34.

14.6 SUMMARY

A regression analysis begins with an explanatory variable X and a dependent variable Y. A scatter diagram is a useful first step to see whether or not there is any relationship between these two variables. We may find that there is a close linear relationship, a nonlinear relationship, or no relationship at all. A scatter diagram can also help identify outliers in the data. These outliers may have been recorded incorrectly, or they may reflect very atypical circumstances. In either case, outliers warrant closer inspection. If there is a recording error, then it can be corrected and we can be thankful that we found it before it ruined our regression. If the numbers are not in error, then we must decide whether the outlier is a misleading curiosity that can be discarded or an important piece of information that should be retained.

The next step is to fit a line to our data by computing the least squares estimates of the equation $Y = \alpha + \beta X + \epsilon$. We will be pleased if the estimate of β is plausible—that it has the right sign and a reasonable magnitude (taking into account the width of the confidence interval). If it has the wrong sign or an implausible magnitude, then we have a puzzle to unravel: perhaps our theory is wrong, or maybe there is an important omitted factor in the error term that is distorting our estimate.

The closeness of the fit can be gauged by the R-squared. If R-squared is close to 1, then most of the variation in the dependent variable is explained by the explanatory variable. If R-squared is close to 0, then the explanatory variable is of little predictive value; the regression equation's squared forecasting errors are not much smaller than those that can be obtained by discarding X and just using the average value of Y. For a more formal measure, we can calculate the t-value. If the t-value is greater than (approximately) 2, then the data reject the null hypothesis that $\beta = 0$: the improvement in predictive ability from using X as an explanatory variable is "statistically significant," which means that it is too great to be plausibly explained by chance alone. If the t-value is less than 2, then 0 is inside our confidence interval for β and cannot be rejected as a possible value for β.

A test of the null hypothesis $\beta = 0$ is interesting, but it is not the only question to be asked. We want our estimate of β (and the associated confidence interval) to be substantial and plausible. Suppose that we are on a business school admissions committee and that we believe GMAT scores are a useful predictor of success at our school, as measured by grade-point average (GPA). Using data for last year's class, we estimate the equation $Y = \alpha + \beta X + \epsilon$, where Y is GPA and X is GMAT score. If our estimate of β is -0.010, with a standard deviation of 0.002, then our t-value is a pleasing 5.0, but our estimate has the wrong sign! According to this estimate, those who do well on the GMATs do poorly in business school. We had better reexamine our data or rethink our theory, as there may be some other factor that is correlated with GMAT scores and is really affecting grades.

Another possible result is that our estimate of β turns out to be 0.00010 with a standard deviation of 0.00002. Again, the t-value is 5.0 and, this time, the estimate has the right sign. But still we would be disappointed in these results. A β of 0.00010 $\pm$ 0.00004 means that a 100-point difference in GMAT scores is associated with only a 0.006 to 0.014 difference in GPAs—hardly enough to make GMAT scores of any real value. A third possibility is that our estimate of β is 0.010 with a standard deviation of 0.006. This t-value is less than 2, but the estimate is substantial and suggestive. The 0.01 value of β means that a 100-point difference in GMAT scores is associated with a resounding 1.0 difference in GPAs. Our 95 percent confidence interval (from -0.2 to $+2.2$) is too wide to rule out no effect at all, but it is also centered at a place that encourages us to gather more data and keep watching GMAT scores. A fourth possibility, and the most satisfying of all, is a β estimate of 0.005 with a standard deviation of 0.001. Now, we've got an estimate with the right sign, a substantial and plausible magnitude, and a t-value that we can brag about.

APPENDIX: THE DURBIN-WATSON d-STATISTIC

Two statisticians, James Durbin and Geoffrey Watson, developed a powerful and widely used test for first-order serial correlation. To apply the Durbin-Watson test, we first estimate the regression equation $Y_t = \alpha + \beta X_t + \epsilon_t$ and then calculate the residuals $e_t = Y_t - (a + bX_t)$. The Durbin-Watson d-statistic is

$$d = \frac{\sum\limits_{t=2}^{n} (e_t - e_{t-1})^2}{\sum\limits_{t=1}^{n} e_t^2}$$

Notice that the numerator has one fewer observation than the denominator, since there is no value of e_{t-1} for the first value of e_t. In practice, the Durbin-Watson d-value is not calculated by hand but is instead part of a computer regression package. (If the lagged value of the Y is used as an explanatory variable, then the Durbin h-statistic should be used in place of the Durbin-Watson d-statistic.)

Since d is a ratio of sums of squared values, it must be positive. The largest possible value of d is 4. Equation 14.5 ($\epsilon_t = \rho \epsilon_{t-1} + u_t$) tells us when to expect the d-value to be close to 0 and when to expect it to be close to 4. If $\rho = 1$, then ϵ_t will, on average, equal ϵ_{t-1}, and d should be close to 0. If $\rho = -1$, on the other hand, then ϵ_t will, on average, equal $-\epsilon_{t-1}$ and $(\epsilon_t - \epsilon_{t-1})^2$ will be approximately equal to $(2\epsilon_t)^2 = 4\epsilon_t^2$, giving a d-value close to 4. If there is no serial correlation, $\rho = 0$, then it can be shown that the expected value of d is equal to 2.

Durbin and Watson computed tables, such as Table 7 at the end of this book, that can be used to determine whether the d-value is sufficiently far from 2 to reject the null hypothesis $\rho = 0$. Their test is unusual in that there are two critical values—d_L and d_U—and the test is sometimes inconclusive.

In the typical case, negative serial correlation is ruled out in advance and the null and alternative hypotheses are

$$H_0: \rho \leq 0$$

$$H_1: \rho > 0$$

As already explained, values of d that are close to 0 provide evidence of positive serial correlation. The upper and lower critical values in Table 7 at the back of this book are used to divide the possible values of d into three regions:

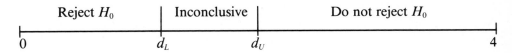

If negative serial correlation is not ruled out in advance, the null and alternative hypotheses are

$$H_0: \rho = 0$$

$$H_1: \rho \neq 0$$

Now values of d that are close to 0 or 4 provide evidence of serial correlation, and the upper and lower critical values in Table 7 are used to divide the possible values of d into five regions:

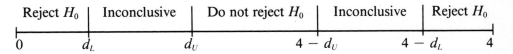

Suppose, for instance, that $n = 50$ observations are used to estimate a regression equation with $k = 1$ explanatory variable. For a one-tailed test of the null hypothesis $H_0: \rho \leq 0$ at the 5 percent level, Table 7 tells us that $d_L = 1.50$ and $d_U = 1.59$. Therefore, the null hypothesis is rejected at the 5 percent level if the Durbin-Watson d-value turns out to be less than 1.50, is inconclusive for a d-value between 1.50 and 1.59, and does not reject the null hypothesis if the d-value is larger than 1.59.

In contrast, for a two-tailed test at the 5 percent level with $n = 50$ and $k = 1$, Table 7 (with .025 in each tail) gives $d_L = 1.42$ and $d_U = 1.50$. Now the null hypothesis is rejected at the 5 percent level if the Durbin-Watson d-value turns out to be less than 1.42 or larger than $4 - 1.42 = 2.58$, is inconclusive for a d-value between 1.42 and 1.50 or between 2.50 and 2.58, and does not reject the null hypothesis if the d-value is between 1.50 and 2.50.

REVIEW EXERCISES

14.36 A rope manufacturer believes that the breaking strength of its ropes depends on what fraction of the cord consists of nylon, rather than cotton. Ten ropes are constructed with varying percents of nylon to study this relationship:

% Nylon X	Breaking Strength (Pounds) Y
0	260
10	360
20	490
20	510
30	600
30	600
40	680
50	820
60	910
70	990

Plot these data and then find the least squares estimates of the relationship $Y = \alpha + \beta X + \epsilon$. Draw this estimated line on your scatter diagram. **a.** What is the predicted breaking strength of a 33 percent nylon rope? **b.** What is the predicted breaking strength of a 66 percent nylon rope?

• 14.37 *(continuation)* Answer these questions that were posed by the company's management: **a.** Why are the observed breaking strengths of the two 20 percent nylon ropes not the same? **b.** What is the equation's predicted breaking strength for a 0 percent nylon rope? Why isn't this prediction 260 pounds, the same as the measured breaking strength for a 0 percent nylon rope? **c.** What is a 95 percent confidence interval for the breaking strength of a 0 percent nylon rope? **d.** Would you discard your results if it turned out that the measured breaking strength of a 0 percent nylon rope was outside of this 95 percent confidence interval?

•• 14.38 *(continuation)* The cost C ($/foot) of manufacturing rope depends on the nylon content (X). The price P ($/foot) that the rope can be sold for depends on its breaking strength. If the cost of manufacturing the rope is given by the equation $C = 0.02 + 0.10X$, then use your least squares estimates to find the minimum price that the firm must charge to make a profit on a rope with a predicted breaking strength

of 500 pounds. What about a rope with an 800-pound predicted breaking strength?

14.39 The Ace Software Company has developed a new computer game. To determine the most profitable price to charge, the company decides to sell this game at different prices in five similar cities. The results are

City	Price	Sales
A	$20	1010
B	$25	813
C	$30	721
D	$35	542
E	$50	305

Plot these data with price (X) on the horizontal axis and sales (Y) on the vertical axis. Find the least squares estimates of the equation $Y = \alpha + \beta X + \epsilon$. **a.** Is there a statistically significant relationship at the 5 percent level between price and sales? **b.** According to the estimated equation, what will be the effect on sales of a $1 increase in price? **c.** Is this estimated effect substantial?

•• 14.40 *(continuation)* Use this estimated equation to find the price that maximizes profits (revenue less costs). The revenue equals the price multiplied by sales. Assume that the cost is $2 per game, so that the total cost is $2 multiplied by sales.

14.41 You are considering two risky investments, each with a .5 probability of a 100 percent gain and a .5 probability of a 50 percent loss. If you put half of your wealth in each investment, what are the possible overall returns (and associated probabilities) if the two outcomes are **a.** independent? **b.** perfectly positively correlated, so that the outcome of the second investment will be the same at the first? **c.** perfectly negatively correlated, so that the outcome of the second investment will be the opposite of the first? Which of these three scenarios seems the safest? Which seems the riskiest?

14.42 You are considering n investments, each with a .5 probability of a 100 percent gain and a .5 probability of a 50 percent loss. The returns are independent. If you divide your wealth equally among these n investments, what are the possible overall returns (and associated probabilities) if n is **a.** 1?

b. 2? **c.** 10? Which of these three strategies is safest? Which is riskiest? Explain.

• **14.43** You are the middleman between two eager gamblers. One wants to bet that some interesting event will happen, while the other wants to bet that it won't. The first will put up $1 against your $1. The second will put up $1.30 against your $0.70. If you have $10,000 available, how should you divide your wagers between these two gamblers to assure yourself a fixed profit, regardless of whether or not the event happens?

14.44 An agronomist has used ordinary least squares to estimate the effect of fertilizer dosage X on crop yield Y:

$$\hat{Y} = 50.5 + 0.05\, X$$
$$\quad\; [1.7] \quad [2.5]$$

The t-values are in brackets. The fertilizer dosage data had a mean value of 210 gallons, with a standard deviation of 20 gallons. **a.** Does fertilizer dosage have a statistically significant effect (at the 5 percent level) on crop yield? **b.** What would be a 95 percent confidence interval for the estimate of the effect of fertilizer dosage on crop yield? **c.** If we were to use this estimated equation to predict Y for different values of X, for which value of X would we be most confident about our predictions? **d.** What is our predicted value of Y for $X = 300$? For $X = 500$? Is there any reason for extra caution about our forecasts of Y for $X = 500$?

14.45 Is there any reason why a stock's beta coefficient and expected return should be related? Calculate the average returns and estimated beta coefficients for the four stocks in Table 14.2 to see if there is any apparent relationship.

• **14.46** Investment advisory services routinely adjust their estimated beta coefficients. For instance, a firm with an estimated β of 1.7 might be assigned a β of 1.5, while a firm with an estimated β of 0.3 might be given a β of 0.5. Can you think of any logical reason for this peculiar practice?

14.47 Baseball players who are batting close to .400 at the halfway point in the season almost always find that their batting average declines during the second half of the season. This slump has been blamed on such factors as fatigue, pressure, and too much media attention. Is there any possible statistical explanation? In view of these data, if you were a major

league manager, would you take a .400 hitter out of the lineup in order to play someone who was batting only .250 at the halfway point?

14.48 The National Football League (NFL) reported that most teams that won more than half their games in 1972 did not do as well in 1973 as they had done in 1972. Conversely, most teams that lost more than half of their games in 1972 did better in 1973 than they had done in 1972.[19] The NFL claims that these data show that teams are balanced by the college player draft, in which the teams with the poorest records choose first. What alternative interpretation is provided by statistical theory?

14.49 In the first 21 Super Bowls (through 1987), only 4 teams repeated as champs. Sid Gillman attributes this falling off to human psychology:

> In this game, you are dealing with the human mind and nothing else.
>
> You know, it's a study for psychologists, what winning Super Bowls does to people. Coaches who win it say they'll keep things the same the following season, but they never do. Players don't either. Everything becomes deaccelerated. Humans just cannot deal with prosperity.[20]

Even if people could deal with prosperity, why might we expect most Super Bowl champions not to be repeat winners?

14.50 Albert Ando and Franco Modigliani collected the following data on the income and consumption of homeowners who are not self-employed.[21]

Income Bracket ($)	Average Income ($)	Average Consumption ($)
0–999	556	2,760
1,000–1,999	1,622	1,930
2,000–2,999	2,664	2,740
3,000–3,999	3,587	3,515
4,000–4,999	4,535	4,350
5,000–5,999	5,538	5,320
6,000–7,499	6,585	6,250
7,500–9,999	8,582	7,460
10,000–above	14,033	11,500

Use least squares to estimate the equation $Y = \alpha + \beta X + \epsilon$, where Y is consumption and X is income. What is the estimated effect on consumption of $1,000 higher income?

14.51 *(continuation)* Ando and Modigliani were concerned about the regression-toward-the-mean phenomenon. People with unusually high income in these data were more likely to have had atypically good years than bad years. Their observed income is consequently an overestimate of the typical income on which their consumption is based. In the same way, those with very low incomes are more likely to have had a bad year than a good year. Because their observed income understates the typical income on which their spending is based, they will seem to be spending a very high fraction of their income (indeed, more than 100 percent!).

Ando and Modigliani regrouped these families by the values of their homes, reasoning that the homes people buy are related to their typical incomes, rather than income in any particular year.

Home Value ($)	Average Income ($)	Average Consumption ($)
7,500–9,999	4,649	4,600
10,000–12,499	5,191	5,140
12,500–14,999	5,729	5,610
15,000–17,499	5,948	5,949
17,500–19,999	7,547	7,170
20,000–24,999	9,607	9,500
25,000–above	11,267	11,100

Because there is no reason to expect the people in any particular home-value bracket to have had either atypically good or bad years, average income should be a good estimate of typical income, which is what influences consumption.

Use these consumption and income data to estimate $Y = \alpha + \beta X + \epsilon$. Compare these estimates with those obtained in the previous exercise. Is there evidence of regression toward the mean?

14.52 A study of U.S. males ages 21 to 54 found a positive correlation between income and blood pressure, a positive correlation between income and age, and a positive correlation between blood pressure and age. What do you think is causing what here?

14.53 A doctor notices that more than half of the young patients are male, while more than half of the older patients are female. Does this indicate that elderly females are less healthy than elderly males?

14.54 Let Y be the first 8 numbers in the second column of Table 3 at the end of the book, and let X be the twenty-fourth through thirty-first numbers in the seventh column. Run a regression to see if there is a statistically significant relationship (at the 5 percent level) between Y and X. If so, does this statistical significance convince you that the numbers in Table 3 aren't really random as advertised?

14.55 To investigate the correlation between two unrelated variables that are both growing over time, use 20 coin flips to generate 10 observations of Y and X:

$$Y = 2t + \epsilon_1 \quad X = 2t + \epsilon_2$$

where $t = 1, 2, \ldots, 10$ and the ϵs are equal to $+1$ if heads and -1 if tails. **a.** Carefully record the data you construct. **b.** Plot these data with X and Y on the vertical axis and t on the horizontal axis. **c.** Use least squares to estimate $Y = \alpha + \beta X + \epsilon$. **d.** Is there a statistically significant relationship between Y and X at the 5 percent level?

14.56 Add a cautionary note to this statement: "The combination of the regression coefficient and the intercept allows us to predict the value of the dependent variable for any value of the independent variable."[22]

14.57 Many economists believe that the faster the rate of growth of a nation's money supply, the faster the rate of inflation—if not in the short run, at least in the long run. The following table shows the annual percentage changes in M1 (a common measure of money) and consumer prices for ten developed countries during the years 1980–1990. Use these data to estimate the equation $Y = \alpha + \beta X + \epsilon$, making the appropriate choice about which variable is X and which is Y. Do your results confirm the theory? What criteria did you use to judge whether these data are consistent with the theory?

Country	Annual Percentage Change in M1	Annual Percentage Change in Consumer Prices
Britain	16.54	7.55
Canada	11.47	6.65
France	6.83	7.4
Italy	9.67	11.34

Country	Annual Percentage Change in M1	Annual Percentage Change in Consumer Prices
Japan	4.47	2.58
South Korea	12.88	8.30
Spain	11.79	10.46
Switzerland	1.71	3.53
United States	7.14	5.67
West Germany	5.52	2.99

14.58 *(continuation)* Make a scatter diagram of Y and X, labeling each point with the country's name. Does the relationship look linear to you? Do any of the countries seem to be obvious outliers, in that if that country were removed from the data, the shape of the least squares line would change dramatically?

14.59 *(continuation)* Calculate the prediction errors $e = Y - (a + bX)$ for each of these ten countries and make a scatter diagram with e on the vertical axis and X on the horizontal axis. Draw a horizontal line at $e = 0$. If the average value of e does not seem to be zero, explain why it is positive or negative. If there seems to be a pattern in the values of e, identify it. Would a nonzero average value or a pattern for e suggest a violation of the least squares assumptions about ϵ?

14.60 *(continuation)* Construct a 95 percent confidence interval for β. Do these data reject at the 5 percent level the null hypothesis that $\beta = 1$? At the 1 percent level? Explain why a failure to reject $\beta = 1$ does not prove that $\beta = 1$. Give two hypothetical examples of regression results: In the first, the data reject $\beta = 1$; in the second, the data do not reject $\beta = 1$. Yet, in the first case, you are more certain that the value of β is close to 1.

14.61 The following table gives data on the height X of U.S. presidents who died of natural causes and their age at death Y.[23] Use these data to estimate the regression equation $Y = \alpha + \beta X + \epsilon$. Do your estimates indicate a relationship that is positive or negative? Statistically significant? Substantial?

U.S. President	Height	Lifespan
Thomas Jefferson	75	83
Lyndon Johnson	74	64
Franklin Roosevelt	74	63

U.S. President	Height	Lifespan
Chester Arthur	74	56
George Washington	74	67
Andrew Jackson	73	78
Warren Harding	72	57
William Taft	72	72
James Buchanan	72	77
John Tyler	72	71
James Monroe	72	73
Herbert Hoover	71	90
Woodrow Wilson	71	67
Grover Cleveland	71	71
Dwight Eisenhower	70	78
Calvin Coolidge	70	60
Theodore Roosevelt	70	60
Andrew Johnson	70	66
Franklin Pierce	70	64
Millard Fillmore	69	74
Harry Truman	69	88
Rutherford Hayes	69	70
Ulysses Hayes	69	63
Zachary Taylor	68	65
James Polk	68	53
William Harrison	68	68
John Quincy Adams	67	80
John Adams	67	90
Benjamin Harrison	66	67
Martin Van Buren	66	79
James Madison	64	85

14.62 *(continuation)* What has happened to heights and longevity of the general population during the past 200 years? Explain how these trends could create a misleading impression of a causal relationship between the height and longevity of U.S. presidents.

14.63 An investment consultant has phoned to tell you that he has identified the top 0.5 percent of the all mutual funds, funds that averaged a 16 percent annual return during the past five years. Why might you hang up without paying to learn their names?

14.64 Wiesenberger Services monitors the performance of many mutual funds to help investors select an appropriate fund. At the end of 1986 they identified the five best-performing mutual funds in each of twelve fund categories. In 1987, 27 of these 60 funds did better than average in their category and 33 did worse than average.[24] How do you explain this disappointing performance?

14.65 Explain the flaw in this advice: "The problem of picking a mutual fund seems to be simplicity itself: the investor should select whichever fund will give him the largest percentage return, year after year."[25]

14.66 In the 1960s the following data were collected for 21 countries on[26]

Y = deaths from coronary heart disease per 100,000 persons of age 35–64

X = annual per capita cigarette consumption

Use these data to estimate the regression equation $Y = \alpha + \beta X + \epsilon$. Do your estimates indicate a relationship that is positive or negative? Statistically significant? Substantial?

Country	Cigarette Consumption	Coronary Heart Disease
United States	3,900	259.9
Canada	3,350	211.6
Australia	3,220	238.1
New Zealand	3,220	211.8
United Kingdom	2,790	194.1
Switzerland	2,780	124.5
Ireland	2,770	187.3
Iceland	2,290	110.5
Finland	2,160	233.1
West Germany	1,890	150.3
Netherlands	1,810	124.7
Greece	1,800	41.2
Austria	1,770	182.1
Belgium	1,700	118.1
Mexico	1,680	31.9
Italy	1,510	114.3
Denmark	1,500	144.9
France	1,410	144.9
Sweden	1,270	126.9
Spain	1,200	43.9
Norway	1,090	136.3

14.67 In 1990 a Wharton business professor argued that baseball team owners should give in to the players' request for salary arbitration after two years in the majors (instead of three) because "players who are 'overpaid' [compared to peers with similar performance statistics] perform better than their previous career standard. . . . If owners acquiesce . . . the result may well be a higher performing team."[27] For instance, according to the professor's calculations, Bo Jackson was paid 63.4 percent more in 1988 than was justified by his seniority and performance before 1988 and he played better in 1988 than in 1987. As an owner, how could you counter this argument?

14.68 *U.S. News & World Report* once reported: "When a family gets a second or third car, you might expect that miles of driving per car would be reduced—but it doesn't work out that way. A study by economists of the Bureau of Public Roads indicates that families with only one car average 9,900 miles in a year but those with two or more average 10,000 miles for each car."[28] Does the purchase of a car make people feel they must drive more? Give a plausible explanation for the observed positive correlation between the number of cars a family has and the number of miles it drives.

14.69 Explain why we could be skeptical of the implication that driver-training courses make people worse drivers: "A study comparing 34 drivers who had taken courses with 466 who had not revealed that the driver-educated graduates actually had an accident rate 15% higher than their untrained counterparts."[29]

14.70 "Statistics show that more tickets are given to drivers of red cars than to any other car color. This proves that the police pick on people who drive red cars." Give two possible alternative explanations.

14.71 In June 1989 *Sports Illustrated* reported that for Kevin Mitchell, who then had 23 home runs, to threaten the single-season home run record, he "will have to avoid the traditional second-half power outage that plagues most fast starters."[30] According to *Sports Illustrated,* of the 39 batters who hit 20 or more home runs during the first half of the baseball season, only 2 hit 20 or more in the second half. How would a statistician explain this traditional power outage?

14.72 *Sports Illustrated*'s November 18, 1957, cover had a picture of the Oklahoma football team, which hadn't lost in 47 games, with the caption "Why Oklahoma Is Unbeatable." The Saturday after this issue appeared, Oklahoma lost to Notre Dame 7–0, starting the legend of the *Sports Illustrated* cover jinx—that the performance of an individual or team usually declines after they are pictured on the cover of *Sports Illustrated.* Explain how regression toward the mean might be used to explain the *Sports Illustrated* cover jinx.

14.73 The quantity theory equation is

$$MV = PQ$$

where M is a measure of the money supply (such as M1), P is the price level, Q is real annual GNP, and the product PQ is nominal annual gross national product (GNP). $V = PQ/M$ is the velocity of money, the average number of times per year that money is used to pay for GNP transactions. The English economist John Maynard Keynes argued that the velocity of money is positively related to interest rates—that when interest rates increase, money will turn over faster as individuals and businesses try to avoid holding idle money that could be invested earning high interest rates. To test this theory, use the data for 1970–1989 in Table 8 at the end of the book to estimate the equation $Y = \alpha + \beta X + \epsilon$, where $Y =$ velocity (nominal GNP/M1) and $X =$ interest rate on Treasury bills. Does your estimated equation confirm Keynes's theory?

14.74 *(continuation)* Calculate the prediction error $e = Y - (a + bX)$ for each of these 19 years, and make a scatter diagram with e on the vertical axis and time on the horizontal axis. Draw a horizontal line at $e = 0$. Does the average value of e seem to be zero? If not, how do you explain this? Do the values of e seem to be randomly scattered about the line $e = 0$, or does there seem to be a pattern? If there is a pattern, how do you explain it?

14.75 *(continuation)* Make a scatter diagram with the prediction error $e = Y - (a + bX)$ on the vertical axis and the Treasury bill rate (X) on the horizontal axis. Draw a horizontal line at $e = 0$. Does the average value of e seem to be zero? If not, how do you explain this? Do the values of e seem to be randomly scattered about the line $e = 0$, or does there seem to be a pattern? If there is a pattern, explain it.

14.76 *(continuation)* Make a scatter diagram of Y and X. Does the relationship look linear to you? Reuse the data in Exercise 14.73 to estimate the equation

$$\ln[Y] = \alpha + \beta\ln[X] + \epsilon$$

where the dependent variable is the natural logarithm of velocity (the natural logarithm of the ratio nominal GNP/M1) and the explanatory variable is the natural logarithm of the interest rate on three-month Treasury bills. Does this estimated equation support Keynes's theory about interest rates and velocity?

14.77 *(continuation)* The Federal Reserve monitors velocity and uses a regression equation to predict velocity for different values of the Treasury bill rate. Imagine that you work for the Fed and have estimated the equation in Exercise 14.73 and the one in Exercise 14.76. Which would you recommend to the Fed's Board of Governors? Defend your choice.

14.78 The quantity theory equation $MV = PQ$ described in Exercise 14.73 can also be interpreted as a statement that the demand for money is proportional to nominal GNP,

$$M = \frac{1}{V}(PQ)$$

Rearranging, we have money demand relative to nominal GNP,

$$\frac{M}{PQ} = \frac{1}{V}$$

If, as Keynes argued, a rise in interest rates increases velocity V, how does a rise in interest rates affect money demand? Use the 1970–1989 data defined in Exercise 14.73 to estimate the regression equation

$$\frac{M}{PQ} = \alpha + \beta X + \epsilon$$

where X is the interest rate on three-month Treasury bills. Identify consistencies and inconsistencies between this estimated equation and the equation estimated in Exercise 14.73.

14.79 *(continuation)* Now use these 1970–1989 data to estimate the equation

$$\ln\left[\frac{M}{PQ}\right] = \alpha + \beta\ln[X] + \epsilon$$

Identify any consistencies and inconsistencies between this estimated equation and the equation estimated in Exercise 14.76.

14.80 Use regression toward the mean to explain why movie sequels are usually not as good as the original.

14.81 Why do you suppose that the consumption function estimated in Chapter 13 used real per capita data on consumption and income? What regression assumption might have been violated by the use of the aggregate dollar values of consumption and income?

14.82 When in junior high school, Elizabeth Likens considered listening to classical music after she was told that "studies show that students who listen to classical music get better grades." What would you advise?

14.83 A meteorologist once observed that the price of corn in the fall is correlated with severity of hay fever (which is caused by ragweed). Does this mean that corn prices affect ragweed growth or that ragweed growth affects corn prices? Or is there another plausible explanation? Do you suppose that corn prices and hay fever are positively or negatively related?

14.84 The annual returns on a company's stock (r) and on the entire market r_M for the years 1972–1980 are shown in the following table. Use these data to estimate the company's beta coefficient. Does this stock have an above-average or below-average beta coefficient?

Year	Stock Return	Market Return
1972	23.9%	18.98%
1973	0.5	−14.66
1974	−4.5	−26.47
1975	21.6	37.20
1976	32.3	23.84
1977	1.9	−7.18
1978	7.6	6.56
1979	−5.6	18.44
1980	1.4	32.42

14.85 Identify a portfolio of stocks that has a beta of 1 and an R^2 of 1.

• **14.86** A researcher obtained the Scholastic Aptitude Test (SAT) (X) scores and grade-point averages (Y) of 396 college students and reported these regression results:

$$Y = \begin{matrix} 1146.0 \\ [5.86] \end{matrix} + \begin{matrix} 1.4312\, X, \\ [0.02] \end{matrix} \qquad R^2 = 0$$

[ ]: t-values

There is an error in these results. What is it?

14.87 *(continuation)* The researcher concluded, "Therefore I reject the null hypothesis and conclude that SAT scores and grades are not related." What is the statistical basis for this conclusion, and why is it wrong? What other important explanatory variables do you think affect grade-point averages?

14.88 Two Dutch researchers found that non-drinkers "are 50% more likely to get hurt on the [ski] slopes than skiers who imbibe heavily, and that people who sleep long and deeply are more prone to injury than those who are late-to-bed and early-to-rise."[31] Provide an explanation other than that the route to safer skiing is to drink heavily and sleep little.

14.89 Crime in downtown Houston increased by 40 percent during the first month after the Houston police launched a 1981 anticrime campaign, leading *Texas Monthly* to quip that "they should have launched a pro-crime campaign."[32] What other logical explanation is there for these data?

14.90 Table 9.2 gives the returns from a sample of 25 mutual funds for the years 1986–1989. Let Y be the mutual fund's return in the current year (either 1987, 1988, or 1989) and let X be the fund's return in the preceding year. Use these 75 observations on X and Y to estimate the regression equation $Y = \alpha + \beta X + \epsilon$. Do your estimates indicate a relationship that is positive or negative? Statistically significant? Substantial? What conclusion regarding mutual fund performance do you draw from this exercise?

14.91 The *Economic Report of the President* contains some economic data for seven major industrial countries, including the change in the civilian unemployment rate (Y) and the annual percentage growth rate of real gross national product (X). Do you think that these variables are positively or negatively related. Use the following data for 1984–1989 to estimate the equation $Y = \alpha + \beta X + \epsilon$ and see if you are right.

Country	Annual Percentage Growth in Real GNP	Change in Civilian Unemployment Rate
Canada	4.4%	−3.6%
France	2.4	0.1
Italy	3.0	2.0
Japan	4.6	−0.5
United Kingdom	3.4	−5.3
United States	4.0	−2.2
West Germany	2.8	−1.7

14.92 *(continuation)* Use the data in Exercise 14.91 to make a scatter diagram, labeling each point with the country's name. Which one of these points seems to be an outlier? Reestimate the equation $Y = \alpha + \beta X + \epsilon$, using the data for the six other countries. Did the removal of this one country make much different to your results?

14.93 The equations estimated in Exercises 14.91 and 14.92 were cross-sectional, comparing countries during a specified time period. Following are time series data for the United States during this same period. Use these data to estimate the equation $Y = \alpha = \beta X + \epsilon$, where Y is the annual change in the civilian unemployment rate and X is the annual percentage growth rate of real GNP.

Year	Percentage Growth in Real GNP	Change in Civilian Unemployment Rate
1984	6.8%	−2.1%
1985	3.4	−0.3
1986	2.7	−0.2
1987	3.7	−0.8
1988	4.4	−0.7
1989	3.0	−0.2

14.94 *(continuation)* Draw a scatter diagram with the data in Exercise 14.93, and label each year. Which year seems to be an outlier? Reestimate the equation $Y = \alpha + \beta X + \epsilon$, using the data for the five other years. Did the removal of this one year affect your results substantially?

14.95 The Boston Snow (B.S.) indicator uses the presence or absence of snow in Boston on Christmas Eve to predict the direction that stock prices will move the following year: "The average gain following snow on the ground was about 80% greater than for years in which there was no snow."[33] How do you explain the success of the B.S. indicator: Does snow on Christmas Eve affect the stock market, or does the stock market affect the weather?

14.96 Give a logical statistical explanation for the Cy Young jinx:

Chicago White Sox pitcher LaMarr Hoyt thinks his 1983 American League Cy Young Award [for the best pitcher] may have been a jinx, and the four previous winners might agree with him. "I'll tell you, there have been a lot of times this season I've felt jinxed," said

Hoyt. "A lot happens to me that is unexplainable.". . . Since Mike Flanagan of Baltimore won the award in 1979, all the AL winners have suffered disappointing seasons the following years. Flanagan has never won more than 16 games in a season since winning 23 in 1979. Steve Stone, the 1980 winner, retired less than two years later because of tendinitis. Rollie Fingers, the 1981 winner, and Pete Vuckovich, the 1982 winner, have been hampered with injuries since their big seasons.[34]

14.97 When people with high IQ scores are retested, most get somewhat lower scores. When people with low IQ scores are retested, most get somewhat higher scores. Carefully explain this phenomenon.

14.98 Rosenblatt and Cunningham found a strong positive correlation between family tension and the number of hours spent watching television.[35] Give a logical explanation other than television shows tend to increase family tension.

14.99 A petition urging schools to offer, and students to take, Latin noted: "Seventy per cent of English words are of Latin origin, hence Latin students have larger English vocabularies. In fact, Latin students score 150 points higher on the verbal section of the SAT than their non-Latin peers." How might a skeptic challenge this argument?

14.100 A Club Med advertisement was titled "Classical Music Linked to Movement of Tides":

"I was sitting near the water after a busy day of tennis and windsurfing," said a Philadelphia stockbroker vacationing here, "listening to classical music with a bunch of other people—it was Handel's 'Water Music'—when I looked out at the ocean. I couldn't believe my eyes.

"The tide, which had been moving in for hours, actually stopped in its tracks, turned around and started rolling out . . ."

At these villages, vacationers gather by the water's edge at the end of each day to listen to classical music—which has reportedly had other unusual effects as well.

"The sun had been shining brightly all day long," according to an art director from New York. "But when the music began, the sun started to disappear behind the ocean. Before long," she continued, "it was dark."

What is the statistical fallacy here?

15

Multiple Regression

The mouse is an animal which, killed in sufficient numbers under carefully controlled conditions, will produce a Ph.D. thesis.

Anonymous

TOPICS

In the natural sciences, researchers can often conduct controlled laboratory experiments to isolate the effects of one variable on another. For instance, an agronomist may be interested in the effects of nitrogen on the growth of a certain plant. This agronomist believes that the plant's growth depends on how much light, water, and nutrients (such as nitrogen, phosphorus, potassium, and boron) it receives. To isolate the effects of nitrogen on plant growth, the researcher can conduct a controlled experiment involving 50 similar plants, and try to give each plant the same amounts of light, water, and important nutrients (other than nitrogen), in order to control for these effects on plant growth. The 50 plants will be given different amounts of nitrogen and their growth will be recorded. Because the other important growth factors have been held constant, the researcher can attribute the observed variations in plant growth to the differing applications of nitrogen.

There will, of course, still be some inevitable chance variation in observed plant growth, because (a) not all plants are created equal; (b) there are errors in measuring growth; (c) the applications of light, water, and non-nitrogen nutrients cannot be precisely identical for each plant; and (d) there are always other growth influences over which the researcher had no control. However, if the most important influences have been identified and the controls imposed carefully, then the experiment should yield useful data for estimating the equation

$$Y = \alpha + \beta X + \epsilon$$

where Y is a measure of plant growth, X is the amount of nitrogen applied, and ϵ allows for the slight imperfections in the experiment.

In the social sciences, researchers can seldom set up these sorts of controlled laboratory experiments. Instead, they must make do with "nature's experiments"—and try to unravel the effects of one variable on another in a world in which many variables are changing at the same time. Did X_1 cause Y to go up, or was it caused by X_2 or maybe X_3? There are always lots of other Xs.

For instance, economists think that consumer spending depends both on household income and the value of household wealth. Normally, household wealth changes very little and thus has little effect on consumer spending. Therefore, little harm is done by relegating wealth to the error term and only taking into account the effect of income on consumer spending. However, there are other occasions when wealth changes dramatically due to stock market booms and busts, and economists suspect that consumer spending may be affected substantially during these occasions. The stock market crash in the 1930s has long been given some of the blame for the Great Depression. Although not as catastrophic, there have been wild swings in the stock market in recent years, too. In 1973–1974, the market fell by some $500 billion, which most likely contributed to the 1974–1975 recession; between 1984 and 1986, the market rose by more than a trillion dollars, giving the economy a welcome boost.

How can we estimate the effect of stock prices on consumer spending? When both income and stock prices collapsed in the 1930s, how much of the drop in consumer spending was due to the decline in income and how much was due to the decline in stock prices? If economists could follow the example of the agrono-mist's controlled experiments, they would hold the nation's income constant while varying stock prices to see how much consumer spending was affected. But econ-omists can't use the economy as their laboratory! They can't satisfy their intellec-tual curiosity by freezing everybody's income and manipulating the stock market.*

Instead, they have to make do with analyzing whatever data nature provides, and what they have come up with is a multiple regression procedure.

> In **multiple regression,** an equation with several explanatory variables is estimated, in an attempt to isolate the separate effect of each on the depen-dent variable.

15.1   THE LOGIC AND A FEW DETAILS

In general, a multiple regression equation can be written as

$$Y = \alpha + \beta_1 X_1 + \beta_2 X_2 + \cdots + \beta_k X_k + \epsilon$$ (15.1)

This model says that the dependent variable Y depends on k explanatory variables, plus an error term that encompasses various unspecified omitted factors. It doesn't matter in what order the explanatory variables are listed. What does matter is which k variables the researcher chooses to identify and which are left unspecified in the error term. Because there are, in principle, virtually an unlimited number of pos-sible explanatory variables, the art of regression analysis is choosing those with important effects and ignoring those with negligible effects.

The parameter β_1 gauges the effect of the first explanatory variable X_1 on the dependent variable Y, *holding the other explanatory variables constant.* Similarly, β_2 gives the effect of X_2 on Y, holding the remaining explanatory variables constant. These parameters measure the same thing that controlled laboratory experiments seek, which is the effect on Y of an isolated change in one explanatory variable. But while a controlled experiment can estimate a parameter β_i by actually holding the other variables constant, a multiple regression procedure estimates β_i by taking into account how uncontrolled changes in the other variables influence Y.

The mechanics for computing multiple regression estimates are quite compli-cated. They are best done with matrix algebra, and a thorough understanding requires at least one entire course advertised as *regression analysis, econometrics,* or some similar label. However, the logic is actually quite simple, because multiple regression is a straightforward extension of the least-squares estimation of an equa-tion with just one explanatory variable. Because researchers invariably use com-

---

*And even if they did, the results would be artificial, because people's behavior would be influenced by their knowledge of this temporary economic manipulation. For instance, they would rush to sell if they knew that an increase in a stock's price was part of an experiment, and therefore unrelated to its actual value.

puters to calculate their estimates, I will explain the logic and show you some computer results, without getting us bogged down in computational details.

Equation 15.1 is our model of how Y is determined. We want to obtain parameter estimates a, b_1, b_2, . . . , b_k, in order to make accurate predictions $\hat{Y}$, for selected values of the explanatory variables:

$$\hat{Y} = a + b_1X_1 + b_2X_2 + \cdots + b_kX_k \qquad (15.2)$$

The least squares criterion says that we should compute parameter estimates a, b_1, b_2, . . . , b_k so as to minimize the sum of squared prediction errors for the data we have. This can be done, although the actual calculations are usually left to a computer.

The standard error of estimate (SEE) is an estimate of the standard deviation of ϵ:

$$\text{SEE} = \sqrt{\frac{\Sigma(Y_i - \hat{Y}_i)^2}{n - (k + 1)}} \qquad (15.3)$$

Because we are using n observations to estimate $k + 1$ parameters, we divide by $n - (k + 1)$, the degrees of freedom. R-squared gauges predictive accuracy by comparing the model's errors to the errors of a benchmark model in which all of the β_i are set equal to zero and the sample mean alone is used to predict Y:

$$R^2 = 1 - \frac{\Sigma(Y_i - \hat{Y}_i)^2}{\Sigma(Y_i - \overline{Y})^2} \qquad (15.4)$$

The adjusted (for degrees of freedom) R-squared is

$$R_a^2 = 1 - \frac{\Sigma(Y_i - \hat{Y}_i)^2/(n - k - 1)}{\Sigma(Y_i - \overline{Y})^2/(n - 1)} \qquad (15.5)$$

We can also compute a standard deviation s_{b_i} for each estimate b_i (and for the intercept too, although this is usually the least interesting parameter). The actual formulas are complicated, and again it is best to let a computer do the work. Once a computer has done its job and revealed the estimated standard deviations, we can easily calculate a "plus or minus two standard deviations" 95 percent confidence interval for any parameter:

$$b_i \pm 2s_{b_i}$$

Hypothesis tests can be conducted directly, or they can be done implicitly by seeing whether or not the hypothesized value is inside the confidence interval. For the particular hypothesis

$$H_0: \beta_i = 0$$

we can see whether the value of the t-statistic

$$t_{b_i} = |b_i|/s_{b_i}$$

is greater than or less than 2. (This is, of course, still just a rule of thumb; the exact cutoff, using Table 5 at the end of the book with $n - (k + 1)$ degrees of freedom, will be either a bit larger or smaller than 2.0.) To test the hypothesis that *all* of the β_i equal zero, we can use the F-test discussed in the next chapter.

An Expanded Consumption Function

To illustrate the multiple regression procedure, let's include both income and wealth as explanatory variables in our consumption function. The expanded equation is

$$Y = \alpha + \beta_1 X_1 + \beta_2 X_2 + \epsilon$$

where Y = real per capita consumption
   X_1 = real per capita disposable income
   X_2 = real per capita value of corporate stocks

I entered annual data for 1977 through 1981 for consumption, income, and wealth into a multiple regression computer program. The results are shown in Figure 15.1. These results could be reported either as

$$\hat{Y} = -176 + 0.935\ X_1 + 0.047\ X_2, \quad SEE = 7.935, \quad R^2 = .993$$
$$\phantom{\hat{Y} = -176}\ (190)\quad (0.057)\quad\ (0.024)$$

( ): standard deviations

or as

$$\hat{Y} = -176 + 0.935\ X_1 + 0.047\ X_2, \quad SEE = 7.935, \quad R^2 = .993$$
$$\phantom{\hat{Y} = -176}\ [0.9]\quad\ [16.5]\quad\ \ [1.9]$$

[ ]: t-values

Notice first that the inclusion of wealth in the consumption function does not substantially alter the estimated coefficient for income. In the simple regression reported in Table 13.1, with only income as an explanatory variable, we estimated that a \$1 rise in income raises spending by $0.947 \pm 2(0.078)$. Our multiple-regression estimate is that, holding wealth constant, a \$1 rise in income increases spending by $0.935 \pm 2(0.057)$. Our estimate is little affected by taking into account the possible effects of wealth on spending.

It might have been the case that consumption depends only on wealth, but that income and wealth tend to go up and down together. If so, then a simple regression

```
  MTB > read     c1      c2     c3
DATA >         3042    3319   2481
DATA >         3124    3421   2229
DATA >         3108    3404   2125
DATA >         2994    3276   2133
DATA >         2971    3271   2064
DATA > end

    5 ROWS READ

MTB > name c1 'CONSUMPT' c2 'INCOME' c3 'WEALTH'
MTB > regress c1 2 c2 c3


The regression equation is
CONSUMPT = -175.9+0.935 INCOME+0.047 WEALTH

Predictor       Coef        Stdev      t-ratio         P
Constant      -176.09      190.27        0.93       0.235
INCOME          0.9347       0.0566     16.52       0.000
WEALTH          0.0470       0.0243      1.94       0.095


 s = 7.934     R-sq = 99.3%     R-sq (adj) = 98.6%
```

(b) **MYSTAT**

```
>edit
  'CONSUMPT'    'INCOME'    'WEALTH'
       3042        3319        2481
       3124        3421        2229
       3108        3404        2125
       2994        3276        2133
       2971        3271        2064
>save CONSUMPT
>quit


>use CONSUMPT
>model CONSUMPT = constant + INCOME + WEALTH
>estimate


DEP VAR:CONSUMPT  N: 5  MULTIPLE R: .996  SQUARED MULTIPLE R:    .993
ADJUSTED SQUARED MULTIPLE R: .986  STANDARD ERROR OF ESTIMATE: 7.934

VARIABLE     COEFFICIENT    STD ERROR      T       P(2 TAIL)
CONSTANT     -176.087        190.268      0.925       0.235
  INCOME        0.935          0.056     16.514       0.000
  WEALTH        0.047          0.024      1.935       0.095
```

*Enter
Data*

*Specify
Regression*

*Regression
Results*

*Enter and
Save Data*

*Specify
Regression*

*Regression
Results*

Figure 15.1 *Excerpts from a Multiple Regression of Consumption on Income and Wealth* of consumption on income would indicate that extra income increases spending when, in fact, it is the concurrent rise in wealth that increases spending. This is apparently not the case. Even when we allow for the influence of wealth on spending, we still find that $1 of additional income raises spending by close to $1.

The second thing to notice is that wealth, too, has a separate influence on spending. Even when we hold income constant, a $1 increase in stock prices tends to raise spending by about 5 cents. A $1 drop in stock prices tends to reduce spending by about 5 cents. Five cents may not seem like much, but the stock market is so volatile that these nickels can add up. A $100 billion swing in stock prices is not at all unusual, and at 5 cents to the dollar, a $100 billion swing in stock prices means a $5 billion change in consumer spending—which can be the difference between prosperity and recession. Notice, though, that our estimate of the effect of wealth on spending is not statistically significant. Because of the unrealistically small number of observations, there are only $n - 3 = 2$ degrees of freedom, the t-value for a 95 percent confidence interval is 4.303, and the confidence interval is $0.047 \pm 4.303(0.024)$, that is, between -0.056 and 0.150. We cannot rule out zero as the true value, nor can we rule out the possibility that the effect may be almost 15 cents of additional spending for every $1 of additional wealth.

Third, compare the standard error of estimate and R-squared for our multiple regression with those for our simple regression. The sum of squared prediction errors is always reduced by the inclusion of additional explanatory variables. Consider the equation

$$Y = \alpha + \beta_1 X_1 + \beta_2 X_2 + \epsilon$$

The least squares procedure selects those parameter estimates that minimize the sum of squared prediction errors. A simple regression with X_2 omitted assumes that β_2 is zero. A multiple regression with X_2 included can use a zero estimate for β_2 if that minimizes the sum of squared prediction errors and can use a nonzero estimate if that does even better. Thus, a simple regression can never have a smaller sum of squared prediction errors than a corresponding multiple regression. By the same reasoning, a simple regression can never have a higher R-squared than a corresponding multiple regression. The Standard Error of Estimate (SEE) and adjusted R-squared are a bit different because both take into account the degrees of freedom, and every extra parameter that is estimated uses up a degree of freedom. Thus, a simple regression can have a smaller SEE and a larger adjusted R-squared than a corresponding multiple regression, if the increase in the sum of the squared prediction errors is smaller than the accompanying increase in degrees of freedom.

In our consumption function example, it turns out that the multiple regression reduces the SEE (from 10.98 to 7.94) and increases the R-squared (from .980 to .993). The 30 percent reduction in the SEE does seem substantial. The R-squared increase is small, because we are already so close to 1.0. We can also ask whether these improvements are statistically significant: Is the improvement in predictive accuracy sufficiently impressive to convince us that wealth belongs in the consumption function? If we rephrase the question slightly, the answer is immediately apparent. We want to know if the improvement in predictive accuracy is sufficient to warrant the rejection of the null hypothesis that wealth has no effect on consumption,

$$H_0: \beta_2 = 0$$

Our t-statistic for b_2 answers this question! Here, it turns out that the t-value is 1.939, which is not high enough to show convincingly that wealth does indeed belong in the consumption function. The decrease in the sum of squared prediction errors and corresponding increase in R-squared are not quite large enough to rule out chance as the explanation.

Is History Math or English?

Exercise 14.4 examined the relationship between grades in a freshman history course (Y) and math (X_1) and verbal (X_2) SAT scores. Simple regression yielded these results:

$$\hat{Y} = 29.3 + 0.08 \ X_1, \qquad R^2 = .34$$
$$\quad [1.8] \quad [3.1]$$

$$\hat{Y} = 14.1 \quad 0.11 \ X_2, \qquad R^2 = .85$$
$$\quad [2.2] \quad [10.2]$$

[   ]: t-values

History grades seem to be positively related to math scores and verbal scores. These statistical relationships are substantial and statistically significant—a 100-point advantage on the math SAT test is worth 8 (± 5) extra points on the history test, while 100 points on the verbal SAT test is worth 11 (± 2) history points.

However, if we now run a multiple regression, using both math and verbal scores to predict history performance, we get

$$\hat{Y} = 16.5 - 0.01 \ X_1 + 0.11 \ X_2, \qquad R^2 = .85$$
$$\quad [2.1] \quad [0.5] \qquad [7.8]$$

Now the estimated effect of math scores on history scores is negative and not statistically significant! What's going on here?

This example illustrates the difference between simple and multiple regression, and the need for care in interpreting their parameters. The multiple regression coefficients measure the effect on Y of an increase in the associated explanatory variable, *holding the other explanatory variables constant*. Thus, the coefficient of math scores asks, if Moe and Joe both have 570 verbal SATs, but Moe has a 740 math SAT and Joe only a 520, should we expect Moe to get the higher grades in history courses? The answer, according to these data, is "No." Similarly, the coefficient of

Multiple Regression and the Law

In recent years, multiple regression estimates have been admitted as evidence in a wide variety of legal proceedings, including estimates of the effect of the death penalty on murder rates, of train firemen on railroad safety, and of cable television systems on network profits.[1] In each case, there are other important influences as well and controlled laboratory experiments are impractical; so multiple regression is used to estimate the effects that are of interest, taking into account the other influences, too.

Multiple regression has been used especially often in cases alleging discrimination by sex or race.[2] Consider, for instance, the claim that a company's wage rates discriminate against women. To support this claim, the plaintiff might produce data showing that, on average, the female employees of this company earn less than the males. In response,

the company might argue that its wage rates are based on experience, training, and skill, and that its female employees happen, on average, to be younger and less experienced than the male employees. It might even argue that, ironically, this situation is due to its recent, successful efforts to hire more women. How might a jury evaluate such claims statistically? As one econometrician who has often appeared as an expert legal witness says,

Multiple regression is well suited to answer this sort of question fairly precisely. Moreover, without multiple regression, it is difficult to see how it could be decided.[3]

A multiple regression equation explaining wages that takes into account the experience, training, skill, and sex of each employee can evaluate the firm's claim that its wages depend on the first three factors, but not the fourth.

the verbal SAT scores asks, if two students have the same math score, should the one with the higher verbal score be expected to do better in history courses? This answer is "Yes."

The simple regression equations ask somewhat different questions: If all we know about two students is that one had the higher verbal SAT score, would we predict a difference in history scores? If all we know is that one had the higher math SAT score, would we predict a difference in history scores? The answer to both of these questions is "Yes," because verbal and math SAT scores are positively related and verbal scores are apparently a good predictor of success in history courses. That is, students with high math SAT scores tend to do well in history classes, not because of their math skills, but because they also tend to have high verbal SAT scores and these do matter.

The apparent relationship between math and history scores is an example of a spurious correlation, in that one variable (history scores) seems to be related to another (math scores) but is, in fact, related to a third (verbal scores), which happens to be correlated with the second. One of the great appeals of multiple regression analysis is that it can sort out these interrelationships.

The Effect of Air Pollution on Life Expectancy

Many scientists believe that air pollution is hazardous to human health, yet they cannot run laboratory experiments on humans to confirm this theory. Instead, they must make do with "nature's data," statistics showing that life expectancies are lower and incidence of respiratory disease is higher for people living in cities with substantial air pollution. But perhaps these grim statistics can be explained by other factors—that the population of cities tends to be older, gets little exercise, and can be aggravated by a rushed lifestyle.

Multiple regression techniques can be used in place of laboratory experiments to control for these other factors. One study used biweekly data from 117 Standard Metropolitan Statistical Areas to estimate the following equation:[4]

$$\hat{Y} = 19.61 + 0.041X_1 + 0.71X_2 + 0.001X_3 + 0.41X_4 + 6.87X_5$$
$$\quad\quad\quad\;\; [2.5] \quad\quad [3.2] \quad\quad [1.7] \quad\quad [5.8] \quad\quad [18.9]$$

where $\hat{Y}$ = mortality rate, per 10,000 population (average = 91.4)
X_1 = average suspended particulate reading (average = 118.1)
X_2 = minimum sulfate reading (average = 4.7)
X_3 = population density per square mile (average = 756.2)
X_4 = fraction of population that is nonwhite (average = 12.5)
X_5 = fraction of population 65 and older (average = 8.4)
[ ] = t-values

The inclusion of the last three explanatory variables is intended to accomplish the same objectives as a laboratory experiment, in which these potential influences on mortality are held constant to isolate the effects of pollution on mortality. The two pollution measures both have substantial and statistically significant effects on mortality. For the average city, a 10 percent increase in the average suspended particulate reading, from 118.1 to 129.8, increases the mortality rate by 11.81(0.041) = 0.48. A 10 percent increase in the minimum sulfate reading increases mortality by .47(0.71) = 0.34.

Dummy Variables

So far, our explanatory variables have all had numerical values. Income was $10,000, $20,000, or $30,000; SAT scores were 400, 500, or 650. However, as illustrated by the preceding boxed discussion "Multiple Regression and the Law," some explanatory variables are qualitative rather than quantitative: male or female, white or nonwhite, and Republican or Democrat.

We can handle such qualitative variables by using **0-1 dummy variables,** variables whose values are equal to 0 or 1, depending on whether or not a certain characteristic is true. Suppose, for instance, that a certain firm claims that its salaries are based solely on experience and we want to investigate whether, taking experience into account, there are, in fact, systematic differences in the salaries that this firm pays male and female employees. We could collect a sample of male and female salaries and estimate the equation

$$Y = \alpha + \beta_1 D + \beta_2 X + \epsilon$$

where Y = annual salary
         X = experience (number of years)
         D = 1 if male, 0 if female

The parameters of this equation can be interpreted by considering the possible values of the dummy variable:

male: $D = 1$ and $Y = (\alpha + \beta_1) + \beta_2 X + \epsilon$

female: $D = 0$ and $Y = \alpha + \beta_2 X + \epsilon$

The parameter β_1 gauges the extra salary received by males, for any given experience level. The employer's claim is that $\beta_1 = 0$ (and presumably $\beta_2 > 0$). This equation can be estimated in the usual way, using the t-value for b_1 to test the employer's claim.

Perhaps we survey 47 people and obtain these estimates:

$$\hat{Y} = 19,960 + 500\,D + 1,020\,X, \quad R^2 = .62$$
$$(200.0) \quad (159.4)$$
$$[2.5] \quad\ [6.4]$$

( ): standard deviations
[ ]: t-values

These empirical results indicate that salary increases by about $1,020 with each additional year of experience and that, for any given level of experience, males earn $500 more than females. With $n = 47$ observations and $k = 2$ explanatory variables, there are $n - (k + 1) = 44$ degrees of freedom. Using the t distribution with 44 degrees of freedom, there is a .05 probability of a t-value with an absolute value larger than 2.1. Therefore, a 95 percent confidence interval for β_1 is $500 \pm 2.1($200) = $500 \pm $420 = $80 to $920. Since 0 is outside this 95 percent confidence interval (and the 2.5 t-value is larger than the 2.1 cutoff), these estimates reject at the 5 percent level the firm's claim that $\beta_1 = 0$.

Instead of using the dummy variable $D = 1$ if male, 0 if female, we could have specified the model using $D' = 1$ if female, 0 if male, and estimated this equation:

$$Y = \alpha' + \beta_1' D' + \beta_2' X + \epsilon$$

Since there is no logical reason for preferring one specification to the other, it would be disconcerting if it made a difference in our results. Fortunately, it doesn't.

The two alternative dummy variables, D and D', are related as follows:

	D	D'	$D + D'$
Male	1	0	1
Female	0	1	1

Thus $D = 1 - D'$, implying

$$Y = \alpha + \beta_1(1 - D') + \beta_2 X + \epsilon$$
$$= (\alpha + \beta_1) + (-\beta_1)X + \beta_2 D' + \epsilon$$

Therefore,

$$\alpha' = \alpha + \beta_1$$
$$\beta_1' = -\beta_1$$
$$\beta_2' = \beta_2$$

These theoretical relationships are, in fact, borne out by a reestimation of our model,

$$\hat{Y} = 20{,}460 - 500\,X + 1{,}020\,D', \qquad R^2 = .62$$
$$\phantom{\hat{Y} = 20{,}460} (200.0) \quad (159.4)$$
$$\phantom{\hat{Y} = 20{,}460} [2.5] \quad\;\; [6.4]$$

( ): standard deviations
[ ]: t-values

The coefficient of X is unchanged. The coefficient of D in our first specification measured the additional salary earned by a male, relative to a female with the same experience; a positive value indicates discrimination in favor of males. The coefficient of D' in our second specification measures the additional salary earned by a female relative to a male with the same experience; the negative value indicates discrimination in favor of males. Thus, if we are careful in our interpretation, it doesn't matter whether the dummy variable is 1 for male and 0 for female, or the other way around.

So far in our example, we have assumed that gender affects only the level of salaries, not the size of the increase that accompanies experience. We can handle the more general case with this specification:

$$Y = \alpha + \beta_1 D + \beta_2 X + \beta_3 DX + \epsilon$$

Did the *Washington Post* Profit from the Watergate Scandal?

It has long been believed that sensational stories sell newspapers, ranging from those headed by William Randolph Hearst and Joseph Pulitzer (labeled yellow journalists because they favored yellow paper) to today's supermarket tabloids. Clearly, many newspapers believe that emphasizing the sensational (or the bizarre) is more profitable than restricting their attention to straight news. A different question is whether a legitimate newspaper's temporary coverage of a sensational event, such as a war or scandal, boosts its profits.

It seems self-evident that a newspaper will increase its profits if it can increase its sales by covering an event that captures public attention. Yet Fred McChesney, a professor of law, business, and economics at Emory University, argues persuasively that this is not necessarily so.[5] He points out that newspaper revenue comes from both sales of the paper and from advertising revenue. In the United States, on average, advertisements take up more than half of newspaper space and advertising revenue accounts for 60 to 80 percent of a newspaper's total revenue. All-advertising papers succeed, while all-news papers fail.

News stories are, in effect, a loss-leader in that their costs are not covered by circulation revenue. News stories help persuade people to read newspapers, where their attention is attracted to the advertising whose revenue is essential to the newspaper's profitability. In fact, the news hole—the newspaper space containing news—is determined, not by the number of newsworthy events, but by the size of the advertising space.

In these circumstances, a temporary increase in circulation does not increase a newspaper's profits. In the short run, the price of the newspaper is fixed and so is its advertising revenue. (To increase advertising revenue, a newspaper must persuade clients that there has been a permanent increase in circulation.) An increase in sales reduces a newspaper's profits because the price charged for a newspaper does not cover the cost of printing it.

To test these ideas, McChesney examined the effect of the Watergate scandal, arguably the most sensational political event of this century, on the circulation and profitability of newspapers overall and of the *Washington Post* in particular, which provided the earliest, most original, and most thorough coverage. The Watergate scandal began with the *Washington Post*'s October 1972 disclosure that the June 1972 break-in at the Democratic party's Watergate headquarters was part of "a massive campaign of political spying and sabotage" and ended either in August 1973 with the Senate's public hearings and the disclosure of the existence of President Richard Nixon's taped conversations or in August 1974, when Nixon resigned. McChesney estimated several equations for the *Post* and for all newspapers, using these starting and stopping dates, and obtained similar results, indicating that his conclusions were robust. Two of his estimated equations are shown here.

The only available newspaper circulation data are annual. These data were used to compare the *Washington Post* with the *Wall Street Journal,* which provided relatively little coverage of Watergate. McChesney used a dummy variable to reflect the fact that the *Washington Post*'s principal competitor, the

Washington Star, folded in August 1981, and used another dummy variable for the Watergate years:

$$\hat{Y} = 290.10 + 0.761\ X + .325\ D_1 + .058\ D_2$$
$$\qquad\qquad\ [14.3]\qquad [6.1]\qquad\ [1.3]$$

$$R^2 = .97$$

where

Y = *Washington Post* circulation

X = *Wall Street Journal* circulation

D_1 = number of months that year that *Star* did not publish

D_2 = 1 in Watergate years, 0 otherwise

[ ] = *t*-values

The positive, statistically significant coefficient of X indicates that the *Washington Post*'s circulation increased over time, along with that of the *Wall Street Journal* (and other newspapers). The positive, statistically significant coefficient of D_1 indicates that, as expected, the demise of the *Star* had a favorable effect on the *Post*'s circulation. The positive coefficient of D_2 suggests that the Watergate scandal boosted the *Post*'s circulation, but this coefficient is not statistically significantly different from 0 at the 5 percent level.

To gauge the effect of Watergate on the *Post's* profitability, McChesney looked at the monthly rates of return on *Washington Post* stock, in comparison with the returns on all stocks and the returns on other newspaper stocks:

$$\hat{Y} = 0.014 + .947\ X_1 + 0.931\ X_2 - .018\ D$$
$$\qquad\qquad [9.1]\qquad\quad [5.3]\qquad\quad [0.8]$$

$$R^2 = .61$$

where

Y = *Washington Post* stock return

X_1 = overall market stock return

X_2 = index of other newspaper stock returns

D = 1 in Watergate months, 0 otherwise

[ ] = *t*-values

The positive, statistically significant coefficients of X_1 and X_2 indicate that, as expected, the *Washington Post*'s stock return went up and down with the overall market and with other newspaper stocks. The negative coefficient of the dummy variable suggests that the Watergate scandal had a negative impact on the *Washington Post*'s stock, but this coefficient is not statistically significant at the 5 percent level.

Based on these and similar regressions, McChesney concluded that there was no convincing statistical evidence that the reporting of the Watergate scandal was profitable for the *Washington Post* or for newspapers in general. If anything, the results suggest that the scandal may have temporarily increased circulation but reduced profits.

Again, our interpretation of the parameters of this equation can be aided by considering the possible values of the dummy variable:

male: $D = 1$ and $Y = (\alpha + \beta_1) + (\beta_2 + \beta_3)X + \epsilon$

female: $D = 0$ and $Y = \alpha + \beta_2 X + \epsilon$

A positive value for β_1 indicates that, independent of experience, males earn an additional amount β_1. A positive value for β_3 indicates that, for every additional year of experience, males earn an amount β_3 more than females.

For practice, interpret these results on your own:

$$\hat{Y} = 19,750 + 200\, D + 1,010\, X + 92\, DX, \qquad R^2 = .68$$
$$\qquad\quad (63) \qquad (185) \qquad (57)$$
$$\qquad\quad [3.2] \qquad [5.5] \qquad [1.6]$$

( ): standard deviations
[ ]: t-values

EXERCISES

15.1 An academic economist believes that consumer spending depends not only on income (positively) but also on interest rates (negatively). Having only read Chapters 13 and 14, he regresses consumption on interest rates and forgets about income. Are there any perils to this procedure? As it turns out, he obtains a positive slope estimate—high interest rates apparently encourage spending. Is there any possible statistical explanation of this result that is consistent with his theory?

15.2 *(continuation)* His sister, who is a chemist, proposes a controlled experiment in which incomes are held constant while interest rates are varied: "In this way, you will be able to see the effects of interest rate changes alone on spending." Why might the economist resist this sisterly advice?

15.3 In theory, a strong economy is good for the stock market and high interest rates are bad for the market. Here are some December data on the Standard & Poor's 500 stock price index, the unemployment rate, and long-term Treasury bond rates:

Year	S&P 500 Y	Unemployment X_1	Interest Rate X_2
1980	133.48	7.2	11.89
1981	123.79	8.6	12.88
1982	139.37	10.7	10.33
1983	164.36	8.2	11.44
1984	164.48	7.2	11.21
1985	207.26	6.9	9.60
1986	249.78	6.9	7.65

Estimate the equation

$$Y = \alpha + \beta_1 X_1 + \beta_2 X_2 + \epsilon$$

and write a brief summary of your results, telling whether these data are consistent with the theory. How does this estimated equation explain the market surge in 1985 and 1986?

15.4 *(continuation)* The reason a strong economy is good for the stock market is that profits rise, enabling firms to pay higher dividends in the future. Here are some fourth-quarter data on corporate after-tax profits:

Year	Profits X_1
1980	146.7
1981	135.0
1982	100.8
1983	141.1
1984	141.0
1985	140.0
1986	140.0

Estimate the equation

$$Y = \alpha + \beta_1 X_1 + \beta_2 X_2 + \epsilon$$

again, this time using profits as the first explanatory variable. Are these estimates consistent with the theory? How does this equation explain the 1985–1986 market rally?

15.5 A government economist has estimated the following consumption function

$$\hat{Y} = -25.0 + 0.90\, X_1 - 2.5\, X_2, \qquad R^2 = .98$$
$$\quad\;\;[0.8] \qquad [9.0] \qquad\; [1.5]$$

where $Y =$ consumption ($ billions), average 2,200

       $X_1 =$ income ($ billions), average 2,500

       $X_2 =$ interest rate (percent), average 10

       [ ] = t-values

What is the predicted value of consumption if **a.** income is 2,500 and the interest rate is 10? **b.** income is 2,600 and the interest rate is 10? **c.** income is 2,500 and the interest rate is 11? **d.** income is 2,600 and the interest rate is 11?

15.6 *(continuation)* In this estimated consumption function, explain what each of these numbers measures: **a.** 0.90 **b.** −2.5 **c.** −25.0 **d.** 9.0 **e.** 0.98

15.7 *(continuation)* Economists have long debated whether a higher interest rate encourages consumption, discourages consumption, or has no effect at all. Does this estimated equation settle that debate? What does R-squared tell us about this estimated equation? Would the R-squared be significantly affected if we omitted the interest rate and just regressed consumer spending on income?

15.8 Data collected by a county agricultural agency led to the following multiple regression:

$$Y = -120 + 0.10\, X_1 + 5.33\, X_2, \qquad R^2 = .50$$
$$\quad\;\;[13.4] \quad\; [5.1] \qquad\;\; [3.7]$$

where $Y =$ corn yield (bushels/acre), average value 30

       $X_1 =$ fertilizer (pounds/acre), average value 100

       $X_2 =$ rainfall (inches), average value 30

       [ ] = t-values

a. What does the coefficient of X_1 measure? If the population value of this parameter really is 0.10, what is the implication (in simple English)? **b.** What does the coefficient of X_2 measure? If the population value really is 5.33, what is the implication? **c.** What does the estimate −120 measure? If this is the correct population value, should we conclude that Y will equal −120 if there is no rainfall or fertilizer?

15.9 *(continuation)* The *Farmer's Almanac* predicts 2 inches less rain this year than usual. If the previous estimates are the correct population values, then how much additional fertilizer must farmers apply if they want to offset this rain shortfall? Corn is expected to sell for $2 per bushel, and an extra pound of fertilizer costs $0.25. If the previous estimates are correct, should farmers apply more fertilizer?

15.10 Throughout the 1960s and much of the 1970s, the Federal Reserve generally "leaned against the wind," using open market operations to stabilize interest rates. In October 1979, concerned about double-digit inflation, the Federal Reserve made a dramatic decision to pay less attention to stabilizing interest rates and to focus more on its monetary aggregate targets. Interest rate volatility subsequently increased substantially. To see if this had an effect on the relationship between stock dividend yields and long-term interest rates, use the 1970–1989 data in Table 8 at the end of the book to estimate the equation $Y = \alpha + \beta_1 D + \beta_2 X + \epsilon$, where $Y =$ stock dividend yield (S&P 500 dividend as a percentage of S&P 500 price; for example, the 1970 value is 3.83 percent), $X =$ long-term Treasury bond rate, and $D = 0$ for 1970–1979 and $D = 1$ for 1980–1989. What does the parameter β_1 measure? Does the sign of your estimate of β_1 make sense? Do your results indicate a statistically significant difference between pre-1979 and post-1979?

15.11 *(continuation)* Redo the preceding exercise, using data for 1950–1989, with $D = 0$ for 1950–1979 and $D = 1$ for 1980–1989. Are your results substantially affected by the inclusion of these pre-1970 data?

15.2   SEVERAL APPLICATIONS

To see multiple regression in action, we'll now look at a variety of applications. In each case, we are trying to explain some phenomena, but a simple regression model with one explanatory variable just won't do. We have good reason to suspect that there are several important influences, and relegating some to the error term leaves the model impotent. Instead, we will use a multiple regression with multiple explanatory variables.

The Effect of Auto Speeds on Fuel Consumption

OPEC's manipulation of oil prices and supplies in the 1970s created an energy crisis in the United States. The enormous U.S. consumption of fuel is expensive and makes us dependent on uncertain foreign oil supplies. A primary component of this fuel consumption is our 150 million motor vehicles—cars, trucks, buses, and motorcycles. Each year, these vehicles use about 800 gallons of fuel apiece, more than 100 billion gallons of fuel in all. The amount of fuel consumed by our vehicles depends on how much we travel, how much horsepower our vehicles have, and how fast we drive.

Some data are shown in Table 15.1. The speed data are based on a government sample of several hundred thousand cars traveling on rural interstate highways during offpeak hours. You can see how travel, horsepower, and speed all declined during the 1973–1974 oil embargo. So did fuel consumption. In the late 1970s, speed and travel both crept back up, while horsepower kept falling. Fuel consumption went up, too. Then in 1979–1980, oil prices were jacked up again, and travel and speed fell once more.

Table 15.1   *Some Data on U.S. Motor Vehicle Fuel Consumption*

Year	Fuel Used per Vehicle (Gallons) Y	Travel per Vehicle (Miles) X_1	Horsepower per Vehicle X_2	Average Speed (MPH) X_3
1970	851.5	10,341	178.3	63.8
1971	863.4	10,496	183.5	64.7
1972	884.7	10,673	183.0	64.9
1973	879.1	10,414	183.2	65.0
1974	818.3	9,900	178.8	57.6
1975	820.2	10,008	178.7	57.6
1976	835.4	10,195	175.7	58.2
1977	839.9	10,372	175.7	58.8
1978	843.0	10,431	174.5	58.8
1979	804.3	10,072	175.3	58.3
1980	738.1	9,763	175.6	57.5

Table 15.2 *Midterm, Homework, and Final Exam Scores*

Student Initials	Midterm X_1	Homework X_2	Final Exam Y
LC	82	94	77
CC	94	96	90
JC	80	84	70
MC	78	94	94
DD	83	84	87
MG	86	88	69
BH	81	92	77
PH	88	84	82
LK	86	76	75
DL	81	72	72
BM	80	68	62
LM	97	86	88
PM	90	90	86
LO	88	80	88
JP	89	86	83
MR	79	78	77
NR	83	84	84
AS	90	96	83
FS	91	94	86
SS	88	92	83
AT	80	78	81
BT	86	88	90
NT	83	96	77
HV	84	82	87
JV	86	96	87
RW	80	66	73
SW	83	68	73
VW	93	88	90
WW	86	94	87

The *t*-values indicate that midterm and homework scores are both useful in predicting final exam scores. The inclusion of a second explanatory variable causes a statistically significant drop in the standard error of estimate (and corresponding increase in R-squared).

The relationship is still far from perfect. Table 15.3 compares the final exam scores with those predicted by the estimated equation. In one-third of the cases, the actual score differs from the prediction by more than 5 points. The two largest surprises are MC, who did 13.1 points better than predicted, and MG, who did 13.7 points worse than predicted. Apparently the final exam is a necessary evil, in that it contains information that is not obtainable from the midterm and homework scores.

Table 15.3   *Final Exam Scores Are Not Predicted Perfectly from Midterm and Homework Scores*

Name	Final Score	Predicted Score	Prediction Error
LC	77	82.9	−5.9
CC	90	89.7	.3
JC	70	78.1	−8.1
MC	94	80.9	13.1
DD	87	79.6	7.4
MG	69	82.7	−13.7
BH	77	81.6	−4.6
PH	82	82.2	−.2
LK	75	78.1	−3.1
DL	72	74.1	−2.1
BM	62	72.1	−10.1
LM	88	87.5	.5
PM	86	85.4	.6
LO	88	80.7	7.3
JP	83	83.4	−.4
MR	77	75.3	1.7
NR	84	79.6	4.4
AS	83	87.7	−4.7
FS	86	87.5	−1.5
SS	83	85.2	−2.2
AT	81	75.8	5.2
BT	90	82.7	7.3
NT	77	84.1	−7.1
HV	87	79.4	7.6
JV	87	85.7	1.3
RW	73	71.3	1.7
SW	73	73.6	−.6
VW	90	86.2	3.8
WW	87	84.9	2.1

Can College Success Be Predicted from SAT Scores?

Colleges use a variety of criteria in deciding which applicants to accept for admission—including high school grades, SAT or ACT test scores, and letters of recommendation. Their primary objective is to identify those who are most likely to be successful college students. A college that routinely admits students who do poorly or drop out is not doing itself or the students any favors.

Many colleges have tried to estimate formal regression models using admissions criteria to predict college success. The first thing needed for such an endeavor is some measure of college success. College Grade Point Average (GPA) is a natural yardstick. We must be careful, though, in that courses are not all equally difficult. It may be harder to get an A in nuclear physics than in physical education, and harder to get an A in nonlinear differential equations than in introductory statistics. Because freshmen usually take very similar courses, we can reduce the differences between departments and between advanced and beginning courses by using the freshman-year GPA to gauge whether or not the admitted student is succeeding in college.

For explanatory variables, the obvious choices are those very factors that admissions committees look at. It may be a bit of work, though, to quantify some factors. For example, committee members must read carefully (and often between the lines) in deciding whether a letter of recommendation is enthusiastic, strong, lukewarm, or negative. Committee members also commonly adjust high school GPAs to take into account whether or not the applicant took challenging courses at a demanding high school.

Once data have been assembled on college success and admissions criteria, then a multiple regression equation can be used to see whether or not useful predictions can be made. Some colleges do this on their own. Others use the data analysis provided by the College Entrance Examination Board, which administers the SAT tests. To illustrate the procedure, I've obtained the actual regression equation estimated for 359 freshmen who enrolled at a certain very selective liberal arts college in the fall of 1977. The data are:

Y = freshman year GPA, average 3.0

X_1 = high school GPA, average 3.7

X_2 = verbal SAT score, average 591

X_3 = math SAT score, average 629

The least squares estimates are

$$\hat{Y} = 0.437 + 0.361\ X_1 + 0.0014\ X_2 + 0.0007\ X_3, \qquad R^2 = .26$$
$$\quad\ \ [20.3] \quad\ [6.23] \qquad\quad [5.50] \qquad\qquad [2.64]$$

[ ]: t-values

Each of the three estimated coefficients of the admissions variables is positive, substantial, and statistically significant. A student with a 3.7 high school GPA and with 600 verbal and math SAT scores is predicted to have a 3.03 freshman GPA at this college. If the high school GPA was a point lower (2.7 instead of 3.7), the predicted college GPA would fall to 2.67, all other things being equal. A 100-point decline in the verbal SAT score reduces the predicted freshman GPA by 0.14. So does a 200-point drop in the math SAT. A student with a 3.0 high school GPA and 500 SAT scores has a 2.46 predicted freshman GPA at this college.

This college has used its regression equation in a variety of ways. In some years, it calculates a predicted freshman GPA for each applicant and then admits some

The Risk of Coronary Heart Disease

Researchers are often interested in predicting the value of a probability—for example, the probability of producing a defective television, passing a licensing examination, or dying from coronary heart disease. If a regression equation such as

$$Y = \alpha + \beta_1 X_1 + \beta_2 X_2 + \beta_3 X_3 + \epsilon$$

is used with the dependent variable Y equal to a probability, then the value of Y must be between 0 and 1. However, a normally distributed error term ϵ has no minimum or maximum. In addition, there will inevitably be values of the estimated coefficients for which the predicted value of Y is less than 0 or larger than 1—neither of which can be accurate.

A variety of procedures have been developed to maintain the advantages of the regression model but not violate the 0–1 constraints on probabilities. One popular procedure is the logistic risk function, in which the dependent variable Y is a nonlinear function of the probability P that is of interest:

$$Y = \ln \left[\frac{P}{1 - P} \right]$$

where ln denotes the natural logarithm (base $e = 2.71828$). For $P = 0$, $Y = -\infty$; for $P = 1$, $Y = +\infty$. The probability P is given by

$$P = \frac{\ln[Y]}{\ln[Y] + 1}$$

and, as desired, can never be negative or larger than 1.

The logistic risk function was used in an analysis of data gathered over several years, beginning in the late 1950s, on the residents of Framingham, Massachusetts.[7] One of the estimated equations was

$$\ln \left[\frac{P}{1 - P} \right] = -9.7512 + 0.01575\, X_1 + 0.00649\, X_2 + 0.03315\, X_3$$

applicants solely on the basis of their high predicted GPAs. The remaining admissions are based not only on predicted GPA, but also on other criteria that are more difficult to quantify, such as letters of recommendation and special talents. In some years, this college does not formally calculate a predicted GPA, but it adds together the math SAT score plus twice the verbal SAT score and uses this together with other, more subjective criteria to select students. The math-plus-twice-verbal rule of thumb is based on the relative weights in the estimated regression equation. Because this same approximate two-to-one weighting holds for many college regression equations, the College Board also computes a selection index by adding the math SAT score to twice the verbal SAT score. The selection index for the PSAT tests taken by high school juniors is used to determine National Merit Scholarship winners.

where P = probability of dying from coronary heart disease

X_1 = systolic blood pressure

X_2 = cholesterol

X_3 = daily cigarette consumption

The estimated coefficients show the effects of the explanatory variables on the logarithmic function, which must be transformed to determine the effects on the probability of dying from coronary heart disease. Table 1 uses a variety of values for the explanatory variables to illustrate the estimated effects of blood pressure, cholesterol, and smoking on the risk of coronary heart disease. A systolic blood pressure of 170 rather than 130 roughly doubles the risk; so does smoking 20 cigarettes a day. An increase in cholesterol from 180 to 240 increases the risk by about 50 percent. The combination of all three risk factors increases the probability of coronary heart disease by 450 percent, from 14 per 10,000 people to 77 per 10,000 people.

Table 1

Blood Pressure	Cholesterol	Daily Cigarettes	Probability of Coronary Heart Disease
130	180	0	.0014
170	240	0	.0027
130	240	0	.0021
130	180	20	.0028
170	240	20	.0077

How accurate are these predictions? One measure is provided by a 95 percent confidence-interval calculation. Here it turns out that, for the average student, the predicted freshman GPA is 3.0 ± 0.8—about what one might expect. Admissions committees have a rough idea about how well a student will do, but there are always some surprises. Figure 15.2 makes this observation a little more concrete. Of these 359 freshmen, 198 had predicted freshman GPAs in the range of 3.0 to 3.33. Figure 15.2 shows that some of these students did very well and some did poorly, but most did as well as expected.

Another way to examine the accuracy of this prediction equation is to compare the performance of students with different predicted GPAs. In this freshman class, 21 students had predicted first-year GPAs in the range 3.33 to 3.67, while another 29 students were between 2.33 and 2.67. Table 15.4 shows that some students in

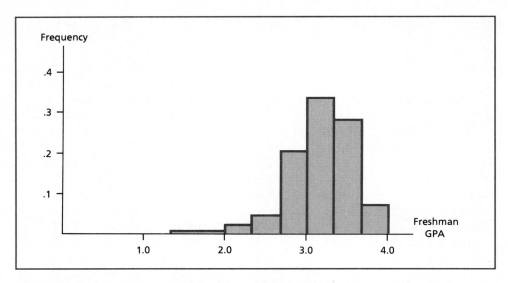

Figure 15.2   *Distribution of Freshman Grades for 198 Students Predicted to Get Between 3.00 and 3.33 GPA*

the second group did better than some students in the first group, but that, again, most students in the first group did better than most in the second group. The average freshman GPA was 3.4 for those in the first group and 2.5 for those in the second group. Six students from the first group got 3.67–4.0; no one in the second group did this well. Three students from the second group got 1.33–1.67; no one in the first group did this poorly.

This GPA prediction equation is not perfect, but it does have some value, and that's why this college, like many others, uses a multiple-regression equation as part of its admissions process; so do many business schools and law schools.

Table 15.4   *The College Performance of Two Groups Predicted to Do Differently*

Freshman Grade	21 Students with Predicted GPA 3.33–3.67	29 Students with Predicted GPA 2.33–2.67
3.67–4.00	6	
3.33–3.67	6	1
3.00–3.33	5	2
2.67–3.00	3	6
2.33–2.67		10
2.00–2.33	1	7
1.67–2.00		
1.33–1.67		3

Mechanical Rules Versus Subjective Evaluation

An interesting study of the University of Oregon's graduate program in psychology considered how well the following three variables explain the applicant's subjective rating by the admissions committee (Y_1) and, for those who choose to enroll in Oregon's program, their success as gauged by subsequent faculty ratings (Y_2):[8]

$$Y_1 = a_1 + 1.02\ X_1 + .0032\ X_2 + .0791\ X_3$$
$$Y_2 = a_1 + 0.76\ X_1 + .0006\ X_2 + .2518\ X_3$$

where

X_1 = undergraduate grade point average

X_2 = verbal plus quantitative GRE scores

X_3 = rating of applicant's undergraduate college

The author concluded that "the admissions committee does not place sufficient weight on the quality of the undergraduate academic institution." The author also found, paradoxically, that the regression equation estimated to explain the ratings of the admissions committee predicted the success of those admitted better than did the ratings themselves! Apparently, the admissions committee ratings tended to be useful, but were too often led astray by subjective assessments of the applicants.

Using the Stock Market to Predict Unemployment

The National Bureau of Economic Research uses the stock market as one of its leading indicators of the course of the economy. Stock prices often go down a few months before an economic recession begins, and then rise a few months before a recession ends. The stock market collapse in late 1929, shortly before the Great Depression, is the most dramatic example. But Figure 15.3 shows that this pattern also holds in recent, if less severe, times. Between 1952 and 1981, there were six recessions, which are indicated by shaded bars in the figure. Each was preceded by a slump in stock prices and, in each case, stock prices turned back up before the end of the recession. A closer inspection of Figure 15.3 reveals that the record is not perfect, in that stock prices sometimes go down without having a recession occur. To paraphrase Paul Samuelson, the stock market has predicted ten out of the last six recessions.

Still, there does seem to be a potentially useful empirical relationship between stock prices and the economy, and there are logical, if multiple, explanations of why this should be so. First, the stock market may influence the economy because wealth affects spending. Second, the economy may influence stock prices—because profitable companies are more attractive than unprofitable ones—and stock prices may dip before a recession because farsighted security analysts can see the recession coming. Third, it may be that some other factor, such as the money supply, pushes

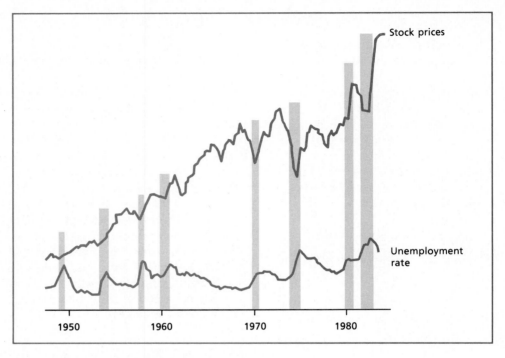

Figure 15.3   *The Stock Market and the Economy*

stock prices and the economy in the same direction, but it affects stock prices more quickly than the economy.

We won't try to sort through these competing explanations here. Instead, let's just use regression analysis to see if we can find confirmation of the idea that stock prices are a useful predictor of the economy's health. Our model might be

$$Y = \alpha + \beta X + \epsilon$$

where X is the current level of stock prices and Y is the unemployment rate one quarter hence.

A least squares regression, using 120 quarterly observations for 1952 through 1981, yielded these results:

$$\hat{Y} = 4.02 + 0.018\,X, \qquad R^2 = .12$$
$$[11.1] \quad [4.1]$$

[ ]: *t*-values

There does seem to be a statistically significant relationship between stock prices and the unemployment rate three months hence. But wait; it has the wrong sign! Our results say that an increase in stock prices presages an increase in the unemployment rate. Yet our intuition and a visual inspection of Figure 15.3 both say that unemployment goes up after a *slump* in stock prices. Where have we gone wrong?

When common sense and empirical results disagree, we should reexamine both to see where the error might lie. Here our common sense seems sound enough, and it is buttressed by an inspection of Figure 15.3. Yet the regression definitely says that there is a positive relationship between stock prices and the unemployment rate. The culprit here is that both sets of data have tended to increase over this 30-year period. Stock prices have risen over time as our economy has grown. The unemployment rate has also increased for a variety of reasons, including the gradual influx of more teenagers and women into the labor force—two demographic groups with relatively high unemployment rates. As I discussed at the end of Chapter 14, when two sets of data increase over time, there will often seem to be a positive relationship between them, even when there is really no relationship at all or even, as is the case here, a negative relationship.

One way to take into account this upward growth over time in both data sets is to include time itself as an explanatory variable.* We can run the multiple regression

$$Y = \alpha + \beta_1 X_1 + \beta_2 X_2 + \epsilon$$

where Y is the unemployment rate one quarter hence, X_1 is the current level of stock prices, and X_2 is a measure of the passage of time. I used the following values for X_2:

$$
\begin{aligned}
X_2 &= 1 \text{ in the first quarter of 1952}\\
&= 2 \text{ in the second quarter of 1952}\\
&= 3 \text{ in the third quarter of 1952}\\
&= 4 \text{ in the fourth quarter of 1952}\\
&= 5 \text{ in the first quarter of 1953}\\
&\quad\quad\quad\quad\vdots\\
&= 120 \text{ in the fourth quarter of 1981}
\end{aligned}
$$

---

*Our underlying model might be that deviations from trend values of stock prices presage deviations from trend in the unemployment rate,

$$Y - Y_T = \alpha_1 + \beta_1(X - X_T) + \epsilon$$

where the trend values are given by

$$Y_T = \alpha_2 + \gamma_2 t$$
$$X_T = \alpha_3 + \gamma_3 t$$

and t is a measure of time that increases by one each quarter. Substitution yields the equation estimated in the text.

This multiple regression yields the following results:

$$\hat{Y} = 5.663 - 0.052\ X_1 + 0.062\ X_2, \qquad R^2 = .46$$
$$\qquad [16.3] \quad [5.76] \qquad\quad [8.43]$$

[ ]: t-values

Now that's more like it. As anticipated, an increase in stock prices (X_1) signals a coming decline in the unemployment rate, while a decrease in stock prices presages an increase in the unemployment rate. This effect is not only very statistically significant; it is also substantial. The -0.052 estimate implies that a 10-point increase in stock prices (about 10 percent in 1980) signals a drop in the unemployment rate of about half a percentage point (from 8 percent to 7.5 percent). The estimated coefficient of X_2 (time) is positive and statistically significant. This estimate indicates that if the stock market does not change in any given quarter, then the unemployment rate can be expected to increase by about 0.062 point. All in all, a multiple regression including time as an explanatory variable provides a worthwhile clarification of some initially puzzling results.

Another way of handling the problem of two variables that both grow over time is to run a regression using percentage rates of changes, instead of levels. Here we could regress

$$Y = \alpha + \beta X + \epsilon$$

with Y being the percentage change in the unemployment rate next quarter and X the percentage change in stock prices this quarter. The results are

$$\hat{Y} = 0.02 - 0.51\ X \qquad R^2 = .11$$
$$\qquad [2.5] \quad [3.7]$$

[ ]: t-values

These estimates are consistent with our multiple regression. A 10 percent increase in stock prices presages about a 5 percent decrease in the unemployment rate, say from 8 percent to 7.6 percent. This estimate is both substantial and statistically significant.

Which Statistics Matter to Investors?

The outcomes of most financial investments are uncertain. The return may turn out to be pleasingly high or painfully low, and we don't know beforehand which will be the case. We may, however, be able to quantify this uncertainty by specifying a rough probability distribution for the return. Such a specification could reflect a subjective assessment or it might be based on a historical investigation of past returns.

The Determinants of Price-Earnings Ratios

The price-earnings ratio is a traditional yardstick for comparing different stocks; it shows how highly a company's shares are valued relative to its current profits. Suppose, for instance, that company XYZ earns $5 a share, while company MBI earns $10 a share. If XYZ stock sells for $50 a share and MBI for $100 a share, then they both have price-earnings ratios of 10. Their earnings are valued equally in that for a $1,000 investment, we can buy 20 shares of XYZ or 10 shares of MBI, representing $100 of earnings in either case. If, instead, XYZ sells for $75 a share while MBI sells for $100, then XYZ's earnings are more highly valued because it has the higher price-earnings ratio. There are several possible explanations for differences in price-earnings ratios:

1. Reported earnings may be misleading because of unusual circumstances or creative accounting. The price-earnings ratio will be low if investors believe that reported earnings are inflated.
2. Firms with high anticipated growth rates have high price-earnings ratios, because they will have higher future earnings.
3. Firms paying large dividends have high price-earnings ratios, because their future growth is not at the expense of current dividends.
4. Riskier firms have low price-earnings ratios, because risk-averse investors won't pay high prices for stock with a very uncertain return.

A classic study of the determinants of price-earnings ratios asked 17 professional security analysts for their opinions of 178 well-known stocks.[9]

P/E = price/"normal" earnings
 g = anticipated growth rate
 d = anticipated dividends relative to earnings
 β = risk, as measured by the firm's beta coefficient (the study also used standard deviations to measure risk, and obtained similar results)

A cross-section multiple regression was then used to explain differences in price-earnings ratios among firms. The results for 1963 were

$$P/E = 3.47 + 2.57\ g + 7.17\ d - 0.84\ \beta$$
$$\qquad\qquad [21.4]\quad\ [2.9]\qquad [1.4]$$

$R^2 = .75$

[   ]: t-values

The signs of the estimated parameters agree with theory; high growth, high dividends, and low risk are the ingredients for a high price-earnings ratio. The first two coefficients, for growth and dividends, are substantial and statistically significant. The third coefficient, for risk, is not. This does not mean that risk is unimportant, only that these data do not show conclusively that it is important.

It is difficult to digest a complete probability distribution, but, earlier in this book, we looked at several summary measures, including the mean, variance, and skewness coefficient. The mean is the expected value of the return, a measure of the long-run average return for such an investment. The variance (or standard deviation) gauges the dispersion in the possible returns, and is thus a measure of uncertainty. The higher the variance, the less certain we are of what the return will turn out to be. If investors are risk-averse, then they will only acquire a high-variance asset if the expected return is also high.* The skewness coefficient measures whether the large deviations from the expected return are more likely to be positive or negative. A symmetrical probability distribution has a zero skewness coefficient. A positive skewness coefficient reflects a "long-shot" probability distribution that has some chance of a very high return (which is balanced by a substantial chance of a return that is slightly below the expected return). A negative skewness coefficient means that there is some chance of a very low return (usually a large loss). If investors prefer positive skewness to negative skewness, then they will only make negatively skewed investments if the expected return is high.

Haim Levy and Marshall Sarnat made an intriguing empirical investigation of these ideas.[10] Using annual data covering the years 1946 through 1967, they calculated the mean, variance, and skewness coefficients for each of 58 mutual funds. They then asked whether, as anticipated, those funds with high variances also had high means and if those funds with negative skewness had higher mean returns than those with positive skewness. To separate the effects of variance and skewness on mean returns, they used a multiple regression:†

$$\hat{Y} = 7.205 + 0.019\ X_1 - 0.000064\ X_2, \qquad R^2 = .86$$
$$[16.2] \quad [9.9] \qquad\quad [8.4]$$

where Y = mutual fund's average return
        X_1 = variance of mutual fund's return (average squared deviation from mean)
        X_2 = mutual fund's skewness (average cubed deviation from mean)
    [   ] = t-values

These empirical results corroborate the theory in that there is a statistically significant relationship between a fund's average return and both its variance and skewness. Funds with high variances tend to have high means, and so do funds with negative skewness. Unfortunately, because Levy and Sarnat did not report the average values and ranges of their explanatory variables, we cannot gauge whether these effects are substantial as well as statistically significant.

---

*For individual securities that will be part of a diversified portfolio, we've seen that the beta coefficient is a better measure of risk. But here, we are going to look at mutual funds, which are often virtually the entire portfolio for small investors.

†Notice that Levy and Sarnat based their probability distributions on historical data, rather than subjective assessments. They also included 18 other summary measures of the probability distributions in their regression, but found that these did not have a statistically significant effect on mean returns.

EXERCISES

15.12 Use the data in Table 15.1 to estimate the equation

$$Y = \alpha + \beta_3 X_3 + \epsilon$$

where Y is fuel consumption and X_3 is average speed. Is this estimate of β_3 the same as the estimate $\beta_3 = -3.57$ given in the text? Explain why they either are or are not the same.

15.13 The manager of a fleet of identical trucks wants to estimate the effect of driver speeds on fuel consumption. Monthly data are collected and the following equation estimated:

$$\hat{Y} = 0.0 + 0.20\, X_1 + 1,000.0\, X_2, \qquad R^2 = .86$$
$$\quad\ \ (11.0)\ (0.04) \qquad (400.0)$$

where Y = fuel consumption (gallons), average value 71,000

X_1 = miles traveled, average value 80,000

X_2 = speed (miles per hour), average value 55

( ) = standard deviations

Is each of these explanatory variables statistically significant at the 5 percent level? What does the estimated coefficient of X_1 measure? If the population value actually is 0.20, explain the implication (in simple English). What does the estimated coefficient of X_2 measure? If the population value is 1,000.0, explain the implication (in simple English).

15.14 *(continuation)* Can we conclude from the much larger value of the second coefficient (1,000 versus 0.2) that X_2 is more important than X_1 in determining fuel consumption?

15.15 *(continuation)* This firm has been offered a 1,000-mile delivery that will net $200 after paying the driver and other nonfuel expenses. To complete the job on time, the truck will have to average 55 miles per hour. Does the equation in Exercise 15.13 indicate that they will make a profit or a loss on this job?

15.16 The text uses the data in Table 15.2 to analyze the relationship between midterm, homework, and final exam scores. Interpret the estimated coefficients in the simple and multiple regression equations. In particular, how can the coefficient of midterm scores be 0.80 in the simple regression and 0.51 in the multiple regression?

15.17 *(continuation)* The data in Table 15.2 were used in this chapter to see how accurately final exam grades could be predicted from midterm and homework scores. Do you think that we would make better or worse predictions if we simply added the midterm and homework scores together and divided by 2? Explain your reasoning. Check your logic by calculating the errors in predicting the final exam score by taking the average of the midterm and homework scores. Compute the standard error of estimate for this prediction method. Compare this to the standard error of estimate for a multiple regression.

15.18 When semiconductors are manufactured, an electrical film is created by the deposit of a gaseous material on a special alloy base; the thickness of the deposit depends on the gas's temperature and concentration. Here are data from one experiment:[11]

Rate of Film Deposition Y	Gas Temperature (in 100°C) X_1	Concentration (Percent) X_2
14.5	7.4	16.0
27.6	8.4	21.0
20.9	7.7	20.0
9.1	6.3	18.0
11.2	7.2	14.0
10.2	7.4	12.0
24.7	8.5	17.5
14.3	7.0	18.0
33.9	10.0	16.0
32.6	8.0	28.2
16.5	8.2	12.2
11.2	7.8	9.8
20.8	7.9	18.0
4.8	6.6	12.0
5.1	6.6	12.0
39.8	10.0	21.5
5.2	6.6	13.0
15.8	8.0	12.5
22.2	8.0	19.0
39.4	10.1	20.0

Estimate this linear equation and interpret your results:

$$Y = \alpha + \beta_1 X_1 + \beta_2 X_2 + \epsilon$$

15.19 A business wants to see how sensitive its sales are to the price it charges. Nine different prices are charged in nine different cities:

City	Per Capita Income ($) X_1	Price Charged ($) X_2	Sales (1000s) Y
A	6,200	10	56
B	6,900	20	57
C	7,100	25	59
D	7,700	20	68
E	6,700	15	57
F	8,000	30	65
G	6,500	20	57
H	6,100	15	52
I	7,400	25	59

To control for the variations in per capita income, a multiple regression is used, with both price and income as explanatory variables:

$$Y = \alpha + \beta_1 X_1 + \beta_2 X_2 + \epsilon$$

Estimate the parameters of this equation. Do both price and income have statistically significant effects (at the 5 percent level) on sales? Do both seem to have important effects on sales? Give a 95 percent confidence interval for the effect on sales of a $5 price increase.

15.20 A business is trying to estimate the contributions of labor and machines to output. Its data are

Output Y	Labor X_1	Machines X_2
34	40	20
42	60	15
33	50	10
60	60	60
37	50	15

Use these data to estimate the model

$$Y = \alpha + \beta_1 X_1 + \beta_2 X_2 + \epsilon$$

Do labor and machines both have statistically significant effects on output at the 5 percent level? Give 95 percent confidence intervals for their coefficients.

15.21 *(continuation)* It costs this firm $25,000 to employ a person for a year and $40,000 to rent a machine for a year. Would you recommend that they substitute labor for machines or machines for labor?

15.22 *(continuation)* Can you think of any reasons for caution in using this estimated equation to justify a complete substitution of labor for machines, or vice versa?

15.3   MULTICOLLINEARITY

In these various examples, multiple regression has handled, with varying degrees of success, the estimation of models using "nature's data" rather than experimental data. The regression model

$$Y = \alpha + \beta_1 X_1 + \beta_2 X_2 + \cdots + \beta_k X_k + \epsilon$$

asks the effects on the dependent variable Y of changes in each of the explanatory variables X_i, holding constant the other explanatory variables. This is the sort of question that could be answered by a laboratory experiment, but often it isn't because such an experiment is impractical. Instead, nature's data are sifted through a multiple regression, which provides the statistical counterpart to laboratory controls.

The statistical logic can be explained roughly as follows. Suppose we are interested in the effects of two variables, income and wealth, on spending, but cannot hold either constant while we vary the other. The available data might show the following information:

1. One year, income went up by 50, wealth fell by 100, and spending increased by 40.
2. Another year, income went up by 100, wealth went up by 100, and spending increased by 95.

What can we infer from these data? If wealth had no effect on spending, then spending, like income, should have increased by twice as much in the second year as in the first year. In fact, spending increased by more than this, indicating that the increase in wealth stimulated spending. Logically, then, an increase in wealth apparently has a positive effect on spending. For the effects of income changes, notice that wealth went down by 100 in the first year and up by 100 in the second. If income were unimportant, then spending also should have gone down in the first year and up equally in the second. In fact, spending, like income, increased in both years—indicating that increases in income have a positive effect on spending.*

It's not nearly this simple, though, because there are other factors—the troublemaking error term—that influence spending. Spending in either year may have been unusually high or low because of omitted factors, leading us to draw incorrect inferences. What we must do, statistically, to overcome this random variation is use more than two observations. With lots of data, we can hope that these annoying disturbances will average out and allow us to draw correct inferences. Multiple regression does this averaging for us by finding the parameter estimates that minimize the sum of squared prediction errors.

Our statistical reasoning can be thwarted, however, not only by too few data, but also by data that don't show the variation needed to allow inferences. If income never changes, then we obviously cannot estimate the effect of income changes on spending. A similar difficulty arises if income and wealth always change in unison, because we can't tell whether it is income or wealth that influences spending if we never see them move separately.

Suppose, for instance, that wealth is always equal to twice income. This substitution into the regression model yields

$$Y = \alpha + \beta_1 X_1 + \beta_2 X_2 + \epsilon$$

$$= \alpha + \beta_1 X_1 + \beta_2 (2X_1) + \epsilon$$

$$= \alpha + (\beta_1 + 2\beta_2)X_1 + \epsilon$$

*These inferences can be quantified, as follows, by letting income be the first explanatory variable and wealth the second:

$$(+95) = \beta_1(+100) + \beta_2(+100)$$
$$(+40) = \beta_1(+50) + \beta_2(-100)$$

The addition of these two equations implies $135 = 150 \beta_1$, so that $\beta_1 = 0.9$ and, hence, $\beta_2 = 0.05$.

Whenever income goes up \$1, wealth goes up \$2, and spending goes up $\beta_1 + 2\beta_2$. With enough data, we can obtain a useful estimate of $\beta_1 + 2\beta_2$. But no matter how plentiful our data is, we'll never be able to obtain separate estimates of β_1 and β_2 as long as income and wealth move in unison. If we try, our computer program will come to an abrupt stop and print a nasty error message, such as INVALID DATA–DIVISION BY ZERO.

Fortunately, matters are seldom this extreme, because few explanatory variables move in perfect unison. But the bad news is that many variables are highly correlated, so that they move in close, although imperfect, unison. The consequence is that, while we can obtain a pretty good estimate of $\beta_1 + 2\beta_2$, we will not get accurate estimates of the separate parameters β_1 and β_2. Our regression may yield a satisfyingly high R-squared, but discomfortingly large standard deviations and low t-values for the parameters β_1 and β_2. This malaise is known as the multicollinearity problem.

> In the **multicollinearity problem,** high correlations among the explanatory variables prevent accurate estimates of the individual coefficients.

The symptoms of the problem are disappointingly low t-values. A multicollinearity problem can be diagnosed by regressing each explanatory variable on the others to see if these R-squareds reveal high intercorrelations among the explanatory variables. Unfortunately, there are no simple cures for the multicollinearity problem. The builder can't make solid bricks without clay and the statistician can't make accurate estimates without information.

If the researcher is willing, a priori, to assign values to some of the parameters, then conditional estimates of the remainder can be computed. For instance, if the estimate is that

$$\beta_1 + 2\beta_2 = 1.03$$

then an a priori belief that $\beta_2 = 0.05$ implies a 0.93 estimate for β_1. More commonly, the desperate researcher will assign zero values to some of the parameters by dropping variables out of the equation. This practice isn't recommended unless zero is really believed to be the most plausible parameter value. Lacking such a priori information, the only alternative is to use additional data in which the explanatory variables are not so highly correlated.

Fortunately, multicollinearity does not undermine predictive accuracy as long as the intercorrelations persist. If wealth continues to be equal to twice the income, then we can forecast spending perfectly well without ever knowing the separate effects of income and wealth. But that's a big "if" and we do need to know the separate effects of the explanatory variables if the intercorrelations evaporate during the prediction period. The only silver lining to this dark cloud on the horizon is that, while these new data will be difficult to predict, at least they will provide the basis for better estimates next time around.

The Coleman Report

In response to the 1964 Civil Rights Act, the U.S. government sponsored the 1966 Coleman Report, which was an ambitious effort to gauge how student scholastic performance is affected by student backgrounds and school resources. One of the more provocative findings involved the estimation of the multiple regression equation

$$Y = \alpha + \beta_1 X_1 + \beta_2 X_2 + \cdots + \beta_k X_k + \epsilon$$

The dependent variable was student performance, as measured by scores on tests of verbal ability. Some of the explanatory variables were measures of school resources (budgets, books in the library, science facilities, and so on) and others were measures of student backgrounds (family income, occupation, education, and the like).

The Coleman Report did not report the estimated coefficients. Instead, they regressed background explanatory variables and, second, using both student background and school resources variables. Because R-squared only improved slightly with the inclusion of the school resources variables, they concluded that these "show very little relation to achievement." The coefficients of these school resources variables were, on the whole, apparently not statistically significant; however, not being able to reject zero as a possible value for a parameter is not at all the same thing as proving that the value is zero. Several critics pointed out the multicollinearity problem inherent in these nonexperimental data. In the 1950s and early 1960s, school resources depended heavily on the local tax base and parents' interest in quality education for their children:

The family background characteristics of a set of students determine not only the advantages with which they come to school; they are also associated closely with the amount and quality of resources which are invested in the schools. As a result, higher status children have two distinct advantages over lower status ones: First, the combination of material advantages and strong educational interests provided by their parents stimulate high achievement and educational motivation; and second, their parents' relatively high incomes and interest in education leads to stronger financial support for and greater participation in the schools that their children attend.[12]

This is not to say that the data show that resources matter, only that the high multicollinearity between resources and backgrounds made it possible to get an accurate estimate of the effect of resources alone on student achievement.

Another Look at Fuel Usage

One of the early examples in this chapter investigated the effects of vehicle travel (X_1), horsepower (X_2), and speed (X_3) on the nation's fuel consumption. The results were a bit disappointing, in that two of the three coefficients were statistically insignificant and one had an unanticipated sign. Perhaps multicollinearity was the culprit! To check, I regressed each of the three explanatory variables on the other two:

$$\hat{X}_1 = 10,534.5 - 33.15 \, X_2 + 93.0 \, X_3, \qquad R^2 = .65$$
$$\qquad\quad [2.7] \qquad\;\; [1.1] \qquad\quad [3.0]$$

$$\hat{X}_2 = 152.3 - 0.004 \, X_1 + 1.11 \, X_3, \qquad R^2 = .72$$
$$\qquad [5.9] \qquad\; [1.1] \qquad\quad [3.7]$$

$$\hat{X}_3 = -98.3 + 0.006 \, X_1 + 0.56 \, X_2, \qquad R^2 = .85$$
$$\qquad\quad [4.0] \qquad\;\; [3.0] \qquad\quad [3.7]$$

[   ]: t-values

There are substantial intercorrelations among these explanatory variables. The third variable, speed, has a particularly high multiple correlation coefficient and this is why its coefficient in the original regression equation was not estimated very accurately. That estimate was -3.6 with a standard deviation of 4.4, giving a 95 percent confidence interval of $-3.6 \pm 2(4.4)$, from -12.4 to $+5.2$.

One of the unsatisfying aspects of our earlier fuel consumption estimates was the large standard deviations and wide confidence intervals on the individual parameter estimates, especially the estimated effect of speed on fuel usage. Now we see that the basic reason for this inadequacy is that the three explanatory variables are intercorrelated. Speed, in particular, has been highly correlated with travel and horsepower over the years covered by this study. In those years when the average speed has fallen, horsepower and travel have slumped, too. When speed has increased, horsepower and travel have also tended to increase. This multicollinearity makes it difficult to estimate how speed alone affects fuel usage. There is not much that can be done about it except to note that the data are not sufficiently uncorrelated to decisively answer all of the questions we have posed.

EXERCISES

15.23 It has been argued that consumer spending depends not only on current income, but also on some notion of average of "permanent" income. One way to take this possibility into account is to include income for both the current and previous year in the consumption function:

$$Y = \alpha + \beta_1 X_1 + \beta_2 X_2 + \epsilon$$

where Y is this year's consumption, X_1 is this year's income, and X_2 is the previous year's income.

Use the consumption and income data in Table 13.1 (or Figure 15.1) to estimate this equation. Interpret your estimates, including their standard deviations and t-values.

15.24 *(continuation)* Is there a multicollinearity problem here?

15.25 The data in Table 15.2 were used in the text to see how accurately final exam scores can be predicted from midterm and homework scores. What sort of multicollinearity problem might cause what sort of problems for this investigation? Is there any evidence of a multicollinearity problem here?

15.26 The preceding section included three regressions that were intended to investigate the interrelations among the data on vehicle travel, horsepower, and speed. Use these three reported regressions to describe these empirical interrelations.

15.27 An economist once attempted to estimate an asset-demand equation that included the following three explanatory variables: current wealth X_1, wealth in the previous quarter X_2, and the change in wealth $X_3 = X_1 - X_2$. Can you guess what problem he encountered?

• **15.28** *(continuation)* Frankly, he was befuddled about how to solve his problem. (I know, because I reviewed his paper!) What would you suggest?

• **15.29** A college admissions office wants to estimate the relationship between math SAT scores, verbal SAT scores, and freshman Grade Point Average (GPA). Larry wants to regress GPAs (Y) on math SAT scores (X) and verbal SAT scores (Z). Moe wants to regress GPA on the total SAT score ($X + Z$) and the verbal score (Z). Joe wants to regress GPA on the total SAT score ($X + Z$) and the amount by which the math score exceeds 500 ($X - 500$). They each run their preferred regressions and obtain these results:

$$\text{Larry: } \hat{Y} = 0.10 + 0.003\ X + 0.002\ Z$$
$$[3.0] \qquad\quad [2.0]$$

$$R^2 = .74$$

$$\text{Moe: } \hat{Y} = 0.10 + 0.002\ (X + Z) + 0.001\ X$$
$$[2.0] \qquad\qquad\quad [0.5]$$

$$R^2 = .74$$

$$\text{Joe: } \hat{Y} = 0.60 + 0.002\ (X + Z) + 0.001\ (X - 500),$$
$$[2.0] \qquad\qquad\quad [0.5]$$

$$R^2 = .74$$

[ ]: *t*-values

Are their estimates consistent with each other? For example, what college GPA would each predict for a student with a 500 math score and a 600 verbal score? What can you learn from Moe's regression that is not apparent from Larry's regression?

• **15.30** *(continuation)* Joanna thinks that GPA should be regressed on total SAT score ($X + Z$) and verbal SAT score (Z). Joshua wants to regress GPA on the amounts by which the math and verbal scores exceed 500: on ($X - 500$) and ($Z - 500$). If they do this, can you predict what their coefficient estimates will be?

15.4 SUMMARY

Multiple regression is the statistician's alternative to controlled laboratory experiments. The regression model is

$$Y = \alpha + \beta_1 X_1 + \beta_2 X_2 + \cdots + \beta_k X_k + \epsilon$$

Each coefficient β_i gauges the effects on the dependent variable Y of changes in each of the explanatory variables X_i, holding constant the other explanatory variables. This question could be investigated in a controlled experiment in which the other explanatory variables really are held constant, while X_i is varied and the effects on Y are recorded. However, many interesting questions, particularly in business and economics, do not lend themselves to laboratory experiments. Humans are reluctant to participate in such experiments and, even if they do, the artificial environment makes the results of dubious value.

Multiple regression uses "nature's data"—a recording of what transpires as people live their lives. If there happens to be sufficient variation in the explanatory variables, then accurate estimates of the parameters may be possible. The specific criterion used by multiple regression is to determine those estimates of the parameters that best explain the observed variation in Y, specifically those values that minimize the sum of squared prediction errors. Because of omitted influences, measurement error, and sampling error, there is an inevitable error term ϵ, which precludes certainty about the values of the parameters. This random variation in ϵ causes random variation in the recorded values of Y and, hence, random variation in the parameter estimates. We must remember, then, that the particular estimates we obtain come from a probability distribution. Assuming normality, this uncertainty can be quantified by computing standard deviations for the estimates. In addition, t-values (the ratio of an estimate to its standard deviation) can be used to test the null hypothesis that a parameter is equal to zero, that is, that the associated explanatory variable doesn't really affect Y after all. The overall predictive success of the model can be measured by the standard error of estimate or, relative to the benchmark of omitting all of the explanatory variables and using only the sample mean, by R-squared.

Although multiple regression doesn't require controlled experiments, it does require some observed variation in the explanatory variables. If one variable never varies, then its effects cannot be judged. The same is true if one variable never varies independently of the others—if, for instance, one variable only goes up when either a second or a third goes down—because then we cannot tell whether it is this variable or the others that is affecting Y. The consequences are imprecise estimates, as reflected in their large standard deviations. This multicollinearity problem can be diagnosed by calculating the correlations among the explanatory variables, but can only be solved by using additional information.

REVIEW EXERCISES

15.31 Jane Daylily computes a total score for each of her students by adding together the midterm score, homework score, and twice the final score. Table 15.5 gives some sample scores. Use these data to estimate the equation

$$Y = \alpha + \beta_1 X_1 + \beta_2 X_2 + \beta_3 X_3 + \epsilon$$

Briefly interpret your results. Do you think that homework, midterm, and final scores are a good predictor of a student's total score?

15.32 An empirical study of professors' salaries obtained the following multiple regression estimates:[13]

$$\hat{Y} = 25,000 + 200\,X_1 + 250\,X_2 + 50\,X_3$$
$$(500) \quad\;\; (20) \qquad (100) \qquad (20)$$
$$+\; 500\,X_4 + \;50\,X_5$$
$$(50) \qquad\quad (400)$$

where Y = annual salary
         X_1 = years teaching
         X_2 = published books

Table 15.5 *Data for Exercise 15.31*

Student	Midterm X_1	Homework X_2	Final X_3	Total Y
1	74	92	83	332
2	94	87	90	361
3	82	78	88	336
4	80	78	66	290
5	89	90	97	373
6	65	73	74	286

X_3 = published articles
X_4 = Ph.D.s supervised
X_5 = teacher rating (1 = above average, 0 = below average)
( ) = standard deviations

Calculate 95 percent confidence intervals and t-statistics for each of the estimated coefficients. Which of these estimates are statistically significant at the 5 percent level? Which seem substantial?

15.33 *(continuation)* According to these estimates: **a.** What is the predicted salary of a professor who has been teaching for 15 years, written 3 books and 20 articles, supervised 6 Ph.D. theses, and is a below-average teacher? **b.** How much higher would this professor's salary be if the teacher rating was above average? **c.** How much is another published article worth, in terms of annual salary?

15.34 *(continuation)* Professors have a finite amount of time that must be allocated to these various academic activities and to a bit of leisure as well. Planning ahead over the next two years, a certain professor can either write a book, write six articles, or work on his teaching (to bring his rating up to above average). Which would you advise?

15.35 The text reports the estimation of a regression equation using travel, horsepower, and speed to explain motor vehicle fuel usage:

$$\hat{Y} = -1201.1 + 0.132\ X_1 + 5.07\ X_2 - 3.57\ X_3$$

What is the predicted fuel usage if we stop using motor vehicles ($X_1 = X_2 = X_3 = 0$)? Do you believe this prediction? If you don't believe the equation's prediction, then should the estimates be discarded?

15.36 A professional football team is trying to develop a model to predict which college players will be successful linemen in the pros. They decide that an objective measure of success is the number of games that the player starts during his professional career. Table 15.6 gives the data collected on 16 former NFL players. "Anger" is the player's score on a psychological test of aggressiveness. The higher this score, the more aggressive the person. Use these data to estimate the equation

$$Y = \alpha + \beta_1 X_1 + \beta_2 X_2 + \beta_3 X_3 + \beta_4 X_4 + \beta_5 X_5 + \epsilon$$

For each of these five explanatory variables, answer these questions: **a.** What is the estimated coefficient? **b.** Is the sign of this estimate plausible? **c.** What is a 95 percent confidence interval for the

Table 15.6 *Data for Exercise 15.36*

Player	Games Started Y	Height (Inches) X_1	Weight (Pounds) X_2	Speed (40 Yds) X_3	IQ X_4	Anger X_5
CA	22	73	240	4.8	105	91
MA	0	75	220	4.5	120	89
DB	104	79	290	5.7	84	93
FB	13	74	210	4.6	110	95
ED	97	78	260	5.3	95	92
FD	142	76	260	5.2	86	96
RG	120	78	280	5.4	92	94
TH	32	80	230	4.7	115	88
UH	9	76	245	5.2	100	92
DJ	120	75	220	4.6	105	95
EJ	172	82	250	4.9	85	94
JN	24	75	240	5.0	110	93
RN	117	71	245	5.5	82	99
MO	35	78	250	4.8	108	89
BS	45	78	300	5.7	86	86
RW	130	71	240	5.3	80	98

estimate? **d.** What is the value of the *t*-statistic? **e.** Does this variable have a statistically significant effect on *Y* at the 5 percent level?

15.37 *(continuation)* This team is considering drafting either BG or HS. BG is 78 inches tall and weighs 280 pounds, with a 5.0 speed, a 102 IQ, and a 93 anger score. HS is 76 inches tall and weighs 260 pounds, with a 4.8 speed, a 98 IQ, and a 98 anger score. If the estimated equation is to be believed, which football player should they draft? Are there any reasons why they should not use this estimated equation alone to make their draft picks?

• **15.38** *(continuation)* Do you have any reason to suspect that some of the five explanatory variables might be correlated with the others? Regress each explanatory variable on the other four to see if you are right. If there is a high correlation, what disease have you diagnosed? If one of these explanatory variables is highly correlated with the other four, does this invalidate the equation estimated in 15.36?

15.39 A nationwide magazine contest assigned each letter of the alphabet a secret value from 1 to 26.[14] The value of a state or city is then the sum of the values of the letters in the name; for example, if $R = 7$, $E = 10$, $N = 3$, and $O = 5$, then R-E-N-O would equal $7 + 10 + 3 + 5 = 25$. The object of the contest is to use the following values to determine the value of ALBUQUERQUE. Can you do it?

Alaska	= 73	Jamestown	= 102
Arizona	= 73	Kansas	= 56
Atlanta	= 81	Maine	= 65
Boston	= 56	Monterey	= 91
Buffalo	= 91	Nome	= 36
Chicago	= 81	Ohio	= 47
Columbia	= 109	Oregon	= 61
Denver	= 72	Salem	= 64
Detroit	= 93	Tampa	= 77
Hawaii	= 106	Texas	= 49
Houston	= 56	Toledo	= 61
Idaho	= 64	Tulsa	= 58
Iowa	= 64	Utah	= 44

15.40 In the IBM antitrust case, Franklin Fisher estimated multiple regression equations explaining computer prices as a function of memory, speed, and other characteristics.

> Despite the fact that *t*-statistics on the order of 20 were obtained for all of the regression coefficients, Alan K. McAdams, appearing as an expert for the government, testified that collinearity made it impossible reliably to separate the effects of the different independent variables and hence that little reliance could be placed on the result.[15]

Explain why you either agree or disagree with McAdams's logic.

15.41 The admissions officer at a very selective law school is interested in how well law school performance can be predicted from college grade-point averages (GPAs) and LMAT test scores. A random sample of 20 students yields these data:

Law School GPA Y	College GPA X_1	LMAT Percentile X_2
3.42	3.28	.96
3.60	3.18	.97
3.28	2.89	.93
3.75	3.72	.99
3.36	3.18	.95
3.96	3.50	.98
3.31	3.04	.94
3.33	3.87	.95
3.60	3.54	.96
4.00	3.27	.99
3.28	3.30	.95
3.44	3.29	.91
3.25	3.17	.93
3.75	3.62	.97
3.30	3.34	.96
3.20	3.08	.90
3.50	3.37	.96
3.28	3.16	.94
3.17	3.20	.95
3.31	3.10	.94

Use these data to estimate the parameters of the prediction equation

$$Y = \alpha + \beta_1 X_1 + \beta_2 X_2 + \epsilon$$

What are the standard deviations of these parameter estimates? What are the *t*-values and (approximate) 95 percent confidence intervals for these parameters? Do both college GPA and LSAT scores have substantial and (at the 5 percent level) statistically significant predictive abilities?

15.42 *(continuation)* If you were the admissions officer at this school, is there any reason why you would not use this estimated equation alone for admitting students?

15.43 *(continuation)* An applicant scored in the 99th percentile on the LMAT test, but has only a 2.48 grade point average at Cal Tech. Is there any reason for caution in using the estimated equation to predict this student's performance in law school?

15.44 Exercise 14.4 gives some data on math and verbal SAT scores and history course scores, which were analyzed in this chapter. Is there any evidence of a multicollinearity problem here?

15.45 As an academic exercise, you want to examine the influence of interest rates X_1, and unemployment rates X_2 on stock prices Y. Lacking a suitable laboratory, you must rely on nature's experiments, using the available historical data to estimate your regression model

$$Y = \alpha + \beta_1 X_1 + \beta_2 X_2 + \epsilon$$

If these data are accurate measurements, why is there an error term ϵ? Why are there prediction errors $e = Y - \hat{Y}$? What is the difference between ϵ and e?

15.46 Is the multicollinearity problem a population or sample characteristic? For example, if you are trying to estimate the effect of interest rates and unemployment on stock prices, do you have a multicollinearity problem if interest rates and unemployment tend to be correlated in general or if they happen to be correlated by historical accident in the particular data you have?

• 15.47 Table 15.7 gives some data on the average weights (in pounds) of U.S. males, by height. Can you see any reason for wariness in fitting the model

$$Y = \alpha + \beta_1 X_1 + \beta_2 X_2 + \epsilon$$

where Y is average weight, X_1 is height, and X_2 is age (the midpoint of each age interval). Use these data to estimate this equation and see if your fears are confirmed.

•• 15.48 To analyze the risk premiums on corporate bonds, the following regression equation was estimated.[16]

$$Y = 0.987 + 0.307\,X_1 - 0.253\,X_2 - 0.537\,X_3$$
$$\quad\quad\quad (0.032)\quad\quad (0.036)\quad\quad (0.031)$$
$$\quad - 0.275\,X_4$$
$$\quad\quad (0.021)$$
$$R^2 = .75$$

Table 15.7 *Data for Exercise 15.47*

Height (Inches)	Age					
	18–24	25–34	35–44	45–54	55–64	65–74
62	130	139	146	148	147	143
63	135	145	149	154	151	148
64	139	151	155	158	156	152
65	143	155	159	163	160	156
66	148	159	164	167	165	161
67	152	164	169	171	170	165
68	157	168	174	176	174	169
69	162	173	178	180	178	174
70	166	177	183	185	183	178
71	171	182	188	190	187	182
72	175	186	192	194	192	187
73	180	191	197	198	197	192
74	185	196	202	204	201	195

where Y = corporate bond return minus return on safe U.S. Treasury bond

X_1 = standard deviation of firm's profits, divided by the average level of profits

X_2 = length of time since firm had defaulted on any bond

X_3 = ratio of the market value of the firm's stock to value of its debt

X_4 = market value of the firm's debt

( ) = standard deviations

The first three explanatory variables are intended to measure the riskiness of the bonds. (For instance, a large value of X_3 indicates a safe bond.) The fourth explanatory variable is intended to measure how easily the bond can be bought and sold. (It's difficult to market small issues by obscure firms.) Interpret the results. In particular, do the signs of the estimates make sense?

•• **15.49** Arthur wants to study the effects of temperature, water, and music on plant growth. A lack of funding has forced him to study just three plants:

Plant	Temperature X_1	Water X_2	Music X_3	Height Y
Left	60	1.5	20	22
Right	70	1.0	0	28
Middle	80	0.5	10	24

The temperature is in °F, the water is in inches per day, the music is in record albums per day, and the height is in inches. Use these data to estimate the equation

$$Y = \alpha + \beta_1 X_1 + \beta_2 X_2 + \beta_3 X_3 + \epsilon$$

What can we conclude from this exercise? Be sure to explain your reasoning.

•• **15.50** An economist claims that high interest rates are inflationary. To prove this, she regresses inflation on interest rates and unemployment (to gauge the state of the economy) and obtains

$$\hat{P} = 10.6 + 0.30\,I - 1.30\,U, \qquad R^2 = .985$$
$$\quad\;\;[4.5] \quad\;\; [1.6] \quad\;\;\; [6.8]$$

where P = inflation rate (%)

I = interest rate (%)

U = unemployment rate (%)

[ ] = t values

After a colleague suggests that perhaps it is the other way around, that it is inflation that raises interest rates, this economist tries it this way and obtains

$$\hat{I} = -9.5 + 1.5\,P + 1.5\,U, \qquad R^2 = .818$$
$$\quad\;\;[0.7] \quad\;\; [1.6] \quad\;\;\; [1.0]$$

Would you conclude from the decline in R-squared that her original formulation is right?

15.51 Some observers attributed an increase in major league home runs in the 1980s to a reluctance of pitchers to brush back batters with inside pitches and to a proclivity by batters to swing from their heels, risking a strikeout but increasing their chances for the fame and fortune lavished on home run hitters. To test these explanations, annual home run data for the National League from 1969 to 1985 and for the American League from 1977 (when two teams were added to the league) to 1985 were regressed on the number of batters hit by the pitch (X_1) and strikeouts (X_2).[17] Write a paragraph interpreting and evaluating the results.:

$$\widehat{NL}: Y = -364.6 + 0.537\,X_1 + 0.150\,X_2$$
$$\qquad\quad (612.2) \quad (0.701) \quad (0.064)$$

$$R^2 = .45$$

$$\widehat{AL}: Y = 888.9 - 1.746\,X_1 + 0.166\;X_2$$
$$\qquad\quad (1051.7) \;(1.727) \qquad (0.072)$$

$$R^2 = .55$$

( ): standard deviations

15.52 Exercise 14.19 investigated the relationship between corporate dividends (Y) and earnings (X) using the 1950–1989 data in Table 8 at the end of the book to estimate the regression equation $Y = \alpha + \beta X + \epsilon$. To consider whether there was a change in corporate dividend policy after 1970, when inflation increased dramatically, estimate the equation $Y = \beta + \beta_1 D + \beta_2 X + \epsilon$, where $D = 0$ for 1950–1969 and $D = 1$ for 1970–1989. Report your results and write a paragraph summarizing your conclusions. Be sure to interpret β_1 carefully.

15.53 (continuation) How does the more general equation $Y = \lambda + \gamma_1 D + \gamma_2 X + \gamma_3 DX + \epsilon$ differ from the equation that was estimated in the preceding exercise? Interpret the parameters γ_1, γ_2, and γ_3 carefully. Now estimate the more general equation, and compare your results with the equation estimated in the previous exercise.

15.54 Exercise 14.31 asked for the estimation of the relationship between the unemployment rate (X) and the rate of inflation (Y), using data for 1960–1969. Most economists believe that the increase in inflation expectations in the 1970s caused a shift in this relationship. To investigate this possibility, use the 1960–1973 data in Table 8 at the end of the book to estimate the equation $Y = \alpha + \beta_1 D + \beta_2 X + \epsilon$, where $D = 0$ for 1960–1969 and $D = 1$ for 1970–1973. Report your results and write a paragraph summarizing your conclusions. Be sure to interpret β_1 carefully.

15.55 *(continuation)* The rate of inflation generally increased between 1970 and 1982, and then declined. Inflation expectations probably did likewise. To consider whether the relationship between unemployment and inflation continued to shift as inflation expectations changed, use the 1960–1989 data in Table 8 to estimate this equation:

$$Y = \alpha + \beta_1 D_1 + \beta_2 D_2 + \beta_3 D_3 + \beta_4 X + \epsilon$$

where $D_1 = 1$ for 1970–1973 and 1985–1989, 0 otherwise
$D_2 = 1$ for 1974–1979 and 1983–1984, 0 otherwise
$D_3 = $ for 1980–1982, 0 otherwise

Provide a rationale for this particular definition of the dummy variables and interpret the parameters β_1, β_2, and β_3.

15.56 *(continuation)* Using the same data and definition of variables as in the preceding exercise to estimate the equation

$$Y = \alpha + \beta_1 D_1 + \beta_2 D_2 + \beta_3 D_3 + \beta_4 X + \beta_5 D_1 X$$
$$+ \beta_6 D_2 X + \beta_7 D_3 X + \epsilon$$

Interpret your results. Make a scatter diagram of X and Y, label each point, and sketch in the four fitted equations corresponding to (1) $D_1 = D_2 = D_3 = 0$; (2) $D_1 = 1$, $D_2 = D_3 = 0$; (3) $D_2 = 1$, $D_1 = D_3 = 0$; and (4) $D_3 = 1$, $D_1 = D_2 = 0$.

15.57 Arbitrage pricing theory (APT) assumes that an asset's return R is related to underlying factors, such as $F_1 = $ change in the unemployment rate and $F_2 = $ percentage change in the interest rate on long-term Treasury bonds, and an idiosyncratic error term ϵ,

$$R = \alpha + \beta_1 F_1 + \beta_2 F_2 + \epsilon$$

Use the data in Table 15.8 to estimate this equation for AT&T and for IBM. Compare these two estimated equations.

15.58 *(continuation)* Could an asset possibly have $\beta_1 = \beta_2 = 0$? Why or why not?

15.59 Data on the selling prices of single-family houses in the northeast section of Fort Wayne, Indi-

Table 15.8 *Data for Exercise 15.57*

	Change in Unemployment	*Percentage Change in Interest Rate*	*AT&T Return*	*IBM Return*
1972	−0.8	0.36	23.9%	21.1
1973	0.5	12.79	0.5	−21.9
1974	3.4	6.61	−4.5	−29.7
1975	−1.2	5.91	21.6	37.4
1976	−0.4	−10.18	32.3	28.1
1977	−1.3	13.48	1.9	1.3
1978	−0.2	15.47	7.6	13.6
1979	1.8	14.59	−5.6	−9.1
1980	−0.1	24.11	1.4	10.8

ana, in 1977–1978 were used to estimate the following regression equation:[18]

$$Y = 5067 + 14.96\,X_1 - 218\,X_2 + 16.68\,X_3$$
$$\quad\ (2206)\quad (170)\qquad\ (79)\qquad\ (3.70)$$

$$\qquad + 3835\,X_4 + 5730\,X_5 + 2302\,X_6 - 5225\,X_7$$
$$\qquad\quad (793)\qquad (1119)\qquad (957)\qquad (1187)$$

$$\qquad + 3396\,X_8 + 1818\,X_9 - 1435\,X_{10}$$
$$\qquad\quad (1235)\qquad (1079)\qquad (868)$$

$$R^2 = .874$$

where
X_1 = total living area, square feet
X_2 = age of house, years
X_3 = garage, square feet
X_4 = central air; 1 if yes, 0 if no
X_5 = basement; 1 if yes, 0 if no
X_6 = fireplace; 1 if yes, 0 if no
X_7 = style; 1 if unusual, 0 if conventional
X_8 = baths, number
X_9 = new; 1 if yes, 0 if no
X_{10} = two-story; 1 if yes, 0 if no
( ) = standard deviation

What is the predicted sale price of a conventional two-story, 10-year-old house with 2,000 square feet of living space, 2.5 baths, a 400-square-foot garage, central air, a basement, and a fireplace? Does this predicted sale price seem reasonable to you? Which of the estimated coefficients in the regression equation are statistically significant at the 5 percent level? How can you tell?

15.60 *(continuation)* Look at each of the ten estimated coefficients (other than the intercept) in the estimated housing price equation. Does the sign of each coefficient seem reasonable? Does the size of each coefficient seem plausible? Do you think that the parameters of housing regression equations change over time? Why or why not? Be specific.

15.61 Table 1.9 in Chapter 1 gives the annual percentage returns for four assets from 1950 through 1988. Estimate the equation

$$Y = \alpha + \beta_1 X_1 + \beta_2 X_2 + \beta_3 X_3 + \epsilon$$

where Y = return on New York Stock Exchange (NYSE) stocks, X_1 = return on Treasury bills, X_2 = return on Treasury bonds, and X_3 = return on corporate bonds. Is there a multicollinearity problem here? How can you tell? Is the presence or absence of

a multicollinearity problem something that might be expected with this model, or is it an unfortunate coincidence? Can you think of any potentially important variables that have been omitted from this model?

15.62 In Exercise 14.79, 1970–1989 data from Table 8 at the end of the book were used to estimate the demand for money equation

$$\ln\!\left[\frac{M}{PQ}\right] = \alpha + \beta\ln[X] + \epsilon$$

where M is M1 (a measure of money), PQ is nominal GNP, and X is the interest rate on Treasury bills. Since $\ln[A/B] = \ln[A] - \ln[B]$, this equation can be rearranged as

$$\ln[M] = \alpha + \beta\ln[X] + \ln[PQ] + \epsilon$$

To consider the possibility that the coefficient of nominal GNP is not 1.0, estimate the equation

$$\ln[M] = \alpha + \beta_1\ln[X] + \beta_2\ln[PQ] = \epsilon$$

using the same data as in Exercise 14.79. Is the estimated value of β_2 statistically significantly different from 1.0 at the 5 percent level? Compare this estimated equation with the one estimated in Exercise 14.79. Is there a substantial change in R^2? Is there a substantial change in the interest-rate coefficient?

15.63 *(continuation)* Plot the regression equation's prediction errors (residuals) versus the year. Do the residuals appear to show heteroskedasticity or serial correlation? If your computer program calculates the Durbin-Watson d-statistic, report this value and test at the 5 percent level the null hypothesis that $\rho \leq 0$.

15.64 Annual data for 1950–1987 were used to estimate the relationship among the amount of airline travel, the gross national product (GNP), and the real cost of airline transportation:[19]

$$Y = \alpha + \beta_1 X_1 + \beta_2 X_2 + \epsilon$$

where
Y = annual passenger miles per capita, average value 493.6
X_1 = real GNP per capita in 1982 dollars, average value \$11,517
X_2 = average price of air transportation per passenger mile, deflated by consumer price index, average value 9.1 cents.

The results were

$$\hat{Y} = -2035.02 + 0.018\ X_1 + 53.26\ X_2$$
$$\qquad\qquad (0.006) \qquad (8.34)$$

$$R^2 = .972$$

( ): standard deviation

a. Which is the dependent variable? **b.** Are the signs of the estimated parameters plausible? **c.** Are the sizes of the estimated parameters plausible? **d.** Are the estimates significantly different from 0 at the 5 percent level?

15.65 Use the data in Table 9.2 to estimate the equation $Y = \alpha + \beta_1 X_1 + \beta_2 X_2 + \epsilon$, where $Y =$ mutual fund return for current year (50 observations using 1988 and 1989 data), $X_1 =$ mutual fund return one year earlier, and $X_2 =$ mutual fund return two years earlier. For values of Y using the 1988 data, X_1 will use 1987 data and X_2 will use 1986 data; for values of Y using the 1989 data, X_1 will use 1988 data and X_2 will use 1987 data. Imagine that you are researching an article for the *Wall Street Journal* and write a paragraph summarizing your results.

15.66 A person's hat size is equal to the circumference of the head, divided by three. Thus a person with a 21-inch head circumference has a hat size of 7. If the equation $Y = \alpha + \beta X + \epsilon$ is estimated, where $Y =$ hat size and $X =$ head circumference, what results do you predict? Is there a multicollinearity problem?

15.67 Housing starts may be affected by the strength of the economy and real mortgage rates (adjusted for inflation). To investigate this relationship, use the 1950–1989 data in Table 8 at the end of the book to estimate the equation

$$Y = \alpha + \beta_1 X_1 + \beta_2 X_2 + \epsilon$$

where $Y =$ housing starts, $X_1 =$ unemployment rate, and $X_2 =$ the long-term Treasury bond rate (a proxy for mortgage rates) minus the rate of inflation. Do your estimates of β_1 and β_2 have plausible signs? Are these estimates statistically significant at the 5 percent level?

15.68 *(continuation)* Plot the residuals $e = Y - (a + b_1 X_1 + b_2 X_2)$ versus the year. Do the residuals seem to display heteroskedasticity or serial correlation? If your computer program gives the value of the

Durbin-Watson d-statistic, report this and test at the 5 percent level the null hypothesis that $\rho \leq 0$.

15.69 *(continuation)* Redo Exercise 15.68, using housing starts per capita, rather than housing starts, as the explanatory variable. Which formulation seems more successful to you? What criteria did you use? Which equation does a better job of predicting housing starts for the years 1950–1989?

15.70 In an IBM antitrust case, an economist, Franklin Fisher, estimated multiple regression equations explaining computer prices as a function of memory, speed, and other characteristics.

> Despite the fact that t-statistics on the order of 20 were obtained for all of the regression coefficients, Alan K. McAdams, appearing as an expert for the government, testified that collinearity made it impossible reliably to separate the effects of the different independent variables and hence that little reliance could be placed on the results.[20]

Explain why you either agree or disagree with McAdams's logic.

15.71 A survey of 137 law professors during the 1972–1973 academic years obtained data on their total salary and 15 explanatory variables:[21]

$Y =$ total salary, thousands of dollars
$X_1 =$ hours worked per week
$X_2 =$ number of published books
$X_3 =$ number of articles published in last two years
$X_4 =$ 1 if have research support, 0 otherwise
$X_5 =$ number of students per semester
$X_6 =$ 1 if have received teaching award, 0 otherwise
$X_7 =$ 1 if hold administrative position, 0 otherwise
$X_8 =$ 1 if do public service consulting, 0 otherwise
$X_9 =$ number of years since degree
$X_{10} =$ 1 if have received recent job offer, 0 otherwise
$X_{11} =$ 1 if have another professional position, 0 otherwise
$X_{12} =$ 1 if also have a nonlaw degree, 0 otherwise
$X_{13} =$ other professional noninstitutional income, thousands of dollars
$X_{14} =$ 1 if are politically liberal, 0 otherwise
$X_{15} =$ 1 if work at a major research university, 0 otherwise

The estimated equation is

$$Y = 23.29 + 0.039\ X_1 + 0.113\ X_2 + 0.598\ X_3$$
$$\qquad\qquad\ \ [1.12]\qquad\ \ [0.92]\qquad\ \ [4.09]$$

$$+\ 0.194\ X_4 + 0.010\ X_5 - 0.527\ X_6 + 1.336\ X_7$$
$$\ \ [0.26]\qquad\ [2.50]\qquad\ \ [0.55]\qquad\ \ [1.59]$$

$$-\ 0.479\ X_8 + 0.364\ X_9 + 0.301\ X_{10} - 0.789\ X_{11}$$
$$\ \ [0.58]\qquad\ [8.10]\qquad\ \ [0.42]\qquad\ \ [1.07]$$

$$+\ 2.233\ X_{12} + 0.031\ X_{13} + 2.005\ X_{14} + 2.363\ X_{15}$$
$$\ \ [2.96]\qquad\ \ [1.11]\qquad\ \ [2.63]\qquad\ \ [3.03]$$

[ ]: t-values

a. Which of the explanatory variables are dummy variables? **b.** Which, if any, of the estimated signs do you find surprising, perhaps implausible? **c.** How many degrees of freedom are there? **d.** Which of the estimated coefficients are statistically significantly different from 0 at the 5 percent level?

15.72 *(continuation)* Give a 95 percent confidence interval for the effect on a law professor's annual salary of teaching an additional 20 students a semester? Use a two-standard-deviations rule of thumb. Does your answer seem like a lot of money or a little?

15.73 *(continuation)* According to the estimated equation, would a law professor's annual salary be increased more **a.** by writing a book or by writing two articles? **b.** by writing an article or by improving one's teaching sufficiently to win a teaching award?

15.74 *(continuation)* According to the estimated equation, how much will a law professor's annual salary be increased by working an extra 10 hours a week? How much is this salary increase per hour of additional work, if the professor works 40 weeks a year? 50 weeks a year?

15.75 *(continuation)* According to the estimated equation, how would the annual salary be affected by a professor working an additional 20 hours a week, publishing two books, obtaining outside research support, teaching an additional 100 students a semester, and receiving a job offer? Do they do it for the money?

15.76 A dean claims that the salaries of humanities professors Y are determined solely by the number of years of teaching experience X. To see if there is a statistically significant difference in the salaries of male and female humanities professors, the following equation was estimated, using the variable $D = 0$ if male, 1 if female. Interpret each of the coefficients.

$$Y = \alpha + \beta_1 D + \beta_2 X + \beta_3 DX + \epsilon$$

15.77 *(continuation)* What advantages does a regression equation have over a difference-in-means test that compares average male and female salaries? Describe a specific situation in which a difference-in-means test could show statistically significant discrimination against females while a regression equation does not. Describe a specific situation in which a regression equation could show discrimination against females while a difference-in-means test does not.

15.78 The following regression equation was estimated to see if male and female professors earn different salaries:[22]

$$Y = \alpha + \beta_1 X + \beta_2 D + \epsilon$$

where Y = annual salary, X = rank (0 if assistant professor, 1 if associate professor, and 2 if full professor), and D = sex (0 if male, 1 if female). Explain why you either agree or disagree with this interpretation:

> We are trying to estimate the separate influence of the variable, "sex." If sex is unrelated to [rank], then the coefficient of sex will show the difference between male and female salaries when [rank] is accounted for. But if sex affects rank and rank has some influence on salary, then all the effect of a change from male to female in the sex variable will not be captured in the coefficient on sex; some of the effect will be embodied in the associated change in rank. Therefore the regression coefficient on sex is misleading.

• 15.79 *(continuation)* The person quoted in the preceding exercise goes on to say that the regression

> . . . showed a statistically insignificant coefficient on sex. This is typical of a problem of multicollinearity. The true effect of sex is obscured by the small regression coefficient and the relatively large associated standard error on that coefficient. The combination of these two yields a low calculated t statistic.
>
> It was in order to avoid this inaccuracy that Dr. deCani omitted rank. The court's discussion of the propriety of excluding rank was essentially one of determining whether rank and sex were correlated. The court concluded that they were not and that therefore excluding the rank was improper.

Explain why you either agree or disagree with this reasoning.

15.80 Table 15.9 lists 20 annual observations on corn yield and 8 weather variables measuring precipitation and temperature. Use multiple regression to estimate the effect of each of these weather variables on corn yield. Do all of the estimated coefficients have plausible signs and magnitudes? Are they each statistically significant at the 5 percent level?

15.81 *(continuation)* Give a 95 percent confidence interval for the effect on corn yield of an additional inch of preseason precipitation, with other factors unchanged. Give a 95 percent confidence interval for the effect on corn yield of an additional two inches of preseason precipitation, with other factors unchanged. Do not use a two-standard-deviations rule of thumb.

15.82 *(continuation)* Is there any reason to suspect that such data might have a multicollinearity problem? Be specific. Is there evidence of a multicollinearity problem in these particular data?

15.83 *(continuation)* What is this equation's predicted corn yield when $X_1 = X_2 = X_3 = X_4 = X_5 = X_6 = X_7 = X_8 = 0$? Why are you skeptical about the accuracy of this prediction? What is this equation's predicted corn yield when $X_1 = 20$, $X_2 = 60$, $X_3 = 5$, $X_4 = 70$, $X_5 = 4$, $X_6 = 75$, $X_7 = 4$, and $X_8 = 72.5$? Give a 95 percent confidence interval for this prediction. Do not use a two-standard-deviations rule of thumb.

15.84 *(continuation)* Plot the estimated equation's prediction errors (residuals) versus the year. Do the residuals display heteroskedasticity or serial correlation? If your computer program reports the value of the Durbin-Watson d-statistic, test at the 5 percent level the null hypothesis that $\rho = 0$.

15.85 A researcher commented on the estimated equation that follows: "Multicollinearity could be a factor as indicated by the very high standard deviation and the small t-value." How does multicollinearity

Table 15.9 *Data for Exercise 15.80*

Year	Corn Yield (Bu/Acre) Y	Preseason Precip. (Inches) X_1	May Temp. (°F) X_2	June Precip. (inches) X_3	June Temp. (°F) X_4	July Precip. (Inches) X_5	July Temp. (°F) X_6	Aug Precip. (Inches) X_7	Aug Temp. (°F) X_8
1943	54.7	19.05	57.5	6.16	71.6	4.56	75.4	5.07	74.0
1944	52.0	20.79	64.6	5.88	71.7	3.73	72.6	5.88	71.8
1945	43.5	21.88	55.1	4.70	64.1	2.96	72.1	3.43	72.5
1946	56.7	20.02	56.5	6.41	69.8	2.45	73.8	3.56	68.9
1947	30.5	23.17	55.6	10.39	66.3	1.72	72.8	1.49	80.6
1948	60.5	19.15	59.2	3.42	68.6	4.14	75.0	2.54	73.9
1949	46.1	18.28	63.5	5.51	72.4	3.47	76.2	2.34	73.0
1950	48.2	18.45	59.8	5.70	68.4	4.65	69.7	2.39	67.7
1951	43.1	22.00	62.2	6.11	65.2	4.45	72.1	6.21	70.5
1952	62.2	19.05	59.6	5.40	74.2	3.84	74.7	4.78	70.0
1953	52.9	15.67	60.0	5.31	73.2	3.28	74.6	2.33	73.2
1954	53.9	15.92	55.6	6.36	72.9	1.79	77.4	7.10	72.1
1955	48.4	16.75	63.6	3.07	67.2	3.29	79.8	1.79	77.2
1956	52.8	12.34	62.4	2.56	74.7	4.51	72.7	4.42	73.0
1957	62.1	15.82	59.0	4.84	68.9	3.54	77.9	3.76	72.9
1958	66.0	15.24	62.5	3.80	66.4	7.55	70.5	2.55	73.0
1959	64.2	21.72	62.8	4.11	71.5	2.29	72.3	4.92	76.3
1960	63.2	25.08	59.7	4.43	67.4	2.76	72.6	5.36	73.2
1961	75.4	17.79	57.4	3.36	69.4	5.51	72.6	3.04	72.4
1962	76.0	26.61	66.6	3.12	69.1	6.27	71.6	4.31	72.5

affect standard deviations and t-values? Why is it clear that multicollinearity is not a problem here?

$$Y = 1146.0 + 1.4312 X, \qquad R^2 = 0$$
$$(195.6) \quad (71.6)$$
$$[5.86] \quad [0.02]$$

( ): standard deviations
[ ]: t-values

15.86 Five plots of land in Georgia were studied over a three-year period to see how the tobacco yield (Y) is affected by the percentage of nitrogen in the fertilizer (X). Use the following data to estimate the equation $Y = \alpha + \beta X + \epsilon$. Does there seem to be a positive or negative relationship between nitrogen and yield? Is this relationship substantial? Is it statistically significant at the 1 percent level?

Year	Plot	Percent Nitrogen	Yield (Pounds)
1924	1	0	867
	2	2	1,094
	3	3	1,206
	4	4	1,281
	5	5	1,235
1925	1	0	889
	2	2	1,101
	3	3	1,180
	4	4	1,238
	5	5	1,237
1926	1	0	914
	2	2	1,096
	3	3	1,157
	4	4	1,224
	5	5	1,219

15.87 *(continuation)* Calculate the prediction errors $e = Y - (a + bX)$ for each of these fifteen observations and make a scatter diagram with e on the vertical axis and X on the horizontal axis. Draw a line $e = 0$. What pattern do you see and how do you explain it?

15.88 *(continuation)* Draw a scatter diagram with Y on the vertical axis and X on the horizontal axis. Does the relationship between X and Y appear to be linear or nonlinear? If you were to fit a quadratic equation, $Y = \alpha + \beta_1 X + \beta_2 X^2 + \epsilon$, to these data, what signs would you anticipate for your estimates of β_1 and β_2? Use multiple regression to estimate this

equation and see whether you are right. Are the estimates of β_1 and β_2 both statistically significant at the 1 percent level?

15.89 *(continuation)* On a scatter diagram of X and Y, sketch the linear and quadratic equations estimated in Exercises 15.86 and 15.88. Does the quadratic equation appear to fit the data substantially better?

15.90 *(continuation)* Looking at the graph drawn in Exercise 15.88, how do you think that a cubic equation, $Y = \alpha + \beta_1 X + \beta_2 X^2 + \beta_3 X^3 + \epsilon$, might differ from a quadratic equation, $Y = \alpha + \beta_1 X + \beta_2 X^2 + \epsilon$? Estimate this cubic equation and sketch the estimated quadratic and cubic equations on a scatter diagram of X and Y. How does the cubic equation differ from the quadratic? Does the cubic equation appear to fit the data substantially better than the quadratic equation?

•• **15.91** *(Requires calculus)* Use the equation estimated in Exercise 15.90 to determine the nitrogen percentage that maximizes yield. What crucial information do you need to know before deciding that this nitrogen percentage maximizes profits?

15.92 In a 1980 court case a federal judge explained how the application of regression analysis to estimate male and female salary equations could be used to determine if an employer discriminates against females.[23] Explain his conclusion that the following estimated equations would make it "clear that females are being discriminated against."

males: $Y = 10,000 + 500 X_1 + 350 X_2$

females: $Y = 9,000 + 400 X_1 + 300 X_2$

where Y = annual earnings
          X_1 = number of years of schooling
          X_2 = number of years of relevant job experience

15.93 *(continuation)* Why, if the values of R^2 do not equal 1, is the conclusion not as ironclad as the judge assumes?

15.94 *(continuation)* Explain why the following regression equations cited by the judge would represent "a more difficult case":

males: $Y = 10,000 + 500 X_1 + 350 X_2$

females: $Y = 10,000 + 400 X_1 + 1,000 X_2$

15.95 *(continuation)* The judge noted that in place of two separate equations, one for males and one for females, the following single equation could be estimated:

$$Y = \alpha + \beta_1 X_1 + \beta_2 X_2 + \beta_3 X_3$$

where Y, X_1, and X_2 are as defined in Exercise 15.92 and $X_3 = 0$ if the person is a male and $X_3 = 1$ if the person is female. Interpret the coefficients of this equation. In what ways is this single equation less general than the two separate equations?

15.96 In a job discrimination court case, a professor estimated salary equations using education, years experience, race, and sex as explanatory variables:[24]

$$Y = \alpha + \beta_1 X_1 + \beta_2 X_2 + \beta_3 X_3 + \beta_4 X_4$$

where Y = annual earnings, X_1 = number of years of schooling, X_2 = number of years of relevant job experience, $X_3 = 0$ if the person is a male and $X_3 = 1$ if the person is female, and $X_4 = 0$ if the person is a white and $X_4 = 1$ if the person is nonwhite. What results would you consider to be evidence of discrimination against blacks in comparison to whites? Against females in comparison to males?

15.97 *(continuation)* In this formulation, is it possible to obtain results showing, in comparison to white men, discrimination against black men and against black women but in favor of white women? Can the estimated equation show, in comparison to white men, discrimination against black men and against white women but in favor of black women?

• **15.98** In the court case discussed in Exercise 15.96, the professor who estimated the salary equation was asked if his equation would have been more valid if he had included additional explanatory variables that measured worker productivity. He responded, "To the extent that [these omitted explanatory variables are] not correlated with race or sex [it] makes no

difference whatsoever." Using your intuitive understanding of multiple regression, explain his reasoning. It may be helpful to begin with a specific example of the contrary case, where an important omitted productivity variable is correlated with race or sex.

• **15.99** *(continuation)* To explore the professor's argument, consider the following hypothetical model. Salaries are determined by the following equation (with no error term):

$$Y = 20{,}000 + 1{,}000\, X_1 - 5{,}000\, X_2 + 10{,}000\, X_3$$

where Y = annual salary
X_1 = number of years of experience
$X_2 = 0$ if the person is male, 1 if female
$X_3 = 1$ if the person had a course in statistics, 0 otherwise

The data are

Y	X_1	X_2	X_3
25,000	5	0	0
35,000	5	0	1
20,000	5	1	0
30,000	5	1	1
30,000	10	0	0
40,000	10	0	1
25,000	10	1	0
30,000	10	1	1

Is X_3 correlated with either X_1 or X_2? Use these data to estimate the equation $Y = \alpha + \beta_1 X_1 + \beta_2 X_2 + \beta_3 X_3 + \epsilon$, and then to estimate the equation $Y = \alpha + \beta_1 X_1 + \beta_2 X_2 + \epsilon$. Do your results confirm the professor's assertion that the exclusion of X_3 "makes no difference whatsoever"?

15.100 A multiple regression equation was constructed with 12 explanatory variables that correctly predicts the winner of all 11 presidential elections from 1932–1972.[25] Give an important reason for skepticism about this equation.

PART

III

Selected Topics

16

Analysis of Variance

Nothing is good or bad but by comparison.

Thomas Fuller

TOPICS

There are many situations in which we suspect that certain equipment or procedures make a difference. Secretaries may be more productive with certain word processors; students may learn more from certain teachers; and the ill may get well sooner with certain medicines. Observed variations in the outcomes can be caused by other factors, though. Some secretaries are more productive than others, and some students will do better than others, regardless of the teacher. An illness may last only a short while, even without medication, or it may last a long time, even with the best treatment.

The trick is to use a variety of data to separate the other influences from the treatments of interest. If we look at only a single secretary who is working on a single machine, we will have no way of judging the machine. Even if we observe one secretary on one machine and another secretary on a second machine, we still won't know whether the observed differences in performance are caused by the secretaries or the machines. Another possible experiment is to have a single secretary do one assignment on two machines, but again, variations in performance may be caused by variations in the secretary's productivity, rather than differences between the two machines. The key is to get a handle on the variation in performances on a single machine, and for this, we need several assignments done on each machine. These could be single assignments done by many secretaries or multiple assignments completed by the same secretary. One way or another, we must get an estimate of how much variation can be expected on a single machine, and then we can compare this to the amount of variation between the two machines. If the two-machine variation is about the same as the variation on a single machine, then the two machines are apparently very similar. If, on the other hand, the variation between the two machines is greater than that on a single machine, then we have identified a statistically significant difference between the machines.

It sounds plausible in theory, but how do we do it in practice? Which statistics should we calculate and what is the appropriate test? The great British statistician, Sir Ronald A. Fisher (1890–1962), wrestled with this question and came up with an **analysis-of-variance** procedure, which is an extension of the *t*-test for the differences between two means. In this chapter, you will learn how Fisher's procedure works. In turns out that Fisher's analysis-of-variance model is a special case of regression analysis, but it is most easily motivated by following Fisher's reasoning.

16.1   VARIATION BETWEEN AND WITHIN TREATMENTS

Let's use a new example to explain Fisher's logic. Several years ago, some researchers found that plants grew better when they were watered with a solution containing fluorescein.[1] In their experiment, these agronomists gave the plants on one plot of land a mixture of fluorescein and water, while the plants on a second plot, the *control group,* were just given water.

The researchers found that the plants that were given fluorescein did substantially better than those that were given plain water. However, other agronomists tried this experiment elsewhere and found that fluorescein made no difference.

What went wrong? The problem with the original experiment was that the two plots of land were not identical—as they seldom are! The composition of the soil (loam, clay, or sand), the nutrients present, and the amount of shade and sunshine can all vary from one plot of land to the next. It so happened that in this experiment, plot 1 had better soil than plot 2. The plants grew more, not because of the fluorescein, but because of the richer soil.

An ideal experiment would use identical soil for both the plants given fluorescein and those given plain water. However, because growing conditions can never really be identical, the next best thing is to estimate the amount of natural variation in growing conditions. By estimating the random variation in plant growth attributable to varying plant and soil conditions, we can determine statistically whether the observed variation between the plants given fluorescein and those given plain water is of the same order of magnitude.

How do we actually decide which plants in which locations are given which treatment? The logic is very similar to public-opinion sampling. We don't want to construct the samples by trying to choose two typical plots of land, because the results will then be too dependent on our subjective assessment of what is "representative" soil. If the plots aren't identical (and they seldom will be), then our results will be biased in ways that we can't take into account. We won't know whether the observed differences in plant growth are caused by the treatments or by our errors in choosing typical plots. This is the trap that the original fluorescein researchers fell into.

We also do not want convenience sampling. It is easiest for pollsters to interview friendly people who live nearby and it is easiest for doctors to give polio vaccines to children whose parents volunteer permission. But such easy samples are, as we have seen, also biased samples. Similarly, it is undoubtedly easiest to put all the plants that are receiving fluorescein together and all of the plants that are receiving plain water together; then all of the plants in each group can be treated at once. But again, that leads to two separate, and perhaps very different, plots of soil. Similarly, if the plants in the experiment are all new plants that arrive in several shipments, it is easiest to label all of the plants in one shipment "fluorescein" or "plain water." But the shipments may well be different, especially if they come from different nurseries or from different parts of a large nursery.

It takes some extra effort to choose random samples, but this effort is well worth it. As in opinion polling, a small random sample is much more reliable than a large biased sample. What we want to do in this experiment is spread the fluorescein and water around on randomly selected plants in randomly selected locations. If the plants are already in the ground, then this randomization will be done in one fell swoop. If the plants are new, then they can be assigned to different locations that are randomly selected to receive either fluorescein or plain water.

We do want to enforce some balance in our experiment to avoid the possibility that, by the luck of the draw, the fluorescein is used on all of the plants or on none of them. If we are using eight plants, for example, we might require that half be given fluorescein and half plain water. The randomness comes in deciding which is which. A simple procedure would be to prepare eight tags, with four marked "fluorescein" (F) and four "plain water" (W). These tags could be placed in a farmer's hat and, then, a drawing made for each of the eight plants. For a more scientific appearance, a table of random numbers could be used. I used eight playing cards to determine the following locations:

$$F \quad W \quad W \quad F$$
$$F \quad W \quad F \quad W$$

Our hope is that the other factors that influence plant growth will balance each other out, that some of the fluorescein plants will be healthy and some weak, and that some will be planted in the most fertile soil and others in less favorable locations. But, by the luck of the draw, this may not happen. By chance, fluorescein may be put only on the healthiest plants in the choicest locations. However, our statistical procedures can take this possibility into account by calculating the probability that random variation (sampling error) is responsible for the results. In the absence of random sampling, no such allowances can be made.

Perhaps the plant heights (in inches) turn out as follows:*

	Treatment	
	Fluorescein	*Plain Water*
	28	27
	33	30
	24	23
	27	24
Mean	28	26
Variance	14	10

First, notice that there is a difference in the means. The average height is 28 inches with fluorescein and only 26 inches with water. Second, notice that, for either treatment, the heights vary from one plant to the next. Because of this inevitable variation, we cannot be certain that the observed difference between the plants that were given fluorescein and those that were given just water is caused by the fluorescein. It may be that, by the luck of the draw, fluorescein was given to healthier plants in more productive soil.

There are two very different explanations, as shown in Figure 16.1, for the observed difference in the sample means. It may be that the expected height with fluorescein is greater than the expected height with plain water; thus, the observed means of 28 and 26 came from two different populations. Or it may be that the

*The variance calculations are

$$\text{fluorescein: } \{(28 - 28)^2 + (33 - 28)^2 + (24 - 28)^2 + (27 - 28)^2\}/3$$
$$= (0 + 25 + 16 + 1)/3 = 42/3 = 14$$
$$\text{plain water: } \{(27 - 26)^2 + (30 - 26)^2 + (23 - 26)^2 + (24 - 26)^2\}/3$$
$$= (1 + 16 + 9 + 4)/3 = 30/3 = 10$$

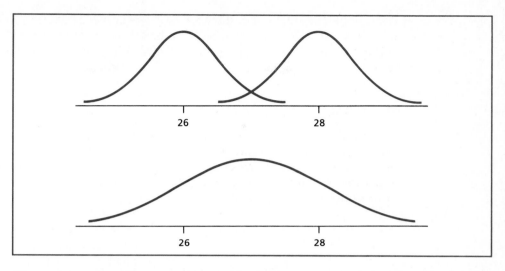

Figure 16.1 *Two Different Sample Means May Have Come from Two Populations with Different Means or from a Single Population with a Large Standard Deviation*

samples came from a single population with a large variance: if there are large height differences from one plant to the next, then it is not surprising that one mean happened to be 28 and the other only 26.

> The observed difference between two sample means can be caused by either the different means of two distributions or the large variance of a single distribution.

A formal model of plant heights, Y, would be

$$\text{plain water: } Y = \alpha_1 + \epsilon$$

$$\text{fluorescein: } Y = \alpha_2 + \epsilon$$

The population parameter α is the expected height. The error term ϵ is a random variable that encompasses all of the many reasons why, either with plain water or fluorescein, heights vary from one plant to another. If fluorescein makes no difference, then α_1 equals α_2 and the heights come from a single distribution, so that observed height variations are simply due to the random error ϵ. If fluorescein does affect plant heights, then the inequality of α_1 and α_2 is partly responsible for observed height differences.

How can we choose between these competing explanations for the observed difference in the sample means? The key to a comparison here is the difference in heights from one plant to the next. If plant heights hardly vary, then the difference between 28 and 26 inches is persuasive evidence that fluorescein makes a difference. If, on the other hand, there are large height variations from one plant to the next, then the difference between 28 and 26 inches is not convincing. So, we must compare the observed variance across plants with the observed difference in sample means.

The Mean of the Sample Variances

The variance in heights across plants is 14 with fluorescein and 10 with plain water. We can view these numbers as two estimates of the true population variance of plant heights—the variance of ϵ, which we assume is the same for both fluorescein and water. A natural estimate of this common variance across plants is the average of these two separate estimates:

$$\begin{matrix} \text{estimated common} \\ \text{variance} \\ \text{across plants} \end{matrix} = \frac{\begin{matrix} \text{fluorescein} & & \text{plain water} \\ \text{variance} & + & \text{variance} \\ \text{across plants} & & \text{across plants} \end{matrix}}{2}$$

$$= \frac{14 + 10}{2}$$

$$= 12$$

The Variance of the Sample Means

Now, with this estimate in our hip pocket, we can gauge whether the observed difference in the means, 28 versus 26, is large or small. If we were drawing two random samples of size 4 from a single population with mean μ and variance σ^2, then the sample means would come from a distribution with expected value μ and variance $\sigma^2/4$.

If the population variance of ϵ is $\sigma^2 = 12$, then the sample means come from a distribution with a variance $12/4 = 3$ (and a standard deviation of $\sqrt{3} = 1.73$). In this light, as Figure 16.2 shows, the observed difference between 28 and 26 is not very convincing evidence against the null hypothesis that the heights come from a single distribution. More concretely, the observed variance of the sample means is

$$\text{variance of the sample means} = \frac{(28 - 27)^2 + (26 - 27)^2}{2 - 1}$$

$$= 2$$

Thus, the observed variance of 12 across plants implies a variance of 3 for sample means drawn from a single distribution. The observed variance of the sample means is, in fact, only 2. If fluorescein really did affect plant growth, the variance in sample means should be larger than 3, and not smaller.

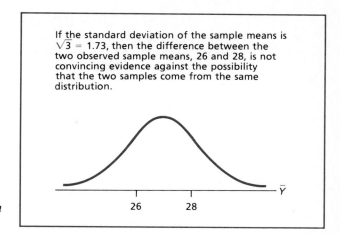

If the standard deviation of the sample means is $\sqrt{3} = 1.73$, then the difference between the two observed sample means, 26 and 28, is not convincing evidence against the possibility that the two samples come from the same distribution.

26   28

$\bar{Y}$

Figure 16.2   *The Fluorescein Single-Distribution Theory*

The *F*-Statistic

Let's retrace our steps before we forget what we did. First, we estimated the variance σ^2 across plants by calculating the average value of the two sample variances. Dividing this by n gives s^2/n, an estimate of the variance of the sample means. Second, we compared this estimate to the observed variance of the sample means. In ratio form, we have

$$F = \frac{\text{(variance of sample means)}}{\text{(mean of sample variances)}/n} \tag{16.1}$$

or, equivalently,

$$F = \frac{n(\text{variance of sample means})}{(\text{mean of sample variances})} \tag{16.2}$$

This is called the **F-statistic** in honor of Fisher's pioneering work. A calculated *F*-value near 1.0 is consistent with the hypothesis that the data came from a single population with a relatively large variance. If, on the other hand, the *F*-value is much larger than 1.0, then the observed difference in the sample means is too large to be explained plausibly by sampling variation from one plant to the next.

According to the previous data, the samples are of size $n = 4$, the mean of the sample variances is 12, the variance of the sample means is 2, and the value of the *F*-statistic,

$$F = (4)(2)/(12) = 0.67$$

is not convincing evidence that fluorescein makes a difference.

For practice, let's run through the calculations again, using somewhat different data this time. This agricultural experiment was conducted on a different crop, using a special solution and plain water. The results (in inches) were

	Treatment	
	Solution	*Water*
	33	31
	32	30
	30	25
	32	27
	36	31
	29	25
	32	27
Mean	32	28
Variance	5	7

The sample variances of the heights are 5 for the solution and 7 for the plain water. Thus, the average variance is 6:

$$\text{mean of sample variances} = (5 + 7)/2 = 6$$

There is relatively little variation in heights with this second crop. If the heights are, in fact, from a single population with a common expected value, then the estimated variance of the sample means is $s^2/n = 6/7 = 0.857$. However, the observed variance of two sample means (32 and 28) is much larger than this:

$$\text{variance of sample means} = \frac{(32 - 30)^2 + (28 - 30)^2}{2 - 1} = 8$$

Thus, the *F*-value is

$$F = \frac{n(\text{variance of sample means})}{(\text{mean of sample variances})}$$
$$= (7)(8)/(6) = 9.33$$

Unlike the fluorescein data, this difference in the sample means is striking, as Figure 16.3 illustrates. There is relatively little variation in the heights within each sample, and so, the observed variation in the sample means is more than nine times larger than might be expected if the solution were no different than plain water. Thus, we are led to the conclusion that the solution does matter.

With the fluorescein data, we got an unimpressive *F*-value of 0.67. With the second batch of data, we got a striking 9.33 *F*-value. Where do we draw the line between unimpressive and striking? Fisher answered this question by figuring out the probability distribution for the *F*-statistic so that, as with any hypothesis test, we can ask whether the observed value is sufficiently unlikely to be persuasive.

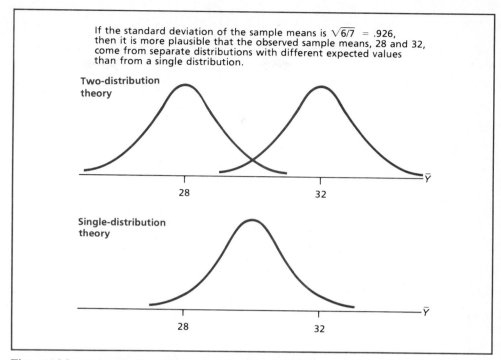

If the standard deviation of the sample means is $\sqrt{6/7} = .926$, then it is more plausible that the observed sample means, 28 and 32, come from separate distributions with different expected values than from a single distribution.

Figure 16.3 *A Convincing Difference in the Sample Means*

The *F*-Distribution

Fisher assumed that the variable of interest is normally distributed, and so will we throughout this chapter. Under the null hypothesis that the data come from a single distribution (that the treatments are unimportant), he derived the distribution of the *F*-statistic.* Figure 16.4 shows one such *F*-distribution. Remember, that an *F*-value that is close to 1.0 is consistent with this hypothesis. However, as is shown in Figure 16.4, the *F*-value will probably be a bit larger or smaller than 1.0 because of sampling variability. The *F*-value could even be as small as 0 or much larger than 1.0. As Equation 16.1 shows, a large *F*-value reflects large observed differences in the sample means, which is evidence against the null hypothesis. For the usual hypothesis-testing procedure, we choose a probability of Type I error, say 5 percent, and then find the corresponding cutoff value of *F*. If it turns out that the *F*-value is this large, then the improbability of this outcome persuades us that the null hypothesis is wrong, and the treatments do matter.

*Formally, if the independent random variables X_1 and X_2 have chi-square distributions with d_1 and d_2 degrees of freedom, then

$$\frac{X_1/d_1}{X_2/d_2}$$

has an *F*-distribution with d_1 and d_2 degrees of freedom.

Figure 16.4   *An F-*
Distribution

The exact distribution of F depends on two considerations: the number of treatments (k) and the number of observations (n) of each treatment. There are essentially k samples; each is size n. The numerator of the F-statistic is the variance of the sample means and, because there are k sample means, the numerator is said to have $k - 1$ degrees of freedom. The denominator of the F-statistic is the mean of the sample variances and, because each of these k variances has $n - 1$ degrees of freedom, the denominator is said to have $k(n - 1)$ degrees of freedom.

These two degrees-of-freedom calculations

| numerator: $k - 1$ = number of treatments $- 1$ |
| denominator: $k(n - 1)$ = (number of treatments)(sample sizes $- 1$) |

determine the exact distribution of the F-statistic.

Figure 16.5 shows three very different shapes for an F-distribution. Notice, though, how the F-distribution becomes bell-shaped as both degrees of freedom increase.

In the fluorescein example, there were $k = 2$ treatments and $n = 4$ observations per treatment, giving the numerator $k - 1 = 1$ degree of freedom and the denominator $k(n - 1) = 2(3) = 6$ degrees of freedom. Table 6 in the back of this book shows that for these degrees of freedom there is a 5 percent chance that the value of the F-statistic will be greater than 5.99, if there really is no difference between the treatments. The complete F-distribution is shown in Figure 16.6. Because the observed value of 0.67 is nowhere near this large, we cannot reject at the 5 percent level the null hypothesis that the data come from a single distribution.

In the second example, there were $k = 2$ treatments and $n = 7$ observations, giving 1 degree of freedom for the numerator and 12 for the denominator. Table 6 says that there is a 5 percent chance of an F-value greater than 4.75, if the null hypothesis is true. The observed 9.33 F-value is greater than this, which leads to the rejection of the null hypothesis at a 5 percent significance level. If the data came from a single distribution, there would be less than a 5 percent chance of observing this large an F-value, and thus, the improbability of this outcome persuades us that the data did not come from a single distribution.

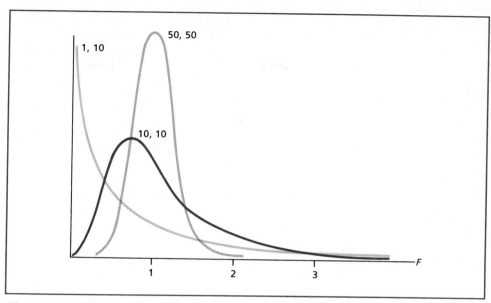

Figure 16.5 *Three Varied F-Distributions, with Differing Degrees of Freedom k — 1, k(n — 1)*

Does the Teacher Make a Difference?

Fisher's F-test is actually an extension of the t-test for the difference between two means. With two treatments, the F-distribution has 1 and $n - 1$ degrees of freedom; a comparison of this column of Table 6 with the t-distribution in Table 4 shows that F is equal to t-squared! This is no accident, because with $k = 2$, the test values of the F-statistic and the square of the t-statistic are identical (see Exercise 6.13). The major advantage of Fisher's F-test is that it can be applied to more than two treatments. For an example, consider the question of whether or not student

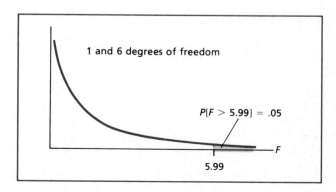

Figure 16.6 *The F-Distribution for the Fluorescein Example*

performance is affected by the teacher. At a certain large university, students who enroll in an introductory statistics course are randomly assigned to sections that are taught by different people. Each section uses the same textbook, does the same homework, and is given the same final examination. At the end of the semester, the final exam is written by all of the teachers, as a group, and is graded by all the teachers, too. To simplify the calculations, I've randomly selected four sections from the fall 1983 semester—one taught by a full professor, one by an associate professor, one by an assistant professor, and one by a graduate student. The final exam scores for these four sections are shown in Table 16.1.

Table 16.1   *Final Exam Scores, for Four Different Teachers*

	Graduate Student	Assistant Professor	Associate Professor	Full Professor
	71	84	79	92
	72	94	92	70
	80	77	73	74
	70	84	86	70
	85	96	82	74
	77	84	98	85
	79	86	98	70
	95	99	82	75
	63	96	91	62
	80	86	64	90
	88	93	71	87
	72	82	80	89
	63	78	65	85
	71	80	97	73
	63	98	94	81
	61	80	75	83
	64	82	59	75
	77	79	87	73
	70	85	78	59
	60	86	81	83
	87	71	70	82
	79	68	71	77
	83	69	91	87
	68	95	73	63
	54	65	92	64
	79	87	82	78
	81	82	59	76
	58	82	75	99
	88	84	71	77
	73	80	61	80
Mean	73.70	83.73	79.23	77.77
Variance	103.25	76.41	136.67	88.81

We can think of these scores as being a sample from the population of prospective students that might be taught by these professors. The professors are the "treatments." We want to know if the expected student score varies from professor to professor, and so the null hypothesis is that the teacher doesn't matter. The first thing to do in testing this blasphemy is to calculate the variance in student scores within each class. These four variances are shown at the bottom of Table 16.1, and their average is about 101:

$$\begin{matrix} \text{mean of} \\ \text{sample} \\ \text{variances} \end{matrix} = \frac{103.25 + 76.41 + 136.67 + 88.81}{4}$$

$$= 101.285$$

This number gives us an idea of how much variance to expect in the sample means. If the teacher is unimportant, so that all of the scores come from a single distribution, then, with 30 students per class, the variance of the sample means should be $\sigma^2/n = 101.285/30 = 3.376$.

In fact, the observed variance of the sample means is more than five times larger than this:

$$\begin{matrix} \text{variance} \\ \text{of sample} \\ \text{means} \end{matrix} = \frac{(73.3 - 78.6)^2 + (83.7 - 78.6)^2 + (79.2 - 78.6)^2 + (77.8 - 78.6)^2}{(4 - 1)}$$

$$= 17.137$$

Thus the value of the F-statistic is about 5:

$$F = \frac{n(\text{variance of sample means})}{(\text{mean of sample variances})}$$

$$= \frac{30(17.137)}{(101.285)}$$

$$= 5.076$$

For $k = 4$ samples of size $n = 30$, the degrees of freedom are $k - 1 = 3$ for the numerator and $k(n - 1) = 4(29) = 116$ for the denominator. Table 6 tells us that if the teacher did not make a difference, there would be a 5 percent chance of observing an F-value of 2.7 and, thus (as is illustrated in Figure 16.7) the data in Table 16.1 are decisive evidence against the hypothesis that the teacher doesn't matter.

Are There Monthly Patterns in Stock Prices?

There is considerable Wall Street folklore about monthly patterns in stock prices; for example, it is said that prices go up with our spirits in the spring, fall in the dog

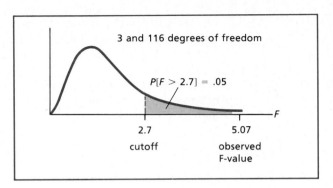

Figure 16.7   *The Data Reject*
the Hypothesis that All
Teachers Are Equal

days of summer, and drop again with end-of-the-year tax sales. If these tales were true, this would be valuable information. If stock prices went up in April and down in August, we could profitably buy in March and sell in July.

A contrary argument is that there are so many people looking for such patterns that the patterns could not persist. If stock prices could be counted on to go up in April, then people would stop buying stocks in April and start buying in March. Then, this March surge would persuade people to buy in February instead (or January, for people anticipating the February surge). The only situation that could persist is chaos. Any profitable pattern would self-destruct.

We can use Fisher's F-test to see whether or not the behavior of the stock market varies with the month of the year. Stock prices go up and down because any of a large number of possible events may happen. Because it is not known in advance what will happen, stock returns are random variables whose uncertainties can be described by probability distributions. The null hypothesis here is that each month's return is drawn from a single probability distribution. The actual data are a sample of what might have happened each month.

Table 16.2 shows the actual monthly percentage returns (dividends plus capital gains) for the Standard & Poor's 500 stocks over the eleven-year period 1970–1980. There are some differences in the average monthly returns, although not those suggested by the folklore cited earlier. April was no better than August over these eleven years, and December was the best month of all—no sign of a tax selloff!

The statistical question is whether these observed differences from one month to the next are large enough to impress us. Or, are they just the sort of averages we would expect if we just picked eleven returns at random, without paying attention to the month in which they occurred? To answer this question, we turn to the variance of the returns and find that the average variance is 23.20. If we use this as an estimate of σ, then we expect the variance of the sample means to be $23.20/n = 23.20/11 = 2.11$ if the months are unimportant. If, instead, the months are important, then the observed variance of the sample means should be larger than this. As it turns out, the observed variance of the sample means is a minuscule 0.634, giving

They Call It Stormy Monday

Mondays also have a reputation on Wall Street—for being wild and often downright nasty. An empirical study of daily percentage stocks returns over the period July 3, 1963 to December 18, 1978 (4132 days in all) obtained the data shown in Table 1.[2]

Over this period, the average daily return was 0.0152. Only Monday had a negative average return, and it works out to be a staggering -33.5 percent on an annual basis. To see if the differences in the daily returns are statistically significant, Fisher's F-statistic can be calculated from*

$$\text{average variance} = (0.670 + 0.551 + 0.644$$
$$+ 0.483 + 0.479)/5$$
$$= 0.5654$$

$$\text{variance of means} = \{(-0.134 - 0.0152)^2 + \cdots$$
$$+ (0.084 - 0.0152)^2\}/(5 - 1)$$
$$= 0.008465$$

so that

$$F = \frac{n(\text{variance of sample means})}{(\text{mean of sample variances})}$$
$$= \frac{(4132/5)(0.008465)}{(0.5654)} = 12.4$$

The numerator has 4 degrees of freedom, while the denominator has so many that we may as well use the ∞ in Table 6 at the back of this book, giving a 2.37 cutoff for a test at the 5 percent level. The observed 12.4 F-value shatters this cutoff, indicating decisively that the days of the week do matter. More complete tables of the F-distribution show that, if the day of the week were unimportant, the probability of such a high F-value is less than .0001.

Why is Monday so stormy? The researchers who compiled these data tried to find a reasonable explanation and came up empty, concluding that it is "an obvious and challenging empirical anomaly." Until proven otherwise, my advice is, "Be cautious on Monday."

Table 1

	Monday	Tuesday	Wednesday	Thursday	Friday
Mean	−0.134	0.002	0.096	0.028	0.084
Variance	0.670	0.551	0.644	0.483	0.479

*These calculations aren't quite right, because holidays left an unequal number of observations for each day of the week. The exact way to handle such data will be discussed in the next section.

Table 16.2   *Eleven Years of Monthly Stock Returns, Percent*

Year	Jan	Feb	Mar	Apr	May	June	July	Aug	Sept	Oct	Nov	Dec
1970	−7.43	5.85	0.30	−8.89	−5.47	−4.82	7.52	5.09	3.47	−0.97	5.35	5.84
1971	4.19	1.41	3.82	3.77	−3.67	0.21	−3.99	4.12	−0.56	−4.04	0.27	8.77
1972	1.94	2.99	0.72	0.57	2.19	−2.05	0.36	3.91	−0.36	1.07	5.05	1.31
1973	−1.59	−3.32	−0.02	−3.95	−1.39	−0.51	3.94	−3.18	4.15	0.03	−10.82	1.83
1974	−0.85	0.19	−2.17	−3.73	−2.72	−1.29	−7.59	−8.28	−11.70	16.57	−4.48	−1.77
1975	12.51	6.74	2.37	4.93	5.09	4.62	−6.59	−1.44	−3.28	6.37	3.13	−0.96
1976	11.99	−0.58	3.26	−0.99	−0.73	4.27	−0.68	0.14	2.47	−2.06	−0.09	5.40
1977	−4.89	−1.51	−1.19	0.14	−1.50	4.75	−1.51	−1.33	0.00	−4.15	3.70	0.48
1978	−5.96	−1.61	2.76	8.70	1.36	−1.52	5.60	3.40	−0.48	−8.91	2.60	1.72
1979	4.21	−2.84	5.75	0.36	−1.68	4.10	1.10	6.11	0.25	−6.56	5.14	1.92
1980	6.10	0.31	−9.87	4.29	5.62	2.96	6.76	1.31	2.81	1.87	10.95	−3.15
Mean	1.84	0.73	0.52	0.47	−0.26	0.98	0.45	0.90	−0.29	−0.07	1.89	1.95
Variance	45.72	10.83	17.27	24.02	12.17	10.90	26.80	18.25	19.07	48.09	32.86	12.45

Source: Roger G. Ibbotson and Rex A. Sinquefield, *Stocks, Bonds, Bills, and Inflation,* Charlottesville, Virginia: Financial Analysts Research Federation, 1979 (and updates).

an *F*-value of

$$F = \frac{n(\text{variance of sample means})}{(\text{mean of sample variances})}$$

$$= \frac{(11)(0.634)}{23.20}$$

$$= 0.30$$

Because the observed *F*-value is less than 1, we do not need to look at the *F*-table to know that these data do not show a statistically significant variation in monthly stock returns.

EXERCISES

16.1 You have been hired to design a sampling procedure for testing a purported "miracle" fertilizer. One hundred plants will be used; half will be given this new fertilizer and half will be given a standard fertilizer. Tell how you would decide which plants get which fertilizer. Be specific!

16.2 An agricultural researcher tests the effects of two different fertilizers on soybean yields from four different plots of land:

Treatment	
Fertilizer 1	*Fertilizer 2*
44	48
37	32
41	40
34	48

Determine the following: **a.** the mean yields for each fertilizer **b.** the variances of the yields for each fertilizer **c.** the mean of the variances **d.** the variance of the means **e.** the value of Fisher's F-statistic **f.** whether or not there is a statistically significant difference at the 1 percent level in the yields with these two fertilizers

16.3 A firm that tries to find customers by phone has two alternative approaches: the "soft sell" and the "hard sell." The spirit of the soft sell is a low-key friendly manner that wins the customer's confidence. The hard sell attacks by making wild claims, promising excessive benefits, and exaggerating the dangers of not buying. A consultant is hired to decide which approach is the more effective. Half of the calls use one approach, and half the other; a shuffled deck of cards is used to determine which potential customer gets which sales approach. (A black card means soft sell, a red card hard sell.) The results can be summarized as follows:

	Sales/Day	
	Soft Sell	Hard Sell
Mean	80	96
Variance	100	120

Number of days tested: $n = 20$

a. Why does each approach have a variance in its sales? **b.** What is the mean of the sample variances? **c.** If the two approaches were equally effective, what would be your estimate of the variance of the sample means, based on the mean of the sample variances? **d.** What is the actual variance of the sample means? **e.** What is the value of Fisher's F-statistic? **f.** Do these data confirm or reject the null hypothesis that each approach is equally effective? Use a 1 percent significance level.

16.4 In the monthly stock returns example discussed in the text, I didn't bother to look up the value of F for which we would reject the null hypothesis that the expected stock return does not depend on the month. Why? What is the appropriate value of F for a 5 percent probability of Type I error?

16.5 Can the value of the F-statistic ever be negative? When? Can the value of the F-statistic ever be zero? When?

16.6 A 1986 newspaper headline read "Barnstable auto fatalities drop 600 percent," referring to the

fact that there had been 12 fatalities in 1980 and 2 in 1985. How large a percentage drop would you report? The accompanying article also noted that

> Of particular significance is that last year neither of the two fatalities involved alcohol, [the selectman] said. That compares with alcohol being a factor in three of the 12 deaths in 1980.[3]

The selectman's daughter was killed in an alcohol-related auto accident in 1982 and there has subsequently been a community-wide campaign to reduce drunken driving. This article used these data to show the effects of this campaign:

Year	Fatalities	Alcohol-Related
1980	12	3
1981	12	4
1982	6	2
1983	4	0
1984	5	1
1985	2	0

Use Fisher's F-test to see if the difference in fatalities in 1980–1982 and in 1983–1985 is statistically significant at the 5 percent level. Apply a similar test to the data on alcohol-related fatalities.

• 16.7 Fisher's F-test uses the ratio of the variance of the sample means to the mean of the sample variances. Why is this ratio multiplied by n, the sample size?

16.8 A chemical firm tested the durability of four kinds of rubber-covered fabric. The following data show the weight loss after some rubbing tests on four samples of each fabric:[4]

	Fabric		
1	2	3	4
268	218	235	236
251	241	229	227
274	226	273	234
270	195	230	225

Use Fisher's F-test to see if the observed differences in durability are statistically significant at the 1 percent level.

16.9 Professor Smith claims that stock prices are no more likely to go up in even-numbered years than in odd-numbered years. To test this theory, he gathers some data on percentage stock returns over the period 1960–1979:

Even Years	Odd Years
0.47	26.89
−8.73	22.80
16.48	12.45
−10.06	23.98
11.06	−8.50
4.01	14.31
18.98	−14.66
−26.47	37.20
23.84	−7.18
6.56	18.44

The null hypothesis is that the expected return in an even year is the same as in an odd year. If Smith's null hypothesis is right, then why aren't the sample means equal? Use these data to conduct an F-test of Smith's theory with a 1 percent probability of Type I error.

16.10 Before selecting an investment advisory service, I compiled a record of the recommendations of three firms over a ten-year period. These ten annual returns for each firm's recommendations can be summarized as follows:

	Firm 1	Firm 2	Firm 3
Mean	2.0	10.8	14.4
Variance	515.4	381.1	303.5

Do these data show a statistically significant difference (at the 5 percent level) in the firms' performance?

16.11 A highway department conducted an experiment with three different brands of paints for traffic lanes. Each paint was used on parts of five different streets and then measured for brightness twelve months later. The results were

	Paint	
Everbright	Neverfade	Longlast
72	80	65
67	71	59
58	65	58
72	71	63
65	66	58

a. Why do the results for any single paint vary from one street to the next? **b.** What are the mean and variance for each brand of paint? **c.** What is the average variance? **d.** If these results come from a single distribution, then what is your estimate of the variance of the means from one brand to the next? **e.** Is the actual variance of the sample means consistent with this estimate? **f.** What is the value of Fisher's F-statistic? **g.** Do these data show any statistically significant differences at the 5 percent level among the three brands?

• **16.12** In Exercise 16.9, you made an F-test of the null hypothesis that the expected stock return in an even year is the same as in an odd year. Make a t-test of this same hypothesis in the manner described in Chapter 12. Do your answers agree?

•• **16.13** *(continuation)* Is the agreement or disagreement found in the preceding exercise a fluke or a lesson? To see which is the case, compare the formulas for the F-statistic and t-statistic.

16.2   THE MEAN SUM OF SQUARES

There is an alternative way to calculate the value of the F-statistic that is simpler, although the underlying logic is not quite as apparent. Let's number the k treatments 1, 2, . . . , k. There are n observations of the effects of each of these treatments. These effects might be plant heights, test scores, or something else. We'll label the measured effects Y_i, as shown in Table 16.3. Because there are n observations on each of k treatments, there are nk values of Y in all. For the first treatment, the average value of Y is $\overline{Y}_1$ and the variance is s_1^2. Similarly, for the second treatment, Y has a mean $\overline{Y}_2$ and a variance of s_2^2. The overall average value of Y

Table 16.3 *A General Format for the Analysis of Variance*

	k Treatments			
	1	*2*	. . .	*k*
n	Y_1	Y_{n+1}		
Observed	Y_2	Y_{n+2}		
effects of	.	.		
each	.	.		
treatment	.	.		
	Y_n	Y_{2n}	. . .	Y_{kn}
Mean	$\overline{Y}_1$	$\overline{Y}_2$		$\overline{Y}_k$
Variance	s_1^2	s_2^2		s_k^2

will be labeled $\overline{Y}$:

$$\overline{Y} = \frac{Y_1 + Y_2 + \cdots + Y_{kn}}{kn}$$

Now let's look at the total sum of squared deviations of Y about its mean $\overline{Y}$:

$$\text{total sum of squares} = \sum_{i=1}^{kn} (Y_i - \overline{Y})^2$$

It can be shown that this total sum of squares consists of two factors:

$$n(\text{squared deviations of means}) = n \sum_{i=1}^{k} (\overline{Y}_i - \overline{Y})^2$$

$$\begin{array}{c} \text{squared deviations} \\ \text{about the means} \end{array} = \left\{ \sum_{i=1}^{n} (Y_i - \overline{Y}_1)^2 + \cdots + \sum_{i=(k-1)n+1}^{kn} (Y_i - \overline{Y}_k)^2 \right\}$$

In particular,*

$$\begin{array}{c} \text{total sum} \\ \text{of squares} \end{array} = n \left(\begin{array}{c} \text{squared deviations} \\ \text{of means} \end{array} \right) + \begin{array}{c} \text{squared deviations} \\ \text{about the means} \end{array}$$

---

*Here's a sketch of a proof. The first of the nk squared deviations can be rewritten as

$$(Y_1 - \overline{Y})^2 = \{(\overline{Y}_1 - \overline{Y}) + (Y_1 - \overline{Y}_1)\}^2$$
$$= (\overline{Y}_1 - \overline{Y})^2 + (Y_1 - \overline{Y}_1)^2 + 2(\overline{Y}_1 - \overline{Y})(Y_1 - \overline{Y}_1)$$

The total sum of squares is the sum of nk such expressions. The sum of the first terms gives n sums of the k squared deviations of the means, and the sum of the second terms is the squared deviations about the k means. The sum of the third terms is zero, because each $\overline{Y}_i$ is the average of nY_i.

For instance, in the fluorescein example, the overall average height is 27 inches and the total sum of squared deviations about this mean is

$$\text{total sum of squares} = (28 - 27)^2 + (33 - 27)^2 + (24 - 27)^2 + (27 - 27)^2$$
$$+ (27 - 27)^2 + (30 - 27)^2 + (23 - 27)^2 + (24 - 27)^2$$
$$= 1 + 36 + 9 + 0 + 0 + 9 + 16 + 9$$
$$= 80$$

The sample size ($n = 4$) multiplied by the squared deviations of the two sample means about the overall mean comes to

$$n \left(\begin{array}{c} \text{squared} \\ \text{deviations} \\ \text{of means} \end{array} \right) = 4\{(28 - 27)^2 + (26 - 27)\}^2$$
$$= 8$$

Within each sample, the squared deviations about that sample's mean sum to

$$\begin{array}{c} \text{squared} \\ \text{deviations} \\ \text{about the means} \end{array} = (28 - 28)^2 + (33 - 28)^2 + (24 - 28)^2 + (27 - 28)^2$$
$$+ (27 - 26)^2 + (30 - 26)^2 + (23 - 26)^2 + (24 - 26)^2$$
$$= 0 + 25 + 16 + 1 + 1 + 16 + 9 + 4$$
$$= 72$$

Thus, as advertised, the squared deviations of the means plus the squared deviations about the means equals the total sum of squares.

The **total sum of squares** is a measure of the variation in Y. As shown, this variation can be split into two parts. The first part, the squared deviations of the means, gauges how much the average value of Y varies from one treatment to the next; this is the variation **between treatments.** The second part, the squared deviations about the treatment means, tells us how much the individual values of Y vary about the mean of each treatment; this is the variation **within treatments.** Thus, another way to describe the total sum of squares is:

$$\boxed{\text{total variation} = \begin{array}{c} \text{variation between} \\ \text{treatments} \end{array} + \begin{array}{c} \text{variation within} \\ \text{treatments} \end{array}}$$

These words are reminiscent of our discussion of whether or not treatments matter, and indeed, as you will soon see, they provide another path to the F-statistic. Before we go down that road, however, let's look at one more set of labels that statisticians apply to this fundamental algebraic identity.

There are two reasons why Y varies. First, each treatment may have a different expected value for Y. Second, for any given treatment, there are other (unidentified) factors that cause Y to vary about its expected value. We could say that part of the variation in Y is "explained" by the different treatments and part is "unexplained," because it is caused by unknown factors. This logic leads to the following description:

$$\text{total variation in } Y = \begin{array}{c}\text{variation explained}\\ \text{by different treatments}\end{array} + \begin{array}{c}\text{unexplained}\\ \text{variation}\end{array}$$

The ANOVA Table

The relationship between these various labels is shown in Table 16.4. This format is called an **ANOVA table**—an acronym for *AN*alysis *O*f *VA*riance. The first column says that we are separating the variation of Y into two parts: the between-

Table 16.4   *The ANOVA Format*

Source of Variation	Sum of Squares	Degrees of Freedom	Mean Sum of Squares
Explained (between treatments)	$n\left(\begin{array}{c}\text{squared}\\ \text{deviations}\\ \text{of means}\end{array}\right)$	$k-1$	$n\dfrac{\left(\begin{array}{c}\text{squared}\\ \text{deviations}\\ \text{of means}\end{array}\right)}{k-1} = n\left(\begin{array}{c}\text{variance}\\ \text{of sample}\\ \text{means}\end{array}\right)$
Unexplained (within treatments)	Squared deviations about the means	$k(n-1)$	$\dfrac{\begin{array}{c}\text{squared}\\ \text{deviations}\\ \text{about the means}\end{array}}{k(n-1)} = \begin{array}{c}\text{mean of}\\ \text{sample}\\ \text{variances}\end{array}$
	Total sum of squares	$nk-1$	

$$F = \frac{n(\text{variance of sample means})}{(\text{mean of sample variances})}$$

$$= \frac{(\text{squared deviations of means})/(k-1)}{(\text{squared deviations about the means})/k(n-1)}$$

$$= \frac{\text{explained variance}}{\text{unexplained variance}}$$

Taste Tests of New Food Product

In developing a new "easy-to-prepare, nutritious, on-the run" food product, General Foods compared the taste of two versions of this product—one liquid and the other solid—to a competitive product that was already being sold successfully.[5] For this taste test, General Foods paid 75 males and 75 females to taste the products and be interviewed for their reaction. These 150 people were divided randomly into three groups of 50; each group was half male and half female. One group tasted the new liquid product, the second group tasted the new solid product, and the third group tasted the established product. Each person was asked to show his

or her reaction to the product they tasted by marking a ballot showing faces ranging from delight to disgust. See Figure 1.

General Foods then assigned numerical scores to each face, going from $+3$ for delight to -3 for disgust (results are shown in Table 1).* General Foods was naturally delighted to find its product more highly rated, on the average, than the established competitor. But the statistical question is whether this difference in the mean ratings can be explained by taste variations from one person to the next. Perhaps, by the luck of the draw, the established product was tasted by very fussy people who automatically give low ratings.

That sort of bad luck is always possible, but what we can do statistically is gauge the

Table 1

		Product	
Score	New Liquid	New Solid	Established
$+3$	7	5	1
$+2$	13	11	5
$+1$	16	19	15
0	11	13	17
-1	2	1	8
-2	1	1	3
-3	0	0	1
Mean	1.18	1.06	0.22
Variance	1.38	1.16	1.44

*For the "new liquid," the mean is the sum of the scores divided by the number of observations,

$$\frac{7(+3) + 13(+2) + 16(+1) + 11(0) + 2(-1) + 1(-2) + 0(-3)}{50} = \frac{59}{50} = 1.18$$

while the variance is the sum of the squared deviations about the mean, divided by the number of observations minus one:

$$\frac{7(3 - 1.18)^2 + 13(2 - 1.18)^2 + \cdots + 0(-3 - 1.18)^2}{50 - 1} = \frac{67.38}{49} = 1.38$$

The other means and variances are calculated similarly.

Please check the box under the picture which expresses how you feel toward the product which you have just tasted.

Figure 1 *Pictorial Taste-Test Ballot (the scores +3 to −3 were assigned from left to right)*

probability of such bad luck. The 1.33 average variance in the ratings for each product gauges the variation in people's ratings. If the ratings of these three products came from a single distribution with that variance, the variance of the means for samples of size 50 would be 1.33/50 = 0.027. The observed differences in the sample means, particularly for the established product, are much larger than this. The variance of these three sample means is 0.274 and the value of Fisher's *F*-statistic is

$$F = \frac{(50)(0.274)}{(1.33)} = 10.3$$

In an ANOVA format, we could report the results as shown in Table 2.

With $k = 3$ and $n = 50$, the degrees of freedom are $k - 1 = 2$ and $k(n - 1) = 147$, so that, according to Table 1, if the ratings came from a single distribution, then there would be about a 5 percent chance of observing an *F*-value greater than 3.0. The chance of an *F*-value greater than 10 is less than .001. Bad luck is always possible, but here that explanation for the differences in the ratings is not very plausible.

Table 2 *ANOVA Table for General Foods Taste Test*

Source of Variation	Sum of Squares	Degrees of Freedom	Mean Sum of Squares	F
Between ratings	27.36	2	13.68	10.3
Within ratings	194.76	147	1.32	
Total	222.12	149		

Figure 1 is from Elisabeth Street and Mavis G. Carroll, "Preliminary Evolution of a New Food Product," in Judith M. Tanur et al., *Statistics: A Guide to the Unknown*, San Francisco: Holden-Day. 1972.

treatment variation that is "explained" by the existence of the different treatments and the within-treatment variation that is left "unexplained." The second column gives the corresponding separation of the total sum of squares. In the third column, the degrees of freedom are separated, too. The k-squared deviations of the means have $k - 1$ degrees of freedom. The n-squared deviations about each mean have $n - 1$ degrees of freedom; with k means, this gives $k(n - 1)$ degrees of freedom. In total, there are nk-squared deviations about the overall mean, giving $nk - 1$ degrees of freedom.

In the fourth column, the sum of squares is divided by the appropriate degrees of freedom. As shown, the average "explained" sum of squares is equal to n times the variances of the sample means; the average "unexplained" sum of squares is equal to the mean of the sample variances.

The F-statistic can be calculated by computing the explained and unexplained sums of squares and dividing each by the appropriate degrees of freedom:

$$F = \frac{n(\text{sum of squared deviations of means})/(k - 1)}{(\text{sum of squared deviations about the means})/k(n - 1)}$$

There are at least four advantages of an ANOVA table. First, it provides a standard reporting format. Second, it gives a needed check on some tricky calculations. The explained and unexplained sums of squares must add up to the total sum of squares. If they don't, then there has been a computational error. Third, the sums of squares are easier to calculate and are often subject to less rounding error than the variances.* Fourth, it can be used to extend Fisher's F-test to data with unequal sample sizes.†

To illustrate an ANOVA table in action, let's take one more look at the fluorescein example. Earlier, we found that the total sum of squares is 80, the squared deviations of the means is 8, and the squared deviations about the means is 72. These calculations can be reported in an ANOVA format like Table 16.5. The vari-

---

*If you do the calculations by hand (ugh!), then you should use the fact that the sum of the squared deviations about a mean is always equal to the sum of the squared values minus the number of observations times the mean squared. For example, the total sum of squares is

$$\sum_{i=1}^{kn} (Y_i - \overline{Y})^2 = \sum_{i=1}^{kn} (Y_i^2) - kn\overline{Y}^2$$

This shortcut was first given in Chapter 1 for calculating variances and standard deviations.

†Let's say that there are k samples of size $n_1, n_2, \ldots, n_k$. Two modifications are needed in our ANOVA calculations. First, the

$$n\begin{pmatrix} \text{squared} \\ \text{deviations} \\ \text{of means} \end{pmatrix} = n_1(\overline{Y}_1 - \overline{Y})^2 + n_2(\overline{Y}_2 - \overline{Y})^2 + \cdots + n_k(\overline{Y}_k - \overline{Y})^2$$

Second, the "within" degrees of freedom is

$$(n_1 - 1) + (n_2 - 1) + \cdots + (n_k - 1)$$

Table 16.5 *ANOVA Table for Fluorescein and Plain Water*

Source of Variation	Sum of Squares	Degrees of Freedom	Mean Sum of Squares	F
Between treatments	8	1	8	
				0.67
Within treatments	<u>72</u>	<u>6</u>	12	
Total	80	7		

ation is separated into "between treatment" and "within treatment," and it is up to the reader to recognize that the relative size of the between-treatment variation indicates whether or not there is a difference in the effects of the treatments; the within-treatment variation is the benchmark by which this between-treatment variation is judged. The ANOVA table shows the corresponding separation of the sum of squares and the degrees of freedom. The two sums of squares divided by the respective degrees of freedom give the mean sums of squares, and the ratio of these gives Fisher's *F*-statistic. Figure 16.8 shows how MINITAB and MYSTAT calculate ANOVA tables with these data.

Figure 16.8 *Excerpts from an ANOVA Calculation*

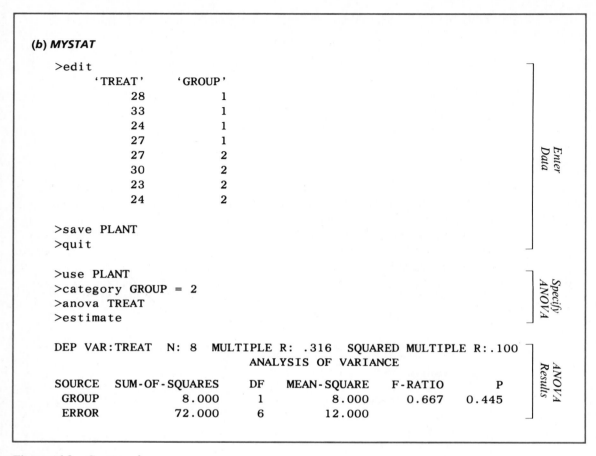

Figure 16.8   *Continued*

EXERCISES

16.14 The *F*-Statistic section in the text gives some data on plant heights for seven plants that were given a solution and seven plants that were given plain water. See if the total sum of squares is equal to the squared deviations of the means plus the squared deviations about the means for these data.

16.15 Use the data in Table 16.1 to see if the total sum of squares is equal to the squared deviations of the means plus the squared deviations about the means.

• **16.16** Show that, for $k = 2$ and $n = 3$, the total sum of squares is equal to the squared deviations of the means plus the squared deviations about the means for any values of *Y*.

16.17 A college consumer-protection group sponsored a beer-tasting, in which 30 students each drank a randomly assigned, unmarked pitcher of beer. Ten of these pitchers contained Spud beer, ten Filler, and ten Schultz. Each of the students was asked to rate the beer they drank on a scale from 1 to 10. The ratings turned out to be as follows:

Spud	Filler	Schultz
7.4	7.6	5.8
6.6	4.4	3.4
5.8	3.3	1.2
7.8	0.8	4.0
9.0	7.8	1.6
6.0	0.0	1.8

Spud	Filler	Schultz
5.0	4.0	0.8
8.4	6.6	3.6
10.0	4.8	0.0
8.0	3.2	3.8

Why might we conclude that differences in ratings such as these are not statistically significant? If one beer gets higher ratings than another, doesn't that prove that it is the more popular beer? Are the differences in these particular ratings statistically significant at the 1 percent level? Use an ANOVA table to summarize your results.

16.18 The Educational Testing Service claims that its SAT tests are a useful predictor of college performance. To test this theory, data are collected on 30 students—10 from each of the following predicted freshman grade-point average (GPA) categories: 2.67–3.00, 3.00–3.33, and 3.33–3.67. The actual freshman-year GPAs are shown in the chart:

Predicted GPA		
2.33–2.67	*2.67–3.00*	*3.00–3.33*
2.53	2.43	2.40
2.90	3.20	3.07
2.43	2.57	3.40
2.37	2.80	2.76
2.60	2.10	3.27
2.16	3.50	3.83
2.03	3.00	2.86
2.80	2.70	3.60
3.17	3.30	3.17
2.70	2.90	3.50

What are the average freshman GPAs for each of these three groups? Are these differences statistically significant at the 5 percent level? What is the null hypothesis that you are testing? Use an ANOVA table to summarize your results.

16.19 Consumer Sports magazine tested the gasoline mileage of three subcompact cars by driving five of each make from Claremont, California to Brewster, Massachusetts. Their results were

Miles per Gallon (MPG)		
Spiffy	*Jiffy*	*Biffy*
32.8	36.2	39.0
31.0	41.4	36.6
30.6	43.0	33.2
26.8	41.0	34.8
30.8	40.6	31.8

What are the average MPGs for each car? Use an ANOVA table to show whether the differences in MPGs are statistically significant at the 5 percent level.

16.20 A wholesaler makes deliveries at 20 retail stores each day. There are five different routes that look like they might be the fastest. Over a ten-week period, the firm tries each route ten times and finds that the travel times (in hours) are as shown in Table 16.6. Why isn't one route consistently faster than the rest? What are the average travel times for each of these five routes? Are the differences in travel times statistically significant at the 1 percent level? What is the null hypothesis that is being tested? Use an ANOVA table to show your results.

Table 16.6 *Data for Exercise 16.20*

Route 1	Route 2	Route 3	Route 4	Route 5
5.8	6.5	6.1	5.0	4.2
7.2	6.2	3.4	7.9	5.8
5.6	4.9	6.2	5.8	4.8
6.1	5.6	5.4	5.1	6.9
6.8	5.4	4.4	7.4	7.3
7.7	7.2	5.2	5.9	5.8
6.7	6.2	5.3	6.7	3.9
4.2	6.4	4.3	7.6	6.6
9.3	6.0	4.7	7.1	6.8
6.4	6.2	5.1	7.9	6.1

16.3   A REGRESSION INTERPRETATION

The multiple regression model discussed in Chapter 15 is sufficiently general to encompass analysis of variance as a special case. However, unlike such quantitative variables as income and the unemployment rate, the explanatory variables here will not have numerical values, since we can't say that one treatment is twice the size of another. All that we can say is that the treatments are different, and that as a consequence, their effects on Y may be different, too.

The treatments model introduced at the beginning of this chapter is

$$\text{treatment 1:} \quad Y = \alpha_1 + \epsilon$$
$$\text{treatment 2:} \quad Y = \alpha_2 + \epsilon$$
$$\cdot$$
$$\cdot$$
$$\cdot$$
$$\text{treatment } k: \quad Y = \alpha_k + \epsilon$$

We can't use the treatments themselves as typical explanatory variables, because the distinction between treatments is qualitative—"Treatment 4 was used, instead of the other treatments." To handle differences that are qualitative rather than quantitative, statisticians invented **0–1 dummy variables,** which are variables that are equal to 0 or 1, depending on whether or not a particular characteristic is true. Here, with k treatments, our regression model can be

$$Y = \alpha + \beta_1 D_1 + \beta_2 D_2 + \cdots + \beta_{k-1} D_{k-1} + \epsilon \qquad (16.3)$$

where $D_1 = 1$ if treatment 1 is used, 0 otherwise
$D_2 = 1$ if treatment 2 is used, 0 otherwise
$\cdot$
$\cdot$
$\cdot$
$D_{k-1} = 1$ if treatment $k - 1$ is used, 0 otherwise

The parameters are

$$\alpha = \alpha_k$$
$$\beta_1 = \alpha_1 - \alpha_k$$
$$\beta_2 = \alpha_2 - \alpha_k$$
$$\cdot$$
$$\cdot$$
$$\cdot$$
$$\beta_{k-1} = \alpha_{k-1} - \alpha_k$$

Notice the clever slight of hand played by these 0–1 dummy variables. If the first treatment is used, then only D_1 is not equal to zero and Y will be

$$Y = \alpha + \beta_1 + \epsilon$$
$$= \alpha_k + (\alpha_1 - \alpha_k) + \epsilon$$
$$= \alpha_1 + \epsilon$$

Similarly, if the second treatment is used, then only D_2 is not equal to zero and Y will be

$$Y = \alpha + \beta_2 + \epsilon$$
$$= \alpha_k + (\alpha_2 - \alpha_k) + \epsilon$$
$$= \alpha_2 + \epsilon$$

and so on. Thus, the multiple-variable Equation 16.3 has all of the k equations embedded in it.

The parameter α in Equation 16.3 gives the value of Y for the kth treatment, the one without a dummy variable. The $k - 1$ parameters β_i represent between-group differences—the extra effect on Y of using the ith treatment rather than treatment k. For instance, β_2 gives the difference in the expected values of Y for the second and kth treatments.

The k parameters α_i in Equation 16.3 can be estimated, as with any multiple regression; the only peculiarity is that the values of the explanatory variables are either 0 or 1. If, for instance, there were three treatments and four observations of the effects of each treatment, the data for the two explanatory variables would look like this:

Treatment	D_1	D_2
1	1	0
1	1	0
1	1	0
1	1	0
2	0	1
2	0	1
2	0	1
2	0	1
3	0	0
3	0	0
3	0	0
3	0	0

It turns out, naturally enough, that such a multiple regression yields the estimates

$$a = \overline{Y}_k$$
$$b_1 = \overline{Y}_1 - \overline{Y}_k$$
$$b_2 = \overline{Y}_2 - \overline{Y}_k$$

and so on, giving the implicit estimates $a_i = \overline{Y}_i$. That is, the estimate of the expected value for each treatment is the observed sample mean for that treatment.

The null hypothesis that the treatment doesn't matter ($\alpha_1 = \alpha_2 = \cdots = \alpha_k$) can be written as

$$H_0 : \beta_1 = \beta_2 = \cdots = \beta_{k-1} = 0$$

Fisher has shown us that this composite hypothesis can be tested with his F-statistic. The lesson is, in fact, much more general: *Fisher's F-test can be used with any regression model to test any composite hypothesis!*

A Generalized *F*-Test

Fisher's F-test can be applied to a wide variety of null hypotheses. A very general statement is as follows: Null hypotheses restrict the values of the parameters, so that some parameters are equal to each other or are equal to zero. To test such a hypothesis, the regression equation can be estimated twice, once with no restrictions and once imposing these restrictions. The resulting sums of the squared prediction errors are called "unrestricted unexplained" and "restricted unexplained," respectively. (Equation 15.3 shows that each can be calculated by multiplying the standard error of estimate (SEE) squared times the degrees of freedom.) Because parameter restrictions necessarily increase the prediction errors, the question is whether this worsened fit is statistically significant—whether the parameter restrictions imposed by the null hypothesis increase the prediction errors sufficiently to persuade us to reject that hypothesis.

The null hypothesis can be tested by computing

$$F = \frac{\left(\begin{array}{c} \text{restricted} \\ \text{unexplained} \end{array} - \begin{array}{c} \text{unrestricted} \\ \text{unexplained} \end{array} \right) \Big/ \text{NUM}}{\left(\begin{array}{c} \text{unrestricted} \\ \text{unexplained} \end{array} \right) \Big/ \text{DEN}} \tag{16.4}$$

where the degrees of freedom for the numerator and denominator are

NUM = number of restrictions imposed by null hypothesis

DEN = degrees of freedom for unrestricted regression

For instance, in the treatments examples emphasized in this chapter, the unrestricted unexplained sum of squares is the squared deviations about the k means. With the $k - 1$ restrictions that the treatments have the same effects, the least squares predictor is the overall mean $\overline{Y}$ and the restricted unexplained sum of squares is equal to the total sum of the squared deviations about $\overline{Y}$. The degrees of freedom for the numerator is the number of restrictions, $k - 1$, and for the denominator it is the total number of observations kn less the number of parameters to be estimated k, giving $k(n - 1)$. Thus

$$
F = \frac{\left(\begin{array}{ccc} \text{total} & & \text{squared} \\ \text{sum of} & - & \text{deviations} \\ \text{squares} & & \text{about means} \end{array} \right) \Big/ (k - 1)}{\left(\begin{array}{c} \text{squared} \\ \text{deviations} \\ \text{about means} \end{array} \right) \Big/ k(n - 1)}
$$

the same as was derived earlier. The beauty of the present description of the F-test is its generality. It can handle an incredible number of models and null hypotheses. The chapter will close with four brief examples.

First, let's analyze the fluorescein data once more, this time using the regression model

$$
Y = \alpha + \beta D + \epsilon
$$

where the dummy variable D is equal to 1 if plain water is used and equal to 0 if the fluorescein solution is applied. The parameter interpretation

$$
\alpha = \alpha_2
$$

$$
\beta = \alpha_1 - \alpha_2
$$

shows that this is indeed the original model

$$
\text{plain water: } Y = \alpha_1 + \epsilon
$$

$$
\text{fluorescein: } Y = \alpha_2 + \epsilon
$$

To estimate the regression model, we use the following data for Y and D:

Treatment	Y	D
F	28	0
F	33	0
F	24	0
F	27	0
W	27	1
W	30	1
W	23	1
W	24	1

Ordinary least squares yields

$$\hat{Y} = 28.0 - 2.0\, D, \qquad R^2 = .1$$
$$[16.2] \quad [0.82]$$

[   ]: t-values

and the sum of the squared prediction errors turn out to be

$$\begin{matrix} \text{unrestricted} \\ \text{unexplained} \end{matrix} = 72$$

The restriction that fluorescein doesn't matter, $\beta = 0$, simplifies the model to

$$\hat{Y} = \overline{Y} = 27$$

with a sum of squared prediction errors

$$\begin{matrix} \text{restricted} \\ \text{unexplained} \end{matrix} = 80$$

The unrestricted regression has DEN $= 8 - 2$ degrees of freedom since 8 observations are used to estimate 2 parameters. The null hypothesis $\beta = 0$ imposes NUM $= 1$ restriction on this regression.

The substitution of these various pieces of information into equation 16.4 gives

$$F = \frac{(80 - 72)/1}{72/(8 - 2)}$$

$$= .667$$

the same F-value obtained earlier in this chapter by a very different route. Thus, as advertised, analysis of variance is a special case of regression analysis.*

*With a single restriction, the hypothesis $\beta = 0$ can also be tested by the 0.8 t-value. Exercises 16.29 and 16.30 show that the t- and F-tests coincide in this case.

Now, to show the generality of Fisher's F-test, let's go back to the example of using midterm scores X_1 and homework grades X_2 to predict final exam scores Y. The data in Table 15.2 yielded this regression equation

$$\hat{Y} = 5.80 + 0.51\ X_1 + 0.38\ X_2, \qquad SEE = 6.1, \qquad R^2 = .42$$
$$[0.3] \quad [2.0] \qquad\quad [2.7]$$

[ ]: t-values

The reported t-values test the hypothesis $\beta_1 = 0$ or $\beta_2 = 0$, but what about the combined hypothesis, that both β_1 and β_2 are equal to 0? For this, we need an F-test.

If β_1 and β_2 are zero, then allowing nonzero estimates shouldn't improve the predictions of Y very much. If, on the other hand, β_1 and β_2 are not equal to zero, then constraining their estimates to be zero should undermine the predictions. So, the relevant question is how much difference these two restrictions make to the predictive accuracy. In the unrestricted equation, the sum of squared prediction errors happens to be 970.0. If we constrain β_1 and β_2 to be zero, then the estimated equation is just the average final exam score:

$$\hat{Y} = \overline{Y} = 81.34$$

and the sum of squared prediction errors is 1658.2. Thus, the appropriate F-statistic is

$$F = \frac{\left(\begin{array}{c}\text{restricted} \\ \text{unexplained}\end{array} - \begin{array}{c}\text{unrestricted} \\ \text{unexplained}\end{array}\right)\Big/ \text{NUM}}{\left(\begin{array}{c}\text{unrestricted} \\ \text{unexplained}\end{array}\right)\Big/ \text{DEN}}$$

$$= \frac{(1658.2 - 970.0)/2}{970.0/(29 - 3)}$$

$$= 9.22$$

The numerator has 2 degrees of freedom, because we're imposing two restrictions. The denominator has $29 - 3 = 26$ degrees of freedom, because the unrestricted equation has 29 observations and 3 estimated parameters. Table 6 at the end of the book tells us that, were the null hypothesis true, there would be a .05 probability of F exceeding 3.37, and so the observed 9.22 F-value leads us to reject at the 5 percent level the hypothesis that both midterm scores and homework grades are of no value in predicting final exam scores. These results are displayed in the ANOVA Table 16.7.

For a third example, let's look again at using high school GPA X_1, verbal SAT score X_2, and math SAT score X_3 to predict freshman-year GPA Y. The estimated

Table 16.7 *ANOVA Table for Predicting Final Exam Scores*

Source of Variation	Sum of Squares	Degrees of Freedom	Mean Sum of Squares	F
Explained by regression	688.2	2	344.1	
				9.22
Unexplained by regression	970.0	26	37.3	
Total	1658.2	28		

equation reported in Chapter 15 is

$$\hat{Y} = 0.437 + 0.361\ X_1 + 0.0014\ X_2 + 0.0007\ X_3, \qquad R^2 = .26$$
$$\quad [20.3] \quad [16.23] \quad\quad [5.50] \quad\quad\quad [2.64]$$

[ ]: *t*-values

What if we want to test the hypothesis that verbal and math SAT scores affect freshman GPA equally: $\beta_2 = \beta_3$? Again, an *F*-test can be devised. If we impose this restriction, then the equation to be estimated is

$$Y = \alpha + \beta_1 X_1 + \beta_2 (X_2 + X_3) + \epsilon$$

Under this null hypothesis, what matters is the total SAT score, not the division between verbal and math, because verbal and math scores count equally. Least squares estimation of this restricted equation yields

$$\hat{Y} = 0.407 + 0.358\ X_1 + 0.0021\ (X_2 + X_3), \qquad R^2 = .25$$

Using the unrestricted and restricted sums of squared prediction errors, the value of the *F*-statistic turns out to be

$$F = \frac{(59.31 - 58.91)/1}{58.91/(359 - 4)}$$
$$= 2.39$$

The numerator has 1 degree of freedom because the null hypothesis imposes one restriction, and the denominator has $359 - 4 = 355$ degrees of freedom because the unrestricted form uses data on 359 students to estimate 4 coefficients. Table 6 says that there is a .05 probability of $F > 3.9$, and thus, at the 5 percent level, we cannot reject the null hypothesis that math and verbal SAT scores have similar implications for college success.

Stock Prices of Problem Banks

Dummy variables can be used in a wide variety of ways to handle data that are qualitative rather than quantitative. For example, bank examiners label those loans considered excessively risky as "other loans especially mentioned," "substandard," "doubtful," or "loss." Examiners also rate each bank's capital position, asset quality, and management on a scale of 1 to 4 (1 being the highest). Based on these bank examinations, supervisory agencies place some banks on a secret "problem bank list."

Wu and Helms[6] were given access to confidential examination reports for 1975 and used a cross-section of 24 banks to see if stock prices incorporated the information in these reports; that is, did stock analysts know, either by surreptitious access to the reports or from other publicly available data, which banks were having problems? In their basic regression, the dependent variable was the price-earnings ratio and the independent variables were

X_1 = net income/total assets

X_2 = average yield on loans

X_3 = 4th-quarter earnings/1st-quarter earnings

X_4 = equity/total assets

X_5 = beta coefficient

X_6 = total deposits

Each of the first three explanatory variables is intended to gauge profitability and is thought to have a positive influence on the price-earnings ratio. The next three explanatory variables are intended to gauge risk, with large values of X_4 and X_6 indicating less risk and a large value of X_5 more risk. Thus, X_4 and X_6 have a positive influence on the price-earnings ratio and X_5 a negative influence.

In addition, Wu and Helms included a variety of data from the confidential bank examinations. If these data are not known by investors, then there should be no observed relationship to share prices. Here is one of their estimated equations:

$$\hat{Y} = -28.2215 + 4.1927\ X_1 + 5.2238\ X_2 + 6.7390\ X_3 + 0.04054\ X_4$$
$$[1.7027] \qquad [1.3810] \qquad [2.6176] \qquad [0.9433] \qquad [2.4477]$$

$$-10.4133\ X_5 + .0000005\ X_6 - 2.4735\ X_{12}, \qquad R^2 = .6891$$
$$[1.3636] \qquad\quad [2.5609] \qquad\quad [1.5844]$$

[ ]: t-values

where X_{12} is a dummy variable equal to 1 if the bank is on the problem list and 0 if it is not. In theory, such information, if available, would have a negative effect on the price of the bank's stock.

The reporting of results to four decimal places gives a spurious aura of accuracy. Most of these data are not measured accurately enough to justify such precise coefficient estimates, and for the t-values and R-squared such accuracy is unnecessary. (It doesn't really matter whether the t-value is 1.7027 or 1.7026.)

All of the estimated coefficients have the anticipated signs, and several are statistically significant at the 5 percent level. But here, as in all eight of the reported regressions, the coefficient of the examination report variable is *not* statistically significant at the 5 percent level. (With $24 - 8 = 16$ degrees of freedom, the critical t-values are 1.7 for a one-tailed test and 2.1 for a two-tailed test.) Wu and Helms conclude

> None of the confidential information variables with regard to loan classification, bank ratings and the problem list is statistically significant. There is thus no evidence that these variables have any impact on bank stock prices. This indicates that the confidential information in the examination reports is indeed confidential and not known to investors.
>
> As noted, previous studies have found that examiners' judgments of the quality of bank loans are extremely accurate, hence should, if known, affect stock prices. Our results however, indicate that current or potential shareholders have no access to this significant information. It would appear that the market for bank stocks cannot be efficient, at least in the strong sense.

One overlooked possibility is that the information in these confidential reports is redundant; that is, the reports, though known surreptitiously, did not add to what was already known about these banks from the other explanatory variables—the yield and risk data. Notice, also, how the authors slide from "cannot reject the null hypothesis" to "the null hypothesis is proved true." Their first sentence in this quotation makes the correct observation that the estimate is not statistically significantly different from zero; that is, not so far from zero that the null hypothesis of zero can be rejected at the usual 5 percent level. The second sentence then converts "insufficient evidence to reject zero" into "no evidence of any impact" and, from there, the authors proceed as if the null hypothesis has been proven true, with their definitive conclusions "is indeed confidential and not known," "no access," and "cannot be."

There is, in fact, some evidence of some impact. According to the least squares estimate, the placement of a bank on the problem list reduces its price-earnings ratio by 2.47, a substantial amount. But the standard deviation for this estimate is 1.56, sufficiently large so that we are reluctant to dismiss the possibility that there is no effect at all. A 95 percent confidence interval ranges from -5.78 to $+0.84$, which, sure enough, includes zero but does not prove zero to be the correct value.

EXERCISES

16.21 Use the data in Table 15.1 to estimate the equation

$$Y = \alpha + \beta_1 X_1 + \beta_2 X_2 + \beta_3 X_3 + \epsilon$$

where the dependent variable is fuel usage and the three independent variables are travel, horsepower, and speed. Make an F-test with a 1 percent significance level of the hypothesis that $\beta_1 = \beta_2 = \beta_3 = 0$. Use an ANOVA table to show your results.

16.22 A certain technical analyst agrees that there is no simple relationship between stock returns from one month to the next, but argues that stock returns can be predicted from returns over the preceding two months. To test this theory, use the data in Table 16.2 to estimate the equation

$$Y = \alpha + \beta_1 X_1 + \beta_2 X_2 + \epsilon$$

where Y is the return this month, X_1 is the return the preceding month, and X_2 is the return the month before that. For example, the first observation is $Y = 0.30$, $X_1 = 5.85$, and $X_2 = -7.43$. Apply an F-test with a 1 percent significance level to the null hypothesis that $\beta_1 = \beta_2 = 0$.

16.23 In Exercise 15.47, you investigated how weight depends on age and height. Now use the data given in Table 15.7 to estimate the equation

$$Y = \alpha_1 + \alpha_2 D_2 + \alpha_3 D_3 + \cdots + \alpha_6 D_6 + \beta_1 X + \epsilon$$

where Y is weight, X is height, and the D_i are five 0–1 dummy variables for age:

$D_2 = 1$ if 25–34, 0 otherwise

$D_3 = 1$ if 35–44, 0 otherwise

$D_4 = 1$ if 45–54, 0 otherwise

$D_5 = 1$ if 55–64, 0 otherwise

$D_6 = 1$ if 65–74, 0 otherwise

Interpret your estimates. What theoretical and empirical differences are there between this model and the model estimated in Exercise 15.47?

16.24 *(continuation)* Make an F-test (with a 1 percent significance level) of the null hypothesis that all five of these dummy-variable coefficients equal zero. In plain English, what is the null hypothesis and what is your conclusion?

16.25 David Harrington estimated this equation using 1973–1984 data (413 observations) on the amount of time network news programs devoted to the monthly unemployment data:[7]

$$\hat{Y} = -59.9 + 184.1 X_1 + 19.6 X_2$$
$$\quad\ (23.8)\quad\ (29.2)\qquad (3.1)$$
$$\quad + 25.9 X_3 + 20.2 X_4$$
$$\qquad (10.7)\quad\ (15.3)$$

where
Y = seconds of news coverage given to unemployment rate

X_1 = absolute value of the change in unemployment rate

X_2 = level of the unemployment rate

X_3 = 1 when unemployment up, 0 otherwise

X_4 = 1 when unemployment constant, 0 otherwise

( ) = standard deviations

Interpret each estimated coefficient. Which, if any, of these estimates are statistically significant at the 5 percent level?

16.26 A political scientist claims that the relationship between the economy and election results has changed since World War II. Use the data in Table 14.5 for the years 1928–1984 to estimate the equation

$$Y = \alpha + \beta X + \epsilon$$

where Y is the incumbent percentage of the vote and X is the change in the unemployment rate. Now use 0–1 dummy variables to reestimate this equation, allowing for the possibility that the intercept α for 1928–1944 is not the same as the intercept for 1948–1984. What, in plain English, is this hypothesis? What are the results of an F-test of the hypothesis with a 5 percent probability of Type I error?

● **16.27** *(continuation)* Now use 0–1 dummy variables to test the hypothesis that the slope β for 1928–1944 is not the same as the slope for 1948–1984. (Assume that the intercept is constant.) What, in plain English, is this hypothesis? What are the results of an F-test with a 5 percent probability of Type I error?

● **16.28** *(continuation)* Now make an F-test at the 5 percent level of the hypothesis that both the intercept and slope are different for 1928–1944 and 1948–1984.

16.29 Use the data given in Table 13.1 to estimate the equation $Y = \alpha + \beta X + \epsilon$, where Y is household spending and X is disposable income. To test the null hypothesis $\beta = 0$, do the following: **a.** Calculate the t-value and compare it to the critical values given in Table 4 at the end of the book for both a 5 percent and 1 percent probability of Type I error. **b.** Now calculate the F-value and compare it to the

critical values given in Table 6 at the end of the book for both a 5 percent and 1 percent probability of Type I error.   **c.** Do these test results agree?

• **16.30** *(continuation)* Do you see any relationship between the numbers used for these t- and F-tests? To see if this is a quirk, use Tables 4 and 6 to compare the t-values with $n - 1$ degrees of freedom and the F-values with 1 degree of freedom for the numerator and $n - 1$ degrees of freedom for the denominator. What do you conclude?

16.4   SUMMARY

The analysis of variance is a powerful, yet quite general procedure. Entire books have been written on the details,[8] but the logic is relatively straightforward. We have a null hypothesis that the expected values of the effects of k treatments are equal. If the n observations of each treatment come from a single distribution, then the variance of the sample means should be σ^2/n. If the observed variance of the sample means is much greater than this, then the data indicate that the observed differences in the means are not reasonably attributed to chance variation, but must instead be caused by different expected values for the various treatments. To implement this procedure, we use the variation within each treatment to estimate σ. Fisher's F-statistic is

$$F = \frac{n(\text{variance of sample means})}{(\text{mean of sample variances})}$$

The F-value can also be calculated from sums of squared deviations,

$$F = \frac{n(\text{sum of squared deviations of } k \text{ means})/(k - 1)}{(\text{sum of squared deviations about the } k \text{ means})/k(n - 1)}$$

Part of its appeal is that, unlike many of our earlier statistical tests, analysis of variance can handle differences in more than two means. It can also handle more than the "one-way" ANOVA that I have shown you. For example, crop yields may depend not only on the watering solution, but also on the varieties planted. ANOVA can deal with this question and much more, but this Pandora's box is best left to a specialized course.

Fisher's F-distribution does assume normality and equal variances for each treatment. Formal statistical tests for these assumptions are covered in more advanced courses; for now, all we can do is make a casual check for obvious violations. For normality, at least, we can appeal to the central limit theorem and hope that the variety of reasons for random variation cause Y to follow a normal distribution.

REVIEW EXERCISES

16.31 A new executive claims that female employees spend more time on the phone than male workers. To test this theory, 10 males and 10 females are selected at random and their calls are monitored for a week (without their knowledge). The results (in minutes) are

Females	Males
960	540
420	640
280	500
540	220
260	140
240	480
220	400
560	700
140	780
500	300

Determine the following: **a.** the mean times for females and males **b.** the variances of the times for females and males **c.** the mean of the variances **d.** the variance of the means **e.** the value of Fisher's F-statistic **f.** if there is a statistically significant difference at the 5 percent level in the time females and males spend on the phone.

16.32 General Stuff is introducing a new breakfast cereal and trying to decide whether to sell it in a package that measures $10 \times 6 \times 3$ inches or $10 \times 9 \times 2$ inches. The product is test marketed in 16 different cities, with the following sales results:

$10'' \times 6'' \times 3''$	$10'' \times 9'' \times 2''$
830	850
490	710
630	820
560	930
480	670
550	810
620	860
640	750

a. What are the mean sales for each package size? **b.** What are the variances of sales for each size? **c.** What is the mean of the variances? **d.** What is the variance of the means? **e.** What is the value of Fisher's F-statistic? **f.** Is there a statistically significant

difference at the 5 percent level in the sales of these two package sizes?

16.33 Farmer Jones claims that Gulp Feed makes pigs plumper than Glup Feed. To prove his claim, he raises nine pigs on Gulp and nine pigs on Glup. After two years, their weights are recorded.

Gulp	Glup
130	128
162	152
130	140
180	124
134	110
156	124
134	114
132	148
108	122

a. If Jones is right, why don't all of the pigs raised on Gulp weigh more than those on Glup? **b.** What are the mean weights for each feed? **c.** What are the variances of weights for each feed? **d.** What is the mean of the variances? **e.** What is the variance of the means? **f.** What is the value of Fisher's F-statistic? **g.** Is there a statistically significant difference at the 1 percent level in the pig weights?

16.34 A researcher wants to see if IQs are related to hair color. A random sample of 30 people yields the following data:

Hair Color		
Dark	Light	Red
120	80	120
85	105	80
100	115	115
95	100	110
115	115	105
105	105	100
105	110	85
115	80	85
90	95	125
110	85	115

a. What are the mean IQs for each hair color? **b.** What are the variances of IQs for each hair color? **c.** What is the mean of the variances? **d.** What is the variance of the means?   **e.** What is the value of Fisher's *F*-statistic?   **f.** Is there a statistically significant difference (at the 5 percent level) in the IQs of the people with different hair colors?

16.35 A marketing executive has been asked to select the color for a new soap box: red, yellow, green, or blue. This executive selects 20 stores at random (5 per color) to test market the product. The sales results are

Red	Yellow	Green	Blue
510	525	320	365
605	500	280	350
580	360	300	325
590	460	260	335
580	450	230	355

Are these differences in sales statistically significant at the 5 percent level? What is the null hypothesis being tested?

16.36 Professor Jones is teaching a statistics class with 150 students. For her final examination, she prepares 50 copies of 3 different tests. These 150 exams are then shuffled and passed out randomly to her students. Their scores can be summarized as follows:

	Test 1	Test 2	Test 3
Mean	82.90	84.70	83.05
Variance	89.87	102.41	107.72

Do the data indicate that the tests are equally difficult?

16.37 A large business wants to know if the number of employee absences varies with the days of the week. Data on the number of people absent each day for a 50-week period can be summarized as shown in Table 16.8. Is there a statistically significant difference in the number of people that are absent on different days of the week? Use a 1 percent probability of Type I error.

16.38 "Bite Back!," a consumer-protection television show, compared three brands of sandwich bags. Each brand was used to wrap ten different sandwiches, and the weight loss after two days was recorded to gauge the moisture retention. The results were as follows:

	Brands	
Plad	Clad	Blad
98	106	105
97	102	98
117	118	106
89	96	99
88	97	95
79	79	96
102	103	101
83	80	91
130	113	112
91	97	98

Why do the results vary from one sandwich to the next? If one bag were better, wouldn't it always have less weight loss? What is the value of Fisher's *F*-statistic here? Is there a statistically significant difference (at the 5 percent level) in the bags' moisture retention?

16.39 Three long jumpers at a track meet are each given five jumps. Their jumps were recorded.

Fred	Ned	Ed
23′ 6″	24′ 9″	25′ 3″
22′ 10″	24′ 7″	26′ 8″
24′ 2″	23′ 5″	25′ 8″
24′ 4″	25′ 6″	24′ 9″
23′ 11″	25′ 0″	26′ 8″

Do these data indicate a statistically significant difference (at the 5 percent level) in their long-jumping abilities? What is the null hypothesis being tested here?

Table 16.8   *Data for Exercise 16.37*

	Monday	Tuesday	Wednesday	Thursday	Friday
Mean	730	620	635	615	670
Variance	3690	4520	5300	4130	3870

16.40 The investigation of daily stock patterns reported in the text was based on the Standard & Poor's 500 stock index. These researchers also looked at an index compiled by the Center for Research in Security Prices (CRSP), which is shown in Table 16.9. Does the F-value for these data reject the null hypothesis (at the 1 percent level) that the expected stock return is the same for each day of the week? (For simplicity, assume that the 4132 days reported are equally divided among these 5 days of the week.)

16.41 *(continuation)* The researchers used these data to estimate the equation

$$Y = \alpha_1 D_1 + \alpha_2 D_2 + \alpha_3 D_3 + \alpha_4 D_4 + \alpha_5 D_5 + \epsilon$$

where D_1 = 1 if Monday, 0 otherwise
D_2 = 1 if Tuesday, 0 otherwise
D_3 = 1 if Wednesday, 0 otherwise
D_4 = 1 if Thursday, 0 otherwise
D_5 = 1 if Friday, 0 otherwise

Can you guess what the values of their least-squares estimates turned out to be?

16.42 *(continuation)* To investigate whether Monday returns were significantly different from the other days of the week, they also regressed the equation

$$Y = \alpha + \beta_2 D_2 + \beta_3 D_3 + \beta_4 D_4 + \beta_5 D_5 + \epsilon$$

where the dummy variables are as defined in the preceding exercise. Can you guess what the values of these parameter estimates turned out to be? How would you test the hypothesis that the expected values of Monday and Tuesday returns are not the same?

16.43 Use a computer program to generate random stock price changes by simulating the repeated flipping of a fair coin. Assume that stock prices go up a point if a head is flipped and down a point if a tail comes up, and that 100 flips represent a day's trading. Set up a calendar of ten 5-day weeks—Monday, Tuesday, Wednesday, Thursday, and Friday—and record the daily price changes for these 50 days. (Don't record the 100 flips each day, just the net price increase that day.) Now calculate the average price change on Monday, on Tuesday, and so on. Which day of the week was the most profitable? The least profitable? Apply an F-test (with a 1 percent significance level) to the null hypothesis that the expected stock return is the same for each day of the week.

16.44 A statistics instructor believes that midterm scores are twice as important as homework in predicting final exam scores. Use the data in Table 15.2 to estimate the equation

$$Y = \alpha + \beta_1 X_1 + \beta_2 X_2 + \epsilon$$

where Y is the final exam score and the two explanatory variables are midterm and homework scores. Make an F-test with a 5 percent significance level of the hypothesis that $\beta_1 = 2\beta_2$.

16.45 Section 14.3 discussed the idea of a political business cycle and the possibility that, as an outlier, the 1932 election may have an undue influence on an estimated regression equation. How would you interpret the coefficients in the following equation?

$$Y = \alpha + \beta_1 X + \beta_2 D + \epsilon$$

where Y = incumbent vote; X = unemployment change; and D = 1 in 1932, 0 otherwise. Use the data in Table 14.5 to estimate this equation and compare your results with the estimated equation $Y = \alpha + \beta X + \epsilon$, excluding the 1932 observation.

16.46 The value of F needed to reject the null hypothesis that the treatments are unimportant depends on the selected probability α of Type I error. Is this requisite F larger for α = .05 or .01? Why?

•• **16.47** The value of F needed to reject the null hypothesis that the treatments are unimportant depends on the size of the sample n. Does this requisite F-value go up or down as n increases? Why?

Table 16.9 *Data for Exercise 16.40*

	Monday	Tuesday	Wednesday	Thursday	Friday
Mean	−.117	.010	.105	.047	.106
Variance	.660	.518	.626	.469	.456

•• **16.48** What is the relationship between Fisher's F-statistic and R-squared? See if you can find a simple equation relating the two. What is the value of F when R-squared $= 0$? When R-squared $= 1$?

• **16.49** Two economists wanted to see if the relationship between the growth rates of money and GNP changed after the Federal Reserve changed its operating procedures in October 1979.[9] They used quarterly data for the second quarter of 1960 through the fourth quarter of 1982 to estimate the equation

$$Y = \alpha_0 + a_1 D + \sum_{i=1}^{5} \beta_i X_i + \sum_{i=1}^{5} \gamma_i DX_i$$
$$+ \sum_{i=6}^{10} \beta_i X_i + \sum_{i=6}^{10} \gamma_i DX_i + \epsilon$$

where Y is the rate of growth of GNP, X_i for $i = 1$ to 5 are the rates of growth of M1 during the current and past four quarters, X_i for $i = 6$ to 10 are the rates of growth of high-employment government spending during the current and past four quarters, and D is a dummy variable equal to one up until the fourth quarter of 1979 and then equal to zero thereafter. How would you go about testing the null hypotheses that **a.** $\gamma_1 = \gamma_2 = \gamma_3 = \gamma_4 = \gamma_5 = 0$? **b.** α_1 and all ten γ_i equal zero?

•• **16.50** *(continuation)* Critique their statistical conclusion that "The stability test for the coefficients on M1 growth is conducted by testing the joint hypothesis that all of the estimates [of γ_1, γ_2, γ_3, γ_4, and γ_5] are simultaneously equal to zero. The calculated F-statistic for this test is 2.01; the critical F-value is 2.72 for the 5 percent significance level. Consequently, the hypothesis that the coefficients on M1 growth have changed [since the fourth quarter of 1979] can be rejected at the 5% significance level."[10]

16.51 Exercise 15.86 shows the tobacco yields in 1924, 1925, and 1926 on five plots of land that were subject to different nitrogen percentages in the fertilizer. Group these data by plot and use Fisher's F-test to see whether the observed yield differences are statistically significant at the 1 percent level.

16.52 *(continuation)* What advantages does multiple regression have, in comparison with analysis of variance, in analyzing the effects of nitrogen on crop yields? Be specific.

16.53 College students in various math classes were given a standard logical reasoning test the first day of class, with the scores kept secret from the instructors and students until the term was over and course grades had been determined.[11] If the students are divided by course grade, there would be approximately 90 A students, 90 B students, and 90 C students. The mean and standard deviation of their logical reasoning scores are given in the following table. Use Fisher's F-test to see whether these observed differences are statistically significant at the 1 percent level. What is the null hypothesis?

Grade	Mean Score	Standard Deviation
A	16.90	2.53
B	15.65	3.26
C	14.01	3.27

16.54 To compare the QWERTY and Dvorak keyboards, 20 randomly selected high school freshmen with no previous typing experience were enrolled in a seven-week course, 10 with each method. The following table shows their typing speeds at the end of the course. Make an F-test at the 5 percent level of the null hypothesis that there is no difference in these keyboards.

QWERTY	Dvorak
25	30
35	45
50	25
40	60
25	40
25	35
30	40
35	40
30	50
40	50

16.55 A man claims that he can tell whether orange juice has been made with oranges from California, Florida, or Texas: "The California orange is clearly superior, and the Texas orange inferior." To test this claim, he samples 75 unmarked glasses of orange juice, 25 from each of these three states, and rates each drink on a scale of 1 to 10. When the rating are separated by state, the mean and variances turn out as shown in the following table. Use an F-test to determine whether these observed differences are statistically significant at the 5 percent level. What is the null hypothesis?

	California	Florida	Texas
Mean	7.8	7.0	6.8
Variance	2.2	3.2	3.6

16.56 A company will choose between two word processing systems. One consideration is the length of time it will take users to learn the system. Five randomly selected users were given instruction with one system and five with the other. The lengths of time (in hours) it took users to achieve an adequate level of competence are listed in the following table. Use an F-test to determine whether these observed differences are statistically significant at the 1 percent level.

System 1	System 2
4	6
7	8
6	9
5	11
3	11

16.57 Table 8 at the end of the book shows annual stock returns for 1950–1989. Separate these returns into two groups, 1950–1979 and 1980–1989, reflecting an October 1979 decision by the Federal Reserve to deemphasize the objective of stable interest rates. Make an F-test at the 5 percent level of the null hypothesis that these two sets of data come from a single distribution.

16.58 (continuation) Use the 1950–1989 annual stock returns (Y) to estimate the regression equation $Y = \alpha + \beta_1 D_1 + \epsilon$, where $D_1 = 1$ if 1980–1989, 0 otherwise. What does the parameter β_1 measure? Make a t-test at the 5 percent level of the null hypothesis that $\beta_1 = 0$, and compare your answer with the results of the F-test in the preceding exercise.

16.59 (continuation) Now use the t distribution with a pooled variance to make a difference-in-means test of the null hypothesis that $\mu_1 = \mu_2$, where μ_1 is the mean of the distribution from which the 1950–1979 data are drawn and μ_2 is the mean of the distribution from which the 1980–1989 data are taken. Compare your results with the preceding two exercises.

16.60 Separate the annual stock return data in Table 8 at the end of the book into four decades: 1950–1959, 1960–1969, 1970–1979, and 1980–1989. Treating these groupings as four samples of ten observations each, make an F-test at the 1 percent level of the null hypothesis that these data come from a single distribution.

16.61 (continuation) Use the 40 annual stock returns (Y) in the preceding exercise to estimate the regression equation $Y = \alpha + \beta_1 D_1 + \beta_2 D_2 + \beta_3 D_3 + \epsilon$, where $D_1 = 1$ if 1960–1969, 0 otherwise; $D_2 = 1$ if 1970–1979, 0 otherwise; and $D_3 = 1$ if 1980–1989, 0 otherwise. Interpret your estimates of β_1, β_2, and β_3 carefully. Make an F-test at the 1 percent level of the null hypothesis that $\beta_1 = \beta_2 = \beta_3 = 0$. Compare your results with the preceding exercise.

16.62 Table 16.10 shows the 1973–1985 annual growth rates of real per capital income for 41 developing countries classified by the degree to which they restricted international trade.[12] Make an F-test at the 1 percent level of the null hypothesis that these 41 growth rates are random variables drawn from a single distribution. Use an ANOVA table to summarize your results.

16.63 (continuation) Redo the preceding exercise, this time combining the strong and moderate free-trade countries into a single group and similarly combining the strong and moderate restricted-trade countries into a single group. Are your conclusions affected by this use of two groups rather than four?

16.64 (continuation) Use the data for the two groups in Exercise 16.63 to make a difference-in-means test at the 1 percent level, using the t distribution and a pooled variance. What is your null hypothesis? Is your conclusion similar to the preceding exercise?

16.65 Velocity is defined as nominal gross national product (GNP) divided by M1, M2, or some other measure of the money supply. In 1980 the Federal Reserve substantially revised its definitions of M1, M2, and other monetary aggregates; the annual data in Table 8 at the end of the book are for the current definitions. Calculate the value of M1 velocity (GNP/M1) for each of these years and then compute the percentage change in M1 velocity each year.

Table 16.10   *Data for Exercise 16.62*

Free-Trade Countries				Restricted-Trade Countries			
Strong		*Moderate*		*Moderate*		*Strong*	
Singapore	6.5%	Malaysia	4.1%	Cameroon	5.6%	Bangladesh	2.0%
Hong Kong	6.3	Thailand	3.8	Indonesia	4.0	India	2.0
South Korea	5.4	Tunisia	2.9	Sri Lanka	3.3	Burundi	1.2
		Brazil	1.5	Pakistan	3.1	Domin. Repub.	0.5
		Turkey	1.4	Yugoslavia	2.7	Ethiopia	0.5
		Israel	0.4	Colombia	1.8	Sudan	−0.4
		Uruguay	0.4	Mexico	1.3	Peru	−1.1
		Chile	0.1	Philippines	1.1	Tanzania	−1.6
				Kenya	0.3	Argentina	−2.0
				Honduras	−0.1	Zambia	−2.3
				Senegal	−0.8	Nigeria	−2.5
				Costa Rica	−1.0	Bolivia	−3.1
				Guatemala	−1.0	Ghana	−3.2
				Ivory Coast	−1.2	Madagascar	−3.4
				El Salvador	−3.5		
				Nicaragua	−3.9		

Divide your data into two groups—pre-1980 and 1980–1989—and make an *F*-test at the 1 percent level of the null hypothesis that these annual percentage changes are drawn from a single distribution.

16.66 *(continuation)* Use the data calculated in the preceding exercise to estimate the regression equation $Y = \alpha + \beta D + \epsilon$, where Y = annual percentage change in M1 velocity and $D = 1$ for 1980–1989, 0 otherwise. Interpret the parameter β and report whether your estimate of β is statistically significantly different from 0 at the 1 percent level.

16.67 Repeat Exercise 16.65, this time working with M2 velocity (GNP/M2).

16.68 Since velocity may increase with interest rates, use the M1 velocity data described in Exercise 16.65 to estimate the equation $Y = \alpha + \beta_1 D + \beta_2 X + \epsilon$, where Y = M1 velocity; X = Treasury bill rate; and $D = 1$ for 1980–1989, 0 otherwise. Is your estimate of β_2 plausibly signed and statistically significantly different from 0 at the 5 percent level? Is your estimate of β_1 statistically significantly different from 0 at the 5 percent level?

16.69 Repeat Exercise 16.66, this time letting Y = M2 velocity.

16.70 An air quality engineer is trying to determine which of three smokestack designs produces fewer contaminants on average. Ten measurements of contaminants per hour at randomly selected times yield the following data. Are the observed differences statistically significant at the 1 percent level?

Design A	*Design B*	*Design C*
152	163	148
143	158	154
148	172	157
169	162	166
163	167	150
157	144	140
145	156	152
149	170	170
171	163	165
160	168	153

16.71 The Dow Jones Industrial Average and the Standard and Poor's 500 (S&P 500) are two widely followed stock indexes. Use the data in Table 8 at the end of the book to calculate the percentage increase in each of these indexes for each year 1980–1989. Now make an F-test at the 5 percent level of the null hypothesis that these two sets of data are drawn from the same distribution.

16.72 Repeat Exercise 16.71, using data for the entire 40-year period 1950–1989.

16.73 Some labor economists believe that shifting composition of the U.S. labor force increased the full-employment unemployment rate in the 1960s. Separate the 1950–1989 unemployment data in Table 8 at the end of the book into two groups: 1950–1969 and 1970–1989 and make an F-test at the 5 percent level of the null hypothesis that these data are drawn from a single distribution. Use an ANOVA table to summarize your findings.

16.74 Use a t distribution with a pooled variance to apply a difference-in-means test at the 5 percent level to the two sets of unemployment data identified in Exercise 16.73.

16.75 Use the data in Table 8 at the end of the book to calculate real corporate dividends and real corporate earnings for each year from 1950 through 1989. Now calculate the annual percentage change in real corporate dividends and real corporate earnings for each year from 1951 through 1989. (Not knowing the 1949 values, you won't be able to calculte the 1950 changes.) Make an F-test with a 1 percent significance level of the null hypothesis that these two sets of percentage change data were drawn from the same normal distribution.

16.76 Use the 1959–1989 data in Table 8 at the end of the book to calculate the percentage changes in the interest rates on one-year Treasury bills and on long-term Treasury bonds for each of the 30 years 1960–1989. Make an F-test at the 5 percent level of the null hypothesis that these two sets of data come from the same normal distribution. Repeat this F-test, this time using percentage changes for the 39 years 1951–1989. Is your conclusion affected by the inclusion of the 1950s data? Does the P-valve increase or decrease?

16.77 (continuation) Combine your 1951–1989 data on the percentage changes in these two interest

rates into a single variable Z, with 39 observations. Estimate the regression equation $Z = \alpha + \beta D + \epsilon$, where $D = 0$ if Z is the percentage change in a Treasury bill rate and $D = 1$ if Z is the percentage change in a Treasury bond rate. Compare your results with the preceding exercise.

16.78 Use the 1950–1989 data in Table 8 at the end of the book to calculate, for each of the 39 years 1951–1989, X = the percentage change in the interest rate on one-year Treasury bills and Y = the percentage change in the interest rate on long-term Treasury bonds. Use these data to estimate the equation $Y = \alpha + \beta X + \epsilon$. Is there a statistically significant relationship (at the 5 percent level) between X and Y? Is the estimate of β statistically significantly different from 1.0 at the 5 percent level? Explain carefully how this regression equation is conceptually different from the regression equation specified in Exercise 16.77.

16.79 (continuation) Use the data in the preceding exercise to estimate the equation $Y = \alpha + \beta_1 D + \beta_2 X + \epsilon$, where the new variable $D = 1$ for the years 1971–1989 and $D = 0$ otherwise. Interpret your estimate of β_2 and test the null hypothesis $\beta_2 = 0$ at the 5 percent level. Make an F-test at the 5 percent level of the null hypothesis $\beta_1 = \beta_2 = 0$. Make an F-test at the 5 percent level of the null hypothesis $\beta_1 = 0$, $\beta_2 = 1$.

16.80 Use the 1950–1989 data in Table 8 at the end of the book to calculate the percentage change in the interest rate on one-year Treasury bills for each of the 39 years 1951–1989. Table 8 also shows the annual rate of inflation for these years. Make an F-test at the 1 percent level of the null hypothesis that these two sets of data come from the same probability distribution.

16.81 Use the 1950–1989 data in Table 8 at the end of the book to calculate Y = the percentage change in the interest rate on one-year Treasury bills for each of the 39 years 1951–1989. Table 8 also shows X = annual rate of inflation for these years. Use these data to estimate the equation $Y = \alpha + \beta X + \epsilon$. Do these data reject at the 5 percent level the null hypothesis that $\beta = 0$? That $\beta = 1$? Explain why using this regression equation to test the null hypothesis that $\beta = 0$ is very different from the use of an F-test in Exercise 16.80.

16.82 *(continuation)* Exercise 16.81 compares the average interest rate on Treasury bills during the year with the percentage change in the consumer price index (CPI) during the year (from December to December). At any point in time, the interest rate on one-year Treasury bills (Y) presumably reflects the rate of inflation that is anticipated during the coming 12 months. For comparison with the average T-bill rate during the year, a better proxy for the anticipated rate of inflation might be the percentage rate of change in the average level of the CPI from this year to the next—for example, a comparison of the Treasury bill rate in 1960 with the percentage change in prices between 1960 and 1961. The CPI levels shown in Table 8 at the back of the book can be used to calculate these rates of inflation (X). Use these data to estimate the equation $Y = \alpha + \beta X + \epsilon$. (Be sure that you match the T-bill rate in year t with the rate of inflation between year t and year $t + 1$.) Do these data reject at the 5 percent level the null hypothesis that $\beta = 0$? That $\beta = 1$? Does the use of different inflation data make much difference to the regression results? Which set of inflation data does a better job of explaining percentage changes in the Treasury bill rate?

16.83 Use the 1950–1989 data in Table 8 at the end of the book to calculate the percentage change in the interest rate on one-year Treasury bills and on long-term Treasury bonds for each of the 39 years 1951–1989. Table 8 also shows the annual rate of inflation for these 39 years. Make an F-test at the 1 percent level of the null hypothesis that these three sets of data came from the same probability distribution.

16.84 Calculate the F-test described in the preceding exercise, this time using data for only the 19 years 1951–1969. Then use only the 20 years 1970–

1989. Is there much difference in the results for these two different time periods? Which has the higher P-value?

16.85 Three girls scout troops operating in very similar areas recorded the following annual sales of cookie boxes. Use these data to make an F-test at the 1 percent level of the null hypothesis that all these data were drawn from the same normal distribution.

Year	Troop 287	Troop 319	Troop 403
1980	1238	815	1436
1981	1159	1478	1152
1982	906	1253	1282
1983	1246	1208	1414
1984	1648	1314	1234
1985	1280	1337	1543
1986	1113	948	1438
1987	1114	1215	1282
1988	852	1207	1548
1989	1197	992	1502

16.86 *(continuation)* If we omit the data for Troop 403 and simply compare Troops 287 and 319, will we necessarily get the same F-value as in Exercise 16.85?

16.87 *(continuation)* How might the assumption of normality be justified here? Why might the mean and standard deviation of the distribution vary among troops and over time?

16.88 A new product was introduced in five very similar test markets at five different prices. Assuming that the weekly sales figures shown in Table 16.11 are random samples, make an F-test at the 5 percent level of the null hypothesis that they are drawn from the same normal distribution.

16.89 *(continuation)* At the company's request, the marketing tests were extended for an additional

Table 16.11   *Data for Exercise 16.88*

P = $9.99	P = $12.99	P = $14.99	P = $19.99	P = $29.99
67	53	51	49	51
62	67	51	29	28
76	56	40	34	38
39	49	47	44	34

Table 16.12 *Data for Exercise 16.89*

P = $9.99	P = $12.99	P = $14.99	P = $19.99	P = $29.99
69	40	48	51	29
33	31	42	38	31
59	54	63	36	40
48	52	52	36	33

four weeks, with the sales results shown in Table 16.12. Combine these data with the data in Table 16.11 so that you have eight observations at five prices. Again, make an *F*-test at the 5 percent level of the null hypothesis that the sales data are drawn from the same normal distribution.

16.90 *(continuation)* The preceding two exercises give eight weekly sales figures at each of five different prices. Unfortunately, during the last week of this marketing experiment, there was a snowstorm in the cities with prices of $9.99 and $12.99. Omit these two observations (48 and 52), leaving seven observations for the first two cities and eight observations for the other three cities. Make an *F*-test at the 5 percent level of the null hypothesis that these sales data are drawn from the same normal distribution.

16.91 Exercises 16.88 and 16.89 describe a marketing experiment that was extended from four weeks to eight weeks. How does such an extension affect the probability of Type I error? Of Type II error? Explain in simple English what these revised probabilities mean.

16.92 A company's production of ball bearings is monitored with hourly random samples. Table 16.13

gives the data on ball bearing diameters (in centimeters) in six such samples. Use these data to make an *F*-test at the 1 percent level of the null hypothesis that these data come from the same normal distribution.

16.93 Use the data in Table 16.13 to calculate the hourly means. Plot these six observations on a control chart with a process mean of 0.8 centimeter and upper and lower control limits at 0.8 ± 0.005 centimeter. Do any of these six samples means violate the control limits?

16.94 Exercises 16.92 and 16.93 apply an *F*-test and a control chart to data that a company used to monitor its production of ball bearings. Explain why these tests could sometimes come to different conclusions regarding a change in the production process. Be specific, giving a numerical example if helpful.

16.95 Many large companies monitor the profitability of each division. If a division is not sufficiently profitable, its management is replaced or the division is liquidated. The following table gives the annual profit figures (percentage return on capital) for four divisions of one company, which are split along product lines. Assuming that these data are random samples, make an *F*-test at the 5 percent level of the

Table 16.13 *Data for Exercise 16.92*

8:30 A.M.	9:30 A.M.	10:30 A.M.	11:30 A.M.	12:30 A.M.	1:30 A.M.
.8025	.7952	.8030	.8002	.8053	.7926
.7961	.8006	.8011	.8040	.7996	.8003
.7990	.7988	.8035	.8033	.7988	.7972
.8053	.7997	.8038	.8021	.7948	.7953
.8007	.8011	.8061	.7981	.7994	.7985

null hypothesis that they are drawn from the same normal distribution.

Kitchenware	Gardening	Toys	Novelty
11.4	−1.5	19.9	−17.5
25.5	−12.3	28.7	12.4
31.9	16.1	23.4	−39.9
−10.0	43.0	33.6	0.0
− 1.0	7.8	38.5	−3.4

16.96 The preceding exercise describes the use of annual data to compare the profitability of four divisions. Why might monthly data be less suitable for such a test? Be specific.

16.97 Following are the annual profits (percentage return on capital) for three divisions of a company. Assuming that these data are random samples, make an F-test at the 5 percent level of the null hypothesis that they are drawn from the same normal distribution.

Hardware	IBM Software	Mac Software
6.79	20.23	62.92
12.35	61.97	1.96
− 4.11	7.04	−10.34

Hardware	IBM Software	Mac Software
38.96	49.52	40.06
28.16	38.46	−12.78

16.98 Since the capital in the IBM and in Mac software divisions in Exercise 16.97 are of approximately equal size, the IBM and Mac software data can be combined by simply averaging the annual returns. Make these calculations so that you have five annual observations on two divisions, hardware and software. Make an F-test at the 5 percent level of the null hypothesis that the returns for these two divisions are drawn from the same normal distribution.

• **16.99** Construct five hypothetical observations for three divisions, such that the F-test described in Exercise 16.97 rejects the null hypothesis decisively, but the F-test in Exercise 16.98 does not reject the null hypothesis at the 5 percent level.

• **16.100** Construct five hypothetical observations for three divisions, such that the F-test described in Exercise 16.98 rejects the null hypothesis decisively but the F-test in Exercise 16.97 does not reject the null hypothesis at the 5 percent level.

17

Some Nonparametric Tests

Statistics, frankly, drive me nuts.

Larry Merchant
The National Football Lottery

TOPICS

Almost all of the many different statistical tests we have considered so far—involving *t*-statistics, *F*-statistics, and so on—assume that the random variation in the data conforms to a normal distribution. For tests about the value of a population mean, we assumed that the sample mean came from a normal distribution. In statistical tests of the parameters of the regression model, the error term is assumed to be normally distributed. In analysis-of-variance tests, the random variation within each treatment is assumed to follow a normal distribution.

We can appeal to the central limit theorem and a wide variety of empirical studies for theoretical and empirical support for this convenient normality assumption, but there are situations in which the samples are small and the data are very obviously not normally distributed. And even if there is a hint of normality, our enthusiasm may be restrained by a bit of prudent caution.

Statisticians have devised several alternative procedures for the researcher who is reluctant to assume normality. These procedures are intended to be appropriate *no matter what* the distribution of the data. They work for normal distributions, of course, but they also work for nonnormal distributions. For this reason, they are called **distribution-free** or **nonparametric** statistics.

If the data do, in fact, follow a normal distribution, then these nonparametric tests are not as powerful as *t*-tests or *F*-tests—which use the known implications of normal distributions. For any given probability of Type I error, the nonparametric tests will have higher probability of Type II error. A test that ignores information about the data, such as their normality, won't be as good as a test that uses this information. On the other hand, if the data do not come from a normal distribution, then nonparametric tests have a clear advantage over tests that make an erroneous assumption of normality. The seriousness of the error, and hence, the accuracy of *t*- and *F*-tests depends, of course, on how incorrect the normality assumption is. Because nonparametric tests are not sensitive to this assumption, they are said to be "robust."

There are many nonparametric tests that are used in widely varying degrees. We will look at a few popular tests in this chapter.

17.1   A SIGN TEST FOR THE MEDIAN

The mean can be a very misleading descriptive statistic. If 99 workers make $10,000 a year, while the boss makes $4,010,000 a year, then the $50,000 mean income does not accurately describe the average worker's pay. In such situations, the median is a popular alternative to the mean. Yet, most statistical tests apply to the mean, because the central limit theorem gives us the convenient result that a sample mean follows a normal distribution. There is no corresponding theorem for the median. With a bit of ingenuity, though, we can concoct a hypothesis test and even a confidence interval for the median.

Let's say that we are professors at a certain small college, which, like most small colleges, does not make faculty salaries public. The president of this college claims that the median salary is $50,000. How could we test this claim, short of finding out everyone's salary? Well, if $50,000 is the median, then half the salaries should be higher than this figure and half lower, so that if a faculty member is randomly selected, there should be a 50 percent chance that the salary will be above $50,000 and a 50 percent chance that it will be below $50,000. A random sample of 20 faculty members should turn up about half above the median and half below. If instead, we find that 17 out of 20 make less than $50,000, then the improbability of this outcome discredits the president's claim. The exact probabilities come from a binomial distribution with a success probability $\pi = .5$.

> For hypothesis tests of the median, the probability that a certain number of sample observations will be above the median and the rest below is given by the binomial distribution with $\pi = .5$.

If we sample 20 faculty members at random, Table 3 at the end of the book tells us that there is a .021 chance that at least 15 out of 20 will be below the population median and a .021 chance that at least 15 out of 20 will be above the median. These numbers can provide the cutoffs for a two-tailed test. If the number below the purported median is greater than 14 or less than 6, we reject this claimed median; if the number below the claimed median is between 6 and 14, we accept the null hypothesis. The probability of Type I error is .042; this is not exactly 5 percent, but close enough.

Perhaps the sampled salaries (in $1,000s) turn out as follows:

Above $50,000	*Below $50,000*		
68.0	49.0	49.0	48.5
63.5	48.5	48.0	48.0
58.5	47.5	47.5	47.0
54.0	47.0	46.5	46.0
51.0	45.5	44.0	43.0

Five are above $50,000; fifteen are below $50,000. This result is sufficiently improbable to reject the null hypothesis that the median salary is $50,000.

This clever procedure is called a **sign test** because it focuses on the signs of the deviations from the purported median. Some data are above the median $(+)$ and some are below $(-)$. If the null hypothesis was true, the number above and below the claimed median should be divided about 50–50. (If an observation happens to be exactly on the claimed median, then it is discarded, because it doesn't provide any information about whether the number above is equal to the number below.)

With a little ingenuity, these sorts of median tests can be applied to a variety of situations. For instance, a farmer may claim that he can predict whether rainfall will be above or below historical average each year. If we are willing to assume that above and below average are equally likely, then a sign test can be constructed readily.

Similarly, a company's claim that a certain diet increases most people's IQ scores would be confirmed statistically if the data reject the null hypothesis that the diet has no effect on IQ scores. Assuming that, under the null hypothesis of "no effect," IQ scores are equally likely to go up or down, a sign test can be used. Perhaps 25 people are given an IQ test, placed on this diet, and then given an IQ test again. The null hypothesis is that scores are as likely to go down as up, and the competing one-sided alternative hypothesis is that scores are more likely to go up than down. If the null hypothesis is true, then the appropriate probabilities are given by the binomial distribution with $\pi = .5$. Table 3 at the end of the book tells us that there is a .054 probability of more than 16 successes in 25 trials. Hence, this is an appropriate cutoff for a test with approximately a 5 percent probability of Type I error. If 16 or fewer scores increase, we'll conclude that these data are consistent with the null hypothesis that scores are equally likely to increase or decrease. If more than 16 out of 25 scores increase, this is sufficiently improbable under the null hypothesis to call for its rejection.

Sign tests also can be used to compare two groups that are not necessarily before-and-after some event. For instance, the null hypothesis that men and women are paid equally well could be tested by a sample of pairs of men and women. If the null hypothesis is true, then the higher-paid person in each pair is equally likely to be the male or female. This is clearly a fitting candidate for a sign test, and you can set one up in Exercise 17.5.

A Confidence Interval

We can even use the logic underlying the sign test to construct a confidence interval for a median. To illustrate the reasoning, let's look again at the 20 faculty salaries listed earlier. We would like to use these data to pin down the median salary for the entire faculty. Although we can't know the exact median for certain unless we know all the salaries, we can calculate confidence intervals with a .95 chance of encompassing the population median.

Remember, a hypothesis test at the 5 percent level can be conducted by constructing a 95 percent confidence interval and seeing whether it includes the value specified by the null hypothesis. Any number inside the confidence interval would be accepted as a possible value for the parameter, while any value outside the interval would be rejected. Turning the logic around, we can construct a 95 percent confidence interval by seeing which parameter values would be accepted as null hypotheses at the 5 percent level.

For the median salary example, the table of binomial probabilities revealed that there is a .021 chance that 15 out of 20 sampled salaries will turn out to be below the median and a .021 chance that 15 out of 20 will be above the median. These probabilities provided a test with a .042 significance level. If the number of salaries below the purported median is greater than 14 or less than 6, the claimed median should be rejected; if the number below the median is between 6 and 14, the null hypothesis should be accepted. The rule implies that we can identify accepted and rejected hypotheses by finding the specific incomes in our data that are sixth from

These values would be rejected since fewer than six observations are below the assumed median

These values for the median would be rejected by a hypothesis test at the 5% level, since fewer than six observations fall above the **assumed** median

95% confidence interval for median ($47,000 – $49,000)

Income ($1000s)

45 50 55 60 65

Figure 17.1 *A Confidence Interval for the Median*

the bottom and sixth from the top. These turn out to be $47,000 and $49,000. If we tested the null hypothesis that the median income is $46,999, we would reject it, because only 5 of the 20 observed salaries are lower than this. In contrast, $47,001 would be accepted as a possible value for the population median, because there are 6 salaries lower than this. At the other end of the scale, $48,999 would be accepted and $49,001 would be rejected. The usual practice is to draw the boundaries right on the borderline observations: $47,000 and $49,000. Thus, our 95 percent confidence interval for the median salary is $47,000 to $49,000. This logic is illustrated in Figure 17.1.

> With 20 observations, the boundaries for a 95 percent confidence interval for the median are the sixth observation from the bottom and the sixth from the top, because any values outside this range would be rejected as null hypotheses at the 5 percent level.

With a different number of observations, the table of binomial probabilities can be used to find how many observations we should count in from the bottom and top.

EXERCISES

17.1 A college dean claims that the median salary for associate professors is $42,000. Cautious by nature, you use a random sample of 17 associate professors to test this claim. For a two-tailed test with approximately a .05 probability of Type I error, what outcomes would you consider persuasive evidence against the dean's claim? The data turn out to be (in $1,000s): 45.0, 38.5, 41.5, 39.0, 43.0, 44.5, 37.5, 34.0, 54.0, 36.0, 37.5, 62.0, 35.5, 43.5, 38.0, 41.0, and 38.5. Do these data confirm or reject the dean's claim? Use a 5 percent significane level.

17.2 At graduation, a college president claims that the median starting salary for those seniors who have accepted jobs is $25,000. A dubious senior uses a random sample of 17 classmates to test this claim. Ruling out the possibility that this claimed median is too low, the senior constructs a one-tailed test at the 5 percent level. For a sample of size 17, what outcomes should be considered evidence against the president's claim? These 17 salaries (in $1,000s) turn out to be: 22.5, 19.0, 28.0, 16.5, 23.5, 24.0, 18.5, 23.5, 29.0, 26.5, 24.5, 32.0, 26.0, 24.0, 26.5, 23.5, and 24.0. Do these data support or refute the president's claim? Use a 5 percent significance level.

17.3 *(continuation)* Is there any reason to think that the answers in this sort of survey may be biased one way or another? If there is a bias, will it increase the chances of incorrectly accepting or rejecting the president's claim? Some people decline to answer salary surveys. Could these nonresponses bias the results or can they be safely ignored?

17.4 An NFL executive claims that the median percentage increase in the salaries of veterans (5 or more years experience) was 24.5 percent last year. How would you structure a sign test of this claim? If you were to sample 50 veterans and wanted approximately a 5 percent Type I error probability, what would be your criterion for accepting or rejecting this claim?

17.5 A large corporation claims that there is no systematic distinction between the salaries of its male and female executives. If you were to use a sign test based on a sample 20 males and 20 females and wanted approximately a 5 percent significance level, what would be your criterion for accepting or rejecting this claim?

17.6 A certain stockbroker claims that, in most years, her clients make more in the stock market than they would have made in a passbook savings account. How would you make a sign test of her claim? If you have 5 years of data, what would her record have to be to convince you that her claim is correct? Allow approximately a 1 percent probability of Type I error.

• 17.7 The text uses 20 faculty salaries to test a college president's claim that the median salary is $50,000. Use these data to make a *t*-test at the 5 percent level of the null hypothesis that the mean salary is $50,000. Does your conclusion coincide with that given in the text? If there is any difference, explain how you would reconcile the results of these two tests.

•• 17.8 *(continuation)* If salaries are actually normally distributed with a mean of $48,000 and a known standard deviation of $5,200, what are the probabilities of Type II error for   **a.** a sign test at the 5 percent level of the null hypothesis that the median equals $50,000?   **b.** a *t*-test at the 5 percent level of the null hypothesis that the mean is $50,000?

17.2   THE WILCOXON-MANN-WHITNEY RANK SUM TEST

Sometimes, we want to compare two populations, such as the income distributions in California and Texas. A random sample of 8 Texans and 10 Californians might give these data:

Texans (Yearly Income $)	Californians (Yearly Income $)
16,000	28,000
3,000	41,000
21,000	12,000
8,000	72,000
65,000	4,000
27,000	32,000
12,000	13,000
32,000	9,000
	22,000
	19,000

To put these numbers in perspective, let's arrange them in order, identifying which are from Texas (T) and California (C):

Rank	Income $
1	72,000 (C)
2	65,000 (T)
3	41,000 (C)
4.5	32,000 (T)
4.5	32,000 (C)
6	28,000 (C)
7	27,000 (T)
8	22,000 (C)
9	21,000 (T)
10	19,000 (C)
11	16,000 (T)
12	13,000 (C)
13.5	12,000 (T)
13.5	12,000 (C)
15	9,000 (C)
16	8,000 (T)
17	4,000 (C)
18	3,000 (T)

Notice that when there is a tie, we use the average of the ranks they would have had if they hadn't tied. Because the two $32,000 incomes tie for fourth and fifth, both are given ranks of 4.5.

One way to compare the rankings of the Californians and Texans is to compute their average rankings. The average rank of the Californians is

$$AR_1 = \frac{1 + 3 + 4.5 + 6 + 8 + 10 + 12 + 13.5 + 15 + 17}{10}$$

$$= \frac{90}{10}$$

$$= 9.0$$

while the average rank of the Texans is

$$AR_2 = \frac{2 + 4.5 + 7 + 9 + 11 + 13.5 + 16 + 18}{8}$$

$$= \frac{81}{8}$$

$$= 10.125$$

On the average, these 10 Californians tend to have higher incomes than these 8 Texans. The statistical question is whether the observed difference in their average ranks is sufficiently improbable to discredit the null hypothesis that they came from similar distributions. The answer requires a knowledge of the distribution of the average ranks under the null hypothesis.

These probabilities have been tabulated and can be found in statistical reference books.[1] If the smaller of the two samples has at least 8 to 10 observations, then the average rank is approximately normally distributed,* and we can use our two-standard-deviations rule of thumb. The expected value of the average rank is

$$\mu_{AR} = \frac{n_1 + n_2 + 1}{2} \qquad (17.1)$$

where n_1 and n_2 are the respective sizes of the two samples. In the California-Texas example, there are 18 observations in all and the expected value of the average rank is 9.5 if the data come from identical distributions. The standard deviation of the average rank depends on the relative sizes of the two samples. With a bit of work, it can be shown that the standard deviation of AR_1 is

$$\sigma_{AR_1} = \sqrt{\frac{(n_1 + n_2 + 1)n_2}{12 n_1}} \qquad (17.2)$$

An analogous formula holds for the standard deviation of AR_2, but we need only look at one of the average ranks (it doesn't matter which one) to make our hypothesis test.

If the observed average rank is within (approximately) two standard deviations of its expected value, then we accept the null hypothesis that the data come from similar distributions; if the observed average rank is more than two standard deviations away, we reject the null hypothesis. This is called the Mann-Whitney, Wilcoxon, or **Wilcoxon-Mann-Whitney rank sum test** (or some other permutation).†

In our California-Texas example, if we let California be number 1, then $n_1 = 10$, $n_2 = 8$, and

$$\mu_{AR} = \frac{(10 + 8 + 1)}{2} = 9.5$$

$$\sigma_{AR_1} = \sqrt{\frac{(10 + 8 + 1)8}{12(10)}} = 1.125$$

---

*Interestingly, while nonparametric statistics are designed for data that do not come from a normal distribution, the statistics themselves usually are normally distributed!

†Equivalently, we can work with the total sum of ranks, SR, instead of the average value AR. The expected value and standard deviation are

$$\mu_{SR_1} = \frac{n_1(n_1 + n_2 + 1)}{2}$$

$$\sigma_{SR_1} = \sqrt{\frac{n_1 n_2 (n_1 + n_2 + 1)}{12}}$$

Because two standard deviations is 2.25, if the income distributions in California and Texas were the same, then California's average rank should be (approximately) within 2.25 of 9.5 about 95 percent of the time. The observed average rank of 9.0 is well within this range, and so it does not provide persuasive evidence for rejecting the null hypothesis that the income distributions are the same.

Matched-Pair Samples

In Chapter 12, we observed that sampling error in comparison tests can be reduced if some of the reasons for random variation can be controlled by choosing matched pairs—essentially twins who receive two different treatments. This sampling design is common in medical experiments, and can sometimes be used in business situations, too. Suppose we want to compare two word processors. We could select two people who seem equally competent and give each person the same ten assignments, with one person using one machine and the second person using the other machine. In this way, we have tried to control for some of the reasons for random variation—that some people are more productive and that some tasks are easier.

Perhaps the data turn out as shown in the first four columns of Table 17.1. If we are willing to assume normality, we can use these data for a matched-pair test, as was described in Chapter 12.* If, however, we are reluctant to assume normality, then we can examine the ranks rather than the magnitudes of the differences. The fifth column gives the absolute values of the differences in working times, so that

Table 17.1 *A Comparison of Two Word Processors*

Task	Number of Minutes to Complete Task		Difference in Working Times	Absolute Value of Difference	Rank of Absolute Value	Signed Rank
	Machine 1	*Machine 2*				
1	34	38	−4	4	5	−5
2	24	27	−3	3	4	−4
3	42	50	−8	8	8	−8
4	72	84	−12	12	10	−10
5	20	18	+2	2	2.5	+2.5
6	46	55	−9	9	9	−9
7	5	4	+1	1	1	+1
8	27	34	−7	7	7	−7
9	32	38	−6	6	6	−6
10	18	16	+2	2	2.5	+2.5
Total			−44			$S = -43$

*Under the null hypothesis that there is no difference in the machines, the mean of a sample of n differences has an expected value of zero and a standard deviation $\sigma/\sqrt{n}$. The differences in the fourth column of Table 17.1 have a mean of 4.4 and a standard deviation of 4.88, giving an estimate for $\sigma/\sqrt{n}$ of $4.88/\sqrt{10} = 1.54$. Thus, the observed sample mean is nearly three standard deviations away from zero, leading to a rejection of the null hypothesis.

we can rank the size of the differences, as is shown in the sixth column. The seventh column then signs the rank, showing us that the second machine only beat the first in three cases, and each was a small margin of victory (a low rank).

If the machines were identical, the sum of the signed ranks, S, would come from an (approximate) normal distribution, with a zero expected value and the standard deviation shown here:

$$\mu_S = 0 \tag{17.3}$$

$$\sigma_S = \sqrt{\frac{(n(n+1)(2n+1)}{6}} \tag{17.4}$$

A statistical test based on these facts is called a **Wilcoxon signed-rank sum test.** For $n = 10$, the standard deviation works out to be 19.6, so that the observed value $S = -43$ is more than two standard deviations away from zero and leads to the rejection at the 5 percent level of the null hypothesis that the machines are equal.

The Battle for Supermarket Shelves

A supermarket has a limited amount of shelf space and thousands upon thousands of products that it might stock. The supermarket manager must decide which products to carry and which to ignore. This is an important decision, because a store that carries the wrong products can lose sales and even customers. It is, of course, an even more important issue for the product manufacturers. A company can't sell its product if potential customers can't find it.

A second consideration is the particular shelves the products are displayed on. You've surely noticed that supermarkets put batteries, cigarettes, candy, magazines, and assorted specials by the checkout counters, where these are certain to be spotted by customers. With batteries and cigarettes, the store's intent is to remind us that we may be running out. Candy, magazines, and specials are "impulse" items—things that may not be on our shopping lists, but that we might buy if these catch our eye.

The same visibility issue arises on every aisle of the store: Which items are put at the end of the aisle, where every passing shopper sees them, and which are put in the middle of the aisle, where they will only be seen by shoppers who walk down the aisle? Another issue is which products get put at eye level, to be easily spotted by hurried shoppers, and which products are put down low or up high, where it takes some effort to find them?

Two researchers made a formal study of the effect of shelf location on sales.[2] In one of their experiments, they arranged for certain breakfast cereals to be displayed at different shelf heights in 24 stores. In 12 of the stores, the selected brand was displayed at eye level; in the other 12 stores, the brand was either above or below eye level. The stores were then ranked according to sales of this brand:

Rank	Sales	Shelf
1	154	eye level
2	150	eye level
3	133	
4	130	eye level
5	126	
6	123	eye level
7	121	
8	112	eye level
9	111	eye level
10	109	
11	96	
12	93	
13	84	eye level
14.5	71	eye level
14.5	71	
16	67	eye level
17	62	eye level
18	58	
19	51	eye level
20	49	
21	38	eye level
22	37	
23	36	
24	27	

The average rank is 10.875 for the displays at eye level and 14.125 for the displays at other levels.

If the level of the display is unimportant, the expected value of the average rank is

$$\mu_{AR} = \frac{n_1 + n_2 + 1}{2}$$

$$= \frac{(12 + 12 + 1)}{2} = 12.5$$

and the standard deviation is

$$\sigma_{AR_1} = \sqrt{\frac{(n_1 + n_2 + 1)n_2}{12n_1}}$$

$$= \sqrt{\frac{(12 + 12 + 1)12}{12(12)}} = 1.44$$

Two standard deviations is 2.88. If the shelf level were unimportant, the average rank for eye-level displays would be 12.5 ± 2.88 in 95 percent of these experiments. Because the observed average rank of 10.875 is inside this range, this test, at least, does not reject at the 5 percent level the hypothesis that shelf height is unimportant.

Now, in fact, the 24 stores were not two independent random samples, but rather 12 matched pairs that were chosen for the similarity in their weekly sales figures. Thus, if the stores really are twins, the test just used overstates the necessary margin for sampling error. With matched pairs, the signed-rank test can be used, as is shown in Table 17.2. Using Equation 17.4, the standard deviation works out to be 25.5, so that the sum of the signed ranks, $S = 52$, is slightly more than two standard deviations away from zero.

This intriguing study is one of those borderline calls. If the stores were matched twins, as intended, then the data reject the null hypothesis that shelf height is unimportant at the 5 percent level. If, at the other extreme, the pairs are unrelated draws, then the observed sales differences are not sufficient to reject the null hypothesis. The conclusion then hinges on whether or not we believe the stores are matched pairs. The safest course, undoubtedly, is to gather some more data.

Table 17.2   *Matched-Pair Analysis of Shelf Height*

Sales in Matched Stores		Difference in Sales	Absolute Value of Difference	Rank of Absolute Value	Signed Rank
Eye-level Shelf	*Other Shelf*				
111	71	+40	40	10	+10
150	121	+29	29	9	+9
130	133	−3	3	2	−2
154	126	+28	28	8	+8
67	93	−26	26	6	−6
112	49	+63	63	12	+12
84	109	−25	25	5	−5
123	96	+27	27	7	+7
71	27	+44	44	11	+11
62	58	+4	4	3	+3
38	36	+2	2	1	+1
51	37	+14	14	4	+4
Total		+197			$S = +52$

Under the null hypothesis that shelf height is unimportant.

$$\mu_S = 0 \quad \text{and} \quad \sigma_S = \sqrt{n(n+1)(2n+1)/6} = \sqrt{12(13)(25)/6} = 25.5$$

EXERCISES

17.9 The following income data have been collected for ten randomly selected families in California and Florida:

California (Yearly Income $)	Florida (Yearly Income $)
9,000	14,000
64,000	26,000
22,000	7,000
42,000	23,000
6,000	48,000
27,000	11,000
17,000	31,000
31,000	5,000
12,000	19,000
24,000	15,000

Make a rank sum test at the 5 percent level of the null hypothesis that the income distributions in California and Florida are the same.

17.10 The text applies a rank sum to test to the incomes of ten California families and eight Texas families. The average ranks of the California families were used in the actual test. Do the test calculations again, this time using the average ranks of the Texas families. Is the result exactly the same? Does it matter, then, which group is used for a rank sum test?

17.11 A tire maker is experimenting with a new, less expensive manufacturing process. Tests of the durability (in miles) of the old and new processes turned up these data:

Old	New
39,000	45,000
46,000	33,500
38,500	38,500
56,500	41,000
46,500	37,500
46,000	33,000
47,000	44,000
39,500	40,000

Make a rank sum test at the 5 percent level of the hypothesis that the new process has no effect on tire durability.

17.12 Birth-weight data on a random sample of black and white babies are given below.

Black	White
7 lb, 3 oz	7 lb, 10 oz
5 lb, 15 oz	6 lb, 14 oz
6 lb, 12 oz	5 lb, 8 oz
7 lb, 7 oz	8 lb, 12 oz
5 lb, 5 oz	7 lb, 2 oz
8 lb, 2 oz	7 lb, 8 oz
5 lb, 10 oz	7 lb, 10 oz
6 lb, 15 oz	6 lb, 2 oz
7 lb, 3 oz	7 lb, 13 oz
5 lb, 7 oz	7 lb, 5 oz
6 lb, 8 oz	6 lb, 5 oz
7 lb, 12 oz	8 lb, 9 oz
6 lb, 5 oz	6 lb, 12 oz
7 lb, 2 oz	8 lb, 6 oz
7 lb, 9 oz	7 lb, 14 oz

Make a rank sum test at the 5 percent level of the null hypothesis that these weights come from the same distribution.

17.13 A home inventor claims to have perfected a gizmo that increases gasoline mileage. Tests are conducted involving two similar cars, one with the gizmo and one without, that are driven over the same road at the same time and at approximately the same speeds. Ten such tests are conducted, with a variety of car models, roads, weather conditions, and speeds.

Test	Miles per Gallon	
	With Gizmo	Without
1	27	24
2	32	30
3	26	29
4	18	22
5	19	26
6	38	34
7	33	37
8	45	43
9	32	28
10	36	35

Assuming these are matched pairs, calculate the sum of signed ranks and test (at the 5 percent level) the null hypothesis that the gizmo has no effect on gasoline mileage.

• **17.14** In Exercise 17.11, a rank sum test was used to compare the durability of tires manufactured by two different processes. Make an analysis-of-variance test using these same data. Do you get the same conclusion?

• **17.15** *(continuation)* Explain why you think that an analysis-of-variance test is either more or less appropriate than a rank sum test.

•• **17.16** *(continuation)* What third type of test could you use? Explain the advantages and disadvantages of the third test you chose and then apply this test to these tire data.

17.3   RANK CORRELATION TESTS

You have now seen several nonparametric counterparts to statistical tests that are based on the assumption of normality. Regression analysis provides yet another example. For the regression model

$$Y = \alpha + \beta X + \epsilon$$

the statistical test of the null hypothesis $\beta = 0$ assumes that the error term ϵ comes from a normal distribution. The reasonableness of this convenient approximation is based on the central limit theorem and the belief that the random variation in ϵ is the result of a great many separate influences.

As in the preceding section, the researcher who is reluctant to assume normality can work instead with rankings of the data. Here, for instance, are some midterm and final exam scores plucked from Table 14.1:

Student	Midterm Exam Score	Midterm Exam Rank	Final Exam Score	Final Exam Rank
CC	75	3	84	1
JG	56	4	69	4
JL	25	5	49	5
SR	78	1	65	3
JV	76	2	83	2

Instead of estimating the parameters of a regression equation using the scores, we can use the ranks:

Midterm Rank (X)	Final Rank (Y)
3	1
4	4
5	5
1	3
2	2

The correlation coefficient R that comes from such a regression is called **Spearman's rank correlation coefficient.** As with any correlation coefficient, R is zero if there is no correlation and close to $+1$ or -1 if there is a close association; that is, R-squared is close to 1.

With ranked data, the average value of each variable must be $(1 + 2 + \cdots + n)/n$ and, if there are no ties, the standard deviation can be figured out in advance, too. Because of these constraints on the data, Spearman's correlation simplifies to

$$R = 1 - \frac{6\Sigma(X_i - Y_i)^2}{n(n^2 - 1)}$$

where $\Sigma(X_i - Y_i)^2$ is the sum of the squared deviations between X and Y. For the midterm and final exam data,

$$\Sigma(X_i - Y_i)^2 = (3 - 1)^2 + (4 - 4)^2 + (5 - 5)^2 + (1 - 3)^2 + (2 - 2)^2 = 8$$

so that

$$R = 1 - \frac{6(8)}{5(25 - 1)} = .6$$

and R-squared is .36. There is a positive association between the rankings based on the midterm and final exam scores, but it is hardly a perfect correlation.

For small samples such as this one, the exact probability distribution for R can be worked out under the null hypothesis that the rankings are randomly determined independently of one another, and such tabulations are given in statistical reference books.[3] For $n = 5$, it turns out that R would have to exceed .9 for the correlation between the rankings to be statistically significant at the 5 percent level.

For larger samples of around size 20, a statistical test can be based on the fact that the distribution of R is—you guessed it—approximately normal, with expected value of zero under the null hypothesis of no correlation and standard deviation

$$\sigma_R = \sqrt{\frac{1}{n - 1}}$$

Normality is so plausible for the regression model that researchers are seldom reluctant to assume it. As a consequence, Spearman's rank correlation coefficient is infrequently used, unless the only data that are available are rankings. For instance, instead of giving students numerical scores, a professor may rank them from best to worst. This is commonly done, by the way, by graduate school admissions committees. When two people rank the applicants, we can use Spearman's rank correlation coefficient to see there is a correlation between their rankings. If there isn't a strong positive correlation, then these two people are using different criteria, or else, the admissions process is pretty arbitrary!

Sports Illustrated Baseball Predictions

Before the start of the major league baseball season, *Sports Illustrated* ranks the teams from best to worst. Here are their 1986 rankings, compared with the actual finish (as gauged by each team's winning percentage):[4]

Team	SI Preseason Ranking	Actual Winning %	Actual Rank
Mets	1	.667	1
Royals	2	.469	16.5
Blue Jays	3	.531	9.5
Yankees	4	.556	5
Reds	5	.531	9.5
Cardinals	6	.491	13
Tigers	7	.537	6.5
Dodgers	8	.451	19.5
Orioles	9	.451	19.5
Cubs	10	.438	23.5
A's	11	.469	16.5
Padres	12	.457	18
Twins	13	.438	23.5
Expos	14	.484	14
Braves	15	.447	21
Red Sox	16	.590	3
Mariners	17	.414	25
White Sox	18	.444	22
Phillies	19	.534	8
Angels	20	.568	4
Rangers	21	.537	6.5
Astros	22	.593	2
Brewers	23	.478	15
Indians	24	.519	11
Giants	25	.512	12
Pirates	26	.395	26

The rank correlation coefficient works out to be 0.11, confirming the visual impression conveyed by a scatter diagram (Figure 17.2). In 1986, there wasn't much correlation between the *Sports Illustrated* predictions and the actual results. For a formal statistical test, the standard deviation is

$$\sigma_R = 1/\sqrt{26 - 1}$$

$$= .2$$

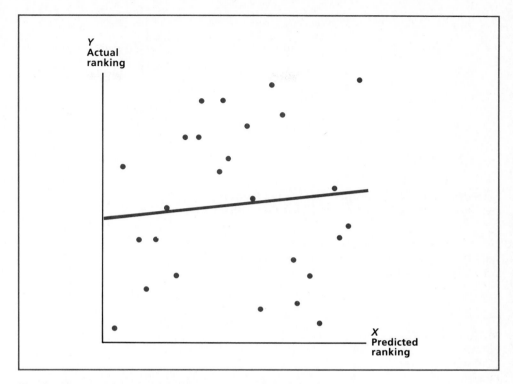

Figure 17.2   *Predicted and Actual Baseball Rankings for 1986*

and the Z-value

$$Z = (0 - 0.11)/.2 = 0.55$$

does not reject at the 5 percent level the null hypothesis that the predictions and results are independent.

EXERCISES

17.17 Rank all 25 of the midterm and final exam scores in Table 14.1 and then estimate the parameters of the regression equation $Y = \alpha + \beta X + \epsilon$, where Y is rank on the final exam and X is rank on the midterm. What is the value of Spearman's rank correlation coefficient? Interpret your results. Is this value of R statistically significant at the 5 percent level? What is the null hypothesis being tested?

17.18 *(continuation)* Compare and contrast these results with the regression equation estimated in

Chapter 14. Why might the results disagree? If they did disagree, how would you choose between them?

17.19 Two vice-presidents ranked ten prospective ventures:

| | Ranking | |
Venture	Bob	Ray
1	4	2
2	2	1

	Ranking	
Venture	Bob	Ray
3	9	8
4	7	7
5	8	9
6	3	4
7	10	10
8	5	5
9	1	3
10	6	6

Gauge the correlation between these rankings by computing Spearman's rank correlation coefficient. Interpret your result.

17.20 Here are some data ranking the 50 states (and Washington, D.C.) on per capita consumption of wine, beer, and soft drinks:[5]

State	Wine	Beer	Soft Drinks
Washington, D.C.	1	15	11
Nevada	2	1	16
California	3	24	25
Oregon	4	29	47
Vermont	5	14	43
Alaska	6	10	32
Washington	7	20	46
New York	8	34	26
Rhode Island	9	17	40
New Jersey	10	35	38.5
New Hampshire	11	2	41
Colorado	12	11	31
Massachusetts	13	23	27
Connecticut	14	41	28.5
New Mexico	15	7	38.5
Hawaii	16	21	28.5
Arizona	17	6	34
Illinois	18	28	21
Florida	19	18	2
Maryland	20	19	15
Michigan	21	12	19
Delaware	22	31	24
Wisconsin	23	3	36.5
Idaho	24	9	50
Maine	25	13	33
Virginia	26	36	5

State	Wine	Beer	Soft Drinks
Pennsylvania	27	25	44
Louisiana	28	26	8
Wyoming	29	5	51
North Carolina	30	47	1
South Carolina	31	40	4
Texas	32	8	12.5
Minnesota	33	27	22
Ohio	34	32	18
Missouri	35	33	9.5
Georgia	36	45	3
Montana	37	4	48
South Dakota	38	38	45
Alabama	39	51	7
Utah	40	49	42
Nebraska	41	16	23
North Dakota	42	22	49
Oklahoma	43	46	30
Indiana	44	44	36.5
Arkansas	45	50	20
Kansas	46	37	12.5
Tennessee	47	39	9.5
Mississippi	48	42	6
Kentucky	49	43	14
West Virginia	50	48	17
Iowa	51	30	35

Is there a statistically significant relationship (at the 1 percent level) between the wine and beer rankings? Between the wine and soft drink rankings? Between the beer and soft drink rankings?

17.21 Rank the income and consumption data in Table 13.1 and compute Spearman's rank correlation coefficient. Interpret your result.

17.22 Rank the unemployment and voting data in Table 14.5 and compute Spearman's rank correlation coefficient. Is the correlation between these rankings statistically significant at the 5 percent level?

17.23 *(continuation)* Write a brief paragraph comparing and contrasting the results of Exercise 17.22 with the equation estimated on page 590 in Chapter 14.

17.4   RUNS TESTS

One of the foundations of probability and statistics is the Bernoulli-trial model, with its assumption that each outcome is independent of the others. The usual analogy is a coin flip, in which the probability of a head or tail does not depend on what happens on other flips. This independence is assumed in the central limit theorem, which underlies the normal distribution, and in the concept of a random sample, which underlies most statistical estimation, hypothesis tests, and confidence intervals.

There are many circumstances, though, in which we may suspect that the outcomes are not really independent. Statisticians have devised some tests for independence and I will show you one here—a **runs test.** Think of the coin-flip analogy again. I flipped a coin 20 times and here are the results:

T H H T H H T H T H H T T T T T H H T H

Heads happened to come up exactly 10 times, which is, of course, far from guaranteed. (The binomial distribution tells us the chances of exactly 10 heads in 20 flips is .1762.) Ten heads in 20 flips are quite consistent with the natural assumption that, when a fair coin is flipped, heads and tails are equally likely. The assumption of interest here is that the outcomes are independent, in that the chances of a head are not influenced by the outcomes of other flips.

One thing to examine would be what happened after each flip of tails. Here, 6 of the tails were followed by heads and 4 were followed by another tail—nothing surprising there. We might also look at the heads: 5 were followed by tails, 4 by heads, and 1 by the end of the experiment. Again, there are no surprises. We could also look for more complicated patterns, such as the second flip after a flip of heads, or the third flip after a double heads. The possibilities are endless; this is not a pleasant prospect for researchers with better things to do. In addition, we know that determined data mining will lead eventually to some sort of pattern. So, if we look until we find a pattern, we will have shown nothing at all but a stubborn determination.

What we would like is a single test that covers the most common kinds of statistical dependence. To obtain one, let's think about the question from the other side. What are the most common kinds of dependence? There may be *momentum* in that when a head appears, this increases the chances of another head. If there were momentum in coin flips, the data might look like this:

H H H H H H T T T T T T T T T T H H H H H

We start out with a head, which leads to heads on the next 5 throws. Then a tail appears, shifting momentum to the tails side. After 9 straight tails, momentum shifts back to heads for 5 more tosses. At the other extreme—*reaction*—coin flips might obey an extreme version of the naive law of averages: Whenever a head comes up, then a tail must follow, to balance things out:

$$\text{H T H T H T H T H T H T H T H T H T H T}$$

Notice the difference between these two contrasting kinds of statistical dependence. With momentum, a run of heads is followed by a long run of tails, and then back to a run of heads—3 runs in all in 20 flips. With reaction, on the other hand, each run of heads or tails lasts but 1 flip, and then the coin comes up on the opposite side; in 20 tosses, there are 20 runs.

With *momentum,* the outcomes are characterized by a small number of long runs. With *reaction,* there are a large number of short runs.

Thus, a possible test for independence is to count the number of runs. If there are very few runs, then the data exhibit a surprising amount of momentum; if there are a large number of runs, then a surprising amount of reaction is shown.

With independence, the exact number of runs is a random variable that will vary from one experiment to the next. We could concoct a formal statistical test of independence, if only we could figure out the probability distribution for the number of runs. This is a formidable task, but some clever statisticians have already figured out these probabilities for us, and it turns out that, like so many things, the number of runs follows a normal distribution.

If there are n independent trials, then the probability distribution for the number of runs R is approximately normal:

$$R \sim N[1 + n/2, (\sqrt{n - 1})/2] \qquad (17.5)$$

(Both the expected value and standard deviation given here are approximations, which are more accurate the larger the sample.)

In the case of 20 independent coin flips, the expected number of runs is $1 + 20/2 = 11$ and the standard deviation is $(\sqrt{20 - 1})/2 = 2.18$. Ninety-five percent of the time, the number of runs in 20 coin flips should be 11 ± 4.36. More than 15 runs indicates reaction; fewer than 7 suggests momentum. In the actual data reported previously, there are 12 runs, providing no reason to question the assumption of independence.

Now, as you may have guessed, the runs test was not concocted for testing coin flips. There are lots of other, much more interesting situations in which there is a very serious question of whether or not independence holds. In each case, we want to phrase the question so that it parallels the coin-flip model and then apply a runs test. For instance, it has been argued that stock prices follow a random walk, in that, at any moment, they are equally likely to go up or down, regardless of their history. If we think of a price increase as "heads" and a decrease as "tails," then the runs test can be applied.

A slightly more sophisticated model recognizes that stocks, like almost all investments, do not have a zero expected return. Stock prices grow with the businesses they represent, and indeed they must for stocks to be an attractive investment. But still, we can postulate independence in the sense that whether the return on a stock is unusually high or low is unrelated to past performance. The coin-flip scenario assumes that heads and tails are equally likely. To apply a runs test to the annual returns on a stock, we need to determine the *median* annual return over the period studied, so that there is an equal chance that a return in any randomly selected year is higher or lower than this median. An above-average return is a "head" and a below-average return is a "tail." Now we can use a runs test to see whether there is momentum or reaction in stock prices.

Runs in Stock Prices

Technical analysts claim that they can find patterns in stock prices, and that by looking at where prices have been, they can tell where prices are going. Economic theorists dispute these claims. They argue that if any profitable pattern begins to appear, it will be exploited out of existence by technical analysts. If stock prices could be counted on to go up at any point in time, then eager buyers would jump the gun and bid prices up in advance, and others would anticipate this jumping the gun and buy even farther in advance. Still others would anticipate this anticipation, and so on. Chaos is the only sustainable pattern.

Part of the evidence against technical analysts is that they seem to make more money by selling their advice than by using it. There are so many hopeful technicians and so few who make money in the market, that good luck and bad luck seem the most plausible explanation of any success they enjoy. As a class, they don't seem any more successful than other investors.

Some other evidence is that technicians so often disagree in their advice. If there were recognizable patterns, you would think that people would at least agree on what these patterns were. Yet, whenever the market goes up, many technicians (those in the momentum camp) advise, "Buy. A body in motion tends to remain in motion." There are just as many reactionaries advising, "Sell. What goes up, must come down."

For some statistical evidence, here are the annual stock returns over the 20-year period 1961–1980:

Year	% Return	Year	% Return
1961	26.89	1971	14.31
1962	−8.73	1972	18.98
1963	22.80	1973	−14.66
1964	16.48	1974	−26.47
1965	12.45	1975	37.20
1966	−10.06	1976	23.84
1967	23.98	1977	−7.18
1968	11.06	1978	6.56
1969	−8.50	1979	18.44
1970	4.01	1980	32.42

Notice that stock returns were positive about twice as often as they were negative. Stocks usually do have a positive return.

The two middle returns are 12.45 and 14.31 percent. If we split the difference, then the median return is 13.38 percent. Now let's relabel these data in terms of above the median ($+$) or below the median ($-$):

Year	% Return	Year	% Return
1961	+	1971	+
1962	−	1972	+
1963	+	1973	−
1964	+	1974	−
1965	−	1975	+
1966	−	1976	+
1967	+	1977	−
1968	−	1978	−
1969	−	1979	+
1970	−	1980	+

As a check, you should always count the number of positive and negative signs to see that these are, as intended, equal. Here there are 10 positive and 10 negative signs.

The question of interest is whether the number of runs seems unusually high or low. Equation 17.5 tells us that for $n = 20$, the number of runs is normally distributed with an expected value of 11 and a standard deviation of 2.18, just as in the case of 20 coin flips. If annual stock returns are independent, then there is a .95 probability that the number of runs over a 20-year period will be 11 ± 4.36, using our two-standard-deviations rule of thumb. If the number of runs is less than 7 or greater than 15, then the improbability of such an outcome casts doubt on the independence hypothesis. As it turns out, there are exactly 11 runs in the data (count them yourself), providing no evidence whatsoever against independence.

On the other hand, other empirical studies have found some evidence of momentum over very short time periods, although this momentum has been too slight to pay taxes, brokerage fees, and still make any money. (Or, as the joke goes, the government made money, the broker made money, and two out of three ain't bad.) For some evidence of brief momentum, reconsider Table 16.2, which shows 132 monthly stock returns over the 11-year period 1970 through 1980. The median return is 0.34 percent. Table 17.3 shows the months in which the returns were above and below this median. Using Equation 17.5, the expected number of runs is $1 + 132/2 = 67$ and the standard deviation works out to be 5.7, so that a two-standard-deviations rule gives 67 ± 11.4 runs. If the monthly returns are independent, then there is only a 5 percent chance that the number of runs over a 132-month period will be fewer than 56 or more than 78. It turns out that a close inspection of Table 17.3 reveals only 55 runs, which is borderline evidence of momentum. It is left as an exercise for the student to determine whether or not this is profitable information.

Table 17.3 *Eleven Years of Monthly Stock Returns, Above (+) or Below (−) Median*

Year	Jan	Feb	Mar	Apr	May	June	July	Aug	Sept	Oct	Nov	Dec
1970	−	+	−	−	−	−	+	+	+	−	+	+
1971	+	+	+	+	−	−	−	+	−	−	−	+
1972	+	+	+	+	+	−	+	+	−	+	+	+
1973	−	−	−	−	−	−	+	−	+	−	−	+
1974	−	−	−	−	−	−	−	−	−	+	−	−
1975	+	+	+	+	+	+	−	−	−	+	+	−
1976	+	−	+	−	−	+	−	−	+	−	−	+
1977	−	−	−	−	−	+	−	−	−	−	+	+
1978	−	−	+	+	+	−	+	+	−	−	+	+
1979	+	−	+	+	−	+	+	+	−	−	+	+
1980	+	−	−	+	+	+	+	+	+	+	+	−

EXERCISES

17.24 If a sequence contains a large number of runs, does this indicate momentum or reaction?

17.25 For each of the following sets of data, indicate whether you would expect to find a large or small number of runs: **a.** whether the cars passing a certain point are traveling faster than 55 MPH **b.** weekly win or loss by a college football team **c.** which team scores in a college basketball game **d.** daily pollution readings in Los Angeles **e.** the monthly unemployment rate.

17.26 With ten observations and a 5 percent significance level, how many runs would be considered evidence against the null hypothesis of independence? Flip a coin 10 times, record your results, and count the number of runs. Do you accept or reject the hypothesis of independence?

17.27 Write down the results of 20 imaginary flips of a coin. Now use a runs test with a 5 percent significance level to see if these imaginary flips really were random.

17.28 Do you think that the success of political parties exhibits momentum, reaction, or randomness? The political parties of the 16 presidential election winners for 1920–1980 are shown here:

Year	Winning Party
1920	Warren Harding, Republican
1924	Calvin Coolidge, Republican
1928	Herbert Hoover, Republican
1932	Franklin Roosevelt, Democrat
1936	Franklin Roosevelt, Democrat
1940	Franklin Roosevelt, Democrat
1944	Franklin Roosevelt, Democrat
1948	Harry Truman, Democrat
1952	Dwight Eisenhower, Republican
1956	Dwight Eisenhower, Republican
1960	John F. Kennedy, Democrat
1964	Lyndon Johnson, Democrat
1968	Richard Nixon, Republican
1972	Richard Nixon, Republican
1976	Jimmy Carter, Democrat
1980	Ronald Reagan, Republican

Use a runs test with a 5 percent significance level to see if these party results are random, or else show either significant momentum or reaction.

17.29 Table 17.4 gives some data on the interest rate on three-month Treasury bills. Find the median interest rate and then identify the monthly levels that are above (+) or below (−) this median. Count the

Table 17.4 *Data for Exercise 17.29*

Year	Jan	Feb	Mar	Apr	May	June	July	Aug	Sept	Oct	Nov	Dec
1975	6.49	5.58	5.54	5.69	5.32	5.19	6.16	6.46	6.38	6.08	5.47	5.50
1976	4.96	4.85	5.05	4.88	5.18	5.44	5.28	5.15	5.08	4.93	4.81	4.35
1977	4.60	4.66	4.61	4.54	4.94	5.00	5.15	5.50	5.77	6.19	6.16	6.06
1978	6.45	6.46	6.32	6.31	6.43	6.71	7.07	7.04	7.84	8.13	8.79	9.12

number of runs and make a runs test with a 5 percent significance level of the hypothesis that the monthly interest rate levels are independent.

17.30 *(continuation)* Now compute the monthly changes in the Treasury bill rate. Find the median and record the monthly changes that are above (+) or below (−) this median. Make a runs test (with a 5 percent significance level) of the hypothesis that the monthly changes in the Treasury bill rate are independent.

17.5   SUMMARY

Many statistical tests assume that the random variation in the data follows a normal distribution. If there are doubts about the appropriateness of the assumption, then a nonparametric test can be used. These tests are "robust" in that they do not assume a particular probability distribution, but of course, they cannot be as powerful as tests that take advantage of (correct) knowledge about underlying probabilities.

The sign test for the median is based on the definition of the median: a random draw is equally likely to be above or below the population median. For a random sample of size n, the binomial distribution with a .5 success probability gives the chances that x of the outcomes will be above the median and $n - x$ below. A claimed value for the population median can be tested by seeing whether there is an improbable imbalance of outcomes, either above or below this median value. A confidence interval for the median can be constructed by locating the range of observed values that would be accepted as null hypotheses.

The Wilcoxon-Mann-Whitney rank sum test compares two population distributions by ranking the outcomes of samples from both distributions. If the populations are identical, then the average rank for each sample should be similar. For reasonably sized samples (at least 8 to 10 observations in the smaller sample), the average rank is approximately normally distributed with an expected value

$$\mu_{AR} = \frac{n_1 + n_2 + 1}{2}$$

The standard deviation of the average rank for the first sample is

$$\sigma_{AR} = \sqrt{\frac{(n_1 + n_2 + 1)n_2}{12 n_1}}$$

with a similar formula holding for the second sample. Either of the observed average ranks can be compared to its expected value and standard deviation to see if the deviation is large enough to cast doubt on the null hypothesis that the two populations are identical.

If the two samples are not independent, but are matched pairs, then the sum of the signed ranks, S, can be calculated as demonstrated in Table 17.1. Under the null hypothesis of identical populations (and a reasonably sized sample), S comes from an approximately normal distribution.

$$S \sim N[0, \sqrt{n(n + 1)(n + 2)/6}]$$

If the observed value of S is more than two standard deviations away from zero, then this discredits the null hypothesis.

Spearman's rank correlation coefficient R is used in regression models when the data are ranks. It can be computed in the usual way, by estimating the regression equation and taking the square root of R-squared, or by the shortcut formula

$$\text{rank correlation:} \quad R = 1 - \frac{6\Sigma(X_i - Y_i)^2}{n(n^2 - 1)}$$

where the $X_i - Y_i$ are the observed differences in the ranks. The null hypothesis that the ranks are unrelated implies that R is normally distributed for large samples,

$$\text{rank correlation:} \quad R \sim N[0, 1/\sqrt{n - 1}]$$

and can be tested by seeing whether the observed value of R is more than two standard deviations from zero.

Runs tests are designed to test the null hypothesis that a sequence of outcomes is independent. There will be a few long runs if the outcomes exhibit momentum and many short runs if there is reaction. Under the null hypothesis of independence, the number of runs is approximately normal,

$$\text{runs:} \quad R \sim N[1 + n/2, (\sqrt{n - 1})/2]$$

If the observed value of R is more than two standard deviations below $1 + n/2$, this indicates momentum; two standard deviations above indicates reaction.

REVIEW EXERCISES

17.31 An admissions officer at a small liberal arts college claims that the median parental income of its students is $26,100. To test this claim, a sample of 20 students yields the following parental income (in $1,000s): 10.5, 14.5, 12.5, 48.5, 250.0, 23.5, 26.0, 34.0, 29.0, 30.3, 16.0, 21.5, 6.2, 18.0, 125.0, 65.0, 27.0, 32.0, 37.5, and 75.0.

Is the claim supported or refuted at the five percent level? What are your cutoffs for rejecting the admission officer's claim? For these data, what is a 95 percent confidence interval for the median?

17.32 An admissions officer at a small liberal arts college claims that the median parental contribution to the student's college expenses is $3,010. To test this claim, a sample of 15 students yields the following data (in $1,000s): 3.5, 0.4, 7.5, 12.5, 0.1, 9.0, 2.5, 12.5, 10.0, 1.0, 0.7, 0.0, 3.6, 8.0, and 3.0.

Is this claim supported or refuted at the 5 percent level? What is your decision rule for rejecting the admission officer's claim? Find a 95 percent confidence interval for the median parental contribution.

17.33 Dr. Quakk claims that older men weigh no more than older women; in particular, if a 65-year-old male and female are chosen at random, the female is, as often as not, the heavier of the two. Set up a sign test of this hypothesis that you could use if you had data for 25 males and 25 females and wanted a 5 percent Type I error probability.

17.34 *(continuation)* Use these weights (in pounds) to test Quakk's hypothesis:

Males	Females	Males	Females
155	161	225	135
155	145	142	151
127	151	147	129
137	123	155	175
175	149	155	159
177	123	131	157
161	141	177	141
204	175	163	141
145	155	153	173
181	161	141	167
181	137	155	125
141	151	147	123
		191	137

• **17.35** *(continuation)* In Exercise 17.34, you made a sign test of Dr. Quakk's theory that older men weigh no more than older women. How would you make an *F* test (with a 5 percent significance level) of Quakk's theory using these same data? Do these two tests give the same conclusion? If they didn't, how would you choose between them?

17.36 A new all-potatoes diet claims to reduce most people's weight by 5 pounds within a week. How would you construct a sign test of this claim with a 5 percent significance level? If 30 people are tested, what would be your criterion for accepting the company's claim?

17.37 The nationwide median salary of full professors in a certain discipline is said to be $40,000. A

random sample of 50 professors in a large state university system found that 30 earned more than $40,000. Does this show (at the 5 percent level) that the median salary in the system is greater than $40,000?

17.38 Here are the IQ scores of some randomly selected school children:

Female	Male
87	87
92	122
112	101
80	103
109	92
82	106
127	121
85	82
97	95
108	106

Make a rank sum test of the hypothesis that the distributions of male and female IQs are identical.

• **17.39** *(continuation)* In Exercise 17.38, a rank sum test was used to test the hypothesis that the distributions of male and female IQs are the same. Make an analysis-of-variance test (at the 5 percent level) using these same data. What are the differences in these tests' assumptions and/or objectives? If the test results disagreed, which test would you find more persuasive?

17.40 A coach claims that the heights of soccer players are no different from the heights of nonplayers. A random sample of 21-year-old males yielded the following heights (in inches):

Soccer Players	Nonplayers
72	66
70	74
67	70
63	68
75	72
71	76
66	73
71	70
69	67

Does a rank sum test at the 5 percent level using these data support or refute the coach's claim?

17.41 A coach claims that the weights of soccer players are no different from the weights of nonplayers. A random sample of 18-year-old males yielded the following weights (in pounds):

Soccer Players	Nonplayers
145	182
164	171
158	146
174	160
155	205
167	115
149	120
170	175
138	133

Does a rank sum test (at the 5 percent level) using these data confirm or refute the coach?

17.42 Consumer Car magazine tested two new car models: a Sleek and a Boate. They purchased eight Sleeks and eight Boates, and calculated the miles per gallon for each in a 1000-mile test run:

Sleeks	Boates
43.5	34.5
41.0	29.5
37.5	35.5
36.5	31.0
44.0	38.5
41.0	35.0
43.0	41.0
40.0	37.5

Use a rank sum test at the 5 percent level to see if the observed differences in miles per gallon are statistically significant.

• **17.43** Is it possible to have a rank sum test find a difference in two distributions that is statistically significant but not substantial? Substantial, but not statistically significant? Make up some data to illustrate your arguments.

17.44 A technical analyst claims that the price of XYZ stock follows a regular cycle; it rises every day for a week, then falls every day the next week, and then rises again the week after that. If so, would a runs test using daily data show a large or small number of runs? (Be careful. This one is tricky.)

17.45 A traffic study recorded the speeds of cars passing a certain point on an expressway between 8:00 and 8:30 A.M. one morning. These speeds are listed horizontally in the chart; the first two speeds are 68 and 67:

68	67	62	54	59	64	65	58	65	54
60	80	70	70	70	65	60	65	55	55
50	48	52	50	46	46	45	42	40	42
40	40	38	40	40	41	38	40	42	42
42	42	40	42	42	43	45	40	40	39
40	41	42	42	42	38	40	40	42	42
42	40	42	44	42	41	43	45	48	50
44	46	44	44	44	43	45	45	50	45
44	47	48	47	50	46	46	48	46	48

Use a runs test with a 5 percent significance level to see if the speeds above and below the median are random.

17.46 Table 17.5 gives some monthly data on the unemployment rate. Find the median unemployment rate and then identify the monthly levels that are above $(+)$ or below $(-)$ this median. Count the number of runs and make a runs test with a 5 percent significance level of the hypothesis that the monthly levels of unemployment are independent.

Table 17.5 *Data for Exercise 17.46*

Year	Jan	Feb	Mar	Apr	May	June	July	Aug	Sept	Oct	Nov	Dec
1975	8.1	8.1	8.6	8.8	9.0	8.8	8.6	8.4	8.4	8.4	8.3	8.2
1976	7.9	7.7	7.6	7.7	7.4	7.6	7.8	7.8	7.6	7.7	7.8	7.8
1977	7.5	7.6	7.4	7.2	7.0	7.2	6.9	7.0	6.8	6.8	6.8	6.4
1978	6.4	6.3	6.3	6.1	7.0	5.9	6.2	5.9	6.0	5.8	5.9	6.0
1979	5.9	5.9	5.8	5.8	5.7	5.7	5.7	6.0	5.8	6.0	5.9	6.0
1980	6.3	6.2	6.3	6.9	7.5	7.5	7.8	7.7	7.5	7.5	7.5	7.3

17.47 *(continuation)* Now compute the monthly changes in the unemployment rate from Exercise 17.46: for example, 0.0 in February 1975 and 0.5 in March 1975. (You won't have an observation for January 1975.) Find the median change in the unemployment rate and record the monthly changes that are above (+) or below (−) this median. Make a runs test with a 5 percent significance level of the hypothesis that the monthly changes in the unemployment rate are independent.

•• **17.48** *(continuation)* Carefully explain the different questions asked by the tests in Exercises 17.46 and 17.47. Write down some hypothetical monthly figures to show how the levels could be independent while the changes are not, and how the changes could be independent while the levels are not.

• **17.49** The text applied a runs test to the monthly stock returns shown in Table 16.2. Use these same data to estimate the equation $Y = \alpha + \beta X + \epsilon$, where Y is the percent return this month and X is the percent return the previous month. (You will not be able to use the January 1970 observation for Y, because you don't know the return the previous month.)

If you find a statistically significant relationship, what does this indicate? If the relationship is not statistically significant, what does this indicate? What differences are there in the assumptions and/or objectives of this regression model and the runs test?

•• **17.50** The text applied a runs test to the monthly stock return data in Table 17.3 and found some statistically significant momentum. Let's see if this momentum can be profitable. If momentum is present, then savvy investors may want to buy after a month in which the market has gone up and sell after a down month. First, compute the average return for those months in which the market went up the preceding month and the average return for those months in which the market went down the preceding month. Compare these two average returns. Second, compute the compounded return if an investor starts with $10,000 and buys stocks after the market has gone up, holds these stocks until after the market drops, and then buys in again after the market has gone up. Ignore taxes, but assume a 2 percent brokerage commission on each transaction.

17.51 Exercise 14.73 asks you to use the data for 1970–1989 in Table 8 at the end of the book to estimate the equation $Y = \alpha + \beta X + \epsilon$, where $Y =$ velocity (nominal GNP/M1) and $X =$ interest rate on Treasury bills. Calculate the regression residuals $Y - (a + bX)$ and make a runs test at the 5 percent level of the null hypothesis that the values of ϵ are independent of each other. If your computer program reports the value of Durbin-Watson's d-statistic, make a test at the 5 percent level of the null hypothesis that $\rho = 0$ and compare your answer with the results of the runs test.

17.52 Exercise 14.32 asks you to use the 1950–1989 data in Table 8 at the end of the book on stock returns (Y) and the rate of inflation (X) to estimate the equation $Y = \alpha + \beta X + \epsilon$. Calculate the regression residuals $Y - (a + bX)$ and make a runs test at the 5 percent level of the null hypothesis that the values of ϵ are independent of each other. If the computer program that you are using reports the value of Durbin-Watson's d-statistic, use the d-value to test at the 5 percent level the null hypothesis that $\rho = 0$. Do your two tests agree? Must they agreee?

17.53 Exercise 14.34 asks you to estimate the consumption function $Y = \alpha + \beta X + \epsilon$, where $Y =$ real per capita consumption and $X =$ real per capita disposable income, using the 1950–1989 data in Table 8 at the end of the book to calculate the values of Y and X. Calculate the regression residuals $Y - (a + bX)$ and make a runs test at the 5 percent level of the null hypothesis that the values of ϵ are independent of each other. If your computer program reports Durbin-Watson's d-value, make a test at the 5 percent level of the null hypothesis that $\rho = 0$.

17.54 A count of the number of runs in the daily percentage changes in the Dow Jones Industrial Average during two great bull markets yielded the data in the following table.[6] In each case, compute the expected number of runs under the null hypothesis that the daily percentage changes are independent and use the difference between the actual and expected number of runs to test this null hypothesis at the 5 percent level. If this null hypothesis is not rejected, can we conclude that daily price changes were independent during these two bull markets?

	Number of Observations	Number of Runs
January 3, 1928– September 3, 1929	495	233
January 2, 1986– August 25, 1987	417	220

17.55 Twenty random digits generated by a computer follow. Letting the even numbers be + and the odd numbers −, make a runs test of their independence at the 1 percent level.

$$5 \quad 2 \quad 4 \quad 8 \quad 1 \quad 7 \quad 6 \quad 6 \quad 8 \quad 5$$
$$3 \quad 9 \quad 0 \quad 1 \quad 4 \quad 9 \quad 8 \quad 4 \quad 7 \quad 7$$

17.56 *(continuation)* Letting the numbers larger that 4 be + and the numbers less than 5 be −, make a runs test of their independence at the 1 percent level.

17.57 Redo Exercise 17.55, this time using a computer random-number generator to generate 20 random digits.

17.58 Redo Exercise 17.56, obtaining your 20 random digits from a computer random-number generator.

17.59 Redo Exercise 17.55, this time using a computer random-number generator to generate 50 random digits. If the computer's random-number generator is functioning as intended, is a test at the 1 percent level more likely to reject independence if 20 digits are tested or if 50 are tested?

17.60 A taxi manufacturer claims that the median number of breakdowns per year for its cars is two. A random sample of 84 taxis finds

Number of Breakdowns	Number of Taxis
0	6
1	11
2	34
3	29
>3	4
	84

Use these data for a sign test at the 5 percent level of the null hypothesis that the median is two.

17.61 Following are some data on annual income (in $1,000) and education (highest year completed). Why might a study of the relationship between income and education examine the rank correlation coefficient? Calculate the rank correlation coefficient. Is its value statistically significant at the 5 percent level? What is the null hypothesis?

Income	Education
8	8
24	9
10	10
30	12
16	12
17	12
14	12
50	12
18	13
20	16
15	16
40	16
25	18
35	20

17.62 The following are scores that Soviet and American judges gave an American gymnast. Make a rank sum test at the 5 percent level of the hypothesis that these scores come from the same distribution.

Soviet	American
7.5	9.0
9.0	9.5
8.0	9.5
8.5	10.0
7.5	9.5
8.0	9.5

17.63 A chief executive officer feels that sales are "stop and go," with intermittent lulls and hectic activity. Here are some daily sales data, beginning on a randomly selected day and continuing for 20 days (in this order, $900, $950, . . .):

900	950	840	1010	1090	1030	1210
1050	1110	970	850	930	980	910
820	940	1040	1180	1230	1060	

a. Find the median and identify which days are above and which are below the median. **b.** How many runs are there? Does this suggest momentum or reaction? **c.** Make a runs test at the 5 percent level of the null hypothesis that the outcomes are independent.

17.64 From 1949 through 1984 American League baseball teams won 18 World Series titles and National League teams won 18. Here are the year-to-year results reading from left to right:

```
A A A A A N N A N A N N
A A N N N A N A N A N A
A A N N A A N N N A A
```

Calculate the number of runs and the expected number of runs, assuming they were 36 independent trials. Does the actual number of runs suggest momentum or reaction in World Series results? Is the difference between the actual and expected number of runs statistically significant at the 5 percent level?

17.65 Table 15.9 gives annual weather data for 20 years, from 1943 through 1962. Calculate the median preseason precipitation and then identify the years that had preseason precipitation above (+) or below (−) this median. What is the expected value of the number of runs if the annual data are independent. Count the actual number of runs and test the independence hypothesis at the 5 percent level.

17.66 *(continuation)* Why might annual June precipitation data not be independent? If your theory is correct, would you expect to find an unusually small or large number of runs. Repeat the calculations in Exercise 17.65, using the June precipitation data in Table 15.9.

17.67 Redo Exercise 17.66, this time using the July temperature data in Table 15.9.

17.68 Redo Exercise 17.66, this time using the corn yield data in Table 15.9.

17.69 *(continuation)* Do you think that daily or annual weather data are more likely to reject the null hypothesis of independence. Explain your reasoning, with an example if possible.

17.70 A researcher examined the Friday prices of ten stocks over the 153-week period from August 8, 1986 to July 7, 1989.[7] For each of these ten stocks, 152 weekly prices changes were computed, the median price change was determined, and each week was labeled + or − depending on whether its price change was above or below the median. The number of statistical runs were then determined, with the fol-

lowing results. What is the expected value of the number of runs for each stock if the weekly price changes are independent? Do any of these stocks show a statistically significant (at the 1 percent level) departure from independence?

Stock	Number of Runs
American South African	72
Burlington Northern	74
Dow Chemical	81
Georgia Pacific	80
Greyhound	68
Louisiana Land and Exploration	81
Pittston	80
Southern Company	73
Tandy	78
Unocal	75

17.71 Exercise 10.97 gives 10 annual ozone readings in the Inland Empire. Calculate the median reading and use a runs test for independence, with a 5 percent significance level.

17.72 *(continuation)* Daily noon readings for August 1990 follow. Calculate the median reading for these data and again make a runs test for independence with a 5 percent significance level. Why might daily pollution data be dependent even though annual readings are independent?

```
139  196  179  149  165  174  161  158
179  223  226  227  241  224  181  124
160  187  165   98  120  113  178  120
153  139   84  141  189  154  173
```

17.73 Table 17.6 shows several 1989 preseason and postseason ratings of college football teams.[8] (In the *New York Times* postseason rankings, Alabama and Southern Cal are tied and so are Brigham Young and Michigan State.) Compare the *New York Times* and Associated Press preseason rank ordering of the 20 teams ranked preseason by the *New York Times*. (Ignore the 5 teams that are ranked by the Associated Press but not by the *Times* so that in each case the ranks range from 1 to 20.) Gauge the correlation between these 20 rank orderings by calculating Spearman's rank correlation coefficient. Is this value of R statistically significant at the 5 percent level? At the 1 percent level?

Table 17.6 *Data for Exercise 17.73*

Rank	Preseason		Postseason	
	New York Times	*Associated Press*	*New York Times*	*Associated Press*
1	Michigan	Michigan	Miami (Fla.)	Miami (Fla.)
2	Nebraska	Notre Dame	Notre Dame	Notre Dame
3	Miami (Fla.)	Nebraska	Florida State	Florida State
4	Notre Dame	Miami (Fla.)	Auburn	Colorado
5	Southern Cal	Southern Cal	Colorado	Tennessee
6	Louisiana State	Florida State	Houston	Auburn
7	Brigham Young	Louisiana State	Tennessee	Michigan
8	Florida State	Auburn	Michigan	Southern Cal
9	Auburn	UCLA	Clemson	Alabama
10	UCLA	Arkansas	Alabama	Illinois
11	Arizona	Penn State	Southern Cal	Nebraska
12	Ohio State	Clemson	Penn State	Clemson
13	Arkansas	Syracuse	Nebraska	Arkansas
14	Clemson	Colorado	Illinois	Houston
15	Alabama	Oklahoma	Arkansas	Penn State
16	Syracuse	Alabama	Texas Tech	Michigan State
17	Colorado	West Virginia	Washington	Pittsburgh
18	Illinois	Arizona	Virginia	Virginia
19	Penn State	Brigham Young	Brigham Young	Texas Tech
20	Oklahama	Pittsburgh	Michigan State	Texas A&M
21		Houston		West Virginia
22		Illinois		Brigham Young
23		Iowa		Washington
24		North Carolina		Ohio State
25		Ohio State		Arizona

17.74 *(continuation)* Now determine whether there is a higher correlation among the first 10 or second 10 rankings. First compare the *New York Times* and Associated Press preseason rank ordering of the 10 teams ranked 1 to 10 preseason by the *New York Times*. (Ignore the other teams; in each case, the ranks should range from 1 to 10.) Gauge the correlation between these 10 rank orderings by calculating Spearman's rank correlation coefficient. Is this value of R statistically significant at the 5 percent level? Repeat these calculations, using the rank orderings for the 10 teams ranked 11 to 20 by the *New York Times*. (Again, the ranks should range from 1 to 10.) Is this value of R statistically significant at the 5 percent level? Is the value of R higher for the first 10 or the second 10?

17.75 Repeat Exercise 17.73, this time comparing the postseason *New York Times* and Associated Press rank orderings of the 20 teams ranked by the *New York Times*. Remember that the *Times* rankings have two ties.

17.76 Repeat Exercise 17.74, this time comparing the postseason *New York Times* and Associated Press rank orderings of the first 10 and the second 10 teams ranked by the *New York Times*.

17.77 Use the data in Table 17.6 to compare the preseason and postseason *New York Times* rank orderings of the top 10 teams in the *New York Times* preseason rankings. Assume that UCLA had the lowest postseason ranking among these 10 teams and that

Louisiana State had the next lowest ranking. (Ignore the other teams, so that in each case the ranks range from 1 to 10.) Calculate Spearman's rank correlation coefficient for these two sets of 10 rankings. Is this value of R statistically significant at the 5 percent level?

17.78 *(continuation)* Repeat Exercise 17.77, this time comparing the preseason and post-season Associated Press rank orderings of the top 10 teams in the Associated Press preseason rankings. Again assume that UCLA had the lowest postseason ranking among these 10 teams and that Louisiana State had the next lowest ranking. Is the preseason/postseason value of R higher for the *New York Times* or the Associated Press?

17.79 *(continuation)* Why is the procedure used in Exercises 17.77 and 17.78, which focuses on the top 10 teams in the preseason poll, an imperfect measure of the correlation between preseason and postseason polls? Give an example of preseason and postseason poll results for which the Spearman rank correlation coefficient is 1.0, yet most would consider the preseason poll a poor predictor of the postseason poll?

17.80 Each of the students in an investments class used newspaper and magazine articles to pick a stock to follow from September 9 to December 4, 1987. During this period, the stock market as a whole fell 28.4 percent. Twelve of these 38 students picked stocks that did better than this, while 26 picked stocks that did worse. Overall, the student stocks had an average return of -34.7 percent with a standard deviation of 13.9 percent.

Use the sample mean to test at the 1 percent level the null hypothesis that the student stocks were randomly selected from a distribution with a mean of -28.4 percent and standard deviation of 13.9 percent. Use the sample median to test at the 1 percent level the null hypothesis that the student stocks were randomly selected from a distribution with a median of -28.4 percent.

17.81 The following are the rankings of 10 large cities on per capita sales of two types of novels.[9] Compute Spearman's rank correlation coefficient and interpret your result.

City	Historical	Occult
New York	1	2
Chicago	5	8
Los Angeles	2	1
Philadelphia	9.5	4
Houston	3	5.5
Detroit	7.5	10
Dallas	6	8
San Diego	9.5	5.5
Phoenix	7.5	3
Baltimore	4	8

17.82 When a cereal maker's production process is under control, the net weights of cereal boxes are normally distributed with a mean of 48 ounces. Each of the following 24 observations, taken at one-hour intervals, is the average of the weight of five boxes. Use these data to estimate the standard deviation of these sample means and to locate the control limits at 48 ounces plus or minus three standard deviations. Sketch a control chart with these control limits and plot the observations. Do any of these 24 observations violate the control limits?

49.53	47.74	49.56	46.58	49.39	47.59
46.76	48.51	49.93	48.10	48.13	46.60
46.62	50.27	47.66	48.36	48.48	49.61
48.35	47.78	47.51	48.12	48.70	48.37

17.83 Use the data in Exercise 17.82 to test at the 5 percent level the null hypothesis that sample means are drawn from a distribution with a median weight of 48 ounces. Compare your results with a t-test at the 5 percent level of the null hypothesis that sample means are drawn from a distribution with a mean weight of 48 ounces.

17.84 The data in the Exercise 17.82 are 24 sequential observations, reading left to right. Use these data to calculate the median value and then make a runs test of independence with a 5 percent significance level.

17.85 A large mail-order firm monitors the number of clerical errors made each week, using a control chart with a process mean of 50 errors, an upper control limit of 65 errors, and a lower control limit of 35 errors. The following are the weekly errors over a two-month period. Sketch a control chart with these 10 observations. Do any of these observations violate the firm's control limits?

50 49 48 51 48 44 44 46 43 47

17.86 Use the data in Exercise 17.85 to make a binomial test of the null hypothesis that these data are from a distribution with a median of 50, and to make a t-test of the null hypothesis that these data are from a normal distribution with a mean of 50. Use a 5 percent significance level for each test.

17.87 Use the sequential data in Exercise 17.82 to make a runs test of independence with a 5 percent significance level.

17.88 A supermarket's sales of a certain brand of soup have historically averaged 200 cans a week. Draw a control chart with an upper control limit at 215 and a lower control limit at 185 and plot the following weekly sales data. Do any of these observations violate the control limits? Why, nonetheless, might the supermarket be concerned? Make a runs test of independence with a 5 percent significance level.

209 203 200 208 211 201 202 194
203 204 211 201 212 204 206 209

17.89 Compare the results of a binomial test of the null hypothesis that the data in Exercise 17.88 are from a distribution with a median of 50 with the results of a t-test of the null hypothesis that these data are from a normal distribution with a mean of 50. Are the P-values similar or quite different?

17.90 Table 8 at the end of the book contains data on the monetary aggregates M1 for each of the 31 years 1959–1989. Calculate the percentage change in M1 for each of the 30 years 1960–1989 and divide these data into three 10-year periods: 1960–1969, 1970–1979, and 1980–1989. For each of these 10-year periods, rank the 10 observations from the largest percentage change in M1 to the smallest. Now do the very same calculations for the 1959–1989 consumer price index data in Table 8. When you are done, for each of these three decades, you will have 10 rankings of the percentage changes in M1 and the percentage changes in the consumer price index. Calculate Spearman's rank correlation coefficient for each of these

three decades. Assuming that the correlation coefficients are normally distributed, how many are statistically significant from zero at the 5 percent level? At the 1 percent level?

17.91 Use the data in Table 8 at the end of the book to calculate the percentage changes in the monetary aggregate M1 and in the consumer price index for each of the 30 years 1960–1989. For each of these two variables, rank the 30 observations from the largest percentage change to the smallest. Now calculate Spearman's rank correlation coefficient and, assuming it to be normally distributed, test at the 5 percent level the null hypothesis that these ranks are uncorrelated.

17.92 Follow the instructions in the preceding exercise, this time comparing the rankings of the percentage changes in M2 and the consumer price index over the 30-year period 1960–1989.

17.93 Use the step-by-step procedure described in Exercise 17.91 to calculate and compare the rankings of the percentage changes in M1 and M2 over the 30-year period 1960–1989.

17.94 Use the data in Table 8 at the end of the book to calculate the percentage changes in the monetary aggregates M1 and M2 for each of the 10 years 1980–1989. Now make a rank sum test at the 5 percent level of the null hypothesis that these percentage changes are from the same probability distribution.

17.95 Use the data in Table 8 at the end of the book to calculate the percentage changes in the Dow Jones and S&P 500 stock price indexes for each of the 10 years 1980–1989. Now make a rank sum test at the 5 percent level of the null hypothesis that these percentage changes come from the same probability distribution.

17.96 (continuation) Apply an analysis of variance test to the 1980–1989 annual percentage changes in the Dow Jones and S&P 500 stock price indexes calculated in the preceding exercise, again testing at the 5 percent level the null hypothesis that these percentage changes come from the same probability distribution. Do these two tests come to the same conclusion?

■ **17.97** *(continuation)* Now apply a difference-in-means *t*-test (with a pooled variance) to these 1980–1989 data to test at the 5 percent level the null hypothesis that the annual percentage changes in the Dow Jones and S&P 500 stock price indexes are from the same probability distribution.

■ **17.98** Use the 1959–1989 data in Table 8 at the end of the book to calculate the percentage changes in the interest rates on one-year Treasury bills and on long-term Treasury bonds for each of the 30 years 1960–1989. Make a rank sum test at the 1 percent level of the null hypothesis that these percentage changes came from the same probability distribution.

■ **17.99** *(continuation)* Use the percentage returns described in Exercise 17.98 to make a rank correlation test at the 5 percent level of the null hypothesis that there is no correlation between the annual percentage changes in the interest rates on one-year Treasury bills and on long-term Treasury bonds.

■ **17.100** Use the percentage returns described in Exercise 17.98 to make a runs tests at the 5 percent level of the null hypothesis that the annual percentage changes in the interest rate on one-year Treasury bills are independent of each other. Do the same for the annual percentage changes in the interest rate on long-term Treasury bonds.

18

Time Series Analysis and Indexes

The age of chivalry is gone; that of sophisters, economists, and calculators has succeeded.

Edmund Burke
Reflections on the Revolution in France

TOPICS

In some tests, data are collected at virtually identical times, or else, the time differences are unimportant; examples are the test scores of different students, the percentage returns this year on several different stocks, and product sales at somewhat different times in several different stores. In such cases, our interest is centered on how test scores vary across students, how returns differ from one stock to the next, and how sales vary from one store to another. The particular day, month, or year in which the data are collected is not of any special interest. There are other situations, though, in which our attention is focused on time, and in particular, how behavior changes as time passes. We might wonder how SAT scores have varied over the past 20 years, if a stock's return this year is correlated with its return in previous years, or how sales vary with the seasons of the year. These questions of how behavior varies as time passes can be answered with **time series analysis** and, if needed, **index numbers.**

18.1 A PICTURE IS WORTH 1,000 WORDS

The most basic tool in time series analysis is a graph, with time on the horizontal axis and the variable of interest on the vertical axis. Figure 18.1 provides an example. Time is marked off in years, from 1970 through 1980. There are 11 annual observations of some hypothetical variable Y. The story told by this picture is that Y has been steadily increasing as time passes.

A persistent movement of a variable, either upward or downward, is called its **trend.**

Figure 18.2 shows some quarterly data on another hypothetical variable over the period 1978–1980. In this case, there is no apparent trend, but there is an obvious pattern to the fluctuations. This variable tends to go up in the summer and down in the winter; perhaps it is a graph of ice cream sales. Some variables show weekly patterns; for instance, there may be increased spending during the first week

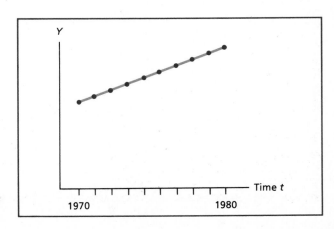

Figure 18.1 *A Variable that Grows over Time*

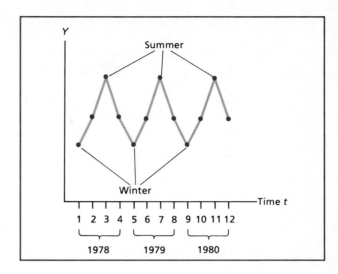

Figure 18.2 *A Variable that Goes Up in the Summer and Down in the Winter*

of the month, after the monthly paycheck is received, and decreased spending in the last week of the month, as funds run out. There are also daily patterns (movie attendance on Friday and job absenteeism on Monday). Other variables tend to be abnormally high or low on holidays, such as the sales of turkeys before Thanksgiving and Christmas.

> Short-term, regular fluctuations with the months, weeks, or days of the year are known as **seasonal variations.**

A variable can, of course, exhibit both a trend and seasonal patterns. Figure 18.3 shows a hypothetical example. This variable has been growing over time, but it also

Figure 18.3 *A Variable with Trend and Seasonality*

regularly increases during the last quarter of each year. This might be the sales of toys, books, or jewelry, which grow with the economy, and also predictably increase shortly before Christmas.

Some fluctuations are unrelated to time, and are just the haphazard consequence of a variety of events. Sales go up because a business decides to reduce the price it charges; sales go down because the Surgeon General declares that the product is hazardous to one's health. Other sales fluctuations may be due to no more than when people happen to run out of a product or when they happen to find the time to go shopping. If sales are graphed versus time, as shown in Figure 18.4, these variations will seem to be just random wiggles and bumps, with no apparent trend or seasonal pattern.

> Fluctuations in a variable that are not related to time are called **random disturbances.**

A variable may have trend, seasonality, and random fluctuations, too. Figure 18.5 shows an example. The trick is to see through the random fluctuations and discover the trend and seasonal patterns.

Some Data

The graphs have so far been hypothetical, to ensure that the patterns are clear. Real data are seldom so well behaved, and the regularities are a bit harder to find, so that the untrained eye cannot always separate the trend, seasonal, and random influences. Let's try a few and then we will see how statistical procedures can be used to unravel time series data.

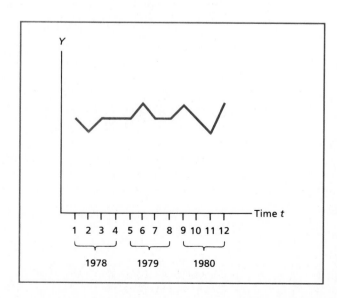

Figure 18.4 *A Randomly Fluctuating Variable*

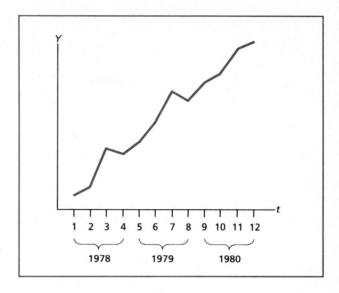

Figure 18.5 *A Variable with Trend, Seasonal, and Random Fluctuations*

Figure 18.6 shows some annual data on U.S. Gross National Product (GNP), going back to the beginning of the century.* This graph was prepared by the Federal Reserve Board and they fitted a trend line, showing that GNP has tended to increase by about 3.1 percent a year over the long run. This long-run growth of the U.S. economy is caused by an expanding population, the accumulation of tools and equipment, and increasing productivity. The fluctuations about this trend line are booms and recessions, which are the business cycles that economists try to explain by looking at the behavior of consumers, businesses, and government. These business cycles appear to have some persistence to them, in that GNP seems to stay below the trend line for an extended period, and then go back above trend for a while.

Figure 18.7 shows U.S. output per hour over this same historical period. Here, the Federal Reserve Board fitted two trend lines, because the rate of growth of productivity seems to have increased since the 1930s. One of the advantages of a graph is that it may uncover such changes. Figure 18.8 shows an even more dramatic example of how a single fitted trend line can be very misleading. In the first third of the century, the average work week dropped dramatically, from 53 hours in 1900 to 40 hours in 1930. The work week collapsed in the Great Depression, when many people had to settle for part-time work, and then jumped to 45 hours during World War II, as people worked overtime to arm the nation. After World War II, the average work week went back to 40 hours and fluctuated around that level for the next 40 years. These differing historical periods would be masked by a single trend line.

*Figure 18.6 uses a **ratio scale,** which is equivalent to graphing the logarithm of GNP. In the next section, you will see that if a variable grows at a constant percentage rate, then a graph of its logarithm versus time will be a straight line.

Figure 18.6 *Gross National Product, Billions of 1972 Dollars*

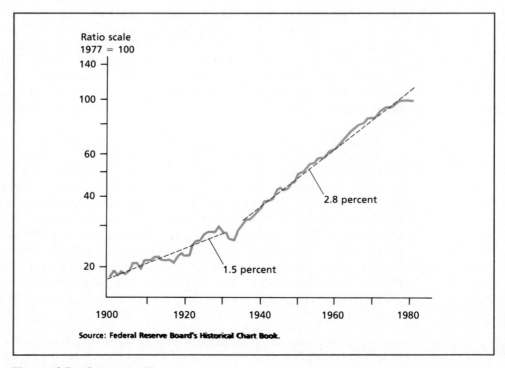

Figure 18.7 *Output per Hour*

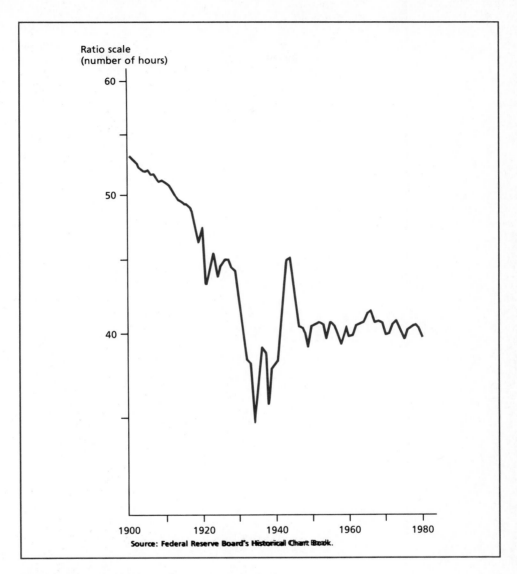

Figure 18.8   *Average Manufacturing Work Week*

These graphs use annual data going back to 1900 to show long-run trends. Short-term seasonal fluctuations would be lost in these graphs, in that any quarterly bumps and wiggles would be so small as to be indiscernible. Figure 18.9 puts a magnifying glass on quarterly GNP fluctuations by showing the shorter period, 1976–1982. It is apparent that output regularly increases in the fourth quarter and slumps in the first quarter of each year. This pattern may be related to the holiday season and to inhospitable weather during the first three months of the year.

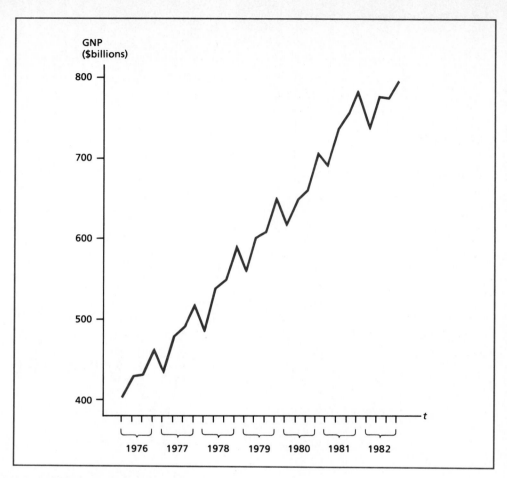

Figure 18.9 *Quarterly GNP, 1976–1982*

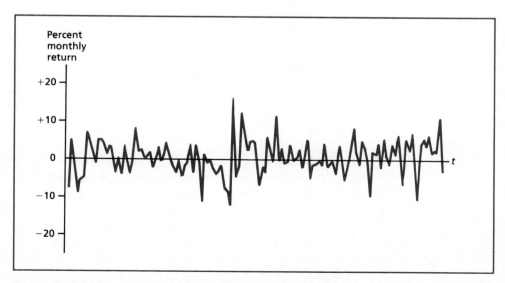

Figure 18.10 *Eleven Years of Monthly Stock Returns, 1970–1980*

Figure 18.10 shows some different data: monthly stock returns over the eleven-year period 1970–1980. Neither trend nor seasonality is very apparent here. Perhaps you can spot something, but I can't.

EXERCISES

18.1 Does the following graph show any trend or seasonality?

18.2 Does the following graph show any trend or seasonality?

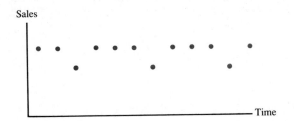

18.3 Does the following graph show any trend or seasonality?

18.4 Does the following graph show any trend or seasonality?

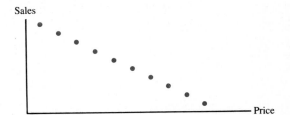

18.2 MODELING A LONG-RUN TREND

The logic of a trend line is that a variable steadily grows or declines as time passes. The simplest model that captures this idea is to let the variable Y be a linear function of some measure of time t,

$$Y_t = \alpha + \beta t \tag{18.1}$$

The subscript t on the variable Y shows that this is the value of the dependent variable at time period t. With annual data, the natural units for t are years. We could let

$$t = 0 \text{ in } 1960 \quad \text{or} \quad t = 1960$$
$$t = 1 \text{ in } 1961 \qquad\qquad t = 1961 \qquad\qquad \textit{(continues)}$$

$$t = 2 \text{ in } 1962 \qquad t = 1962$$

.   .

.   .

.   .

The important point is that t increases by 1 each year, so that the annual increase in t records the passage of time. The annual increase in Y as time passes is given by β,

$$Y_t - Y_{t-1} = \beta$$

With monthly data, the most natural scheme is for t to increase by 1 each month; for weekly data, t could increase by 1 each week. In general, the units for t can be set to match the units for our data. The parameter α scales the data by giving the value of Y at $t = 0$,

$$Y_0 = \alpha$$

In this **linear model,** $Y_t = \alpha + \beta t$, the increase in Y each period is a constant β. An alternative assumption is that the *percentage* increase in Y is the same each period:

$$\frac{Y_t - Y_{t-1}}{Y_{t-1}} = g$$

If so,

$$Y_t - Y_{t-1} = gY_{t-1}$$

or

$$Y_t = Y_{t-1}(1 + g)$$

Looking back t periods implies*

$$Y_t = Y_0(1 + g)^t \tag{18.2}$$

where Y_0 is the value of Y at $t = 0$. (Remember that any number, such as $1 + g$, which is raised to the power 0 is equal to 1.) The parameter g is the constant percentage growth rate of Y as time passes.

---

*The substitution of $Y_{t-1}(1 + g)$ into $Y_t = Y_{t-1}(1 + g)$ gives $Y_t = Y_{t-2}(1 + g)^2$. Repeated substitutions yield Equation 18.2.

Another way of writing this model is to take the logarithms of both sides of Equation 18.2:

$$\ln[Y_t] = \ln[Y_0] + \ln[1 + g]t$$

These could either be common logarithms (base 10) or, as shown here, natural logarithms (base $e = 2.71828$) denoted by ln.* Because $\ln[Y_0]$ and $\ln[1 + g]$ are constants that do not vary with t, we can rewrite the equation as

$$\ln[Y_t] = \alpha + \beta t \qquad\qquad (18.3)$$

Now compare this equation with the linear model Equation 18.1. In Equation 18.1, Y is a linear function of time t. If we plot Y versus t, we will get a straight line with slope β and intercept α. The slope is the constant *amount* by which Y increases each period. Equation 18.3 is said to be a **log-linear model,** because the logarithm of Y is a linear function of t. If the **log-linear model** is appropriate, then a graph of the logarithm of Y versus time t will yield a straight line. The slope of this line, β, is the logarithm of $(1 + g)$, where g is the constant *percentage* by which Y increases each period. For small values of g, the natural log of $1 + g$ is approximately equal to g:

g	$\ln[1 + g]$
.00	0
.01	.0010
.02	.0198
.03	.0296
.04	.0392
.05	.0488
.06	.0583
.07	.0677
.08	.0770
.09	.0862
.10	.0953
.15	.1398
.20	.1823
.25	.2231

Thus, if natural logs are used, the slope of a fitted line is approximately equal to the percentage increase in Y each period.

The easiest way to make a log-linear graph is to buy some "semi-log" graph paper from your college bookstore. This special paper has a ratio scale, so that when you plot a number Y, you automatically get the logarithm $\ln[Y]$ (multiplied by

*The logarithm, base 10, of Y is that number X such that $10^X = Y$. The log of Y, base e, is the number Z such that $e^Z = Y$. The one property we use here is that $\ln[AB] = \ln[A] + \ln[B]$ and, hence, $\ln[(1 + g)^t] = t \ln[1 + g]$.

some unimportant constant). Thus when you plot Y versus t on semi-log paper, you magically obtain a graph of $\ln[Y]$ versus t.

We can draw two conclusions so far:

If Y increases by a constant *amount* each period, then we can model this behavior by expressing Y as a linear function of time t.

If Y increases by a constant *percentage* each period, then the logarithm of Y is a linear function of t.

In practice, matters are never this simple. There are almost always factors other than the simple passage of time that affect Y. If we put these *deviations from trend* into an error term ϵ, then our two trend models become

$$Y_t = \alpha + \beta t + \epsilon \qquad (18.4)$$

$$\ln[Y_t] = \alpha + \beta t + \epsilon \qquad (18.5)$$

The parameters α and β can be estimated by a simple least squares regression.

To illustrate, let's use the annual GNP data for 1951–1981 that are shown in Table 18.1. (The 1982 and 1983 data will be used later in a forecasting test.) If I graph GNP versus time, as is shown in Figure 18.11, the relationship does not seem to be linear. Instead of a straight line, GNP curves upward, in that the annual *dollar* increase in GNP during the 1970s is generally larger than the *dollar* increase during the 1950s. Perhaps the annual *percentage* increases are comparable. To check this, Figure 18.12 uses semi-log paper to graph the log of GNP versus time. Now the relationship could be described by a straight line. Because the log-linear model seems the most appropriate, we'll use the 1951–1981 data to calculate the least

Table 18.1   *Gross National Product, in Billions of 1972 Dollars*

Year	GNP	Year	GNP	Year	GNP
1951	579.4	1962	800.3	1973	1,254.3
1952	600.8	1963	832.5	1974	1,246.3
1953	623.6	1964	876.4	1975	1,231.6
1954	616.1	1965	929.3	1976	1,298.2
1955	657.5	1966	984.8	1977	1,369.7
1956	671.6	1967	1,011.4	1978	1,438.6
1957	683.8	1968	1,058.1	1979	1,479.4
1958	680.9	1969	1,087.6	1980	1,475.0
1959	721.7	1970	1,085.6	1981	1,513.8
1960	737.2	1971	1,122.4	1982	1,485.4
1961	756.6	1972	1,185.9	1983	1,534.8

Figure 18.11 *A Graph of Annual GNP for 1951–1981 Seems to Curve Upward*

squares estimates of Equation 18.5:

$$\ln[\hat{Y}_t] = -60.028 + 0.034t, \qquad R^2 = .991$$

where $t = 1951$ in 1951. The slope 0.034 is an estimate of $\ln[1 + g]$, and the implicit estimate of the annual growth rate g works out to be 0.035, or 3.5 percent per year:

$$g = e^{0.034} - 1 = 0.035$$

This is slightly different than the Federal Reserve's estimate because we used a slightly different time period.

The implicit fitted values of Y can be computed by taking the antilogarithm of $\ln[\hat{Y}_t]$,

$$Y_t = e^{\ln[\hat{Y}_t]}$$

In practice, this would be done with a computer program, or else by a hand calculator with a button marked e^x, exp, or something similar. To illustrate, for $t =$

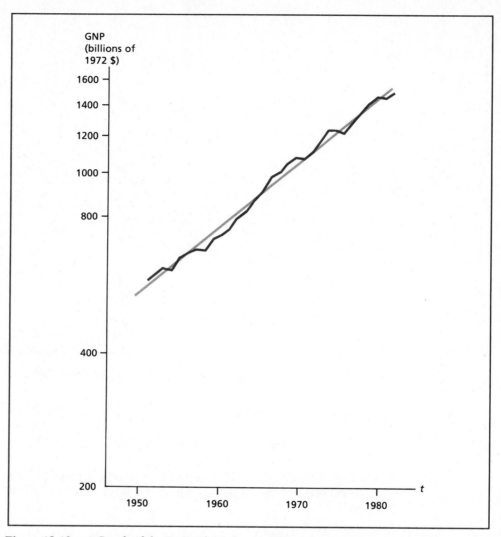

Figure 18.12   *A Graph of the Logarithm of Annual GNP for 1951–1981 Seems More Nearly Linear*

1960, the predicted value of ln[Y] is $-60.027840 + 0.03401671(1970) = 6.985074$, which implies a prediction for Y of $e^{6.985074} = 1080.4$, as compared to the actual value, $Y = 1,085.6$ (from Table 18.1).

Misleading Trends

One of the most bitter criticisms of statisticians is that, "Figures don't lie, but liars figure." The complaint is that an unscrupulous statistician can prove anything by carefully choosing favorable data and conveniently ignoring conflicting evidence. Trend lines are a favorite vehicle for such chicanery. By a clever selection of the years to start and end a time-series analysis, a dishonest person may be able to force the trend line up or down, in whichever direction is desired.

If we are fitting a trend to a long series of relatively stable data, the specific years used in the regression won't matter very much, but if we use a short series of volatile data, the chosen years can make all the difference in the world. Figure 18.13 illustrates this point with some data on stock prices. If we fit a trend line to the data for 1981 through mid-1982, then it appears that stock prices tend to fall by about 10 percent a year, but if we analyze the data for mid-1982 through 1983, stock prices seem to increase by about 30 percent a year. Which is correct? Well, both time periods are too short to show convincing trends for something as volatile as the stock market: A much longer perspective is needed; for example, over the past 60 years, stock prices have increased by around 4.0 percent per year.[1]

Would anyone be so naive as to extrapolate stock prices on the basis of only one or two years of data? Investors do it all the time, over even shorter periods,

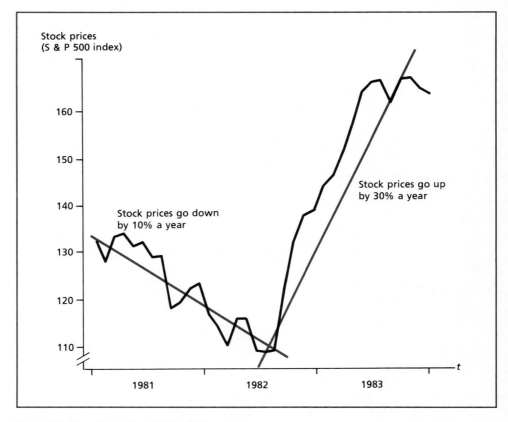

Figure 18.13 *Some Trend Line Chicanery*

hoping that the latest upward blip is the prelude to many happy profits or fearing that a recent slip is the start of a crash. The dedicated choose their data carefully, to give the trend they want. In December 1985, an investment advisory service touting gold and other tangible assets fit a line to two points—an 11 percent inflation in 1974 and a 13.5 percent inflation in 1979—and concluded that we will have at least 23 percent inflation by 1995!

Another source of misleading trends is fundamental changes in the underlying behavior. Recall Figure 18.8 on the average work week. A line fitted to pre-World War II data would show a steep downward trend in the average work week; a trend line fitted to the entire period 1900–1980 would show a less drastic downward trend; a line fitted to data for the 1930s and 1940s would show a steep upward trend; a line for 1940–1980 would show a steep downward trend; and a line for 1950–1980 data would show no trend at all. Any single trend line is misleading, because it ignores the fact that behavior during the years 1950–1980 was very different from 1900–1930 and that the 1930s and 1940s were exceptional periods with very unusual changes in the work week.

The best antidotes for misleading trends are a visual inspection of the data and some common sense. A graph may reveal obvious breaks in the data, such as a change in behavior or an extraordinary deviation. Also, if the data appear to be volatile, then a bit of caution (and a longer perspective) may be in order. Finally, common sense can sometimes catch the problem. You may know that behavior now is much different than it used to be or that some unusual events may have occurred that give the appearance of a trend where there is none.

Time series analysis is essentially descriptive; it is a picture that is meant to tell a story. As with any story, bumblers may forget the punch line and dishonest people may lie. The conscientious person, though, can use time series analysis to describe, in a simple and yet accurate way, the behavior of a variable as time passes. There is room for mistakes and even chicanery, but there is also room for enlightenment. As with all statistics, time series analysis is an art as well as a science.

EXERCISES

18.5 In 1924, Computing-Tabulating-Recording Company became International Business Machines (IBM). Here are some data on its net after-tax profits (in millions of dollars):

Year	Profits
1925	2.8
1935	7.1
1945	10.9
1955	55.9
1965	476.9
1975	1,979.9
1985	6,555.0

Use the equation $Y_t = Y_0(1 + g)^t$ to calculate IBM's annual rate of growth of profits for each ten-year period 1925–1935, 1935–1945, and so on, and for the periods **a.** 1925–1975 **b.** 1975–1985 **c.** 1925–1985. Explain why we should be cautious about estimating the equation

$$\ln[\text{profits}] = \alpha + \beta t + \epsilon$$

18.6 Table 14.2 gives the annual returns for some selected stocks and for the stock market as a whole. See if these data show any trend in the market return by estimating the equation $Y = \alpha + \beta t + \epsilon$, where Y is the market rate of return and t is the year ($t = 1972$ in 1972, and so on).

18.7 Here are some data on the size of the U.S. population (in millions) from the U.S. census taken every ten years:

Year t	Population Y
1940	132
1950	151
1960	179
1970	203
1980	227

Plot these data, with Y on the vertical axis and t on the horizontal axis. Does there appear to be a linear relationship? Now, plot $\ln[Y]$ versus t; if you have semi-log paper, plot Y versus t. Does this relationship appear to be linear? Determine the least squares estimates of α and β in the model $\ln[Y] = \alpha + \beta t + \epsilon$. What is your estimate of the percentage growth rate g of population over this forty-year period?

18.8 Table 18.1 shows annual GNP for the years 1951–1983. Estimate the trend line $\ln[Y] = \alpha + \beta t + \epsilon$, where Y is annual GNP and $t = 1951$ in 1951, using data for **a.** 1951–1960 **b.** 1961–1970 **c.** 1971–1980. In each case, what is the estimated growth rate of GNP? Are these three sets of estimates roughly consistent with each other? If there are substantial differences, explain why these specific differences occur.

18.9 *(continuation)* Acting out the role of unscrupulous data massager, select some consecutive years for which a trend-line regression gives a negative estimate of β.

18.10 The height of Lake Michigan varies from month to month and year to year. Here are the highest monthly mean levels (in feet) every ten years, for a hundred years:

Year t	Height Y
1860	583.3
1870	582.7
1880	582.1
1890	581.6
1900	580.7
1910	580.5
1920	581.0
1930	581.2
1940	579.3
1950	580.0

Find the least squares estimates of the equation $Y = \alpha + \beta t + \epsilon$, in order to determine if there was (at the 5 percent level) a statistically significant trend in the height of Lake Michigan. Why might you be cautious about extrapolating the trend equation forward 20,000 years or back 20,000 years?

18.11 Table 13.1 gives some data on household disposable income X and spending Y, for the five years 1977–1981. Use these data to see if there are (at the 5 percent level) any statistically significant trends in income, spending, and the fraction Y/X of income that is spent. In particular, make three graphs, plotting X, Y, and Y/X versus time t and then find the least squares estimates of the equations

$$X = \alpha_1 + \beta_1 t + \epsilon_1$$
$$Y = \alpha_2 + \beta_2 t + \epsilon_2$$
$$Y/X = \alpha_3 + \beta_3 t + \epsilon_3$$

(Set $t = 1977$ in 1977.) Which, if any, of these three variables shows a statistically significant trend? Explain.

• **18.12** In Exercise 18.7, you estimated the model

$$\ln[Y] = \alpha + \beta t + \epsilon, \text{ where } t = 1940 \text{ in } 1940$$

An alternative formulation is

$$\ln[Y] = \gamma + \delta T + \epsilon, \text{ where } T = 0 \text{ in } 1940$$

That is, $T = t - 1940$. What is the relationship between the parameters in these two formulations? Check your logic by using the data in Exercise 18.7 to estimate both equations.

18.3 MODELING SEASONAL REGULARITIES

One of the most important reasons for deviations about a trend line is seasonal fluctuations. The easiest way to account for these seasonal patterns is to use 0–1 dummy variables. Perhaps we believe that Y varies with the four quarters of the year. If so, we can use the model

$$Y_t = \alpha_1 + \alpha_2 D_2 + \alpha_3 D_3 + \alpha_4 D_4 + \epsilon \qquad (18.6)$$

where Y_t = quarterly value of Y
           D_2 = 1 in second quarter of each year, 0 otherwise
           D_3 = 1 in third quarter of each year, 0 otherwise
           D_4 = 1 in fourth quarter of each year, 0 otherwise

The first quarter is selected arbitrarily as the reference quarter. In the first quarter, the three dummy variables are all equal to zero, and the equation simplifies to

first quarter: $Y_t = \alpha_1 + \epsilon$

In the other three quarters, the equation changes to

second quarter: $Y_t = (\alpha_1 + \alpha_2) + \epsilon$

third quarter: $Y_t = (\alpha_1 + \alpha_3) + \epsilon$

fourth quarter: $Y_t = (\alpha_1 + \alpha_4) + \epsilon$

The parameters α_2, α_3, and α_4 gauge the quarterly variations in Y. If α_2 is positive, then Y tends to be higher in the second quarter than in the first quarter, by an amount α_2. If α_2 is negative, then Y tends to be lower in the second quarter than the first, by an amount α_2. The parameters α_3 and α_4 similarly compare the third and fourth quarters with the first quarter. If these three parameters are not significantly different from zero, then there is no apparent systematic variation in Y from quarter to quarter.

The parameters of our seasonal dummy model, Equation 18.6 can be estimated by a least squares computer program. The procedure, called *multiple regression,* is explained in Chapter 15 and the actual calculations would be left to the computer. With monthly data, we can use 11 monthly dummy variables; for weekly data, 51 weekly dummies could be used.

If there is both trend and seasonality, then a time variable can be included along with the seasonal dummy variables. For example, if we believe that Y grows by a constant percentage rate and also varies with the quarters of the year, then we can use least squares to estimate the model

$$\ln[Y_t] = \alpha_1 + \alpha_2 D_2 + \alpha_3 D_3 + \alpha_4 D_4 + \beta t + \epsilon \qquad (18.7)$$

Seasonal Adjustment

It is again important to recognize that these models are merely descriptive. They do not attempt to explain *why* there are long-run trends and short-run seasonality;

they just try to identify the existence of such patterns. Similarly, they do not explain why there are fluctuations in addition to trend and seasonality, even though these "disturbances" are often fuel for the most interesting of questions—for example, why the economy has economic booms and recessions. The answer to such questions requires good theory and a regression model with true explanatory variables in place of the descriptive time variable. All that time series analysis offers are methods for describing the data.

The way these methods are often used in practice is to excise the trend and seasonality, so that we can more easily see the fluctuations that need to be explained by something else. For instance, the movements of GNP above and below the trend lines in Figures 18.6 and 18.12 allow us to identify periods of economic boom and recession. Most economic data are also **seasonally adjusted** to remove seasonal patterns.

The actual seasonal-adjustment calculations are quite complicated and are best left to a computer program. However, for a simple example of the logic involved, let's look at the quarterly GNP shown in Figure 18.9. (The data pictured in this graph are given in Table 18.2.) The graph seems to show a regular increase in output in the fourth quarter of each year, followed by a decline in the first quarter. To confirm and quantify this impression, I've estimated the equation

$$Y = \alpha_1 + \alpha_2 D_2 + \alpha_3 D_3 + \alpha_4 D_4 + \beta t + \epsilon$$

where Y = quarterly GNP
           D_2 = 1 in second quarter, 0 otherwise

Table 18.2   *Quarterly GNP, in Billions of Dollars, 1976–1982*

Year	Quarter	GNP	Year	Quarter	GNP
1976	1	398.9	1980	1	616.1
	2	426.5		2	648.8
	3	430.8		3	661.0
	4	461.8		4	705.7
1977	1	436.9	1981	1	689.8
	2	476.6		2	733.7
	3	488.8		3	752.8
	4	515.6		4	777.7
1978	1	484.7	1982	1	733.6
	2	536.0		2	773.4
	3	547.6		3	772.5
	4	587.9		4	793.5
1979	1	559.8			
	2	598.5			
	3	607.1			
	4	649.8			

$D_3 = 1$ in third quarter, 0 otherwise

$D_4 = 1$ in fourth quarter, 0 otherwise

$t = 0$ in the first quarter of 1976, and then increases by 1 each quarter

The least squares estimates for 1976–1982 are

$$\hat{Y} = 382.85 + 24.34D_2 + 19.17D_3 + 37.46D_4 + 14.76t, \qquad R^2 = .992$$
$$\quad\ [68.7] \qquad [3.9] \qquad\ [3.0] \qquad\ \ [5.9] \qquad\ \ [53.0]$$

[   ]: t-values

The coefficients of the trend variable t and the three seasonal dummies are all statistically significant at the 5 percent level. According to the seasonal estimates, GNP tends to be about $24 billion higher in the second quarter than in the first quarter, $19 billion higher in the third quarter than in the first, and $37 billion higher in the fourth quarter than the first.

The trend term means that GNP increases by about $15 billion each quarter. The seasonal regularities imply that we shouldn't be surprised by seasonal fluctuations about this trend. Between the fourth quarter of each year and the first quarter of the next, we can expect quarterly GNP to drop by $22 billion (the usual $37 billion seasonal decline less the expected $15 billion trend increase). This is the expected pattern. If GNP drops by more than $22 billion, then we might become worried, because something bad may have occurred, beyond the usual first-quarter dip.

A more formal way of taking these expected seasonal patterns into account is to seasonally adjust the data. For any given year, let's label the unadjusted quarterly data

$$Y_1 = \text{first-quarter GNP}$$

$$Y_2 = \text{second-quarter GNP}$$

$$Y_3 = \text{third-quarter GNP}$$

$$Y_4 = \text{fourth-quarter GNP}$$

The seasonally adjusted data will be

$$S_1 = Y_1 + A_1$$

$$S_2 = Y_2 + A_2$$

$$S_3 = Y_3 + A_3$$

$$S_4 = Y_4 + A_4$$

where the A_i are the additive seasonal adjustments. (Multiplicative adjustments can also be used.)

Now, it would be nice if the annual values of the adjusted data coincided with the annual values of the unadjusted data:

$$S_1 + S_2 + S_3 + S_4 = Y_1 + Y_2 + Y_3 + Y_4$$

Thus, annual GNP for 1978 will be the same, regardless of whether we add up the seasonally adjusted or unadjusted quarterly data. For this to be true, the four seasonal adjustments must sum to zero:

$$A_1 + A_2 + A_3 + A_4 = 0$$

That is, if we add a few billion dollars here or there to make a seasonal adjustment, then we must take them away from somewhere else, to get the annual numbers to add up.

Our regression estimate is that seasonality causes GNP to be higher by a_2 (our estimate of α_2) if it is the second quarter rather than the first. Thus, if we add some amount A_1 to first-quarter GNP, then we want to add $A_1 - a_2$ to second-quarter GNP. In this way, we can take into account the fact that we expect GNP to be higher in the second quarter than in the first. Similar adjustments for the third and fourth quarters give

$$A_2 = A_1 - a_2$$
$$A_3 = A_1 - a_3$$
$$A_4 = A_1 - a_4$$

This logic, along with the desire to have the four adjustments add up to zero, gives us our rules:

$$0 = A_1 + A_2 + A_3 + A_4$$
$$= A_1 + (A_1 - a_2) + (A_1 - a_3) + (A_1 - a_4)$$
$$= 4(A_1) - (a_2 + a_3 + a_4)$$

implies that

$$
\begin{aligned}
A_1 &= (a_2 + a_3 + a_4)/4 \\
A_2 &= (a_2 + a_3 + a_4)/4 - a_2 \\
A_3 &= (a_2 + a_3 + a_4)/4 - a_3 \\
A_4 &= (a_2 + a_3 + a_4)/4 - a_4
\end{aligned}
\tag{18.8}
$$

Thus, we want our seasonal adjustment to reduce: second-quarter GNP by a_2 relative to first-quarter GNP; third-quarter GNP by a_3 relative to first-quarter GNP;

and fourth-quarter GNP by a_4 relative to first-quarter GNP. But we also want the adjustments to have no effect on annual GNP, so we make these three subtractions, but then add $(a_2 + a_3 + a_4)/4$ to each quarter.

The least squares estimates a_i for the quarterly GNP example yield

$$A_1 = 20.24$$

$$A_2 = -4.10$$

$$A_3 = 1.08$$

$$A_4 = -17.22$$

The seasonal regularity is that GNP typically rises by a little in the second quarter and by a lot in the fourth quarter, and falls by a lot in the first quarter and by a little in the third quarter. To offset this pattern, we add several billion dollars to first- and third-quarter GNP and subtract several billion dollars from second- and fourth-quarter GNP. As planned, the seasonal adjustments sum to zero and the difference between the first adjustment and each of the others is equal to the estimated coefficient of the relevant dummy variable.

Table 18.3 shows the seasonally adjusted data for the three-year period 1980–1982, with the adjusted and unadjusted data graphed together in Figure 18.14. Notice how the seasonally adjusted data fluctuate less than the unadjusted data. This is because the seasonal adjustment is an attempt to correct for seasonal fluctuations, so that we can see other changes in GNP. Look, for example, at Table 18.3 at the fourth quarter of 1980 and the first quarter of 1981. In the unadjusted data, there is a $15.9 billion drop in GNP. But there is usually a $37.46 billion seasonal drop in GNP between the fourth and first quarters. (Remember, 37.46 is

Table 18.3 *Seasonally Adjusted Quarterly GNP, in Billions of Dollars, 1980–1982*

Year	Quarter	GNP	Seasonal Adjustment	Adjusted GNP
1980	1	616.1	20.2	636.3
	2	648.8	−4.1	644.7
	3	661.0	1.1	662.1
	4	705.7	−17.2	688.5
1981	1	689.8	20.2	710.0
	2	733.7	−4.1	729.6
	3	752.8	1.1	753.9
	4	777.7	−17.2	760.5
1982	1	733.6	20.2	753.8
	2	773.4	−4.1	769.3
	3	772.5	1.1	773.6
	4	793.5	−17.2	776.3

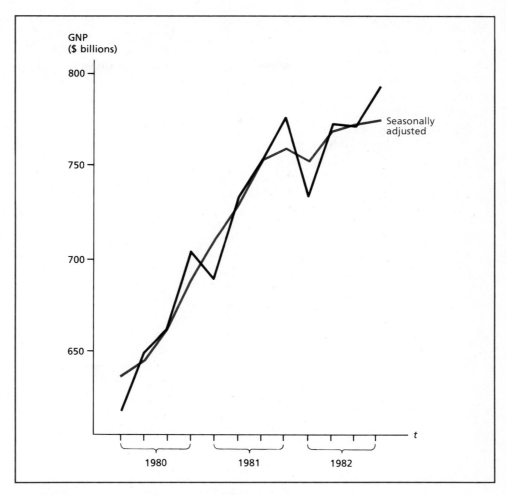

Figure 18.14 *Seasonally Adjusted and Unadjusted Quarterly GNP*

the estimated coefficient of the fourth-quarter dummy variable.) Because the observed $15.9 billion drop is $21.56 billion less than might be expected, the seasonally adjusted data show a $21.56 billion increase in GNP. The first quarter of 1981 was actually a relatively good quarter, because GNP did not drop as much as usual.

It has been said that, "With seasonally adjusted temperatures, you could eliminate winter in Canada."[2] That's funny, but not true unfortunately. Seasonal adjustment doesn't eliminate seasonal patterns; it merely takes them into account so that we can see if anything unusual is happening. Seasonally adjusted temperatures wouldn't eliminate winter, but they would tell us whether or not a particular winter was unusually cold.

Toys "R" Us

Toys "R" Us is the world's largest toy retailer, which operates some 150 toy supermarkets throughout the United States. Since 1978, their sales have increased by 30 percent per year, while their profits have grown by 40 percent per year. However, because they are in the toy business, these sales and profits are not evenly distributed throughout the year, as indicated in Table 1.

As you can see, more than half of their sales and almost all of their profits come dur-

ing the Christmas season. We can formally describe this extreme seasonal pattern with the model*

$$Y = \alpha_1 + \alpha_2 D_2 + \alpha_3 D_3 + \alpha_4 D_4 + \beta t + \epsilon$$

where

$D_2 = 1$ in May–July quarter, 0 otherwise
$D_3 = 1$ in Aug–Oct quarter, 0 otherwise
$D_4 = 1$ in Nov–Jan quarter, 0 otherwise

Table 1

Fiscal Year Begins	Quarter	Sales ($ Millions)	Earnings per Share
1978	Feb 1–Apr 30	38.0	−0.08
	May 1–July 31	53.6	0.00
	Aug 1–Oct 31	57.5	0.02
	Nov 1–Jan 31	200.0	0.87
1979	Feb 1–Apr 30	56.5	−0.02
	May 1–July 31	75.8	0.08
	Aug 1–Oct 31	78.3	0.03
	Nov 1–Jan 31	269.7	1.13
1980	Feb 1–Apr 30	70.2	−0.02
	May 1–July 31	92.7	0.01
	Aug 1–Oct 31	101.8	0.00
	Nov 1–Jan 31	332.6	1.30
1981	Feb 1–Apr 30	97.3	0.03
	May 1–July 31	123.7	0.14
	Aug 1–Oct 31	132.9	0.14
	Nov 1–Jan 31	429.4	1.81
1982	Feb 1–Apr 30	138.3	0.11
	May 1–July 31	167.6	0.22
	Aug 1–Oct 31	189.9	0.25
	Nov 1–Jan 31	545.9	2.27

*Their growth is not really linear, but logarithms of profits can't be used, because of the zero and negative values. The linear trend model was used as a rough approximation, because I couldn't resist showing you the exceptional seasonality that buffets the world's largest toy seller.

Here, we will look at Y = earnings per share. The sales data are left as an exercise.

The least squares estimation of this time-series equation gives

$$\hat{Y} = -0.28 + 0.05D_2 + 0.02D_3$$
$$\quad\; [2.0] \quad\; [0.4] \quad\quad [0.2]$$
$$\quad + 1.38D_4 + 0.03t$$
$$\quad\quad [9.7] \quad\quad [3.5]$$

[ ]: t-values

There is (at the 5 percent level) a statistically significant growth trend and a statistically sig-nificant difference between the fourth and first fiscal-quarter earnings. On average, fourth-quarter earnings tend to be about $1.38 higher than first-quarter earnings. Earnings also seem to be a bit higher in the second and third quarters than in the hapless first quarter, but these differences are small and not statistically significant, because the t-statistics are less than 2.

Table 2 shows the seasonal-adjustment calculations and Figure 1 depicts the season-ally adjusted and unadjusted data. Notice how the seasonal adjustment smooths out the data by abstracting from the regular seasonal

Table 2   *Seasonal Adjustment of Toys "R" Us Earnings*

Fiscal Year Begins	*Quarter*	*Unadjusted Earnings per Share*	*Seasonal Adjustment*	*Adjusted Earnings per Share*
1978	Feb 1–Apr 30	−0.08	0.36	0.28
	May 1–July 31	0.00	0.31	0.31
	Aug 1–Oct 31	0.02	0.34	0.36
	Nov 1–Jan 31	0.87	−1.01	−0.14
1979	Feb 1–Apr 30	−0.02	0.36	0.34
	May 1–July 31	0.08	0.31	0.39
	Aug 1–Oct 31	0.03	0.34	0.37
	Nov 1–Jan 31	1.13	−1.01	0.12
1980	Feb 1–Apr 30	−0.02	0.36	0.34
	May 1–July 31	0.01	0.31	0.32
	Aug 1–Oct 31	0.00	0.34	0.34
	Nov 1–Jan 31	1.30	−1.01	0.29
1981	Feb 1–Apr 30	0.03	0.36	0.39
	May 1–July 31	0.14	0.31	0.45
	Aug 1–Oct 31	0.14	0.34	0.48
	Nov 1–Jan 31	1.81	−1.01	0.80
1982	Feb 1–Apr 30	0.11	0.36	0.47
	May 1–July 31	0.22	0.31	0.53
	Aug 1–Oct 31	0.25	0.34	0.59
	Nov 1–Jan 31	2.27	−1.01	1.26

(continued)

fluctuations. Notice also, how the seasonally adjusted numbers sometimes tell a different story than the unadjusted data. Look at the fourth quarter of 1980, for instance. Earnings per share jumped from $0.0 to $1.30; however, taking into account the expected growth and the traditionally strong fourth-quarter results, this was actually a disappointing quar-

ter. During the next three months, the first quarter of fiscal 1981, earnings slumped to $0.03 a share, but this was a much smaller drop than usual. On a seasonally adjusted basis, the $0.03 earnings in the first quarter of 1981 was actually more encouraging news than the $1.30 earnings in the fourth quarter of 1980.

Figure 1   *Seasonally Adjusted and Unadjusted Quarterly Earnings per Share for Toys "R" Us*

"*Yes, I'm somewhat depressed,*
but seasonally adjusted I'm probably happy enough."

Drawing by Mankoff; © 1984 The New Yorker Magazine, Inc.

These GNP data were deseasonalized, so that we could focus on GNP fluctuations that weren't simply due to seasonal patterns. In the same way, data can also be detrended, so that we abstract from increases in GNP that just represent normal growth. By looking at deviations about the trend, we can identify periods in which GNP is increasing faster or slower than usual. Thus, when GNP increases, but not by as much as usual, this is a sign of economic weakness. The details of detrending are left as Exercise 18.20.

Monthly Stock Returns

Table 16.2 contains monthly stock returns over the eleven-year period 1970–1980. To see if there is any trend or seasonality in these data, I estimated the equation

$$Y = \alpha + \alpha_2 D_2 + \alpha_3 D_3 + \cdots + \alpha_{12} D_{12} + \beta t + \epsilon$$

where Y = the monthly percentage return
D_i = 11 monthly dummy variables, with January as the reference month
t = time, with $t = 0$ in January 1970 and increased by 1 each month

A least squares regression yielded these estimates:

$$\hat{Y} = 1.19 - 1.16D_2 - 1.34D_3 - 1.40D_4 - 2.15D_5 - 0.92D_6 - 1.46D_7$$
$$\quad [0.7] \quad [0.6] \quad\quad [0.7] \quad\quad [0.7] \quad\quad\quad [1.0] \quad\quad [0.4] \quad\quad\quad [0.7]$$
$$\quad - 1.02D_8 - 2.22D_9 - 2.01D_{10} + 0.06D_{11} + 0.01D_{12} + 0.01t$$
$$\quad\quad [0.5] \quad\quad [1.1] \quad\quad [1.0] \quad\quad\quad [0.0] \quad\quad\quad [0.0] \quad\quad\quad [1.0]$$

$R^2 = .03$

[ ]: t-values

Not one of the seasonal dummies is statistically significant at the 5 percent level; nor is the time trend. There is no convincing evidence in these data of trend or seasonal patterns in stock returns. Government data collectors agree, in that securities prices are almost the only economic data that are not seasonally adjusted. The government statisticians can't find any seasonal patterns either!

If there were seasonality in stock returns, extraordinary profits would be available to those who bought low and sold high. It is most likely that such profit incentives would cause the seasonality to disappear, in that the pressure of people trying to buy low would push prices up in the low quarters, while the pressure of sellers would pull prices down in the high quarters. In this way, any regular patterns would be destroyed by speculators. Such a self-destructive mechanism does not exist for quarterly GNP, because seasonality in GNP does not offer speculative profits. If consumers want to spend in the fourth quarter of each year, then businesses will produce more to meet this demand. There is no reason for them to produce three months early. If the weather is lousy in February, then home builders will wait for better weather. There is no reason for them to stop working in November in anticipation of February's dip. If anything, they will produce more in the fourth quarter, in a rush to finish before the weather turns nasty.

EXERCISES

18.13 Table 18.3 shows that in the first quarter of 1982 GNP fell by $44.1 billion, while seasonally adjusted GNP fell by only $6.7 billion. How can this be?

18.14 A business has estimated the following equation to identify seasonal fluctuations in its profits (in millions of dollars):

$$\hat{Y} = 20 + 10D_2 + 10D_3 + 10D_4$$

where Y = quarterly profits
            D_2 = 1 in second quarter, 0 otherwise
            D_3 = 1 in third quarter, 0 otherwise
            D_4 = 1 in fourth quarter, 0 otherwise

How should we interpret these estimates? Do profits tend to be higher in the first quarter or second quarter? Are profits usually higher in the second or third quarter? Make a rough sketch to show the usual seasonal pattern.

18.15 The text estimates the trend and seasonality in earnings per share for *Toys "R" Us*. Use the data given in this example to estimate the trend and seasonality in their sales:

$$\ln(Y) = \alpha_1 + \alpha_2 D_2 + \alpha_3 D_3 + \alpha_4 D_4 + \beta t + \epsilon$$

where Y = quarter sales (in $ millions)
            D_2 = 1 in May–July quarter, 0 otherwise

$D_3 = 1$ in Aug–Oct quarter, 0 otherwise
$D_4 = 1$ in Nov–Jan quarter, 0 otherwise

What is your estimate of the quarterly growth rate of their sales? What, if any, seasonal pattern is there in their sales?

18.16 *(continuation)* Use your estimated *Toys "R" Us* sales equation to calculate a seasonally adjusted sales series. Plot the adjusted and unadjusted sales data on the same graph, with time on the horizontal axis. (Use semi-log paper, if you have some.)

18.17 Determine the quarterly earnings per share over a ten-year period for a mythical Las Vegas, Inc. by rolling a pair of dice 40 times. For each of these 40 quarters, the earnings per share are equal to $9 minus the number that is rolled. Now see if there is any trend or seasonality in these data by estimating the equation

$$Y = \alpha_1 + \alpha_2 D_2 + \alpha_3 D_3 + \alpha_4 D_4 + \beta t + \epsilon$$

where

Y = quarterly profits
$D_2 = 1$ in second quarter, 0 otherwise
$D_3 = 1$ in third quarter, 0 otherwise
$D_4 = 1$ in fourth quarter, 0 otherwise
$t = 1$ to 40, up by 1 each quarter

18.18 *(continuation)* Do these calculations again using the same rolled numbers, but a different formula for calculating earnings. Let earnings at $t = 0$ be $2. Now, let the *change* in earnings each quarter equal $9 minus the number rolled.

18.19 Here are some monthly data on U.S. beer production (in thousands of barrels):

	Year			
Month	1975	1976	1977	1978
Jan	12,549	12,441	11,978	12,871
Feb	11,181	11,890	11,482	12,713
Mar	12,412	11,855	16,200	15,860
Apr	14,496	13,688	16,027	15,624
May	14,343	15,175	16,788	16,567
June	15,762	15,758	16,904	16,878
July	16,076	16,539	15,921	16,741
Aug	14,719	16,096	15,309	17,614
Sept	13,345	14,313	13,255	14,625
Oct	12,350	13,422	12,606	14,014
Nov	11,217	11,288	12,021	12,714
Dec	12,150	11,191	12,014	12,872

Do you discern any seasonal patterns in these data? Estimate the model

$$Y = \alpha_1 + \alpha_2 D_2 + \alpha_3 D_3 + \cdots + \alpha_{12} D_{12} + \epsilon$$

where the D_i are eleven seasonal dummies (with January as the reference month). In which, if any, months are there (at the 5 percent level) statistically significant differences in production, relative to January? Does this apparent seasonality make sense?

• 18.20 The text uses the 1951–1981 GNP data shown in Table 18.1 to estimate a trend equation for GNP. Use these estimates to calculate a *detrended* GNP for 1978–1981. What does this detrended GNP measure?

18.4 FORECASTING

Many decisions by households, businesses, and government are based on forecasts of one sort or another. Your choice of a college major involves forecasts of how well you will do in your courses, what you want to do when you graduate, and what your career opportunities will be. You won't major in physics if you think you can't pass physics courses; you probably won't major in education if you want to go to medical school; and you may be dissuaded from majoring in Latin by glum job prospects. Businesses routinely make forecasts of their sales, materials costs, interest rates, and a dizzying array of other things. Government policymakers forecast output, unemployment, inflation, interest rates, and so on.

Businesses and government often use time series analyses to make these forecasts, hoping that extrapolations of the past will yield accurate predictions of the future. For instance, the annual GNP data shown in Table 18.1 was used earlier in this chapter to estimate a trend equation for GNP,

$$\ln[\hat{Y}] = -60.0 + 0.0340t$$

where $t = 1,951$ in 1951, $t = 1,952$ in 1952, and so on. A projection of this estimated equation gives annual GNP forecasts into the future. Here is a sampling:

Year	Projected GNP Billions of 1972 $
1982	1,625.0
1983	1,681.2
1990	2,133.3
2000	2,997.6

Of course, we know from Chapter 13 that least squares predictions have confidence intervals associated with them, and that these confidence intervals widen the farther one gets from the center of the available data. The computer program that I used to estimate this equation gave these confidence intervals:

Year	Projected GNP Billions of 1972 $	95% Confidence Interval
1982	1,625.0	1,523.0 to 1,733.9
1983	1,681.2	1,575.1 to 1,794.5
1990	2,133.3	1,992.2 to 2,284.3
2000	2,997.6	2,782.5 to 3,229.4

As it turns out, the 1982 and 1983 projections were overly optimistic. In 1982, the Federal Reserve Board used a tight monetary policy to wring inflation out of the economy, and the United States suffered its most severe recession since the Great Depression in the 1930s. The unemployment rate averaged 9.7 percent and GNP came to $1,485.4 billion, some $40 billion outside our 95 percent confidence interval. In 1983, it was more of the same, with a 9.6 percent unemployment rate and $1,534.8 billion GNP.

It is important to recognize that time series forecasts are just mechanical extrapolations. Their justification is their simplicity and usefulness in describing the implications of the past. They provide no underlying rhyme or reason why things should continue as they have, just a computation of what will occur if recent trends continue. Time series analysis takes no account of real or imagined changes in behavior that may be occurring. To take such changes into account, we either have to make a subjective adjustment of our forecast or else construct a formal model that explains the effects of such changes.

For instance, an extrapolation may indicate that the unemployment rate will drop by 0.2 percent next month and by 0.5 percent next year. However, if we know

A Projection Gone Awry

International Business Machines (IBM) is the world's largest producer of computers, and in early 1984, it had 600 million shares of stock outstanding that were valued at slightly over $100 a share, giving its stock a staggering aggregate market value of $60 billion. Stockholders have liked IBM for a long time because it earns a very healthy profit, pays a good dividend, and grows relentlessly. Between 1945 and 1965, IBM profits grew by 21 percent per year. Some analysts cautioned that this phenomenal growth could not continue forever. The U.S. economy, as a whole, was only growing by about 3 percent per year as its population, equipment, and productivity slowly increased. If these respective growth rates continued, by the year 2000, IBM would be producing almost all of the nation's gross national product. At that point, something would have to give: either the U.S. economy would start growing by 21 percent per year, or else IBM would slow to a more normal 3 percent growth rate.

Let's imagine that we are security analysts in 1978, and are trying to project IBM's profits per share over the next three years. A peek back over the previous thirteen years gives the data opposite.

If we fit a time series trend model by ordinary least squares, we get

$$\ln[\hat{Y}] = -226.7 + 0.1357t, \quad R^2 = .988$$
$$\qquad\quad [30.1] \qquad [30.2]$$

[ ]: t-values

The coefficient of t is an estimate of $\ln[1 + g]$, and implies an annual growth rate $g = 0.145$ (14.5 percent per year). An extrapolation of this trend equation gives the following forecasts and associated 95 percent confidence intervals:

Year	Predicted IBM Earnings per Share	Confidence Interval
1979	$5.99	$5.20 to $6.90
1980	$6.86	$5.94 to $7.93
1981	$7.86	$6.77 to $9.13

Well, in 1979, IBM's uninterrupted growth was interrupted by a nasty combination of a weak world economy and increased competition in the computer industry. IBM's earnings per share fell to $5.16 in 1979, rebounded to $6.10 in 1980, and then fell again to $5.63 in 1981. The disappointing 1981 result was more than $2 below predicted earnings per share and well outside the computed confidence interval.

Year t	Earnings per Share Y
1966	$0.94
1967	1.16
1968	1.54
1969	1.64
1970	1.78
1971	1.88
1972	2.21
1973	2.70
1974	3.12
1975	3.34
1976	3.99
1977	4.58
1978	5.32

that the Federal Reserve Board intends to tighten up the nation's money supply, then we may predict that unemployment won't decline at all and may even rise somewhat. The exact unemployment prediction that we come up with will be a subjective shading of the extrapolation, an adjustment based on our knowledge of the economy and the Federal Reserve Board. As another example, a business extrapolation of the sales of a product introduced two years ago may indicate a 20 percent increase in sales this year, but we know that, after two years of success, other firms usually start flooding the market with imitations that take away sales. Based on past experience, then, we may predict a slower growth or even a decline in sales.

In both of these cases, the alternative to subjective adjustments is a formal model that attempts to explain behavior, rather than merely extrapolate it. For their unemployment forecasts, government economists build regression models—several equations describing how unemployment depends on the state of the economy and how the state of the economy depends on the Federal Reserve's monetary policy. Each equation can be estimated by the least-squares procedures described in Chapters 13, 14, and 15. For sales forecasts, business economists build regression models that explain how sales depend on a product's price, on the existence of competing products, and on the number of years that a product has been on the market. Using data on experiences with similar products, these equations also can be estimated by least squares.

Those who build elaborate behavioral models can use a time-series model as a benchmark standard of comparison. The time-series model is a "naive extrapolation" that the person hopes to outforecast with a behavioral model that takes into account *why* things change. Time series models do best when the projections are short term and nothing dramatic occurs—when the future is just a simple extension of the past. Forecasts for one quarter ahead normally can't go too far off, but predictions five or ten years into a foggy future can be very wrong. The farther ahead we extrapolate, the more danger there is that a major turn of events will make the future unlike the past. In such circumstances, a time series model is bound to fail. A behavioral model may fail, too, but at least it has a chance.

EXERCISES

18.21 Explain why you either agree or disagree with this conclusion: "Because sales have increased steadily over the past five years, we can be certain that the trend will continue."

18.22 Use your results from Exercise 18.7 to predict the U.S. population in the years 2000, 2020, and 2030. Give a 95 percent confidence interval for each of these extrapolations.

18.23 In the text, IBM earnings-per-share data for 1966–1978 are extrapolated to predict earnings for 1979–1981, and these predictions are then compared

with the actual values. Reestimate the trend equation, using all of the data for 1966–1981. Does the inclusion of the 1979–1981 data raise or lower the estimated growth rate? Why? Now extrapolate your equation to predict IBM's earnings per share for 1982, 1983, and 1984. Give 95 percent confidence intervals for each of these three predictions.

• **18.24** *(continuation)* Go to a library and find out what IBM's earnings per share actually were in 1982, 1983, and 1984. How accurate were your predictions? Now reestimate the equation using data for 1966–

1984, and compare the predicted and actual values for 1985, 1986, and 1987.

18.25 The Democrats and Republicans are the two major political parties in the United States. Here is the ratio of the popular vote received by the Democratic party's presidential candidate to that of the Republican candidate:

Election Year t	Democrat Republican Y
1928	0.702
1932	1.447
1936	1.664
1940	1.222
1944	1.163
1948	1.099
1952	0.805

Election Year t	Democrat Republican Y
1956	0.733
1960	1.003
1964	1.587
1968	0.984
1972	0.618
1976	1.043
1980	0.808
1984	0.690

See if there is any statistically significant trend (at the 5 percent level) in these data by estimating the model $Y = \alpha + \beta t + \epsilon$. Extrapolate your estimated equation to predict the 1988 election outcome. Is your extrapolation accurate? Why might this sort of extrapolation be an unsuccessful predictor?

18.5 INDEXES

We live in a complex society with more detailed statistical data than anyone could absorb or appreciate. Indexes are an attempt to summarize masses of data and show the general changes that take place as time passes. They are a statistical average, with the specific details of the averaging depending on the particular data being summarized. To illustrate the principles involved, we will look at two well-known indexes: the Consumer Price Index and the Dow Jones Average.

The Consumer Price Index

The consumer price index (CPI) was started during World War I to determine whether workers' wages were staying ahead of prices, and today, it is still used by both sides in wage negotiations. In addition, many union contracts include formal escalator clauses, which automatically increase wages when the CPI goes up. Social Security benefits are also linked to the CPI, as are many rental leases, alimony agreements, and other contracts. The Congressional Budget Office reported in 1981 that

> ... almost a third of federal expenditures is directly linked to the CPI or related price measures, and over half of the federal budget is affected if indirectly linked expenditures are added. A one percent increase in the CPI will automatically trigger nearly $2 billion of additional federal expenditures, at 1981 program levels.[3]

In addition to Social Security payments, federal CPI-indexed programs include military pay, food stamps, student loans, and dairy price supports. Government economic policies are also influenced by the rate of inflation, as measured by the CPI.

If inflation seems rapid, Congress may increase taxes and the Federal Reserve is almost certain to step on the monetary brakes, slowing the growth of the money supply. The last three economic recessions in the United States resulted from deliberate efforts to combat inflation by cooling off the economy.[4]

This is certainly a most important index, but how is the CPI actually computed? The CPI is an ambitious attempt to keep track of the cost of living for typical U.S. households—the cost of buying food, clothing, shelter, VCRs, and the other necessities and luxuries we consume. Every ten years or so, the Bureau of Labor Statistics (BLS) makes an intensive survey of thousands of households to learn the details of their buying habits—most recently, in 1982–1984, when they interviewed more than 140,000 households nationwide. Some cities, such as New York and Los Angeles, were included automatically, because the BLS keeps separate price indexes for the nation's largest cities. Other cities were randomly selected from size and geographic groups. The specific people surveyed in these cities were also picked by random selection methods. Half were interviewed every three months for five consecutive quarters to monitor their major purchases (cars, televisions, and video recorders) and their regular expenses (rent, utility bills, insurance premiums). The other half recorded their daily expenses in diaries for two weeks.

Based on this survey, the BLS constructs a market basket of 2,000 goods and services. Each month thereafter, 250 BLS agents call or visit 20,000 stores in these 56 cities to collect current price data on about 400 of these goods and services. The 20,000 stores, 56 cities, 400 items, and the time of visit are all samples from much larger populations.

The resultant price data are then combined into price indexes, which measure the current cost of the market basket relative to the cost in a base period,

$$P = (100) \frac{\text{current cost of market basket}}{\text{cost of market basket in base period}} \tag{18.9}$$

The logic can best be demonstrated by an example. Table 18.4 shows some hypothetical data for a market basket of three items. This market basket cost $20 in

Table 18.4 *A Price-Index Calculation*

	Quantity	*1975*		*1980*	
		Price of One	*Cost of Basket*	*Price of One*	*Cost of Basket*
Loaf of bread	10	$0.90	$9.00	$1.10	$11.00
Pound of meat	6	$1.50	$9.00	$2.50	$15.00
Gallon of milk	2	$1.00	$2.00	$2.00	$4.00
Total			$20.00		$30.00

This market basket cost $20 in 1975 and $30 in 1980, indicating a 50 percent increase.

1975 and $30 in 1980, indicating a 50 percent increase ($30/$20 = 1.5). The CPI and most other price indexes are scaled to equal 100 in the base year; this is why there is a 100 in Equation 18.9. Thus, if 1975 is the base year for the data in Table 18.4, the price index would be set at $P = 100$ in 1975 and $P = 150$ in 1980. These numbers, 100 and 150, don't mean anything by themselves. The market basket didn't cost $100 or contain 150 loaves of bread. A price index only means something in comparison to its value in another period. Thus, a comparison of $P = 150$ in 1980 with $P = 100$ in 1975 tells us that prices increased by 50 percent over this five-year period, in that this market basket cost 50 percent more in 1980 than in 1975.

Real Income

The most commonplace usage of price indexes is to see whether or not one's income is keeping ahead of prices. If prices increase by 50 percent and your income does too, then you are running fast but staying in the same place. You have 50 percent more dollars to spend but, since prices are up 50 percent, you can't buy any more now than you could before. Economists make such comparisons by calculating *real income:* nominal income divided by a price index.

Let's say that a person earned $20,000 in 1975 and $33,000 in 1980. Purchasing power could be calculated by dividing these incomes by the cost of the bundles shown in Table 18.4:

Year	Dollar Income	Cost of Bundle	Number of Bundles
1975	$20,000	$20	$20,000/$20 = 1000
1980	$33,000	$30	$33,000/$30 = 1100

In terms of purchasing power, this person's real income went up by 10%—he could purchase 10% more in 1980 than in 1975.

Economists usually express real income data relative to some base year. If 1975 is selected as the base year, we calculate prices relative to what they were in 1975 and then divide dollar income by this relative price calculation:

Year	Dollar Income	Price Index	Prices Relative to 1975	Real Income (1975$)
1975	$20,000	100	100/100 = 1.0	$20,000/1.0 = $20,000
1980	$33,000	150	150/100 = 1.5	$33,000/1.5 = $22,000

The division of dollar incomes by prices relative to 1975 gives real income in terms of 1975 dollars, indicated by the shorthand 1975$. Thus this person earned enough in 1980 to purchase what a $22,000 income bought in 1975. Since, in fact, $20,000 were earned in 1975, real income was 10 percent higher in 1980 than in 1975.

Alternatively, real incomes can be expressed in terms of 1980 dollars by using prices relative to 1980:

Year	Dollar Income	Price Index	Prices Relative to 1980	Real Income (1980$)
1975	$20,000	100	100/150 = 0.67	$20,000/0.67 = $30,000
1980	$33,000	150	150/150 = 1.0	$33,000/1.0 = $33,000

This person earned $20,000 in 1975, enough to buy what a $30,000 income in 1980 would purchase. The actual 1980 income of $33,000 was 10 percent higher than this.

Thus there are a variety of ways of telling the same story: this person's real income, taking into account the diminished purchasing power of the dollar, increased by 10 percent between 1975 and 1980.

Price-Index Weights

As is always the case, individual prices change by varying amounts. Some prices go up a lot, some go up a little, and some even go down. A price index is, implicitly, a weighted average of individual prices, where the weights are determined by consumer buying habits. Here are the weights used in 1986 for the seven major expenditure categories, along with the level of each index in July 1986:

Expenses	Weights	Level (1967 = 100)
Food & beverages	21%	312.2
Housing	35%	361.5
Apparel and upkeep	8%	203.2
Transportation	22%	304.7
Medical care	5%	434.6
Entertainment	5%	274.4
Other	5%	344.9
Overall	100%	328.0

The overall value of the CPI is a weighted average of the price indexes for these seven categories,

$$312.2(.21) + 361.5(.35) + \cdots + 344.9(.05) = 328.0$$

The cost of medical care increased the fastest over this 20-year period and the price of entertainment the least (admit it, you still think $7 is a lot to pay to see a mediocre movie).

Within each expenditure category are dozens of individual items and the BLS keeps price indexes for these too. Here are some more detailed data on the July

1986 prices for several food groups (all relative to 100 in 1967):

Roasted coffee	513.0
Apples	420.5
Food away from home	360.8
Cold drinks	317.3
Overall food and beverages	312.2
White bread	272.9
Ground beef	242.2
Fresh whole milk	226.7
Eggs	175.2

Thus we can see at a glance that the first four groups increased more than average, and the last four less so.

The BLS also maintains data for several cities and metropolitan areas. Here are some for July 1986 (all equal to 100 in 1967):

San Diego	383.1
Denver	358.4
Milwaukee	331.3
Los Angeles	330.9
Overall average	328.0
New York	325.1
Philadelphia	323.0
Detroit	318.4
Chicago	311.1

Remember, these numbers do not necessarily mean that San Diego is a more expensive place to live than New York City, only that prices of a typical market basket increased somewhat faster there between 1967 and 1986.

There are two inherent difficulties with all price indexes. The first is that purchases adjust to price changes. For example, gasoline consumption fell after its price soared. Should a price index's typical market basket reflect the quantities purchased before or after the price change? If before, then the price index overstates the change in the cost of living, because it ignores the fact that people will substitute less expensive items for more expensive ones. If after, then the price index understates the ill effects of the higher prices, because it ignores the lifestyle that people were forced to give up.

There is no good answer to this before-or-after question. Some indexes, called **Laspeyres indexes,** use the "before" quantities purchased in the base period. Others, called **Paasche indexes,** use the "after" quantities purchased each period.*

---

*Because the Paasche weights vary from year to year, changes in the index reflect quantity as well as price changes. A Paasche index can go up or down even if all prices are constant! An intermediate approach is a *chain index,* in which each period's computations use the market basket from the immediately preceding period. In an annual chain index, for example, a calculation of the average change in prices in 1980 would use 1979 spending habits, while the 1981 calculation would use 1980 spending habits.

It Might Be Cheaper to Switch to Another Song

If you received the items listed in the song "The 12 Days of Christmas" today, be impressed—they cost a bundle.

Each year at this time Hugh Gee, publisher of *Money Power Confidential,* a San Francisco-based investor newsletter, issues his Christmas Index—the cost of the items in the song—and compares them with the year before. His findings:

A partridge costs $6 this year, compared with $5 in 1982, a 20% gain. Pear trees go for $15 vs. $10, up 50%. Two turtle doves sell for $20, unchanged from a year ago. Three French hens (a fancy name for chickens) can be purchased for $4.25 wholesale in Lodi, compared with $3.69 last year, up 15%.

Four calling birds (Gee picked rare grahlas) cost $2,000, unchanged from 1982. Five golden rings are a bargain this year at $1,875 vs. $2,125, a 14% drop because the price of gold has eased. Six geese-a-laying (Canadian honkers) sell for $360 vs. $300, up 20%.

Seven swans-a-swimming cost $2,100, compared with $1,750, up 20% for the birds. And a 15-foot by 30-foot pool for them to do their swimming sells for $22,000 vs. $19,000, a 16% increase. For 8 maids-a-milking, Gee checked with several temporary help firms and found a place he could get them for $369.60, compared with $320, up 16%. Uni-

forms for the maids rent for $320 vs. $280, up 14%. Gee didn't list what the maids were supposed to milk, so you might assume they took turns milking the chickens.

Nine dancers dancing (at union rates) cost $1,485, compared with $1,323, up 12%. And because nude dancing is gauche at Christmas, it cost an extra $216 for costumes vs. $166.50, a 29% increase. Ten lords-a-leaping (real lords weren't available so Gee had to substitute union dancers willing to leap) cost $1,650, compared with $1,470, up 12%. Their lord suits rented for $400 vs. $350, a 14% gain.

Eleven pipers piping (at union rates) cost $1,045, compared with $891.88, a 17% increase. The fee for renting 11 flutes totaled $495 vs. $275, up 80% (apparently 1983 was a good year to be in the flute-rental business). And 12 drummers drumming cost $1,140 (no Shelley Manne at these rates), compared with $972.96, up 17%. Drum rentals added another $960 vs. $624, a 54% hike.

The total cost this year came to $36,460.85, up from last year's $31,887.03. In 1981 the total was a paltry $29,184.23. It seems that Christmas goods have bigger price mark-ups than everyday items because although the Christmas Index surged 14%, the government's Consumer Price Index rose a mere 3.5%.

© Ron W. Heinzel, *Los Angeles Times,* December 25, 1983. Reprinted by permission.

Thus, the Laspeyres weights are fixed by base period spending habits, while the Paasche weights vary from year to year with spending habits. The CPI is a Laspeyres index, while the *consumption deflator* (another government index of consumer prices) is a Paasche index.

This distinction can make a big difference in the behavior of price indexes, particularly when relative prices have changed greatly and the base-period survey of spending habits was made several years ago. The major CPI surveys were made in 1934–1939, 1952, 1963, 1972–1973, and 1982–1984. Many economists believe that the validity of the CPI in the late 1970s was undermined by significant changes in consumer buying patterns, such as the decrease in gasoline usage.

The Commerce Department calculates a personal consumption deflator, which differs from the CPI in that it is a Paasche index (with weights that change with spending habits). From the fourth quarter of 1969 to the fourth quarter of 1980, the CPI increased by about 130 percent, while the consumption deflator went up by 100 percent. The average annual percentage increases over these eleven years works out to be 7.8 percent for the CPI and 6.7 percent for the consumption deflator. There was inflation, no matter how we look at it, but how we look at it does affect our estimate of the severity of the inflation.

Quality Adjustments

The second inherent difficulty with price indexes is that the quality of many items varies as time passes. When the price goes up 10 percent, but the box says "new and improved," should we believe it? If the quality has in fact improved 10 percent, then there hasn't really been any inflation. It's like buying a 10 percent larger size; you simply paid more to get more. But how do you measure quality? Volkswagen Rabbits are very different from Beetles; they are roomier, they get better gas mileage, and they have front-wheel drive, water-cooled engines, fuel injection, and disc brakes. What is the percentage change in their quality? Many consumer goods, like personal computers, didn't even exist a short while ago. Another example is that of medical doctors; they have drugs and equipment that didn't exist 20 years ago, their fees are also much higher, and they don't make house calls. What is the percentage change in the quality of medical services?

The other side of this coin is "candy bar inflation." For years, candy bar prices were constant, while the bars gradually became smaller and smaller. Many other products and services have similarly deteriorated. Their quality has gone down rather than up. Statisticians try to make quality adjustments in the price indexes, but this task is difficult and subjective. In the 1960s, many economists argued that quality improvements were underestimated to the extent that the reported annual increases in the CPI were about 1 to 2 percent too high.

In the 1970s, another quality issue emerged. The state and federal governments pushed for breathable air, drinkable water, and safe working environments. The costs of moving toward these objectives showed up in higher prices for automobiles, electricity, and other products, but the benefits weren't counted in any market basket. We paid more, but the price indexes didn't show that we also were getting more—better air, water, and so on. It is difficult to say how we could reasonably adjust price indexes for these sorts of changes in the quality of our living and working environments. All we can really say is that the neglect of environmental quality changes caused reported price indexes to understate inflation in the 1960s and to overstate inflation in the 1970s.

The Dow Jones Average

In 1880, Charles Dow and Edward Jones started a financial news service that they called Dow-Jones. Today, the most visible results are the *Wall Street Journal*, the most widely read newspaper in the United States, and the Dow Jones Industrial Average, the most widely reported stock market index.

The Dow Jones averages began as a barometer of economic activity. Dow was convinced that stock prices reflected informed opinion about future business conditions, and thus, that stock prices provided advance information about the economy. Two indexes were constructed: the Industrial Average would reflect production, while the Railroad Average would monitor commerce. The Railroad Average has always been the less prominent of the two indexes, an imbalance exacerbated by the decline of railroads in the United States. In 1969, 8 of the 20 railroad stocks were replaced by 6 airlines and 2 trucking firms, and it was renamed the Transportation Average; however, the Industrial Average still gets the headlines.

The Industrial Average began with 12 stocks in 1896 and grew to its present size, 30 stocks, by 1928. These select stocks represent the nation's industrial giants, the bluest of blue chips. It is a measure of economic evolution in the United States that only one company, General Electric, has been in the Dow Jones Industrial Average from the beginning (and even GE was dropped for a while). The other 29 stocks have been added over time as other, once powerful companies faded. The Great Depression was, of course, the blow that killed many of the mightiest. Twenty-five of the 30 blue chips in the Dow Jones Industrial Average were dropped in the 1930s and replaced with healthier companies. Another substitution was made in 1956, 4 more occurred in 1959, and 3-M was added in 1976 after the Anaconda copper company was taken over by Atlantic Richfield. In 1979, an ailing Chrysler and an unglamorous Esmark (formerly Swift & Company, a meatpacking firm) were replaced by Merck, a leading producer of health care products, and by IBM, the company with the largest market value of any company ($43 billion at the time and even more now). In 1982, American Express replaced troubled Manville, and in 1985, Philip Morris (manufacturer of such products as Marlboro, Benson & Hedges, Miller beer and 7-Up) acquired General Foods and took its place in the Dow, while McDonald's was substituted for American Brands (manufacturer of Pall Mall, Tareyton, and Sunshine Biscuits) to avoid overweighting the Dow with tobacco and packaged foods. In 1987, Boeing and Coca Cola replaced Owens-Illinois (which disappeared in a leveraged buyout) and Inco (a producer of nickel and other metals). The current list of the "biggest and best" is shown in Table 18.5.

Originally, the Dow Jones Average was computed simply by adding together the prices of the 12 stocks and dividing by 12. This computation had to be modified as other companies were added or substitutions were made; otherwise, the Dow Jones Average would have jumped or plummeted for no real reason. Suppose, for instance, that all 12 of the original stocks were selling for $50 a share, giving a Dow Jones Average of 50, and another stock is added that happens to be selling for $10

Table 18.5 *The Dow Jones Industrial Average, 1989*

Stock	Price per Share (As of May 1, 1989)	Number of Shares (Millions)	Total Market Value (Billions of Dollars)
Allied Signal	33 1/2	148.7	5.0
Alcoa	62 3/4	88.3	5.5
American Express	32 1/2	417.1	13.6
AT&T	34 1/2	1073.7	37.0
Bethlehem Steel	22 1/2	74.5	1.7
Boeing	77 1/8	153.2	11.8
Chevron	53 5/8	342.1	18.3
Coca Cola	54 3/4	359.0	19.7
DuPont	109 1/4	239.4	26.2
Eastman Kodak	47 1/8	324.2	15.3
Exxon	42 7/8	1289.0	55.3
General Electric	48 3/4	900.5	43.9
General Motors	41 1/4	612.9	25.3
Goodyear	52 1/4	57.3	3.0
IBM	113 3/8	591.4	67.0
International Paper	49 1/8	111.1	5.5
McDonald's	55 3/8	188.0	10.4
Merck	67 3/8	396.4	26.7
MMM	71 1/2	226.8	16.2
Navistar	5 3/8	252.8	1.4
Philip Morris	126 3/8	231.0	29.2
Primerica	21 1/8	96.0	2.0
Procter & Gamble	97 5/8	169.4	16.5
Sears Roebuck	45 3/4	380.0	17.4
Texaco	54 1/4	244.3	13.3
Union Carbide	31 5/8	135.6	4.3
United Technologies	51 3/8	130.7	6.7
USX	34 1/2	260.8	9.0
Westinghouse	57 1/2	144.0	8.3
Woolworth	52 1/4	63.9	3.4

a share. If we add up the 13 prices and divide by 13, we get a Dow Jones Average of

$$\frac{50 + 50 + \cdots + 50 + 10}{13} = \frac{610}{13} = 46.92$$

indicating that the average stock price dropped by more than 7 percent, when in fact, all that happened was the index was broadened from 12 stocks to 13. The same sort of false signal would be given if one stock was substituted for another that happened to be selling for a different price.

The same thing would happen whenever a stock splits. In a two-for-one split, each original share is replaced with two new shares, so that every owner of 100 old shares now has 200 new shares. For the company as a whole, the number of shares outstanding is doubled; this is a simple bookkeeping change that promptly halves the value of each share.*

The Dow Jones Averages allow for these cosmetic changes by adjusting the divisor. In our first example, if we don't want the inclusion of a thirteenth stock to shock the average, we can divide by some number k that keeps the average at 50:

$$\frac{50 + 50 + \cdots + 50 + 10}{k} = \frac{610}{k} = 50$$

implies $k = 610/50 = 12.2$, rather than 13. By the same logic, the divisor is adjusted when a substitution is made, and every time a stock splits. The cumulative effect of these adjustments has been to reduce the Dow Jones Industrial divisor to 0.682 in May 1989. (The current divisor is published in the *Wall Street Journal* every day.)

> The Dow Jones Industrial Average is computed by adding up the per-share prices of 30 prominent companies and dividing by a divisor that changes from time to time in order to maintain a logically consistent daily index of stock prices running all the way back to 1896.

There are two primary criticisms of the Dow Jones Industrial Average. One is that it only includes 30 stocks. Admittedly, these 30 are among the biggest and the best, but still, every other stock is ignored completely. At present, these 30 companies are so prominent that they represent one-fourth of the total market value of the nearly 2000 stocks traded on the New York Stock Exchange. It is impressive that 2 percent of the companies have 25 percent of the market value, but it may be misleading to ignore the other 75 percent—particularly because the excluded companies are quite different from those that are included.

The 30 stocks in the Dow Jones Industrial Average are not a random sample. The Dow seems top heavy with "smokestack America," industrial giants from the past, whereas America's growth now seems to be with high-technology and service companies. In addition, stocks are chosen because they are large, well-established companies, the bluest of the blue chips, and as such, their performance may not be

---

*A shareholder with 100 out of 1 million old shares owned .01 percent of the company and received .01 percent of all the dividends distributed to shareholders. After a two-for-one split, this shareholder will have 200 out of 2 million new shares, still .01 percent of the company and its dividends. The market value of these 200 new shares should be the same as the value of the 100 old shares, implying a halving of the price per share. If the old shares were worth $4,000 (100 shares at $40 per share), then the new shares should also be worth $4,000 (200 shares at $20 per share).

representative of other, smaller and less prominent companies. The excluded companies may grow faster and are certainly riskier than the giants, and there may well be systematic differences in the behavior of their prices. Indeed, some security analysts compare movements in the Dow Jones Industrial Average with broader averages, such as the New York Stock Exchange (NYSE) Composite Index, to gauge the mood of the market. If the broader averages are soaring while the Dow is slumbering, this is interpreted as a signal that the speculators are loose again, gambling on risky stocks.

In addition, in the long run, it can make all the difference in the world which particular 30 are anointed the biggest and the best. IBM was originally added to the big 30 in 1932, replacing National Cash Register, but then was dropped mysteriously in 1939 in favor of AT&T. If IBM hadn't been left out for forty years, from 1939 to 1979, the Dow would have grown much faster and the stock market would have seemed much healthier. By 1979, the Dow would have been twice as high, at 1,700 rather than 850.

The second criticism is that the Dow Jones Averages reflect per-share prices instead of market values. We've seen how per-share prices are dramatically affected by stock splits. The more general lesson is that the per-share price depends on how many shares the company happens to have issued. If one of two equally profitable companies has twice as many shares as the other, it will also have a per-share price that's half the size of the other. This is unimportant for the company and its shareholders, but it makes a big difference to the Dow Averages, in that a company's influence on the Dow Average depends on how many shares it happens to have issued!

Look at Table 18.5 again. In May 1988 Merck was the highest-priced stock among the Dow industrials, selling for $150 a share. But on May 26, 1988, Merck split three-for-one and its price dropped to $50 a share, reducing its importance to the Dow. At that time, the Dow divisor was about 0.754, so that, before the split, a 10 percent drop in the price of Merck (from 150 to 135) would have reduced the numerator of the Dow average by 15, causing the Dow to drop by $15/0.754 = 20$ points. After the split, Merck lost its premiere position; a 10 percent drop in the price of Merck would have reduced the Dow by only 7 points.

In May 1989 Philip Morris at 126 3/4 had the highest price and Navistar at 5 3/8 the lowest, so that a 4 percent drop in the price of Philip Morris would have reduced the Dow by 8 points, the same as if the price of Navistar had fallen by 100 percent.

The Dow Jones Industrial Average treats a $1 change in any of the 30 stock prices the same, regardless of whether it is a $1 change in a stock selling for $6, $40, or $120 a share.

Nor does the Dow average pay attention to how many shareholders are affected by price changes. In May 1989 General Electric (GE) and Goodyear were both selling for about $50 a share, so a $5 increase in either price would mean a $5/0.682 = 7.3$-point increase in the Dow and almost a 10 percent gain for shareholders.

But GE had sixteen times as many shares outstanding as Goodyear. A $5 per share increase means $286 million in capital gains to Goodyear shareholders, but $4.5 billion in capital gains to GE shareholders. The Dow treats the two events as identical while they are, in fact, quite different—both as a barometer of the economy and in their implications for consumer wealth.

It is for just these reasons that most analysts prefer an index of stock prices based on market values

$$\text{market value index} = \sum \begin{pmatrix} \text{number} \\ \text{of} \\ \text{shares} \end{pmatrix} \begin{pmatrix} \text{price} \\ \text{per} \\ \text{share} \end{pmatrix}$$

For an example, consider an index based on just two stocks—GE and Goodyear:

$$(48\ 3/4)(900.5) + (52\ 1/4)(57.3) = 46,893$$

(The index would, of course, be scaled to equal 100 in a chosen base period.) In a market value index, a 10 percent change in the value of the index represents a 10 percent change in the aggregate market value of the companies included in the index. The Standard and Poor's 500, the NYSE Composite, and most other stock indexes are based on market value, but the Dow has its long tradition and an entrenched spot in the newspaper headlines. When the Dow is up 60 points, people can interpret that statistic readily as meaning very good news about the stock market, simply because they are used to hearing the Dow reported and remember that the Dow seldom goes up as much as 60 points in a single day. If they were to read that the NYSE Composite was up 6 points, they wouldn't know what to make of that statistic, even though it is describing the same events and probably describing them more accurately. As one securities analyst put it,

> Statistically [the Dow] isn't very good, but I follow it avidly. It's a kind of love-hate relationship. When you get down to it, it's the most convenient shorthand for the market.[5]

EXERCISES

18.26 Mr. Bunker lives on beer and pretzels. In 1970, he bought 1,000 six-packs of beer at $1.00 each and 500 bags of pretzels at $0.50 each. In 1980, Bunker's beer cost $1.50 and his pretzels cost $1.00. **a.** One way to calculate Bunker's rate of inflation is

Bunker's inflation
from 1970 to 1980

$$= \frac{\text{cost of 1,000 6-packs and 500 bags in 1980}}{\text{cost of 1,000 6-packs and 500 bags in 1970}}$$

What was the percentage increase in Bunker's cost of living from 1970 to 1980? If we calculated a Bunker

Price Index (BPI) and scaled this BPI to equal 100 in 1970, what would be the 1980 value of the BPI? **b.** Another way to calculate Bunker's rate of inflation is

Bunker's inflation
from 1970 to 1980

$$= \begin{pmatrix} \% \text{ increase in} \\ \text{price of beer} \end{pmatrix} \begin{pmatrix} \text{fraction of 1970 budget} \\ \text{spent on beer} \end{pmatrix}$$

$$+ \begin{pmatrix} \% \text{ increase in} \\ \text{pretzel prices} \end{pmatrix} \begin{pmatrix} \text{fraction of 1970 budget} \\ \text{spent on pretzels} \end{pmatrix}$$

What was Bunker's inflation rate using this approach? **c.** Bunker earned $2 an hour in 1970 and $3 an hour in 1980. Did his wages win or lose the race against prices?

18.27 Here are data on the consumer price index and the median family income in the United States:

Year	Median Income	CPI (1967 = 100)
1950	$ 3,319	72.1
1960	5,620	88.7
1970	9,867	116.3
1980	21,028	246.8

Calculate the real median income for each of these four years (in 1967 dollars) and the percentage increases from 1950 to 1960, 1960 to 1970, and 1970 to 1980.

18.28 The consumer price index increased by 112.4 percent between 1970 and 1980. Here are the average retail prices of some items in the CPI:

CPI Items	1970	1980
1 dozen large eggs	$ .61	$1.00
1 lb sirloin steak	1.35	2.93

CPI Items	1970	1980
1 lb can of coffee	.91	2.82
1/2 gallon ice cream	.85	1.92

Calculate the percentage increase in each of these four items. Joanna's market basket contains 1 dozen large eggs, 3 pounds of sirloin steak, 1 pound of coffee, and 2 gallons of ice cream. What did her basket cost in 1970? In 1980? What would be the 1980 value of a Joanna Price Index set equal to 100 in 1970?

18.29 In July 1979, IBM (at $70 a share) and Merck (at $68) replaced Chrysler and Esmark (at $9 and $26, respectively) in the Dow Jones Industrial Average. To keep the Dow Jones index at 850, did the divisor of 1.443 have to be adjusted up or down? What was the new value of the divisor?

18.30 Use the data in Table 18.5 to analyze the effects of a $3 decrease in the price of AT&T and a $3 decrease in the price of Allied Signal, roughly 9 percent in each case. **a.** Using a Dow divisor of 0.682, how much does each of these price decreases affect the Dow Industrial Average? **b.** How much does each of these price decreases affect shareholder wealth?

18.6   SUMMARY

Time series analysis and index numbers are statistical efforts to describe behavior as time passes. Time series analysis focuses on the identification of trend and seasonal patterns, while indexes are concerned with aggregating excessive detail. The two most popular trend models are linear and log-linear

$$Y_t = \alpha + \beta t + \epsilon$$

$$\ln[Y_t] = \alpha + \beta t + \epsilon$$

Seasonality can be taken into account by adding seasonal dummies. The parameter estimates provide a statistical description of the behavior of Y, and also predictions of how Y will behave if past trend and seasonal patterns continue. In practice, time series models are used for two primary purposes. First, they can be used to detrend and deseasonalize data, so that we can identify events of interest. Periods when GNP is below trend indicate an economic recession; a January when *Toys "R" Us* sales didn't collapse would be phenomenal. The second usage is to provide a benchmark forecast that can be adjusted for other information that, it is hoped, will improve the forecasts.

Time series analysis is descriptive. It identifies trends and seasonal patterns in the data and tells what will happen if these continue. It does not explain why these

patterns occur, why they should continue, or, most interesting, why there are fluctuations about these patterns.

Indexes are attempts to summarize masses of data and describe "average"
behavior as time passes. The CPI, for instance, shows changes in the cost of buying
a market basket purchased by typical households. Although the Dow Jones Industrial Average is an important exception, most stock indexes show changes in the
market value of stocks.

REVIEW EXERCISES

18.31 United States per capita gross national
product (GNP) was $205 in 1885 and $16,704 in
1985. Prices in 1985 were 14.17 times 1885 prices
(i.e., on average, something costing $1 in 1885 cost
$14.17 in 1985). Calculate real 1885 GNP, in terms of
1985 prices, and determine the annual rate of growth
of real per capita GNP over this period.

18.32 A business has estimated the following
equation to identify seasonal fluctuations in its sales:

$$\hat{Y} = 450 + 50D_2 + 50D_3 + 150D_4$$

where Y = quarterly sales
         D_2 = 1 in second quarter, 0 otherwise
         D_3 = 1 in third quarter, 0 otherwise
         D_4 = 1 in fourth quarter, 0 otherwise

How should we interpret these estimates? Do sales
tend to be higher in the first quarter or second quarter? In the second or third quarter? In the third or
fourth quarter? Make a rough sketch to show the
usual seasonal pattern.

18.33 Gordon Jewelry is a leading retail jeweler.
Its earnings per share during the fiscal years 1978–
1982 are shown in Table 18.6. (Often, the *fiscal* year
that a particular business uses for accounting purposes
does not begin on January 1.) Plot these data versus
time and see if any seasonal trend is apparent. Now
find the least squares estimates of the equation

$$Y = \alpha_1 + \alpha_2 D_2 + \alpha_3 D_3 + \alpha_4 D_4 + \beta t + \epsilon$$

where Y = quarterly earnings per share
         D_2 = 1 in Dec–Feb quarter, 0 otherwise
         D_3 = 1 in Mar–May quarter, 0 otherwise
         D_4 = 1 in June–Aug quarter, 0 otherwise

Interpret your estimates. Using a 5 percent significance level, are there any statistically significant sea-

sonal or trend patterns? If so, carefully explain them
in language that a nonstatistician can understand.

18.34 Table 18.7 gives some data on quarterly
earnings per share for Carter Hawley, which operates
a variety of department stores and specialty shops,
including Broadway, Neiman-Marcus, and Walden-
books. Plot these data versus time and see if any seasonal trend is apparent. Now find the least squares
estimates of the equation

$$Y = \alpha_1 + \alpha_2 D_2 + \alpha_3 D_3 + \alpha_4 D_4 + \beta t + \epsilon$$

where Y = quarterly earnings per share
         D_2 = 1 in May–July quarter, 0 otherwise
         D_3 = 1 in Aug–Oct quarter, 0 otherwise
         D_4 = 1 in Nov–Jan quarter, 0 otherwise

Summarize the trend and seasonal patterns in these
earnings data. Extrapolate your estimated equation to
predict Carter Hawley earnings per share for the fiscal
years 1983–1985.

18.35 *(continuation)* Table 18.8 gives some data
on quarterly earnings per share for the R.H. Macy
company, which operates Macy's department stores.
Plot these data versus time and see if any seasonal
trend is apparent. Now find the least squares estimates
of the equation

$$Y = \alpha_1 + \alpha_2 D_2 + \alpha_3 D_3 + \alpha_4 D_4 + \beta t + \epsilon$$

where Y = quarterly earnings per share
         D_2 = 1 in May–July quarter, 0 otherwise
         D_3 = 1 in Aug–Oct quarter, 0 otherwise
         D_4 = 1 in Nov–Jan quarter, 0 otherwise

Summarize the trend and seasonal patterns in these
earnings data. Write a brief paragraph comparing the
seasonal and trend patterns in the earnings of Carter
Hawley and R.H. Macy over this period.

Table 18.6 *Data for Exercise 18.33*

			Fiscal Year Ends		
Quarter	1978	1979	1980	1981	1982
Sept 1–Nov 30	0.32	0.34	0.45	0.57	0.21
Dec 1–Feb 28	1.29	1.77	2.15	2.19	1.97
Mar 1–May 31	0.37	0.56	0.55	0.54	0.25
June 1–Aug 31	0.32	0.60	−0.40	−0.15	0.32

Table 18.7 *Data for Exercise 18.34*

			Fiscal Year Beginning		
Quarter	1978	1979	1980	1981	1982
Feb 1–Apr 30	0.19	0.21	0.21	0.21	0.17
May 1–July 31	0.29	0.31	0.29	0.23	0.08
Aug 1–Oct 31	0.39	0.45	0.37	0.24	0.20
Nov 1–Jan 31	1.65	1.70	1.24	0.87	1.20

Table 18.8 *Data for Exercise 18.35*

			Fiscal Year Ends		
Quarter	1978	1979	1980	1981	1982
Aug 1–Oct 31	0.41	0.42	0.72	0.72	0.75
Nov 1–Jan 31	1.21	1.29	1.74	1.84	1.89
Feb 1–Apr 30	0.12	0.17	0.23	0.42	0.48
May 1–July 31	0.31	0.55	0.55	0.70	0.78

Table 18.9 *Data for Exercise 18.36*

			Fiscal Year Beginning		
Quarter	1979	1980	1981	1982	1983
Apr 1–June 30	1.12	1.14	1.16	1.38	1.35
July 1–Sept 30	1.24	0.78	1.40	1.50	1.65
Oct 1–Dec 31	0.98	0.66	0.80	0.95	1.10
Jan 1–Mar 31	0.31	0.26	0.39	0.42	0.55

18.36 Table 18.9 gives some data on quarterly earnings per share for Molson, one of Canada's two largest brewers. Plot these data versus time and see if any seasonal trend is apparent. Now find the least squares estimates of the equation

$$Y = \alpha_1 + \alpha_2 D_2 + \alpha_3 D_3 + \alpha_4 D_4 + \epsilon$$

where Y = quarterly earnings per share
D_2 = 1 in July–Sept quarter, 0 otherwise
D_3 = 1 in Oct–Dec quarter, 0 otherwise
D_4 = 1 in Jan–Mar quarter, 0 otherwise

Use these estimates to calculate seasonally adjusted values for Molson's quarterly earnings per share over this period, 1979–1983.

18.37 *(continuation)* What do you think will happen if you reestimate the model

$$Y = \alpha_1 + \alpha_2 D_2 + \alpha_3 D_3 + \alpha_4 D_4 + \epsilon$$

this time using your seasonally adjusted values of Y? Check your logic by doing just that.

18.38 Table 18.10 gives some data on quarterly earnings per share for General Cinema, the largest U.S. theatre chain. Plot these data versus time and see if any trend or seasonal pattern is apparent. Now find the least squares estimates of the equation

$$Y = \alpha_1 + \alpha_2 D_2 + \alpha_3 D_3 + \alpha_4 D_4 + \beta t + \epsilon$$

where Y = quarterly earnings per share
D_2 = 1 in Feb–Apr quarter, 0 otherwise
D_3 = 1 in May–July quarter, 0 otherwise
D_4 = 1 in Aug–Oct quarter, 0 otherwise

Use these estimates to calculate seasonally adjusted values for General Cinema's quarterly earnings per share over this period, 1979–1983. Now extrapolate your estimated equation to predict General Cinema's (seasonally unadjusted) earnings per share for the fiscal years 1984–1985.

18.39 Here are some monthly data on the residential use of electric power (in millions of kilowatt-hours):

Month	1975	1976	1977	1978
Jan	53,299	59,088	64,516	64,624
Feb	50,716	54,530	61,705	64,283
Mar	48,595	48,656	52,686	59,283
Apr	46,036	45,365	47,118	49,722
May	42,424	42,786	44,086	46,764
June	45,741	45,262	49,481	51,533
July	52,275	53,312	59,748	60,266
Aug	55,310	57,556	61,541	62,366
Sept	53,057	53,746	57,687	60,883
Oct	44,430	47,296	50,599	52,656
Nov	43,824	48,582	47,568	49,440
Dec	50,442	56,893	55,611	57,458

Use these data to estimate the trend and seasonal patterns in residential use of electricity,

$$Y = \alpha_1 + \alpha_2 D_2 + \alpha_3 D_3 + \cdots + \alpha_{12} D_{12} + \beta t + \epsilon$$

where the D_i are eleven seasonal dummies (with January as the reference month). In which, if any, months are there statistically significant differences (at the 5 percent level) in electric power usage, relative to January? Do these seasonal patterns make sense to you?

•• 18.40 In the text, estimates are reported for the following time series model of quarterly GNP:

$$\ln[Y] = \alpha_1 + \alpha_2 D_2 + \alpha_3 D_3 + \alpha_4 D_4 + \beta t + \epsilon$$

where Y = quarterly GNP
D_2 = 1 in second quarter, 0 otherwise
D_3 = 1 in third quarter, 0 otherwise
D_4 = 1 in fourth quarter, 0 otherwise
t = 0 in first quarter of 1976, then increases by 1 each quarter

Table 18.10 *Data for Exercise 18.38*

	Fiscal Year Ends				
Quarter	1979	1980	1981	1982	1983
Nov 1–Jan 31	0.45	0.50	0.60	0.78	1.00
Feb 1–Apr 30	0.57	0.56	0.69	1.19	1.05
May 1–July 31	0.95	0.98	1.45	1.70	2.10
Aug 1–Oct 31	0.27	0.68	1.25	1.48	1.75

An alternative formulation is

$$\ln[Y] = \beta_4 + \beta_1 D_1 + \beta_2 D_2 + \beta_3 D_3 + \beta t + \epsilon$$

What is the relationship between the parameters in these two formulations? Check your logic by using the data in Table 18.2 to estimate both equations.

•• **18.41** *(continuation)* Yet another formulation is

$$\ln[Y] = \gamma_1 D_1 + \gamma_2 D_2 + \gamma_3 D_3 + \gamma_4 D_4 + \epsilon$$

What is the relationship between the parameters in this equation and the two equations in the preceding exercise? Check your logic by using the data in Table 18.2 to estimate the third equation and to compare your results with the first two equations.

18.42 Use the GNP data for 1951–1983 shown in Table 18.1 to estimate the trend equation $\ln[Y] = \alpha + \beta t + \epsilon$. Now extrapolate this equation to predict GNP for the years 1984, 1985, and 1986. Give 95 percent confidence intervals for each of these predictions. Go to a library and find out what the actual GNP (in 1972 $) was in 1984. Explain why your prediction was too low, too high, or just about right.

18.43 The annual production of electrical power (in billions of kilowatt-hours) by U.S. electric utilities is as follows:

Year	Production	Year	Production
1948	283	1961	792
1949	291	1962	852
1950	329	1963	914
1951	371	1964	984
1952	399	1965	1,055
1953	443	1966	1,144
1954	472	1967	1,214
1955	547	1968	1,329
1956	601	1969	1,442
1957	632	1970	1,492
1958	645	1971	1,612
1959	710	1972	1,750
1960	753	1973	1,861

Use these data to estimate the trend equation $\ln[Y] = \alpha + \beta t + \epsilon$, where Y is electricity production and t is time.

Now extrapolate this estimated equation to predict production in 1974. Actual production in 1974 turned out to be 1,867 billion kilowatt-hours. What is the size of your prediction error? Why is there such a prediction error?

18.44 It has been alleged that statisticians do not fully take into account quality improvements in what we buy. If so, will their estimates of inflation and real income be too high or too low?

18.45 Critically evaluate the following economic commentary:

> When it comes to measuring inflation, the average consumer can do a far better job than the economics experts.... Over the years I have been using a system which is infallible.... The Phindex [short for the Phillips index] merely requires you to divide the total dollar cost of a biweekly shopping trip by the number of brown paper bags into which the purchases are crammed. You thus arrive at the average cost per bagful.
>
> When I started this system some 10 years ago, we would walk out of the store with about six bags of groceries costing approximately $30—or an average of $5 per bag. ...
>
> On our most recent shopping trip, we emerged with nine bagsful of stuff and nonsense, totaling the staggering sum of $114.... the Phindex shows a rise from the initial $5 to almost $13, a whopping 153 percent.
>
> Why is the government worrying about double-digit inflation when the Phindex clearly shows that over a relatively short time we have already escalated into triple digits?[6]

18.46 Explain the error in this interpretation of inflation data:

> In the 12-month period ending in December of 1980, consumer prices rose by 12.4 percent—after a 13.3 percent increase the year before. Similar measures of inflation over the next three years were 8.9 percent, 3.9 percent, and 3.8 percent.... We are certainly paying less for what we buy than we were at the end of the Carter years.[7]

18.47 Under what circumstances would you expect to find a significant difference between Laspeyres and Paasche price indexes? Which index would increase more under these circumstances?

18.48 The Standard & Poor's 500 (S&P 500) is an index of the prices of 500 stocks. Could the Dow Jones Industrial Average and the S&P 500 ever move in opposite directions? If so, how would you tell the story in words that is being told by these indexes in numbers?

18.49 Here are some Sotheby Index data, based on auctions affiliated with Sotheby Parke Bernet:[8]

	1975	*1982*	*1983*
Old master paintings	100	199	217
Chinese ceramics	100	460	445
Continental silver	100	134	156

a. Do Chinese ceramics cost more than continental silver? **b.** Which of these three would have been the best investment between 1975 and 1982? Between 1982 and 1983?

• **18.50** Monthly CPI data can be found in *Business Conditions Digest.* (Check their end pages to find the issue that has historical data.) Use these data for 1971–1980 to see whether consumer prices display trend and/or seasonality by estimating the equation

$$\ln[Y] = \alpha + \beta_1 t + \sum_{i=2}^{12} \beta_i D_i + \epsilon$$

where the D_i are 11 seasonal dummies for the months February through December. Write a one-paragraph report summarizing your findings.

18.51 What's wrong with the following table of consumer prices, which uses 1967 as a base year?[9]

Year	*% Change in Prices from Previous Year*	*% Change from 1967*	*Price Index*
1970	6%	116%	116
1971	4%	121%	121
1972	3%	125%	125
1973	6%	133%	133

18.52 To identify patterns in daily school absences, a sample of five schools were studied over a two-year period and the following model estimated:

$$\hat{Y} = 234 - 41.8\, D_1 - 62.7\, D_2 - 71.4\, D_3 - 53.2\, D_4$$
$$\quad\quad [3.7] \quad\quad [4.2] \quad\quad [5.1] \quad\quad [2.8]$$

where Y = number of students absent
  D_1 = 1 if Tuesday, 0 otherwise
  D_2 = 1 if Wednesday, 0 otherwise
  D_3 = 1 if Thursday, 0 otherwise
  D_4 = 1 if Friday, 0 otherwise
  [ ] = t-values

Do there tend to be more absences on **a.** Monday or Tuesday? **b.** Monday or Wednesday? **c.** Tuesday or Wednesday? Which day of the week has the most absences?

18.53 Data for 1976–1981 were used to examine the growth of federal Social Security payments. Here are the data:

t (Year)	*Y ($ Million)*
1976	125.6
1977	136.3
1978	146.7
1979	165.2
1980	197.6
1981	228.6

Estimate the least squares equation $Y = \alpha + \beta t + \epsilon$ and answer the following questions: **a.** Why is t used as an explanatory variable, if time itself does not *cause* changes in these payments? **b.** What is the predicted value of Y in 1990? **c.** What is the predicted value of Y in 1930? In 2050? Is there any reason for caution in making such extrapolations?

18.54 In a 1978 Pennsylvania lawsuit, evidence was presented on the percentage of new entrants to an operating engineers union who were minorities.[10] Estimate the equation $Y = a + bT + e$, where Y = minority percentage and T = year, to see whether these data show a statistically significant trend. According to your estimated equation, in what year will the value of Y be equal to 11.5 percent, the minority percentage of the local labor pool from which these operating engineers were hired?

Year	*Minority Percentage of New Entrants*
1966	5.5%
1967	2.8%
1968	5.8%
1969	9.8%
1970	4.8%
1971	8.7%

18.55 Use the 1950–1989 U.S. population data in Table 8 at the end of the book to estimate the equation $Y = \alpha + \beta t + \epsilon$, where Y population and t = year (for example, $t = 1950$ in 1950). Interpret the null hypothesis $\beta = 0$. Do these data reject that null hypothesis at the 1 percent level? Plot the prediction errors (regression residuals) $e = Y - (a + bT)$ versus time. Are there any apparent nonrandom patterns? If so, how would you transform the regression equation?

18.56 *(continuation)* Use the 1950–1989 U.S. population data in Table 8 at the end of the book to estimate the equation $\ln[Y] = \alpha + \beta t + \epsilon$, where $\ln$ = natural logarithm, Y population, and t = year (for example, $t = 1950$ in 1950). Do these data reject that null hypothesis at the 1 percent level? If you had to choose between this estimated equation and the one estimated in Exercise 18.55, which would you choose? Why?

18.57 In a court case concerned with depreciation schedules, the Federal Energy Regulatory Commission (FERC) argued that the depletion of two major fields would cause much of Northern Natural Gas's pipeline network to become useless before it deteriorated physically.[11] The FERC used the following data to extrapolate the annual discovery of reserves in these two fields. Use these data to estimate the equation $Y = \alpha + \beta t + \epsilon$. Is the relationship statistically significant at the 5 percent level? (Do *not* use a two-standard-deviations rule of thumb.) According to your estimated equation, in what year will the annual discovery of reserves hit zero? (Round off to the nearest year.)

Year t	Reserves Discovered Y (Billions of Cubic Feet)
1967	3,307
1968	984
1969	1,107
1970	2,012
1971	1,871
1972	1,883
1973	1,567
1974	1,139
1975	791
1976	650

18.58 *(continuation)* Give a 95 percent prediction interval for the reserves discovered in the year that discovered reserves are predicted to hit zero. (Do *not* use a two-standard-deviations rule of thumb.) If you were the judge in this case, would you be persuaded that discovered reserves will be approximately zero in this year?

• **18.59** *(continuation)* Find the approximate year for which the upper limit of an 80 percent prediction interval for discovered reserves is equal to 0.

18.60 Table 15.9 gives annual weather data for 20 years, from 1943 through 1962. See if there was a trend in preseason precipitation (Y) by estimating the equation $Y = \alpha + \beta t + \epsilon$, where $t = 1943$ in 1943. Do these data indicate an upward or downward trend? Does the trend seem substantial? Is it statistically significant at the 5 percent level?

18.61 *(continuation)* Redo Exercise 18.60, using the June precipitation data in Table 15.9.

18.62 Redo Exercise 18.60, this time using the July temperature data in Table 15.9.

18.63 Redo Exercise 18.60, this time using the corn yield data in Table 15.9

18.64 *(continuation)* Does it seem more plausible to you that there would be a trend in temperature, rain, or corn yield? Explain your reasoning.

18.65 *(continuation)* Why might policymakers be keenly interested in weather trends?

18.66 The Dow Jones Industrial Average and the Standard and Poor's 500 (S&P 500) are two widely followed stock indexes. Use the data in Table 8 at the end of the book to estimate the equation $Y = \alpha + \beta X + \epsilon$, where Y = S&P 500 index and X = Dow Jones Index. Interpret the estimate of β and report whether it is statistically significantly different from 0 at the 5 percent level. Does your estimate reject at the 5 percent level the null hypothesis that $\beta = 1$?

18.67 *(continuation)* Plot the regression residuals versus time. Does there appear to be a heteroskedasticity or serial correlation problem? If your computer program gives the value of the Durbin-Watson d-statistic, test at the 5 percent level the null hypothesis that $\rho \leq 0$.

18.68 *(continuation)* Calculate the annual percentage changes in the Dow Jones Industrial Average and the Standard and Poor's 500 (S&P 500) for each year 1950–1989 and estimate the equation $Y = \alpha + \beta X + \epsilon$, where Y = percentage change in the S&P 500 index and X = the percentage change in the Dow Jones Index. Is the estimate of β statistically significantly different from 0 at the 5 percent level? From 1 at the 5 percent level?

18.69 *(continuation)* Is the use of percentage changes rather than levels a potential cure for heteroskedasticity or serial correlation? Plot the regression residuals from Exercise 18.68 versus time. In comparison with the regression residuals from the equation estimated in Exercise 18.66, does there appear to be more or less of a heteroskedasticity problem? Of a serial correlation problem? If your computer program gives the value of the Durbin-Watson d-statistic, compare this value with the d-value for the residuals from the equation estimated in Exercise 18.66.

18.70 Some economists believe that the changing composition of the U.S. labor force increased the full-employment unemployment rate in the 1960s. Use the 1950–1989 unemployment data in Table 8 at the end of the book to estimate the equation $Y = \alpha + \beta D + \epsilon$, where Y = unemployment rate and D = 1 in 1970–1989, 0 otherwise. Interpret your estimates of α and β carefully. Do these data indicate a statistically significant (at the 5 percent level) change in the unemployment rate after 1970?

18.71 *(continuation)* Instead of a sharp change in the unemployment rate in 1970, there may have been a gradual upward evolution. To consider this possibility, estimate the equation $Y = \alpha + \beta t + \epsilon$, where Y = unemployment rate and t = year (t = 1950 in 1950, and so on). Does this estimated equation or the one estimated in Exercise 18.70 seem to be a better description of the unemployment rate? What criteria did you use to make your choice?

18.72 *(continuation)* Is there any plausible reason to think that the equation estimated in Exercise 18.71 might have a heteroskedasticity or serial correlation problem? Plot the regression residuals versus time and see whether there is evidence of either. If your computer program gives the value of the Durbin-Watson d-statistic, test at the 5 percent level the null hypothesis that $\rho = 0$.

18.73 The following table gives some U.S. Department of Agriculture data on an index of the prices received by U.S. farmers and an index of the prices paid for producing farm products. Both indexes are scaled to equal 100 in 1977. Calculate the ratio Y of prices received to prices paid and estimate the equation $Y = \alpha + \beta t + \epsilon$, where t = 1950, 1955, and so on. Interpret your estimate of β. Using a 1 percent significance level, is there a statistically significant trend in this ratio?

Year	Prices Received	Prices paid
1950	56	37
1955	51	43
1960	52	46

Year	Prices Received	Prices paid
1965	54	47
1970	60	55
1975	101	89
1980	134	138
1985	128	162
1990	150	180

18.74 *(continuation)* Plot the regression residuals $e = Y - (a + bt)$ versus time and see whether there appears to be a heteroskedasticity or serial correlation problem. If your computer program reports the Durbin-Watson d-value, test at the 5 percent level the null hypothesis that $\rho = 0$.

18.75 Use the data in Table 8 at the end of this book to calculate real corporate dividends (X) and real corporate earnings (Y) for the years 1950–1989. Use these data to estimate the equations $\ln[X] = \alpha + \beta t + \epsilon$ and $\ln[Y] = \gamma + \delta t + \upsilon$, where t = year (for example, t = 1950 in 1950). Interpret your estimates of β and δ and calculate 95 percent confidence intervals for each of these parameters.

18.76 *(continuation)* Examine the residuals from each of these regressions and indicate whether there appears to be a heteroskedasticity or serial correlation problem.

18.77 Use the Dow Jones price index data in Table 8 at the end of this book to calculate real stock prices (X) and real corporate earnings (Y) for the years 1950–1989. Use these data to estimate the equations $\ln[X] = \alpha + \beta t + \epsilon$ and $\ln[Y] = \gamma + \delta t + \upsilon$, where t = year (for example, t = 1950 in 1950). Are your estimates of β and δ statistically significantly different from 0 at the 5 percent level?

18.78 *(continuation)* Do the residuals from your regressions indicate either a heteroskedasticity or serial correlation problem? If so, does this seem plausible?

18.79 Calculate the percentage change in the price of beer at Dodger Stadium between 1977 and 1985, using the data in this letter that Bruce Alan wrote to the *Los Angeles Times* (April 17, 1985).

Did you ever wonder how the Dodgers keep from raising their ticket prices every year? Here's a brief history of beer prices the last nine seasons:
1977—65 cents for a 16-oz. cup
1978—85 cents for a 16-oz. cup

1979—$1.00 for a 16-oz. cup
1980—$1.00 for a 14-oz. cup
1981—$1.00 for a 12-oz. cup
1982—$1.25 for a 12-oz. cup
1983—$1.50 for a 12-oz. cup
1984—$1.75 for a 12-oz. cup
1985—$2.00 for a 12-oz. cup

18.80 If you were a government CPI statistician, how would you answer this June 4, 1986, letter to the editor of the *Los Angeles Times?*

The headline in your paper (May 22) "Living Costs Decline for 3rd Month" made me feel as if someone was lying about me. I'm 80 years old and have not driven a car for 16 years because I cannot afford it.

Even with careful shopping in three different stores my food costs have gone up 10% or more since the first of the year.

18.81 Use the data in Table 8 at the end of this book to calculate real per capita gross national product (Y) for the years 1950–1989. Use these data to calculate $\ln[Y]$ for the years 1950–1989 and to estimate the equation $\ln[Y] = \alpha + \beta t + \epsilon$, where t = year (for example, t = 1950 in 1950). Calculate a 95 percent confidence interval for β. Do these data reject $\beta = 0$ at the 5% level?

18.82 *(continuation)* Give a logical reason why the equation estimated in the preceding exercise might have a heteroskedasticity or serial correlation problem. Do the residuals from your regression indicate the presence of either problem?

18.83 Use the data in Exercise 18.81 to estimate the equation $\ln[Y] = \alpha + \beta_2 D + \beta_1 t + \epsilon$, where D = 0 for 1950–1969 and D = 1 for 1970–1989 and t = year (for example, t = 1950 in 1950). Calculate 95 percent confidence intervals for β_1 and β_2 and interpret your estimates of these two parameters.

18.84 *(continuation)* Examine the residuals from this regression for evidence of a heteroskedasticity or serial correlation problem.

18.85 Here are some data on aggregate personal income ($ billion) in the United States and the consumer price index:

	Income	CPI
December 1983	2,860.4	301.5
December 1984	3,113.6	312.2

What is the value of real December 1984 income in terms of December 1983 dollars? What was the percentage change in real income during this year? Did nominal income increase faster or slower than prices in 1984?

• 18.86 Here are some data for three price indexes, the consumer price index and the prices of two of the items included in the CPI:

	CPI	Food	Medical Care
December 1983	301.5	293.9	357.3
December 1984	312.2	205.1	388.5

What can we conclude from these data about the price of food relative to medical care, and about the price of medical care relative to everything else in the CPI?

• 18.87 *Business Conditions Digest* reported the following CPI data:

Year	CPI (P = 100 in 1975)
1975	161.2
1976	170.5
1977	181.5

The *United Nations Statistical Bulletin* reported these CPI data:

Year	CPI (P = 100 in 1975)
1978	121.2
1979	135.1
1980	153.1

Convert the UN data to a 1967 base so that they are comparable to the *Business Conditions Digest* data. What was the percentage increase in prices between 1977 and 1978?

• 18.88 The federal government estimates a "poverty level" of income, that amount needed to maintain a decent standard of living, depending on the number of people in the family and their ages. The index is based on the cost of an economy food plan devised by the Department of Agriculture and the varied needs of children, the elderly, and so on. (There are different poverty-level incomes for a single person under 65 years old, a single person over 65, a family of four, and so on.) Here are some data on the consumer price index and on the estimated poverty levels for a family of four and for a family of five:

		Poverty Level	
Year	CPI	Family of 4	Family of 5
1970	116.3	$3,968	$4,680
1980	246.8	$8,414	$9,966

What are the percentage changes in the CPI and in these two poverty levels? Is it just a coincidence that the percentage changes in the poverty level are roughly equal to the percentage change in the CPI? Why aren't the percentage changes exactly equal?

18.80 Clear up the confusions in this newspaper editorial, which appeared in *The Cape Codder,* August 24, 1984.

> Inflation was once in double figures. Now it is in single figures, and we are the beneficiaries.
>
> But, glad as we are, we're sort of puzzled, too. . . . Where do these single digit figures from from? . . . Why are the chocolate bars that used to cost a nickel now about half the size, and priced at 50 cents; and why is that considered four per cent? . . .
>
> We know, of course, that statistics can be diddled.
>
> One of the simplest ways of diddling statistics, to make the results support the one who ordered the statistics, is to be selective. The costs of some things rise faster than others; and costs rise faster in tourist areas than in downtown Peoria. So it is possible, statistically, to measure the change in the cost of woolen blankets, rubber-soled shoes, and the wages of upstairs maids and get one result; and then measure the cost of breakfast cereal, bread, and baby sitters and get another. Somebody chooses the items to be included in the inflation figures. And whoever it is hasn't been to the grocer's lately. . . .
>
> Perhaps we ought to spend our money on Rolls Royces and caviar, and take advantage of the lower inflation rate.
>
> Or just ignore the propaganda.

18.90 In January 1985, the seasonally unadjusted unemployment rate was 8.0 percent while the seasonally adjusted unemployment rate was 7.4 percent. Is the unemployment rate generally higher or lower in January than in other months? Explain your reasoning. Of what use is a seasonally adjusted unemployment rate?

18.91 M1 and M2 are two monetary aggregates that the Federal Reserve uses to help monitor U.S. financial markets. M1 includes cash outside banks, balances in checkable accounts, and travelers checks. M2 is a broader index that includes not only the assets in M1 but also such items as savings deposits, small time deposits, money market deposit accounts, and money market mutual fund shares. Use the data in Table 8 at the end of this book to calculate the ratio of M2/M1 for the years 1959–1984 and then estimate the equation M2/M1 $= \alpha + \beta t + \epsilon$, where t is the year (1959, 1960, and so on). Calculate a 95 percent confidence interval for β. Using a 5 percent significance level, is the estimate of β statistically significantly different from 0? Do these data provide evidence of either an upward or downward trend in the ratio of M2 to M1?

18.92 *(continuation)* Use your equation estimated from 1959–1984 data to predict the values of M2/M1 in each of the five years 1985–1989. Do your predictions seem reasonably accurate? Reestimate the equation M2/M1 $= \alpha + \beta t + \epsilon$, using data for 1959–1989. Are the results much affected by the inclusion of the 1985–1989 data?

18.93 *(continuation)* Examine the residuals $e =$ M2/M1 $- (a + bt)$ for evidence of either a heteroskedasticity or serial correlation problem. If your computer program gives the value of the Durbin-Watson d-statistic, test at the 5 percent level the null hypothesis that $\rho = 0$.

18.94 Use the data in Table 8 at the end of this book to calculate the ratio M2/M1 for the years 1959–1989 and then estimate the equations M2/M1 $= \alpha_1 + \beta_1 t + \epsilon$ and ln[M2/M1] $= \alpha_2 + \beta_2 t + \epsilon$, where t is the year (1959, 1960, and so on). Are the estimates of β_1 and β_2 both statistically significantly different from 0 at the 5 percent level? Are your estimates of β_1 and β_2 consisent with each other regarding evidence of an upward or downward trend in the ratio of M2 to M1?

18.95 *(continuation)* Use two scatter diagrams of prediction errors versus time to compare the residuals from the two equations estimated in Exercise 18.94. Do the residuals in either scatter diagram appear to have more of a nonrandom pattern than in the other diagram? If your computer program gives the value of the Durbin-Watson d-statistic, compare the d-values for these two equations.

18.96 *(continuation)* The two equations specified in Exercise 18.94 are both of the form $Y_t = \alpha + \beta t +$

ϵ_t, where the subscript t identifies the year of the observation. For each of these two specifications of Y_t, estimate the equation $(Y_t - Y_{t-1}) = \gamma + v_t$ using 1960–1989 data, where γ is a constant and v_t is the error term. (You won't have a 1959 observation for $Y_t - Y_{t-1}$.) What logical basis is there for transforming a regression model in this way? Interpret your estimates of γ and state where the estimates in this exercise are consistent with those in Exercises 18.94 regarding evidence of an upward or downward trend in the ratio of M2 to M1.

18.97 *(continuation)* Imagine that you work for the Federal Reserve's Board of Governors and must recommend one of the four equations estimated in Exercises 18.94 and 18.96 for predicting the ratio M2/M1 for the next few years. Which equation do you recommend? Write a paragraph defending your choice.

18.98 Because of financial innovations in the 1970s, the Federal Reserve redefined M1, M2 and other monetary aggregates in 1980. Use the data in Table 8 at the end of this book to calculate the ratio M2/M1 for the years 1959–1989 and then use the data for the years 1959–1984 (omitting 1985–1989) to estimate the equation M2/M1 $= \alpha + \beta_1 D + \beta_2 t +$ $\beta_3 Dt + \epsilon$, where $D = 0$ for 1959–1975, $D = 1$ for 1976–1984, and t is the year (1959, 1960, and so on). Calculate 95 percent confidence intervals for β_1, β_2, and β_3 and interpret your estimates of these three parameters. Do these data provide evidence of either an upward or downward trend in the ratio of M2 to M1? Of a change in the behavior of M2/M1 after 1975?

18.99 *(continuation)* Plot a scatter diagram with M2/M1 on the vertical axis and t on the horizontal axis, showing each data point and the regression lines estimated in Exercise 18.98 (one line for pre-1975, one for post-1975). Does there seem to be a break in the ratio M2/M1 after 1975?

18.100 *(continuation)* Add the years 1985–1989 to the scatter diagram constructed in Exercise 18.99. Does there appear to be a change in M1/M2 after 1984? Use the data for the years 1959–1989 to estimate the equation M2/M1 $= \alpha + \beta_1 D_1 + \beta_2 D_2 + \beta_3 t$ $+ \beta_4 D_1 t + \beta_5 D_2 t + \epsilon$, where $D_1 = 1$ for 1975–1984, 0 otherwise; $D_2 = 1$ for 1985–1989, 0 otherwise; and t is the year (1959, 1960, and so on). Interpret each of your estimated coefficients and identify the ones that are statistically significantly different from 0 at the 5 percent level.

19

Decision Theory Analysis

A pinch of probability is worth a pound of perhaps.

James Thurber

TOPICS

Decision theory is an internally consistent framework for making rational decisions in an uncertain world. Should a movie studio price a new home video at $29.95 or $79.95? Should a real estate developer build a one- or two-story apartment complex? Should the Federal Reserve raise or lower interest rates? Should a doctor prescribe drugs for a patient when a test suggests heart disease? Should a suspected criminal be acquitted or convicted? Should a military unit attack or hold? Decision theory provides a coherent, consistent framework for making such decisions.

19.1   DECISION TREES AND PAYOFF MATRIXES

Our starting point is the observations that a decision is a choice of one of many possible courses of action and a list of these alternative actions may be helpful in reaching a decision.

Imagine that we are the managers of a small firm that has introduced a new breakfast cereal, which has met with some initial success. We are now considering expanding our production facilities because we feel that our cereal may be on the verge of a sudden wave of popularity. We could list our alternatives as

The box indicates a decision point, and the two branches out of the box are the choices. We will choose one of these branches, or paths (or forks, analogous to the choice faced by a traveler confronted with a fork in the road).

Rational decision making requires not only an enumeration of the options, but also the possible consequences. In this case, the product may prove to be popular (with heavy demand) or unpopular (with only slight demand). As shown in Figure 19.1, these branches can be added to our graph with circles, indicating that the path taken depends on fate (chance, rather than choice). The dollar amount shown on the branches indicates the net profit (the payoff) if that decision is made and that outcome occurs. This figure is called a *decision tree*.

> A **decision tree** uses branches to show the possible decisions and, for each decision, uses branches to show the possible outcomes. Some outcomes possibly lead to additional situations.

Figure 19.1 says that if we expand our production capacity, we will make $1,100,000 if the product is popular, and lose $500,000 if it is not. If we stay at the same level, we will make only $200,000 (because of our limited capacity) if the product is popular, and $100,000 if it is unpopular.

Another framework for displaying the possible actions and consequences is a **payoff matrix,** as in Table 19.1. Here, I've labeled what might happen as events;

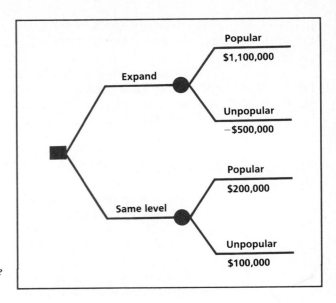

Figure 19.1 *A Decision Tree for the Cereal Manufacturer*

Table 19.1 *A Payoff Matrix*

Actions	Events Popular	Unpopular
Expand	$1,100,000	−$500,000
Same Level	$200,000	$100,000

alternatively, these can be called outcomes, situations, states of nature, or states of the world. Each entry in a payoff matrix shows the payoff if the decision maker chooses the action indicated by that row and the event in that column happens. For instance, the −$500,000 entry in the first row, second column, shows that if the manufacturer expands and the cereal proves unpopular, there will be a $500,000 loss.

EXERCISES

19.1 You own a personal computer and are considering buying theft insurance. The insurance costs $100 and will reimburse you $2,000 if your computer is stolen. Use a decision tree and payoff matrix to show your options and the consequences.

19.2 A movie studio is about to release a home video version of its hit movie "Teenage Vampire," and is considering pricing it at either $29.95 or

$79.95. Here are its subjective sales probabilities

	Probabilities	
Sales	*$29.95*	*$79.95*
100,000	.0	.3
500,000	.3	.4
1,000,000	.5	.2
3,000,000	.2	.1

Either way, the cost (for production, distribution, and so on) is $10 per video. Calculate the profits for each sales event and show these using both a decision tree and payoff matrix.

19.2 MAXIMAX AND MAXIMIN

If we know in advance whether or not the product will be popular, our decision is easy: Expand only if the product is going to be popular. But as with most decisions, we don't know in advance. One rule, suitable for the incurable optimist, is a best-case scenario: For each action, assume that the best event that can happen will happen, and then choose the action that does best. In Table 19.1, for each action, the largest payoff occurs if the cereal is popular: $1,100,000 if we expand and $200,000 if we don't. Expansion, with its $1,100,000 payoff, maximizes these maximum payoffs and would be the choice of the blithe optimist. Since this rule maximizes the maximum payoffs, the best-case scenario is known as *maximax*.

> The **maximax strategy** looks at the maximum gain possible with each action (i.e., assume the best possible outcome), and chooses the action that maximizes this maximum gain.

(If the problem is stated in terms of minimizing losses—for example, minimize the costs of producing a product, managing a payroll, or disposing of surplus material—then the best-case strategy is called *minimin*.)

A polar opposite strategy, suitable for a conservative or a risk-averse pessimist, is the "worst-case scenario": As Murphy's Law says, "If anything can go wrong, it will." For the worst-cast strategy, you choose the action that does best in trying circumstances.[1] In Table 19.1, the minimum payoff with either action occurs if the cereal is unpopular—a $500,000 loss if we expand or a $100,000 profit if we don't. The latter action, sticking with the same level of production, gives the better worst-case payoff and would be the choice of a gloomy pessimist. The worst-case rule is known as the *maximin* or *minimax* criterion, depending on whether the problem is stated in terms of maximizing gain or minimizing losses. Here we are maximizing gains, and the maximin strategy says:

> Find the minimum gain resulting from each action (i.e., assume the worst possible outcome), and choose the action that maximizes this minimum gain—hence the label **maximin**.

Many people do behave in these ways; they routinely assume that things will always work out for the best or for the worst. One advantage of maximax and maximin strategies is that there is no need to specify subjective probabilities for the various events that may occur. But this is also its most glaring weakness: Maximax and maximin strategies ignore all of the payoffs, except the most extreme, and pay no attention whatsoever to how probable or improbable these extreme outcomes are. Why should a decision depend solely on the worst possible outcome, which may have only a 1-in-1000 chance of occurring, and ignore the payoffs that have a 999-in-1000 chance of occurring?

Military Precautions

Before the Shah's overthrow in 1979, the United States supplied Iran with about 80 F-14 fighter planes equipped with Phoenix missiles. The Phoenix system can track 24 separate targets 50 miles away and fire missiles simultaneously at 6 of these targets: It has been called "the most sophisticated combat weaponry known to the free world."[2] When the Shah fled and Ayatollah Khomeini took power, the U.S. military made the maximin assumption that all of the weapons in Iran would be examined thoroughly by the Soviet military. In the words of one official, we "assumed that every weapon in Iran was compromised. We assumed the worst case and we made fixes." The U.S. modified all of its Phoenix missiles to offset any defensive measures that the Soviets might have uncovered in a thorough examination of the weapons lost in Iran.

In 1985, the U.S. military feared that Iranian agents had broken into weapons stockpiles in California and stolen 60 cartons of replacement parts for the F-14. Once again, U.S. military experts assumed the worst possible scenario—that, although Iran now labels both the U.S. and Soviet Union as Satans, these thefts allowed the Soviets to study the modified Phoenix system. Again, the U.S. modified all of its Phoenix missiles to guard against this possibility.

With a large budget and national security on the line, the military can and often does follow a maximin strategy.

EXERCISES

19.3 There have been many investor lawsuits recently alleging that because brokerage firms did not exercise adequate supervision, stock brokers churned these investors' accounts with a flurry of transactions, which generated large commissions and painful losses. One customer who lost $30,000 this way has been offered a $30,000 settlement but is considering suing for $1 million for mental anguish and punitive damages. If she loses this suit, she will have to pay substantial court costs and lawyer fees. The relevant events are whether or not she would win her suit. Here is her payoff matrix:

	Events	
Actions	Would Win	Would Lose
Settle	$30,000	$30,000
Sue	$1,000,000	−$100,000

Which action is maximin? Which is maximax? Draw a decision tree.

19.4 A car manufacturer is introducing a new car, which can be small, medium, or large. How well it sells depends on whether the price of gasoline goes up or down. Here is the payoff matrix (in millions):

	Events		
Actions	Prices Up	Stable	Prices Down
Small	+$130	+$100	−$20
Medium	+$70	+$100	+$110
Large	−$80	+$100	+$210

Does any of these three actions dominate another? Which action is maximax? Which is maximin?

19.5 Dirty Harry is running for mayor and a week before the election he is considering starting some nasty rumors to smear his opponent's reputation. If it works, he figures to gain 100,000 votes; if it backfires, Harry thinks it will cost him 100,000 votes.

Table 19.2 *Payoff Matrix for Exercise 19.7*

Actions	Events		
	Economic Boom	*Muddle Through*	*Recession*
Two-Story	$30,000,000	$15,000,000	−$5,000,000
One-Story	$10,000,000	$10,000,000	$5,000,000
Nothing	$0	$0	$0

Draw a payoff matrix using votes and identify the maximin action. What factors might influence Harry's decision to take the maximin action?

19.6 It's Valentine's Day and you have decided to earn some extra money by selling roses on a street corner for $1 each. The roses cost you $.25 each in quantities of 100. You will either buy 100 or 200, and demand will either be weak (exactly 100 sold) or strong (demand for 200), but there is no way of knowing which will be the case until sometime after you have taken up your position on the street corner.

Draw a payoff matrix for your net profits. Which decision is maximin?

19.7 A developer must choose between building a two-story apartment complex, a one-story complex, or nothing at all. The profits depend on the state of the economy after the construction is completed. The payoff matrix is shown in Table 19.2. Which decision is maximax? Which is maximin? Why might you resist making the maximax or maximin decision? Are there any actions you can rule out, based on the payoff matrix?

19.3 EXPECTED VALUE MAXIMIZATION

Many decision makers focus on the event considered the most likely to occur, and choose the action that does the best in this most likely scenario. Theorists call this a *maximum likelihood strategy*.

> The **maximum likelihood strategy** maximizes the payoff when the most probable event occurs.

Look at the payoff matrix in Table 19.1 again. A maximum likelihood strategist would ask whether the cereal is more likely to be popular or unpopular. If popular, then expand; if unpopular, don't expand.

This strategy appeals to common sense and that is why it is used so often. Yet, a closer examination reveals that a myopic focus on the most likely event is not always sensible. Paul Samuelson, the Nobel-winning economist, put it even more strongly: "I think the greatest error in forecasting is not realizing how important are the probabilities of events other than those everyone is agreeing upon."[3]

Let's change the payoff matrix drastically to make the point clearly:

Actions	Events	
	Popular	*Unpopular*
Expand	$100,000	−$5,000,000
Same Level	$50,000	$50,000

Expansion is expensive and gives a staggering $5,000,000 loss if the cereal proves unpopular, or recovers the cost and makes a small profit if the cereal turns out to be popular. Now suppose we believe there is a .51 probability that the cereal will be popular and a .49 probability that it won't be popular—essentially a flip of a slightly bent coin. The maximum likelihood strategist focuses on the (slightly) more likely outcome—a popular cereal—and decides to expand, completely ignoring the fact that it is almost equally likely that the cereal will prove unpopular, inflicting catastrophic losses. A sensible person would look not only at the most likely outcome, but also at what happens if less likely events occur.

Even more forcefully, imagine that there are eleven possible events, representing different degrees of popularity, with one event having a 10 percent chance of occurring and the other ten events having 9 percent chances. If we focus on the single most likely event, the one with a .10 probability, then we are unreasonably ignoring the other possible outcomes, one of which is almost certain to occur.

An attractive alternative is to take into account both the best and worst cases (and those in between) by using probabilities to quantify the relative likelihood of all possible outcomes, and then using these probabilities to calculate a (weighted) average payoff; that is, an expected value. For our initial payoff matrix in Table 19.1, perhaps we believe that the product is as likely to be popular as unpopular and, so, assign .5 probabilities to each outcome. The expected values of the profits from each decision are

$$\text{expand: } (\$1,100,000)(.5) + (-\$500,000)(.5) = \$300,000$$

$$\text{same level: }\quad (\$200,000)(.5) + (\$100,000)(.5) = \$150,000$$

By this criterion, expansion would be the choice. A convenient way to display these expected values, together with the underlying data, is to add another column to the payoff matrix, as shown in Table 19.3.

> The **expected-value maximization** rule is to choose the option with the largest expected value of profits (or smallest expected value of costs).

A somewhat different way to pose the issue is to let P be the probability that the product will be popular and then find the borderline value of P such that each decision has exactly the same expected value:

$$\$1,100,000(P) - \$500,000(1 - P) = \$200,000(P) + \$100,000(1 - P)$$

$$\$1,600,000(P) - \$500,000 = \$100,000(P) + \$100,000$$

$$\$1,500,000(P) = \$600,000$$

$$P = .40$$

If we believe that there is more than a .4 probability that the product will be popular, then we should expand.

Table 19.3 *Payoff Matrix with Expected Values*

Actions	Events		Expected Value
	Popular	Unpopular	
Expand	$1,100,000	−$500,000	($1,100,000)(.5) + (−$500,000)(.5) = $300,000
Same Level	$200,000	$100,000	($200,000)(.5) + ($100,000)(.5) = $150,000

The determination of a borderline probability relieves some of the anxiety accompanying the choice of a subjective probability and reveals how sensitive our choice is to the probability value. If we are having a difficult time deciding whether the product has a .6 or .7 probability of being popular, it is reassuring to know that any probability over .4 favors expansion. Unfortunately, this simple procedure cannot be used when there are more than two possible outcomes or when the probabilities depend on the decision—for instance, in designing a product, whether demand is strong or weak may depend on the design chosen. In this case, more complicated sensitivity analyses are needed.

Expected Opportunity Loss

After an event occurs, with 20-20 hindsight we can see which action would have had the largest payoff and, in comparison, how much we lost by taking the action we chose. This difference is called the *opportunity loss*.

> For each event, the **opportunity loss** is the amount lost by not doing what, in retrospect, would have been the best possible action.

Look again at the payoff matrix in Table 19.1 and then at the opportunity losses shown in Table 19.4. If the cereal turns out to be popular, expansion would have been the most profitable action; the opportunity loss from this action is zero, since we did the best we could. A decision to stay at the same level has an opportunity loss of

$$\$1,100,000 - \$200,000 = \$900,000$$

Table 19.4 *Opportunity Losses*

Actions	Events	
	Popular	Unpopular
Expand	$0	$600,000
Same Level	$900,000	$0

Notice that we don't actually lose $900,000 out of our pocket; in fact, we make $200,000. The $900,000 figure is an opportunity loss in the sense that we had an opportunity to make an additional $900,000 and didn't. For this reason, the opportunity loss is sometimes called a "regret."

If, on the other hand, the cereal turns out to be unpopular, no expansion would have been the best choice and has a zero opportunity loss. By comparison, the amount lost by expansion will be

$$\$100,000 - (-\$500,000) = \$600,000$$

Again, this $600,000 opportunity loss represents the difference between the best and actual payoffs.

The expected values of the opportunity loss are

$$\text{expand: } (\$0)(.5) + (\$600,000)(.5) = \$300,000$$

$$\text{same level: } (\$900,000)(.5) + (\$0)(.5) = \$450,000$$

Because these are losses, we want to minimize the expected value and, so, choose to expand, the same action chosen by a maximization of the expected value of the payoff. This equivalence always applies.

> The action that maximizes the expected value of the payoff also minimizes the expected value of the opportunity loss.

The advantages of recasting the payoffs in terms of opportunity losses are primarily expositional. Compare Tables 19.1 and 19.4 one more time. The opportunity loss format in Table 19.4 shows at a glance which actions fare best for each event and how much is lost by inferior actions.

EXERCISES

19.8 A computer company must decide whether its new computer, Granny Smith, will have a closed architecture or an open one, allowing users to add enhancements. Here are the estimated profits, in millions of dollars, depending on whether potential customers prefer an open or closed architecture, events believed to be equally likely.

	Events	
Actions	Prefer Open	Prefer Closed
Open	500	300
Closed	200	800

a. Which action is maximax? **b.** Which is minimax? **c.** Which action maximizes the expected value of the payoff? **d.** For what range of probability values does an open architecture have the higher expected payoff?

19.9 Tricia is going to invest in one of five sectoral mutual funds:

Smokestack: manufacturers do well when the economy booms

Utilities: regulated utilities do somewhat better when interest rates fall

S&Ls: savings and loan associations do well when interest rates fall

Table 19.5 *Payoff Matrix for Exercise 19.9*

Actions	Events			
	Y and R Up (P = .20)	Y Up, R Down (P = .40)	Y Down, R Up (P = .20)	Y and R Down (P = .20)
Smokestack	+20%	+40%	−30%	0%
Utilities	−5%	+25%	−10%	+20%
S&Ls	−20%	+80%	−30%	+60%
Housing	0%	+90%	−60%	+20%
T-Bills	+6%	+4%	+6%	+4%

Housing: home-building companies do well when the economy is up and interest rates are down

T-bills: a money market fund, buys short-term bonds

Table 19.5 gives her estimated returns over the next year, depending on whether the economy (*Y*) and interest rates (*R*) go up or down. (The respective probabilities are in parentheses under each event.) Which action **a.** is maximax? **b.** is minimax? **c.** is maximum likelihood? **d.** maximizes the expected value of the payoff?

19.10 A classic study by Howard, Matheson, and North considered the question of whether the federal government should seed hurricanes to weaken them.[4] They allowed for five possible percentage changes in maximum wind speed over a 12-hour period. Table 19.6 gives some of their data for a representative hurricane. Notice how seeding makes it more likely that the wind speed decreases and less likely that it will increase. The costs (in $ millions) for unseeded hurricanes were estimated from past hurricane damage. With seeding, these damages were increased by the cost of seeding, and by estimates of the increased gov-

ernment legal and cleanup costs if the government were to try to weaken hurricanes and fail.

Set up a decision tree showing the government's choices—to seed or not to seed—and the possible outcomes, with their respective costs and probabilities. Which action would be made by a minimaxer who minimizes the maximum cost? Which is the maximum likelihood action? Which action minimizes the expected value of the cost?

19.11 Look at the data in Exercise 19.2 again. Which decision is maximin? Which is maximum likelihood? Which maximizes the expected value of profits? Now construct an opportunity loss matrix. Which decision minimizes the expected value of the opportunity loss?

19.12 Using the data in Table 19.1, which is the maximin action? Using the corresponding opportunity losses in Table 19.3, which action is minimax, minimizing the worst possible opportunity loss? Do these two criteria agree on which action is recommended? What if the payoff when the manufacturer stays at the same level and the cereal turns out to be popular is changed from $200,000 to $700,000?

Table 19.6 *Data for Exercise 19.10*

Change in Wind Speed	If Unseeded		If Seeded	
	Probability	Cost	Probability	Cost
> +25%	.054	335.8	.038	504.0
+10% to +25%	.206	191.1	.143	248.7
−10% to +10%	.480	100.0	.392	105.3
−25% to −10%	.206	46.7	.255	47.0
< −25%	.054	16.3	.172	16.6

19.4   EXPECTED UTILITY

In Chapter 4, several examples were used to show that people do not invariably maximize expected returns: Many purchase lottery tickets or fire insurance despite the negative expected returns; most prefer $1 million to a .5 chance at $2.1 million; and all prefer a few dollars to the St. Petersburg gamble with its infinite expected return. In such cases, something other than expected return is needed to describe decision making accurately.

The modern approach to decision making under uncertainty began with the brilliant British logician Frank Ramsey (who died at age 26) in a paper published posthumously in 1931.[5] Its widespread acceptance was due, however, to John von Neumann and Oskar Morgenstern's classic 1944 book *Theory of Games and Economic Behavior.*[6]

These pioneers showed that if a person's decisions conform to certain plausible axioms,* then a utility function can be constructed such that decisions can be described as maximizing expected utility. This construction of a personal utility function begins with the assignment of two arbitrary values for the function, subject only to the restriction that the higher utility number is assigned to the preferred event. For example, we could let the utility of $10 be 1 and let the utility of $0 be 0:

$$U[\$10] = 1, \; U[\$0] = 0.$$

Utility values can now be assigned to other payoffs by learning this person's preferences. Consider the utility of $5, for instance. First, the question is posed, "Would you prefer $5 or a gamble in which you have a .50 probability of receiving $10 and a .50 probability of receiving nothing?" Perhaps the safe $5 is chosen. Then it is asked, "Would you prefer $5 or a 60 percent chance at $10?" Proceeding in this fashion, we eventually find some probability P such that the person is indifferent between a safe $5 and a gamble with this probability P of winning $10. At this point of indifference, the person's expected utilities for the two alternatives are equal:

expected utility of safe $5 = expected utility of $10 gamble

$$U[\$5] = U[\$10](P) + U[\$0](1 - P)$$

Substituting $U[\$10] = 1$ and $U[\$0] = 0$, we see that $U[\$5] = P$. Continuing in a similar fashion, we can derive utility values for all possible payoffs.

The precise utility values depend on the preferences of the person being inter-

---

*The axioms are (1) completeness: A person either prefers A to B, B to A, or is indifferent between A and B; (2) transitivity: If A is preferred to B and B to C, then A is preferred to C; (3) dominance: A person prefers a larger chance of winning A to a smaller chance; (4) continuity: If A is preferred to B and B to C, for some value of P the person is indifferent between B and a lottery with probability P of winning A and $1 - P$ of winning C; (5) independence: A person indifferent between A and B is indifferent between a lottery with probability P of winning A and one with probability P of winning B.

rogated. One important and interesting characteristic of the resulting utility function is its concavity or convexity. As shown in Figure 19.2, a person whose utility function is concave is said to be *risk averse,* while a person whose utility function is convex is *risk seeking.* A *risk-neutral* person has a linear utility function.

These shapes coincide with the discussion in Chapter 4. A person who is risk neutral simply maximizes expected return and, thus, is indifferent between a safe $5 and a 50 percent chance at $10; the utility of $5 is consequently 0.5. A risk-averse person will take the sure $5 unless there is more than a 50% chance at $10; thus, the utility of $5 is greater than 0.5, and the utility function is concave. A risk-seeker will take the gamble even when there is less than a 50 percent chance of winning and, consequently, has a convex utility function.

Risk Bearing

If a person is not risk neutral, then the appropriate amounts to enter in a payoff matrix are not dollars, but utility values, and the preferred action is the one that maximizes the expected value of utility. Consider an individual, named Joe, who owns a $50,000 house and little else. This house is near an earthquake fault and Joe is afraid that he will lose his house in a quake. To construct his utility function,

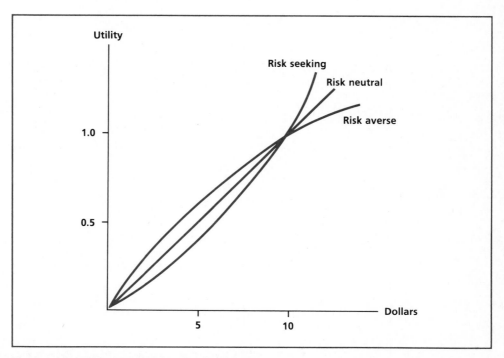

Figure 19.2 *Three Kinds of Utility Functions*

we'll set

$$U[-\$100] = -100 \qquad U[\$0] = 0$$

If Joe were risk neutral, his utility function would be linear, and $U[-\$50,000]$ would equal $-50,000$. But he is not. Joe is risk averse, as shown in Figure 19.3 and, I assume,

$$U[-\$50,000] = -150,000$$

Joe knows that $50,000 is 500 times as much money as $100, but he values its loss 1500 times as highly.

Jane, on the other hand, is very rich and winning or losing $50,000 to her is

Figure 19.3 *Joe Is Risk Averse; Jane Isn't*

like Joe flipping a coin to see who pays for beer. Her utility function in Figure 19.3 is a straight line, because she is risk neutral over this range of monetary values.

Now let's see what happens if Jane offers to insure Joe against an earthquake for, say, a $100 fee. Table 19.7 shows the payoff matrixes, both in terms of dollars and utility. Because Jane is risk neutral, nothing is changed by using utility in place of money. For Joe, in contrast, there is a big difference, because he values a $50,000 loss so highly.

Whether Jane is willing to sell such insurance for $100, and whether Joe is willing to buy, depend on the probability that the house will be destroyed by an earthquake. Let's assume that this probability is .001. The expected value of the insurance payoff is

$$.001(\$50,000) = \$50$$

only half the $100 cost of the policy. Thus a risk neutral person, like Jane, would be willing to take this gamble (selling for $100 something with an expected value of $50). We can confirm this by applying the .001 probability to her (utility) payoff matrix

$$\text{sell ins.: } E[U] = (-49,900)(.001) + (100)(.999) = +50$$

$$\text{don't: } E[U] = (0)(.001) + (0)(.999) = 0$$

Her expected utility is higher if she sells the insurance.

Is Joe willing to spend $100 to avoid a loss with an expected value of $50? Yes, if he is sufficiently risk averse. Here are his expected utilities:

$$\text{insure: } E[U] = (-100)(.001) + (-100)(.999) = -100$$

$$\text{don't: } E[U] = (-150,000)(0.001) + (0)(.999) = -150$$

There is room for a deal, in that he is willing to buy what she is willing to sell.

Table 19.7 *Payoff Matrixes for Joe and Jane*

Joe:	*Payoff in Dollars*		*Payoff in Utility*	
Actions	*Earthquake*	*None*	*Earthquake*	*None*
Insure	−$100	−$100	−100	−100
Don't Insure	−$50,000	$0	−150,000	0

Jane:	*Payoff in Dollars*		*Payoff in Utility*	
Actions	*Earthquake*	*None*	*Earthquake*	*None*
Sell Insurance	−$49,900	$100	−49,900	−100
Don't Insure	$0	$0	0	0

An earthquake is a real physical danger, a natural risk that has to be borne by someone. Those who are the least risk averse are the most willing to bear such risks. Those who are more risk averse can, for a price, use financial contracts to pass such risks on to others.

EXERCISES

19.13 Which would you prefer? **A.** $200,000 **B.** a lottery with a 20 percent chance of winning $1,200,000 and an 80 percent chance of losing $50,000? How would you characterize a person who prefers A to B? Prefers B to A? Is indifferent between A and B?

19.14 Refer back to Exercise 19.8. Which action maximizes expected utility if the company is risk neutral? In fact, the company's utility function has these values: $U[0] = 0$, $U[200] = 2$, $U[300] = 2.9$, $U[500] = 4.6$, $U[800] = 6.8$. How would you characterize its risk preferences? Which action maximizes expected utility?

19.15 You have been offered $2 or a raffle ticket with a 1 percent chance of winning $100. Your utility function is $U[x] = x + .02x^2$, where x is your monetary gain. Which choice maximizes your expected return? Which maximizes your expected utility? Explain any difference.

19.16 The text shows how to find the utility of $5 when $U[\$0] = 0$ and $U[\$10] = 1$. How would you next find the utility of $20? What will be true of $U[\$20]$ if the person being interrogated is risk neutral? risk averse?

19.17 Jill's wealth consists entirely of stock in company NEW, which may be taken over by company GLOM. If so, her stock will be worth $360,000; if not, it will be worth $160,000. She considers the two possible outcomes to be equally likely. Her utility function is $U = \sqrt{W}$, where W is her wealth. **a.** What is the expected value of her wealth? **b.** What is the expected value of her utility? **c.** What is the minimum amount for which she would be willing to sell her stock? Why is this minimum amount larger or smaller than the expected value of her stock? **d.** Who could she sell to?

● **19.18** Bernoulli used this logarithmic utility function to argue that the poor are more likely to buy insurance than the rich:

$$U = \log (W/W_o)$$

where W_o is the initial wealth, before the gamble, and W is the wealth after the gamble. Consider now a $100,000 boat with a .01 probability of sinking and insurance that costs $2,000. If Rich has $W_o = \$1,000,000$, will she buy insurance? If Poor has $W_o = \$200,000$, will he buy insurance? Does Bernoulli's utility function reflect risk-averse or risk-seeking behavior? Can you think of any reasons why the poor might be *less* likely to buy insurance?

19.5 POSTERIOR ANALYSIS

So far, our probabilities have been purely subjective, revealing prior beliefs about the relative likelihood of possible outcomes. Often, businesses are able to use sampling to revise and, it is hoped, improve these probabilities. In our example, we might conduct a marketing survey to learn more about the potential demand for our cereal, before committing the firm's resources to expensive expansion plans. Perhaps it is known from past experience that when a product is popular, there is a .8 probability that the survey results will be positive and that, when unpopular, such marketing surveys have a .6 probability of being negative; that is, using obvious notation, these conditional probabilities are

$$P[\text{pos}|P] = .8 \text{ and } P[\text{neg}|U] = .6$$

What we are interested in, of course, is the reverse conditional probabilities—that demand will be strong if the survey results are positive, that demand will be weak if the survey results are negative—and Bayes' theorem using our prior probabilities $P[P] = P[U] = .5$ is the way to reverse these conditional probabilities:

$$P[P|\text{pos}] = \frac{P[P]\,P[\text{pos}|P]}{P[P]\,P[\text{pos}|P] + P[U]\,P[\text{pos}|U]}$$

$$= \frac{.5(.8)}{.5(.8) + .5(.4)}$$

$$= 2/3$$

$$P[U|\text{neg}] = \frac{P[U]\,P[\text{neg}|U]}{P[U]\,P[\text{neg}|U] + P[P]\,P[\text{neg}|P]}$$

$$= \frac{.5(.6)}{.5(.6) + .5(.2)}$$

$$= 3/4$$

These revised probabilities, based on the survey results, are called *posterior probabilities*. With these, we can calculate the expected payoff from each decision if the survey is positive:

expand: ($1,100,000)(2/3) + (−$500,000)(1/3) = $566,667

same level: ($200,000)(2/3) + ($100,000)(1/3) = $166,667

and if it is negative:

expand: ($1,100,000)(1/4) + (−$500,000)(3/4) = − $100,000

same level: ($200,000)(1/4) + ($100,000)(3/4) = $125,000

We maximize expected value by expanding if the marketing survey is encouraging and maintaining the same level if the survey is discouraging.

The Value of Information

We can calculate the value of the survey information, even before the survey is taken. Early on, we saw that with no survey and nothing to go on but our prior subjective probabilities, we choose to expand and the expected value of our profits is $350,000. At the other extreme is perfect information. If we know the cereal will be popular, we expand and make $1,100,000; knowing the cereal will be unpopular, we maintain the same level of production and earn $100,000. What is the most we would pay to find out which is the case, before we have to decide whether or not to expand? Based on our prior probabilities, there is a .5 probability that we will learn the cereal is popular and a .5 probability that it is unpopular. The expected value of our profits is

$$\$1,100,000(.5) + (\$100,000)(.5) = \$600,000$$

This is an improvement over $350,000, because now we will not make the mistake of expanding when there is no demand, or not expanding when there is. The expected value of perfect information is $600,000 − $350,000 = $250,000, the difference between the expected value of profits with perfect information and the expected value without.

One more question. What is the most we would pay for the marketing survey described above? It must be less than $250,000, since that is the most that we would pay for infallible information and the survey is, unfortunately, fallible. The probabilities of encouraging and discouraging survey results are

$$P[\text{pos}] = P[P]P[\text{pos}|P] + P[U]P[\text{pos}|U] = .5(.8) + .5(.4) = .6$$

$$P[\text{neg}] = P[U]P[\text{neg}|U] + P[P]P[\text{neg}|P] = .5(.6) + .5(.2) = .4$$

There is .6 probability that the results will be encouraging and that we will consequently expand, with a $566,667 expected value of profits. There is a .4 probability of a discouraging survey, persuading us not to expand, with a $125,000 expected value of profits. Overall, the expected value of profits is

$$\$566,667(.6) + \$125,000(.4) = \$390,000$$

more than the $350,000 without the survey, but less than the $600,000 with perfect information. The expected value of the survey is $390,000 − $350,000 = $40,000 and we should pay no more than this for it.

This anticipatory reasoning is called *preposterior analysis* and, here, comes to the conclusion that (1) if the survey costs less than $40,000, it should be acquired and, then, we expand if the results are encouraging and not if they are discouraging; and (2) if the survey costs more than $40,000, we forego it and expand solely based on our prior probabilities. This can all be laid out in a many-branched decision tree, but, mercifully, I have not done so.

Another Look at Acceptance Sampling

In Chapter 5 we used the example of bomb shipments to discuss acceptance sampling. The Army wanted at least 90 percent of its bombs to explode properly and we examined the rule of rejecting a shipment if a sample of ten bombs turns up

Table 19.8 *Prior and Posterior Probabilities for a Bomb Shipment*

		10-Bomb Test		100-Bomb Test	
π	$P[\pi]$	$P[x = 1\|\pi]$	$P[\pi\|x = 1]$	$P[x = 10\|\pi]$	$P[\pi\|x = 10]$
.05	.4	.3151	.4092	.0167	.1425
.10	.3	.3874	.3773	.1319	.8432
.20	.2	.2684	.1743	.0034	.0143
.30	.1	.1211	.0393	.0000	.0000

The payoff matrix:

	Events				Expected Payoff	
Actions	$\pi = .05$	$\pi = .10$	$\pi = .20$	$\pi = .30$	1 of 10	10 of 100
Reject	−1	−1	−1	−1	−1	−1
Accept	0	0	−10	−30	−2.92	−.143

more than one dud. Our analysis was from a classical perspective, looking at the probability that a particular sample will pass or fail the test if, in fact, a fraction π (.05, .10, .20, or .30) of all the bombs in the shipment are duds. Decision theory looks at the question the other way around, calculating based on the test results, the posterior probability that a fraction π of the bombs are duds, and then uses an explicit payoff matrix to determine acceptance or rejection.

To implement the decision theory approach, we first must assign prior probabilities to the shipment proportions π. Suppose that, based on past experience with this supplier, we consider four representative values of π with the prior probabilities shown in the second column of Table 19.8. Second, we obtain the test results—let's say that a test of ten bombs turns up one dud. The binomial table gives the probability of one dud in ten trials for various assumed values of π, and these are shown in the third column of Table 19.8. Bayes' theorem then combines the prior probabilities and the test results to give the posterior probabilities shown in the table's fourth column.* Notice how the test result, one dud in a sample of ten, makes it more likely that 5 percent or 10 percent of the total shipment are duds and less likely that 20 percent or 30 percent are duds. Still, there is more than a 20 percent chance that the proportion of defectives is unacceptably high.

*The numerator is

$$P[x = 1] = P[\pi = .05]\, P[x = 1|\pi = .05] + P[\pi = .1]\, P[x = 1|\pi = .1]$$
$$+ P[\pi = .2]\, P[x = 1|\pi = .2] + P[\pi = .3]\, P[x = 1|\pi = .3]$$
$$= .4(.3151) + .3(.3874) + .2(.2684) + .1(.1211) = .30805$$

so that, for instance,

$$P[\pi = .05|x = 1] = \frac{P[\pi = .05]\, P[x = 1|\pi = .05]}{P[x = 1]} = \frac{.4(.3151)}{.30805} = .4092$$

To decide whether that 20 percent chance is sufficient reason to reject the shipment, we need a payoff matrix, specifying the relative costs of rejecting a shipment and of accepting an unsatisfactory one. This, too, is shown in Table 19.8. The cost of rejecting a shipment is assumed to be a constant amount, reflecting the disposal of this shipment and replacement with a new one. The cost of accepting a shipment with 20 percent defectives is assumed to be ten times this amount and, with 30 percent defectives thirty times as large, representing the damage to the nation's security. There is no cost to accepting a satisfactory shipment, with π equal to .05 or .10. It turns out that, as shown, rejection of the shipment has the higher expected value (lower expected cost).

For comparison, Table 19.8 also shows the results if 100 bombs are tested and 10 are found to be duds. Notice that the larger sample dramatically reduces the posterior probabilities that the shipment is 30 percent, 20 percent, or even 5 percent defective. This illustrates the general principle that a substantial amount of sample data will overwhelm prior probabilities. The payoff matrix shows that acceptance of the shipment now has the higher expected payoff (lower expected cost).

As this example indicates, a formal decision theory analysis does require the specification of prior probabilities and a payoff matrix, and does necessitate some additional calculations. The overriding benefit is that it makes explicit the sort of considerations that should influence a rational decision.

Medical Decision Theory

Medical diagnoses are sometimes clearcut—the x-rays reveal a broken bone—and, at other times, ambiguous—this patient may or may not have heart disease. One way of handling an uncertain diagnosis is with probabilities. Two doctors give this example:

> A 55-year-old asymptomatic man underwent a submaximal electrocardiographic stress test before undertaking a jogging program. . . . Two minutes into the recovery period, 1 mm of horizontal S-T segment depression developed in his lateral precordial leads. . . . When told of these findings, the patient asked a straightforward question: "Do I have coronary heart disease?"[7]

A definite answer of either "yes" or "no" is not warranted since this test is far from perfect; yet, the patient should, in some way, be told that the results suggest the possibility of disease. Words alone are inadequate because the doctor and patient may interpret them very differently. When sixteen physicians were asked to assign a numerical probability corresponding to the diagnosis "cannot be excluded," the answers ranged from a 5 percent probability to 95 percent, with a mean of 47 percent. When they were asked to interpret "likely," the probabilities ranged from 20 percent to 95 percent, with a mean of 75 percent. Even the phrase "low probability" elicited answers ranging from 0 percent to 80 percent (with a mean of 18 percent).[8] If, by "low probability," one person means an 80 percent chance and another means no chance at all, then it is better to state the probability one has in mind than to risk a severe misinterpretation of ambiguous words.

Where do these numerical probabilities come from? Most likely, the values reflect a subjective weighing of the available evidence. Bayes' theorem provides a formal method for incorporating test results. The two doctors cited earlier answered the patient's question in this way, using a computer program called CADENZA, which pools published medical data. The prior probability that a 55-year-old man with no symptoms of disease has coronary heart disease is

$$P[D] = .10$$

The probability of a 1 mm horizontal S-T depression is .077 in diseased patients and .020 in healthy patients:

$$P[1 \text{ mm} | D] = .077 \qquad P[1 \text{ mm} | H] = .020$$

By Bayes' rule, the posterior probability is

$$P[D | 1 \text{ mm}] = \frac{P[D] \, P[1 \text{ mm} | D]}{P[D] \, P[1 \text{ mm} | D] + P[H] \, P[1 \text{ mm} | H]}$$

$$= \frac{(.10)(.077)}{(.10)(.077) + (.90)(.020)}$$

$$= .303$$

The probability of heart disease has increased threefold, from .10 to .30, but is still far from a certainty. Indeed, it is still more likely than not that the patient does not have heart disease. The most effective and unambiguous way of communicating this diagnosis accurately—that the test is suggestive, but far from conclusive—is with a probability.

How the doctor and patient should use this probability depends on the treatment options and, implicitly, on a payoff matrix. Suppose the payoff matrix looks like this:

	Events	
Actions	Disease	No Disease
Treat	−50,000	−100,000
Don't Treat	−1,000,000	0

In this case, treatment is relatively inexpensive, compared to the cost of leaving the disease untreated. The expected payoffs are

$$\text{treat: } .303(-50,000) + .697(-100,000) = -84,850$$

$$\text{don't treat: } \quad .303(-1,000,000) + .697(0) = -303,000$$

making treatment the more attractive option.

If, in contrast, the treatment of a nondiseased person is very expensive—perhaps a complicated, dangerous operation or a drug with serious side effects—the payoff matrix might look like this

	Events	
Actions	Disease	No Disease
Treat	−300,000	−500,000
Don't Treat	−1,000,000	0

Now, no treatment has the lower expected cost:

$$\text{treat: } .303(-300,000) + .697(-500,000) = -439,400$$

$$\text{don't treat: } .303(-1,000,000) + .697(0) \quad = -303,000$$

These are, to be sure, complex and difficult issues. But decision theory ensures that the doctor and patient are asking the right questions and provides a formal way to organize the answers.

EXERCISES

19.19 Major Oil has obtained a lease to drill for oil on some as-yet-unexplored federal land. The probability of a significant oil find is thought to be .15 and the payoff matrix (in millions of dollars) is

	Events	
Actions	Oil	No Oil
Drill	100	−10
Don't Drill	0	0

a. Which action is maximax? Which is minimax? **b.** Which action maximizes the expected payoff? **c.** A $1 million seismic test has these probabilities of positive and negative readings: $P[+|\text{oil}] = .6$ and $P[-|\text{no oil}] = .9$. What is the posterior probability of oil if the test turns out to be positive? Negative? **d.** Which action maximizes the expected payoff if the seismic test is positive? If it is negative?

19.20 *(continuation)* What is the expected value of perfect information? What is the probability that the test result will be positive? That it will be negative? If the seismic test costs $1 million, would an expected value maximizer choose to test?

19.21 Five percent of all 60-year-olds have a certain medical disease, which can only be cured by

major surgery: $P[D] = .05$. The accepted test gives a positive reading in 70 percent of the diseased patients and in 30 percent of the healthy ones:

$$P[\text{pos}|D] = .70 \qquad P[\text{pos}|H] = .30$$

Assume that the payoff matrix looks like this:

	Events	
Actions	Disease	No Disease
Surgery	−200,000	−600,000
No surgery	−1,000,000	0

a. What action would be chosen by a minimaxer? By a miniminer? **b.** Before a test is conducted, which action has the higher expected payoff (lower expected cost)? What is the expected value of perfect information? **c.** After a test, what is the posterior probability of disease if the test is positive? If it is negative? Which action has the higher expected payoff in each case? **d.** What is the probability that the test result will be positive? That it will be negative? What is the most that an expected value maximizer would pay for a test?

• **19.22** *(continuation)* If a test costs $1,000, would you recommend a second, independent test if the results of the first are positive? If they are negative?

19.23 Apply the bomb acceptance sampling analysis in the text to the case of 15 duds in a 100-bomb test.

19.24 Sunspots Incorporated makes the following claim: "We can predict with 90 percent accuracy whether the Dow Jones Average will be higher or lower one week from the date of our prediction." You obtain a trial subscription to this service for two weeks (two predictions). You consider only two hypotheses—Sunspots can forecast with either .5 or .9 accuracy—and believe the first hypothesis to be three times as likely as the second. Your payoff matrix is:

	Events	
Actions	$\pi = .5$	$\pi = .9$
Accept $\pi = .5$	0	0
Accept $\pi = .9$	$-\$1,000$	$+\$10,000$

a. Without the benefit of the trial subscription, which action has the higher expected value? **b.** What is the most that an expected value maximizer would pay for perfect information about this firm's accuracy? **c.** List the possible number of correct forecasts for this trial period and the probabilities of these outcomes for each assumption about Sunspots' accuracy. **d.** Both of the firm's predictions turn out to be correct! What are your revised probabilities of .9 and .5 accuracy? **e.** Using these posterior probabilities, does accepting .9 or .5 accuracy give a larger expected return?

•• **19.25** *(continuation)* Before knowing the results, what is the most that an expected value maximizer would be willing to pay for a trial subscription?

• **19.26** *(continuation)* Compare the approach in Exercise 19.24 with a classical hypothesis test of H_0: π = .5. What advantages and disadvantages do you see for each procedure?

19.6 SEQUENTIAL DECISIONS (OPTIONAL)

A decision tree shows how decisions and fate interact to determine payoffs. The first insight facilitated by this apparatus is that an intelligent decision requires a consideration of the options and their consequences. The second insight, to be explained now, is that actions taken today may well affect the consequences of future decisions. If so, then anticipated future decisions should influence present decisions.

In our original cereal example, let's say that we are also considering the introduction of a new pet food next year. We will either have to build a new plant to produce this pet food or, if we expand now and find our cereal to be unpopular, this excess capacity can be converted to pet food production. An expanded decision tree is shown in Figure 19.4.

The payoffs are below each event line and the probabilities are on top. The logic reads from left to right: We will expand or maintain the same level of production; the cereal will be popular or unpopular; we will or will not introduce a pet food; and this pet food will be popular or unpopular. The most convenient analysis reads backwards, from right to left. This is *backward induction:* To determine the best decision today, first anticipate what decisions will be made tomorrow.

In each of the four cases, the pet food has a .4 probability of being popular and earning a $400,000 profit. In three of the cases, the pet food will lose $300,000 if unpopular. In the fourth, the loss will be only $100,000 because we convert the unused cereal facility, instead of building a new plant. In this case, the expected value is positive:

$$(\$400,000)(.4) + (-\$100,000)(.6) = \$100,000$$

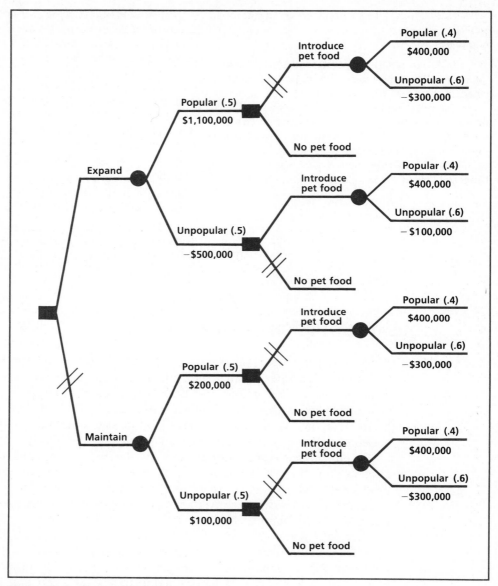

Figure 19.4   *Sequential Decisions*

In the other three cases, it is negative:

$$(\$400,000)(.4) + (-\$300,000)(.6) = -\$20,000$$

Therefore, to maximize expected value, we will not introduce the pet food unless we decide to expand cereal production and find the cereal unpopular, leaving excess production capacity. I have drawn in slash marks on the decision tree in Figure 19.4 to show the branches that will not be followed (that have been "pruned").

Now that we know what we will do tomorrow for each of today's possible actions, we can choose between today's options. If we expand cereal production, there is a .5 probability that we will earn $1,100,000, and a .5 probability that we will lose $500,000 and then convert to pet food with an expected gain of $100,000; the overall expected value is

$$(\$1,100,000)(.5) + (-\$500,000 + \$100,000)(.5) = \$350,000$$

If we instead maintain the same level of production, there is a .5 probability of earning $200,000, and a .5 probability of earning $100,000, for an expected value of $150,000. Expansion has the higher expected value and thus the same production level branch is pruned.

Now we can read Figure 19.4 again from left to right, avoiding the pruned branches. We choose to expand cereal production. If the cereal proves popular, we will not introduce the pet food; if the cereal is unpopular, our excess capacity will be converted to pet food production.

The nice thing about decision trees is that they make explicit the consideration of alternatives and consequences that should at least be implicit in rational decision making. The difficulties are the elicitation of subjective probabilities and the formidable complexity when several sequential decisions with a variety of outcomes are analyzed. Although subjective probabilities may be challenging, they are implicit in most decisions; decision theory just brings them out in the open, where they can be scrutinized and revised. As for the complexity of realistic decision trees, a leading corporate finance book responded this way:

> Life is complex, and there is very little we can do about it. It is therefore unfair to criticize decision trees because they can become complex. Our criticism is reserved for analysts who let the complexity become overwhelming. The point of decision trees is to allow explicit analysis of possible future events and decisions. They should be judged not on their comprehensiveness but on whether they show the most important links between today's and tomorrow's decisions. Decision trees . . . display only a small fraction of possible future events and decisions. Decision trees are like grapevines: they are productive only if they are vigorously pruned.[9]

EXERCISES

19.27 In 1974, an inventor offered to sell Cutler-Hammer the marketing rights to a flight-safety system patent.[10] The ultimate value of this patent depended not only on the quality of the system, but also on pending legislation and the attractiveness of alternative systems. Cutler-Hammer's immediate decision was whether or not to spend $125,000 on a six-month option to the patent rights. They figured that at the end of six months, there was a .29 probability that they would abandon the project due to discouraging developments and a .71 probability that they would exercise their option and bid for a defense contract, in which case they would either lose an additional $575,000 with .85 probability or earn $5,375,000 (not counting the cost of the option) with a probability of .15. Draw a decision tree and determine whether an expected-value maximizer would buy the option.

19.28 Many business inventories are perishable—for instance, the raspberries stocked by a grocery or the daily newspapers at a newsstand. Imagine that you are the buyer for a department store deciding how many Calvin Keen winter coats to order, at a price of $100 each. You will price these coats at $200, but any that are unsold at the end of the season will be disposed of for $60. You will order either one or two dozen coats, and believe that there is a .75 probability that people will buy one dozen, and a .25 probability that they will buy two dozen. (Of course, if you order one dozen, that is the most they can buy.) Sketch a decision tree, showing your two options and the two possible levels of sales. Show the probability of each sales level and the net profit, including the proceeds from disposing of unsold coats. **a.** Which is the maximin decision? **b.** Which decision maximizes the expected value of profits? **c.** How high does the probability of two dozen sales have to be to persuade an expected-value maximizer to order two dozen coats?

19.29 *(continuation)* Now extend your decision tree to show the option, if coats are unsold, of either disposing of them for $60 apiece or trying to sell them for $100. Assume that there is a .5 probability that you will be able to sell these surplus half-price coats and that, if unsuccessful, you will be forced to dispose of them for $40. Use backward induction to show the decision sequences that maximize expected return.

19.30 *(continuation)* What is the most that an expected-value maximizer would pay to know in advance whether one or two dozen coats can be sold for $200 apiece?

19.31 *(continuation)* A marketing survey has a 70 percent success rate; that is, if people will buy one dozen coats, there is a .70 probability that the survey will predict this; if people will buy two dozen, there is a .70 probability that the survey will predict this. Would an expected-value maximizer pay $50 for such a survey? (For simplicity, assume that all unsold coats are disposed of for $60.)

19.7   GAME THEORY (OPTIONAL)

In earlier chapters (and so far in this chapter), we have assumed that a single decision maker calculated probabilities, established confidence intervals, and tested hypotheses. Now we consider a situation with more than one decision maker, where the actions, real or perceived, of each influence the actions of others. These decision makers, players in current jargon, can cooperate or compete with each other.

There are many situations in which the consequences of our actions—the payoffs—depend on the actions of adversaries. In poker, for example, my chances of winning against you depend not only on the cards you and I will be dealt, but on whether you, as my opponent, will bet or fold. In business, the demand for my product depends not only on the quality of my product and the price I charge, but also on the quality and price of the competing product that you sell as my competitor. In war, the success of my advance depends on whether you, as my foe, defend or retreat. In these situations, there is a conflict between our interests, and a rational strategy entails an anticipation of what one's opponent will do. An analysis of such situations is called the *theory of games of strategy* or, more briefly, *game theory,* and such theories have been applied successfully to poker, business, and war.

If You Win, I Lose

In a game of poker, each person's winnings are at the expense of others. Every dollar that goes into one person's pocket comes out of another's. This is called a *zero-sum game.*

> In a **zero-sum game,** the sum of the payoffs to all of the players is zero.

In such games, the conflict among the players' own interests is evident. I can only win if you lose, and vice versa.

In a similar spirit, Lester Thurow, an influential economist, wrote a book called *The Zero-Sum Society,* the title referring to the author's argument that in a world with a fixed amount of resources ("a fixed pie"), distributional issues are paramount. If one sector of society gets more of the available resources, there is necessarily less for others. A central issue for a zero-sum society is how to slice the pie.

Many situations are not zero-sum games. In games of chance, if we reckon the payoffs in terms of personal utility rather than dollars, the sum need not be zero. In business, my sales don't have to be at your expense; we may both sell more if we cut prices or improve product quality—a *positive-sum game.* War is a *negative-sum game,* in that lives are lost and property is destroyed. We can have a positive-sum society if educated and motivated people use technology to increase aggregate output—allowing more pie for everyone. Two-person zero-sum games, however, are by far the easiest to analyze and will be used here to show some important lessons of game theory without getting bogged down in unnecessarily messy details.

First, consider the following game. You and I are both bakers, who have been coexisting by selling donuts and coffee cake in the morning. Now we have both learned to bake croissants, but mine are better than yours. Table 19.9 shows my payoff matrix (often called the game matrix), where the dollar amounts represent daily profits. The strategies, no and yes, describe each of our decisions whether or not to introduce croissants. The dollar entries are my additional daily profits, for each combination of decisions. We assume that this is a zero-sum game, so that any sales of croissants are at the expense of donuts and coffee rolls, just a reshuffling

Table 19.9 *A Bakery Game*

		You		
		No	*Yes*	*Row Minima*
Me	*No*	$0	−$100	−$100
	Yes	$200	$100	$100—*Maximin*
Column Maxima		$200	$100	*Noncooperative Equilibrium*
			Minimax	

of a fixed amount of bakery revenue. Thus, my dollar gains shown in the game matrix represent dollar losses to you.

If neither of us introduces croissants, our revenues are unchanged. If you do but I don't, I will lose $100 in profits to you. If I do but you don't, you will lose $200 in profits to me. If we both introduce croissants, on balance, my profits will go up $100 and your profits will go down $100.

Our interests are obviously antithetical. I am trying to maximize my payoff and you are trying to minimize my payoff (because every additional dollar I make is at your expense). From my standpoint, I would prefer you to not introduce croissants, so that I can make $200. But I cannot count on you to be so generous. Similarly, you would prefer that I not introduce croissants, but must anticipate that if I am sensible, I will.

The row minima show my payoffs if I assume the worst—that whichever action I take, you will make the best possible response. My maximin action (maximizing the minimum amount I will receive) is yes, ensuring a payoff of at least $100. The column maxima show the payoffs you must make, assuming that I make the best possible response. Your minimax strategy (minimizing the maximum you must pay me) is yes, ensuring a payoff of no more than $100.

If we are both rational and anticipate the other's actions, we will both introduce croissants, giving me a $100 gain. This is a *noncooperative equilibrium* in the sense that we are both competing fiercely with each other (not cooperating) and, knowing what the other will do, we both have no incentive to change our actions (equilibrium). If I change from yes to no, my $100 gain will turn into a $100 loss; if you change from yes to no, your $100 loss will grow to $200.

Madness in Our Method

Now let's look at a somewhat different payoff matrix with a very different outcome. We are going to play a simple game of odd or even. At the count of three, I will stick out 1 or 2 fingers and you will do the same. I win if they match; you win if they don't. The payoff matrix is shown in Table 19.10. Now my maximin ($-\$100$) doesn't coincide with your minimax ($+\$100$), and there is no noncooperative equi-

Table 19.10   *Odd or Even*

		You		
		1	*2*	*Row Minima*
Me	*1*	$100	−$200	−$200
	2	−$100	$200	−$100—*Maximin*
Column Maxima		$100	$200	
		Minimax		

librium. If I know that you will show 1 finger, then I will show 1 also. If you know what I will do, you will do the opposite.

In this sort of game, the timing of our actions is very critical. If you must act first, I have the advantage and can counter with a winning move. If I go first, you will win by exploiting knowledge of my move. If we act simultaneously, there is no single, dominant action for either side. What we can say is that if either's action is predictable, then it is as if that person acted first, allowing the other to counter profitably. Therefore, an unpredictable, probabilistic mixture of actions is preferable.

Let's say that I choose 1 with probability p and 2 with probability $1 - p$, while you choose 1 with probability q and 2 with probability $1 - q$. Assuming independence, the probability of a 1–1 match is pq, the probability of 2–2 is $(1 - p)(1 - q)$, and so on, giving an expected payoff of

$$\mu = (\$100)pq + (-\$100)(1 - p)q + (-\$200)p(1 - q) + (\$200)(1 - p)(1 - q)$$

It turns out* that I maximize my expected payoff by choosing $p = 1/2$, and you minimize my expected payoff by choosing $q = 2/3$. With these values the expected payoff is exactly zero, a result that cannot be improved on systematically without knowing the other's action in advance.

To implement these strategies, I could flip a coin to determine my action; you could shuffle three cards: two red and one black. The interesting, perhaps surprising, feature of this analysis is that the most rational strategy is based on the seemingly arbitrary flip of a coin or deal of a card. The reason is easy to understand in the context of this simple, familiar game. To be effective, one's actions must be unpredictable. To ensure that your opponent does not know your action in advance, you use a random strategy to guarantee that you do not know your action in advance either!

This is one of the basic arguments of the great mathematician, John von Neumann, who laid the foundation for modern game theory.[11] As a child, von Neumann had been fascinated by how poker combined both chance (in the deal of the cards) and strategy (when to raise, call, or fold). As an adult, he provided a rigorous proof that a randomized strategy is far better than a fixed (and, hence, predictable) strategy. The poker player whose bets depend solely on the strength of his hand will, by his bets, reveal that strength just as surely as the fool who always smiles when dealt a good hand and frowns when holding a bad one. It is much better to

---

*Maxima and minima can be found by setting the derivatives equal to zero:

$$\partial\mu/\partial p = \$100q + \$100q - \$200(1 - q) - \$200(1 - q)$$
$$= -\$400 + \$600q$$
$$= 0 \text{ if } q = 2/3$$
$$\partial\mu/\partial q = \$100p - \$100(1 - p) + \$200p - \$200(1 - p)$$
$$= -\$300 + \$600p$$
$$= 0 \text{ if } p = 1/2$$

wear a "poker face" and to mix up one's bets to disguise one's hand. A successful poker player must occasionally bluff, so that the opponents will not know for sure whether a large bet is backed by a strong hand.

For similar reasons, successful businesses keep competitors guessing about new products, manufacturing plans, and advertising campaigns. The military must keep its plans secret, too. Even more dramatically, game theory suggests that a successful military strategy should have irrational (i.e., unpredictable) elements. If a nation details its responses to all enemy provocations, then a rational enemy will nibble away at the areas where the responses are weak or ineffective. A more successful strategy may be to keep the opponent guessing. In May 1987, President Reagan was asked what the United States would do if Iran fired missiles at U.S. ships in the Persian Gulf. He refused to be specific, explaining that "It's far better if the Iranians go to bed every night wondering what we might do than us telling them in advance." Similarly, President Richard Nixon reportedly believed in the "mad-man theory" of diplomacy—that foreign policy goals can be accomplished by an apparent irrationality that paralyzes other governments. Germany's Adolf Hitler, Iran's Ayatollah Khomeini, and Libya's Colonel Qaddafi are role models for this theory.

Even more ominously, consider the paradoxes (or contradictions) inherent in mutual deterrence—the doctrine that because the Soviet Union and the United States both have the means to annihilate each other, neither will ever use these ultimate weapons. How credible is a threat to use nuclear weapons, if its purpose is to avoid the use of such weapons? More specifically,

> picture the circumstances of leaders whose country has just been annihilated in a first strike. Now their country is on its way to becoming a radioactive desert, but the retaliatory nuclear force survives in its silos, bombers, and submarines. These leaders of nobody, living in underground shelters or in "doomsday" planes that could not land, would possess the means of national defense but no nation to defend. What rational purpose could they have in launching the retaliatory strike? Since there was no longer a nation, "national security" could not be the purpose. Nor could the defense of other peoples be the purpose, since the retaliatory strike might be the action that would finally break the back of the ecosphere and extinguish the species.[12]

If a retaliatory strike is irrational, then how can the threat of a retaliatory strike be credible enough to deter a first strike? One answer is to persuade the other side that the nation's leaders, motivated by revenge or other emotions, would not behave rationally—creating what Herman Kahn called the "rationality of irrationality." Similar thinking underlies brinkmanship, the threat to launch a first strike if a nation's vital interests are threatened—if the Soviet Union were to invade West Germany, Japan, or Saudi Arabia; if the United States were to invade East Germany, North Korea, or Iran. Such threats are only credible if the other side believes that the leaders are crazy enough to instigate a nuclear war. Irrational actions in a nuclear age involve extraordinarily dangerous bluffing.

Prisoner Dilemmas

The possible gains from unpredictable actions is one interesting and well-known implication of game theory. Another involves what is known as the prisoner's dilemma. This stylized story is as follows. Two people have committed a felony punishable by a twenty-year prison sentence, and the police have arrested two suspects; but so far they have only circumstantial evidence—enough to convict each of a lesser crime, carrying a one-year prison sentence. The suspects are placed in separate rooms and each is offered this deal: "If you provide incriminating evidence against your fellow suspect, we will persuade the judge to knock a year off of your prison sentence, whatever it turns out to be." This is not a zero-sum game, in that talking reduces each suspect's prison time by 1 year, but adds 20 years to the other's time. We can write the payoff matrix as follows, using the notation x, y, with x being suspect 1's prison time and y being suspect 2's time:

	Suspect 2	
Suspect 1	Quiet	Talk
Quiet	1,1	20,0
Talk	0,20	19,19

The objective of each suspect is to minimize his own prison time. The prisoner's dilemma is that each sees an apparent advantage of squealing, but if both talk, both will be made worse off. Actions that seem beneficial to the individual are detrimental to the group.

There is no obvious strategy. The cooperative solution (with both keeping quiet) is far better than both talking; but if there is no mechanism for enforcing the cooperative solution, each feels compelled to talk. Each reasons, "The other will choose to talk or not, regardless of what I do. And no matter what he does, I am better off if I talk." So they are drawn inevitably to mutually destructive squealing.

Analogous situations abound. Collusive businesses agree to charge similar high prices, feeling that price wars are self-destructive. If one firm cuts prices, others quickly follow, and market shares remain unchanged. If aggregate sales are not very responsive to price, then an industry-wide price cut simply reduces profits. For instance, in the aftermath of the banking collapse in the 1930s, bankers appeared before Congress pleading for protection from ruinous deposit rate wars. In the 1920s, these rate wars led to ever higher rates on deposits and left banks no alternative but to make speculative investments, hoping to earn enough to pay depositors the interest that had been promised. (I'm not making this up!) The cooperative solution is for all banks to agree to pay depositors little or no interest, ensuring large profits for all banks.

Responding to these pleas, Congress enforced the cooperative solution with Regulation Q, which prohibited banks from paying any interest on checking accounts and more than a ceiling rate set by the Federal Reserve on savings

Sherlock Holmes Tries to Escape

Von Neumann and Morgenstern use a Sherlock Holmes episode to illustrate the application of game theory.[13] Attempting to shake his nemesis, Professor Moriarty, Sherlock Holmes boards a train going from London to Dover, where he can take a boat to the Continent and make good his escape. As the train leaves the London station, Holmes sees Moriarty on the platform and reasons, correctly, that Moriarty will hire a faster private train and soon catch up with Holmes. Holmes has two options: to continue to Dover or to disembark at the train's only other stop, at Canterbury. Likewise, Moriarty can leave his train at either of these two stops. If Moriarty guesses correctly, Holmes will be killed. If Holmes chooses Dover and Moriarty chooses Canterbury, Holmes will escape to safety, a major victory for Holmes. If Holmes chooses Canterbury and Moriarty chooses Dover,

Holmes will be alive, but still not have escaped to the Continent—essentially a tie game. What should each do?

Von Neumann and Morgenstern model this life-or-death choice as a zero-sum game with the payoff matrix shown in Table 1. There is no noncooperative equilibrium. Either, knowing what the other will do, could win the game. So, a probabilistic mixed strategy is recommended. Holmes maximizes his expected payoff by using a .6 probability of going to Canterbury and .4 to Dover; Moriarty should use a .4 probability of Canterbury and .6 of Dover. The author, Conan Doyle, did not have the benefit of von Neumann-Morgenstern's game theory analysis and ("excusably" according to von Neumann-Morgenstern) does not have his fictional characters base their actions on dice rolls. But he does choose the most likely outcome—Holmes gets off at Canterbury, while Moriarty continues on to Dover.

Table 1   *The Game Between Holmes and Moriarty*

		Moriarty	
		Dover	*Canterbury*
Holmes	*Dover*	−100	+50
	Canterbury	0	−100

accounts. This collusive, cooperative solution held together for nearly 50 years, protecting the profits of banks, large and small, efficient and inefficient, until nonbanks (and a few banks, too) found ways around Regulation Q, and lured depositors with above-average rates. Congress legalized deposit rate wars by abolishing Regulation Q, effective in 1986, and the consequent erosion of profits has been a major factor in the subsequent shakeout in banking.

College Athletics

Many colleges collude on financial aid packages to applicants. At professional meetings, department chairs exchange tentative information on the salaries that will be offered new Ph.D.s, an exchange that can be interpreted as a way of encouraging uniform salaries. Colleges in close proximity often agree not to raid each other's faculties, which short-circuits potential bidding wars. In college athletics, with big money at stake, collusion is out in the open and enforced vigorously.

The recruiting of college athletes involves an apparent prisoner's dilemma. The cooperative solution is for colleges not to pay athletes, since they will play anyway in preparation for later professional careers. Yet each college, seeing how much ticket sales, TV revenue, and alumni contributions can be generated by a winning team, has a strong incentive to recruit athletes with financial inducements. But what happens if all colleges pay student athletes? There is little or no increase in the total number of athletes available and, if colleges offer comparable wages, there will be little or no change in the distribution of athletes among colleges. All that will happen is that college athletes get some of the revenue they are earning for colleges—a most unsatisfactory solution from the colleges' standpoint.

So how do the colleges enforce the more profitable cooperative solution? The National Collegiate Athletic Association (NCAA) enforces strict rules on financial inducements to college athletes (and professional football and basketball cooperate, with a few "financial hardship" exceptions, by not hiring an athlete until his college class has graduated). The intentions are usually stated in terms of morality—to avoid corrupting college athletics and athletes. But the real objective may just be a collusion to enforce a cooperative solution.

In 1987, the *Wall Street Journal* reported that a similar rate war was undermining the cooperative solution on credit card rates:

> The great credit-card war may finally be breaking out.... Issuers have long held to a fragile truce—and charged interest rates averaging 17.8%, far above the cost of other forms of credit....
>
> But the truce is in serious jeopardy: American Express Co. has launched a new revolving charge card, called Optima, that will undercut most Visa and Master-Cards by initially offering a comparatively low 13.5% rate on outstanding balances. Distressed, Visa USA Inc. urged some of its bank clients to stop selling American Express traveler's checks....
>
> Even a modest rate war could cost banks dearly. Charles Russell, Visa USA's president, estimates that even if American Express gets only a quarter of the 10 million holders of upscale premium cards to throw out their bank cards in favor of Optima, the banks would lose $1 billion in interest income and $125 million in fees. In a full-scale rate war, the lost profits would be even higher.[14]

In military affairs, the cooperative solution is for each nation to devote its energies and resources to producing enjoyable goods and services, rather than ever more terrifying weapons that, it is hoped, will never be used. Yet if every nation were unarmed, it would be in the interest of each to produce weapons that enable it, through threat or actual force, to gain control over other nations. And, so, the cooperative solution of peaceful coexistence gives way to a wasteful and frightening arms race.

EXERCISES

19.32 In the game Rock, Scissors, Paper (also known as Roshambo), two players simultaneously show hand signals representing a rock (closed fist), scissors (two fingers), or paper (open palm). Rock beats scissors, scissors beats paper, and paper beats rock. Show the payoff matrix, using $+1$ to represent a win, -1 a loss, and 0 a tie. Is there a noncooperative equilibrium?

19.33 *(continuation)* What mixed strategy would you recommend?

•• 19.34 *(continuation)* How would you counter an opponent who has a .4 probability of showing rock, .3 scissors, and .3 paper?

19.35 The text analyzes a game of odd or even. How would the analysis be altered if the payoffs were $100 to me if we match, $100 to you if we don't?

• 19.36 Outland can attack Inland through either of two routes, but not both, and Inland can choose to mass its defenses at either route, but not both. Assume that route 1 is the more valuable, giving this payoff matrix:

| | *Inland* | |
Outland	*1*	*2*
1	100	-100
2	-200	100

What strategies do you recommend for each side?

19.8 SUMMARY

Decision trees and payoff matrixes offer ways of organizing our plans, recognizing explicitly the options that are available and how each action might be affected by uncertain events. The maximax and maximin strategies assume that, whatever action we take, the best, or worst, possible event will occur. A maximum likelihood strategy assumes that the most likely event will occur. Expected value maximization weighs the possible events by their respective probabilities. For those who are not risk neutral, we can assign utility values to the various possible monetary payoffs and choose the action that maximizes expected utility.

Posterior probabilities use Bayes' theorem to combine prior probabilities and sample data—perhaps a marketing survey, tests for defective products, or a medical examination. These posterior probabilities can be used to figure the expected value of each possible action and, before the data are collected, to calculate the expected value of the data themselves. A sequential analysis encourages us to plan ahead, to take into account tomorrow's options when selecting today's actions.

Game theory analyzes choices when the consequences depend not on mere chance, but on competitors made hostile by the fact that the payoff may, to some extent, be at their expense. The simplest case involves zero-sum games, where the payoff to one person is entirely at the expense of others. Among the lessons of game theory are that some situations have a noncooperative equilibrium (where, even knowing the others' actions, no one has an incentive to behave differently); other situations require a randomized strategy (to ensure that others cannot anticipate your actions), and some involve a prisoner's dilemma (where the players are lured to individually beneficial, but mutually destructive actions).

REVIEW EXERCISES

19.37 Ira Betaman is considering investing $100,000 in either Treasury bills, with a guaranteed 5 percent return, or in a Bynight Inc. "junk" bond, a security considered so speculative that it is not rated by the major bond services. Because of its speculative nature, this bond is inexpensive and the prospective return high, if there is no default. The payoff matrix is given in Table 19.11. Which should Ira buy if he follows a maximax strategy? If he follows a maximin strategy? What other information might be of interest to Ira?

19.38 *(continuation)* Historical data for junk bonds issued by similar companies suggest that 5 percent have a total default and another 20 percent a partial default. If Ira uses these as prior probabilities for the Bynight bond, which action is favored by maximum likelihood? By maximum expected value?

19.39 *(continuation)* Ira's utility function is $U = .20R - .001R^2$, where R is the rate of return on this investment ($R = 5, 20$, etc.). Which action maximizes his expected utility? Explain why Ira does not always choose the action that maximizes the expected value of his return.

• **19.40** *(continuation)* Ira's brokerage firm has rated Bynight's bond relative to other junk bonds as "above average risk." Half of all junk bonds are so

rated, including 100 percent of those that (later) have a total default, 75 percent of those that have a partial default, and 40 percent of those that never default. Calculate the posterior probabilities and the expected value of the return, using these revised probabilities.

19.41 Charlie Hustle earns a living by typing and by giving windsurfing lessons. On any given day, he can either stay home and type papers or drive to the beach and set up his surfboards, with a payoff dependent on the weather:

	Events	
Actions	*Rain*	*No Rain*
Surf	$0	$120
Type	$40	$40

Which action is maximax? Which is maximin? At this time of year, it rains 40 percent of the days. If Charlie uses this historical frequency as his prior probability, which action is maximum likelihood? Which maximizes the expected value of his payoff?

19.42 *(continuation)* A barometer correctly predicts "rain" for 80 percent of the days when it does rain and correctly predicts "no rain" for 70 percent of the days without rain. When this barometer predicts "rain," what are the posterior probabilities of rain and

Table 19.11 *Payoff Matrix for Exercise 19.37*

	Events		
Actions	*No Default*	*Partial Default*	*Total Default*
Junk	+20%	−10%	−100%
T-Bills	+5%	+5%	+5%

no rain? Which action is maximum likelihood? Which maximizes the expected value of the payoff? Answer these same three questions for the case where the barometer predicts "no rain."

19.43 *(continuation)* What is the expected value of perfect information? Calculate the most that an expected value maximizer would pay to use the barometer; explain this answer.

19.44 An *Esquire* column discussed how a businessperson should react after a "humongous breakdown in which your malfeasance, laziness, or stupidity creates a problem for the whole company."[15] One possibility is to keep quiet, hoping that unexpected events will make the mistake irrelevant and it will go undiscovered. However, if you do not admit your mistake and it is discovered, you will be treated harshly. The columnist recommends that "When the possibility of discovery reaches 60 percent, blow the whistle on yourself."

If the payoff matrix looks like this, for what value of X will it be true that admitting a mistake maximizes the expected payoff for discovery probabilities larger than .60?

	Events	
Action	*Discovered*	*Not Discovered*
Admit Mistake	$-X$	$-X$
Keep Quiet	-100	0

19.45 It has been estimated that 10 percent of American workers use illegal drugs and that a urine test has anywhere from a 30 percent to 95 percent probability of correctly detecting such drugs, depending on the drug and the skill of the person conducting the test. Drug tests also often give false positive readings, especially if certain medications or particular enzymes are present in the urine. For this exercise, let's suppose that

$$P[\text{drug}] = .10 \quad P[\text{"pos"}|\text{drug}] = .70$$
$$P[\text{"pos"}|\text{no drug}] = .20$$

What are the posterior probabilities of drugs and no drugs if a test gives a positive reading? If a test gives a negative reading?

Here is an assumed payoff matrix, depending on whether the employee is fired or not and whether the employee really does or does not have a drug problem:

	Events	
Actions	*Drug*	*No Drug*
Fire	0	-50
Keep	-100	0

Which action would be taken by a minimaxer? Explain. Which action maximizes the expected value of the payoff if there is no test? If the test is positive? If the test is negative? Explain why your answers change if the payoff for firing someone who is drug-free is reduced from -50 to -20.

19.46 Traditionally, security guards have walked their appointed rounds along fixed routes, punching time clocks at assigned times to show that they are not loafing or, even worse, sleeping on the job. What improvement does game theory suggest?

19.47 It is one week before a presidential election and one of the candidates, Brett Blueyes, must decide whether to campaign in California (with 47 electoral votes) or New York (with 36 electoral votes). The election is winner-take-all, in that whoever gets the most votes in a state gets all of its electoral votes. Here are his (independent) probabilities:

Campaign Options	*Probability of Winning*	
	California	*New York*
Only California	.70	.10
Only New York	.20	.75
Both States	.40	.50

Which of these three options would an expected-value maximizer choose? Can you think of any rational reason why Blueyes might choose a different option?

19.48 It is the weekend before the presidential election and, as things stand, it is a tossup whether Blueyes or his opponent, Thickhair, will win California's 47 electoral votes and New York's 36 votes. Each is prepared to launch an all-out media blitz, but each can afford to saturate only one of these two states. Here is the game matrix:

	Thickhair	
Blueyes	*California*	*New York*
California	0	$+11$
New York	-11	0

Is there a noncooperative equilibrium? Explain.

19.49 Mr. Smith is a New Yorker who knows nothing about weather forecasting other than it is wet in New York one-third of the time. He decides to construct a "weather spinner" similar to those that many professionals are rumored to use. He will draw a circle and color part of it green ("wet") and the remainder yellow ("dry"). A free-spinning needle is then attached and spun each day to make a weather forecast. If Mr. Smith wants to maximize the probability of making a correct forecast, what fraction of the circle should be colored green? Don't just guess; try to derive a formal answer.

Mr. Smith is going to move to San Juan, Puerto Rico, where there is, on average, precipitation 199 days a year. How should he adjust his weather spinner?

19.50 Good, Bad, and Ugly are to fight a three-cornered pistol duel. All know that Good's probability of hitting a target is .3, Ugly's chance is .5, and Bad never misses. They are to fire at their choice of target in succession—Good, then Bad, then Ugly—until only one is left unhit. What should Good's strategy be? (Give a definite answer.)

19.51 A small company is contractually committed to marketing a new data compression technology for personal computers that increases storage capability and reduces telephonic transmission time. They will either launch an all-out advertising campaign, hoping to grab a large market share before rivals are able to introduce similar products, or else proceed cautiously to avoid an expensive mistake. Table 19.12 shows their payoff matrix and associated probabilities. **a.** Which action is maximax? Which is maximin? **b.** Which action is maximum likelihood? **c.** Which action maximizes expected value?

19.52 *(continuation)* What is the value of perfect information about the product's success?

19.53 *(continuation)* If the company decides to proceed cautiously and is maximizing expected utility, would you characterize this decision as risk neutral, risk averse, or risk seeking? Why?

19.54 *(continuation)* The firm decides to interview five respected potential buyers; their reaction is negative. Historically, these buyers have never been negative for a product that ultimately turned out to be very popular, have been negative on 50 percent of the products that turned out to be mildly popular, and have been negative on 80 percent of the products that proved to be unpopular. What should be the firm's revised probabilities of this product's success in light of this negative reaction? Now which action maximizes expected value?

19.55 In the Indian game Tong, two players simultaneously show 1, 2, or 3 fingers, with one player winning if the sum is even and the other winning if the sum is odd. Assume that you are playing for even, with the following payoff matrix. If your opponent is equally likely to show 1, 2, or 3 fingers, what strategy maximizes your expected payoff?

		Opponent		
		1	*2*	*3*
	1	$+1$	-1	$+1$
You	*2*	-1	$+1$	-1
	3	$+1$	-1	$+1$

19.56 Exercise 9.61 explains the rules of Roshambo (rock-scissors-paper), in which two players simultaneously move their fists up and down three times and then show a fist (rock), two fingers (scissors), or an open hand (paper). Rock beats scissors, scissors beats paper, and paper beats rock. You are going to play once against an opponent, receiving $1 for a win, paying $1 for a loss, and nothing for a tie.

Table 19.12 *Payoff Matrix for Exercise 19.51*

	Event		
Action	*Very Popular* *(P = .1)*	*Mildly Popular* *(P = .4)*	*Unpopular* *(P = .5)*
Caution	$1,000,000	$200,000	$- $50,000
Flood	$5,000,000	$600,000	$- $200,000

(The game will not be replayed, even if there is a tie.) Show the payoff matrix for this game. Is this a zero-sum game?

19.57 *(continuation)* Is there a noncooperative equilibrium?

19.58 *(continuation)* In 50 games of Roshambo, the student mentioned in Exercise 9.61 played rock 26 times, scissors 9 times, and paper 15 times. If this student is your opponent and we assume that these empirical frequencies are the probabilities that this student's first move will be rock, scissors, or paper, what strategy maximizes the expected value of your payoff?

19.59 *(continuation)* If one player plays rock with probability P_R, scissors with probability P_S, and paper with probability $(1 - P_R - P_S)$, and the opponent plays rock with probability Q_R, scissors with probability Q_S, and paper with probability $(1 - Q_R - Q_S)$, what values of these probabilities do you suppose maximizes the expected value of the payoff for each side? (Use your intuition—do not attempt a formal proof.) For these expected-value-maximizing probabilities, what is the expected value of each player's payoff?

•• 19.60 *(continuation)* Use calculus to determine formally the probability values that maximize the expected value of the payoff for each side.

19.61 In 1989 it was estimated that Activase, which cost \$2,500 per treatment, had a 70 percent chance of dissolving blood clots in heart-attack patients, while Streptokinase, which cost \$250, had a 35 percent chance of success. If we let X be the amount one would pay for a drug that is certain to dissolve clots, then a risk-neutral person would be willing to pay PX for a drug with a probability P of dissolving clots. For what values of X would such a person choose Activase? Choose Streptokinase? Choose neither?

19.62 *(continuation)* In 1990 some studies indicated that Activase was no more effective than Streptokinase. If so, for what values of X would the risk-neutral person described in the preceding exercise choose Activase? Choose Streptokinase?

19.63 A soft-drink company is considering a special summer advertising campaign, with the potential payoffs (in millions of dollars) shown in the following table. Which action is maximax? Which is maximin?

How would you describe a company that consistently follows a maximax strategy? That consistently follows a maximin strategy?

	Success of Ad Campaign	
Actions	Hit	Flop
Advertise	300	−100
Don't	0	0

19.64 *(continuation)* If hit and flop are equally likely, which strategy maximizes the expected value of the payoff? What is the minimal probability of a hit for which advertising has a nonnegative expected value?

19.65 Why do the payoffs in Exercise 19.63 depend critically of what is assumed about the action's of the company's competitors? Do you think that the competitors' actions are affected by whether or not this soft-drink company launches a special summer advertising campaign? Give specific examples to illustrate and bolster your arguments.

19.66 The soft-drink company described in Exercise 19.63 actually considered three, rather than two, possible results of the proposed ad campaign, with the potential payoffs (in millions of dollars) shown below. Which action is maximax? Which is maximin?

	Success of Ad Campaign		
Actions	Great	Modest	Flop
Advertise	300	50	−100
Don't	0	0	0

19.67 *(continuation)* If great, modest, and flop are equally likely, which action has the highest expected payoff? What if there is a .2 probability of great success, a .3 probability of modest success, and a .5 probability of a flop?

19.68 Consider the payoff matrix in Table 19.13. If great, modest, and flop are equally likely, which action will be taken by a risk-neutral person? In this situation, is a risk-averse person more likely or less likely to advertise than a risk-neutral person?

• **19.69** The payoff matrix in Exercise 19.66 was presented to the company's board of directors, but for simplicity, the possible outcomes from the ad campaign had to be reduced from three to two. The modest success category was consequently split in half, as shown in the following table. Explain, using specific examples if helpful, why this procedure is or is not appropriate.

Actions	Success of Ad Campaign	
	Plus	Minus
Advertise	325	−75
Don't	0	0

19.70 (continuation) Alternatively, it was proposed that the most success category be averaged with the two extremes, as shown in the following table. (For example, $(300 + 50)/2 = 175$.) Explain, using specific examples if helpful, why this procedure is or is not appropriate.

Actions	Success of Ad Campaign	
	Plus	Minus
Advertise	175	−25
Don't	0	0

19.71 (continuation) On balance, if you were making the presentation to the board of directors, how would you choose between the payoff matrixes in the two preceding exercises?

19.72 A city in Michigan considered three alternatives for repairing several streets that had been damaged by winter storms: patching, resurfacing, and rebuilding. A rebuilt road might not need any further maintenance for ten years. The maintenance required for a patched or resurfaced road depends on the severity of future winters. The city planners came up with the following cost estimates. (All figures are present values, in millions of dollars.) Which action is minimin? Which is minimax? Does one of these strategies dominate the other two?

Actions	Winter Weather		
	Mild	Normal	Severe
Patch	10	20	50
Resurface	20	30	40

Actions	Winter Weather		
	Mild	Normal	Severe
Rebuild	30	30	30

19.73 (continuation) Label the respective probabilities of mild, moderate, and severe weather during the next several years P_m, P_n, and P_s, where $P_m + P_n + P_s = 1$. If $P_m = P_n = P_s$, which action minimizes the expected value of costs?

19.74 (continuation) Using the standard deviation to gauge risk, which action would be taken by someone who is risk neutral? Risk seeking? Risk averse?

19.75 (continuation) If $P_m = .25$, $P_n = .50$, and $P_s = .25$, which action minimizes the expected value of costs?

19.76 (continuation) If $P_n = .60$, are there any values of P_m and P_s for which rebuilding has a lower expected cost than patching?

19.77 Exercise 19.72 describes the potential dollar cost of various street-repair actions. Taking into account the political repercussions of road costs and inconvenience, the mayor's cost matrix, in terms of utility, is as follows. If $P_m = P_n = P_s$, which action maximizes the mayor's expected utility?

Actions	Winter Weather		
	Mild	Normal	Severe
Patch	−40	−50	−100
Resurface	−50	−50	−80
Rebuild	−60	−60	−60

19.78 (continuation) If $P_m = .25$, $P_n = .50$, and $P_s = .25$, which action maximizes the mayor's expected utility?

19.79 Why do you suppose that when several people are suspected of involvement in a crime, police usually prefer to interrogate each suspect separately? Are there any circumstances in which the police might prefer to interrogate two suspects simultaneously?

19.80 Why do professional football teams keep records of the frequencies with which each of their opponents calls certain plays in certain situations? Why do they keep records of the frequencies with which they themselves call certain plays in certain situations?

Tables

Table 1 *Binomial Probabilities*

These are the probabilities of x successes in n Bernoulli trials, each with probability π of success:
$P[x] = \binom{n}{x} \pi^x (1 - \pi)^{n-x}$. For instance, with a $\pi = .20$ success probability, the probability of $x = 2$ successes in $n = 5$ trials is .2048. (*Note:* For $\pi > .5$, use the probability of $n - x$ successes with success probability $1 - \pi$. For instance, the probability of 4 successes in 6 trials when $\pi = .75$ is the same as the probability of 2 successes when $\pi = .25$, i.e., .2966.)

n	x	.01	.05	.10	.15	.20	π .25	.30	.35	.40	.45	.50
1	0	.9900	.9500	.9000	.8500	.8000	.7500	.7000	.6500	.6000	.5500	.5000
	1	.0100	.0500	.1000	.1500	.2000	.2500	.3000	.3500	.4000	.4500	.5000
2	0	.9801	.9025	.8100	.7225	.6400	.5625	.4900	.4225	.3600	.3025	.2500
	1	.0198	.0950	.1800	.2550	.3200	.3750	.4200	.4550	.4800	.4950	.5000
	2	.0001	.0025	.0100	.0225	.0400	.0625	.0900	.1225	.1600	.2025	.2500
3	0	.9703	.8574	.7290	.6141	.5120	.4219	.3430	.2746	.2160	.1664	.1250
	1	.0294	.1354	.2430	.3251	.3840	.4219	.4410	.4436	.4320	.4084	.3750
	2	.0003	.0071	.0270	.0574	.0960	.1406	.1890	.2389	.2880	.3341	.3750
	3	.0000	.0001	.0010	.0034	.0080	.0156	.0270	.0429	.0640	.0911	.1250
4	0	.9606	.8145	.6561	.5220	.4096	.3164	.2401	.1785	.1296	.0915	.0625
	1	.0388	.1715	.2916	.3685	.4096	.4219	.4116	.3845	.3456	.2995	.2500
	2	.0006	.0135	.0486	.0975	.1536	.2109	.2646	.3105	.3456	.3675	.3750
	3	.0000	.0005	.0036	.0115	.0256	.0469	.0756	.1115	.1536	.2005	.2500
	4	.0000	.0000	.0001	.0005	.0016	.0039	.0081	.0150	.0256	.0410	.0625

Table 1 *Continued*

n	x	.01	.05	.10	.15	.20	π .25	.30	.35	.40	.45	.50
5	0	.9510	.7738	.5905	.4437	.3277	.2373	.1681	.1160	.0778	.0503	.0312
	1	.0480	.2036	.3280	.3915	.4096	.3955	.3602	.3124	.2592	.2059	.1562
	2	.0010	.0214	.0729	.1382	.2048	.2637	.3087	.3364	.3456	.3369	.3125
	3	.0000	.0011	.0081	.0244	.0512	.0879	.1323	.1811	.2304	.2757	.3125
	4	.0000	.0000	.0004	.0022	.0064	.0146	.0284	.0488	.0768	.1128	.1562
	5	.0000	.0000	.0000	.0001	.0003	.0010	.0024	.0053	.0102	.0185	.0312
6	0	.9415	.7351	.5314	.3771	.2621	.1780	.1176	.0754	.0467	.0277	.0156
	1	.0571	.2321	.3543	.3993	.3932	.3560	.3025	.2437	.1866	.1359	.0938
	2	.0014	.0305	.0984	.1762	.2458	.2966	.3241	.3280	.3110	.2780	.2344
	3	.0000	.0021	.0146	.0415	.0819	.1318	.1852	.2355	.2765	.3032	.3125
	4	.0000	.0001	.0012	.0055	.0154	.0330	.0595	.0951	.1382	.1861	.2344
	5	.0000	.0000	.0001	.0004	.0015	.0044	.0102	.0205	.0369	.0609	.0938
	6	.0000	.0000	.0000	.0000	.0001	.0002	.0007	.0018	.0041	.0083	.0156
7	0	.9321	.6983	.4783	.3206	.2097	.1335	.0824	.0490	.0280	.0152	.0078
	1	.0659	.2573	.3720	.3960	.3670	.3115	.2471	.1848	.1306	.0872	.0547
	2	.0020	.0406	.1240	.2097	.2753	.3115	.3177	.2985	.2613	.2140	.1641
	3	.0000	.0036	.0230	.0617	.1147	.1730	.2269	.2679	.2903	.2918	.2734
	4	.0000	.0002	.0026	.0109	.0287	.0577	.0972	.1442	.1935	.2388	.2734
	5	.0000	.0000	.0002	.0012	.0043	.0115	.0250	.0466	.0774	.1172	.1641
	6	.0000	.0000	.0000	.0001	.0004	.0013	.0036	.0084	.0172	.0320	.0547
	7	.0000	.0000	.0000	.0000	.0000	.0001	.0002	.0006	.0016	.0037	.0078
8	0	.9227	.6634	.4305	.2725	.1678	.1002	.0576	.0319	.0168	.0084	.0039
	1	.0746	.2793	.3826	.3847	.3355	.2670	.1977	.1373	.0896	.0548	.0312
	2	.0026	.0515	.1488	.2376	.2936	.3115	.2965	.2587	.2090	.1569	.1094
	3	.0001	.0054	.0331	.0839	.1468	.2076	.2541	.2786	.2787	.2568	.2188
	4	.0000	.0004	.0046	.0185	.0459	.0865	.1361	.1875	.2322	.2627	.2734
	5	.0000	.0000	.0004	.0026	.0092	.0231	.0467	.0808	.1239	.1719	.2188
	6	.0000	.0000	.0000	.0002	.0011	.0038	.0100	.0217	.0413	.0403	.1094
	7	.0000	.0000	.0000	.0000	.0001	.0004	.0012	.0033	.0079	.0164	.0312
	8	.0000	.0000	.0000	.0000	.0000	.0000	.0001	.0002	.0007	.0017	.0039
9	0	.9135	.6302	.3874	.2316	.1342	.0751	.0404	.0207	.0101	.0046	.0020
	1	.0830	.2985	.3874	.3679	.3020	.2253	.1556	.1004	.0605	.0339	.0176
	2	.0034	.0629	.1722	.2597	.3020	.3003	.2668	.2162	.1612	.1110	.0703
	3	.0001	.0077	.0446	.1069	.1762	.2336	.2668	.2716	.2508	.2119	.1641
	4	.0000	.0006	.0074	.0283	.0661	.1168	.1715	.2194	.2508	.2600	.2461
	5	.0000	.0000	.0008	.0050	.0165	.0389	.0735	.1181	.1672	.2128	.2461
	6	.0000	.0000	.0001	.0006	.0028	.0087	.0210	.0424	.0743	.1160	.1641
	7	.0000	.0000	.0000	.0000	.0003	.0012	.0039	.0098	.0212	.0407	.0703
	8	.0000	.0000	.0000	.0000	.0000	.0001	.0004	.0013	.0035	.0083	.0176
	9	.0000	.0000	.0000	.0000	.0000	.0000	.0000	.0001	.0003	.0008	.0020
10	0	.9044	.5987	.3487	.1969	.1074	.0563	.0282	.0135	.0060	.0025	.0010
	1	.0914	.3151	.3874	.3474	.2684	.1877	.1211	.0725	.0403	.0207	.0098
	2	.0042	.0746	.1937	.2759	.3020	.2816	.2335	.1757	.1209	.0763	.0439
	3	.0001	.0105	.0574	.1298	.2013	.2503	.2668	.2522	.2150	.1665	.1172
	4	.0000	.0010	.0112	.0401	.0881	.1460	.2001	.2377	.2508	.2384	.2051

(continues)

Table 1 *Continued*

n	x	.01	.05	.10	.15	.20	π .25	.30	.35	.40	.45	.50
	5	.0000	.0001	.0015	.0085	.0264	.0584	.1029	.1536	.2007	.2340	.2461
	6	.0000	.0000	.0001	.0012	.0055	.0162	.0368	.0689	.1115	.1596	.2051
	7	.0000	.0000	.0000	.0001	.0008	.0031	.0090	.0212	.0425	.0746	.1172
	8	.0000	.0000	.0000	.0000	.0001	.0004	.0014	.0043	.0106	.0229	.0439
	9	.0000	.0000	.0000	.0000	.0000	.0000	.0001	.0005	.0016	.0042	.0098
	10	.0000	.0000	.0000	.0000	.0000	.0000	.0000	.0000	.0001	.0003	.0010
11	0	.8953	.5688	.3138	.1673	.0859	.0422	.0198	.0088	.0036	.0014	.0005
	1	.0995	.3293	.3835	.3248	.2362	.1549	.0932	.0518	.0266	.0125	.0054
	2	.0050	.0867	.2131	.2866	.2953	.2581	.1998	.1395	.0887	.0513	.0269
	3	.0002	.0137	.0710	.1517	.2215	.2581	.2568	.2254	.1774	.1259	.0806
	4	.0000	.0014	.0158	.0536	.1107	.1721	.2201	.2428	.2365	.2060	.1611
	5	.0000	.0001	.0025	.0132	.0388	.0803	.1321	.1830	.2207	.2360	.2256
	6	.0000	.0000	.0003	.0023	.0097	.0268	.0566	.0985	.1471	.1931	.2256
	7	.0000	.0000	.0000	.0003	.0017	.0064	.0173	.0379	.0701	.1128	.1611
	8	.0000	.0000	.0000	.0000	.0002	.0011	.0037	.0102	.0234	.0462	.0806
	9	.0000	.0000	.0000	.0000	.0000	.0001	.0005	.0018	.0052	.0126	.0269
	10	.0000	.0000	.0000	.0000	.0000	.0000	.0000	.0002	.0007	.0021	.0054
	11	.0000	.0000	.0000	.0000	.0000	.0000	.0000	.0000	.0000	.0002	.0005
12	0	.8864	.5404	.2824	.1422	.0687	.0317	.0138	.0057	.0022	.0008	.0002
	1	.1074	.3413	.3766	.3012	.2062	.1267	.0712	.0368	.0174	.0075	.0029
	2	.0060	.0988	.2301	.2924	.2835	.2323	.1678	.1088	.0639	.0339	.0161
	3	.0002	.0173	.0852	.1720	.2362	.2581	.2397	.1954	.1419	.0923	.0537
	4	.0000	.0021	.0213	.0683	.1329	.1936	.2311	.2367	.2128	.1700	.1208
	5	.0000	.0002	.0038	.0193	.0532	.1032	.1585	.2039	.2270	.2225	.1934
	6	.0000	.0000	.0005	.0040	.0155	.0401	.0792	.1281	.1766	.2124	.2256
	7	.0000	.0000	.0000	.0006	.0033	.0115	.0291	.0591	.1009	.1489	.1934
	8	.0000	.0000	.0000	.0001	.0005	.0024	.0078	.0199	.0420	.0762	.1208
	9	.0000	.0000	.0000	.0000	.0001	.0004	.0015	.0048	.0125	.0277	.0537
	10	.0000	.0000	.0000	.0000	.0000	.0000	.0002	.0008	.0025	.0068	.0161
	11	.0000	.0000	.0000	.0000	.0000	.0000	.0000	.0001	.0003	.0010	.0029
	12	.0000	.0000	.0000	.0000	.0000	.0000	.0000	.0000	.0000	.0001	.0002
13	0	.8775	.5133	.2542	.1209	.0550	.0238	.0097	.0037	.0013	.0004	.0001
	1	.1152	.3512	.3672	.2774	.1787	.1029	.0540	.0259	.0113	.0045	.0016
	2	.0070	.1109	.2448	.2937	.2680	.2059	.1388	.0836	.0453	.0220	.0095
	3	.0003	.0214	.0997	.1900	.2457	.2517	.2181	.1651	.1107	.0660	.0349
	4	.0000	.0028	.0277	.0838	.1535	.2097	.2337	.2222	.1845	.1350	.0873
	5	.0000	.0003	.0055	.0266	.0691	.1258	.1803	.2154	.2214	.1989	.1571
	6	.0000	.0000	.0008	.0063	.0230	.0559	.1030	.1546	.1968	.2169	.2095
	7	.0000	.0000	.0001	.0011	.0058	.0186	.0442	.0833	.1312	.1775	.2095
	8	.0000	.0000	.0001	.0001	.0011	.0047	.0142	.0336	.0656	.1089	.1571
	9	.0000	.0000	.0000	.0000	.0001	.0009	.0034	.0101	.0243	.0495	.0873
	10	.0000	.0000	.0000	.0000	.0000	.0001	.0006	.0022	.0065	.0162	.0349
	11	.0000	.0000	.0000	.0000	.0000	.0000	.0001	.0003	.0012	.0036	.0095
	12	.0000	.0000	.0000	.0000	.0000	.0000	.0000	.0000	.0001	.0005	.0016
	13	.0000	.0000	.0000	.0000	.0000	.0000	.0000	.0000	.0000	.0000	.0001

Table 1 *Continued*

n	x	.01	.05	.10	.15	.20	π .25	.30	.35	.40	.45	.50
14	0	.8687	.4877	.2288	.1028	.0440	.0178	.0068	.0024	.0008	.0002	.0001
	1	.1229	.3593	.3559	.2539	.1539	.0832	.0407	.0181	.0073	.0027	.0009
	2	.0081	.1229	.2570	.2912	.2501	.1802	.1134	.0634	.0317	.0141	.0056
	3	.0003	.0259	.1142	.2056	.2501	.2402	.1943	.1366	.0845	.0462	.0222
	4	.0000	.0037	.0349	.0998	.1720	.2202	.2290	.2022	.1549	.1040	.0611
	5	.0000	.0004	.0078	.0352	.0860	.1468	.1963	.2178	.2066	.1701	.1222
	6	.0000	.0000	.0013	.0093	.0322	.0734	.1262	.1759	.2066	.2088	.1833
	7	.0000	.0000	.0002	.0019	.0092	.0280	.0618	.1082	.1574	.1952	.2095
	8	.0000	.0000	.0000	.0003	.0020	.0082	.0232	.0510	.0918	.1398	.1833
	9	.0000	.0000	.0000	.0000	.0003	.0018	.0066	.0183	.0408	.0762	.1222
	10	.0000	.0000	.0000	.0000	.0000	.0003	.0014	.0049	.0136	.0312	.0611
	11	.0000	.0000	.0000	.0000	.0000	.0000	.0002	.0010	.0033	.0093	.0222
	12	.0000	.0000	.0000	.0000	.0000	.0000	.0000	.0001	.0005	.0019	.0056
	13	.0000	.0000	.0000	.0000	.0000	.0000	.0000	.0000	.0001	.0002	.0009
	14	.0000	.0000	.0000	.0000	.0000	.0000	.0000	.0000	.0000	.0000	.0001
15	0	.8601	.4633	.2059	.0874	.0352	.0134	.0047	.0016	.0005	.0001	.0000
	1	.1303	.3658	.3432	.2312	.1319	.0668	.0305	.0126	.0047	.0016	.0005
	2	.0092	.1348	.2669	.2856	.2309	.1559	.0916	.0476	.0219	.0090	.0032
	3	.0004	.0307	.1285	.2184	.2501	.2252	.1700	.1110	.0634	.0318	.0139
	4	.0000	.0049	.0428	.1156	.1876	.2252	.2186	.1792	.1268	.0780	.0417
	5	.0000	.0006	.0105	.0449	.1032	.1651	.2061	.2123	.1859	.1404	.0916
	6	.0000	.0000	.0019	.0132	.0430	.0917	.1472	.1906	.2066	.1914	.1527
	7	.0000	.0000	.0003	.0030	.0138	.0393	.0811	.1319	.1771	.2013	.1964
	8	.0000	.0000	.0000	.0005	.0035	.0131	.0348	.0710	.1181	.1647	.1964
	9	.0000	.0000	.0000	.0001	.0007	.0034	.0116	.0298	.0612	.1048	.1527
	10	.0000	.0000	.0000	.0000	.0001	.0007	.0030	.0096	.0245	.0515	.0916
	11	.0000	.0000	.0000	.0000	.0000	.0001	.0006	.0024	.0074	.0191	.0417
	12	.0000	.0000	.0000	.0000	.0000	.0000	.0001	.0004	.0016	.0052	.0139
	13	.0000	.0000	.0000	.0000	.0000	.0000	.0000	.0001	.0003	.0010	.0032
	14	.0000	.0000	.0000	.0000	.0000	.0000	.0000	.0000	.0000	.0001	.0005
	15	.0000	.0000	.0000	.0000	.0000	.0000	.0000	.0000	.0000	.0000	.0000
16	0	.8515	.4401	.1853	.0743	.0281	.0100	.0033	.0010	.0003	.0001	.0000
	1	.1376	.3706	.3294	.2097	.1126	.0535	.0228	.0087	.0030	.0009	.0002
	2	.0104	.1463	.2745	.2775	.2111	.1336	.0732	.0353	.0150	.0056	.0018
	3	.0005	.0359	.1423	.2285	.2463	.2079	.1465	.0888	.0468	.0215	.0085
	4	.0000	.0061	.0514	.1311	.2001	.2252	.2040	.1553	.1014	.0572	.0278
	5	.0000	.0008	.0137	.0555	.1201	.1802	.2099	.2008	.1623	.1123	.0667
	6	.0000	.0001	.0028	.0180	.0550	.1101	.1649	.1982	.1983	.1684	.1222
	7	.0000	.0000	.0004	.0045	.0197	.0524	.1010	.1524	.1889	.1969	.1746
	8	.0000	.0000	.0001	.0009	.0055	.0197	.0487	.0923	.1417	.1812	.1964
	9	.0000	.0000	.0000	.0001	.0012	.0058	.0185	.0442	.0840	.1318	.1746
	10	.0000	.0000	.0000	.0000	.0002	.0014	.0056	.0167	.0392	.0755	.1222
	11	.0000	.0000	.0000	.0000	.0000	.0002	.0013	.0049	.0142	.0337	.0667

(continues)

Table 1 *Continued*

n	x	.01	.05	.10	.15	.20	π .25	.30	.35	.40	.45	.50
	12	.0000	.0000	.0000	.0000	.0000	.0000	.0002	.0011	.0040	.0115	.0278
	13	.0000	.0000	.0000	.0000	.0000	.0000	.0000	.0002	.0008	.0029	.0085
	14	.0000	.0000	.0000	.0000	.0000	.0000	.0000	.0000	.0001	.0005	.0018
	15	.0000	.0000	.0000	.0000	.0000	.0000	.0000	.0000	.0000	.0001	.0002
	16	.0000	.0000	.0000	.0000	.0000 ·	.0000	.0000	.0000	.0000	.0000	.0000
17	0	.8429	.4181	.1668	.0631	.0225	.0075	.0023	.0007	.0002	.0000	.0000
	1	.1447	.3741	.3150	.1893	.0957	.0426	.0169	.0060	.0019	.0005	.0001
	2	.0117	.1575	.2800	.2673	.1914	.1136	.0581	.0260	.0102	.0035	.0010
	3	.0006	.0415	.1556	.2359	.2393	.1893	.1245	.0701	.0341	.0144	.0052
	4	.0000	.0076	.0605	.1457	.2093	.2209	.1868	.1320	.0796	.0411	.0182
	5	.0000	.0010	.0175	.0668	.1361	.1914	.2081	.1849	.1379	.0875	.0472
	6	.0000	.0001	.0039	.0236	.0680	.1276	.1784	.1991	.1839	.1432	.0944
	7	.0000	.0000	.0007	.0065	.0267	.0668	.1201	.1685	.1927	.1841	.1484
	8	.0000	.0000	.0001	.0014	.0084	.0279	.0644	.1134	.1606	.1883	.1855
	9	.0000	.0000	.0000	.0003	.0021	.0093	.0276	.0611	.1070	.1540	.1855
	10	.0000	.0000	.0000	.0000	.0004	.0025	.0095	.0263	.0571	.1008	.1484
	11	.0000	.0000	.0000	.0000	.0001	.0005	.0026	.0090	.0242	.0525	.0944
	12	.0000	.0000	.0000	.0000	.0000	.0001	.0006	.0024	.0081	.0215	.0472
	13	.0000	.0000	.0000	.0000	.0000	.0000	.0001	.0005	.0021	.0068	.0182
	14	.0000	.0000	.0000	.0000	.0000	.0000	.0000	.0001	.0004	.0016	.0052
	15	.0000	.0000	.0000	.0000	.0000	.0000	.0000	.0000	.0001	.0003	.0010
	16	.0000	.0000	.0000	.0000	.0000	.0000	.0000	.0000	.0000	.0000	.0001
	17	.0000	.0000	.0000	.0000	.0000	.0000	.0000	.0000	.0000	.0000	.0000
18	0	.8345	.3972	.1501	.0536	.0180	.0056	.0016	.0004	.0001	.0000	.0000
	1	.1517	.3763	.3002	.1704	.0811	.0338	.0126	.0042	.0012	.0003	.0001
	2	.0130	.1683	.2835	.2556	.1723	.0958	.0458	.0190	.0069	.0022	.0006
	3	.0007	.0473	.1680	.2406	.2297	.1704	.1046	.0547	.0246	.0095	.0031
	4	.0000	.0093	.0700	.1592	.2153	.2130	.1681	.1104	.0614	.0291	.0117
	5	.0000	.0014	.0218	.0787	.1507	.1988	.2017	.1664	.1146	.0666	.0327
	6	.0000	.0002	.0052	.0301	.0816	.1436	.1873	.1941	.1655	.1181	.0708
	7	.0000	.0000	.0010	.0091	.0350	.0820	.1376	.1792	.1892	.1657	.1214
	8	.0000	.0000	.0002	.0022	.0120	.0376	.0811	.1327	.1734	.1864	.1669
	9	.0000	.0000	.0000	.0004	.0033	.0139	.0386	.0794	.1284	.1694	.1855
	10	.0000	.0000	.0000	.0001	.0008	.0042	.0149	.0385	.0771	.1248	.1669
	11	.0000	.0000	.0000	.0000	.0001	.0010	.0046	.0151	.0374	.0742	.1214
	12	.0000	.0000	.0000	.0000	.0000	.0002	.0012	.0047	.0145	.0354	.0708
	13	.0000	.0000	.0000	.0000	.0000	.0000	.0002	.0012	.0045	.0134	.0327
	14	.0000	.0000	.0000	.0000	.0000	.0000	.0000	.0002	.0011	.0039	.0117
	15	.0000	.0000	.0000	.0000	.0000	.0000	.0000	.0000	.0002	.0009	.0031
	16	.0000	.0000	.0000	.0000	.0000	.0000	.0000	.0000	.0000	.0001	.0006
	17	.0000	.0000	.0000	.0000	.0000	.0000	.0000	.0000	.0000	.0000	.0001
	18	.0000	.0000	.0000	.0000	.0000	.0000	.0000	.0000	.0000	.0000	.0000

Table 1 *Continued*

n	x	.01	.05	.10	.15	.20	π .25	.30	.35	.40	.45	.50
19	0	.8262	.3774	.1351	.0456	.0144	.0042	.0011	.0003	.0001	.0000	.0000
	1	.1586	.3774	.2852	.1529	.0685	.0268	.0093	.0029	.0008	.0002	.0000
	2	.0144	.1787	.2852	.2428	.1540	.0803	.0358	.0138	.0046	.0013	.0003
	3	.0008	.0533	.1796	.2428	.2182	.1517	.0869	.0422	.0175	.0062	.0018
	4	.0000	.0112	.0798	.1714	.2182	.2023	.1491	.0909	.0467	.0203	.0074
	5	.0000	.0018	.0266	.0907	.1636	.2023	.1916	.1468	.0933	.0497	.0222
	6	.0000	.0002	.0069	.0374	.0955	.1574	.1916	.1844	.1451	.0949	.0518
	7	.0000	.0000	.0014	.0122	.0443	.0974	.1525	.1844	.1797	.1443	.0961
	8	.0000	.0000	.0002	.0032	.0166	.0487	.0981	.1489	.1797	.1771	.1442
	9	.0000	.0000	.0000	.0007	.0051	.0198	.0514	.0980	.1464	.1771	.1762
	10	.0000	.0000	.0000	.0001	.0013	.0066	.0220	.0528	.0976	.1449	.1762
	11	.0000	.0000	.0000	.0000	.0003	.0018	.0077	.0233	.0532	.0970	.1442
	12	.0000	.0000	.0000	.0000	.0000	.0004	.0022	.0083	.0237	.0529	.0961
	13	.0000	.0000	.0000	.0000	.0000	.0001	.0005	.0024	.0085	.0233	.0518
	14	.0000	.0000	.0000	.0000	.0000	.0000	.0001	.0006	.0024	.0082	.0222
	15	.0000	.0000	.0000	.0000	.0000	.0000	.0000	.0001	.0005	.0022	.0074
	16	.0000	.0000	.0000	.0000	.0000	.0000	.0000	.0000	.0001	.0005	.0018
	17	.0000	.0000	.0000	.0000	.0000	.0000	.0000	.0000	.0000	.0001	.0003
	18	.0000	.0000	.0000	.0000	.0000	.0000	.0000	.0000	.0000	.0000	.0000
	19	.0000	.0000	.0000	.0000	.0000	.0000	.0000	.0000	.0000	.0000	.0000
20	0	.8179	.3585	.1216	.0388	.0115	.0032	.0008	.0002	.0000	.0000	.0000
	1	.1652	.3774	.2702	.1368	.0576	.0211	.0068	.0020	.0005	.0001	.0000
	2	.0159	.1887	.2852	.2293	.1369	.0669	.0278	.0100	.0031	.0008	.0002
	3	.0010	.0596	.1901	.2428	.2054	.1339	.0716	.0323	.0123	.0040	.0011
	4	.0000	.0133	.0898	.1821	.2182	.1897	.1304	.0738	.0350	.0139	.0046
	5	.0000	.0022	.0319	.1028	.1746	.2023	.1789	.1272	.0746	.0365	.0148
	6	.0000	.0003	.0089	.0454	.1091	.1686	.1916	.1712	.1244	.0746	.0370
	7	.0000	.0000	.0020	.0160	.0545	.1124	.1643	.1844	.1659	.1221	.0739
	8	.0000	.0000	.0004	.0046	.0222	.0609	.1144	.1614	.1797	.1623	.1201
	9	.0000	.0000	.0001	.0011	.0074	.0271	.0654	.1158	.1597	.1771	.1602
	10	.0000	.0000	.0000	.0002	.0020	.0099	.0308	.0686	.1171	.1593	.1762
	11	.0000	.0000	.0000	.0000	.0005	.0030	.0120	.0336	.0710	.1185	.1602
	12	.0000	.0000	.0000	.0000	.0001	.0008	.0039	.0136	.0355	.0727	.1201
	13	.0000	.0000	.0000	.0000	.0000	.0002	.0010	.0045	.0146	.0366	.0739
	14	.0000	.0000	.0000	.0000	.0000	.0000	.0002	.0012	.0049	.0150	.0370
	15	.0000	.0000	.0000	.0000	.0000	.0000	.0000	.0003	.0013	.0049	.0148
	16	.0000	.0000	.0000	.0000	.0000	.0000	.0000	.0000	.0003	.0013	.0046
	17	.0000	.0000	.0000	.0000	.0000	.0000	.0000	.0000	.0000	.0002	.0011
	18	.0000	.0000	.0000	.0000	.0000	.0000	.0000	.0000	.0000	.0000	.0002
	19	.0000	.0000	.0000	.0000	.0000	.0000	.0000	.0000	.0000	.0000	.0000
	20	.0000	.0000	.0000	.0000	.0000	.0000	.0000	.0000	.0000	.0000	.0000

(continues)

Table 1 *Continued*

n	x	.01	.05	.10	.15	.20	π .25	.30	.35	.40	.45	.50
25	0	.7778	.2774	.0718	.0172	.0038	.0008	.0001	.0000	.0000	.0000	.0000
	1	.1964	.3650	.1994	.0759	.0236	.0063	.0014	.0003	.0000	.0000	.0000
	2	.0238	.2305	.2659	.1607	.0708	.0251	.0074	.0018	.0004	.0001	.0000
	3	.0018	.0930	.2265	.2174	.1358	.0641	.0243	.0076	.0019	.0004	.0001
	4	.0001	.0269	.1384	.2110	.1867	.1175	.0572	.0224	.0071	.0018	.0004
	5	.0000	.0060	.0646	.1564	.1960	.1645	.1030	.0506	.0199	.0063	.0016
	6	.0000	.0010	.0239	.0920	.1633	.1828	.1472	.0908	.0442	.0172	.0053
	7	.0000	.0001	.0072	.0441	.1108	.1654	.1712	.1327	.0800	.0381	.0143
	8	.0000	.0000	.0018	.0175	.0623	.1241	.1651	.1607	.1200	.0701	.0322
	9	.0000	.0000	.0004	.0058	.0294	.0781	.1336	.1635	.1511	.1084	.0609
	10	.0000	.0000	.0000	.0016	.0118	.0417	.0916	.1409	.1612	.1419	.0974
	11	.0000	.0000	.0000	.0004	.0040	.0189	.0536	.1034	.1465	.1583	.1328
	12	.0000	.0000	.0000	.0000	.0012	.0074	.0268	.0650	.1140	.1511	.1550
	13	.0000	.0000	.0000	.0000	.0003	.0025	.0115	.0350	.0760	.1236	.1550
	14	.0000	.0000	.0000	.0000	.0000	.0007	.0042	.0161	.0434	.0867	.1328
	15	.0000	.0000	.0000	.0000	.0000	.0002	.0013	.0064	.0212	.0520	.0974
	16	.0000	.0000	.0000	.0000	.0000	.0000	.0004	.0021	.0088	.0266	.0609
	17	.0000	.0000	.0000	.0000	.0000	.0000	.0001	.0006	.0031	.0115	.0322
	18	.0000	.0000	.0000	.0000	.0000	.0000	.0000	.0001	.0009	.0042	.0143
	19	.0000	.0000	.0000	.0000	.0000	.0000	.0000	.0000	.0002	.0013	.0053
	20	.0000	.0000	.0000	.0000	.0000	.0000	.0000	.0000	.0000	.0001	.0016
	21	.0000	.0000	.0000	.0000	.0000	.0000	.0000	.0000	.0000	.0000	.0004
	22	.0000	.0000	.0000	.0000	.0000	.0000	.0000	.0000	.0000	.0000	.0001
30	0	.7397	.2146	.0424	.0076	.0012	.0002	.0000	.0000	.0000	.0000	.0000
	1	.2242	.3389	.1413	.0404	.0093	.0018	.0003	.0000	.0000	.0000	.0000
	2	.0328	.2586	.2277	.1034	.0337	.0086	.0018	.0003	.0000	.0000	.0000
	3	.0031	.1270	.2361	.1703	.0785	.0269	.0072	.0015	.0003	.0000	.0000
	4	.0002	.0451	.1771	.2028	.1325	.0604	.0208	.0056	.0012	.0002	.0000
	5	.0000	.0124	.1023	.1861	.1723	.1047	.0464	.0157	.0041	.0008	.0001
	6	.0000	.0027	.0474	.1368	.1795	.1455	.0829	.0353	.0115	.0029	.0006
	7	.0000	.0005	.0180	.0828	.1538	.1662	.1219	.0652	.0263	.0081	.0019
	8	.0000	.0001	.0058	.0420	.1106	.1593	.1501	.1009	.0505	.0191	.0055
	9	.0000	.0000	.0016	.0181	.0676	.1298	.1573	.1328	.0823	.0382	.0133
	10	.0000	.0000	.0004	.0067	.0355	.0909	.1416	.1502	.1152	.0656	.0280
	11	.0000	.0000	.0001	.0022	.0161	.0551	.1103	.1471	.1396	.0976	.0509
	12	.0000	.0000	.0000	.0006	.0064	.0291	.0749	.1254	.1474	.1265	.0806
	13	.0000	.0000	.0000	.0001	.0022	.0134	.0444	.0935	.1360	.1433	.1115
	14	.0000	.0000	.0000	.0000	.0007	.0054	.0231	.0611	.1101	.1424	.1354
	15	.0000	.0000	.0000	.0000	.0002	.0019	.0106	.0351	.0783	.1242	.1445
	16	.0000	.0000	.0000	.0000	.0000	.0006	.0042	.0177	.0489	.0953	.1354
	17	.0000	.0000	.0000	.0000	.0000	.0002	.0015	.0079	.0269	.0642	.1115
	18	.0000	.0000	.0000	.0000	.0000	.0000	.0005	.0031	.0129	.0379	.0806
	19	.0000	.0000	.0000	.0000	.0000	.0000	.0001	.0010	.0054	.0196	.0509

Table 1 *Continued*

n	x	.01	.05	.10	.15	.20	π .25	.30	.35	.40	.45	.50
	20	.0000	.0000	.0000	.0000	.0000	.0000	.0000	.0003	.0020	.0088	.0280
	21	.0000	.0000	.0000	.0000	.0000	.0000	.0000	.0001	.0006	.0034	.0133
	22	.0000	.0000	.0000	.0000	.0000	.0000	.0000	.0000	.0002	.0012	.0055
	23	.0000	.0000	.0000	.0000	.0000	.0000	.0000	.0000	.0000	.0003	.0019
	24	.0000	.0000	.0000	.0000	.0000	.0000	.0000	.0000	.0000	.0001	.0006
	25	.0000	.0000	.0000	.0000	.0000	.0000	.0000	.0000	.0000	.0000	.0001
40	0	.6690	.1285	.0148	.0015	.0001	.0000	.0000	.0000	.0000	.0000	.0000
	1	.2703	.2706	.0657	.0106	.0013	.0001	.0000	.0000	.0000	.0000	.0000
	2	.0532	.2777	.1423	.0365	.0065	.0009	.0001	.0000	.0000	.0000	.0000
	3	.0068	.1851	.2003	.0816	.0205	.0037	.0005	.0001	.0000	.0000	.0000
	4	.0006	.0901	.2059	.1332	.0475	.0113	.0020	.0003	.0000	.0000	.0000
	5	.0000	.0342	.1647	.1692	.0854	.0272	.0061	.0010	.0001	.0000	.0000
	6	.0000	.0105	.1068	.1742	.1246	.0530	.0151	.0031	.0005	.0000	.0000
	7	.0000	.0027	.0576	.1493	.1513	.0857	.0315	.0080	.0015	.0002	.0000
	8	.0000	.0006	.0264	.1087	.1560	.1179	.0557	.0179	.0040	.0006	.0001
	9	.0000	.0001	.0104	.0682	.1386	.1397	.0849	.0342	.0095	.0018	.0002
	10	.0000	.0000	.0036	.0373	.1075	.1444	.1128	.0571	.0196	.0047	.0008
	11	.0000	.0000	.0011	.0180	.0733	.1312	.1319	.0838	.0357	.0105	.0021
	12	.0000	.0000	.0003	.0077	.0443	.1057	.1366	.1090	.0576	.0207	.0051
	13	.0000	.0000	.0001	.0029	.0238	.0759	.1261	.1265	.0827	.0365	.0109
	14	.0000	.0000	.0000	.0010	.0115	.0488	.1042	.1313	.1063	.0575	.0211
	15	.0000	.0000	.0000	.0003	.0050	.0282	.0774	.1226	.1228	.0816	.0366
	16	.0000	.0000	.0000	.0001	.0019	.0147	.0518	.1031	.1279	.1043	.0572
	17	.0000	.0000	.0000	.0000	.0007	.0069	.0314	.0784	.1204	.1205	.0807
	18	.0000	.0000	.0000	.0000	.0002	.0029	.0172	.0539	.1026	.1260	.1031
	19	.0000	.0000	.0000	.0000	.0001	.0011	.0085	.0336	.0792	.1194	.1194
	20	.0000	.0000	.0000	.0000	.0000	.0004	.0038	.0190	.0554	.1025	.1254
	21	.0000	.0000	.0000	.0000	.0000	.0001	.0016	.0097	.0352	.0799	.1194
	22	.0000	.0000	.0000	.0000	.0000	.0000	.0006	.0045	.0203	.0565	.1031
	23	.0000	.0000	.0000	.0000	.0000	.0000	.0002	.0019	.0106	.0362	.0807
	24	.0000	.0000	.0000	.0000	.0000	.0000	.0001	.0007	.0050	.0210	.0572
	25	.0000	.0000	.0000	.0000	.0000	.0000	.0000	.0003	.0021	.0110	.0366
	26	.0000	.0000	.0000	.0000	.0000	.0000	.0000	.0001	.0008	.0052	.0211
	27	.0000	.0000	.0000	.0000	.0000	.0000	.0000	.0000	.0003	.0022	.0109
	28	.0000	.0000	.0000	.0000	.0000	.0000	.0000	.0000	.0001	.0008	.0051
	29	.0000	.0000	.0000	.0000	.0000	.0000	.0000	.0000	.0000	.0003	.0021
	30	.0000	.0000	.0000	.0000	.0000	.0000	.0000	.0000	.0000	.0001	.0008
	31	.0000	.0000	.0000	.0000	.0000	.0000	.0000	.0000	.0000	.0000	.0002
	32	.0000	.0000	.0000	.0000	.0000	.0000	.0000	.0000	.0000	.0000	.0001
50	0	.6050	.0769	.0052	.0003	.0000	.0000	.0000	.0000	.0000	.0000	.0000
	1	.3056	.2025	.0286	.0026	.0002	.0000	.0000	.0000	.0000	.0000	.0000
	2	.0756	.2611	.0779	.0113	.0011	.0001	.0000	.0000	.0000	.0000	.0000
	3	.0122	.2199	.1386	.0319	.0044	.0004	.0000	.0000	.0000	.0000	.0000
	4	.0015	.1360	.1809	.0661	.0128	.0016	.0001	.0000	.0000	.0000	.0000

(continues)

Table 1 *Continued*

n	x	.01	.05	.10	.15	.20	π .25	.30	.35	.40	.45	.50
	5	.0001	.0658	.1849	.1072	.0295	.0049	.0006	.0000	.0000	.0000	.0000
	6	.0000	.0260	.1541	.1419	.0554	.0123	.0018	.0002	.0000	.0000	.0000
	7	.0000	.0086	.1076	.1575	.0870	.0259	.0048	.0006	.0000	.0000	.0000
	8	.0000	.0024	.0643	.1493	.1169	.0463	.0110	.0017	.0002	.0000	.0000
	9	.0000	.0006	.0333	.1230	.1364	.0721	.0220	.0042	.0005	.0000	.0000
	10	.0000	.0001	.0152	.0890	.1398	.0985	.0386	.0093	.0014	.0001	.0000
	11	.0000	.0000	.0061	.0571	.1271	.1194	.0602	.0182	.0035	.0004	.0000
	12	.0000	.0000	.0022	.0328	.1033	.1294	.0838	.0319	.0076	.0011	.0001
	13	.0000	.0000	.0007	.0169	.0755	.1261	.1050	.0502	.0147	.0027	.0003
	14	.0000	.0000	.0002	.0079	.0499	.1110	.1189	.0714	.0260	.0059	.0008
	15	.0000	.0000	.0001	.0033	.0299	.0888	.1223	.0923	.0415	.0116	.0020
	16	.0000	.0000	.0000	.0013	.0164	.0648	.1147	.1088	.0606	.0207	.0044
	17	.0000	.0000	.0000	.0005	.0082	.0432	.0983	.1171	.0808	.0339	.0087
	18	.0000	.0000	.0000	.0001	.0037	.0264	.0772	.1156	.0987	.0508	.0160
	19	.0000	.0000	.0000	.0000	.0016	.0148	.0558	.1048	.1109	.0700	.0270
	20	.0000	.0000	.0000	.0000	.0006	.0077	.0370	.0875	.1146	.0888	.0419
	21	.0000	.0000	.0000	.0000	.0002	.0036	.0227	.0673	.1091	.1038	.0598
	22	.0000	.0000	.0000	.0000	.0001	.0016	.0128	.0478	.0959	.1119	.0788
	23	.0000	.0000	.0000	.0000	.0000	.0006	.0067	.0313	.0778	.1115	.0960
	24	.0000	.0000	.0000	.0000	.0000	.0002	.0032	.0190	.0584	.1026	.1080
	25	.0000	.0000	.0000	.0000	.0000	.0001	.0014	.0106	.0405	.0873	.1123
	26	.0000	.0000	.0000	.0000	.0000	.0000	.0006	.0055	.0259	.0687	.1080
	27	.0000	.0000	.0000	.0000	.0000	.0000	.0002	.0026	.0154	.0500	.0960
	28	.0000	.0000	.0000	.0000	.0000	.0000	.0001	.0012	.0084	.0336	.0788
	29	.0000	.0000	.0000	.0000	.0000	.0000	.0000	.0005	.0043	.0208	.0598
	30	.0000	.0000	.0000	.0000	.0000	.0000	.0000	.0002	.0020	.0119	.0419
	31	.0000	.0000	.0000	.0000	.0000	.0000	.0000	.0001	.0009	.0063	.0270
	32	.0000	.0000	.0000	.0000	.0000	.0000	.0000	.0000	.0003	.0031	.0160
	33	.0000	.0000	.0000	.0000	.0000	.0000	.0000	.0000	.0001	.0014	.0087
	34	.0000	.0000	.0000	.0000	.0000	.0000	.0000	.0000	.0000	.0006	.0044
	35	.0000	.0000	.0000	.0000	.0000	.0000	.0000	.0000	.0000	.0002	.0020
	36	.0000	.0000	.0000	.0000	.0000	.0000	.0000	.0000	.0000	.0001	.0008
	37	.0000	.0000	.0000	.0000	.0000	.0000	.0000	.0000	.0000	.0000	.0003
	38	.0000	.0000	.0000	.0000	.0000	.0000	.0000	.0000	.0000	.0000	.0001

Table 2 *Standardized Normal Distribution*

$P[Z > \text{cutoff}]$
For instance, $P[Z > 1.96] = .025$

Cutoff	.00	.01	.02	.03	.04	.05	.06	.07	.08	.09
0.0	.5000	.4960	.4920	.4880	.4840	.4801	.4761	.4721	.4681	.4641
0.1	.4602	.4562	.4522	.4483	.4443	.4404	.4364	.4325	.4286	.4247
0.2	.4207	.4168	.4129	.4090	.4052	.4013	.3974	.3936	.3897	.3859
0.3	.3821	.3783	.3745	.3707	.3669	.3632	.3594	.3557	.3520	.3483
0.4	.3446	.3409	.3372	.3336	.3300	.3264	.3228	.3192	.3156	.3121
0.5	.3085	.3050	.3015	.2981	.2946	.2912	.2877	.2843	.2810	.2776
0.6	.2743	.2709	.2676	.2643	.2611	.2578	.2546	.2514	.2483	.2451
0.7	.2420	.2389	.2358	.2327	.2296	.2266	.2236	.2206	.2177	.2148
0.8	.2119	.2090	.2061	.2033	.2005	.1977	.1949	.1922	.1894	.1867
0.9	.1841	.1814	.1788	.1762	.1736	.1711	.1685	.1660	.1635	.1611
1.0	.1587	.1562	.1539	.1515	.1492	.1469	.1446	.1423	.1401	.1379
1.1	.1357	.1335	.1314	.1292	.1271	.1251	.1230	.1210	.1190	.1170
1.2	.1151	.1131	.1112	.1093	.1075	.1056	.1038	.1020	.1003	.0985
1.3	.0968	.0951	.0934	.0918	.0901	.0885	.0869	.0853	.0838	.0823
1.4	.0808	.0793	.0778	.0764	.0749	.0735	.0722	.0708	.0694	.0681
1.5	.0668	.0655	.0643	.0630	.0618	.0606	.0594	.0582	.0571	.0559
1.6	.0548	.0537	.0526	.0516	.0505	.0495	.0485	.0475	.0465	.0455
1.7	.0446	.0436	.0427	.0418	.0409	.0401	.0392	.0384	.0375	.0367
1.8	.0359	.0352	.0344	.0336	.0329	.0322	.0314	.0307	.0301	.0294
1.9	.0287	.0281	.0274	.0268	.0262	.0256	.0250	.0244	.0239	.0233
2.0	.0228	.0222	.0217	.0212	.0207	.0202	.0197	.0192	.0188	.0183
2.1	.0179	.0174	.0170	.0166	.0162	.0158	.0154	.0150	.0146	.0143
2.2	.0139	.0136	.0132	.0129	.0125	.0122	.0119	.0116	.0113	.0110
2.3	.0107	.0104	.0102	.0099	.0096	.0094	.0091	.0089	.0087	.0084
2.4	.0082	.0080	.0078	.0075	.0073	.0071	.0069	.0068	.0066	.0064
2.5	.0062	.0060	.0059	.0057	.0055	.0054	.0052	.0051	.0049	.0048
2.6	.0047	.0045	.0044	.0043	.0041	.0040	.0039	.0038	.0037	.0036
2.7	.0035	.0034	.0033	.0032	.0031	.0030	.0029	.0028	.0027	.0026
2.8	.0026	.0025	.0024	.0023	.0023	.0022	.0021	.0021	.0020	.0019
2.9	.0019	.0018	.0017	.0017	.0016	.0016	.0015	.0015	.0014	.0014
3.0	.0013	.0013	.0013	.0012	.0012	.0011	.0011	.0011	.0010	.0010
4.0	.000 031 7									
5.0	.000 000 287									
6.0	.000 000 000 987									
7.0	.000 000 000 001 28									
8.0	.000 000 000 000 001									

Table 3 *Random Digits*

10 09 73 25 33	76 52 01 35 86	34 67 35 48 76	80 95 90 91 71	39 29 27 49 45
37 54 20 48 05	64 89 47 42 96	24 80 52 40 37	20 63 61 04 02	00 82 29 16 65
08 42 26 89 53	19 64 50 93 03	23 20 90 25 60	15 95 33 47 64	35 08 03 36 06
99 01 90 25 29	09 37 67 07 15	38 31 13 11 65	88 67 67 43 97	04 43 62 76 59
12 80 79 99 70	80 15 73 61 47	64 03 23 66 53	98 95 11 68 77	12 17 17 68 33
66 06 57 47 17	34 07 27 68 50	36 69 73 61 70	65 81 33 98 85	11 19 92 91 70
31 06 01 08 05	45 57 18 24 06	35 30 34 26 14	86 79 90 74 39	23 40 30 97 32
85 26 97 76 02	02 05 16 56 92	68 66 57 48 18	73 05 38 52 47	18 62 38 85 79
63 57 33 21 35	05 32 54 70 48	90 55 35 75 48	28 46 82 87 09	83 49 12 56 24
73 79 64 57 53	03 52 96 47 78	35 80 83 42 82	60 93 52 03 44	35 27 38 84 35
98 52 01 77 67	14 90 56 86 07	22 10 94 05 58	60 97 09 34 33	50 50 07 39 98
11 80 50 54 31	39 80 82 77 32	50 72 56 82 48	29 40 52 42 01	52 77 56 78 51
83 45 29 96 34	06 28 89 80 83	13 74 67 00 78	18 47 54 06 10	68 71 17 78 17
88 68 54 02 00	86 50 75 84 01	36 76 66 79 51	90 36 47 64 93	29 60 91 10 62
99 59 46 73 48	87 51 76 49 69	91 82 60 89 28	93 78 56 13 68	23 47 83 41 13
65 48 11 76 74	17 46 85 09 50	58 04 77 69 74	73 03 95 71 86	40 21 81 65 44
80 12 43 56 35	17 72 70 80 15	45 31 82 23 74	21 11 57 82 53	14 38 55 37 63
74 35 09 98 17	77 40 27 72 14	43 23 60 02 10	45 52 16 42 37	96 28 60 26 55
69 91 62 68 03	66 25 22 91 48	36 93 68 72 03	76 62 11 39 90	94 40 05 64 18
09 89 32 05 05	14 22 56 85 14	46 42 75 67 88	96 29 77 88 22	54 38 21 45 98
91 49 91 45 23	68 47 92 76 86	46 16 28 35 54	94 75 08 99 23	37 08 92 00 48
80 33 69 45 98	26 94 03 68 58	70 29 73 41 35	53 14 03 33 40	42 05 08 23 41
44 10 48 19 49	85 15 74 79 54	32 97 92 65 75	57 60 04 08 81	22 22 20 64 13
12 55 07 37 42	11 10 00 20 40	12 86 07 46 97	96 64 48 94 39	28 70 72 58 15
63 60 64 93 29	16 50 53 44 84	40 21 95 25 63	43 65 17 70 82	07 20 73 17 90
61 19 69 04 46	26 45 74 77 74	51 92 43 37 29	65 39 45 95 93	42 58 26 05 27
15 47 44 52 66	95 27 07 99 53	59 36 78 38 48	82 39 61 01 18	33 21 15 94 66
94 55 72 85 73	67 89 75 43 87	54 62 24 44 31	91 19 04 25 92	92 92 74 59 73
42 48 11 62 13	97 34 40 87 21	16 86 84 87 67	03 07 11 20 59	25 70 14 66 70
23 52 37 83 17	73 20 88 98 37	68 93 59 14 16	26 25 22 96 63	05 52 28 25 62
04 49 35 24 94	75 24 63 38 24	45 86 25 10 25	61 96 27 93 35	65 33 71 24 72
00 54 99 76 54	64 05 18 81 59	96 11 96 38 96	54 69 28 23 91	23 28 72 95 29
35 96 31 53 07	26 89 80 93 54	33 35 13 54 62	77 97 45 00 24	90 10 33 93 33
59 80 80 83 91	45 42 72 68 42	83 60 94 97 00	13 02 12 48 92	78 56 52 01 06
46 05 88 52 36	01 39 09 22 86	77 28 14 40 77	93 91 08 36 47	70 61 74 29 41
32 17 90 05 97	87 37 92 52 41	05 56 70 70 07	86 74 31 71 57	85 39 41 18 38
69 23 46 14 06	20 11 74 52 04	15 95 66 00 00	18 74 39 24 23	97 11 89 63 38
19 56 54 14 30	01 75 87 53 79	40 41 92 15 85	66 67 43 68 06	84 96 28 52 07
45 15 51 49 38	19 47 60 72 46	43 66 79 45 43	59 04 79 00 33	20 82 66 95 41
94 86 43 19 94	36 16 81 08 51	34 88 88 15 53	01 54 03 54 56	05 01 45 11 76
98 08 62 48 26	45 24 02 84 04	44 99 90 88 96	39 09 47 34 07	35 44 13 18 80
33 18 51 62 32	41 94 15 09 49	89 43 54 85 81	88 69 54 19 94	37 54 87 30 43
80 95 10 04 06	96 38 27 07 74	20 15 12 33 87	25 01 62 52 98	94 62 46 11 71
79 75 24 91 40	71 96 12 82 96	69 86 10 25 91	74 85 22 05 39	00 38 75 95 79
18 63 33 25 37	98 14 50 65 71	31 01 02 46 74	05 45 56 14 27	77 93 89 19 36
74 02 94 39 02	77 55 73 22 70	97 79 01 71 19	52 52 75 80 21	80 81 45 17 48
54 17 84 56 11	80 99 33 71 43	05 33 51 29 69	56 12 71 92 55	36 04 09 03 24
11 66 44 98 83	52 07 98 48 27	59 38 17 15 39	09 97 33 34 40	88 46 12 33 56
48 32 47 79 28	31 24 96 47 10	02 29 53 68 70	32 30 75 75 46	15 02 00 99 94
69 07 49 41 38	87 63 79 19 76	35 58 40 44 01	10 51 82 16 15	01 84 87 69 38

Reproduced by permission from The RAND Corporation. "A Million Random Digits." Glencoe, Illinois: 1955.

Table 4 *Student t-Distribution*

$P[t > \text{cutoff}]$
For instance, with 10 degrees of
freedom, $P[t > 1.812] = .05$

Degrees of Freedom	Probability of a t-value larger than indicated cutoff				
	.10	.05	.025	.01	.005
1	3.078	6.314	12.706	31.821	63.657
2	1.886	2.920	4.303	6.965	9.925
3	1.638	2.353	3.182	4.541	5.841
4	1.533	2.132	2.776	3.747	4.604
5	1.476	2.015	2.571	3.365	4.032
6	1.440	1.943	2.447	3.143	3.707
7	1.415	1.895	2.365	2.998	3.499
8	1.397	1.860	2.306	2.896	3.355
9	1.383	1.833	2.262	2.821	3.250
10	1.372	1.812	2.228	2.764	3.169
11	1.363	1.796	2.201	2.718	3.106
12	1.356	1.782	2.179	2.681	3.055
13	1.350	1.771	2.160	2.650	3.012
14	1.345	1.761	2.145	2.624	2.977
15	1.341	1.753	2.131	2.602	2.947
16	1.337	1.746	2.120	2.583	2.921
17	1.333	1.740	2.110	2.567	2.898
18	1.330	1.734	2.101	2.552	2.878
19	1.328	1.729	2.093	2.539	2.861
20	1.325	1.725	2.086	2.528	2.845
21	1.323	1.721	2.080	2.518	2.831
22	1.321	1.717	2.074	2.508	2.819
23	1.319	1.714	2.069	2.500	2.807
24	1.318	1.711	2.064	2.492	2.797
25	1.316	1.708	2.060	2.485	2.787
26	1.315	1.706	2.056	2.479	2.779
27	1.314	1.703	2.052	2.473	2.771
28	1.313	1.701	2.048	2.467	2.763
29	1.311	1.699	2.045	2.462	2.756
30	1.310	1.697	2.042	2.457	2.750
40	1.303	1.684	2.021	2.423	2.704
60	1.296	1.671	2.000	2.390	2.660
120	1.290	1.661	1.984	2.358	2.626
∞	1.282	1.645	1.960	2.326	2.576

Taken from Table III of Fisher and Yates: *Statistical Tables for Biological, Agricultural, and Medical Research*, 6th edition, 1974, published by Longmans Group Ltd. London (previously published by Oliver and Boyd Ltd., Edinburgh) by permission of the authors and publishers.

Table 5 *Chi-Square Distribution*

$P[\chi^2 > \text{cutoff}]$
For instance, with 16 degrees of
freedom, $P[\chi^2 > 26.30] = .05$

Degrees of freedom	Probability of a chi-square value larger than indicated cutoff				
	.10	.05	.025	.01	.005
1	2.71	3.84	5.02	6.63	7.88
2	4.61	5.99	7.38	9.21	10.60
3	6.25	7.81	9.35	11.34	12.84
4	7.78	9.49	11.14	13.28	14.86
5	9.24	11.07	12.83	15.09	16.75
6	10.64	12.59	14.45	16.81	18.55
7	12.02	14.07	16.01	18.48	20.28
8	13.36	15.51	17.53	20.09	21.96
9	14.68	16.92	19.02	21.67	23.59
10	15.99	18.31	20.48	23.21	25.19
11	17.28	19.68	21.92	24.72	26.76
12	18.55	21.03	23.34	26.22	28.30
13	19.81	22.36	24.74	27.69	29.82
14	21.06	23.68	26.12	29.14	31.32
15	22.31	25.00	27.49	30.58	32.80
16	23.54	26.30	28.85	32.00	34.27
17	24.77	27.59	30.19	33.41	35.72
18	25.99	28.87	31.53	34.81	37.16
19	27.20	30.14	32.85	36.19	38.58
20	28.41	31.41	34.17	37.57	40.00
21	29.62	32.67	35.48	38.93	41.40
22	30.81	33.92	36.78	40.29	42.80
23	32.01	35.17	38.08	41.64	44.18
24	33.20	36.42	39.36	42.98	45.56
25	34.38	37.65	40.65	44.31	46.93
26	35.56	38.89	41.92	45.64	48.29
27	36.74	40.11	43.19	46.96	49.64
28	37.92	41.34	44.46	48.28	50.99
29	39.09	42.56	45.72	49.59	52.34
30	40.26	43.77	46.98	50.89	53.67
40	51.81	55.76	59.34	63.69	66.77
50	63.17	67.50	71.42	76.15	79.49
60	74.40	79.08	83.30	88.38	91.95
70	85.53	90.53	95.02	100.43	104.22
80	96.58	101.88	106.63	112.33	116.32
90	107.57	113.14	118.14	124.12	128.30
100	118.50	124.34	129.56	135.81	140.17

Abridged from Catherine M. Thompson, "Table of percentage points of the x^2 distribution," *Biometrika*, Vol. 32, 1942, p. 187. Reprinted by permission of the Biometrika Trust.

Table 6 *F-Distribution*

$P[F > \text{cutoff}]$
For instance, with 4 degrees of freedom
for the numerator and 20 for the
denominator, $P[F > 2.87] = .05$

Cutoffs for 5 percent probabilities

Degrees of freedom for denominator	Degrees of freedom for numerator									
	1	*2*	*3*	*4*	*5*	*6*	*7*	*8*	*9*	*10*
1	161	200	216	225	230	234	237	239	241	242
2	18.5	19.0	19.2	19.2	19.3	19.3	19.4	19.4	19.4	19.4
3	10.1	9.55	9.28	9.12	9.01	8.94	8.89	8.85	8.81	8.79
4	7.71	6.94	6.59	6.39	6.26	6.16	6.09	6.04	6.00	5.96
5	6.61	5.79	5.41	5.19	5.05	4.95	4.88	4.82	4.77	4.74
6	5.99	5.14	4.76	4.53	4.39	4.28	4.21	4.15	4.10	4.06
7	5.59	4.74	4.35	4.12	3.97	3.87	3.79	3.73	3.68	3.64
8	5.32	4.46	4.07	3.84	3.69	3.58	3.50	3.44	3.39	3.35
9	5.12	4.26	3.86	3.63	3.48	3.37	3.29	3.23	3.18	3.14
10	4.96	4.10	3.71	3.48	3.33	3.22	3.14	3.07	3.02	2.98
11	4.84	3.98	3.59	3.36	3.20	3.09	3.01	2.95	2.90	2.85
12	4.75	3.89	3.49	3.26	3.11	3.00	2.91	2.85	2.80	2.75
13	4.67	3.81	3.41	3.18	3.03	2.92	2.83	2.77	2.71	2.67
14	4.60	3.74	3.34	3.11	2.96	2.85	2.76	2.70	2.65	2.60
15	4.54	3.68	3.29	3.06	2.90	2.79	2.71	2.64	2.59	2.54
16	4.49	3.63	3.24	3.01	2.85	2.74	2.66	2.59	2.54	2.49
17	4.45	3.59	3.20	2.96	2.81	2.70	2.61	2.55	2.49	2.45
18	4.41	3.55	3.16	2.93	2.77	2.66	2.58	2.51	2.46	2.41
19	4.38	3.52	3.13	2.90	2.74	2.63	2.54	2.48	2.42	2.38
20	4.35	3.49	3.10	2.87	2.71	2.60	2.51	2.45	2.39	2.35
25	4.24	3.39	2.99	2.76	2.60	2.49	2.40	2.34	2.28	2.24
30	4.17	3.32	2.92	2.69	2.53	2.42	2.33	2.27	2.21	2.16
40	4.08	3.23	2.84	2.61	2.45	2.34	2.25	2.18	2.12	2.08
60	4.00	3.15	2.76	2.53	2.37	2.25	2.17	2.10	2.04	1.99
120	3.92	3.07	2.68	2.45	2.29	2.18	2.09	2.02	1.96	1.91
∞	3.84	3.00	2.60	2.37	2.21	2.10	2.01	1.94	1.88	1.83

(*continues*)

Table 6 *Continued*

Cutoffs for 1 percent probabilities

Degrees of freedom for denominator	Degrees of freedom for numerator									
	1	2	3	4	5	6	7	8	9	10
1	4052	5000	5403	5625	5764	5859	5928	5982	6022	6056
2	98.5	99.0	99.2	99.2	99.3	99.3	99.4	99.4	99.4	99.4
3	34.1	30.8	29.5	28.7	28.2	27.9	27.7	27.5	27.3	27.2
4	21.2	18.0	16.7	16.0	15.5	15.2	15.0	14.8	14.7	14.5
5	16.3	13.3	12.1	11.4	11.0	10.7	10.5	10.3	10.2	10.1
6	13.7	10.9	9.78	9.15	8.75	8.47	8.26	8.10	7.98	7.87
7	12.2	9.55	8.45	7.85	7.46	7.19	6.99	6.84	6.72	6.62
8	11.3	8.65	7.59	7.01	6.63	6.37	6.18	6.03	5.91	5.81
9	10.6	8.02	6.99	6.42	6.06	5.80	5.61	5.47	5.35	5.26
10	10.0	7.56	6.55	5.99	5.64	5.39	5.20	5.06	4.94	4.85
11	9.65	7.21	6.22	5.67	5.32	5.07	4.89	4.74	4.63	4.54
12	9.33	6.93	5.95	5.41	5.06	4.82	4.64	4.50	4.39	4.30
13	9.07	6.70	5.74	5.21	4.86	4.62	4.44	4.30	4.19	4.10
14	8.86	6.51	5.56	5.04	4.70	4.46	4.28	4.14	4.03	3.94
15	8.68	6.36	5.42	4.89	4.56	4.32	4.14	4.00	3.89	3.80
16	8.53	6.23	5.29	4.77	4.44	4.20	4.03	3.89	3.78	3.69
17	8.40	6.11	5.19	4.67	4.34	4.10	3.93	3.79	3.68	3.59
18	8.29	6.01	5.09	4.58	4.25	4.01	3.84	3.71	3.60	3.51
19	8.19	5.93	5.01	4.50	4.17	3.94	3.77	3.63	3.52	3.43
20	8.10	5.85	4.94	4.43	4.10	3.87	3.70	3.56	3.46	3.37
25	7.77	5.57	4.68	4.18	3.86	3.63	3.46	3.32	3.22	3.13
30	7.56	5.39	4.51	4.02	3.70	3.47	3.30	3.17	3.07	2.98
40	7.31	5.18	4.31	3.83	3.51	3.29	3.12	2.99	2.89	2.80
60	7.08	4.98	4.13	3.65	3.34	3.12	2.95	2.82	2.72	2.63
120	6.85	4.79	3.95	3.48	3.17	2.96	2.79	2.66	2.56	2.47
∞	6.63	4.61	3.78	3.32	3.02	2.80	2.64	2.51	2.41	2.32

Extracted from M. Merrington and C.M. Thompson. "Tables of Percentage Points of the Inverted Beta (F) Distribution." *Biometrika*, Vol. 33, 1943, pp. 73–88. Reprinted by permission of the Biometrika Trust.

Table 7 *Values of d_L and d_U for the Durbin-Watson Test*

n = number of observations
k = number of explanatory variables

For a one-sided test of H_0: $\rho \leq 0$ versus H_1: $\rho > 0$,

For a two-sided test of H_0: $\rho = 0$ versus H_1: $\rho \neq 0$,

For example, with $n = 50$ and $k = 1$, a one-sided test of the null hypothesis H_0: $\rho \leq 0$ is rejected at the 5 percent level if the Durbin-Watson d-value turns out to be less than 1.50, is inconclusive for a d-value between 1.50 and 1.59, and does not reject the null hypothesis if the d-value is larger than 1.59.

(continues)

Table 7 *Continued*

	$\alpha = .05$ for a one-sided test									
	$\alpha = .10$ for a two-sided test									
	$k = 1$		$k = 2$		$k = 3$		$k = 4$		$k = 5$	
n	d_L	d_U	d_L	d_U	d_L	d_U	d_L	d_U	d_L	d_U
15	1.08	1.36	0.95	1.54	0.82	1.75	0.69	1.97	0.56	2.21
16	1.10	1.37	0.98	1.54	0.86	1.73	0.74	1.93	0.62	2.15
17	1.13	1.38	1.02	1.54	0.90	1.71	0.78	1.90	0.67	2.10
18	1.16	1.39	1.05	1.53	0.93	1.69	0.82	1.87	0.71	2.06
19	1.18	1.40	1.08	1.53	0.97	1.68	0.86	1.85	0.75	2.02
20	1.20	1.41	1.10	1.54	1.00	1.68	0.90	1.83	0.79	1.99
21	1.22	1.42	1.13	1.54	1.03	1.67	0.93	1.81	0.83	1.96
22	1.24	1.43	1.15	1.54	1.05	1.66	0.96	1.80	0.86	1.94
23	1.26	1.44	1.17	1.54	1.08	1.66	0.99	1.79	0.90	1.92
24	1.27	1.45	1.19	1.55	1.10	1.66	1.01	1.78	0.93	1.90
25	1.29	1.45	1.21	1.55	1.12	1.66	1.04	1.77	0.95	1.89
26	1.30	1.46	1.22	1.55	1.14	1.65	1.06	1.76	0.98	1.88
27	1.32	1.47	1.24	1.56	1.16	1.65	1.08	1.76	1.01	1.86
28	1.33	1.48	1.26	1.56	1.18	1.65	1.10	1.75	1.03	1.85
29	1.34	1.48	1.27	1.56	1.20	1.65	1.12	1.74	1.05	1.84
30	1.35	1.49	1.28	1.57	1.21	1.65	1.14	1.74	1.07	1.83
31	1.36	1.50	1.30	1.57	1.23	1.65	1.16	1.74	1.09	1.83
32	1.37	1.50	1.31	1.57	1.24	1.65	1.18	1.73	1.11	1.82
33	1.38	1.51	1.32	1.58	1.26	1.65	1.19	1.73	1.13	1.81
34	1.39	1.51	1.33	1.58	1.27	1.65	1.21	1.73	1.15	1.81
35	1.40	1.52	1.34	1.58	1.28	1.65	1.22	1.73	1.16	1.80
36	1.41	1.52	1.35	1.59	1.29	1.65	1.24	1.73	1.18	1.80
37	1.42	1.53	1.36	1.59	1.31	1.66	1.25	1.72	1.19	1.80
38	1.43	1.54	1.37	1.59	1.32	1.66	1.26	1.72	1.21	1.79
39	1.43	1.54	1.38	1.60	1.33	1.66	1.27	1.72	1.22	1.79
40	1.44	1.54	1.39	1.60	1.34	1.66	1.29	1.72	1.23	1.79
45	1.48	1.57	1.43	1.62	1.38	1.67	1.34	1.72	1.29	1.78
50	1.50	1.59	1.46	1.63	1.42	1.67	1.38	1.72	1.34	1.77
55	1.53	1.60	1.49	1.64	1.45	1.68	1.41	1.72	1.38	1.77
60	1.55	1.62	1.51	1.65	1.48	1.69	1.44	1.73	1.41	1.77
65	1.57	1.63	1.54	1.66	1.50	1.70	1.47	1.73	1.44	1.77
70	1.58	1.64	1.55	1.67	1.52	1.70	1.49	1.74	1.46	1.77
75	1.60	1.65	1.57	1.68	1.54	1.71	1.51	1.74	1.49	1.77
80	1.61	1.66	1.59	1.69	1.56	1.72	1.53	1.74	1.51	1.77
85	1.62	1.67	1.60	1.70	1.57	1.72	1.55	1.75	1.52	1.77
90	1.63	1.68	1.61	1.70	1.59	1.73	1.57	1.75	1.54	1.78
95	1.64	1.69	1.62	1.71	1.60	1.73	1.58	1.75	1.56	1.78
100	1.65	1.69	1.63	1.72	1.61	1.74	1.59	1.76	1.57	1.78

Table 7 *Continued*

$$\alpha = .025 \text{ for a one-sided test}$$
$$\alpha = .05 \text{ for a two-sided test}$$

	$k = 1$		$k = 2$		$k = 3$		$k = 4$		$k = 5$	
n	d_L	d_U	d_L	d_U	d_L	d_U	d_L	d_U	d_L	d_U
15	0.95	1.23	0.83	1.40	0.71	1.61	0.59	1.84	0.48	2.09
16	0.98	1.24	0.86	1.40	0.75	1.59	0.64	1.80	0.53	2.03
17	1.01	1.25	0.90	1.40	0.79	1.58	0.68	1.77	0.57	1.98
18	1.03	1.26	0.93	1.40	0.82	1.56	0.72	1.74	0.62	1.93
19	1.06	1.28	0.96	1.41	0.86	1.55	0.76	1.72	0.66	1.90
20	1.08	1.28	0.99	1.41	0.89	1.55	0.79	1.70	0.70	1.87
21	1.10	1.30	1.01	1.41	0.92	1.54	0.83	1.69	0.73	1.84
22	1.12	1.31	1.04	1.42	0.95	1.54	0.86	1.68	0.77	1.82
23	1.14	1.32	1.06	1.42	0.97	1.54	0.89	1.67	0.80	1.80
24	1.16	1.33	1.08	1.43	1.00	1.54	0.91	1.66	0.83	1.79
25	1.18	1.34	1.10	1.43	1.02	1.54	0.94	1.65	0.86	1.77
26	1.19	1.35	1.12	1.44	1.04	1.54	0.96	1.65	0.88	1.76
27	1.21	1.36	1.13	1.44	1.06	1.54	0.99	1.64	0.91	1.75
28	1.22	1.37	1.15	1.45	1.08	1.54	1.01	1.64	0.93	1.74
29	1.24	1.38	1.17	1.45	1.10	1.54	1.03	1.63	0.96	1.73
30	1.25	1.38	1.18	1.46	1.12	1.54	1.05	1.63	0.98	1.73
31	1.26	1.39	1.20	1.47	1.13	1.55	1.07	1.63	1.00	1.72
32	1.27	1.40	1.21	1.47	1.15	1.55	1.08	1.63	1.02	1.71
33	1.28	1.41	1.22	1.48	1.16	1.55	1.10	1.63	1.04	1.71
34	1.29	1.41	1.24	1.48	1.17	1.55	1.12	1.63	1.06	1.70
35	1.30	1.42	1.25	1.48	1.19	1.55	1.13	1.63	1.07	1.70
36	1.31	1.43	1.26	1.49	1.20	1.56	1.15	1.63	1.09	1.70
37	1.32	1.43	1.27	1.49	1.21	1.56	1.16	1.62	1.10	1.70
38	1.33	1.44	1.28	1.50	1.23	1.56	1.17	1.62	1.12	1.70
39	1.34	1.44	1.29	1.50	1.24	1.56	1.19	1.63	1.13	1.69
40	1.35	1.45	1.30	1.51	1.25	1.57	1.20	1.63	1.15	1.69
45	1.39	1.48	1.34	1.53	1.30	1.58	1.25	1.63	1.21	1.69
50	1.42	1.50	1.38	1.54	1.34	1.59	1.30	1.64	1.26	1.69
55	1.45	1.52	1.41	1.56	1.37	1.60	1.33	1.64	1.30	1.69
60	1.47	1.54	1.44	1.57	1.40	1.61	1.37	1.65	1.33	1.69
65	1.49	1.55	1.46	1.59	1.43	1.62	1.40	1.66	1.36	1.69
70	1.51	1.57	1.48	1.60	1.45	1.63	1.42	1.66	1.39	1.70
75	1.53	1.58	1.50	1.61	1.47	1.64	1.45	1.67	1.42	1.70
80	1.54	1.59	1.52	1.62	1.49	1.65	1.47	1.67	1.44	1.70
85	1.56	1.60	1.53	1.63	1.51	1.65	1.49	1.68	1.46	1.71
90	1.57	1.61	1.55	1.64	1.53	1.66	1.50	1.69	1.48	1.71
95	1.58	1.62	1.56	1.65	1.54	1.67	1.52	1.69	1.50	1.71
100	1.59	1.63	1.57	1.65	1.55	1.67	1.53	1.70	1.51	1.72

(continues)

Table 7 *Continued*

	$\alpha = .01$ for a one-sided test									
	$\alpha = .02$ for a two-sided test									
	$k = 1$		$k = 2$		$k = 3$		$k = 4$		$k = 5$	
n	d_L	d_U	d_L	d_U	d_L	d_U	d_L	d_U	d_L	d_U
15	0.81	1.07	0.70	1.25	0.59	1.46	0.49	1.70	0.39	1.96
16	0.84	1.09	0.74	1.25	0.63	1.44	0.53	1.66	0.44	1.90
17	0.87	1.10	0.77	1.25	0.67	1.43	0.57	1.63	0.48	1.85
18	0.90	1.12	0.80	1.26	0.71	1.42	0.61	1.60	0.52	1.80
19	0.93	1.13	0.83	1.26	0.74	1.41	0.65	1.58	0.56	1.77
20	0.95	1.15	0.86	1.27	0.77	1.41	0.68	1.57	0.60	1.74
21	0.97	1.16	0.89	1.27	0.80	1.41	0.72	1.55	0.63	1.71
22	1.00	1.17	0.91	1.28	0.83	1.40	0.75	1.54	0.66	1.69
23	1.02	1.19	0.94	1.29	0.86	1.40	0.77	1.53	0.70	1.67
24	1.04	1.20	0.96	1.30	0.88	1.41	0.80	1.53	0.72	1.66
25	1.05	1.21	0.98	1.30	0.90	1.41	0.83	1.52	0.75	1.65
26	1.07	1.22	1.00	1.31	0.93	1.41	0.85	1.52	0.78	1.64
27	1.09	1.23	1.02	1.32	0.95	1.41	0.88	1.51	0.81	1.63
28	1.10	1.24	1.04	1.32	0.97	1.41	0.90	1.51	0.83	1.62
29	1.12	1.25	1.05	1.33	0.99	1.42	0.92	1.51	0.85	1.61
30	1.13	1.26	1.07	1.34	1.01	1.42	0.94	1.51	0.88	1.61
31	1.15	1.27	1.08	1.34	1.02	1.42	0.96	1.51	0.90	1.60
32	1.16	1.28	1.10	1.35	1.04	1.43	0.98	1.51	0.92	1.60
33	1.17	1.29	1.11	1.36	1.05	1.43	1.00	1.51	0.94	1.59
34	1.18	1.30	1.13	1.36	1.07	1.43	1.01	1.51	0.95	1.59
35	1.19	1.31	1.14	1.37	1.08	1.44	1.03	1.51	0.97	1.59
36	1.21	1.32	1.15	1.38	1.10	1.44	1.04	1.51	0.99	1.59
37	1.22	1.32	1.16	1.38	1.11	1.45	1.06	1.51	1.00	1.59
38	1.23	1.33	1.18	1.39	1.12	1.45	1.07	1.52	1.02	1.58
39	1.24	1.34	1.19	1.39	1.14	1.45	1.09	1.52	1.03	1.58
40	1.25	1.34	1.20	1.40	1.15	1.46	1.10	1.52	1.05	1.58
45	1.29	1.38	1.24	1.42	1.20	1.48	1.16	1.53	1.11	1.58
50	1.32	1.40	1.28	1.45	1.24	1.49	1.20	1.54	1.16	1.59
55	1.36	1.43	1.32	1.47	1.28	1.51	1.25	1.55	1.21	1.59
60	1.38	1.45	1.35	1.48	1.32	1.52	1.28	1.56	1.25	1.60
65	1.41	1.47	1.38	1.50	1.35	1.53	1.31	1.57	1.28	1.61
70	1.43	1.49	1.40	1.52	1.37	1.55	1.34	1.58	1.31	1.61
75	1.45	1.50	1.42	1.53	1.39	1.56	1.37	1.59	1.34	1.62
80	1.47	1.52	1.44	1.54	1.42	1.57	1.39	1.60	1.36	1.62
85	1.48	1.53	1.46	1.55	1.43	1.58	1.41	1.60	1.39	1.63
90	1.50	1.54	1.47	1.56	1.45	1.59	1.43	1.61	1.41	1.64
95	1.51	1.55	1.49	1.57	1.47	1.60	1.45	1.62	1.42	1.64
100	1.52	1.56	1.50	1.58	1.48	1.60	1.46	1.63	1.44	1.65

Table 8 *Historical Annual Data, 1950–1989*

U.S. Population: Population in millions, including armed forces overseas

Nominal Gross National Product: Annual dollar value, in billions of dollars, of U.S. gross national product

Nominal Disposable Income: Annual dollar value, in billions of dollars, of U.S. personal after-tax income

Nominal Consumption: Annual dollar value, in billions of dollars, of U.S. personal consumption expenditures

Nominal Private Net Investment: Annual dollar value, in billions of dollars, of U.S. private investment spending, excluding government investment spending, and net of depreciation of existing private capital

Private Housing Starts: Annual new U.S. private housing units on which construction started that year, in millions

Unemployment Rate: Annual average percentage unemployment rate for civilian workers, excluding members of armed forces

Consumer Price Index: Average value of consumer price index during year, scaled so that average value during 1982–1984 = 100

Annual Rate of Inflation: Percentage rate of change of consumer price index during year, comparing December level with December level one year earlier

Interest Rate on One-Year Treasury Bills: Average annual (secondary market) interest rate on one-year U.S. Treasury bills, converted from discount basis (*Source:* monthly *Federal Reserve Bulletin*)

Interest Rate on Long-Term Treasury Bonds: Average annual (secondary market) interest rate on U.S. Treasury bonds that are neither due nor callable in less than 10 years (*Source:* monthly *Federal Reserve Bulletin*)

S&P 500 Stock Returns: Annual percentage return (January 1 to December 31), including dividends and capital gains, on S&P 500 stocks (*Source:* Roger G. Ibbotson and Rex A. Sinquefield, *Stocks, Bonds, Bills, and Inflation 1990 Yearbook,* Ibbotson Associates, Inc., Chicago)

M1: December value, in billions of dollars, of M1 monetary aggregate, consisting of currency outside banks, checkable deposits, and travelers checks

M2: December value, in billions of dollars, of M2 monetary aggregate, consisting of M1, savings accounts, small (less than $100,000) time deposits, money market deposit accounts, money market mutual fund shares, and overnight Eurodollars and repurchase agreements

Dow-Jones Prices: Annual average value of Dow Jones Industrial Index of 30 stock prices

S&P 500 Prices: Annual average value of Standard & Poor's index of 500 stock prices (*Source:* Standard & Poor's)

S&P 500 Earnings: Earnings of Standard & Poor's 500 stocks, scaled so that the ratio of S&P 500 prices to S&P 500 earnings is equal to the (market-value weighted) average price-earnings ratio of the 500 S&P stocks (*Source:* Standard & Poor's)

S&P 500 Dividends: Dividends of Standard & Poor's 500 stocks, scaled so that the ratio of S&P 500 dividends to S&P 500 prices is equal to the (market-value weighted) average dividend-price ratio of the 500 S&P stocks (*Source:* Standard & Poor's)

(continues)

Table 8 *Continued*

Year	U.S. Population	Nominal Gross National Product	Nominal Disposable Income	Nominal Consumption	Nominal Private Net Investment	Private Housing Starts
1950	152.271	288.3	207.5	192.2	31.5	
1951	154.878	333.4	227.6	208.1	33.3	
1952	157.553	351.6	239.8	219.1	24.4	
1953	160.184	371.6	255.1	232.6	24.0	
1954	163.026	372.5	260.5	239.8	21.6	
1955	165.931	405.9	278.8	257.9	35.3	
1956	168.903	428.2	297.5	270.6	34.6	
1957	171.984	451.0	313.9	285.3	29.9	
1958	174.882	456.8	324.9	294.6	20.8	
1959	177.830	495.8	344.6	316.3	35.5	1.517
1960	180.671	515.3	358.9	330.7	31.8	1.252
1961	183.691	533.8	373.8	341.1	29.4	1.313
1962	186.538	574.6	396.2	361.9	38.2	1.463
1963	189.242	606.9	415.8	381.7	41.8	1.603
1964	191.889	649.8	451.4	409.3	45.7	1.529
1965	194.303	705.1	486.8	440.7	58.8	1.473
1966	196.560	772.0	525.9	477.3	66.5	1.165
1967	198.712	816.4	562.1	503.6	58.3	1.292
1968	200.706	892.7	609.6	552.5	63.1	1.508
1969	202.677	963.9	656.7	597.9	71.8	1.467
1970	205.052	1,015.5	715.6	640.0	60.0	1.434
1971	207.661	1,102.7	776.8	691.6	74.9	2.052
1972	209.896	1,212.8	839.9	757.6	94.1	2.356
1973	211.909	1,359.3	949.8	837.2	120.7	2.045
1974	213.854	1,472.8	1,038.4	916.5	103.4	1.338
1975	215.973	1,598.4	1,142.8	1,012.8	57.8	1.160
1976	218.035	1,782.8	1,252.6	1,129.3	98.4	1.538
1977	220.239	1,990.5	1,379.3	1,257.2	142.5	1.987
1978	2,22.585	2,249.7	1,551.2	1,403.5	186.9	2.020
1979	222.055	2,508.2	1,729.3	1,566.8	189.1	1.745
1980	227.757	2,732.0	1,918.0	1,732.6	133.1	1.292
1981	230.138	3,052.6	2,127.6	1,915.1	167.7	1.084
1982	232.520	3,166.0	2,261.4	2,050.7	64.1	1.062
1983	234.799	3,405.7	2,428.1	2,234.5	105.7	1.703
1984	237.001	3,772.2	2,668.6	2,430.5	249.4	1.750
1985	239.279	4,014.9	2,838.7	2,269.0	205.9	1.742
1986	241.625	4,231.6	3,013.3	2,797.4	199.3	1.805
1987	243.934	4,524.3	3,205.9	3,010.8	213.2	1.621
1988	246.329	4,880.6	3,477.8	3,235.1	236.7	1.488
1989	248.777	5,233.2	3,780.0	3,470.3	224.9	1.374

Table 8 *Continued*

Year	Unemployment Rate	Consumer Price Index	Annual Rate of Inflation	Interest Rate on One-Year Treasury Bills	Interest Rate on Long-Term Treasury Bonds	S&P 500 Stock Returns
1950	5.3	24.1	5.9	1.15	2.32	31.7
1951	3.3	26.0	6.0	1.51	2.57	24.0
1952	3.0	26.5	0.8	1.81	2.68	18.3
1953	2.9	26.7	0.7	2.04	2.94	−1.0
1954	5.5	26.9	−0.7	1.37	2.55	52.6
1955	4.4	26.8	0.4	1.40	2.84	31.5
1956	4.1	27.2	3.0	2.60	3.08	6.6
1957	4.3	28.1	2.9	3.32	3.47	−10.8
1958	6.8	28.9	1.8	2.66	3.43	43.4
1959	5.5	29.1	1.7	3.42	4.08	12.0
1960	5.5	29.6	1.4	5.28	4.02	0.5
1961	6.7	29.9	0.7	2.74	3.90	26.9
1962	5.5	30.2	1.3	3.34	3.95	−8.7
1963	5.7	30.6	1.6	3.14	4.00	22.8
1964	5.2	31.0	1.0	3.87	4.15	16.5
1965	4.5	31.5	1.9	4.13	4.21	12.4
1966	3.8	32.4	3.5	4.99	4.65	−10.0
1967	3.8	33.4	3.0	4.90	4.85	24.0
1968	3.6	34.8	4.7	5.67	5.26	11.1
1969	3.5	36.7	6.2	6.53	6.12	−8.4
1970	4.9	38.8	5.6	8.23	6.58	3.9
1971	5.9	40.5	3.3	4.66	5.74	14.30
1972	5.6	41.8	3.4	4.03	5.64	19.0
1973	4.9	44.4	8.7	5.99	6.31	−14.7
1974	5.6	49.3	12.9	7.65	6.98	−26.4
1975	8.5	53.8	6.9	6.79	7.00	37.2
1976	7.7	56.9	4.9	5.84	6.78	23.8
1977	7.1	60.6	6.7	5.34	7.06	−7.2
1978	6.1	65.2	9.0	7.40	7.89	6.6
1979	5.8	72.6	13.3	10.70	8.74	18.4
1980	7.1	82.4	12.5	12.49	10.81	32.4
1981	7.6	90.9	8.9	14.66	12.87	−4.9
1982	9.7	96.5	3.8	14.86	12.23	21.4
1983	9.6	99.6	3.8	8.83	10.84	22.5
1984	7.5	103.9	3.9	10.12	11.99	6.3
1985	7.2	107.6	3.8	9.22	10.75	32.2
1986	7.0	109.6	1.1	7.88	8.14	18.5
1987	6.2	113.6	4.4	5.86	8.64	5.2
1988	5.5	118.3	4.4	7.07	8.98	16.8
1989	5.3	124.0	4.6	9.27	8.58	31.6

(*continues*)

Table 8 *Continued*

Year	M1	M2	Dow-Jones Prices	S&P 500 Prices	S&P 500 Earnings	S&P 500 Dividends
1950			216.31	18.40	2.57	1.21
1951			257.64	22.34	2.64	1.37
1952			270.76	24.50	2.37	1.42
1953			275.97	24.73	2.52	1.43
1954			333.94	29.69	2.70	1.47
1955			442.72	40.49	3.60	1.65
1956			493.01	46.62	3.34	1.90
1957			475.71	44.38	3.37	1.79
1958			491.66	46.24	2.89	1.75
1959	140.0	297.8	632.12	57.38	3.39	1.83
1960	140.7	312.4	618.04	55.85	3.27	1.95
1961	145.2	335.5	691.55	66.27	3.19	2.02
1962	147.9	362.7	639.76	62.38	3.67	2.13
1963	153.4	393.3	714.81	69.87	4.02	2.28
1964	160.4	424.8	834.05	81.37	4.55	2.50
1965	167.9	459.4	910.88	88.17	5.19	2.72
1966	172.1	480.0	873.60	85.26	5.55	2.87
1967	183.3	524.4	879.12	91.93	5.33	2.92
1968	197.5	566.4	906.00	98.70	5.76	3.07
1969	204.0	589.6	876.72	97.84	5.78	3.16
1970	214.5	628.1	753.19	83.22	5.13	3.14
1971	228.4	712.7	884.76	98.29	5.70	3.07
1972	249.4	805.3	950.71	109.20	6.42	3.15
1973	263.0	861.0	923.88	107.43	8.16	3.38
1974	274.4	908.5	759.37	82.85	8.89	3.60
1975	287.6	1,023.2	802.49	86.16	7.96	3.68
1976	306.5	1,163.7	974.92	102.01	9.91	4.05
1977	331.5	1,286.8	894.63	98.20	10.89	4.67
1978	358.8	1,389.2	820.23	96.02	12.33	5.07
1979	386.1	1,500.3	844.40	103.01	14.86	5.65
1980	412.2	1,633.3	891.41	118.78	14.82	6.16
1981	439.1	1,795.9	932.92	128.05	15.36	6.63
1982	476.4	1,954.5	884.36	119.71	12.64	6.87
1983	522.1	2,186.0	1,190.34	160.41	14.03	7.09
1984	551.9	2,367.2	1,178.48	160.46	16.64	7.53
1985	620.5	2,567.4	1,328.23	186.84	14.61	7.90
1986	725.9	2,811.2	1,792.76	236.34	14.48	8.28
1987	752.3	2,909.9	2,275.99	286.83	17.50	8.81
1988	790.3	3,069.6	2,060.82	265.79	21.29	9.67
1989	797.6	3,217.0	2,508.91	322.84	24.17	11.61

Source: Unless otherwise noted, all data are from the *Economic Report of the President,* February 1990.

References

Chapter 1

Opening quotation: Walter Bagehot.

1. Edward O. Thorpe, *Beat the Dealer,* revised, New York: Random House, 1966.
2. For some real numbers, see Maxine Brady, *The Monopoly Book,* New York: David McKay, 1976.
3. W. Allen Wallis and Harry V. Roberts, *Statistics: A New Approach,* New York: The Free Press, 1956, p. 192.
4. Harold R. Jacobs, *Mathematics: A Human Endeavor,* San Francisco: W. H. Freeman, 1982, p. 539.
5. These data are from Barbara Boyle Torrey, "Sharing Increasing Costs on Declining Income: The Visible Dilemma of the Invisible Aged," *Milbank Memorial Fund Quarterly/Health and Society,* 1985, pp. 377–394.
6. Barbara Boyle Torrey, quoted in Richard Lipkin, "Analyzing the Figures That Shape Daily Lives," *Insight,* May 22, 1989, pp. 10–17.
7. This example is from Harold R. Jacobs, p. 562.
8. Frederick Mosteller and David L. Wallace, *Inference and Disputed Authorship: The Federalist,* Reading, Mass.: Addison-Wesley, 1964.
9. Edward R. Tufte, *The Visual Display of Quantitative Information,* Cheshire, Conn.: Graphics Press, 1983. This book is beautiful, in addition to being informative and entertaining. Read it; you'll enjoy it.
10. John Tukey, "Some Graphic and Semigraphic Displays," in T.A. Bancroft, ed., *Statistical Papers in Honor of George W. Snedecor,* Ames, Iowa: Iowa State University Press, 1972, p. 296.
11. For a thorough explanation of stem-and-leaf diagrams and other tools of exploratory data analysis, see John W. Tukey, *Exploratory Data Analysis,* Reading, Mass.: Addison-Wesley, 1977.
12. G. Udny Yule and M.G. Kendall, *An Introduction to the Theory of Statistics,* London: Charles Griffin, 1948, p. 100.
13. These data are from *Forbes* annual mutual fund issue (August 27, 1984; September 16, 1985; September 8, 1986) and in each case refer to the "preceding 12 months," which doesn't coincide with the calendar year.
14. William A. Spurr and Charles P. Bonini, *Statistical Analysis for Business Decisions,* revised edition, Homewood, Ill.: Irwin, 1973, p. 219.
15. Quoted in "Scorecard," Jerry Kirshenbaum, ed., *Sports Illustrated,* February 1984.
16. John J. Gabarro, "When a New Manager Takes Charge," *Harvard Business Review,* May–June 1985, pp. 110–123.
17. Francis Galton, *Natural Inheritance,* London: Macmillan, 1889.
18. *The Way Things Work Book of the Computer: An Illustrated Encyclopedia of Information Science, Cybernetics and Data Processing,* New York: Simon & Schuster, 1974.
19. Pankaj Ghemawat, "Building Strategy on the Experience Curve," *Harvard Business Review,* March–April 1985, pp. 143–149.

20. Mark Kac, "Probability," *Scientific American,* September 1964, pp. 92–108.

21. For the advice of a statistician, see George Lindsey, "An Investigation of Strategies in Baseball," *Operations Research,* 1963, pp. 477–501.

22. *Boston Globe,* August 14, 1976.

23. Gertrude E. Pearse, *Biometrika,* Vol. 20A, 1928, p. 336.

24. S.J. Pretorius, "Skew Bivariate Frequency Surfaces," *Biometrika,* Vol. 22, 1930–1931, p. 210.

25. Robert M. Wood, "Giant Discoveries of Future Science," *Virginia Journal of Science,* 1970, pp. 169–177.

26. A survey by the National Federation of Independent Businesses, reported in Mark Robichaux, "Business First, Family Second," *Wall Street Journal,* May 12, 1989.

27. One percent of those surveyed were under 20 and were included in the category 20–25. Similarly 1 percent were over 60 and were included in the category 50–60. The survey was conducted by the National Federation of Independent Businesses and reported in Mark Robichaux.

28. Morgan Stanley Capital International data published in "Japan, Land of the High P-Es," *Wall Street Journal,* March 22, 1990.

29. Roger G. Ibbotson and Rex A. Sinquefield, *Stocks, Bonds, Bills and Inflation: 1987 Yearbook,* Chicago: Ibbotson Associates, Inc., 1987.

30. Neil Barsky and Pauline Yoshihashi, "Trump Is Betting That Taj Mahal Casino Will Hit Golden Jackpot in Atlantic City," *Wall Street Journal,* March 20, 1990.

31. These data were compiled from *Value Line Investment Survey,* May 5, 1989, pp. 1076–1118.

32. Karen Slater, "Rating Risk," *Wall Street Journal,* October 20, 1989.

33. Editorial, "A Legal Fiction," *Wall Street Journal,* July 18, 1989.

34. *College New Mag,* Special, *Claremont Courier,* November 1989.

35. Mary Ann Mason, *The Equality Trap,* New York: Simon & Schuster, 1988.

36. *Wall Street Journal,* July 6, 1987.

37. Lynn Rapaport, "The Cultural and Material Reconstruction of the Jewish Communities in the Federal Republic of Germany," *Jewish Social Studies,* Spring 1987, pp. 137–154.

Chapter 2

Opening quotation: E.W. Howe, *Sinner Sermons,* Girard, Ka.: Haldeman-Julius, 1926.

1. Here's a sampling: Oswald Jacoby, *Oswald Jacoby on Poker,* New York: Doubleday, 1981; and Maxine Brady, *The Monopoly Book,* New York: David McKay, 1976. In a different spirit, S.W. Erdnase, *The Expert at the Card Table* (or *Artifice, Ruse, and Subterfuge at the Card Table: A Treatise on the Science and Art of Manipulating Cards),* Chicago: Drake, 1905, argues that gamblers can profit simply by knowing that certain cards are missing from the deck, thereby altering probabilities.

2. John Maynard Keynes, *A Treatise on Probability,* London: Macmillan, 1921, pp. 362–363.

3. "U.S. Operators Find 1,004 Fields in 1977," *Oil and Gas Journal,* July 3, 1978, pp. 32–33.

4. Roger G. Ibbotson and Rex A. Sinquefield, *Stocks, Bonds, Bills, and Inflation: Historical Returns (1926–1978),* Charlottesville, Va.: The Financial Analysts Research Foundation, 1979.

5. Richard Scammon, *Odds: On Virtually Everything,* New York: Putnam, 1980, p. 135.

6. Leo Guild, *You Bet Your Life,* Hollywood, Calif.: Marcel Rodd, 1946.

7. Hans Ulrich Balthasar, Roberto A.A. Boschi, and Michael M. Menke, "Calling the Shots in R&D," *Harvard Business Review,* May–June 1978, pp. 151–160.

8. This quotation and the accompanying discussion is based on Irwin Kabus, "You Can Bank on Uncertainty," *Harvard Business Review,* May–June 1976, pp. 95–105.

9. Edward Yardeni, *Money & Business Alert,* February 4, 1987, p. 1.

10. Richard Scammon, p. 8.

11. John Allen Paulos, "Orders of Magnitude," *Newsweek,* November 24, 1986.

12. Norman Dash, *Great Betting Systems,* Los Angeles: Price/Stern/Sloan and Ravenna Books, 1968, pp. 155–156.

13. Norman Dash, pp. 164–165.

14. Leo Guild, p. 41.

15. Clark S. Judge, *The Book of American Rankings,* New York: Facts on File, 1979.

16. Leo Guild, p. 80.

17. Richard Scammon, p. 10.

18. This quotation is from an article by Rose Gutfeld, "Regional Risks: IRS's Audit Rate Shows That All States Aren't Equal," *Wall Street Journal,* July 22, 1987.

19. D. Premack, "Language in Chimpanzees?" *Science,* 1971, pp. 808–822.

Chapter 3

Opening quotation: Charles Sanders Peirce.

1. Unfortunately, many classic works are not in English; e.g., A.N. Kolmogorov, *Grundbegriffe der Wahrscheinlichkeitsrechnung,* Berlin: Springer, 1933. For a modern English discussion, see Paul G. Hoel, *Introduction to Mathematical Statistics,* fourth edition, New York: Wiley, 1971, Ch. 2.

2. Clement McQuaid, ed., *Gambler's Digest,* Northfield, Ill.: Digest Books, 1971, pp. 24–25.

3. Norman Dash, *Great Betting Systems,* Los Angeles: Price/Stern/Sloan and Ravenna Books, 1968, pp. 155–156.

4. A detailed discussion of this case is given in Paul Meier and Sandy Zabell, "Benjamin Peirce and the Howland Will," *Journal of the American Statistical Association,* September 1980, pp. 497–506.

5. See the Supreme Court of California, *People* v. *Collins,* and William B. Fairley and Frederick Mosteller, "A Conversation about Collins," in Fairley and Mosteller, *Statistics and Public Policy,* Reading, Mass.: Addison-Wesley, 1977, pp. 355–379.

6. Gregg Easterbrook, "Troubled Flights Ahead for U.S. Space Program," *Los Angeles Times,* May 12, 1986.

7. Paul Asquith, David W. Mullins, and Eric D. Wolff, "Original Issue High Yield Bonds: Aging Analysis of Defaults, Exchanges, and Calls," *Journal of Finance,* September 1989, pp. 923–952.

8. Richard R. Nelson, "Uncertainty, Learning, and the Economics of Parallel Research and Development Efforts," *Review of Economics and Statistics,* November 1961, pp. 351–364.

9. Jack McCallum, "Who Needs an Airplane?" *Sports Illustrated,* August 26, 1985.

10. P.S. Laplace, *A Philosophical Essay on Probabilities,* 1812, reprinted New York: Dover, 1951, p. 189.

11. T. Bayes, "Thomas Bayes Essay Toward Solving a Problem in the Doctrine of Chances," 1763, reprinted in *Biometrika,* Vol. 45, 1958, p. 297. Also see Karl Pearson, *The History of Statistics in the 17th and 18th Centuries,* London: Charles Griffin, 1978, Chapter X.

12. A 1958 study that estimated these probabilities to be 10 to 1 was cited by the landmark 1964 report of the Surgeon General of the U.S. Public Health Service: E. Cuyler Hammond and Daniel Horn, "Smoking and Death Rates—Report on Forty-Four Months of Follow-Up of 187,783 men, II: Death Rates by Cause," *Journal of the American Medical Association,* 1958, pp. 1294–1308.

13. "Judge Awards $1.2 Million in Fishermen's Deaths," *Cape Cod Times,* August 13, 1985.

14. Michael O. Finkelstein and William B. Fairley, "A Bayesian Approach to Identification Evidence," *Harvard Law Review,* 1970, p. 489.

15. David T. Lykken, "Polygraphic Interrogation," *Nature,* February 23, 1984, pp. 681–684. Another careful study of six polygraph experts found that, on average, 37 percent of 50 innocent people were classified as liars, while 24 percent of 50 confessed thieves were judged truthful; Benjamin Kleinmuntz and Julian J. Szucko, "A Field Study of the Fallibility of Polygraphic Lie Detection," *Nature,* March 29, 1984, pp. 449–450.

16. From the magazine *California Highways,* cited by Darrell Huff in *How to Take a Chance,* New York: Norton, 1959, p. 110.

17. John Fisher, *Never Give a Sucker an Even Break,* New York: Pantheon Books, 1976.

18. Leo Guild, *You Bet Your Life,* Hollywood, Calif.: Marcel Rodd, 1946, pp. 40–41.

19. Hans Zeisel, "Dr. Spock and the Case of the Vanishing Women Jurors," *University of Chicago Law Review,* Fall 1969, p. 12.

20. Jacob W. Ulvila and Rex V. Brown, "Decision Analysis Comes of Age," *Harvard Business Review,* September–October 1982, pp. 130–141.

21. Bob Oates, "Knee Braces: Debate Hinges on Evidence," *Los Angeles Times,* April 11, 1987.

22. John R. Dorfman, "Stock Ace Stumbles; Random Throws Beat Pros," *Wall Street Journal,* June 8, 1989.

23. *Los Angeles Times,* February 8, 1985.
24. J. Yerushalmy, J.T. Harkness, J.H. Cope, and B.R. Kennedy, "The Role of Dual Reading in Mass Radiography," *American Review of Tuberculosis,* 1950, pp. 443–464.
25. *The Rocky Mountain News,* November 27, 1968.

Chapter 4

Opening quotation: James Thurber, "The Fairly Intelligent Fly," in *Fables for Our Time,* New York: Harper & Row, 1939, p. 13.

1. Tyler Hokama, "Fros and Treys: A Study in Shooting," Pomona College, Spring 1990.
2. The details are recounted by I. Nelson Rose in the December 6, 1985, *Los Angeles Times.*
3. John J. Goldman, "Thousands Seek Millions as Jackpot Fever Grips N.Y.," *Los Angeles Times,* May 12, 1984.
4. This example is from Joseph Newman, *Management Applications of Decision Theory,* New York: Harper & Row, 1971, pp. 32–61.
5. Bob Baker, "Door-to-Door Sales—'The Good, the Bad and the Ugly,'" *Los Angeles Times,* March 23, 1986.
6. *Statistisk arsbok for Sverige, 1986,* Stockholm: Statistiska Central Byran, 1986, pp. 55–56.
7. M.E. Bitterman, "The Evolution of Intelligence," *Scientific American,* January 1965.
8. Milton Friedman and Leonard Savage, "The Utility Analysis of Choices Involving Risk," *Journal of Political Economy,* August 1948, pp. 279–304.
9. Roger G. Ibbotson and Rex A. Sinquefield, *Stocks, Bonds, Bills, and Inflation: Historical Returns (1926–1978),* Charlottesville, Va.: The Financial Analysts Research Foundation, 1979. More recent data cited can be found in Kidder, Peabody, & Co., *Investment Strategy,* July 1986, p. 1.
10. Adam Smith, *The Wealth of Nations* (1776), reprinted London: Methuen, 1920, Vol. 1, Ch. 10, p. 109.
11. Frank J. Prial, *New York Times,* February 17, 1976.
12. William J. Peters, "The Psychology of Risk in Consumer Decisions," in George Fisk, ed., *The Frontiers of Management & Psychology,* New York: Harper, 1964.
13. Ibbotson and Sinquefield.
14. Hans R. Stoll and Robert E. Whaley, "Program Trading and Expiration-Day Effects," *Financial Analysts Journal,* March–April 1987, pp. 16–28.
15. John Dorfman, "Company Offers Gold at a Big Discount in Phone Solicitations—but with a Catch," *Wall Street Journal,* October 12, 1987.
16. Thomas C. Noddings, *Advanced Investment Strategies,* Homewood, Ill.: Dow Jones-Irwin, 1978, p. 51.
17. Maria Crawford Scott, "Editor's Note: To the Matophiles," *AAII Journal,* American Association of Individual Investors, March 1984, p. 3.
18. Robert Merton, Myron Scholes, and Mathew Gladstein, "The Returns and Risks of Alternative Put-Option Portfolio Investment Strategies," *Journal of Business,* 1982, pp. 1–55.
19. Alexandra Peers, "Psych 101: Investor Behavior," *Wall Street Journal,* November 13, 1987.
20. Francis C. Brown, III, and Jonathan Dahl, "New Data on Airline Performance May End Up Misleading Travelers," *Wall Street Journal,* November 23, 1987.
21. Paul Kagan Associates, reported in Laura Landro, "Summer Movies Off to Bruising Start," *Wall Street Journal,* June 7, 1990.
22. Steve Phenow, "Getting Your Red Herring Bearings," *Los Angeles Reader,* June 8, 1990, pp. 10–11.

Chapter 5

Opening quotation: J.M. Keynes, *A Tract on Monetary Reform,* London: Macmillan, 1923, p. 65.

1. More elaborate tables can be found in the *Harvard Computation Laboratory Tables of the Cumulative Binomial Distribution,* Cambridge, Mass.: Harvard University Press, 1955, and Sol Weintraub, *Tables of the Cumulative Binomial Probability Distribution for Small Values of p,* New York: Free Press of Glencoe, 1963.
2. For example, Marek Fisz, *Probability Theory and Mathematical Statistics,* third edition, New York: Wiley, 1963, p. 131.
3. Jonathan Dahl and Francis C. Brown, III, "Late Arrivals: New Figures Reveal Airlines' Dismal On-Time Record," *Wall Street Journal,* May 18, 1987.

4. Samuel C. Browstein and Mitchel Weiner, *Barron's How to Prepare for the Graduate Record Examination,* fourth edition, Woodbury, N.Y., 1979.

5. This example is based on Michael D. Intriligator, "Problems of Economic Reform in the Soviet Union," unpublished, April 1990.

6. Quoted in Clement McQuaid, ed., *Gambler's Digest,* Northfield, Ill.: Digest Books, 1971, p. 287.

7. Formal experiments have found a widespread belief in the law of averages; a survey is given by G.S. Tune, "Response Preferences: A Review of Some Relevant Literature," *Psychological Bulletin,* April 1964, pp. 286–302.

8. T. Gilovich, R. Vallone, and A. Tversky, "The Hot Hand in Basketball: On the Misperception of Random Sequences," *Cognitive Psychology,* 1985, pp. 195–314; but see Patrick D. Larkey, Richard A. Smith, and Joseph B. Kadane, "It's Okay to Believe in the 'Hot Hand,'" *Chance,* Fall 1989, pp. 22–30.

9. W. Allen Wallis and Harry V. Roberts, *Statistics: A New Approach,* New York: The Free Press, 1956, p. 322.

10. Some of the details are given in an article by Sheldon E. Haber and Rosedith Sitgreaves. "An Optimal Inventory Model for the Intermediate Echelon When Repair Is Possible," *Management Science,* February 1975, pp. 638–648.

11. Tom Clancy, "In a Frigate's Combat Center, Time and Information Run Out Quickly," *Los Angeles Times,* May 20, 1987.

12. William Simon, "'Back-to-the-Wall Effect': 1976 Perspective," in Shaul P. Ladany and Robert E. Machol, *Optimal Strategies in Sports,* Amsterdam: North-Holland, 1977, pp. 109–111.

13. Clement McQuaid, ed., p. 287.

11. Edgar Allen Poe, "The Mystery of Marie Roget," in *Murders in the Rue Morgue,* Girard, Ka.: Haldeman-Julius.

15. John R. Dorfman, "Stock Ace Stumbles; Random Throws Beat Pros," *Wall Street Journal,* June 8, 1989.

16. Frederick C. Klein, "Mean Streaks: A's Lead Giants," *Wall Street Journal,* October 17, 1989.

17. Ross Newhan, "Dodgers Continue Downfall," *Los Angeles Times,* May 14, 1989.

18. Frederick C. Klein.

19. *Los Angeles Times,* December 2, 1983.

Chapter 6

Opening quotation: Sir Francis Galton.

1. Marek Fisz, *Probability Theory and Mathematical Statistics,* third edition, New York: Wiley, 1963, pp. 196–202.

2. H.P. Bowditch, "The Growth of Children," Report of the Board of Health of Massachusetts, VIII, 1877.

3. For a discussion of several such systems, see Timothy O. Bakke, "Body Language Security Systems," *Popular Science,* June 1986, pp. 76–78, 112–113.

4. A discussion of several limit theorems is given in Fisz, Chap. 6.

5. Mercer and Hall, "The Experimental Error of Field Trials," *Journal of Agricultural Science,* Vol. 4, 1911, p. 107.

6. George Gamow, *One Two Three . . . Infinity,* New York: Viking Press, 1947, pp. 214–215.

7. Edward Kasner and James Newman, *Mathematics and the Imagination,* New York: Simon & Schuster, 1940.

8. J.E. Littlewood, *A Mathematician's Miscellany,* London: Methuen, 1953, p. 110.

9. Eugene F. Fama, "Efficient Capital Markets: A Review of Theory and Empirical Work," *Journal of Finance,* May 1970, pp. 383–417.

10. Paul Samuelson, "Proof that Properly Anticipated Prices Fluctuate Randomly," *Industrial Management Review,* Spring 1965, pp. 41–49.

11. A very readable source of further details is Burton Malkiel, *A Random Walk Down Wall Street,* second edition, New York: Norton, 1978.

12. Roger G. Ibbotson and Rex A. Sinquefield, *Stocks, Bonds, Bills, and Inflation: Historical Returns (1926–1978),* Charlottesville, Va.: The Financial Analysts Research Foundation, 1979.

13. Karl Pearson and Alice Lee, "On the Laws of Inheritance in Man," *Biometrika,* 1903, p. 364.

Chapter 7

Opening quotation: Lord Kelvin.

1. This example is based on D.D. Kosambi, "Scientific Numismatics," *Scientific American,* February 1966.

2. R.A. Howard, J.E. Matheson, and D.W. North, "The Decision to Seed Hurricanes," *Science,* June 16, 1972, pp. 1191–1202.

3. Michael S. Rozeff and William R. Kinney, "Capital Market Seasonability: The Case of Stock Market Returns," *Journal of Financial Economics,* October 1976.

4. Dean C. Coddington, Lowell E. Palmquist, and William V. Trollinger, "Strategies for Survival in the Hospital Industry," *Harvard Business Review,* May–June 1985, pp. 129–138.

5. Final Report of the Anthropometric Committee to the British Association, 1883, p. 256; cited in G. Udny Yule and M.G. Kendall, *An Introduction to the Theory of Statistics,* London: Charles Griffin, 1948, p. 94.

6. Harold Jacobs, *Mathematics: A Human Endeavor,* San Francisco: W.H. Freeman, 1982, p. 570.

7. E.S. Savas, "The Political Properties of Crystalline H_2O: Planning for Snow Emergencies in New York," *Management Science,* October 1973.

8. J.E. Vader, "Mr. Bases Loaded," *Sports Illustrated,* April 24, 1989, p. 73.

9. Andrew Paik, "Haitian Voodoo Baseball Stitchers Cause Uproar as Moon Shots Abound in 1987," Pomona College, December 1989.

10. Niels Diffrient, Alvin R. Tilley, and Joan C. Bardagjy, *Humanscale Manual 1–9,* Cambridge, Mass.: MIT Press, 1974, p. 4. The data in Exercises 7.71–7.79 are from this fascinating handbook.

Chapter 8

Opening quotation: Sir Arthur Conan Doyle, "The Adventure of the Copper Beeches," *Adventures of Sherlock Holmes,* New York: Harper and Brothers, 1892, p. 289.

1. This example is based on Michael Wheeler, *Lies, Damn Lies, and Statistics,* New York: Norton, 1976, pp. 257–259.

2. This example is from Paul Meier, "The Biggest Public Health Experiment Ever: The 1954 Field Trial of the Salk Poliomyelitis Vaccine," in Judith Tanur et al., *Statistics: A Guide to the Unknown,* San Francisco: Holden-Day, 1972, pp. 2–13.

3. Eugene Carlson, "A.C. Nielsen Took the Measure of the Crowd," *Wall Street Journal,* May 16, 1989.

4. Many of the details in this example are from Michael Wheeler, Ch. 10.

5. The RAND Corporation, *A Million Random Digits with 100,000 Normal Deviates,* New York: The Free Press, 1955.

6. The details of this case are from Hans Zeisel, "Dr. Spock and the Case of the Vanishing Women Jurors," *University of Chicago Law Review,* Fall 1969, pp. 1–18.

7. W. Allen Wallis and Harry V. Roberts, *Statistics: A New Approach,* New York: The Free Press, 1956, pp. 479–480.

8. "Coke-Pepsi Slugfest," *Time,* July 26, 1976, pp. 64–65.

9. "Down on the Farm," *Newsweek,* May 2, 1949, pp. 47–48.

10. L.L. Bairds, *The Graduates,* Princeton, N.J.: ETS, 1973.

11. Betsy Morris, "In This Taste Test, the Loser Is the Taste Test," *Wall Street Journal,* June 3, 1987.

12. Michael J. McCarthy, "New Coke Gets a New Look, New Chance," *Wall Street Journal,* March 7, 1990.

13. M.C. Bryson, "The Literary Digest: Making of a Statistical Myth," *The American Statistician,* 1976, pp. 184–185, argues that nonresponse bias was more important than the *Digest's* selection bias.

14. Earl W. Kintner, *A Primer on the Law of Deceptive Practices,* New York: Macmillan, 19712, p. 153.

15. *Sports Illustrated,* August 27, 1984, p. 14.

16. F.F. Stephan and P.J. McCarthy, *Sampling Opinions,* New York: Wiley, 1958, p. 286.

17. Darrell Huff, *How to Lie with Statistics,* New York: Norton, 1954, p. 26.

18. Michael Wheeler, p. 88.

19. Michael Wheeler, p. 90.

20. *New York Times,* November 16, 1980, p. 1.

21. *U.S. News & World Report,* October 21, 1968, p. 28.

22. Darrell Huff, p. 24.

23. The numbers and quotations in this example are from Richard Boudreaux and Marjorie Miller, "Sandinistas Conclude They Lost Touch with Populace," *Los Angeles Times,* March 4, 1990.

24. Cited in Lucy Horwitz and Lou Ferleger, *Statistics for Social Change,* Boston: South End Press, 1980, pp. 181–182.

25. Stanley L. Warner, "Randomized Response: A Survey Technique for Eliminating Evasive Answer Bias," *Journal of the American Statistical Association,* Vol. 60, 1965, pp. 63–69.

26. Michael Wheeler, p. 15.

27. Darrell Huff, *How to Take a Chance,* New York: Norton, 1959, pp. 115–117.

28. Michael Wheeler, p. 289.

29. *University of Chicago Magazine,* April 1952, p. 10.

30. *Civil Liberties,* December 1952.

31. Ben Wattenberg, "Women Are Getting What They Want, Surprisingly Fast," *Los Angeles Times,* May 24, 1983.

32. Joseph R. LaPlante, "Studds Finds Cape Political Climate Not So Temperate," *Cape Cod Times,* July 18, 1984.

33. U.S. Department of Commerce, *Statistical Abstract,* Washington, D.C.: U.S. Government Printing Office, 1981, Table 202, p. 123.

34. *Cape Cod Times,* August 28, 1984.

35. Michael Wheeler, p. 297.

36. Robert Reno, "Poll Shows Strong Work Ethic," *Newsday,* July 1983.

37. David Fay Smith and John Kochevar, "How Computers Are Changing Lives," *Dial,* June 1984, pp. 26–33, 55.

38. "More Children Equal Less Divorce," *Look,* February 13, 1951, p. 80.

39. "Scorecard," *Sports Illustrated,* August 27, 1984, p. 14.

40. Marcia Peoples Halio, "Student Writing: Can the Machine Maim the Message?" *Academic Computing,* January 1990, pp. 16–19, 45.

41. Associated Press, "College Broadens Not Just Mind, Study Says," *Cape Cod Times,* August 10, 1984.

42. *Business Week,* October 26, 1968, p. 97.

43. Sam Blum, "What Makes the Perfect Husband?" *McCall's,* August 1967, p. 61.

44. Ellen Hume and Peter Gumbel, "Gorbachev Impresses Group of Missourians; He Is High in Poll, Too," *Wall Street Journal,* December 4, 1987.

45. Charles Kenney, *Boston Globe Magazine,* August 30, 1987.

46. Robert A. Smith, "Recall and Recognition of Commercials Viewed at a Fast-Forward Speed and at the Normal Broadcast Speed on a Video-cassette Recorder," psychology Ph.D. thesis, U.S. International University, 1989.

47. This example is discussed by John Neter, "How Accountants Save Money By Sampling," in Tanur et al., *Statistics: A Guide to the Unknown,* San Francisco: Holden-Day, 1972, pp. 203–211.

48. *Chicago Daily News,* April 8, 1955.

49. "Selling It," *Consumer Reports,* May 1990, p. 371.

Chapter 9

Opening quotation: Sir Arthur Conan Doyle, *A Study in Scarlet,* Philadelphia: J.B. Lippincott, 1893, p. 136.

1. Leonid F. Maistrov, *Probability Theory,* translated and edited by Samuel Kotz, New York: Academic Press, 1974, p. 153.

2. Arthur Conan Doyle, *The Sign of Four,* copyright 1890 by Sir Arthur Conan Doyle.

3. R. Clay Sprowls, *Elementary Statistics for Students of Social Science and Business,* New York: McGraw-Hill, 1955, p. 104.

4. James T. McClave and P. George Benson, *Statistics for Business and Economics,* second edition, San Francisco: Dellen, 1982, p. 279.

5. Robert J. Samuelson, "The Strange Case of the Missing Jobs," *Los Angeles Times,* October 27, 1983.

6. This example is from W. Allen Wallis and Harry V. Roberts, *Statistics: A New Approach,* New York: The Free Press, 1956, p. 471.

7. Lisa Birnbach, ed., *The Official Preppy Handbook,* New York: Workman, 1980, p. 85.

8. Anthony Lewis, "Chief Justice Bird: Calm at the Center," *New York Times,* October 23, 1986.

9. W. Allen Wallis and Harry V. Roberts, p. 100.

10. *Reserve Mining Co.* v. *EPA* (1975), cited in David W. Barnes, *Statistics As Proof,* Boston: Little, Brown, 1983, p. 244.

11. Ellen Hume and Peter Gumbel, "Gorbachev Impresses Group of Missourians; He Is High in Poll, Too," *Wall Street Journal,* December 4, 1987.

12. Mike Budenholzer, "Underdogs Bomb Away in the NCAAs," Pomona College, Spring 1990.

13. "Inside Pitch," *Sports Illustrated,* June 13, 1983, p. 62.

14. "How They Scored, How They're Eating," *Consumer Reports,* May 1990, p. 325.

15. Todd Austin, "Roshambo," Pomona College, Spring 1990.
16. David W. Barnes, p. 35.
17. Clark Jensen, "Hometown Refs," Pomona College, Spring 1990.
18. "Macworld Survey," *Macworld,* June 1990, p. 127.
19. Peter Joe, "Typo Probabilities," Pomona College, Spring 1990.
20. David Kahn, "Modern Cryptology," *Scientific American,* July 1966, pp. 38–46. Also see Henry Beker and Fred Piper, *Cipher Systems,* New York: Wiley, 1982, esp. p. 25 and Appendix 1.
21. John Allen Paulos, *Innumeracy,* New York: Hill & Wang, 1988, p. 111.
22. Jaime Diaz, "Perils of Putting," *Sports Illustrated,* February 1989, pp. 76–78.
23. Paul Farmer, "Make a Wish! (Your Candle Will Go Out in 715 Seconds.)," Pomona College, Spring 1990.
24. John E. Hervey, "Michael Jordan: Do the Chicago Bulls Elevate as He Does?" Pomona College, Spring 1990.
25. Daniel Matsuda, "Effect of the Backboard on Shooting Accuracy," Pomona College, Spring 1990.
26. Christopher Helwig, "A Statistical Analysis of Stock Mutual Fund Returns," Pomona College, Spring 1990.
27. Charles Kenney, *Boston Globe Magazine,* August 30, 1987.
28. Dina Colman, Pomona College, 1987.
29. Tracy Hobbs, Pomona College, 1987.

Chapter 10

Opening quotation: Sherlock Holmes.
1. Edward Leamer, "Let's Take the Con Out of Econometrics," *American Economic Review,* March 1983, p. 36.
2. H.S. Diehl, A.B. Baker, and D.W. Cowan, "Cold Vaccines," *Journal of the American Medical Association,* September 24, 1938, pp. 1168–1173.
3. H.S. Diehl et al., p. 1173.
4. From Harley Tinkham, "Morning Briefing," *Los Angeles Times,* November 26, 1983.
5. Otto Klineberg, *Negro Intelligence and Selection Migration,* New York: Columbia University Press, 1935.

6. Lester Kauffman, "Statistical Quality Control at the St. Louis Division of American Stove Company," cited in Acheson Duncan, *Quality Control and Industrial Statistics,* Homewood, Ill.: Irwin, 1959.
7. Sheldon Blackman and Don Catalina, "The Moon and the Emergency Room," *Perceptual and Motor Skills,* 1973, pp. 624–626.
8. E. Yehle, "Accuracy in Clerical Work," in Victor Lazzaro, ed., *Systems & Procedures: A Handbook for Business and Industry,* second edition, Englewood Cliffs, N.J.: Prentice-Hall, 1968.
9. William R. Simpson, "A Probabilistic Formulation of Murphy Dynamics as Applied to the Analysis of Operational Research Problems" in *The Best of the Journal of Irreproducible Results,* New York: Workman, 1983, pp. 120–123.
10. R.A. Fisher, *Experiments in Plant Hybridization,* Edinburgh: Oliver and Boyd, 1965, p. 53. This book put Mendel's original paper together with comments by Fisher, based on a 1936 paper.
11. Kenneth B. Clark and Mamie P. Clark, "Racial Identification and Preference in Negro Children" in Eleanor E. Macoby, Theodore M. Newcomb, and Eugene L. Hartley, eds., *Readings in Social Psychology,* third edition, New York: Holt, Rinehart, and Winston, 1958, pp. 602–611.
12. Seymour Siwoff, Steve Hirdt, and Peter Hirdt, *The New Elias Baseball Analyst,* New York: Macmillan, 1985.
13. Martin Gardner, "Great Fakes of Science," in *Science: Good, Bad and Bogus,* New York: Prometheus Books, 1981, p. 123.
14. Roger C. Vergin and Michael Scriabin, "Winning Strategies for Wagering on National Football League Games," *Management Science,* Vol. 24, April 1978, pp. 809–818.
15. P.K. Whelpton and Clyde V. Kiser, *Social and Psychological Factors Affecting Fertility,* Vol. 1, New York: Milbank Memorial Fund, 1950, p. 109.
16. Roger C. Vergin and Michael Scriabin, p. 812.
17. Frederick Mosteller, Robert E.K. Rourke, and George B. Thomas, Jr., *Probability with Statistical Applications,* Reading, Mass.: Addison-Wesley, 1961, p. 17.
18. *A.B.G. Instrument & Engineering, Inc.* v. *United States* (1979), cited in David W. Barnes, *Statistics As Proof,* Boston: Little, Brown, 1983, p. 261.

19. "The FDA Cardio-Renal Committee Replies," *Wall Street Journal,* July 20, 1987.
20. Thomas H. Maugh, II, "Breast Cancer Survival Rate May Double, Study Finds," *Los Angeles Times,* May 14, 1990.
21. Pamela Weintraub, "Preschool?" *Omni,* August 1989, pp. 34–38.
22. Tom Purcell, "Free Throws and Rebounds from Misses," Pomona College, December 1989.
23. Andrew Jacobson, "Is the Mail Finally Here?" Pomona College, Spring 1990.
24. David Muscatel, "Crashing the Boards," Pomona College, December 1989.
25. *Cape Cod Times,* July 24, 1984.
26. *Pennsylvania* v. *Local 542, International Union of Operating Engineers,* cited in David W. Barnes, p. 27.
27. *Los Angeles Times*/Booz-Allen & Hamilton survey, "Who's on Top in Technology?" *Los Angeles Times,* February 21, 1988.
28. *The Skeptical Inquirer,* Spring 1987, pp. 228–230.
29. David W. Barnes, p. 175.

Chapter 11

Opening quotation: Ronald H. Coase.
1. R.A. Fisher, "The Arrangement of Field Experiments," *Journal of the Ministry of Agriculture of Great Britain,* 1926, p. 504.
2. Arthur Melton, ed., *Journal of Experimental Psychology,* Vol. 64, 1962, pp. 553–557.
3. "Science at the EPA," *Wall Street Journal,* October 2, 1985.
4. Darrell Huff, *How to Lie with Statistics,* New York: Norton, 1954, pp. 58–59.
5. F. Arcelus and A.H. Meltzer, "The Effect of Aggregate Economic Variables on Congressional Elections," *American Political Science Review,* Vol. 69, 1975, pp. 1232–1239.
6. F. Mosteller and R. Rourke, *Sturdy Statistics,* Reading, Mass.: Addison-Wesley, 1971, p. 68.
7. *New York Times,* Jan. 18, 1976.
8. Lucy Horwitz, and Lou Ferleger, *Statistics for Social Change,* Boston: South End Press, 1980, p. 177.
9. Francis Iven Nye, *Family Relationships and Delinquent Behavior,* New York: Wiley, 1958, p. 29.
10. Lloyd Shearer in *Parade,* October 8, 1972. For additional evidence, see Dennis D. Miller, "Is It Height or Sex Discrimination?," *Challenge,* September/October 1986, pp. 59–61. On the other hand, an article in the October 1986 *Pediatrics* suggests that height and IQ are related.
11. Sue Avery, "Market Investors Will Be High on Redskins Today;" and "Morning Briefing: Wall Street 'Skinnish' on Big Game," *Los Angeles Times,* January 30, 1983.
12. James Bates, "Reality Wears Loser's Jersey in Super Bowl Stock Theory," *Los Angeles Times,* January 22, 1989.
13. *Sporting Edge,* 1988.
14. Gail Howard, "State Lotteries: How to Get In It . . . and How to Win It!" fifth edition, Ben Buxton, 1986.
15. Cited by Martin Gardner in *Fads and Fallacies in the Name of Science,* New York: Dover, 1957, p. 305.
16. Norman Bloom, *The New World,* July 4, 1973.
17. Anonymous, "Pickles and Humbug," *The Journal of Irreproducible Results: Selected Papers,* 1981, p. 107.
18. "Scorecard," *Sports Illustrated,* January 5, 1987, p. 10.
19. Roger C. Vergin and Michael Scriabin, "Winning Strategies for Wagering on National Football League Games," *Management Science,* Vol. 24, April 1978, p. 814.
20. "News Briefs," *Cape Cod Times,* July 5, 1984.
21. *Newsweek,* February 4, 1974.
22. Dan Shaughnessy, "Pro Basketball," *Boston Globe,* June 24, 1984.
23. David Van Oppen, Pomona College, 1987.
24. Wayne Whaley, "Were They Fab or Just Lucky?" Pomona College, Spring 1990.
25. *Business Week,* April 29, 1985.
26. David W. Barnes, *Statistics As Proof,* Boston: Little, Brown, 1983, p. 162.
27. Floyd Norris, "Predicting Victor in Super Bowl," *New York Times,* January 17, 1989.
28. Don Thornburgh, "Telepathy: A Statistical Evaluation," Pomona College, Spring 1990.
29. R.L. Atkinson, R.C. Atkinson, E.E. Smith, and D.J. Bem, *Introduction to Psychology,* San Diego: Harcourt Brace Jovanovich, 1990, p. 240.
30. Robert W. Creamer, "Scorecard," *Sports Illustrated,* August 13, 1984, p. 13.

31. David W. Barnes, p. 144.

32. Lee Dye, "Proof Positive," *Los Angeles Times,* May 14, 1990.

33. William Feller, *Scientific Research,* February 3, 1969.

34. *United States* v. *Fatico* (1978), cited in David W. Barnes, pp. 146–154.

35. Fred Schwed, Jr., *Where Are the Customers' Yachts?* New York: Simon & Schuster, 1940, p. 47.

36. Beatrice E. Garcia, "Will It Be Friday the 13th, Part 6? Study Ties 3 in Year to Recession," *Wall Street Journal,* November 13, 1987.

37. In *Re Forte-Fairbairn, Inc.,* 62 F.T.C. 1146 (1963), cited in David W. Barnes, p. 132.

38. JS&A advertisement, *Discoveries,* United Airlines, June 1986.

39. *Alabama Nursing Home Association* v. *Califano* (1979), cited in David Barnes, p. 230.

40. John Allen Paulos, *Innumeracy,* New York: Hill & Wang, 1988, p. 99.

41. Quoted by Lee Dye, "Proof Positive," *Los Angeles Times,* May 14, 1990.

Chapter 12

Opening quotation: Maurice G. Kendall, "On the Reconciliation of Theories of Probability," 1949.

1. Charles W. Dunnett, "Drug Screening: The Never-Ending Search for New and Better Drugs," in Judith M. Tanur et al., eds., *Statistics: A Guide to the Unknown,* San Francisco: Holden-Day, 1972, pp. 23–33.

2. Donna Alcosser explained this test to me.

3. *Journal of the American Medical Association,* January 20, 1984.

4. David E. Harrington, "The Network Coverage of Economic News," working paper, July 1985.

5. Roger L. Jenkins, Richard C. Reizenstein, and F.G. Rogers, "Report Cards on the MBA," *Harvard Business Review,* September–October 1984, pp. 20–30.

6. Georg De Leon, "The Baldness Experiment," *Psychology Today,* October 1977, pp. 62–66.

7. John L. Coulehan et al., "Vitamin C Prophylaxis in a Boarding School," *New England Journal of Medicine,* January 1974, pp. 6–10.

8. Frank J. Massey, Jr. et al., "Vasectomy and Health," *Journal of the American Medical Association,* August 24, 1984, pp. 1023–1029.

9. Karl P. Koenig and John Masters, "Experimental Treatment of Habitual Smoking," *Behavior Research and Therapy,* 1965, pp. 235–243.

10. David P. Phillips, "Deathday and Birthday: An Unexpected Connection," in Judith M. Tanur et al., eds., pp. 52–65.

11. Karl Pearson, "On a Certain Double Hypergeometrical Series and Its Representation by Continuous Frequency Surfaces," *Biometrika,* Vol. 16, 1924, pp. 172–188.

12. Christopher Robson and Brendan M. Walsh, "Alphabetical Voting: A Study of the 1973 General Election in the Republic of Ireland," Dublin: The Economic & Social Research Institute, June 1973.

13. P.L. Yu, C. Wrather, and G. Kozmetsky, "Auto Weight and Public Safety, a Statistical Study of Transportation Hazards," Research Report 233, Center for Cybernetic Studies, University of Texas, Austin, 1975.

14. Reported in G. Udny Yule and M.G. Kendall, *An Introduction to the Theory of Statistics,* London: Charles Griffin, 1948, p. 419.

15. Lester Kauffman, "Statistical Quality Control at the St. Louis Division of American Stove Company," cited in Acheson Duncan, *Quality Control and Industrial Statistics,* Homewood, Ill.: Irwin, 1959.

16. Dennis Lendrum, "Should John McEnroe Grunt?," *New Scientist,* July 21, 1983. Professor Lendrum was kind enough to send me his original data.

17. Harriet H. Imrey, "Smoking Cigarettes: A Risk Factor for Sexual Activity Among Adolescent Girls," *Journal of Irreproducible Results,* Nov./Dec. 1983, p. 11.

18. Joan R. Rosenblatt and James J. Filliben, "Randomization and the Draft Lottery," *Science,* Vol. 171 (1971), pp. 306–308. Interestingly, there was also a draft lottery in 1940, in which the capsules apparently were not very well mixed. See the statement by Samuel Stouffer and Walter Bartky in the *Chicago Tribune,* November 2, 1940, p. 4.

19. Gerald A. Hudgens, Victor H. Denenberg, and M.X. Zarrow, "Mice Reared with Rats: Effects of Preweaning and Postweaning Social Interactions

upon Adult Behavior," *Behaviour,* 1968, pp. 259–274.

20. The poll results are summarized "A Look at the Opposite Sex," *Newsweek on Campus,* April 1984, p. 21. The numbers used in the text are approximations, since the article does not give the fractions of the 523 people polled who were male and female.

21. Francis Iven Nye, *Family Relationships and Delinquent Behavior,* New York: Wiley, 1958, p. 37.

22. Dennis McConnell, John A. Haslem, and Virginia R. Gibson, "The President's Letter to Stockholders: A New Look," *Financial Analysts Journal,* September–October 1986, pp. 66–70.

23. Bruno S. Frey, Werner W. Pommerehne, Friedrich Schneider, and Guy Gilbert, "Consensus and Dissension Among Economists: An Empirical Inquiry," *American Economic Review,* December 1984, pp. 986–994.

24. D. Weiss, B. Whitten, and D. Leddy, "Lead Content of Human Hair (1871–1971)," *Science,* 1972, pp. 69–70.

25. Hans R. Stoll and Robert E. Whaley, "Program Trading and Expiration-Day Effects," *Financial Analysts Journal,* March–April 1987, pp. 16–28.

26. Niels Ehlers, "On Corneal Thickness and Intraocular Pressure, II," *Acta Ophthalmologica,* Vol. 48, pp. 1107–1112.

27. Roger L. Faith, Donald R. Leavens, and Robert D. Tollison, "Antitrust Pork Barrel," *Journal of Law and Economics,* October 1982, pp. 329–342.

28. Leo A. Goodman and William H. Kruskal, "Measures of Association for Cross Classifications," *Journal of the American Statistical Association,* 1954, pp. 732–764.

29. Adam Clymer, "Poll Studies Hispanic Party Loyalties," *New York Times,* July 18, 1986.

30. Warren E. Miller, Arthur H. Miller, and Edward J. Schneider, *American National Election Studies Data Sourcebook, 1952–1978.* Cambridge, Mass.: Harvard University Press, 1980, p. 268.

31. H.C. White, "Cause and Effect in Social Mobility Tables," *Behavioral Science,* Vol. 7, 1963, pp. 14–27.

32. Marion Gillim, "Physical Measurements of Mount Holyoke College Freshmen in 1918 and 1943," *Journal of the American Statistical Association,* March 1944, pp. 53–56.

33. Karl Pearson, "On the Laws of Inheritance in Man, II," *Biometrika,* Vol. 3, 1904, pp. 131–190.

34. William E. Fruhan, Jr., "How Fast Should Your Company Grow?" *Harvard Business Review,* January–February 1984, pp. 84–93.

35. H.S. Diehl, A.B. Baker, and D.W. Cowan, "Cold Vaccines," *Journal of the American Medical Association,* September 24, 1938, pp. 1168–1173.

36. G.R. Lindsey, "A Scientific Approach to Strategy in Baseball," in Shaul P. Ladany and Robert E. Machol, *Optimal Strategies in Sports,* Amsterdam: North-Holland, 1977, pp. 1–30.

37. The Steering Committee of the Physicians' Health Study Research Group, "Preliminary Report: Findings from the Aspirin Component of the Ongoing Physicians' Health Study," *New England Journal of Medicine,* January 28, 1988, pp. 262–264.

38. *New England Journal of Medicine,* May 3, 1990.

39. Marilyn Chase, "Old Heart Drug Works as Well as Costly TPA in Study," *Wall Street Journal,* March 8, 1990.

40. Nicola Morf, "Which Students Excel in College: Those from Public or Private Schools?" Pomona College, December 1989.

41. Lucy Hovis, "Statistics and Stock Prices," Pomona College, Spring 1990.

42. Tyson Thomas, "The Magic of Three," Pomona College, December 1989.

43. Wei Hopeman, "47?" Pomona College, Spring 1990.

44. Hugh Chace, Pomona College, December 1989.

45. Daniel Matsuda, "Effect of the Backboard on Shooting Accuracy," Pomona College, Spring 1990.

46. Jennifer D. Sharp, "Gender Differences in Student Major and Course Selection," Pomona College, December 1989.

47. Julie Scheer, Pomona College, 1987.

48. Laura Hubbard, "Women and Commercials," Pomona College, December 1989.

49. G.R. Lindsey, pp. 1–30.

50. Kenneth Bruce Bayley, "A Statistical Analysis of Discrimination in the Alaska Criminal Justice System," Ph.D. dissertation, Claremont Graduate School, 1983.

51. Ben Rollins, "Procrastination. How Far Can It Go?" Pomona College, Spring 1990.

52. Jane Luh, "The Effect of Birth Order on Academic Performance," Pomona College, Spring 1990.

53. Harold Clayton, cited in Burton Crane, *The Sophisticated Investor*, New York: Simon & Schuster, 1959, p. 207.

54. "Germans Become Less Obsessed With Work," *Los Angeles Times*, June 4, 1984.

55. Lawrence S. Wrightsman, "Wallace Supporters and Adherence to 'Law and Order,'" in Robert A. Baron and Robert M. Liebert, eds., *Human Social Behavior*, Homewood, Ill.: Dorsey Press, 1971, pp. 217–225.

56. Michel Hersen, "Personality Characteristics of Nightmare Sufferers," *Journal of Nervous and Mental Diseases*, 1971, pp. 29–31.

57. *New England Journal of Medicine*, May 3, 1990.

58. Owen L. Davies, ed., *The Design and Analysis of Industrial Experiments*, London: Oliver and Boyd, 1960, p. 13.

59. Gloria Lena Vega and Scott M. Grundy, "Treatment of Primary Moderate Hypercholesterolemia with Lovastatin (Mevinolin) and Colestipol," *Journal of the American Medical Association*, January 2, 1987, pp. 33–38.

Chapter 13

Opening quotation: Edward Leamer, "Let's Take the Con Out of Econometrics," *American Economic Review*, March 1983, p. 37.

1. John Maynard Keynes, *The General Theory of Employment, Interest, and Prices* (1936), reprinted New York: Harcourt Brace Jovanovich, 1964, p. 96.

2. See Arthur Okun, *The Political Economy of Prosperity*, New York: Norton, 1970, p. 132, and, for an update, Gary Smith, "Okun's Law Revisited," *Quarterly Review of Economics and Business*, Winter 1975, pp. 37–54.

3. Dorothy S. Brady, "Family Saving, 1888 to 1950," in Raymond W. Goldsmith, Dorothy S. Brady, and Horst Medershausen, *A Study of Saving in the United States*, Vol. 3, Princeton, N.J.: Princeton University Press, 1956, p. 183.

4. Joel Dean, "Statistical Cost Functions of a Hosiery Mill," *Journal of Business*, 1941, Supplement, 2, pp. 1–51.

5. *Re Western Union Tel. Co.*, No. 18935, F.C.C., August 12, 1970.

6. Robert Cunningham Fadeley, "Oregon Malignancy Pattern Physiographically Related to Hanford, Washington, Radioisotope Storage," *Journal of Environmental Health*, 1965, pp. 883–897.

7. These data are from Roger G. Ibbotson and Rex A. Sinquefield, *Stocks, Bonds, Bills, and Inflation: Historical Returns (1926–1978)*, Charlottesville, Va: The Financial Analysts Research Foundation, 1979.

8. Wiesenberger Investment Companies Service, *Investment Companies*, 1987, pp. 154–157.

9. James Shields, *Monozygotic Twins*. London: Oxford University Press, 1962. Three similar, separate studies by Cyril Burt all reported the same value of R^2 ($= .594$)! The most logical explanation is that the data were flawed; see Nicholas Wade, "IQ and Heredity: Suspicion of Fraud Beclouds Classic Experiment," *Science*, 1976, pp. 916–919.

10. "Computerized Investing," Chicago: American Association of Individual Investors, December 1983/January 1984.

11. S. Karelitz, V.R. Fisichelli, J. Costa, R. Kavelitz, and L. Rosenfeld, "Relation of Crying in Early Infancy to Speech and Intellectual Development at Age Three Years," *Child Development*, 1964, pp. 769–777.

12. John R. Dorfman, "Rating Investment-Advice Givers: The Only Constant Is Inconsistency," *Wall Street Journal*, March 27, 1987.

13. J.F. Fraumeni, Jr., "Cigarette Smoking and Cancers of the Urinary Tract: Geographical Variation in the United States," *Journal of the National Cancer Institute*, 1968, pp. 1205–1211.

14. Frederick E. Croxton and Dudley J. Cowdon, *Applied General Statistics*, second edition, Englewood Cliffs, N.J.: Prentice-Hall, 1955, pp. 451–454.

15. George W. Pierce, *The Songs of Insects*, Cambridge, Mass.: Harvard University Press, 1949, pp. 12–21.

16. Gary Smith, "Coping with the Term Structure," in William C. Brainard, ed., *Essays in Honor of James Tobin*, Cambridge, Mass.: MIT Press, 1990.

17. W. Scott Baumann and Richard Dowen, "Growth Projections and Common Stock Re-

turns," *Financial Analysts Journal,* July/August 1988, pp. 79–80.

18. Ulysses Hwang, Pomona College, 1987.

19. Amanda Mondragon, Pomona College, 1987.

20. A.R. Wyler, M. Minoru, and T.H. Holmes, "Magnitude of Life Events and Serious Illness," *Psychomatic Medicine,* 1971, pp. 115–122.

21. John Leusner, Pomona College, 1987.

22. These estimates are reported, using changes rather than levels, in the *Federal Reserve Bank of St. Louis Review,* June/July 1984.

23. Scott Vanourek, "The 1960 Yankee Sluggers—Grace Under Pressure or Luck?" Pomona College, Spring 1990.

24. Charles Richter, *Elementary Seismology,* San Francisco: W.H. Freeman, 1958, p. 366.

Chapter 14

Opening quotation: Lewis Carroll, *Alice in Wonderland,* New York: Heritage Press, 1941.

1. Securities and Exchange Commission (SEC), *Institutional Investor Study Report,* 1971, Vol. 2, pp. 352–347.

2. SEC, *Institutional Investor Study Report,* 1971, Vol. 4, p. 333. For some other studies, see Fischer Black, Michael Jensen, and Myron Scholes, "The Capital Asset Pricing Model: Some Empirical Tests," and Merton Miller and Myron Scholes, "Rates of Return in Relation to Risk: A Re-examination of Some Recent Findings," both in Michael Jensen, ed., *Studies in the Theory of Capital Markets,* New York: Praeger, 1972, pp. 47–127.

3. Robert E. Penn, "The Economics of the Market in Modern Prints," *Journal of Portfolio Management,* Fall 1980, pp. 25–31.

4. James B. Cloonan, American Association of Individual Investors *Journal,* October 1983, p. 37.

5. Ray C. Fair, "The Effect of Economic Events on Votes for President," *The Review of Economics and Statistics,* Vol. 60 (May 1978), pp. 159–173; Gerald H. Kramer, "Short-Term Fluctuations in U.S. Voting Behavior, 1896–1964," *The American Political Science Review,* Vol. 65 (March 1971), pp. 131–143.

6. Charles Osterberg, "Unsafe at Zero Speed," *Journal of Irreproducible Results,* Sept./Oct. 1983, p. 19.

7. Mark Twain, *Life on the Mississippi,* 1874.

8. Karl Pearson and Alice Lee, "On the Laws of Inheritance in Man," *Biometrika,* 1903, p. 362.

9. Christopher Jencks et al., *Inequality,* New York: Basic Books, 1972, p. 59.

10. Amos Tversky and Daniel Kahneman, *Psychological Review,* 1973, p. 237.

11. Harold Hotelling, review of Horace Secrist, "The Triumph of Mediocrity in Business," *Journal of the American Statistical Association,* Vol. 28, 1933, pp. 463–465. Secrist and Hotelling debated this further in the 1934 volume of this journal, pp. 196–199.

12. William F. Sharpe, *Investment,* third edition, Englewood Cliffs, N.J.: Prentice-Hall, 1985, p. 430.

13. These baseball statistics are from *Esquire,* April 1981.

14. Darrell Huff, *How to Take a Chance,* New York: Norton, 1959, p. 141.

15. Lawrence S. Ritter and William F. Silber, *Principles of Money, Banking, and Financial Markets,* New York: Basic Books, 1986, p. 533.

16. John Llewellyn and Roger Witcomb, letters to *The Times,* April 4–6, 1977, and David Hendry, quoted in *The New Statesman,* November 23, 1979, pp. 793–795.

17. Edward Ames and Stanley Reiter, "Distributions of Correlation Coefficients in Economic Time Series," *Journal of the American Statistical Association,* Vol. 56, September 1961, pp. 636–656.

18. R.J. Hoyle, "Decline of Language as a Medium of Communication," in George H. Scherr, ed., *The Best of the Journal of Irreproducible Results,* New York: Workman, 1983, pp. 134–135.

19. "Scorecard," *Sports Illustrated,* March 18, 1974.

20. Gary Pomerantz, "Getting to the Top Is Easier Than Staying There," *Washington Post,* July 1983.

21. Albert Ando and Franco Modigliani, "The 'Permanent Income' and 'Life-Cycle' Hypotheses of Saving Behavior: Comparisons and Tests," in I. Friend and R. Jones, eds., *Consumption and Saving,* Vol. II, 1960, p. 154.

22. David W. Barnes, *Statistics As Proof,* Boston: Little, Brown, 1983, p. 298.

23. R.J. Larson and M.L. Marx, *An Introduction to Mathematical Statistics,* Englewood Cliffs, N.J.: Prentice-Hall, 1981, p. 349.

24. Wiesenberger Services, "Mutual Funds Investment Report," February 1988, p. 7.
25. Michael Laurence, "Playboy's Guide to Mutual Funds," *Playboy,* June 1969, p. 186.
26. R. Mulcahy, J.W. McGilvray, and N. Hickey, "Cigarette Smoking Related to Geographic Variations in Coronary Heart Disease Mortality and to Expectations of Life in the Two Sexes," *American Journal of Public Health,* 1970, p. 1516.
27. Joseph W. Harder, "For Better Ball, Nothing Beats Greed," *Wall Street Journal,* March 12, 1990.
28. *U.S. News & World Report,* March 4, 1968, p. 86.
29. *Time,* November 3, 1967, p. 49.
30. Peter Gammons, "Inside Baseball," *Sports Illustrated,* June 19, 1989, p. 68.
31. Robert Sullivan, "Scorecard," *Sports Illustrated,* February 24, 1986, p. 7.
32. *Texas Monthly,* January 1982, p. 83.
33. David Upshaw of Drexel Burnham Lambert, quoted in John Andrew, "Some of Wall Street's Favorite Stock Theories Failed to Foresee Market's Slight Rise in 1984," *Wall Street Journal,* January 2, 1985.
34. "Hoyt Says Cy Young Award in Jinx," *Cape Cod Times,* August 4, 1984.
35. P.C. Rosenblatt and M.R. Cunningham, "Television Watching and Family Tensions," *Journal of Marriage and the Family,* 1976, pp. 105–111.

Chapter 15

Opening quotation: Anonymous.
1. For a good survey, see Franklin Fisher, "Multiple Regression in Legal Proceedings," *Columbia Law Review,* 1980, pp. 702–736.
2. For a discussion of several cases, see Michael O. Finkelstein, "The Judicial Reception of Multiple Regression Studies in Race and Sex Discrimination Cases," *Columbia Law Review,* 1980, pp. 737–754.
3. Fisher, p. 730.
4. Lester B. Lave and Eugene P. Seskin, "Does Air Pollution Shorten Lives?," in John W. Pratt, ed., *Statistical and Mathematical Aspects of Pollution Problems,* New York: Marcel Dekker, 1974, pp. 223–247.
5. This example is based on Fred S. McChesney, "Sensationalism, Newspaper Profits and the Mar-

ginal Value of Watergate," *Economic Inquiry,* January 1987, pp. 135–144.
6. Frank A. Haight, "Do Speed Limits Reduce Traffic Accidents?," in Judith M. Tanur et al., *Statistics: A Guide to Business and Economics,* San Francisco: Holden-Day, 1976, pp. 130–136.
7. J. Truett et al., "A Multivariate Analysis of the Risk of Coronary Heart Disease in Framingham," *Journal of Chronic Disease,* 1967, pp. 511–524; and J.C. Kleinman et al., "Regional and Urban-Suburban Differentials in Coronary Heart Disease Mortality and Risk Factor Prevalence," *Journal of Chronic Disease,* 1981, pp. 11–19.
8. Robyn M. Dawes, "A Case Study of Graduate Admissions: Application of Three Principles of Human Decision Making," *American Psychologist,* February 1971, pp. 180–188.
9. Burton G. Malkiel and John G. Cragg, "Expectations and the Structure of Share Prices," *American Economic Review,* September 1970, pp. 601–617.
10. Haim Levy and Marshall Sarnat, *Investment and Portfolio Analysis,* New York: Wiley, 1972, pp. 244–248.
11. This example is from Albert Romano, *Applied Statistics for Science and Industry,* Boston: Allyn & Bacon, 1977, p. 128.
12. Samuel Bowles and Henry Levin, "The Determinants of Scholastic Achievement—An Appraisal of Some Recent Evidence," *Journal of Human Resources,* Winter 1968, p. 15. Also see Bowles and Levin, "More on Multicollinearity and the Effectiveness of Schools," *Journal of Human Resources,* Summer 1968, pp. 393–400; and G. Cain and H. Watts, "Problems in Making Policy Inferences from the Coleman Report," *American Sociological Review,* April 1970, pp. 228–249.
13. An economist did estimate this sort of equation! See D.A. Katz, "Faculty Salaries, Promotions, and Productivity at a Large University," *American Economic Review,* June 1973.
14. "How Much is Albuquerque?" *Games,* May 1986, p. 53.
15. Franklin M. Fisher, "Statisticians, Econometricians, and Adversary Proceedings," *Journal of the American Statistical Association,* June 1986, pp. 277–286.

16. Lawrence Fisher, "Determinants of Risk Premiums on Corporate Bonds," *Journal of Political Economy,* Vol. 67, June 1959, pp. 217–237. The actual regression equation uses the logarithms of the variables.

17. Casey Fale, Pomona College, 1987.

18. Chien-Ching Cho and Alan K. Reichert, "An Application of Multiple Regression Analysis for Appraising Single-Family Housing Values," *Business Economics,* January 1980, pp. 47–52.

19. Michael Linenberg, "Regression Analysis and the Airline Industry," Pomona College, Fall 1989.

20. Franklin M. Fisher, "Statisticians, Econometricians, and Adversary Proceedings," *Journal of the American Statistical Association,* June 1986, pp. 277–286.

21. John J. Siegfried and Charles E. Scott, "The Determinants of Academic Lawyer's Salaries and Non-Institutional Professional Income," *Journal of Legal Education,* 1977, pp. 281–287.

22. David W. Barnes, *Statistics As Proof,* Boston: Little, Brown, 1983, p. 340; the actual equation was somewhat more complicated than this.

23. *Vuyanich* v. *Republic National Bank of Texas,* 1980.

24. *Vuyanich* v. *Republic National Bank of Dallas,* 1980.

25. Margaret Cronin Fisk, *The Gambler's Bible,* New York: Drake, 1976, p. 192.

Chapter 16

Opening quotation: Thomas Fuller.

1. W.J. Youden, "Chance, Uncertainty, and Truth in Science," *Journal of Quality Technology,* 1972.

2. Mark Gibbons and Patrick Hess, "Day of the Week Effects and Asset Returns," *Journal of Business,* October 1981, pp. 579–595.

3. John Leaning, "Barnstable Auto Fatalities Drop 600%," *Cape Cod Times,* August 7, 1986.

4. O.L. Davies, *The Design and Analysis of Industrial Experiments,* London: Oliver and Boyd, 1956, p. 164.

5. Elisabeth Street and Mavis B. Carroll, "Preliminary Evaluation of a New Food Product," in Judith M. Tanur et al., *Statistics: A Guide to Business and Economics,* San Francisco: Holden-Day, 1976, pp. 103–112.

6. Hsiu-Kwang Wu and Billy P. Helms, "Confidential Bank Examination Data and the Efficiency of Bank Share Prices," *Financial Analysts Journal,* November–December 1984, pp. 31–33.

7. David E. Harrington, "The Network Coverage of Economic News," working paper, July 1985.

8. Henry Scheffe, *The Analysis of Variance,* New York: Wiley, 1959.

9. Dallas S. Batten and Courtenay C. Stone, "Are Monetarists an Endangered Species?" Federal Reserve Bank of St. Louis *Review,* May 1983.

10. Batten and Stone, p. 9.

11. June Webb, "College Mathematics Achievement and Logical Reasoning," *Perceptual and Motor Skills,* 1985, pp. 15–22.

12. Cletus C. Coughlin, K. Alec Chrystal, and Geoffrey E. Wood, "Protectionist Trade Policies: A Survey of Theory, Evidence, and Rationale," *Federal Reserve Bank of St. Louis Review,* January/February 1988, pp. 12–26.

Chapter 17

Opening quotation: Larry Merchant, *The National Football Lottery,* New York: Holt, Rinehart, and Winston, 1963.

1. For example, William Beyer, ed., *Handbook of Tables for Probability and Statistics,* second edition, Cleveland, Ohio: Chemical Rubber Company, 1972.

2. Ronald Frank and William Massy, "Shelf Position and Space Effects on Sales," *Journals of Marketing Research,* February 1970, pp. 59–66.

3. William Beyer, ed.

4. Peter Gammons, "Inside Baseball," *Sports Illustrated,* October 13, 1986, p. 88.

5. Clark S. Judge, *The Book of American Rankings,* New York: Facts on File, 1979, pp. 236–239.

6. G.J. Santoni, "The Great Bull Markets 1924–1929 and 1982–1987: Speculative Bubbles or Economic Fundamentals?" *Federal Reserve Bank of St. Louis Review,* November 1987, pp. 16–30.

7. Susie Helfrey, "Do Stocks Walk?" Pomona College, Spring 1990.

8. *New York Times,* August 27, 1989; *Los Angeles Times,* August 27, 1989; *New York Times,* January 3, 1990.

9. Clark S. Judge, *The Book of American Rankings,* New York: Facts on File, 1979, pp. 215–221.

Chapter 18

Opening quotation: Edmund Burke, *Reflections on the Revolution in France,* 1790.

1. Roger G. Ibbotson and Rex A. Sinquefield, *Stocks, Bonds, Bills, and Inflation: Historical Returns (1926–1978),* Charlottesville, Va.: The Financial Analysts Research Foundation, 1979.
2. Robert Stansfield, *New York Times,* February 21, 1971.
3. Congressional Budget Office, *Indexing with the Consumer Price Index: Problems and Alternatives,* Washington, D.C.: U.S. Government Printing Office, June 1981.
4. A good reference is Gary Smith, *Macroeconomics,* New York: W.H. Freeman, 1985.
5. David Upshaw, of Drexel Burnham Lambert, quoted in Jack Egan, "Dow and the Law of Averages," *New York,* May 21, 1979, p. 12.
6. Lil Phillips, "Phindex Shows Horrific Inflation," *Cape Cod Times,* July 10, 1984.
7. John A. Johnson, "Sharing Some Ideas," *Cape Cod Times,* July 12, 1984.
8. *Barrons,* July 30, 1984.
9. David W. Barnes, *Statistics As Proof,* Boston: Little, Brown, 1983, p. 18.
10. *Pennsylvania* v. *Local 542, International Union of Operating Engineers,* cited in David W. Barnes, p. 28.
11. *South Dakota Public Utilities Commission* v. *Federal Energy Regulatory Commission,* 1981.

Chapter 19

Opening quotation: James Thurber.

1. This criterion was suggested by Abraham Wald, *Statistical Decision Functions,* New York: Wiley, 1950.
2. The anonymous quotations are from "Soviets May Have Seen Missile Data," *Cape Cod Times,* July 19, 1985.

3. Quoted in "Two Poor Years for the Forecasters," *Business Week,* December 21, 1974, p. 51.
4. R. Howard, J. Matheson, and D. North, "The Decision to Seed Hurricanes," *Science,* June 1972.
5. Frank Ramsey, "Truth and Probability," in R.B. Braithwaite, ed., *The Foundations of Mathematics and Other Logical Essays,* London: K. Paul, Trench, Trubner & Co., 1931, pp. 156–198.
6. John von Neumann and Oskar Morgenstern, *Theory of Games and Economic Behavior,* Princeton, N.J.: Princeton University Press, 1944.
7. George A. Diamond and James S. Forrester, "Metadiagnosis," *The American Journal of Medicine,* July 1983, pp. 129–137. They argue that doctors should not only give their patients numerical probabilities, but also assign probabilites to these probabilities, to show their patients that the answer will vary from doctor to doctor.
8. Geoffrey D. Bryant and Geoffrey R. Norman, "Expressions of Probability: Words and Numbers," letter to the *New England Journal of Medicine,* February 14, 1980, p. 411.
9. Richard Brealey and Stewart Myers, *Principles of Corporate Finance,* New York: McGraw-Hill, 1984, p. 216.
10. This example is from Jacob W. Ulvila and Rex V. Brown, "Decision Analysis Comes of Age," *Harvard Business Review,* September–October 1982, pp. 130–141.
11. His most famous work is the book cited in reference 6.
12. Jonathan Schell, "The Fate of the Earth," *New Yorker,* February 15, 1982, p. 68.
13. Von Neumann and Morgenstern, pp. 176–178; the episode is from *The Final Problem* by Arthur Conan Doyle; copyright 1894 by Sir Arthur Conan Doyle.
14. Charles F. McCoy and Steve Swartz, "Big Credit-Card War May Be Breaking Out, To Detriment of Banks," *Wall Street Journal,* March 19, 1987.
15. Stanley Bing, "The Strategist: When Disaster Strikes," *Esquire,* September 1986, p. 64.

Business
MYSTAT®

An instructional business version of SYSTAT for IBM-PC/ compatibles

Contents

Introduction

This is a real statistics program—it is not just a demonstration. You can use Business MYSTAT to enter, transform, and save data, and to perform a wide range of statistical evaluations. Please use Business MYSTAT to solve *real* problems.

Business MYSTAT is a subset of SYSTAT, our premier statistics package. We've geared Business MYSTAT especially for teaching business statistics, with special forecasting and time series routines all in a single, easy-to-use package.

Business MYSTAT provides descriptive statistics, cross-tabulation, Pearson and Spearman correlation coefficients, multiple regression, time series forecasting, and much more.

Business MYSTAT is available in Macintosh, IBM-PC/compatible, and VAX/VMS versions. Copies are available at a nominal cost.

For more information about SYSTAT, FASTAT, and our other top-rated professional statistics and graphics packages, please call or write.

Installation

Business MYSTAT requires 512K of RAM and a floppy or hard disk drive. It can handle up to 50 variables and up to 32,000 cases.

Your Business MYSTAT disk contains three files: MYSTAT.EXE (the program), MYSTATB.HLP (a file with information for on-line help), and DEMO.CMD (a demonstration that creates a data file and demonstrates some of Business MYSTAT's features).

Hard disk

Set up the CONFIG.SYS file
The CONFIG.SYS file in your root directory must have a line that says FILES=20. If you don't have a CONFIG.SYS file, you can create one with the COPY CON command:

```
>CD \
>COPY CON CONFIG.SYS
FILES=20
[F6]
```

Press the Enter key after each line to tell the computer to start working. Nothing happens until you press Enter. (On some keyboards, the key is marked Return or with a ↵ symbol.)

After typing "FILES=20" and pressing Enter, press the F6 key and then press Enter. F6 signals the computer that you are done writing to CONFIG.SYS, so you get another DOS prompt (>).

Set up the AUTOEXEC.BAT file
The AUTOEXEC.BAT file in the root directory must have the following PATH line. You can create an AUTOEXEC.BAT file with the following commands. Remember to press Enter at the end of each line.

```
>COPY CON AUTOEXEC.BAT
PATH=C:\;C:\SYSTAT
[F6]
```

Reboot your machine

Copy the files on the MYSTAT disk into a \SYSTAT directory
Now, make a \SYSTAT directory, insert the MYSTAT disk in drive A, and copy the Business MYSTAT files into the directory. (You must have the help file in the same directory as MYSTAT.EXE. If you put MYSTAT.EXE in a directory other than \SYSTAT, the help file MYSTATB.HLP must be either in that directory or the \SYSTAT directory.)

```
>MD \SYSTAT
>CD \SYSTAT
>COPY A:*.*
```

You are now ready to begin using Business MYSTAT. From now on, all you need to do to get ready to use Business MYSTAT is boot and move (CD) into the \SYSTAT directory. Save your MYSTAT master disk as a back up copy.

Floppy disk

Boot your machine
Insert a "boot disk" into drive A. Close the door of the disk drive and turn on the machine.

Set up a CONFIG.SYS file on your boot disk
The boot disk must contain a file named CONFIG.SYS with a line FILES=20. If you don't have such a file, type the following lines when you get a DOS prompt (>).

```
>COPY CON CONFIG.SYS
FILES=20
[F6]
```

Press the Enter key at the end of every line to tell the computer to start working. Nothing happens until you press Enter. (On some keyboards this key is marked Return or with a ↵ symbol.)

After typing "FILES=20" and pressing Enter, press the F6 key and then press Enter. F6 signals the computer that you are done writing to CONFIG.SYS, so you get another DOS prompt (>).

Set up an AUTOEXEC.BAT file on your boot disk
The boot disk must also contain a file named AUTOEXEC.BAT with a line PATH=A:\;B:\. You can create one as follows:

```
>COPY CON AUTOEXEC.BAT
PATH=A:\;B:\
[F6]
```

Reboot your machine

Make a copy of the Business MYSTAT disk
When you get a DOS prompt (>), remove the boot disk. Put the Business MYSTAT disk in drive A and a blank, formatted disk in drive B and type the COPY command at the prompt:

```
>COPY A:*.* B:
```

Remove the master disk from drive A and store it. If anything happens to your working copy, use the master to make a new copy.

Use Business MYSTAT
Now, switch your working copy into drive A. If necessary, make drive A the "logged" drive (the drive your machine reads from) by issuing the command A: at the DOS prompt (>).

Business MYSTAT reads and writes its temporary work files to the currently logged drive. Since there is limited room on the Business MYSTAT disk, you should read and write all your data, output, and command files from a data disk in drive B.

From now on, all you need to do to use Business MYSTAT is boot, insert your working copy, and log the A drive.

Getting Started

To start, type MYSTAT and press Enter.

```
>MYSTAT
```

When you see the MYSTAT logo, press Enter. You'll see a command menu listing all the commands you can use in MYSTAT.

Command menu

The command menu shows a list of all the commands that are available for Business MYSTAT. The menu divides the commands into six groups: information, file handling, miscellaneous, graphics, statistics, and forecasting.

DEMO	EDIT	MENU	PLOT	STATS	LOG
HELP		NAMES	BOX	TABULATE	MEAN
SYSTAT	USE	LIST	HISTOGRAM	TTEST	SQUARE
	SAVE	FORMAT	STEM	CORRELATE	TREND
	PUT	NOTE			PCNTCHNG
	SUBMIT				DIFFRNCE
					INDEX
QUIT	OUTPUT	SORT	CHARSET	MODEL	SMOOTH
		RANK		CATEGORY	ADJSEAS
		WEIGHT		ANOVA	EXP
				ESTIMATE	TPLOT
					ACF, PACF
					CLEAR

As you become more experienced with MYSTAT, you might want to turn the menu off. Turn it on and off with the MENU command.

```
>MENU
```

Data Editor

MYSTAT has a built-in Data Editor with its own set of commands. To enter the Editor, use the EDIT command.

```
>EDIT
```

Inside the Editor, you can enter, view, edit, and transform data. When you are done with the Editor, type QUIT to get out of the Editor and back to MYSTAT, where you can do statistical and graphical analyses. To quit MYSTAT itself, type QUIT again.

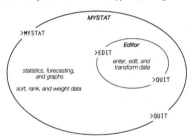

The Data Editor is an independent program inside MYSTAT and has its own commands. Five commands—USE, SAVE, HELP, QUIT, and FORMAT—appear both inside *and* outside the Editor.

Demo

To see an on-line demonstration of MYSTAT, use the DEMO command. (Remember to press Enter after you type the command.)

```
>DEMO
```

When the demonstration is finished, Business MYSTAT returns you to the command menu. After you've seen the demo, you might want to remove the DEMO.CMD file and the CITIES.SYS data file that it creates to save disk space.

Help

The HELP command provides instructions for any command—inside *or* outside the Editor. HELP lists and describes all the commands.

```
>HELP
```

You can get help for any specific command . For example:

```
>HELP EDIT
EDIT starts the MYSTAT full screen editor.

EDIT [filename]

EDIT       (edit a new file)
EDIT CITIES   (edit CITIES.SYS)

For further information, type EDIT [Enter], [ESC], and then type
HELP [Enter] inside the data editor.
```

The second line shows a summary of the command. You see that any EDIT command must begin with the command word EDIT. The brackets indicate that specifying a file is optional. Anything in lowercase, like "filename," is just a placeholder—you should type a real file name (or a real variable name, or whatever).

Customizing DEMO and HELP

All help information is stored in the text file MYSTATB.HLP. You can use a text editor to customize your help information. Teachers can design special demonstrations by editing the file DEMO.CMD.

Information about SYSTAT

For information about SYSTAT and how to order it, use the SYSTAT command.

Data Editor

The Business MYSTAT Data Editor lets you enter and edit data, view data, and transform variables. First, we enter data; later, we show you how to use commands.

To use the Editor to create a new file, type EDIT and press Enter. If you already have a MYSTAT data file that you want to edit, specify a filename with the EDIT command.

```
>EDIT [<filename>]
```

If you do not specify a filename, you get an empty Editor like the one above. MYSTAT stores data in a rectangular worksheet. *Variables* fill vertical columns and each horizontal row represents a *case*.

Entering data

First enter variable names in the top row. Variable names *must* be surrounded by single or double quotation marks, must begin with a letter, and can be no longer than 8 characters.

Character variables (those whose values are words and letters) must have names ending with a dollar sign ($). The quotation marks and dollar sign do not count toward the eight character limit.

Numeric variables (those whose values are numbers) can be named with subscripts; e.g., ITEM(3). Subscripts allow you to specify a range of variables for analyses. For example, STATS ITEM(1-3) does descriptive statistics on the first three ITEM(i) variables.

The cursor is already positioned in the first cell in the top row of the worksheet. Type 'CITY$' or "CITY$" and then press Enter.

Business MYSTAT enters variable names in upper-case whether you enter them in lower- or upper-case.

The cursor automatically moves to the second column. You are now ready to name the rest of the variables, pressing Enter to store each name in the worksheet.

```
'STATE$'
"POP"
'RAINFALL'
```

Now you can enter values. Move the cursor to the first blank cell under CITY$ by pressing Home. (On most machines, Home is the 7 key on the numeric keypad. If pressing the 7 key types a 7 rather than moving the cursor, press the NumLock key and try again.)

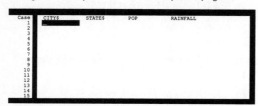

When the cell under CITY$ is selected, enter the first data value:

```
'New York'
```

When you press Enter, Business MYSTAT accepts the value and moves the cursor to the right. You can also use the cursor arrow keys to move the cursor.

Character values can be no longer than twelve characters. Like variable names, character values must be surrounded by single or double quotation marks. Unlike variable names, character values are case sensitive—upper-case is not the same as lower-case (for example, 'TREE' is not the same as 'tree' or 'Tree'). Enter a blank space surrounded by quotation marks for missing character values. To use single or double quotation marks as part of a value, surround the whole value with the opposite marks.

Numeric values can be up to 10^{35} in absolute magnitude. Scientific notation is used for long numbers; e.g., .000000000015 is equivalent to 1.5E–11. Enter a decimal (.) for missing numeric values.

Enter the first few cases: type a value, press Enter, and type the next value. The cursor automatically moves to the beginning of the next case when a row is filled.

"New York"	"NY"	7164742	57.03
"Los Angeles"	"CA"	3096721	7.81
"Chicago"	"IL"	2992472	34

(etc.)

Case	CITY$	STATE$	POP	RAINFALL
1	New York	NY	7164742.000	57.030
2	Los Angeles	CA	3096721.000	7.810
3	Chicago	IL	2992472.000	34.000
4	Dallas	TX	974234.000	33.890
5	Phoenix	AZ	853266.000	14.910
6	Miami	FL	346865.000	60.020
7	Washington	DC	638432.000	37.730
8	Kansas City	MO	448159.000	38.770

Moving around

Use the cursor keys on the numeric keypad to move around in the Editor.

Esc		toggle between Editor and command prompt (>)
Home	7	move to first cell in worksheet
↑	8	move upward one cell
PgUp	9	scroll screen up
←	4	move left one cell
→	6	move right one cell
End	1	move to last case in worksheet
↓	2	move down one cell
PgDn	3	scroll screen down

If these keys type numbers rather than move the cursor, press the NumLock key which toggles the keypad back and forth between typing numbers and performing the special functions. (If your computer does not have a NumLock key or something similar, consult the manual that came with your machine.)

Editing data

To change a value or variable name move to the cell you want, type the new value, and press Enter. Remember to enclose character values and variable names in quotation marks.

Data Editor Commands

When you have entered your data, press the Esc key to move the cursor to the prompt (>) below the worksheet. You can enter editor commands at this prompt. Commands can be typed in upper- or lower-case. Items in <angle brackets> are placeholders; for instance, you should type a specific filename in place of <filename>.

DELETE and DROP

DELETE lets you remove an entire case (row) from the dataset in the Editor. You can specify a range or list of cases to be deleted. The following are valid DELETE commands.

DELETE 3	Deletes third case from dataset.
DELETE 3-10	Deletes cases 3 through 10.
DELETE 3, 5-8, 10	Deletes cases 3, 5, 6, 7, 8, and 10.

DROP removes variables from the dataset in the Editor. You can specify several variables or a range of subscripted variables.

DROP RAINFALL	Drops RAINFALL from dataset.
DROP X(1-3)	Drops subscripted variables X(1-3).
DROP X(1-3), GROUP$	Drops X(1-3) and GROUP$.

Saving files

The SAVE command saves the data in the Editor to a Business MYSTAT data file. You must save data in a data file before you can analyze them with statistical and graphic commands.

```
SAVE <filename>          Saves data in a MYSTAT file.
/DOUBLE | SINGLE         Choose single or double precision.
```

Business MYSTAT filenames can be up to 8 characters long and must begin with a letter. Business MYSTAT adds a ".SYS" extension that labels the file as a Business MYSTAT data file. To specify a path name for a file, enclose the entire file name, including the file extension, in single or double quotation marks.

Business MYSTAT stores data in double-precision by default. You can choose single precision if you prefer: add /SINGLE to the end of the command. Always type a slash before command options.

```
SAVE CITIES/SINGLE       Saves CITIES.SYS in single precision
```

Single precision requires approximately half as much disk space as double precision and is accurate to about 9 decimal places. The storage option (single or double precision) does not affect computations, which always use double precision arithmetic (accurate to about 15 places).

```
SAVE CITIES              Creates data file CITIES.SYS.
SAVE b:new               Creates file NEW.SYS on B drive.
SAVE 'C:\DATA\FIL.SYS'   Creates FIL.SYS in \DATA directory of C.
```

You can save data in text files for exporting to other programs with the PUT command. PUT is not an Editor command, though. First QUIT the Editor, USE the data file, and PUT the data to a text file.

Reading files

USE reads a MYSTAT data file into the Editor.

```
USE [<filename>]         Reads data from MYSTAT data file.
```

Starting new data files

Use the NEW command to clear the worksheet and start editing a new data file.

Importing data from other programs

You can import data from other programs through plain text (ASCII) file. ASCII files contain only plain text and numbers—they have no special characters or formatting commands.

Start the program and save your data in a plain ASCII file according to the instructions given by that program's manual. The ASCII file must have a ".DAT" extension, data values must be separated by blanks or commas, and each case must begin on a new line.

Then, start Business MYSTAT. Use EDIT to get an empty worksheet, and enter variable names in the worksheet for each variable in the ASCII file. Next, use GET to read the text file. Finally, SAVE the data.

```
GET [<filename>]         Reads data from ASCII text file.
```

Finding a case

FIND searches through the Editor starting from the current cursor position, and moves the cursor to the first value that meets the condition you specify. Try this:

```
>FIND POP<1000000
```

This moves the cursor to case 4. After Business MYSTAT finds a value, use the FIND command without an argument (that is, "FIND" is the entire command) to find the next case meeting the same condition. All functions, relations, and operators listed above are available.

Some valid FIND commands:

```
>FIND AGE>45 AND SEX$='MALE'
>FIND INCOME<10000 AND STATE$='NY'
>FIND (TEST1+TEST2+TEST3)>90
```

Decimal places in the Editor

The FORMAT command specifies the number (0–9) of decimal places to be shown in the Editor. The default is 3. Numbers are stored the way you enter them regardless of the FORMAT setting; FORMAT affects the Editor display only.

```
FORMAT=<#>               Sets number of decimal places to <#>.
/UNDERFLOW               Displays tiny numbers in scientific notation.
```

For example, to set a two-place display with scientific notation for tiny numbers use the command:

```
>FORMAT=2/UNDERFLOW
```

Transforming variables

Use LET and IF...THEN to transform variables or create new ones.

```
LET <var>=<exprn>        Transforms <var> according to <exprn>.
IF <exprn> THEN LET      Transforms <var> conditionally according to
<var>=<exprn>               <exprn>.
```

For example, we can use LET to create LOGPOP from POP:

```
>LET LOGPOP=LOG(POP)
```

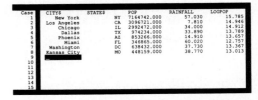

LET labels the last column of the worksheet LOGPOP and sets the values to the natural logs of the POP values. (If LOGPOP had already existed, its values would have been replaced.)

Use IF...THEN for *conditional* transformations. For example:

```
>IF POP>1000000 THEN LET SIZE$='BIG'
```

creates a new character variable, SIZE$, and assigns the value BIG for every city that has population greater than one million.

For both LET and IF...THEN, character values must be enclosed in quotation marks and are case sensitive (i.e., "MALE" is not the same as "male"). Use a period to indicate missing values.

Some valid LET and IF-THEN commands:

```
>LET ALPHA$='abcdef'
>LET LOGIT1=1/(1+EXP(A+B*X))
>LET TRENDY=INCOME>40000 AND CAR$='BMW'
>IF SEX$='Male' THEN LET GROUP=1
>IF group>2 THEN LET NEWGROUP=2
>IF A=-9 AND B<10 OR B>20 THEN LET C=LOG(D)*SQR(E)
```

Functions, relations, and operators for LET and IF...THEN

+	addition	CASE	current case number
-	subtraction	INT	integer truncation
*	multiplication	URN	uniform random number
/	division	ZRN	normal random number
^	exponentiation		
<	less than	AND	logical and
<=	less than or equal to	OR	logical or
=	equal to	SQR	square root
<>	not equal to	LOG	natural log
>=	greater than or equal	EXP	exponential function
>	greater than	ABS	absolute value

Logical expressions

Logical expressions evaluate to one if true and to zero if false. For example, for LET CHILD=AGE<12, a variable AGE would be filled with ones for those cases where AGE is less than 12 and zeros whenever AGE is 12 or greater.

Random data

You can generate random numbers using the REPEAT, LET, and SAVE commands in the Editor. First, enter variable names. Press Esc to move the cursor to the command line. Then, use REPEAT to fill cases with missing values and LET to redefine the values.

```
REPEAT 20              Fills 20 cases with missing values.
LET A=URN              Fills A with uniform random data.
LET B=ZRN              Fills B with normal random data.
SAVE RANDOM            Saves data in file RANDOM.SYS.
```

Distribution functions

MYSTAT can compute the following distribution functions.

Cumulative distribution functions compute the probability that a random value from the specified distribution falls below a given value; that is, it shows the proportion of the distribution below that value. Inverse distribution functions do the same thing backwards: you specify an alpha value (a probability or proportion) between zero and one, and MYSTAT shows the critical value below which lies that proportion.

Distribution	Cumulative	Inverse
Normal	ZCF(z)	ZIF(alpha)
Exponential	ECF(x)	EIF(alpha)
Chi-square	XCF(chisq,df)	XIF(alpha,df)
T	TCF(t,df)	TIF(alpha,df)
F	FCF(F,df1,df2)	FIF(alpha,df1,df2)
Poisson	PCF(k,lambda)	PIF(alpha,k)
Binomial	NCF(p,k,n)	NIF(alpha,k,n)
Logistic	LCF(x)	LIF(alpha)
Studentized range	SCF(x,k,df)	SIF(alpha,k,df)
Weibull	WCF(x,p,q)	WIF(alpha,p,q)

The uniform distribution is uniformly distributed from zero to one. The normal distribution is a standard normal distribution with mean zero and standard deviation one. The exponential distribution has parameter one. The cumulative Poisson function calculates the probability of the number of random events between zero and k when the expected value is *lambda*. The binomial cumulative function provides the probability of k or more occurrences in n trials with the binomial probability p.

Use the cumulative distribution functions to obtain probabilities associated with observed sample statistics. Use the inverse distributions to determine critical values and to construct confidence intervals. Finally, to generate pseudo-random data for the functions, apply the appropriate inverse cumulative distribution function to uniform random data.

Leaving the Data Editor

Use the QUIT command to leave the Editor and return to the main Business MYSTAT menu.

```
>QUIT
```

General MYSTAT Commands

Once your data are in a Business MYSTAT data file, you can use Business MYSTAT's statistical and graphics routines to examine them.

Open a data file

First, you must open the file containing the data you want to analyze:

```
USE <filename>              reads the data in <filename>
```

To analyze the data we entered earlier, type:

```
>USE CITIES
```

MYSTAT responds by listing the variables in the file.

```
VARIABLES IN MYSTAT FILE ARE:
    CITY$     STATE$     POP     RAINFALL   LOGPOP
```

See variable names and data values

The NAMES command shows the variable names in the current file.

```
>NAMES
VARIABLES IN MYSTAT FILE ARE:
    CITY$     STATE$     POP     RAINFALL   LOGPOP
```

The LIST command displays the values of variables you specify. If you specify no variables, all variables are shown.

```
>LIST CITY$
                          CITY$

CASE    1           New York
CASE    2           Los Angeles
CASE    3           Chicago
CASE    4           Dallas
CASE    5           Phoenix
CASE    6           Miami
CASE    7           Washington
CASE    8           Kansas City

  8 CASES AND    5 VARIABLES PROCESSED
```

Decimal places

Use the FORMAT command to specify the number of digits to be displayed after the decimal in statistical output. This FORMAT command has the same syntax and works the same as in the Editor:

```
FORMAT=<#>                 Sets number of decimal places to <#>.
    /UNDERFLOW             Uses scientific notation for tiny numbers.
```

Sorting and ranking data

SORT reorders the cases in a file in ascending order according to the variables you specify. You can specify up to ten numeric or character variables for nested sorts. Use a SAVE command after the USE command to save the sorted data into a MYSTAT file. Then, open the sorted file to do analysis.

```
>USE MYDATA
>SAVE SORTED
>SORT CITY$ POP
>USE SORTED
```

RANK replaces each value of a variable with its rank order within that variable. Specify an output file before ranking.

```
>USE MYDATA
>SAVE RANKED
>RANK RAINFALL
>USE RANKED
```

Weighting data

WEIGHT replicates cases according to the integer parts of the values of the weighting variable you specify.

```
WEIGHT <variable>          Weights according to variable specified.
```

To turn weighting off, use WEIGHT without an argument.

```
>WEIGHT
```

Quitting

When you are done with your analyses, you can end your session with the QUIT command. Remember that the Data Editor also has a QUIT. To quit MYSTAT from the Editor, enter QUIT twice.

```
>QUIT
```

Statistics

STATS produces basic descriptive statistics.

Descriptive statistics

```
>STATS
TOTAL OBSERVATIONS:  8

                 POP      RAINFALL   LOGPOP

N OF CASES         8          8          8
MINIMUM      346865.000     7.810     12.757
MAXIMUM     7164742.000    60.020     15.785
MEAN        2064361.375    35.520     14.028
STANDARD DEV 2335788.226   18.032      1.068
```

Here is a summary of the STATS command. The box on the facing page describes the notation we use for command summaries in this manual.

STATS <var1> <var2>…	Statistics for the variables specified.
MEAN SD SKEWNESS,	Choose which statistics you want.
KURTOSIS MINIMUM,	
MAXIMUM RANGE SUM,	
SEM	
/BY <gvar>	Statistics for each group defined by the grouping variable <gvar>. The data must first be SORTed on the grouping variable.

For example, you can get the mean, standard deviation, and range for RAINFALL with the following command.

```
>STATS RAINFALL / MEAN SD RANGE
TOTAL OBSERVATIONS:  8

                 RAINFALL

N OF CASES          8
MEAN               35.520
STANDARD DEV       18.032
RANGE              52.210
```

Tabulation

TABULATE provides one-way and multi-way frequency tables. For two-way tables, MYSTAT provides the Pearson chi-square statistic. You can produce a table of frequencies, percents, row percents, or column percents. You can tell MYSTAT to ignore missing data with the MISS option and suppress the chi-square statistic with NOSTAT.

TAB <var1>*<var2>…	Tabulates the variables you specify
/LIST	Special list format table.
FREQUENCY PERCENT	Different types of tables
ROWPCT COLPCT MISS	
NOSTAT	

TABULATE	Frequency tables of all numeric variables.
TABULATE AGE/LIST	Frequency table of AGE in list format.
TAB AGE*SEX	Two-way table with chi-square.
TAB AGE*SEX$*STATE$. /ROWPCT	Three-way row percent table.
TAB A.AGE*SEX/FREQ. PERC	Two two-way frequency and cell percent tables (A*SEX and AGE*SEX).
TAB AGE*SEX/MISS	Two-way table excluding missing values.
TAB AGE*SEX/NOSTAT	Two-way table excluding chi-square.

One-way frequency tables show the number of times a distinct value appears in a variable. Two-way and multi-way tables count the appearances of each unique combination of values. Multi-way tables count the appearances of a value in each subgroup. Percent tables convert the frequencies to percentages of the total count; row percent tables show percentages of the total for each row; and column percent tables show percentages of the total for each column.

T-tests

TTEST does dependent and independent *t*-tests. A dependent (paired samples) *t*-test tests whether the means of two continuous variables differ. An independent test tests whether the means of two groups of a single variable differ.

To request an independent (two-sample) *t*-test, specify one or more continuous variables and one grouping variable. Separate the continuous variable(s) from the grouping variable with an asterisk. The grouping variable must have only two values.

To request a dependent (paired) *t*-test, specify two or more continuous variables. MYSTAT does separate dependent *t*-tests for each possible pairing of the variables.

TTEST <var1>…[*<gvar>]	Does *t*-tests of the variables you specify.
TTEST A B	Dependent (paired) t-test of A and B.
TTEST A B C	Paired tests of A and B, A and C, B and C.
TTEST A*SEX$	Independent test.
TTEST A B C*SEX	Three independent tests.

You can also do a one-sample test by adding a variable to your data file that has a constant value corresponding to the population mean of your null hypothesis. Then do a dependent *t*-test on this variable and your data variable.

Correlation

PEARSON computes Pearson product moment correlation coefficients for the variables you specify (or all numerical variables). You can select pairwise or listwise deletion of missing data; pairwise is the default. RANK the variables before correlating to compute Spearman rank-order correlations. CORRELATE is a synonym for PEARSON.

PEARSON <var1> ...	Pearson correlation matrix.
/PAIRWISE | LISTWISE	Pairwise or listwise deletion of missing values.
PEARSON	Correlation matrix of all numeric variables.
CORR HEIGHT IQ AGE	Matrix of three variables.
PEARSON /LISTWISE	Listwise deletion rather than pairwise.

Correlation measures the strength of linear association between two variables. A value of 1 or −1 indicates a perfect linear relationship; a value of 0 indicates that neither variable can be linearly predicted from the other.

Regression and ANOVA

Business MYSTAT computes simple and multiple regression and balanced or unbalanced ANOVA designs. For unbalanced designs, MYSTAT uses the method of weighted squares of means.

The MODEL and ESTIMATE commands provide linear regression. MODEL specifies the regression equation and ESTIMATE tells MYSTAT to start working. Your MODEL should almost always include a CONSTANT term.

```
>MODEL Y=CONSTANT+X        Simple linear regression.
>ESTIMATE
```

```
>MODEL Y=CONSTANT+X+Z      Multiple linear regression.
>ESTIMATE
```

Use CATEGORY and ANOVA commands for fully factorial ANOVA. CATEGORY specifies the number of categories (levels) for one or more variables used as categorical predictors (factors). ANOVA specifies the dependent variable and produces a fully-factorial design from the factors given by CATEGORY.

All CATEGORY variables must have integer values from 1 to *k*, where *k* is the number of categories.

```
>CATEGORY SEX=2            One-way design with independent variable
>ANOVA SALARY             SALARY and one factor (SEX) with 2 levels
>ESTIMATE
```

```
>CATEGORY A=2,B=3          Two-by-three ANOVA
>ANOVA Y
>ESTIMATE
```

Saving residuals

Use a SAVE command before MODEL or ANOVA to save residuals in a file. MYSTAT saves model variables, estimated values, residuals, and standard error of prediction as the variables ESTIMATE, RESIDUAL, and SEPRED. When you use SAVE with a linear model, MYSTAT lists cases with extreme studentized residuals or leverage values and prints the Durbin-Watson and autocorrelation statistics.

```
>SAVE RESIDS              >SAVE RESID2
>MODEL Y=CONSTANT+X+Z     >CATEGORY SEX=2
>ESTIMATE                 >ANOVA SALARY
                         >ESTIMATE
```

You can USE the residuals file to analyze your residuals with MYSTAT's statistical and graphic routines.

Graphics

Use the CHARSET command to choose the type of graphic characters to be used for printing and screen display. If you have IBM screen or printer graphic characters, use GRAPHICS; if not, use GENERIC. The GENERIC setting uses characters like +, −, and l.

```
CHARSET GRAPHICS          For IBM graphic characters.
CHARSET GENERIC           For any screen or printer.
```

Scatterplots

PLOT draws a two-way scatterplot of one or more Y variables on a vertical scale against an X variable on a horizontal scale. Use different plotting symbols to distinguish Y variables.

```
PLOT <yvar1>…*<xvar>      Plots <yvar(s)> against <xvar>.
  /SYMBOL=<var$> |        Use character variable values or character
    '<char>'               string as plotting symbol.
  YMAX=<#> YMIN=<#>       Specify range of X and Y values.
    XMAX=<#> XMIN=<#>
  LINES=<#>              Specify number of screen lines for graph.
PLOT A*B/SYMBOL='*'       Uses asterisk as plotting symbol.
PLOT A*B/SYMBOL=SEX$      Uses SEX$ values for plotting symbol.
PL Y1 Y2*X/SY='1','2'     Plot Y1 points as 1 and Y2 points as 2.
PLOT A*B/LINES=40        Limits graph size to 40 lines on screen.
```

For example, we can plot LOGPOP against RAINFALL using the first letter of the values of CITY$ for plotting symbols.

```
>PLOT LOGPOP*RAINFALL/SYMBOL=CITY$
```

The SYMBOL option is powerful. If you are plotting several Y variables, you can label each variable by specifying its own plotting symbol.

```
>PLOT Y1 Y2*X/SYMBOL='1','2'
```

Or, you can name a character variable to plot each point with the first letter of the variable's value for the corresponding case:

```
>PLOT WEIGHT*AGE/SYMBOL=SEX$
```

Box-and-whisker plots

BOX produces box-and-whisker plots. Include an asterisk and a grouping variable for grouped box plots.

```
BOX <var1>…[*<gvar>]      [Grouped] box plots of the variables.
  /GROUPS=<#>,            Show only the first <#> groups.
  MIN=<#> MAX=<#>         Specify scale limits.
BOX                       Box plots of every numeric variable.
BOX SALARY                Box plot of SALARY only.
BOX SALARY*RANK           Grouped box plots of SALARY by RANK
BOX INCOME*STATE$/GR=10   Box plots of first 10 groups only
```

The center line of the box marks the *median*. The edges of the box show the upper and lower *hinges*. The median splits the ordered batch of numbers in half and the hinges split these halves in half again. The distance between the hinges is called the *Hspread*. The *whiskers* show the range of points within 1.5 Hspreads of the hinges. Points outside this range are marked by asterisks and those more than 3 Hspreads from the hinges are marked with circles.

Histograms

HISTOGRAM displays histograms for one or more variables.

```
HISTOGRAM <var1> …        Histograms of variables specified
  /BARS=<#>               Limits the number of bars used.
  SCALE,                  Forces round cutpoints between bars.
  MIN=<#> MAX=<#>         Specifies scale limits.
HISTOGRAM                 Histograms of every numeric variable.
HISTOGRAM A B/BARS=18     Forces 18 bars for histograms of A and B.
HIST A/MIN=0 MAX=10       Histogram of A with scale from 0 to 20.
```

A histogram shows the *distribution* of a variable. The data are divided into equal-sized intervals along the horizontal axis. The number of values in each interval is represented by a vertical bar. The height of each bar can be measured in two ways: the axis on the right side shows the actual number of cases that have that value, and the left axis shows the same information in proportion per standard unit.

Stem-and-leaf diagrams

STEM plots a stem-and-leaf diagram.

STEM <var1>..	Diagrams of variables specified.
/LINES=<#>	Specify number of lines in diagram.
STEM	Stem-and-leaf of every numeric variable.
STEM TAX	Stem-and-leaf of TAX.
STEM TAX/LINES=20	Stem-and-leaf with 20 lines.

The numbers on the left side are *stems* (the most significant digits in which variation occurs) . The *leaves* (the subsequent digits) are printed on the right. For example, in the following plot, the stems are 10's digits and the leaves are 1's digits.

```
      STEM AND LEAF PLOT OF VARIABLE: RAINFALL, N =    48

MINIMUM IS:        7.000
LOWER HINGE IS:       26.500
MEDIAN IS:        36.000
UPPER HINGE IS:       43.000
MAXIMUM IS:        60.000

             0   78
             1   0114
             1   55556
             2
             2 H 588
             3   00113333
             3 M 5557999
             4 H 001123333
             4   55567999
             5   0
             5   9
             6   0
```

(We added 40 more cases to the dataset for this plot. You will get a different plot if you try this example with the CITIES.SYS data file.)

Hinges are defined under "Box-and-whisker plots," above.

Forecasting

MYSTAT's forecasting routines let you transform, smooth, and plot time series data. All of the forecasting routines operate in the computer's memory; that is, the results are saved in the working memory rather than being saved to a data file. When you execute a transformation or a smooth, the only results you see are messages like "Series is smoothed" or "Series is transformed."

Your usual strategy will be to plot the series with TPLOT, ACF, or PACF; do a transformation or smoothing operation; and then replot

the data to see your progress. You can save a modified series in a new data file by preceding the transformation or smoothing command with a SAVE command. You can use the CLEAR command at any time to restore the variable's original values.

Usually, you will do your work in memory, and when you find some satisfactory results, CLEAR the series, and then repeat the successful transformation using the SAVE command. Suppose, for example, that you wanted to save a demeaned and then smoothed series:

```
>MEAN MYSERIES
>SAVE A:NEWSERIES
>SMOOTH MYSERIES
```

This saves the smoothed, demeaned series in the file NEWSERIES.SYS. If you also want to save the intermediate (demeaned but not smoothed) series, use two SAVE commands:

```
>SAVE A:MEANSERIES
>MEAN MYSERIES
>SAVE A:FINALSERIES
>SMOOTH MYSERIES
```

Transforming

LOG replaces each value in a series with its natural logarithm. Logging removes nonstationary variability, such as increasing variances across time.

LOG <var>	Logs the variable specified.
LOG	Logs first numerical variable in file.
LOG SPEED	Logs SPEED.

MEAN subtracts the variable's mean from each value. Demeaning centers a series vertically on zero.

MEAN <var>	Demeans the variable specified
MEAN	Demeans the first numeric variable.
MEAN SUNSPOTS	Demeans SPEED.

SQUARE squares each value. Squaring can help normalize variance in a series.

SQUARE <var>	
SQUARE	
SQUARE MAGNITUD	

TREND removes linear trend to make a series "level."

TREND <var>	
TREND	
TREND WEATHER	

PCNTCHNG replaces each value with its percentage difference from the immediately preceding value.

PCNTCHNG <var>	
PCNTCHNG	
PCNTCHNG SERIES	

DIFFRNCE replaces each value with its difference from the immediately preceding value; you can specify a different lag to show differences from a greater number of positions previous. Differencing helps make some series "stationary" by removing the dependence of one point on the previous point.

DIFFRNCE	
/LAG=<#>]	Compute differences between values <#> cases apart.
DIFFRNCE	Difference interval of 3 values.
DIFFRNCE TROKES/LAG=3	

INDEX divides each value by the first value of the series, or by the base you specify.

INDEX <var>	Indexes series to its 12th value.
INDEX SERIES/BASE=12	

Smoothing

Smoothing removes local variations from a series and makes it easier to see the general shape of the series. Running smoothers replace each value with the mean or median of that value and its neighbors. The "window" is the number of values considered at a time; 3 is the default. You can optionally specify weights for each value, separated by commas.

```
SMOOTH <var1>...          Smooths the variables specified.
   /MEAN-<#>|MEDIAN-<#>   Mean or median smoothing with window
                             size of <#>
   WT-<#,#,...>           Specifies smoothing weights for each value.
SMOOTH SERIES             Running mean smoother, window of 3.
SMOOTH SERIES/MEAN=5      Running mean smoother with window 5.
SMOOTH SERIES/MEDIAN=4    Running median smoother, window 4
SMOOTH SERIES,            Weighted mean smoother, window 3
   /WT=.1,.2..1
```

Seasonal decomposition removes additive or multiplicative seasonal effects according to the period you specify; the default is 12.

```
ADJSEAS <var>             Smooths specified variable.
   /SEASON-<#>.           Number of observations per period.
   ADDI-<#>|MULT-<#>      Removes additive or multiplicative seasonal
                             effects with weight <#>.
ADJSEAS                   Adjusts first variable with period 12.
ADJSEAS SALES/SEASON=4    Adjusts SALES with period 4.
```

Exponential smoothing forecasts future observations as a weighted average (running smooth) of previous observations. You can optionally specify a linear or percentage growth component and an additive or multiplicative seasonal effects component. Specify the weights for each component and the number or range of cases to be forecast.

```
EXP <var>                 Smooths specified variable.
   /LINEAR-<#>|PERCENT-<  Remove linear or percentage growth trend.
   #>,
   SEASON-<#>,            Number of observations per period.
   ADDI-<#>|MULT-<#>,     Removes additive or multiplicative seasonal
                             effects with weight <#>.
   FORECAST-<#>|<#>-<#>,  Forecasts number or range of cases.
   SMOOTH-<#>             Specifies weight for level component.
EXP/FORE-4                Default model with 4 forecasted cases.
EXP SALES/LINEAR ADD      Holt-Winter's model: linear trend, additive
                             seasonals with period 12.
EXP SERIES/MULT SEAS=4    Multiplicative seasonality with period 4.
```

Plotting

TPLOT produces time series plots, which plot the variable against Case (time). TPLOT's STANDARDIZE option removes the series mean from each value and divides each by the standard deviation. MYSTAT uses the first fifteen cases unless you specify otherwise with LAG.

By default, TPLOT fills the area from the left axis to the plotted point. Include the NOFILL option to specify that the plot not be filled. Use the CENTER option to fill the plot from the observed value to the mean of the series.

```
TPLOT <var>               Case plot of variable specified.
   /LAG-<#>               Plots first <#> cases.
   STANDARDIZE,           Standardizes before plotting.
   CENTER|NOFILL,         Changes the way the plot is filled.
   MIN-<#>,MAX-<#>        Sets scale limits.
TPLOT                     Case plot of first numeric variable.
TPLOT PRICE/LAG=10        First 10 cases of PRICE.
TPLOT PRICE/STAN          Standardizes before plotting.
```

ACF plots show how closely the points in a series relate to the values immediately preceding them.

```
ACF <var>                 Autocorrelation plot of specified variable.
   /LAG-<#>               Plots first <#> cases.
ACF                       Autocorrelation plot of first variable.
ACF PRICE/LAG=10          First 10 cases of PRICE.
```

PACF plots show the relationship of values in a series to preceding points after partialing out the influence of intervening points.

```
PACF <var>                Partial autocorrelation plot of variable.
   /LAG-<#>               Plots first <#> cases.
PACF                      Partial autocorrelation plot of first variable.
PACF PRICE/LAG=10         First 10 cases of PRICE.
```

Clearing the series

CLEAR restores the initial values of the series, removing the effects of smoothings and transformations.

Submitting Files of Commands

You can operate MYSTAT in batch mode, where MYSTAT executes a series of commands from a file and you sit back and watch. (You've already seen a command file in action: the DEMO demonstration uses a file of commands, DEMO.CMD.)

The SUBMIT command reads commands from a file and executes the commands as though they were typed from the keyboard. Command files must have a ".CMD" file extension.

Use a word processor to create a file of commands—one command per line, with no extraneous characters. Save the file as a text (ASCII) file. (Use the command "COPY CON BATCH.CMD" and [F6] to type commands into a file if you have no word processor.)

```
SUBMIT <filename>         Submits file of commands.
SUBMIT COMMANDS           Reads commands from COMMANDS.CMD.
SUBMIT B:NEWJOB           Reads commands from file on drive B.
```

Redirecting Output

Ordinarily, Business MYSTAT sends its results to the screen. OUTPUT routes *subsequent* output to an ASCII file or a printer.

```
OUTPUT *                  Sends output to the screen only
OUTPUT @                  Sends output to the screen and the printer.
OUTPUT <filename>         Sends output to the screen and a text file.
```

MYSTAT adds a .DAT suffix to ASCII files produced by OUTPUT. You must use OUTPUT *or QUIT the program to stop redirecting.

Printing and saving analysis results

To print analysis results, use OUTPUT @ before doing the analysis or analyses. Use OUTPUT <filename> to save analysis results in a file. Use OUTPUT * to turn saving or printing off when you are finished.

Printing data or variable names

You can print your data by using OUTPUT @ and then using the LIST command. Use LIST <var1>... to print only certain variables. Don't forget to turn printing off when you are done.

You can print your variable names by using OUTPUT @ and then NAMES. Don't forget to turn printing off when you are done.

Putting Comments in Output

NOTE allows you to write comments in your output. Surround each line with quotation marks, and issue another NOTE command for additional lines:

```
>NOTE 'Following are descriptive statistics for the POP'
>NOTE "and RAINFALL variables of the CITIES dataset."
>STATS POP RAINFALL
TOTAL OBSERVATIONS:  8

                        POP      RAINFALL

N OF CASES               8           8
MINIMUM          346865.000       7.810
MAXIMUM         7164742.000      60.020
```

```
MEAN        2064361.375  35.520
STANDARD DEV 2335788.226 18.032
>NOTE "Note that the average annual rainfall for these"
>NOTE 'cities is 35.52 inches.'
```

Saving Data in Text Files

You can save datasets to ASCII text files with the PUT command. PUT saves the current dataset in a plain text file suitable for use with most other programs. Text files have a .DAT extension.

USE <filename>   Opens the dataset to be exported
PUT <filename>   Saves the dataset as a text file

Note that the PUT command is *not* an Editor command. To save a newly created dataset in a plain text file, you must QUIT from the Editor, USE the datafile, and finally PUT the data in an ASCII file.

```
>EDIT
[editing session here]
 >SAVE A:NEWSTUFF
 >QUIT
>USE A:NEWSTUFF
>PUT A:NEWTEXT
```

The above commands would create a plain ASCII file called NEWTEXT.DAT on the A disk.

Index of Commands

This index lists all of MYSTAT's commands and the Editor commands, describes each briefly, and shows where to look for a description in this booklet.

Editor commands

DELETE	9	delete a row (case)
DROP	9	drop a column (variable)
Esc key	8	toggle between command line and worksheet
FIND	10	find a particular data value
FORMAT	10	set number of decimal places in Editor
GET	10	read data from an ASCII file
HELP	5	get help for Editor commands
IF...THEN	10–12	conditionally transform or create a variable
LET	10–12	transform or create a variable
NEW	9	create a new data file
QUIT	12	quit the Editor and return to MYSTAT
REPEAT	12	fill cases with missing values
SAVE	9	save data in a data file
USE	9	read a data file into Editor

MYSTAT commands

ACF	23	autocorrelation plot
ADJSEAS	22	seasonal decomposition
ANOVA	17	analysis of variance
BOX	19	box-and-whisker plot
CATEGORY	17	specify factors for ANOVA
CHARSET	18	choose type of characters for graphs
CLEAR	23	restore original values of series
CORRELATE	16	Pearson correlation matrix
DEMO	5	demonstration of Business MYSTAT
DIFFRNCE	22	difference transformation
EDIT	4,6	edit a new or existing data file
ESTIMATE	17	start computations for regression
EXP	23	exponential smoothing
FORMAT	13	set number of decimal places in output
HELP	5	get help for MYSTAT commands
HISTOGRAM	19	draw histogram
INDEX	22	index transformation
LIST	13	display data values
LOG	21	log transformation
MEAN	21	demean a series
MENU	4	turn the command menu on/off
MODEL	17	specify a regression model
NAMES	13	display variable names
NOTE	25	put comment in output
OUTPUT	24	redirect output to printer or text file
PACF	23	partial autocorrelation plot
PCNTCHNG	22	percent transformation
PEARSON	16	Pearson correlation matrix
PLOT	18	scatterplot (X-Y plot)
PUT	25	save data in text file
QUIT	14	quit the MYSTAT program
RANK	14	rank data
SAVE	17, 21	save results in a file
SMOOTH	22	smooth a series
SORT	14	sort data
SQUARE	21	square transformation
STATS	15	descriptive statistics
STEM	20	stem-and-leaf diagram
SUBMIT	24	submit a batch file of commands
SYSTAT	5	get information about SYSTAT
TABULATE	15	one-way or multi-way tables
TPLOT	23	case plot
TREND	22	remove trend from a series
TTEST	16	independent and dependent t-tests
USE	13	read a data file for analysis
WEIGHT	14	weight data

Answers

NUMERICAL ANSWERS TO SELECTED ODD-NUMBERED EXERCISES

Chapter 1

1.3 5000
1.5 .487 males and .513 females
1.9 Using $10,000 units for the histogram widths:

Income Group	Relative Frequency	Histogram Height
Low	.525	.210
Middle	.170	.170
Upper	.261	.065

1.13 1984: -10.25 and -9.53
1.15 $409.95 and $1,229.84
1.17 first sample: 3.55, 3.5, and 3
1.19 mean: 2.09, 1.43, 1.62, 2.10
       std. dev.: .8102, .7906, .8459, 1.1091
1.21 mean: 20,453 and 13,260
       median: 17,412 and 10,269
       mode: 30,000 and 2,500
       avg. abs. dev.: 11,589 and 8,493
       std. dev.: 14,934 and 10,949
1.27 6.11
1.37 $63,389
1.39 a. 30   b. 26.67
1.45 $8,676–$19,374, $25,600, and $23,507

Chapter 2

2.1 .4, .4, .2
2.3 a. .4   b. .4   c. .48   d. .88   e. 0
2.7 .25, .80
2.9 136
2.11 24
2.13 45 semesters
2.15 .1646
2.23 .667
2.29 .25, .667
2.31 .0046
2.35 .0851, .1702
2.37 .444
2.39 .6
2.45 2,600,000; 17,576,000; 17,576,000; and 175,760,000
2.47 a. 39,916,800   b. 3,113,510,400
       c. 239,500,800

Chapter 3

3.3 .0385
3.5 .6
3.7 .5, .5, 1/3, 1/3, 1/6, no
3.13 .000977

3.15 .778, .360, and .072
3.17 .00278, .00833
3.19 .965
3.21 .94235, .99981
3.23 .0586, .2378, .4950
3.25 .99701, .99967
3.27 .5
3.29 .0033, .0596
3.33 .487, .512, .488, .504, .493, .433
3.35 .2857
3.37 a. .6651   b. .6187   c. .5973
3.45 .93, .73
3.47 .25, .0001, .004975

Chapter 4

4.1
x	$p[x]$
1	1/6
2	1/6
3	1/6
4	1/6
5	1/6
6	1/6

4.3 3.5
4.5 mean: 10 and 15
    std. dev.: 15 and 20
4.7 193,999.59 and 4.695
4.11 $P = .001, \mu = .018$
4.13 -16.2% and -13.9%
4.15 .545
4.17 age 20: $190 and $72
4.19 expected value = $1.028, expected cost = $1028
4.21 a. $0.75x$ b. $0.6875x$ c. $0.5x$
4.29 14, 18, and 1
4.33 .00364
4.35 $-33.33\%, +33.33\%, 0$
4.37 20, 40, 160
4.39 .167

Chapter 5

5.3 For $\pi = .95$: .735, .540
5.5 .0156, .0938, .2344
5.7 .1921
5.9 .6047, .0017, .133
5.11 .0625, .1667, .375, 1
5.13 .377, .252, .0284, .0568

5.15 .736
5.25 .16667, .5, .3, .0333
5.27 $P[2 - 2] = .407$
5.29 for $n = 12,000,000$: .3679, .6321, .3679, .2642, .4180
5.31 .0758, .0153
5.33 .0527
5.35 .3874, .6513
5.37 .4040
5.41 -50.1%
5.45 .1, .90483, .09049, .00452, .00016

Chapter 6

6.1 a. discrete   b. continuous
    c. continuous   d. continuous   e. discrete
    f. discrete   g. discrete   h. continuous
6.5 a. 1   b. 0   c. .25   d. .75
    e. .4375
6.7 2.5 and 1.118. $Z = -1.342, -.447, +.447, +1.342$ with mean 0 and standard deviation 1
6.13 a. .5, 1, 5   b. .5, .7071, .158114
    c. $-3.162, -2.530, -1.897, -1.265, -.632, 0, .632, 1.265, 1.897, 2.530, 3.162$
    d. binomial
6.21 each is $0
6.23 $P[d$ feet to right] is given by binomial with $d = 2x - 360$, where x is the number of successful steps to the right.
6.27 The dollar change is $.5x - 2.5$, where x is the number of heads, and the probability distribution of x is binomial.
6.31 a. 1   b. .5   c. .5   d. .375   e. .75

Chapter 7

7.1 a. .5   b. .1587   c. .0228   d. .0013
    e. .1587
7.3 a. 0   b. 1.96   c. -1.96
7.5 a. 50   b. 15.87   c. 97.72
7.7 .0869, .4602, .0594, .0024
7.13 1096, 6216, 164
7.15 .984, .001
7.17 .9437, .9757, $>.999$
7.19 .128
7.21 .136, .001
7.25 a. 867,600   b. 243,333   c. 30,200
    d. 2,000   e. 60

7.29 .3286, .0465, .0018
7.31 .5, .3821, .2420
7.33 .0600
7.35 .2525, .0228
7.39 .0228, .00003
7.41 200, 220, 240, 250
7.45 800, .5, .5

Chapter 8

8.9   ratings: 20%, 18%, and 22%
      shares: 30%, 27%, and 33%
8.11 a. $246 gain   b. $4,246   c. $4,000
8.27 .2
8.49 .24

Chapter 9

9.1   a. .5   b. .0228   c. .5   d. .0000317
9.3   .1587, 16.1165, .1587, 0228, .0000317
9.9   .0062, 2.1 to 2.9
9.11 186,269.552 to 186,270.448
9.15 .000000287, 63.87 to 68.13
9.17 11.91 to 16.45
9.19 a. 15.99   b. 0.1056   c. 15.99 ± .08
9.21 a. .0962
9.23 2401
9.27 .1586, .9544, 1
9.33 a. 4   b. 16   c. 2,213 (using 1.96 std
      dev.)
9.37 9,604 (using 1.96 std. dev.)

Chapter 10

10.7 $Z = 18.8$
10.9   .9938, .5, 93.4, .1977
10.11 555, .560, .75, .2266
10.15 29.5; $Z = 5.33$
10.17 $t = 3.29$
10.21 Jackson: $Z = 2.95$
10.27 .32 ± .174
10.29 a. .0000187   b. .000000000000004
10.33 17.95
10.35 $Z = -2$, $P = .0228$, 1701 to 1799
10.37 .7881
10.39 $Z = -5.37$
10.43 $Z = -.8$
10.45 60 correct, .50, .0228
10.47 $Z = 1.27$, 15.62

Chapter 11

11.1 $t = 2.13$
11.9 P-value $= .006$
11.17 .0228, .9641, .5, $<.001$, virtually 0
11.23 .0918
11.39 $Z = -.679$
11.41 cutoffs: 133, 117
11.43 .59, .172
11.45 .5, .001 P-value
11.47 .012
11.49 a. 2   b. 1   c. 3

Chapter 12

12.1 $Z = 3.28$
12.5 $Z = .35$
12.13 0.155 chi-square value
12.17 35.94 chi-square value
12.19 35.71 chi-square value
12.23 38.6 chi-square value
12.25 19.21 chi-square value
12.27 25.7 chi-square value
12.31 26.97 chi-square value
12.37 $Z = 1.67$
12.39 1,073.5 chi-square value
12.41 $Z = 3.12$
12.45 2.03 chi-square value

Chapter 13

13.7 $a = 23.87$, $b = .709$
13.9 $a = 830.65$, $b = .520$
13.11 cannot be estimated
13.15 a. 2.83   c. $t = 2.07$ e. .09 to 5.57
13.17 $t = 1.22$
13.21 $a = -126.7$, $b = 4.07$, $t = 7.45$
13.23 .63
13.25 $R^2 = .286$ and .014
13.27 $a = 15,907$, $b = -269.71$, SEE $= 230.91$,
       $R^2 = .995$
13.31 a. $b = .039$, $t = 4.7$ b. $b = -.167$,
       $t = 6.9$ c. $b = .008$, $t = 1.0$
13.35 $a = 157.15$, SEE $= 3.89$
13.41 $a = 6.9105$, $b = .0052$

Chapter 14

14.1   a. $a = 15.72$, $b = .109$ b. SEE $= 9.28$,
       $R^2 = .54$
14.5   a. $a = -167.8$, $b = 4.74$ b. SEE $=$

8.25, $R^2 = .77$ c. $t = 5.2$ d. 2.90 to
6.58
14.7 $a = 1.32, b = 2.88$
14.9 $\beta = 1.02$
14.15 $b = 2.53, t = .59$
14.19 -1.67 to $-.65$
14.23 a. 16.81   b. 2.31
14.31 a. 622.0   b. 969.5
14.33 $2.16 for 500-pound breaking strength
14.35 $30.81
14.45 $720
14.49 $t = 5.1$

Chapter 15

15.5 a. 2,200   b. 2,290
15.9 106.6
15.19 est. coeff.: $-1.14, .01, -.42$
15.23 est. coeff.: $-404.3, .947, .086.$
   t-values: 1.5, 16.1, 1.3
15.31 $R^2 = 1$
15.33 a. $32,750   b. $50   c. $50
15.35 -1201.1
15.37 BG: 113.0, HS: 164.7
15.41 est. coeff.: $-3.93, .10, 7.4$
15.47 est. coeff.: $-150, 4.5, .23$
15.49 no unique solution

Chapter 16

16.3 b. 110   c. 5.5   d. 128   e. 23.28
16.9 $F = 1.5$
16.11 $F = 4.8$
16.15 total = 13,291.39
16.17 $F = 14.3$
16.19 means = 30.4, 40.44, 35.08. $F = 19.6$
16.21 $F = 14.3$
16.27 $F = .61$
16.29 a. 12.2   b. 148.8
16.31 a. 412, 470   b. 58,595.56, 43,222.22
   c. 50,908.89   d. 1,682.00
16.33 $F = 1.8$
16.35 $F = 49.4$
16.37 $F = 26.35$
16.39 $F = 9.38$

Chapter 17

17.1 reject if fewer than 5 above or below
   claimed median

17.7 mean = 30, std. dev. = 6.4
17.9 average ranks = 9.55 and 11.45
17.11 average ranks = 6.06 and 10.94
17.13 $Z = .204$
17.17 $R = .544$
17.19 $R = .93$
17.21 $R = 1$
17.29 5 runs
17.31 confidence interval: 18.0 to 34.0
17.39 $F = .313$
17.41 $Z = .14$
17.45 11 runs
17.49 $a = .82, b = -.096$

Chapter 18

18.7 $a = -39.22, b = .0226$
18.11 $b = -24.09, -27.19, 18.94$
18.13 $37.4 billion seasonal difference
18.15 est. coeff.: 3.651, .199, .207, 1.316, .071
18.25 $a = 14.36, b = -6.81$
18.29 1.564
18.31 1.76%
18.33 est. coeff.: .334, 1.491, .066, $-.255, .005$
18.43 prediction = 2,072.53

Chapter 19

19.3 sue; settle
19.7 two stories; one
19.9 housing; T-bills; housing; S&Ls
19.11 $79.95 in each case
19.15 expected utility = 3
19.17 a. $260,000   b. 500   c. $250,000
19.19 c. .514, .073
19.23 $P[\pi = .05 | X = 15] = .0021;$
   $P[\pi = .10 | X = 15] = .5033$
19.25 $160.85
19.27 $100,425
19.29 Ordering two dozen has an expected value
   of $1,230.
19.31 The survey is worth $102.
19.35 $P = .5$
19.37 Junk is maximax; T-bills are maximin.
19.39 .975, .78
19.41 expected value for surfing is $72.
19.43 perfect info is worth $16, a barometer $0.
19.45 $P[\text{drug} | \text{"pos"}] = .28; P[\text{drug} | \text{"neg"}] = .04$

Index

STANDARDIZED NORMAL DISTRIBUTION (Table 2)

Cutoff	.00	.01	.02	.03	.04	.05	.06	.07	.08	.09
0.0	.5000	.4960	.4920	.4880	.4840	.4801	.4761	.4721	.4681	.4641
0.1	.4602	.4562	.4522	.4483	.4443	.4404	.4364	.4325	.4286	.4247
0.2	.4207	.4168	.4129	.4090	.4052	.4013	.3974	.3936	.3897	.3859
0.3	.3821	.3783	.3745	.3707	.3669	.3632	.3594	.3557	.3520	.3483
0.4	.3446	.3409	.3372	.3336	.3300	.3264	.3228	.3192	.3156	.3121
0.5	.3085	.3050	.3015	.2981	.2946	.2912	.2877	.2843	.2810	.2776
0.6	.2743	.2709	.2676	.2643	.2611	.2578	.2546	.2514	.2483	.2451
0.7	.2420	.2389	.2358	.2327	.2296	.2266	.2236	.2206	.2177	.2148
0.8	.2119	.2090	.2061	.2033	.2005	.1977	.1949	.1922	.1894	.1867
0.9	.1841	.1814	.1788	.1762	.1736	.1711	.1685	.1660	.1635	.1611
1.0	.1587	.1562	.1539	.1515	.1492	.1469	.1446	.1423	.1401	.1379
1.1	.1357	.1335	.1314	.1292	.1271	.1251	.1230	.1210	.1190	.1170
1.2	.1151	.1131	.1112	.1093	.1075	.1056	.1038	.1020	.1003	.0985
1.3	.0968	.0951	.0934	.0918	.0901	.0885	.0869	.0853	.0838	.0823
1.4	.0808	.0793	.0778	.0764	.0749	.0735	.0722	.0708	.0694	.0681
1.5	.0668	.0655	.0643	.0630	.0618	.0606	.0594	.0582	.0571	.0559
1.6	.0548	.0537	.0526	.0516	.0505	.0495	.0485	.0475	.0465	.0455
1.7	.0446	.0436	.0427	.0418	.0409	.0401	.0392	.0384	.0375	.0367
1.8	.0359	.0352	.0344	.0336	.0329	.0322	.0314	.0307	.0301	.0294
1.9	.0287	.0281	.0274	.0268	.0262	.0256	.0250	.0244	.0239	.0233
2.0	.0228	.0222	.0217	.0212	.0207	.0202	.0197	.0192	.0188	.0183
2.1	.0179	.0174	.0170	.0166	.0162	.0158	.0154	.0150	.0146	.0143
2.2	.0139	.0136	.0132	.0129	.0125	.0122	.0119	.0116	.0113	.0110
2.3	.0107	.0104	.0102	.0099	.0096	.0094	.0091	.0089	.0087	.0084
2.4	.0082	.0080	.0078	.0075	.0073	.0071	.0069	.0068	.0066	.0064
2.5	.0062	.0060	.0059	.0057	.0055	.0054	.0052	.0051	.0049	.0048
2.6	.0047	.0045	.0044	.0043	.0041	.0040	.0039	.0038	.0037	.0036
2.7	.0035	.0034	.0033	.0032	.0031	.0030	.0029	.0028	.0027	.0026
2.8	.0026	.0025	.0024	.0023	.0023	.0022	.0021	.0021	.0020	.0019
2.9	.0019	.0018	.0017	.0017	.0016	.0016	.0015	.0015	.0014	.0014
3.0	.0013	.0013	.0013	.0012	.0012	.0011	.0011	.0011	.0010	.0010
4.0	.000 031 7									
5.0	.000 000 287									
6.0	.000 000 000 987									
7.0	.000 000 000 001 28									
8.0	.000 000 000 000 001									

STUDENT'S *t* DISTRIBUTION (Table 4)

$P[t > \text{cutoff}]$
For instance, with 10 degrees of freedom, $P[t > 1.812] = .05$

Degrees of Freedom	Probability of a t-value larger than indicated cutoff				
	.10	.05	.025	.01	.005
1	3.078	6.314	12.706	31.821	63.657
2	1.886	2.920	4.303	6.965	9.925
3	1.638	2.353	3.182	4.541	5.841
4	1.533	2.132	2.776	3.747	4.604
5	1.476	2.015	2.571	3.365	4.032
6	1.440	1.943	2.447	3.143	3.707
7	1.415	1.895	2.365	2.998	3.499
8	1.397	1.860	2.306	2.896	3.355
9	1.383	1.833	2.262	2.821	3.250
10	1.372	1.812	2.228	2.764	3.169
11	1.363	1.796	2.201	2.718	3.106
12	1.356	1.782	2.179	2.681	3.055
13	1.350	1.771	2.160	2.650	3.012
14	1.345	1.761	2.145	2.624	2.977
15	1.341	1.753	2.131	2.602	2.947
16	1.337	1.746	2.120	2.583	2.921
17	1.333	1.740	2.110	2.567	2.898
18	1.330	1.734	2.101	2.552	2.878
19	1.328	1.729	2.093	2.539	2.861
20	1.325	1.725	2.086	2.528	2.845
21	1.323	1.721	2.080	2.518	2.831
22	1.321	1.717	2.074	2.508	2.819
23	1.319	1.714	2.069	2.500	2.807
24	1.318	1.711	2.064	2.492	2.797
25	1.316	1.708	2.060	2.485	2.787
26	1.315	1.706	2.056	2.479	2.779
27	1.314	1.703	2.052	2.473	2.771
28	1.313	1.701	2.048	2.467	2.763
29	1.311	1.699	2.045	2.462	2.756
30	1.310	1.697	2.042	2.457	2.750
40	1.303	1.684	2.021	2.423	2.704
60	1.296	1.671	2.000	2.390	2.660
120	1.290	1.661	1.984	2.358	2.626
∞	1.282	1.645	1.960	2.326	2.576

Abridged from Table III of Fisher and Yates, *Statistical Tables for Biological, Agricultural, and Medical Research* (London: Longman Group Ltd.). By permission.